The City
and Its Sciences

Cristoforo S. Bertuglia · Giuliano Bianchi
Alfredo Mela (Eds.)

The City and Its Sciences

With 123 Figures
and 16 Tables

Physica-Verlag
A Springer-Verlag Company

Prof. Cristoforo Sergio Bertuglia
Dipartimento di Scienze e Tecniche per i Processi di Insediamento
Politecnico di Torino
Viale Mattioli 39
10125 Torino
Italia

Prof. Giuliano Bianchi
Rete dell'Alta Tecnologia
Regione Toscana
Via Ciro Menotti 6
50136 Firenze
Italia

Prof. Alfredo Mela
Dipartimento di Scienze e Tecniche per i Processi di Insediamento
Politecnico di Torino
Viale Mattioli 39
10125 Torino
Italia

The editors gratefully acknowledge the financial support received from the Italian National Research Council (CNR) which has permitted the development of the study on "The City and Its Sciences", the results of which are published in this volume. Grants have been provided under the following contracts: Progetto Finalizzato Trasporti 2: no.94.01344.PF74, no.94.01345.PF74, no.96.00015.PF74, no.96.00016.PF74, no.97.00185.PF74, and no.97.00186.PF74; Comitato Nazionale Scienze e Tecnologie dell'Ambiente e Habitat: no.AI.96.00495.13.

ISBN-13: 978-3-642-95931-8 e-ISBN-13: 978-3-642-95929-5
DOI: 10.1007/978-3-642-95929-5

Library of Congress Cataloging-in-Publication Data applied for
Die Deutsche Bibliothek - CIP-Einheitsaufnahme

The City and its sciences / ed.: Cristoforo S. Bertuglia ... - Heidelberg:
Physica-Verl., 1998
ISBN-13: 978-3-642-95931-8

Softcover reprint of the hardcover 1st edition 1998

Coverdesign: Erich Kirchner, Heidelberg

SPIN 10647032 88/2202-5 4 3 2 1 0 – Printed on acid-free paper

To Ludovico Franco who, at three months, was the youngest participant in the seminar.

Preface

This book is the outcome of an ambitious project involving a large number of urban researchers, analysts and practitioners, mainly from Italy, but also other countries. The main focus of the work was the Seminar organised at Perugia in Umbria, from 28-30 September, 1995, which provided the opportunity for three days of intense discussion on the principal themes which had emerged from the preparatory stages. As explained in the Introduction, the project was developed systematically, with close interaction between the contributors both before the seminar and in the two years since then, during which the final versions of the papers have been drawn up.

The aim of the project was to investigate whether and in what way the concept of complexity has already generated, and is likely to generate in the future, changes in (i) how the city is seen, (ii) the sciences which deal with the city, (iii) the way in which city planning is conceived and practised, and (iv) the methods and tools used in the analysis of the city. These four issues correspond in fact to the four sections into which this book has been divided.

The selection of papers published here are those which represent the most compact discussion around each of the main themes. They cover an extremely wide range of topics and points of view, but have as a common thread their desire to contribute to an improved understanding, and hence more effective planning and management of the city.

The whole project was made possible by the support of the following organisations to whom we give our thanks: the Italian Regional Science Association, for their far-sightedness in promoting the Perugia Seminar, the Department of Sciences and Techniques for Settlement Processes of Turin Polytechnic, which served as a logistical base for the network of contacts with and between contributors, the Institute of Environmental Engineering of the University of Perugia, and in particular its Director, Lorenzo Berna, who made an invaluable contribution to the organisation of the Perugia seminar, and finally the Regional Authority for Tuscany, which prepared four volumes containing the initial version of the papers distributed before the Perugia seminar.

The editors would also like to thank Angela Spence, who has been involved in all stages of the work and handled with great professionality the linguistic editing of the book, as well as the translation of several chapters, and Franco Vaio for his competent checking of the technical content of the papers and preparation of the camera ready copy.

Lastly, but certainly not least, we wish to acknowledge the scientific sponsorship of the Italian National Research Council and their financial and practical support, provided in particular by the Transport Project 2 and the National Committee for the Science and Technology of the Environment and Habitat, as well as the encouragement of Lucio Bianco, formerly Director of the Transport Project 2 and currently President of the National Research Council.

Naturally, the whole project would not have been possible without the commitment of all the contributors, and their willingness to present, discuss and develop their personal reflections and experience in the field.

We offer to everyone involved our warmest thanks and gratitude.

Cristoforo S. Bertuglia

Giuliano Bianchi

Alfredo Mela

Contents

SESSION 2: THE SCIENCES OF THE CITY

SESSION 4: THE METHODOLOGIES OF THE URBAN SCIENCES

1. Introduction

Cristoforo S. Bertuglia, Giuliano Bianchi, Alfredo Mela

1.1 The Guiding Idea Behind the Seminar

The international seminar on "*The City and its Sciences*" (Perugia, 28-30 September, 1995) from which this volume originated, arose from the following idea, which served as a guide and also a working hypothesis (and like all working hypotheses, required verification):

"The city, at least since the industrial revolution, has become a highly complex entity and appears to be getting more complex. In fact, it is characterised by a growing number of nonlinear interactions between numerous urban actors, who generate space-time dynamics which are always irreversible, often discontinuous and sometimes chaotic. The sciences of the city, moving from a structural and functionalist paradigm towards an evolutionary paradigm, are beginning to make use of the theory of complexity and the disciplines which provide tools for its application. At the same time, the planning of the city, moving from a rational/ comprehensive concept towards a process of social learning, is once again linking the phases of analysis and design, favouring a synthesis of their methodologies. These methodologies, in order to deal with urban complexity, are increasingly taking the form of data organisation and management techniques which emulate memory processes, techniques of data processing which emulate reasoning processes, and decision-making techniques which emulate processes of social participation" (Bertuglia, Detragiache and Rabino, 1993, p. 172-173, our translation).

It was on the basis of this hypothesis that the seminar discussion was divided into four main themes: "the city as a highly complex entity", "the sciences of the city", "the planning of the city" and "the methodologies of the urban sciences". For each of these themes a main speaker and two discussants were chosen. During the year preceding the seminar, the preparation was organised in the following way:

- the four main speakers prepared drafts of their papers and sent them to the discussants;
- the discussants prepared their own drafts and sent them to the main speakers, who were able to revise their papers accordingly;
- the above set of papers was made available to all the other contributors, who could take into account the contents in completing the preparation of their papers;
- all contributions were distributed to all participants, in the form of four preprint volumes, before the seminar was held.

The above procedure meant that the introductory papers for each session were the result of an interactive process, and that all other contributors could take into consideration the contents in preparing their papers. Finally, every participant had the opportunity to read all the papers before the seminar itself. The seminar benefited considerably, since the presentations could be concise and focus on the key points. It was, in fact, characterised by tight discussion between reciprocally informed participants.

This also influenced the final versions of the papers, prepared after the seminar, which show strong interconnections (evident in the many cross citations). A further contribution to the process of integration was made by the three editors who, with full respect for the autonomy of each author, suggested a number of modifications and clarifications, encouraging comparisons between papers, as well as respect for an overall formal homogeneity.

The editors have no hesitation in saying that, in facing the task of preparing the introduction to this volume, they were aware that they had before them an extremely wide-ranging work, which was compact, but full of rich details, covering a broad field, but containing in-depth investigation. Overall, they feel that it represents a real state-of-the-art review of the city and urban processes, of theories and methods of approach. The work, obviously, bears the stamp of the complexity viewpoint, but this has taken on many facets and meanings, has stimulated different ways of seeing things. This variety has undoubtedly enriched the discipline, but made necessary some attempt at systematic treatment, provided in Section 1.4.

1.2 The Sciences of the City: A Moment of Crisis?

1.2.1 The Debate on Regional Science and the Sciences of the City

As explained above, this volume represents an attempt to reflect on the whole field of the city and its sciences. It begins, in fact, with the more general aspects (the problem of the scientific status of the sciences of the city), moving on to a 'technical' investigation of the problem of representation of urban phenomena, and goes so far as to question the use of scientific knowledge in the solution of urban problems. We could say, to give a classical tone - and making the Aristotelian distinction proposed by Flyvbjerg (1990) with reference to planning - that the book intends to touch on all three levels of knowledge: *episteme* (universal knowledge, that which has to be proved true), *techne* (knowledge of the art of producing something) and lastly the *phronesis* (knowledge of what to do in specific circumstances) (see Peattie, 1994).

But, in proposing such an ambitious project, we must not ignore an aspect which is abundantly clear, that the sciences of the city are at present passing through a conflictual phase, of which the dominant feature seems to be uncertainty. On the one hand, the nineties seem to be ending with renewed interest in town and country planning and, more specifically, the city. As we shall see in Section 1.2.3, this is not in contrast with the concern about the processes of economic and socio-cultural globalisation currently affecting the entire world system. It is in some ways, in fact, complementary. As far as the city is concerned, the interest has been stimulated both by the importance of its role and the social problems by which it is afflicted as we enter what could be defined the 'post-modern' era (Amendola, 1997).

These various stimuli do not, however, for the moment seem to have been translated into an opportunity to advance the sciences of the city, either in relation to university teaching, or in their impact on urban policy. Nor can we say that there appears to be a real commitment to the consolidation of the results, preparing the ground for a more effective theoretical synthesis. On the contrary, it would seem at present that the differences between alternative approaches and paradigms seems to be widening, not only in different disciplines, but also within the same field. The sciences of the city are subject to growing concern, frequently expressed in the desire to make evaluations, often critical, of the state-of-the-art and to assess the prospects, often seen as problematic, for its future evolution.

This debate is well documented, for example in the numerous contributions to *Papers of Regional Science, 73,* 1 (1994), in particular the paper of Bailly and Coffey, and in the *International Regional Science Review, 17,* 3 (1995), especially the paper by Isserman. The debate focusses on two problems which Regional Science (and also the sciences of the city) seem unable to resolve effectively.

The first problem concerns the definition of the epistemological status of the field of research whose aim is the analysis and modelling of the city. This can be broken down into two further aspects. On the one hand, the problem of the definition of the subject of the research, i.e. the city. As many have remarked, in contemporary society the city seems to have lost, at least partially, the characteristics which once made it a clear and distinct entity, self-evident and distinguishable from other spatial systems, and therefore an almost 'natural' term of reference. On the other hand, there is the question of the relations between the points of view of all the various disciplines which contribute, each in its own way, to the analysis of the city and its planning. To what extent are these points of view compatible? Are they linked to fundamentally incompatible assumptions and therefore destined to 'share out' the field of study in a nonconflictual way, exchanging results only *a posteriori*, or is it possible to define common principles which could sustain a process of real disciplinary unification?

The second problem concerns the social role of the city sciences, that is the effectiveness of their application to the development of urban systems, or of tackling specific difficulties in the functioning of such systems. In other words, it regards the capacity of the sciences of the city to offer specialised competence in response to the demand from various actors (mainly public, but also private) operating on the urban scene. This problem, too, can be broken down into further aspects, one concerning the supply, the other the demand. We may ask, therefore, whether the sciences of the city are really socially useful, and able to contribute to the improvement of urban life. We should also enquire whether the bodies with the major decision-making powers affecting the future of the city are really interested in making use of the sciences of the city and, if so, whether they are able to formulate appropriate questions.

In the following part of this introductory chapter, we examine in greater depth the problem areas briefly referred to above. As well as illustrating in more detail the implications, we make some attempts to define a possible response. We also try to establish in what way the papers collected in this volume contribute to the various aspects of the investigation of the city, indicating the directions of research which seem most original and promising.

1.2.2 The Scientific Status of the Sciences of the City

The city already represented a theme of wide-ranging discussion, involving highly diverse aspects of knowledge and expertise, long before the development of the social sciences, in the modern sense of the term. Since the 19th century, under the prevalent influence of the positivist philosophical paradigm, and also, in certain disciplines, of organistic thought, many of the new social and human sciences (from economics to sociology and geography) have adopted human settlement, and in particular the city, as a specific subject of study. In doing so, they have provided common ground for the disciplines concerned with technology and design (civil engineering, architecture and town planning), whose task was to deal with the problems caused by the rapid urban expansion.

But, despite the common subject of interest, the aspects of the city with which these various disciplines are concerned are profoundly different. Spatial and urban economics, for example, while to some extent marginal to the mainstream of economic theory, became far more formalised than other fields of study, such as sociology or geography. This disparity in the degree of formalisation and, more generally, in the conceptualisation and language used for empirical analysis, made dialogue between the various sciences of the city more difficult than might have been expected.

A distinct improvement in this dialogue occurred after the second World War in the climate of optimism about the potential of scientific analysis to resolve social and economic problems. With the formation of the Regional Science Association in 1954, the sciences concerned with human settlement, and among these the urban sciences, seemed to offer, more explicitly than in the past, a possible area of synthesis. According to the main promoter, Walter Isard, Regional Science was something more than the application to territorial analysis of approaches from existing sciences (such as economics, sociology, political science and anthropology). The intention was, in fact, that attention should focus on the spatial dimension of the region, and that spatial analysis should no longer be considered only a marginal aspect, as it had been in the dominant paradigms of the 1950s in both economics and the social sciences. Regional Science was also to promote research aimed at the building and empirical verification of theoretical models, with the use of reliable statistical information (Isserman, 1995). In the construction of such models, the researchers of the 1950s maintained that it was necessary, at least initially, to make use of conceptual tools derived from other approaches, adapting them as necessary: a classic example of this kind of theoretical transposition was the gravity model (Stewart and Warntz, 1958).

The development of this research programme was made possible, above all, by the emergence of a new and powerful unifying paradigm: General Systems Theory (Romanoff, 1955). This paradigm shift appeared to provide the basis not only for greater interdisciplinary collaboration to be achieved, but also the development of a common language, which was previously lacking. Systems theory offered a meta-theoretical point of reference common to many disciplines, making it possible, at least in principle, for the theories developed in the various scientific areas to become comparable and compatible.

In the following decades, the effects of this revolution produced extremely important results. One of these was the extension of the use of mathematical language and modelling procedures to areas outside economics, with the development of mathematical theories and algorithms to tackle the problems raised by the representation of the city as a highly complex system with nonlinear dynamics. During the 1980s, this theoretical basis of the systems paradigm applied to the analysis and modelling of the city was given new vitality from a number of theories and models, largely of biological derivation, i.e. the theories and models often referred to in connection with the 'science of complexity'. In the 1990s, as explained in this volume by Pumain, these new additions became consolidated, allowing developments on numerous fronts, including catastrophe theory, the master equations approaches, dynamic models of chaos, cellular automata, the application of fractals to urban structure, and so on.

At the same time, it was becoming evident that two other factors were acting as stimuli to scientific evolution. The first was the increasing political importance, both at national and international level, being given to the environmental question. This gave rise to the need for many scientific fields to contribute to the definition, in theoretical and operative terms, of the concept of ecologically sustainable development. For the city and territorial sciences, it involved identifying forms of spatial organisation capable of reconciling the processes of transformation and disequilibrium implicit in the idea of 'development' with the need to maintain "The natural eco-biological system of support of life in its various forms, which represents the 'basis' for all human activities" (Fusco Girard and Nijkamp, eds., 1997, p. 21, our translation). The second factor concerned the recognition, in certain disciplines, of conceptions which gave greater importance to the spatial dimension. This occurred in particular in economics, in relation to which Anselin and Rey (1997) claimed that "In recent years, the role of space in general and of spatial externalities in particular, has gained an increasingly prominent position in mainstream economics, in part stimulated by the visibility of Krugman's work on the 'new economic geography' (Krugman, 1991a, 1991b, 1996)" (p. 1). In addition, similar observations could be made in relation to sociology, where recent years have seen the development of a similar orientation, defined as 'spatialist' (Mela, 1996) since it holds that the spatial dimension, like the time dimension, is fundamental to individual social action, as well as to systems which derive from the aggregation of a multiplicity of actions (see the paper in this volume by Preto and Mela).

Beyond these undeniable advances, however, it would be hard to claim that, over the 40 years since its beginnings, the research programme of Regional Science has fully achieved its aims, or that - and this concerns us more closely - it is producing a transdisciplinary understanding of the city. The strong points of the programme, mentioned above, have had positive effects, but have also left areas of shadow. For example, the development of urban modelling seems, in many cases, to have become so highly specialised that the models, rather than representing powerful and flexible tools of urban analysis to help resolve problems, have become an object of scientific interest among a very limited number of exponents, and understood neither by the public at large, nor even by other urban scientists who are 'non modellers'.

The influence of the systemic paradigm too, and its role as a point of reference for dialogue between the various sciences of the city, despite the theoretical advances referred to previously, does not appear to have undergone further consolidation, but rather to be creating perplexity and attracting criticism from many sources, including many branches of geography (especially those of a humanistic imprint, not belonging to the field of quantitative geography) and sociology (especially 'urban political economy' and the radical branches). In this respect, it is worth noting that much of the debate which, in the last ten or fifteen years has involved these scientific sectors, has had a relatively weak echo in the more formalised fields, i.e. the mainstream of city sciences inspired by the original paradigm of Regional Science.

To give just two particularly significant examples, we should like to recall the debate on the post-modern condition and the city (Harvey, 1990), and the debate arising from the critical analyses carried out from the point of view of the various movements inspired by 'differences', above all the feminist movement (Jacobs, 1993). It is interesting to note that in both of these debates, especially the second, radical criticism is made of the conceptual tools used not only by social scientists, but also planners and public authorities, for the interpretation of urban phenomena in contemporary society and the definition of urban policy. It should be stressed that these criticisms are not purely political or ideological, but also touch on the epistemological assumptions of the urban sciences and their methodological procedures. In this connection, it is interesting to note that, since the 1980s, there has been a reappraisal of the qualitative methods of analysis, and also strong vindication of the hermeneutic approach to the city and the variety of meanings found within it (Jacobs, 1993). Much space has been dedicated to these orientations in books and journals about the city, especially in English speaking countries. These have, however, mainly been publications belonging to a literary genre relatively distant from the more formalised sectors of urban science - sufficiently distant for the exponents of each type not to indulge in mutual exchange of views or even, in many cases, adopt the other as a target for their polemics.

Maybe a neutral observer would have the impression that, despite the results achieved and the repeated attempts to create opportunities for discussion through seminars and interdisciplinary debate, as a whole researchers concerned with the city are not much less heterogeneous now than fifty years ago. In fact, they would seem to be divided into two distinct camps, one adopting procedures and methods closer to those of the natural and physical sciences, the other supporting the use of an interpretative approach more typical of the humanities and historical/ literary criticism. It could possibly be added that while the former is advancing with method and precision, but adopting an abstract approach to the city and its problems, the second is attempting to identify the problems in more concrete terms and with greater sensitivity to current issues, but tends to express itself in critical form and with little inclination to build up knowledge cumulatively.

The above represents, however, only one aspect of the present situation. In fact, one of the most significant difficulties faced today by the urban sciences lies not so much in the choice between a more formalised or more hermeneutic approach to analysis, as the disorientation caused by the profound transformation in urban phenomena themselves as a result of the globalisation of society. We mention just a few of the most obvious changes: the process of decentralisation and 'scattering' of development, which is weakening the spatial identity of the city; the new systems of communication and telematics which are further loosening many of the factors which previously constrained whole sets of activities to an urban or metropolitan location. This is having the effect of increasing the freedom of movement of capital, goods, information and a part of the population, and also the importance of what Castells (1989) calls 'the space of flows'. Network relations are being established between the nodes in economic space, even over large distances, and similarly between cities, producing the form of urban development described by Batten (1995) as network cities.

Even though the processes of urbanisation continue to be a central aspect of the post-industrial scene, the single city is tending to lose, at least partly, the 'obviousness' recognised by the social sciences. Does this mean that the sciences of the city are destined to lose their subject of study, or have even perhaps already lost it without realising it? Not at all! It should be said, on the contrary, that it would be meaningless today (as it has always been) to adopt a 'naturalistic' attitude in relation to the city, that is to consider it an identity given by nature and whose laws of functioning and evolution are simply to be discovered. As stated more explicitly later in this volume (in particular in the paper by Tinacci Mossello), the understanding of the city can only be a 'research programme' which presupposes an interaction between the complexity of the reality of the social, economic and physical phenomena and the point of view of an observer, i.e. the objectives of his or her research, the conceptual apparatus used, the tools of observation, and so on.

Having said this, there arises the problem of the definition of guidelines for this programme of research. Should it basically be a unitary programme aiming to find an overall theoretical convergence between the points of view of different observers, or should it maintain a pluralistic nature, simply requiring a guarantee of commensurability between the points of view? In any case, which paradigmatic references would allow dialogue between the various points of view?

These are some of the questions which many of the papers collected in this volume attempt to answer. Both questions and answers will have a substantial influence on the future of the sciences of the city.

1.2.3 The Practical Relevance of the Sciences of the City

Alongside the epistemological questions, there is a second kind of problem which should not be neglected. These concern the practical relevance of the urban disciplines, that is their capacity to affect the social, economic and technological activities which incessantly modify the form of the city, as well as the policies designed to regulate these activities.

This is a point highlighted by Bailly and Coffey (1994) in their proposal to promote - to use an expression employed in their article - "a more open and relevant Regional Science". (The same thing in fact would apply if they had referred exclusively to the sciences of the city.) The first principle established by the two authors for the achievement of this objective, is "to close the gap between the subject of Regional Science and the subject of its practice" (p. 8). In effect, it is undeniable that there exists a wide gap between the increasingly sophisticated theoretical and methodological tools produced by urban studies and the impact of this knowledge on the political, technical and economic practitioners who, through their decisions, influence the evolution of the city and have to resolve the problems arising in the various parts of the urban system. The sensation, common to many researchers, is that the accumulated knowledge belonging to the various disciplines is not being translated into know how which can be made available to urban decision-makers and that, vice versa, the know how used (consciously or otherwise) by the latter is formed in a way which has little to do with the scientific effort put into the urban sciences.

In an optimistic vein, we could say that in the market for scientific knowledge concerning the city there simply exist obstacles to the match between demand and supply. More pessimistically, we could add that this market is in reality very limited. The demand for knowledge about urban analysis or policy is rarely expressed in a form likely to solicit a response from experts or, more precisely, from academic researchers in urban science.

Naturally, the responsibility for this mismatch lies partly on the side of the demand, partly on the supply side. One of the causes may be that the approach has been too abstract, with excessive importance given to internal questions or inter-school disputes. Also, as far as attempts to model the city are concerned, we cannot ignore the fact that, especially in the recent past, many urban models have had operative difficulties which constituted a serious obstacle to their widespread use in defining urban policy. At the same time, the relative lack of interest in urban science by urban decision-makers could be attributed to the way in which the latter, especially those responsible for the formulation and implementation of plans and projects, tend to define their own role and orient their own procedures.

In relation to this second aspect, it is obvious that there are considerable differences between countries regarding the institutional organisation of urban planning and the degree of public intervention accepted in city development. It is not possible here to go into further detail (it is discussed in several papers in this volume, such as those by Faludi and by Mazza), but we might recall some of the major changes which have occurred in recent decades in the demand for knowledge for planning purposes.

The change of paradigm which, in the 1950s and 60s, seemed to have laid the basis for a definitive convergence in the field urban sciences, occurred at a time when in many countries, especially in the Central and Northern Europe, State intervention was being consolidated as an active part of the regulation of social and economic processes, as well as the structuring of urban development. At this time, the influence of forms of thought which put great faith in the potential of scientific reasoning for the solution of development problems was still strong. In these conditions, it was fairly natural for public operators involved in urban planning to see themselves as elements of a 'steering system' (Le Moigne, 1976) for urban systems and to consider the urban sciences as conceptual and operative tools for the overall regulation of urban growth processes. In practice, planning processes did not always conform precisely to this ideal of rational/comprehensive control, but nevertheless its prestige had the effect of legitimising the roles of public planning practitioner and urban scientist, and making them, at least to a certain extent, complementary.

This picture began to change radically after the late 1970s. Firstly, there were changes in the pattern of urbanisation as a result of technological innovation, as mentioned in 1.2.2, and economic and financial globalisation processes. Secondly, the role of the State in the regulation of urban development had changed and new tools were being applied. Economic restructuring processes are now making it difficult for individual countries to control economic variables as these depend on international players and, in any case, are more effectively influenced by private decision-makers (such as the major financial centres) who are able to move fast and have global impact. At the same time, most state and local authorities have fewer resources available to implement urban development policies. In addition, the processes of social tranformation connected with these phenomena mean that in many countries there is a neo-free market tendency which is reducing the consensus given to the whole idea of planning. Finally, the scientific paradigms which previously guided the work of planners and social scientists are now being undermined. There has been strong criticism, also on theoretical grounds, of the very idea that a city plan can be conceived as a tool which overrides the choices of numerous urban decision-makers, able a priori to integrate them and channel the economic, social and physical development along a predefined path.

All in all, town planning during the 1980s, at least at the strategic level, tended to be delegitimised, and was increasingly considered "an annoying obstacle to individual liberty and the functioning of the free market economy, or even considered incompatible with it" (Albrechts, 1996, p. 84, our translation). Scepticism was not only expressed by the neo-free market political orientations of many governments, but also the cultural criticism of post-modernism, which has been spreading in many fields and becoming a pervasive cultural climate (Mela, 1990). Both, in fact, as observed by Healey (1991) tend to retain that progress, if it occurs, cannot be planned.

Given this situation, the relatively straightforward and previously dominant hypothesis of a division of roles between planner and researcher is beginning to seem less and less credible. It is proving difficult, however, to find a new, more complex hypothesis to fill the space left vacant. The 1980s and early 90s were characterised by relative uncertainty, dominated more by criticism (often more than justified) of the past, than constructive attempts to formulate the problems in new ways and find innovative solutions. Inevitably, the situation has increased doubts about the effectiveness of the urban sciences, even pushing some exponents to practice elsewhere and to convince universities that the urban sciences represent a field of only academic interest to which limited space should be dedicated, since they play a subordinate role in the training of technicians who have a real influence on urban planning processes (i.e. architects, engineers, management experts, etc.).

In the late 1990s, this uncertainty about the role of planning and its tools continues to characterise the political and cultural scene. But signs are beginning to emerge, at least in the European context, of a renewed interest in strategic planning, especially as far as the urban and metropolitan scale is concerned.

In recent years, in fact, many factors have contributed to increase the impact of the role of the city, which has played an active part in the definition and implementation of innovative policies. This is largely the result of the growing weight of international factors, both economic and political: the former connected with the phenomenon of globalisation, the latter with the growing influence of international organisations (above all, the European Community) in the definition of political policy at national, regional and even local level (Salone, 1997). In any case, the overall effect has been to give greater significance, in the structuring of economic and political systems, to processes occurring at a scale which goes far beyond the range of action of the individual State.

Due to this jump in scale and weakened by the crisis in public finance, the state is no longer able to fulfill the function adopted after the second World War in most countries (in different forms according to local political and cultural tradition), of co-ordinator and promoter of economic development, capable of mitigating the inequalities continuing to arise between strong and weak regions, and between the various social groups.

The weakening of the nation-state, however, has not been accompanied by a reduction in the problems requiring intervention. On the contrary, the problems linked to uncertainties and disparities between models of development have worsened and been joined by a generalised increase in unemployment, which has also hit the sectors which were the motors of Fordist expansion, by a growth of social marginalisation and increase in the risks linked with environmental degradation (Revelli, 1997). All this is causing a re-orientation of political demand. Whereas in the 1960s and 70s it was directed above all to the central government, it now tends to be directed to a much wider range of decision-making levels - these include of course the international level, but also local government and, above all, urban municipalities. However, with the weakening or transformation of the political idealisms which sustained the great social conflicts of the first three quarters of this century, expectations and worries are focussed increasingly on the conditions which will determine the scenarios and quality of daily life. These scenarios coincide for most people with those of the city and, for most of the urban population, with the suburbs which have grown up around the city, and urban development which more recently has spread through the surrounding region.

This does not mean that there is a restriction in the horizons of political demand, as opposed to the globalisation of economic horizons. The urban scenario in which the various social groups are involved, does not only consist of one's own city or neighbourhood. It is a highly urbanised world in which the processes which occur and problems which arise are common, although they vary according to the context. This world is made up of cities, constituting a dense network in which, thanks to increasingly sophisticated technological development (Graham and Marvin, 1996), proximity and distance, co-operation and competition, centrality and marginality do not depend so much on geographical distances or administrative boundaries as the continually changing relations between the various nodes of the network. The economic, social and urban scenario which is developing is characterised by processes oriented towards globalisation and, at the same time, the recognition of the increasing importance of local phenomena. The main feature is the "growing interpenetration between local and global" due to which "small and large-scale interdependencies are no longer separable" (Veltz, 1998, p. 133, our translation).

The most evident features of the picture described above are its heterogeneity and the close interdependence between the problems facing the city. These problems involve, first of all, the redefinition of the economic and employment base. In fact, for the cities which followed most closely the Fordist model of industrialisation, it is clear that the large manufacturing firms can no longer represent the only motor of the urban economy (Dematteis, 1997). For those towns in areas of less intense industrial development, which in the past concentrated their efforts on trying to 'catch up', there is a need to redimension the growth objectives linked with industrial expansion and to exploit their resources in other ways. At the same time, the illusions cultivated by many in the 1980s concerning the potential of the advanced services sector and high technology industry to substitute the large industrial firms as leading poles of the urban economy have been shattered, or at least become more realistic. The experience gained and the failure of many attempts at urban regeneration based only on 'high tech' and on the rarer services (especially those linked to international finance) emphasise that the role of high level financial centre and innovation pole can be played only by a limited number of metropolises and that, in any case, the employment generated by such sectors does not compensate for the loss of jobs caused by industrial reorganisation and the cutbacks in the public sector.

For the post-Fordist city there is no model of development to serve as a general point of reference. This means that there is an urgent need for each city to reflect on its own qualities, and the advantages and disadvantages connected with the position that it occupies in the network of European and world cities. It implies a complete 'reconnaissance' of the resources available and the social and economic energies which can be activated to stimulate the consolidation of the most diversified economic initiatives: from small and medium manufacturing firms, to the arts, tourism, computers and telecommunications, from personal services to the new environmental professions, and various forms of job creation.

However, the need to mobilitate urban resources of all kinds raises the problem of their distribution between the various social groups and also the different parts of the metropolitan area. In recent years, in fact, the cities of the more advanced nations (especially in America, but also in Europe) have been affected by fragmentation and an increase in internal social imbalance - a phenomenon which sociologists have referred to as 'dualism' (Castells, 1989), or the 'quartered city' (Marcuse, 1989), among other terms. There is also an increase in phenomena such as immigration from underdeveloped countries and the existence of the chronically unemployed (especially among the young people and single parent families) largely dependent on the state. It is significant that these processes do not occur only in declining cities. They can exist alongside upwardly mobile social sectors and successful economic activities. From the spatial point of view, they appear as the tendency of the city to break down into areas which may be adjacent but have strongly contrasting character: declining inner city districts next to gentrified areas, suburban social housing ghettos next to wealthy residential estates.

The activation of the resources of the metropolitan *milieu* (Governa, 1997) cannot be completely separated from a policy which aims to combat social exclusion and the fragmentation of the city. But, at the same time, this cannot be exclusively a policy of subsidy to support specific groups, and achieved through the redistribution of resources drained through tax increases. This in fact is made impossible by the need, now essential in all countries, to cut public expenditure, and would be in contrast with the policy of actively involving weaker social groups in the urban context, and stimulating personal initiative.

Hence the need to find more complex solutions able to adapt realistically to specific situations, and to adopt integrated policies with objectives not limited to single sectors. This is clearly not an easy task and will prove a hard test not only for those countries which until now have had only a fragmented approach to urban policy, but also those with consolidated policy traditions.

The processes and circumstances described above, which imply the need for some form of city and metropolitan management, bring us back to the whole question of planning. They also imply the need to re-examine the question of which conceptual tools and methodologies are required to formulate such plans and strategic projects. It is reasonable to conclude that the current situation of uncertainty in the relationship between the processes of planning and the urban sciences does not mean that the latter are irrelevant. It is simply a symptom of the need to redefine the relationship, radically revising both the policy and scientific assumptions on which it was based in the past. This certainly involves numerous difficulties and will involve experimentation with numerous different paths, without knowing in advance which will prove most fruitful. Nevertheless, we can already state that this experimentation is not only necessary, but is a task of great theoretical and practical relevance.

1.3 Some Questions and Initial Replies Emerging from the Seminar

The seminar and this volume of proceedings have made it possible to attempt a reply to the question: is the hypothesis presented in 1.1 verified?

In our view, they have not only made it possible to reply to this question, but have also highlighted two other problem fields, those fields referred to in 1.2.2 and 1.2.3, which can be summarised in the following questions:

1. has the idea of complexity, which has permeated our examination of the city, become something more than an idea, is it tending to become a paradigm? What does this imply for the tools, including the more formal ones, used in the study of the city?
2. are the sciences of the city useful for the planning and definition of policies for the city?

We begin in this section by considering how the seminar and this book have verified the initial hypothesis, and in 1.4 and 1.5 we shall see whether, and to what extent, it is possible to give an initial timid reply to the two questions above.

First of all, we shall examine what has emerged in relation to the city as a highly complex entity:

- urban multidimensionality (particularly underlined by Pumain): resulting in the large number of descriptions of the city which cannot be reduced to each other (we recall, as an extreme case that made by Socco) and which, together with Casti (1986), lead us to claim that 'the city is a complex system';
- the adaptation, from the spatial point of view, of the city to different situations and its multilevel, and hierarchical, organisational structure;
- the large number of interacting decision-makers;
- the multiplicity of time dimensions of the city and their overlap (examined in detail by Lepetit and Pumain, eds., 1993), which are at the origin of the causal mechanisms in the urban field, and the problem of their identification, as such mechanisms depend on how an observer perceives the time dimension of the urban phenomenon.

These are all considerations which, together with those of La Bella, lead us to state: the city is a complex system and, at least under certain conditions, a self-organising one (a more direct specific reference to this latter quality is made by Mela and Preto).

Urban systems of this kind are not explained only on the basis of competition. The role of co-operation is of increasing importance, as claimed by Allen, (in accord with the studies carried out by the Santa Fe Institute). It is an observation whose significance, as stressed by Tinacci Mossello, goes well beyond the field of interest of urban analysts.

The development of urban systems is a process which is neither completely unpredictable nor a completely deterministic. It lies between these two extremes. Butera suggests that it is a guided self-organisation process, and Lucchi Basili observes that the city, before the industrial revolution, behaved as a self-organising system. It is with the accelerated growth since then the city has required a planned external control.

The planner (designer) must therefore learn to co-operate with the spontaneous processes of urban organisation with a plan which leaves the necessary space for these processes to unfold and which is, at the same time, able to guide them moment by moment, to compensate for the lack of communication which occurs when the system grows too rapidly. We return shortly to the meaning of planning that this view implies.

We should observe that the discovery of complexity, or the complexity approach, has made it possible to introduce more adequate models (illustrated by Pumain, who nevertheless expresses appropriate caution), where catastrophic processes, and hence the possibility of catastrophic jumps in the values of variables, and changes in the nature of solutions are structurally allowed.

In reply to Morin who, in 1977, stressed that the problem now is to transform the discovery of complexity into the 'complexity method', we can say that those studying urban phenomena have accepted his invitation and are proceeding in that direction, leaving behind the determinism of certain models and methodological monism, but also avoiding, as Camagni warns, the delays resulting from the attitude that 'everything goes', which could discourage, rather than promote a new thought adventure.

In the context briefly described above, what emerges is a concept of planning as a process of interactive learning (as suggested by both Faludi and Mazza) between the community of planners and the outside world. It is a process in which grand ideas play a fundamental role; a conception able to produce, as demonstrated by Rabino, a new synthesis of analysis and design.

We should like to add, as argued by Lombardo, but also with reference to the investigation in the architecture field by F. Bertuglia, that planning (or design) if correctly understood, does not involve the reduction of complexity, but increases it. Planning should widen and not narrow the range of choice. In other words, complexity becomes a quality that the plan (or design) should seek to produce.

More generally, we are seeing a shift from attempts to control the urban system to a play of interaction. We could say that evolutive projects depend on the interaction between (i) general mechanisms which operate as constraints, (ii) the variety, individuality and contingency of events, (iii) the strategies and choices adopted by decision-makers, who move between constraints and events to construct new scenarios and new possibilities. For an operator, and especially in the public sector, the achievement of an effective solution is the outcome of a long and tiring process of exploration, in which the random element of discovery coexists with the tension caused by the attempt to go beyond the limits of the existing situation.

To imagine that this way of seeing things means that the action of the operator carries less responsibility and, therefore less need for decision support tools, is a mistake. In reality, the only thing to emerge clearly is the difficulty in which the operator finds himself: the public operator in particular needs tools to support his or her decision but, at the same time, knows that these cannot offer 'the solution', sometimes not even 'a good solution'. Often they only indicate what 'not to do', indicating in some way the decision area. They can however provide appropriately organised information, useful for exploring the alternative choices and potential impacts.

Van Geenhuizen and Nijkamp, as well as Tadei and Dellasette, have focussed on information systems, underlining the links between the 'toolbox' of the urban analyst (containing dynamic nonlinear models, multicriteria methods, performance indicators, etc.) which has been gradually built up and added to over past decades, as illustrated with particular reference to transport in the paper by Cascetta.

The first question we need to ask is: what influence can the use and diffusion of information systems of this kind have on the overall methodological evolution? By methodological, we refer here to the meaning underlined by Occelli, i.e. tools whose use allows us to obtain some kind of 'information gain'.

The second question is: how does the construction, use and diffusion of information systems affect the context of application of methodologies and, more in general, the approach to urban problems? What is the nature of the relationship between information systems, methodologies, theories and society? Here we need to bear in mind the problems, raised by Occelli, concerning: (i) the numerous nonequivalent descriptions of the system, (ii) the difficulty in identifying the objectives, which are growing in number and involve increasingly complicated hierarchies, (iii) the growing number of decision units concerned and, hence, the growing conflictuality, which though maybe less radical is more widespread.

These are not rhetorical questions, and we do not have ready answers. In fact, they are questions to which we admit to being unable to give exhaustive replies. We can simply try to make a modest contribution, referring briefly to some tools of which we have some knowledge: models and performance indicators relating to spatial phenomena.

To do this, we begin at a certain distance, quoting Cini (1990): "Centuries never finish at their precise end date ... The twenty-first century has already begun, somewhere between the second half of the seventies and the first half of the eighties ... The twentieth century was the century of the working class, electricity, the dream of the future ... Suddenly the third millennium is upon us ... the working class has disappeared ... Its language, its ideology, its vision of the future and model of society exist no longer ... it is the passage from a linear way of seeing things ... to the awareness that every part of the world around us is related to all the others through a complex chain of interactions and reciprocal feedbacks ... not by chance does the end of the twentieth century ... coincide with the toppling of physics as the perfect science by biology ... And, naturally, it coincides with the explosion of informatics" (p. 111-113, our translation).

Cini stresses thus the epoch-making nature of this change, beginning with the globality and interdependence of the transformations pervading technology, economics, society, science, ethics, etc. and outlining a cyclical causal system (technology → society → science → technology) which explains the cumulative nature of the phenomenon and, hence, the explosive and radical character of the changes occurring.

It is possible that, at a deductive level, Cini's arguments are not completely convincing, nevertheless it is certain that in focussing attention on the relationship between society, science and technology, he identifies a conceptual structure able to systematically capture all the principal phenomena involved in the change taking place. Exploring this structure, beginning with the relationship between technology and science, there emerges firstly the rapid acceleration in recent years of the speed of transfer of scientific advance into technological products as well as the rising scientific value added in the products, and secondly the ability of technology to provide scientific research with increasingly sophisticated instruments for investigation and measurement.

But it also emerges that it is becoming increasingly difficult to distinguish between science and technology. Technology often has to face problems of such a vast scale and complication that the technological project is not very different from a scientific research activity. And even more mundane projects, when we consider nowadays the enormous number of people likely to be involved, tend to become original scientific problems in themselves. But science, too, is undergoing a profound transformation. Technological progress has radically changed the way in which research is done (it is sufficient to think of the automatic processing of large quantities of data), opening new fields or making it possible to deal with previously untreatable fields, such as those concerning nonlinear, discontinuous, disequilibrium and irreversible phenomena.

It is with this accumulated force of science and technology that modern society (at least the so-called 'developed countries') is confronted. It is a society already profoundly changed, but still changing, largely thanks to science and technology, in respect to:

- its demographic structure;
- the foundation of social organisation;
- and, above all, the cultural dimension.

It is a society more deeply and widely educated and informed, and for this reason more complex and differentiated. Being more educated and informed, modern society is particularly sensitive to the changes taking place and attentive to the signs, inevitably ambiguous and potentially contradictory, indicating the direction of change. Therefore, it is also a society which, reflecting on past changes and aware of the transformations now occurring, looks to the future with great expectations, but also deep uncertainty.

It is evident that the whole process of change outlined above has been reflected in the organisation of human settlements, and in particular the structure of the urban system.

We cannot go into all the implications for cities: from their sudden growth caused by the demographic boom and the increasing standard of urban living, the development of their service function linked to the transformation of activities and work, the congestion of city centres due to the massive use of the private car, and the development of networks of inter and infra urban relations due to the increase in mobility. It is worth observing, however, that the question of the relationship between technology, scientific and social innovation and the city is one of the central problems of urban studies, both at the scientific and project level (see Bertuglia, Fischer and Preto, eds., 1995, Bertuglia, Lombardo and Nijkamp, eds., 1997, Batten *et al.*, eds., 1998). At this point, we should observe that the modelling of urban systems has been particularly receptive to the changes occurring in the city under the effect of technological, social and economic factors, sometimes more so than other disciplines, by virtue of its specific language and the *forma mentis* of urban modellers, more consonent with the science and technology responsible for the changes.

The transition from urban models of the first generation, i.e. models of the Lowry type, to models of the second generation, like those of Allen *et al.* (1978), Mela, Preto and Rabino (1987), Bertuglia, Leonardi and Wilson, eds., (1990) (to limit ourselves to models by authors in this book), can be seen as the passage from models of the 20th century city to models of the 21st century city, to use the language of Cini.

Not that Lowry's model (1964) relates strictly to the 20th century city, in fact it represents a general logical scheme of the city, still largely valid. However, its usual interpretation, in relation to the causal mechanism as well as its operative implementation, represents the link between residence and place of work, for example, as a physical journey, recalling the image of the industrial city. Also linked to this conception is the use of the model for planning (the city seen as a machine which can be guided through exogenously defined actions towards the desired objectives) or the use of the model as a tool for rigorously exploring cause-effect relations.

Similarly, the second generation models are not linked exclusively to the 21st century city, but capture aspects of change towards that new type of city. For example, the model of Allen *et al.* (1978) focusses on the problem of self-organisation and irreversibility of urban dynamics, the model of Mela, Preto and Rabino (1987) underlines the large number of spatial organisation principles present in urban and metropolitan structures, the model of Bertuglia, Leonardi and Wilson (eds.) (1990) examines the synergetic effects operating in the micro-macro relations.

The second tool we wish to briefly discuss are performance indicators (which Fusco Girard and also Tadei and Dellasette have examined in this book). The construction of performance indicators from urban models needs to be based on the vision of the city underlying the model (Bertuglia, Rabino and Tadei, 1991) and in particular on those aspects which the model specifically captures. Models are able to suggest performance indicators which reflect the needs and objectives of our society, as they derive from an understanding and description of the processes occurring in economic, social and urban systems.

A Lowry type model is still therefore an excellent basis for constructing indicators associated with the working of the urban system, especially the spatial distribution of functions and, in particular, those which involve the physical movement of goods or people. Examples are indicators of accessibility and systemic indicators of effectiveness and spatial efficiency (Bertuglia and Occelli, 1995).

As evident also from Bertuglia, Clarke and Wilson (eds.) (1994), models of the second generation suggest and permit the construction of performance indicators:

- focusing on new economic and social features;
- attentive to new forms of production of goods and services and the new relations between them;
- sensitive to the effects induced by the new technologies;
- adequate in their treatment of the time dimension.

From the above, it follows that (i) the recent evolution of methodologies has been not only influenced, but made possible by development of information systems; (ii) at the same time, the recent evolution of methodologies has imposed new developments on information systems, which must incorporate models and indicators (and be based on the organised accumulation of raw and calculated data); (iii) in connection with the previous two points, the approach to urban systems has been reoriented (see Bertuglia, Rabino and Tadei, 1991, 1992). In the urban field, as in other disciplines, we are beginning to see how the techniques of data organisation and management are tending to emulate the memory process, the techniques of data processing the reasoning process, and decision techniques processes of social participation.

From the second generation of models is emerging, above all, the impossibility (in socio-economic and urban systems) of making a simple separation between the system of government (decision-makers) and the system being governed. This leads us to the new connection between analysis and design in the social, economic and urban sciences identified by Rabino. A set of indicators suggested by these models should therefore include those relating to the interrelations between the two subsystems, such as indicators of the conflict between decision-makers and of the degree to which the system can be controlled.

At this point, we can conclude by saying that the hypothesis underlying the seminar has found in the papers presented in the seminar and in this volume not only confirmation, but a great many useful additions and enrichments which have stimulated further discussion, including our own contribution in the concluding part of this section. It is perhaps unnecessary to add that we have been able for reasons of space to include only a limited number of observations.

We now pass on to the questions posed at the beginning of Section 1.3 and some first tentative replies.

1.4 The Concept of Complexity[1]

1.4.1 Introduction

Until a few decades ago, it was generally accepted, by most scientists at least, that simple systems behaved in a simple way, while complex behaviour was explained in terms of complicated causes. Currently, nonlinear dynamic theories are suggesting a different view, according to which even simple models, like that defined by May's logistic map (1976), can give rise to unpredictable behaviour and, vice versa, complicated equilibrium models can generate very simple behaviour.

In this section we discuss these concepts as well as the fundamental, and more general, concept of complexity to which it is linked, making reference to physical and social systems. We examine the transfer of methods and techniques from the natural sciences, of older origin, to the more recent social sciences. In order to gain a clearer idea of the concept of complexity, we make a brief historical review which reveals how the knowledge and representation of reality, in particular of the natural world, have evolved through successive interpretations and models. We then arrive at the heart of our discussion, where we attempt to outline a framework of the meaning of complexity, describing a number of points of view, putting forward some ideas of our own, and suggesting some future directions for research, with the aim of arriving at a more precise and efficient definition. To conclude, with reference to urban complexity, we investigate the analytical tools available and their application.

1.4.2 Physical Systems and Social Systems

The natural sciences have, over the centuries, mostly considered qualitative arguments only as a preliminary phase to theorisation and, as a consequence, have largely used the language of mathematics for the formulation of their theories. The social sciences, on the other hand, with the exception of certain sectors, such as econometrics, have in general preferred qualitative discourse.

The first attempts to describe social sciences in terms of social physics go back to the last century (Weidlich, 1991). They consisted of comparisons, more or less direct, of the behaviour of physical systems and their equations with social behaviour (for example, attempts to describe

[1] This paragraph is partly due to Franco Vaio.

the behaviour of a society by means of the state equation of pure gases, using concepts like pressure, temperature, density, etc.). But an approach of this type, the so-called physicalistic approach, in a strict sense can function only if there is a very close structural isomorphism in the interactions between the constituent elements of one type of system (elementary particles, atoms, molecules, planets, galaxies, etc.) and the other (individuals, families, firms, social groups, peoples, etc.). In general, true isomorphism does not exist, so the analogy can relate only to phenomena. The development of a description of social systems on the lines of physical systems runs into other kinds of difficulty. Firstly, there is the problem of identifying all the relevant variables of social systems, secondly that of quantifying them. More serious is the fact that social systems do not lend themselves to laboratory experiment. This explains why the approaches derived from the biological sciences, especially those inspired by the evolutionist concept, have had a greater influence. Despite the limitations posed by the use of analogy, scientific research has sometimes proceeded successfully in this way, i.e. assuming in one discipline the results obtained in another. But attempts have generally been made to break free of the constraints as soon as possible. The theory of spatial interaction is one such example.

The first mathematical formulation of the theory of spatial interaction was derived, by analogy, from Newton's theory of universal gravity. The result obtained in this way by Reilly (1931) has remained as a special case in successive formulations (see, for example, Wilson, 1970) which have since been used extensively to describe the relationships between flows (e.g. journeys to work, shopping trips etc.) and points of attraction (work places, city centres, services, etc.)

Ever since the 16th century, science has almost always tackled the description of phenomena from the reductionist point of view, isolating the various constituent elements from each other (a simple, but rigorous, definition of reductionism and holism is given in Bertuglia and Rabino, 1990). Scientific progress has derived great benefit from this approach. It is undeniable that despite being the outcome of a partial view of natural phenomena, it has led to considerable progress and permitted fundamental developments, especially in certain branches of physics. On the other hand, most of the social sciences, as well as some of the natural ones, such as biology, have adopted the view that regularities do not always exist, or at least that it is pointless always searching for them. In addition, the mathematical laws able to efficiently describe what was observed in nature were discovered empirically from the study of extremely simple dynamic systems, mainly of the gravity type, as in the case of a falling body, or the orbital movement of a two-body system. But if we consider systems with far less straightforward dynamics, for example a fluid mass in critical conditions near the point of transition to turbulence, any attempt to derive a law of motion from a series of observations fails, even when conducted on a statistical basis. As explained by McCauley, "Reductionism cannot explain *everything* mathematically, but reductionism is required in order to explain the phenomena that *can* be understood mathematically from the human perspective" (McCauley, 1977, p. 26).

During the 20th century, the social sciences sought to establish their foundations independently of the natural sciences, both because of the awareness that the individual and human society require different methods of analysis and survey from the natural sciences, and because of the conviction that the characteristics which are typically 'human' belong to a very different category from those used for the natural sciences (Weidlich, 1991).

Currently, we find ourselves in a situation where it is clear that the reductionist view is insufficient for a vast class of phenomena found in many disciplines. Increasingly, the systemic view is being taken into consideration in the contemporary approach to the natural sciences, including in physics and chemistry, in line that always adopted not only in economics and sociology, but also history and anthropology. At the moment, science is following a path in which it is trying to recapture something of the ancient holistic vision which characterised the science, if we can call it that, of Aristotle, although many elements of his vision have been superseded by the evolution of Western culture.

Moreover, it is accepted that the mathematical concepts used to describe the dynamics of complex systems are universal and therefore applicable to the social sciences, as well as physics and other natural sciences, without scientists feeling the kind of 'discomfort' mentioned previously. In this connection, Benoît Mandelbrot states that: "the variety of natural and social phenomena is infinite, while the mathematical techniques able to 'tame' them are much less numerous, so it is that phenomena without any common aspects share the same mathematical structure" (Mandelbrot, 1997, p. 16, our translation). In this sense, it is completely natural to make use of an apparatus of technical and linguistic tools common to many fields; the only problem is to do it in an appropriate way. As observed by the theoretical physicist Weidlich, who has been investigating questions connected with the quantitative description of social systems for over twenty years, "these analogies are *not* due to a *direct similarity* between physical and social systems. Instead they reflect the fact that, due to the universal applicability of certain mathematical concepts to statistical multicomponent systems, all such systems exhibit an *indirect similarity* on the macroscopic collective level, which is *independent* of their possible comparability on the microscopic level" (Weidlich, 1991, p. 5). In cases where it is difficult, or even questionable, to attempt quantitative modelling, as in the case of social systems, the analysis of laws of evolution of complex physical systems provide not only useful mathematical techniques, but also a series of qualitative concepts, analogies and metaphors.

The natural and social sciences, therefore, are able today to identify an area of comparison different from that identified in the past. It seems natural to extend to biological, economic, social, political and human systems, the interpretation of their 'movement' in terms of systems dynamics, adopting an approach from physics, the first discipline to investigate these dynamics. The physics of complex systems gives particular emphasis to the nonlinear interactions between subsystems as cause of self-organisation processes. In this framework, self-organising dynamics are essential for understanding a system, and are not just exceptional events or 'disturbances'. Self-organisation processes reveal a type of evolution which is quite different from that foreseen by classical physics: open physical systems display reversible processes (microscopic) and an irreversible evolution towards the decay of order (in the macroscopic processes). An 'organisational suggestion' from outside is elaborated by the system, which is capable of evolving towards states of different complexity. The analogy with social and human phenomena is clear: from the growth of the individual, to the development of the species, of societies, economic systems, mass phenomena, etc. Self-organisation provides an important stimulus to establish an analogy between the natural and social sciences: "social changes consist of dynamic processes of self-organisation with spontaneous formation of increasingly subtle and complicated structures. It resembles a turbulent movement of liquid, in which varied and relatively stable forms of current and whirlpools constantly influence each other. The accidental nature and the presence of structural changes like catastrophe and bifurcation which are characteristic of nonlinear systems and whose further trajectories is determined by chance make social dynamics irreversible" (Andersson and Zhang, 1997, p. 112).

Within this panorama, however, there is one aspect of the comparison between the natural and social sciences which deserves particular attention. While the natural scientist considers (or would like to consider) him or herself a detached and objective observer of reality, the social scientist participates in the socio-political system he or she is studying:

making judgements, distinguishing what is desirable from what is not, according to his or her social, political and moral beliefs[2].

In the next section, we shall present a brief historical panorama relating to the evolution of certain aspects of scientific thought: from classical physics and the birth of the first deterministic models, through various crises in the concepts of determinism, to their revision, which has led in the last few decades to the emergence of the concept of deterministic chaos. We shall try to identify the successive elements of innovation in scientific thought which have led to our current view: the concept of complexity, which brings together both the science of nature and of society. We pay particular attention to the evolution of physics, understood in a general sense, since it was the first discipline to develop as a science, and reflects most, if not all, of the developments in the scientific culture and thought encountered in other disciplines.

1.4.3 From Determinism to Deterministic Chaos in the Natural Sciences

Over the last four centuries, the natural sciences have undergone various profound changes, not only in the experimental techniques and the knowledge gradually acquired, but also in the interpretation of description which these provided of nature and natural processes.

The 17th century saw the completion of a representation of the universe which had taken over two centuries to build up (from Copernicus, through Galileo and Kepler, to Newton), which replaced the model proposed by Tolomeus. The new picture made use of a new and specially created mathematical tool: mathematical analysis (Newton and Leibniz). Galileo, in fact, claimed that mathematics is the "language in which the book of nature is written". Already with Galileo, the concept of the 'physical laws' of the universe was born, as expression of the regularity of phenomena, observed experimentally and expressed mathematically. For the whole 18th century, mathematical physics had a prodigious growth, along with mathematical analysis, which was its chief instrument (Euler, the Bernoulli) as far as Laplace and his famous statement contained in *Essai* (1814). It was the era of the triumph of determinism, with clear relations between cause and effect, space and absolute time, reversible phenomena and, at least in principle, a predictable future.

---

[2] On this point see Casti (1986), discussed in 1.4.4, and also the paper by Dendrinos in this volume, in which the author proposes a model of 'social reality-actor-observer' where tasks and functions of the various actors are defined and analysed in detail.

and, at least in principle, a predictable future.

The first signs of the forthcoming crisis in these 'happy certainties' began to appear at the beginning of the 19th century, with the appearance of non-Euclidean geometries. They overturned the two thousand years of certainty in the 'truths' of Euclidean geometry and were accepted by official science only in the second half of the century. Something similar, although in a more rapid and disruptive way, happened around 1860, with Darwin's new conceptions in biology and with the discovery and study of whole new sectors of physics (in particular, thermodynamics, cultivated initially only by people outside the academic environment), which diverged from the ideas and principles of classical physical mathematics.

The real crisis of classical science began in the last decades of the 19th century and lasted until the 1920s. The new conceptions had different origins: (i) statistical thermodynamics (which had became official science with Kelvin and completed with a statistical interpretation of entropy by Boltzmann) was the first physico/mathematical theory in which phenomena were not considered reversible, (ii) the theory of relativity of Einstein demonstrated that the geometry of the universe is not Euclidean, that space and time are strictly linked, that neither absolute space nor absolute time exist, and neither does a privileged observer, (iii) quantum mechanics, finally, questioned and excluded, at least according to the interpretation of the Copenhagen School (Bohr), for the moment that which is officially accepted, the possibility of a complete and detailed knowledge of the state of a physical system (Heisenberg's indetermination principle).

Alongside these fundamental revolutions of thought, and relatively quietly, at the end of the 19th century the idea of deterministic chaos emerged (Poincaré took up the age-old problem of celestial mechanics, involving the description of the evolution of a system of three or more bodies). It does not relate to fundamental principles recognisable in nature, nor to the type of mathematics most suitable for its description, it simply highlights how very small, even imperceptible, changes in the initial state of a dynamic system can cause increasing effects over time, making it totally impossible to make any forecast of its future state after a certain time.

After several decades in which the scientific world paid little attention, the new concept of deterministic chaos was once more investigated in the 60s (Lorenz) and even more fully in the 70s (May, Feigenbaum), both in physics, in numerous other branches of the natural sciences and, finally, in the social sciences.

Science, in other words, after the great success achieved through the development of the analytical-reductionist method, has discovered, gradually other new approaches for the description and interpretation of phenomena. The search for regularity and laws is no longer an exclusive feature of the most advanced sectors of scientific research, increasingly it is accepted that the properties of multicomponent systems (complex systems, as we shall call them) do not depend only on their structure and the laws which describe the mechanisms which act at elementary level, but derive, also in an unrepeatable way, from their individual history, that is their process of evolution.

The approach which attempts to unify various phenomena through the identification of simple common elements has led to the prodigious development of classical science. The role played by this approach is undeniable, it has been and still is fundamental, but has limits which are increasingly evident. As the only method is not sufficient, in fact it can produce misleading results along the path of search for an effective description of natural and social phenomena. Currently, there is a growing tendency to stress the fact that structurally identical systems can display completely different behaviour, as happens for example in phenomena characterised by deterministic chaos (Ruelle, 1991). There is a tendency to give epistemological priority to the categories of 'complexity', 'disorder' and 'chaos' rather than 'simplicity', 'order', and 'regularity'.

Science, at least until recently, has always tried to eliminate the need to resort to chance in explanation (or, rather, description) of phenomena. This was the Galilean tradition, to interrogate the book of nature where it presented aspects of simplicity, regularity and purity, as opposed to change, irreversibility and the irregularity which characterise the phenomena as they appear. Darwin (1859), amidst bitter hostility, was the first scientist to admit chance not as a synonym of ignorance, as Laplace maintained, but as an essential component of a scientific theory. The mix of random generation of a morphological feature and the deterministic selection of features which were more suitable for survival in existing environmental circumstances characterises the evolutionism of Darwin with respect to that of Lamarck. Science, however, for decades still resisted this vision, excluding chance and probability from its methods, witnessed by the hostility encountered by the new ideas of Boltzmann, father of the probabilistic interpretation of the second principle of thermodynamics and entropy, and one of the founders of statistical mechanics.

The essence of mathematics, since its origin in classical Greece, has always been distinguished by regularity, order and harmony. These are not however properties of the reality in which we live, which appears irregular, disordered and chaotic. Mathematics has always, as have all the sciences until fairly recently, looked beyond appearances for the aspects of regularity present in nature which are easily described with mathematical language. Only very recently has there been a reversal of this tendency, which has appeared as a revaluation of those features of natural phenomena considered of minor importance, unessential, a source of disturbance or small blemishes to the elegant aesthetic framework. As we have seen, chance, chaos and irreversibility are now considered uneliminable and general properties, while the phenomena expressible in mathematical terms characterised by continuity and regularity, are seen as particular cases of little importance.

In contrast to Euclid's conception, according to which geometry is the study of the 'perfect' forms to which nature 'tends', since the mid seventies a new geometry has developed, i.e. the geometry of fractals, which studies mechanisms generating certain irregular forms, like those found in nature. The introduction of fractals in science, and in particular mathematics, constitutes an important step in the direction of the search for a geometrical language suitable for the description of irregularities in the real world. Euclidean geometry studies abstract forms (where do we find a straight line or a circumference, if not in the mind?), whereas fractal geometry derives from real forms found in nature. The phenomena of deterministic chaos and, more in general, complexity, makes use of geometrical representations characterised by singularity at each point, like fractals, which have now become a working tool for describing forms impossible to represent in terms of regular geometrical figures, and an example of this new way of thinking.

We can give different images of reality, according to the aspects which seem most important or problematic, and which tools are retained most suitable for tackling them, including the various languages which have been developed to describe or represent these aspects. Between the abstract order of Euclid's geometry and the ungraspable nature of that which seems due to pure chance, there exists an area of irregularity which is, in a certain sense, 'regular', i.e. an area of deterministic chaos, or fractal order.

A more detailed examination of this fractal view is given in Bertuglia and Vaio (1997), where some of the difficulties encountered are pointed out and, discussing certain fractal curves, the structural incapacity of the human mind to understand everything logically is illustrated. It is in this imperfection of human logic that maybe the roots of the incomplete understanding of what we call complexity should be sought. Exploring the concept of deterministic chaos, the authors emphasise that logical categories do not allow us to calculate accurately nor to acquire all the map is given, May, 1976). This recalls in a certain sense what was said in relation to fractals. All this aims to highlight how the concept of complexity has gradually come to the attention in the course of the scientific evolution.

1.4.4 Complexity: Situation and Prospects

In this section, following the historical and methodological discussion above, we examine some aspects of the phenomenology of complexity. In particular, after having highlighted the main problem areas, we shall describe some of the fundamental characteristics of complexity, and present various interpretations. We then attempt to indicate some of the prospects for future research.

The problems posed by complexity

In recent years, in relation to the dynamic nature of complex systems, there has emerged a more general conception than deterministic chaos, of particular importance for multicomponent systems, currently referred to as complexity.

Rather than concept, we should perhaps speak of the category of complexity, since there is as yet no single or generally shared definition, nor even understanding of the concept. In the category of complexity, we find a large number of aspects typical of multicomponent (and multi-interaction) systems. In this context, the use of the term complexity is, perhaps, a little unfortunate because of its existing use in other contexts and with other meanings. We are not referring, for example, to the computational complexity of a mathematical algorithm, for which there exists precise definitions. The category includes certain phenomena which are inaccessible to linear reasoning, unpredictable and incomprehensible from an analytical point of view or through reduction to single components. They can be understood only in a synthetic vision, combining the many components and interactions between them.

The classical method, as explained in the previous section, had always sought simplicity by breaking things down into more elementary parts, and has been highly successful in describing a great many phenomena, both macroscopic and microscopic. But it was recognised that there were difficulties in describing certain macroscopic phenomena, both in the natural sciences (lasers or fluid dynamics for instance) or social sciences (macro-economic processes, psychology of masses, etc.). The conviction, going back to Descartes, that the macroscopic world is more complicated than the microscopic one, was one of the methodological principles fundamental to classical science.

This picture was upset, however, by the discovery of unpredictable difficulties also at microscopic level. The supposition of determinism, i.e. the possibility of knowing precisely and contemporarily the position and speed of a particle was overturned by quantum mechanics which, also attributing fundamental importance to the effects of interaction between observer and observed, profoundly modified the concept of measurement, and hence the meaning of the concept of knowledge.

The whole is not the sum of the parts, it is more. This does not only mean that certain phenomena or certain quantities are typically collective; that the former are not observable and the latter meaningless if considered as single components: in fact, a collective phenomenon is not always complex. As Cini said: "The concept of predictability presupposes the existence of someone, outside the system, who is able to make predictions ... It could be that he does not have sufficient information on the system to be able to predict its evolution, although this is in principle perfectly predictable ... Is it possible therefore to distinguish between unpredictability which is the fruit of ignorance of the person observing the system from that which derives from the complexity of the system? Some, referring to Laplace's tradition deny this. If this were so ... complex systems would be only complicated systems. Others, and I agree with them, sustain that the difference is substantial. The predictability of a complicated system would derive from the possibility of isolating it from the surrounding environment, and identifying those few agents significant in determining its behaviour. The unpredictability of a complex system would, on the other hand, be the consequence of the practical and theoretical impossibility of identifying the set of factors which do not influence its future evolution" (Cini, 1990, p. 93-94, our translation).

We could add that, even though the discoveries of scientific experimental research can be unexpected, or happen by chance, this does not mean that the phenomenon observed is unpredictable, since either it can be included in an existing theory, as an aspect until then not investigated, or the theory is reformulated and extended to include the phenomenon observed, or the theory substituted by a new more general theory. There may be a vast phenomenology, which comes to light gradually in particular situations, but this does not mean that there is not necessarily an interpretative framework which can account for all of these, and hence they are predictable, at least *a posteriori*. A solution to an equation may be incomprehensible in the existing framework, but in a certain sense is already present in the equations, and therefore is not unpredictable.

The deterministic conception of Descartes and Laplace is now superseded; deterministic chaos is one of the phenomena which characterises the category of complexity. The prediction of the evolution of chaotic systems is impossible, but this has nothing to do with the concept of probability. Take, for example, the unpredictable dynamics of the fluid masses of the earth's atmosphere which make the forecasting of the weather impossible more than a few days into the future: in these masses anything can happen. Even if the forces at play at the elementary level are well known and have long been studied, there are so many elements that is impossible to have perfect knowledge of the present state, which is necessary for a deterministic forecast of the future state.

What about social and economic systems then? The case of social systems is more complicated than natural ones due to the absence (or at least, less frequent presence) of regularities that could play a role similar to the role that laws play in natural sciences. Here, we cannot quantify, and often do not even know the forces at play, and it is impossible to recognise invariable or regular features which act as 'laws of motion', since neither the situations, nor the individuals which participate in them, are ever identical[3] . Even the microscopic level is poorly known: as well as uncertainties about the choices of variables and the forms of interaction, there is also uncertainty about the parameters of the equations and initial values of variables.

A typical case of complex phenomena is the evolution of financial markets. Often, the forces involved are too numerous and difficult to grasp, and the opinions of operators too variable to be able to make predictions whose validity goes beyond that provided by the identification of average trends. The technical analysis of the behaviour of financial markets aims to recognise trends *a posteriori,* by means of the identification of statistical elements (supports, resistances, moving averages etc.), with the aim of capturing signs of changes in trends before they occur (see, for example, Edwards and Magee, 1964, Plummer, 1989, Nagurney and Siokos, 1997). The basic idea is that there exist underlying mechanisms in market laws, beyond the level of the decisions of single operators, but relating to the behaviour of operators *en masse*. They are not always comprehensible mechanisms, and not always possible to

[3] "In the natural sciences, the traditional idea was to arrive at certainty associated with a deterministic description, so much so that even quantum mechanics follows this ideal. On the contrary, the notions of uncertainty, choice and risk dominate the human sciences, whether we are dealing with economics or sociology" (Prigogine, 1993, p. 5, our translation).

interpret, but nevertheless, it is assumed that they exist and can be identified through the statistical analysis of certain phenomena. It is, in other words, an example of interpretation of reality which could be read as an application to financial markets of concepts of historical materialism (Secciani, 1997). This is still a long way from scientific prediction in the Galilean sense, of the *ex ante* kind, which is fruit of the identification of something similar to a law of universal motion.

In examples such as those cited, it is often easy to realise with an *a posteriori* examination how processes observed in the past have occurred, to follow the pattern and understand how the present situation has been generated. The spectrum of possibilities in the future is too wide to be able to make forecasts, with too many possibilities which could become true[4].

Characteristics of complexity

It is not only the unpredictability, due to the imperfect knowledge of the current state, typical of deterministic chaos, which can account for complex phenomena. The mathematician, John Casti (1986, 1994) identifies some fundamental characteristics of the phenomena associated with complex systems which he considers typical of human systems and contemporary life. We cite these briefly below:

- While the phenomena typical of simple systems show predictable behaviour from knowledge of external input (or decisions) and laws of motion, conceived from the consideration of numerous similar cases, which permit the definition of invariables, the phenomena typical of complex systems cannot be arrived at through a reasoning process and have a behaviour apparently noncausal and full of surprises. In certain circumstances it can happen, for example, that a reduction of tax and interest rates leads to an increase in unemployment, or policies of rehabilitation of residential areas through the construction of low cost social housing results in worse conditions than previously, or the opening of a new road increases traffic congestion. Allen, in this volume,

---

[4] In a note to the introduction of his book "*The Dynamics of Cities*", Dendrinos observes acutely that the "Statistical verification of chaotic attractors ... in socio-spatial dynamics is not feasible at least at present. Thus, one cannot be certain that chaotic attractors ... do exist in socio-spatial dynamics, and those who try to identify the presence of strange attractors in the stock markets (presumably to make money by trading appropriately) are likely to be disappointed. Consequently, one must talk about 'model chaos', rather than 'actual chaos', particularly since one cannot be sure that what the theory of model chaos produces is what real chaos contains" (Dendrinos, 1992, note 17, p. 5).

discusses problems of this kind, making numerous references to simulations undertaken.

- Complex systems are characterised by the presence of many interactions of different kinds, and numerous feedback and feedforward cycles which allow the system to restructure, or at least modify the pattern of interactions between the variables. For example, while primitive economies based on the barter of few goods functioned according to elementary and easily understandable rules, in modern economies the path from the input of raw materials to the output of the finished product is extremely complicated and involves a huge number of interactions between various intermediate products, capital, labour, also with repercussions from the output towards the input.
- While in simple organisational systems the deciding power is, in general, concentrated in few points (oligarchy type system) in complex systems there is a wide dispersion of authority (democratic type systems). As a consequence, such systems tend to be elastic and more stable than centralised ones, with a greater capacity to resist errors or unexpected environmental fluctuations.
- While simple systems have weak interactions between components, so if some component is separated from the system it retains its own physionomy, behaving more or less as before, complex systems are difficult or even impossible to take apart. A living organism, for example, cannot be divided into parts without one or all of them losing their living characteristics. An emblematic case from the world of physics is the well-known problem of the system composed of three bodies, each in gravitational interaction. This cannot be assimilated, even approximately, as simpler two body system taken as a whole in interaction with a third body, without the dynamics of the system changing radically.
- From the characteristics cited above, we can derive two fundamental aspects of complex phenomenology: synergy and self-organisation, i.e. the capacity of complex systems (physical, economic, social, etc.) to be unstable, but to respond to stimuli from the external environment, redefining their structure and creating organised structures, even in conditions far from equilibrium, in apparently chaotic situations.

Synergy and self-organisation, understood as the capacity of a dynamic system far from equilibrium to make transitions between different regimes, have been introduced, fully discussed and proposed as central elements of the phenomenology of complex systems by Prigogine and others

(Progogine and Stengers, 1984, Nicolis and Prigogine, 1987). They draw attention to the capacity of multicomponent systems to give rise to unexpected organised structures when they find themselves in situations of nonequilibrium as a result of feedback mechanisms. In other words, the traditional view of equilibrium (stable or unstable) as the only organised form of a multicomponent system, outside which there is only instability and chaos, is substituted by a new vision in which multicomponent systems far from equilibrium are able, through a form of feedback, i.e. a return of information, or interaction with itself, to diffuse information among its components which re-organise, giving place to a new but unexpected ordered structure. The emergence of self-organising structures from individual or microscopic interactions is at the base of a vast set of theories and methods in physics, chemistry and molecular biology to explain the formation of complex structures. These methods have also been applied in other fields, such as economics, sociology and urban planning (for a review of these theories and methods see Schweitzer, ed., 1997).

While simple systems, which are predictable (in the above sense), easily describable with mathematical laws assumed to be invariable in time, follow a behaviour which is reversible, other systems of unpredictable behaviour have nonreversible evolution. This poses the question of time, or rather the arrow of time, as central to the behaviour of complex systems. To quote Prigogine: "the introduction of time in the conceptual scheme of classical science signifies immense progress. But it has impoverished the notion of time, as no distinction was made between past and future. On the contrary, in all the phenomena we perceive around us belonging to the macroscopic physics, chemistry, biology or the human sciences, the future and the past play different roles. Everywhere we find the arrow of time. However, this raises the question of how the arrow can emerge from non-time. Is the time we perceive an illusion? This is the question which leads to the 'paradox' of time" (Prigogine, 1993, p. 6, our translation). The systems we call complex self-organise and evolve in an irreversible way. It is the emergence of new types of behaviour or new structures, unpredictable *a priori* which characterise complexity (Cini, 1990).

These questions are discussed in a number of papers, in particular, those of Allen, Butera, Tinacci Mossello and Cavallaro.

Interpretations of complexity

Casti (1986) also puts forward the idea that complexity is a latent property which comes to light only when the system is in interaction with another system. It therefore characterises the observer/observed interaction (note in this idea the reappearance in some way of one of the principles of quantum physics referred to in 1.4.3).

According to this point of view, in our interaction with another system the complexity of that system becomes evident and we have the sensation that, to be able to understand it and manage it, we have to simplify. So we simplify, giving a reduced representation, a model (mental, physical, mathematical, etc.) eliminating variables, aggregating others, ignoring minor interactions, transforming into constants the slow variables, etc. Casti (1986) introduces in this connection, the concept of design complexity and control complexity. The former indicated with $C_O(S)$ is the complexity of a system S as perceived by observer O, the latter, indicated with $C_S(O)$,is the complexity of the observer O perceived by the system S. By the complexity of S or O we mean the number of nonequivalent descriptions which they can provide. The complexity of S therefore depends on O and vice versa.

Therefore, to face the problem of the management of complex social systems, we need to consider explicitly not only the design complexity (the complexity of S seen by O), but also the control complexity (the complexity of O seen by S). Casti proposes the rule $C_O(S) = C_S(O)$, according to which the best situation is that where the two complexities are equivalent In other words, it is necessary to be complex to manage complexity. Which is like saying that a great master is necessary for a great pupil, but conversely only a great pupil is able to understand a great master.

Complexity, therefore, is in imperfect eyes of the observer and is, in fact, intrinsic in things. It is in our eyes because "an important aspect of complexity is certainly our ignorance: that which we do not know perfectly, about which we have little information, which is unfamiliar, appears complex. The boundary of complexity is therefore historically mobile, it changes with changing knowledge" (Serra and Zanarini, 1986, p. 13, our translation). According to this view, we call 'complex' that which we are not able to locate in our usual, limited an imperfect structures of interpretation of reality. The boundary of the complex has continually moved in the course of scientific history and will continue, we imagine, to move in the future, following new ways of thinking, new techniques and new approaches, etc.

But complexity is also intrinsic in systems. There exist systems whose mechanisms are well known, like for example laser emission or cellular automata which, despite this, appear to produce complex phenomena. The latter, in particular, can have extremely simple rules of definition but, through self-organisation mechanisms, give rise to completely unpredictable long term evolution. The heart of the concept of complexity seems to lie in phenomena of self-organisation, which cannot be understood in an exclusively microscopic knowledge of the system. They belong to a macroscopic description which uses a limited number of variables, is based on knowledge of the microscopic interactions, but which goes beyond the microscopic level[5].

According to Casti (1986), in a complex system, characterised by a certain number of variables (or 'observables'), it is possible to distinguish different time scales: slow, intermediate and fast variables. The first are commonly called parameters, the intermediate ones independent variables (decisions, input), while the fast ones represent dependent variables, the output of the system[6]. To this is added a kind of feedback, since the fast variables influence the intermediate ones, and together with these, also the slow ones. In this framework, causality is not an intrinsic property of the system, but is a way in which the observer perceives the various working time scales of the system. In classical science this point is not usually particularly important except at extreme scales: the quantum level and the cosmological level. In the science of society, on the other hand, it is a fundamental question which, at least in part, explains the difficulties found, constructing a predictive model, in deciding what causes what.

In complex phenomenology we can distinguish three levels of description of reality (Serra and Zanarini, 1986):

- the microscopic level, at which all variables are taken into account. In the case of complex systems, a description at this level of detail is impossible, due to the number of variables;
- the mesoscopic level, at which we have few order parameters, fluctuating both due to interactions with the environment and the effect of the degrees of freedom not considered. The dynamic at this level is usually described by means of differential stochastic equations;
- the macroscopic level, at which the fluctuations are limited and are ignored. The description can be of a deterministic type. In correspondence with bifurcation points, i.e. critical situations, there are

---

[5] The two extreme positions were emblematically expressed in the two aphorisms: "you don't understand complexity" and "complexity is what you don't understand".

[6] Casti (1986) sums up the concept calling slow variables 'genotypical observables', intermediate ones 'environmental observables' and fast ones 'phenotypical observables'.

various alternatives for the future evolution. The choice between alternatives is determined by the fluctuations of the variables of the lower level, i.e. it is random.

In this framework, unlike the Cartesian conception, the descent towards the microscopic level does not guarantee in itself a continuous evolution towards greater simplicity, but neither does the macroscopic description, as chaotic systems demonstrate. The possibility of building sufficiently aggregated models of complex systems is limited to specific descriptive levels, to choices of a particular resolution in the study of the system. Despite this, scientific experience shows the possibility of constructing simple models of complex systems, i.e. to identify with few equations the dynamics of the order parameters.

Socio-economic variables have different speeds of evolution in different cultural situations. For examples, prices forms very rapidly in a market economy, while the construction of a communications infrastructure requires longer in developing countries than developed countries. Similarly, the time needed to take decisions, to create new ideas in laboratories and, in general, the speed with which information and new knowledge affect socio-economic systems vary for different peoples, nations and areas. Also the values and social norms, usually considered to be stable in daily life, are in fact changing, as shown for example by the fall of the communist ideology or the processes of secularisation.

The fact that many economic variables, such as prices, change rapidly implies that in economics it is acceptable to 'forget' the slow variables, like institutions, which become constant parameters in the short term dynamics; nevertheless, as we have said, the fast variables have a feedback effect on the slow ones, and therefore also affect the long term evolution.

In the hierarchical picture described, each level has its own complexity characteristics which are not found at other levels, especially the lower ones. The complexity of complex systems can be simplified by reducing their dimension, on an appropriate space-time scales, providing a reduced description of reality by means of a few aggregate macrovariables, i.e. order parameters, which describe the specific characteristics of a higher level. The influence of the ignored lower level variables is manifested in the form of random fluctuations of the macrovariables, causing the phenomena of unpredictability typical of complex systems. With variations in the parameters which describe the interaction of the system with its environment, both the stationary states of the system and its stability properties change.

Although the concepts of reductionism and holism, as defined for example in Bertuglia and Rabino (1990) may seem unreconcilable, according to Weidlich (1991), the holistic and reductionist vision, correctly understood, can find elements of a reciprocal integration within a single framework. Wilson (1981) affirms that whereas the objective must be substantially holistic, the methods employed may be reductionist. In other words, in a holistic approach, the properties of a complex level of organisation, not observable with an exclusively reductionist view, can be seen as the collective result, the aggregate effect of interactions between constituents belonging to a lower level of organisation. In fact, the careful study, in a reductionist approach, of the elementary interactions between the components in a multicomponent system can allow us to clarify the way in which the various levels of structure contained within each other organise themselves and interact.

In this connection, there are two principles, derived from physics, which can be easily extended to all kinds of complex system. The first is the so-called *slaving principle* (Haken, 1978), which states that the dynamics at the macroscopic level are determined by a small number of slow variables, acting at long time scales, which dominate the fast variables acting at short time scales. The second is the principle of self-consistence, which claims that each particle contributes to the formation of a collective field, under whose influence it moves. The field is the collective effect of individual particles in interaction whose states (wave functions) are in turn determined by the field[7]. In a certain sense, the principle of self-consistence expresses the cyclical compatibility between cause and effect. According to this principle, where extended to the social sciences, societies are not seen as systems of interacting individuals, systems in which "the individual members of a society contribute, via their cultural and economic activities to the generation of a general 'field' of civilisation consisting of cultural, political, religious, social and economic components" (Weidlich, 1991, p. 10). All state, religious, economic, legal, etc. institutions belong to this collective field, which can be considered as an order parameter of society seen at the macroscopic level. Vice versa, this collective field acts on individuals, integrating them in its traditions and culture, providing them with or depriving them of information, and preventing them, under

[7] In quantum physics, this field is known as Hartree-Fock's self-consistent field, formulated at the beginning of the 1930s, which states that the interaction between particles is substituted by the interaction between each particle and the collective field generated by all the particles, and hence the particle in question (see, for example, Born, 1935).

pressure of various kinds, from taking completely 'independent' decisions on questions already predetermined by society.

The subject of complexity is relatively new in the history of science and is not yet well delineated as a field of study. It does not fall within any particular field, since complex phenomena are found in many, if not all fields of natural and social science. Mathematics is one of the ways in which the abstract reasoning of the human mind is expressed, but rationality is imperfect, incomplete and not without inconsistencies. Probably, an understanding of complex phenomena needs to be linked with a clearer perception of the area of logical/mathematical thought in which reasoning fails and intuition allows us only a glimpse, an area examined in this book by Angela Spence, and illustrated in the discussion on Mandelbrot set in Bertuglia and Vaio (1997).

Prospects

For all that is said about complexity to become a real science of complexity and not remain a mere phenomenology, it is necessary to have an organic frame of reference, possibly leading to an axiomatisation of complexity, which would allow a step forward, from a simple and nonorganised series of observations to a theory. What is needed is a synthesis similar to that provided by Euclid or Newton, or even only a framework, which is much less than a theory, like that of Ptolemy, or Linneus' classification of species united with Darwin's concepts. In the history of the complexity concept, we are still in the phase of observation, like the Egyptian and Babylonian astronomers and mathematicians before Euclid and Ptolemy. We are in the 'uncomfortable' phase in which we only partly understand, because there is something we cannot yet grasp.

Probably, we need to further develop our mathematical 'tool' to make it more suitable for the description of complexity. It should be, as mathematical analysis was for physics in the 18th century, flexible enough to perceive the more indefinite aspects, and to find its way into the most hidden cracks of the elusive phenomena of complexity. Even with the help of advanced mathematics and the most modern computers, we cannot arrive at a complete description of the behaviour of certain dynamic systems in a three dimensional space, not to speak of what to do with further dimensions. Mathematical analysis, as it is, is perhaps too rigid, rather like the algebra of the Renaissance for the new physics of the 17th and 18th centuries. At that time, the concepts of the derivative, the integral and, above all, differential equations, with the technique of calculus, were invented to provide a description of the kind of motion they wanted to study in new natural philosophy, as algebra was too static. Now maybe we need something more than that which contemporary mathematical analysis can give. Differential equations have solutions: whether it is a general or singular integral, a curve or band of curves, the trajectory is there, stable or not. In theory, it is sufficient to be clever enough to find it. The problem then, as we have said, is the impossible determination of the initial values with perfect precision. A stone under the effect of a gravitational field can do only certain things, not others - the differential equation of movement is given, its behaviour at least theoretically, is given - there is no room for the kind of 'surprise' that Casti (1986) claims is typical of a many-body system.

Surprise, the appearance in certain conditions of turbulence in liquids, a phenomenon typical of self-organising systems, cannot be found in the differential equations. Darwin's mutation in biology cannot at this state of things be expressed mathematically, the disintegration of the Soviet system was not predicted in political science, nor the stock market crash of 1929, or the oil crisis of the 1970s by economic science.

We know how to recognise chaos, but not how to deal with it. We speak of strange attractors in phase space, but this is not enough. Phase space contains the strange attractor, but cannot explain its fractal structure, which is outside the usual logical framework of traditional Euclidean geometry.

Why is it that chaos appears for certain parameter values and not others? What is the meaning theoretically, not only as an empirical given, of that particular value of Feigenbaum's constant (Feigenbaum, 1978) which oversees that passage to chaos for a wide class of maps, among them the logistic map? Does it play a role similar to the number π, at home in many, even apparently very different, sectors of mathematics? Does there exist something to explain all of this, as complex numbers and functions help explain some of the properties of real numbers and functions? Is it possible to introduce surprise and mutation into mathematics? Probably, in replying to this question, developing a new mathematical paradigm, creating a new way of using reasoning and intuition, we could enter into an understanding of complex phenomena. The history of science is full of examples of the creation of paradigms. Now, we see some signs that lead us to think that we are going towards a paradigm of complexity, in a unitary vision of the sciences of nature and of society, which will allow a deeper understanding (*ex-ante* not only *ex-post)* of the dynamics of both.

1.4.5 The Complex Nature of the City

The systemic approach to the city has for long been universally accepted (see, for example, Bertuglia and La Bella, eds., 1991). But urban systems, as we have seen in this introduction, can also be considered to be complex, since they display all the characteristics of complex phenomenology we have identified, as pointed out by numerous authors in this volume. In the light of what has already been discussed so far, we now attempt to take a step forward, trying to see in what way complex phenomenology is manifested in urban systems.

The urbanisation process began thousands of years ago, and has affected all parts of the world, and has not seen interruptions. As observed by Pumain in this book, there is however no general agreement on the cause of urbanisation: among the various causes put forward are religious (see, for example, Wheatley, 1971), political (Duby, ed., 1980) and economic factors (Bairoch, 1985). In this, as in many other respects, the multidimensionality of the city is evident: the co-existence of religious, political and economic dimensions, as well as the social, cultural, geographic ones and so on. For the city, in other words, numerous nonequivalent descriptions are possible, which leads us to conclude that it is an object to which we can apply concepts typical of complex systems.

Who decides in the city? The mayor, the council, industrialists, shopkeepers, households, individuals? To a certain extent they all do, and in any case, many of them. How many and what type of interactions exist between them? The interactions are numerous, different and changing. The large number of centres of power lead us to consider the city as complex. All this can generate unpredictable changes in structure. Even looking at only the second half of this century, we have seen among other changes the transformation from small town centre shops, to large suburban shops, then giant out-of-town stores, from large industrial Fordist factories in urban areas to small post-Fordist firms scattered in various nonurban locations, not to speak of the decline, in Europe, of the population of most large cities, after many decades of growth. We should observe, however, that the bifurcations we find in a city most often involve a single kind of activity, never the entire structure of the city. Probably, to identify bifurcations involving a whole city, we would need to observe a far longer time period.

Where does the city end and the country begin? Are there cities which do not belong to city systems? Is it possible to assign a city with certainty to one system rather than another? Do we know for sure if an agglomeration is a city or a system of cites? Distinguishing between systems and networks of cities, are we referring to different agglomerations or underlining different aspects of a same kind of agglomeration, or even the same agglomeration? Similar kinds of problem are encountered moving in both directions, also passing from the city to smaller agglomeration, showing behaviour similar to fractal structures. We can conclude that as cities are objects to which we can apply the concepts typical of fractal geometry, they are therefore complex.

How many and what are the time scales of the city? They are numerous, and vary according to what we are observing. Journey to work times vary rapidly and adapt quickly to variations in the overall residence-workplace pattern, which changes more slowly. The inevitable consequence is the generation of chronic congestion on the communications network. The life-span of buildings is much longer than the duration of the functions they were built to house, the inevitable consequence is the emergence of contrasts between the 'container' and its function, hence the phenomena of overcrowding, of building disuse and the resulting degradation of the environment (for a more detailed treatment, see Lepetit and Pumain, eds., 1993). The city would therefore appear to be complex also due to the way observers perceive the different orders of time which are behind the causal mechanisms.

The new view of the city is that of a dynamic system not in equilibrium, which seems to be moving towards an ever increasing complexity and therefore, in line with the vision of complexity *à la* Prigogine, may give rise to unpredictable phenomena of self-organisation. The same mechanisms, the same rules of behaviour and the same principles of planning can lead to different, and sometimes unpredictable, structures according to the situation of the specific city and the phase of evolution through which it is passing. The same action (whether an external perturbation or a change generated within the system) may have no effect if the city is in a dynamic stable trajectory, but can profoundly affect structure if it occurs at a moment of instability (see the paper in this volume by Pumain). All of this constitutes a further element of complexity.

The most troublesome aspect of this view of dynamic systems, and hence also of urban systems, is the fact that although the state of the system at a given moment contains the previous trajectory, the fluctuations of the system mean that it is impossible to identify the initial conditions which lead to a given state. This implies that it is impossible, as a theoretical a priori, to make exact predictions. It does not mean however that an analysis of the dynamic behaviour of the system and of its sensitivity to changes in the parameter values does not permit the exploration of possible future states. As we have said already, in the social sciences, it is not useful to always search for regularities. Nevertheless, as frequently pointed out in the papers in this volume, the need to take decisions even in conditions of incomplete knowledge, and hence without being able to make reliable predictions, gives importance to the role of modelling as a way of identifying possible future scenarios.

1.4.6 Analytical Tools for Urban Complexity

As we have emphasised in the previous sections, the issue of complexity is wide open and still far from any coherent formalisation (or an eventual axiomatisation!). There are many possibilities for future developments, relating both to the theory and applications. The city would appear to be one of the most promising fields of applied research, given the evidence reported in 1.4.5 of complex phenomenology. Various aspect of this are examined in the papers in this volume, presenting different points of view, both in relation to their interpretation, and understanding of the concept of complexity itself. The 'paradigm' of complexity is, in other words, a long way from any agreed definition which would make it effectively a paradigm.

Since the mid seventies many novelties have appeared in the analytical tools used for urban research. We could say that a new phase of construction, based on the dynamics of open systems, using catastrophe theory and bifurcations, synergy, self-organisation and chaos theory, has taken shape as a response to the great demand for new theories on the formation and evolution of cities, and on how to control, or at least influence, their evolution (Batty, 1994). New types of modelling, such as systems of nonlinear differential equations, models of competition between species (and more generally ecological models), micro-simulation, cellular automata, neural networks and fractals, have been investigated with the aim of providing appropriate operative tools. There have been some reviews of these theories and models (see, for example, Bertuglia and La Bella, eds., 1991, Batty, 1994, Wegener, 1994), and we should like to stress the importance of making rigorous assessments of the use of these new conceptual and operative tools for tackling urban complexity, such as that undertaken by Pumain in this volume. It is this connection that we should like to add a few observations, without any claim to being exhaustive (in fact excluding this explicitly), limiting ourselves to a series of brief comments on some of the main 'families' of models and concepts which have inspired analytical approaches.

The capacity of systems of nonlinear differential equations to produce a great variety of behaviours and structures (which, deriving from the same equations, derive also from the same type of mechanism, obtained simply by changing the values of certain parameters) has been used in many types of urban model. The basis of the success of such models is that this reflects the fact, already mentioned in 1.4.5, that similar mechanisms, rules of behaviour and planning principles applied in different cities lead to different structures. The equations of models of this type admit many solutions or multiple dynamic equilibria, and present different trajectories corresponding to structures which are qualitatively different. The system may be pushed towards one trajectory or another, towards one form of organisation or another, by the amplification of a small fluctuation. The problem which remains is that of identifying the dynamics which produces a particular sequence of observable structures, i.e. a specific trajectory (Prigogine and Stengers, 1979).

Catastrophe theory (Thom, 1972, 1974) provides some powerful tools of analysis, but to describe the system it considers too few variables and parameters to provide real help in the study of most urban problems. Despite these limitations, there have been worthwhile attempts. The most interesting feature of these models is that they allow a good theoretical understanding of the discontinuous behaviour of certain aggregate variables. It should be said, however, that for the study of the city, some of the most stimulating aspects concern general ideas deriving from catastrophe theory, rather than the models.

The well-known prey-predator model of Volterra-Lotka (Lotka, 1925, Volterra, 1926), describing the competitive evolution of two biological species, has given rise to generalisations (the so-called ecological models) which study the evolution of two interacting populations. This has inspired the construction of models which investigate, for example, the competition for urban space between two types of population for two types of urban use, the cyclical nature of the growth of the centre and suburbs, and urban hierarchies. Urban models deriving from the Volterra-Lotka model are interesting both because of the variety of dynamics they generate from relatively simple equations and because the trajectories and stable states can be calculated analytically. On the other hand, the interpretation to give these parameters is not completely clear and, in any case, spatial interaction does not appear explicitly in these models.

The master equation method (Weidlich and Haag, 1983), deriving from the theoretical principles of synergetics, concerning relations of dependence - and in particular co-operation - between the actors of a system, make it possible to explicitly link the probability of transitions of state at the micro level with the evolution of certain variables which describe the overall structure of some aspect of it at the macro level. It allows the connection of the spatial behaviour of single individuals with the global dynamics of the population, measured through appropriate aggregate variables. The master equation gives the variation over time of the probabilities of possible configurations in space of the state variables. This stochastic formulation is used to obtain a deterministic equation describing the evolution of average values, which then make it possible to make an estimate of the parameters. This method has been used to study, for example, migrations of population, changes in urban hierarchies in relation to the dynamics of urban services, the relationship between movements of population and land values in urban areas, the impact of new communications technology on the structure of interactions in the urban field, and has been used also in comprehensive urban models.

Comprehensive urban models, made up of a large number of state variables and the interactions between them, are more realistic than those referred to above. They focus on the simulation of possible changes in urban areas, and hence the production of a great variety of spatial structures, without paying much attention to the calculation of analytical solutions for stable equilibrium. In general they describe changes in the location of activities and population, and the interactions between them in an urban area divided into many zones. This evolution is induced by both exogenous factors (especially in connection with economic/production mechanisms) and endogenous factors (above all, those associated with residences and services). In models of this kind which deal with the explicit links between the values of certain parameters and urban form (understood as locations interrelated by flows) and having identified the various bifurcations which may appear, they are able to highlight the variety of possible urban forms and explain how a given city's unique trajectories (we could say, in a certain sense, its urban history) can be derived from a single general process of evolution. In fact, in her paper Pumain goes so far as to say that a general theory of the mechanisms governing the evolution of urban structures no longer seems an impossible objective.

Having said this, we should add that, at the operative level, the identification of bifurcations can pose problems. To give just two examples: firstly, it is not always easy to state that two structures are different, and that the passage from one to the other is generated by a bifurcation, secondly, what in a short period may seem like a jump from one trajectory to another, may prove over a longer period to be only a fluctuation in a given trajectory. As Pumain observes, modelling experiments tend to produce far more bifurcations than those found in reality.

'Micro' models originate from the interpretation of the behaviour of single individuals, and both those which make use of the micro-economic approach and those based on random utility are able to generate a large variety of urban forms from different assumptions and constraints relating to the behaviour of urban actors.

Given that the problem is always to generate a variety of urban forms, it is necessary to consider carefully the attempts made with urban micro-simulation, cellular automata, neural networks and fractals, even if they do not, unlike the models described above, have a theoretical foundation.

Under certain conditions, some mathematical urban models can produce chaotic behaviour. But, until now, it has generally been seen that the parameter values at which chaos appears are a long way from values found in a real urban system. Pumain in fact suggests that rather than asking "are our cities chaotic?" we should perhaps ask "why are our cities not more chaotic?". In other words, we should be investigating the mechanisms which prevent chaotic behaviour arising in our cities. Naturally, we make this observation not to discourage interest in this kind of model, but simply to establish the point reached at present.

As can be gathered from these brief observations, the 'new wave' of construction of tools for urban analysis is varied and stimulating, even though the results obtained until now have not always been completely satisfactory. We are still, it should be remembered at an early stage. We wish to point out that, over and above the new types of modelling, which require improvement and, above all, an extensive experimentation programme, the new ideas emerging have already proved to be fruitful. They have profoundly influenced urban analysts, and are beginning to influence the institutional actors in the urban field.

The above relates to the way of conceiving the tools, but we should add, with reference to their use, that the current developments indicate an opportunity to definitively adopt the scenario analysis approach, freeing us from any of the negative aspects remaining from the approach involving the predictive use of models. For this purpose, it is necessary to move the central axis of the approach from a focus on quantitative variables to a focus on the qualitative images, from the emphasis on details to emphasis on trends, from results based on the consideration of the status quo to results based on the influence of future patterns, from a path proceeding from present to future, to a path going from future to present, from a closed future to an open future, from statistical/econometric tests to plausible reasoning, and instead of proceeding from simple to complex, to proceed from complex to simple. Finally, instead of passing from the quantitative to the qualitative, we need to pass from the qualitative to the quantitative (see the paper in this volume by Van Geenhuizen and Nijkamp).

In conclusion, this means (i) integrating the modelling tools developed since the mid seventies within the scenario approach (see, among others, Bertuglia *et al.*, eds., 1987, Bertuglia, Leonardi and Wilson, eds., 1990, Batty, 1994, Wegener, 1994, Pumain, in this volume) in order to test the dynamic robustness of the qualitative changes in the system and (ii) to combine the above with the use of evaluation tools (see, among others, Nijkamp, Rietveld, Voogd, 1985, Bertuglia, Rabino and Tadei, 1991, Bertuglia, Clarke and Wilson, eds., 1994, Van Geenhuizen and Nijkamp, in this volume) with the aim of selecting the general policies to apply in practice.

1.5 The Planning of the City

1.5.1 Some Definitions

Reasoning about the planning of the city (or programming or design of the city - the question of terminology will be taken up later) means reasoning about the practical importance of the sciences of the city (or, more generally, the regional sciences, and here in particular, those which make reference to the concept of complexity), i.e. their use, but also their *usability*.

It is impossible to avoid the question: how does the regional science point of view of city planning express itself? The immediate reply would be obvious: it focuses on the normative approach, that is typical of the regional sciences, which starting from theoretical assumptions frequently imported from other disciplines, develop their theories as generalisations of problem-solving.

To go more deeply into this reply is, however, far from easy. This is witnessed by the world-wide debate on Regional Science, which is able to explain what is happening in a traditional context, but is finds itself in embarrassment *vis-à-vis* the discontinuities resulting, for example, from processes of innovation or the environmental question, which put in difficulty its forecasting and normative apparatus. It is immediately evident that it is a field which cannot be treated hurriedly. Therefore, before proceeding, and in order to avoid both repetition and possible misunderstanding, it would be useful to clear the field of various semantic tangles which infest it.

Firstly, some conventions concerning the terminology used here will be established. In the rest of this section, the terms *system*, *government*, *plan*, and *model* will be used according to the definitions proposed below (even if some discrepancies, in general a certain number of additional specifications, might be found with respect to their use in previous sections).

System: a spatially defined socio-demographic-economic entity (not necessarily institutional), which possesses an (approximate) systemic connotation. In Europe this may range from the region (Land or other similar areas) to the commune, including other intermediate areas such as the province, department, county, etc. The ideal reference is the socio-economic-territorial 'system' in the real sense, i.e. the 'functional region', 'urban system' or 'metropolitan region'. So the term *system* will be used here in an all-inclusive sense.

Government: the political authority (elective assemble or executive board) which governs the system. This may not always have a strong institutional definition, as was the case in Italy in the past for the regional committees for economic planning, and is the case today in Europe for certain commissions and committees responsible for plans of various scales, as well as authorities and consortia for specific sectors (health, transport, water, etc.). The government is responsible for *functions,* which can be divided into *knowledge* (analysis), *decision-making* (choices), *implementation* (management) and *control* (monitoring) functions. The decision, in terms of contents, is expressed in *policies* (general or sectorial; short, medium or long term; for the whole system or parts of it, etc.) over which the government exercises functions of implementation and control.

Plan: the form of policies produced by the decisions, when there is a recognisable minimum, even implicit, *presupposition* deriving from the exercise of the knowledge function and a minimum, even implicit, *prospect* of the function of control. In this sense, the plan is the typical product of the public decision-making process (in which the reference community participates to some degree). In Europe this includes the full range of public decisions, from regional/local development plan to the regional/local plan for technological innovation; from local or strategic plan for a city to the transport plan, as well as the regulatory decisions (on tourism, retailing, services, etc.). The ideal reference is, however, the Plan (regional development programme or Town Plan). So the term *plan* will e used here in an all-inclusive.

Model: any (non banal) technique which produces information or an ordering of information on which the decision (top-down, bottom-up, concerted, etc.) may be based (rationally, maybe discretionally, but not arbitrarily), and also useful for the preceding analytical or successive control stages. The technique is not banal when it admits the possibility of analysis, forecasting (in the sense specified in 1.4) and evaluation of the behaviour of the system (or its parts) or of the effects of policies on the system (or its parts). The definition therefore includes the whole range of techniques from the extrapolation of demographic levels and structure of the system to economic/territorial multi-sectorial models (macro-economic model + structural input/output interdependencies + spatial interaction) for analysis and forecasting, and from socio-economic indicators to multicriteria analysis, including cost/benefit analysis for evaluation. To the extent to which a technique allows analysis, forecasting or evaluation of the type explained above, it is a representation ('model') of the system, with greater or lesser refinement. The ideal reference is, however, the 'model' in the real sense (which may evolve from simple forms of sectorial, territorial or behavioural specifications to more advanced forms). So the term *mdel* wil be aplied here in an al-inclusive sense.

1.5.2 Knowledge and Action: a Real Dichotomy?

Knowability vs. knowledge

Having established the linguistic conventions, we shall now look at some more strictly semantic questions, beginning with the term which is central to our argument: planning.

We accept, together with Faludi in this volume, the definition of planning given by Friedmann (1993, p. 482) "a professional practice that specifically seeks to connect forms of knowledge with forms of action in the public domain" It follows that, as Faludi adds: "planning thought must focus on the relationship between research and design, on the knowledge-action nexus" (p. 521). This nexus is so critical that Faludi uses it as an interpretative canon for the whole history of Dutch planning (one of the most prestigious). He distinguishes a 'classic period', lasting until the end of the 1950s, in which the nexus was simply ignored ("the problem was to obtain knowledge, not how to translate it into action, which was deemed unproblematic" p. 521, the assumption was that after assembling a comprehensive set of data, the requisite course of action would become evident) and a 'modern period', from the 1960s to the present day in which, on the contrary, the focus was on the knowledge-action nexus, which became "the hallmark of modern planning thought" (p. 522), to the point where planning is understood as a learning process: "Attention to how the knowledge-action nexus is being framed is a precondition of planning becoming self-conscious and self-critical" (p. 522). The reasoning is persuasive and reinforced by Faludi's authority. We remain convinced, however, that the theme requires deeper discussion, concerning not only the nexus, but also the associated terms: knowledge and action.

As to knowledge, we know that three forms are recognised (Bara, 1990):

a. explicit knowledge, i.e. the knowledge which we are aware of knowing; the substance of scientific knowledge, formalised or formalisable through paradigms of logic, transmittable through protocols;
b. tacit knowledge, i.e. knowing how to do something, its representation is expressed through procedures and 'learned by doing';
c. modellistic knowledge, which integrates the other two forms, expressed through components and relations; this is the typical form of knowledge used, implicitly or explicitly, in action.

But, as perceptively observed by Rabino (in this book), human action, and therefore necessarily planning, "*inevitably* involves merging the two types of knowledge: tacit (which is the essential, but not exclusive, basis of artistic creation) and explicit (which is specific, but not exclusive, to scientific research)" (p. 583). Nevertheless "society, in its growing cosmo-creativity and consequently *diffused and increasingly aware participation* in planning processes ... necessarily requires explicit (transmittable) knowledge" (p. 583). However, and this is a first tentative conclusion, the possibility of acquiring, accumulating and transmitting explicit knowledge would seem to depend in general, and above all for planning, on the degree of 'knowability' of the object concerned. The accent moves therefore from *knowledge* to *knowability*.

Capacity to act vs. action

The concept of action has no single definition either. We refer here not so much to the contents (see, above, the breadth of the notion of the plan), as to the constraints by which the performance of an action is bound. These constraints may be: institutional, i.e. determined by the powers of the agents; imposed by the context, depending for example on the number of actors involved in the decision-making process; linked to the 'reactivity' or degree to which the object of the action can be modified (in our case, the city, with its well-known characteristics of inertia, irreversibility, etc.). This last aspect is the one which until now has been least explored.

Friedmann's nexus between knowledge and action should now be relocated in a new position between the *knowability* of the object and the *capacity to act* of the subject. Here, it is useful to refer to a diagram, originally suggested by Camagni (1988) and Mela and Preto (1990), reproposed in this volume by Rabino (p. 585), which connects these two aspects (see Fig. 1).

These two variables define the role of knowledge and action in the plan. The first expresses the degree of knowability of the system to be planned, and varies from one extreme at which the system is considered an objective reality, hence perfectly knowable by means of theories and strong interpretations (realism), to the other extreme which considers the object inaccessible to explicit knowledge and therefore expressible only through metaphors, analogies and narrations (nominalism). There are of course many intermediate positions between these two extremes. The second variable expresses the capacity of the agent to act, i.e. his power of command over the actors involved in the decision process. At one extreme, all the actors are autonomous, and the action is entrusted to market mechanisms. At the other extreme, agreement or hierarchy allow actors to make joint decisions in planning the system (the dirigiste approach). Here, too, there are obviously numerous intermediate forms of agreement between actors.

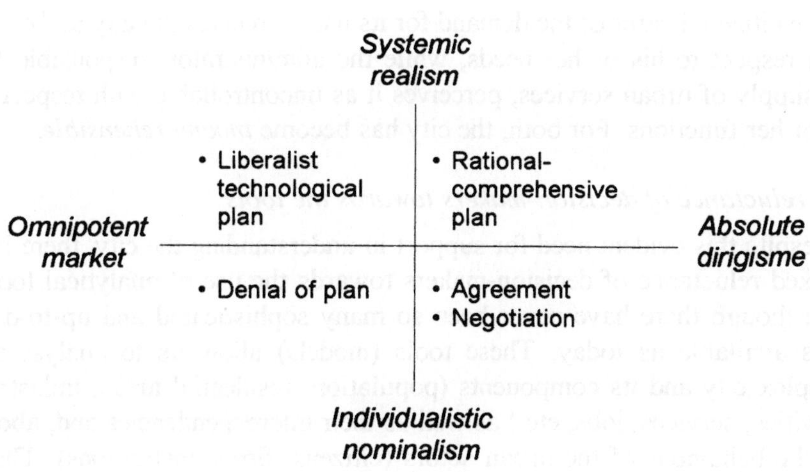

Fig. 1 Knowability, capacity to act, and form of plan

The diagram makes it possible to represent the main approaches to urban planning (or, more generally, programming) as combinations of the two variables (with the usual *caveat* regarding intermediate positions). We therefore go from total negation of the plan to the technocratic plan, from the comprehensive plan to the negotiated or concerted plan.

From theory to practice

This is the theory. Let us now put ourselves in the position of the man in the street and of a city administrator (local politician). The former, frustrated by traffic jams, poisoned by polluted air and afraid to cross his neighbourhood at night, sees the city as an increasingly unliveable place. The latter, continually inundated with requests from the public for more efficient transport and from shopkeepers who oppose traffic restrictions, caught between those who protest about industrial pollution and workers defending their jobs, trying to fight the petty criminality which makes urban life insecure, but not wanting to respond with a state of siege, feels the city to be more and more ungovernable. "We only have to look around us to recognise the existence of dramatic problems, including pollution, traffic, decay, criminality and, more in general, the violence which, although it spares no-one, strikes with untold cruelty the loneliness of the elderly and the impotence of the 'new poor' " (Bertuglia and La Bella, 1991, p. 19-20, our translation).

The citizen, bearer of the demand for its use, perceives the city as *hostile* with respect to his or her needs, while the administrator, responsible for the supply of urban services, perceives it as uncontrollable with respect to his or her functions. For both, the city has become *incomprehensible.*

The reluctance of decision-makers towards the tools

Despite this evident need for support in understanding the city, there is a marked reluctance of decision-makers towards the use of analytical tools, even though there have never been so many sophisticated and up-to-date tools available as today. These tools (models) allow us to analyse the complex city and its components (population, residential areas, industrial activities, services, jobs, etc.) as well as their interdependencies and, above all, the behaviour of the urban actors (citizens, firms, institutions). There are, in addition, refined tools (models, indicators, evaluation methods) which permit reliable predictions of the possible effects of these actors' behaviour and, therefore, help to resolve the problems of equity and effectiveness in the public sector and efficiency in the private sector. These tools have recently made considerable progress, both from the theoretical and applicative points of view, passing from deterministic to probabilistic approaches, and finally to a conception of the city as a living entity (see the papers in this volume, also Bertuglia *et al.*, eds., 1987, Bertuglia, Leonardi and Wilson, eds., 1990, Bertuglia, Clarke and Wilson, eds., 1994, and the exhaustive bibliography provided in these books). Why then is this wide range of tools of analysis and decision support so under-used, in general, but especially in Italy. This is the question.

In 1980, one of the editors presented a brief critical review of the experience of regional planning in Italy (Bianchi, 1982), and ten years later attempted to indicate a viable path to an effective application of regional plans (Bianchi, 1993). Between these two dates, he continued to search in Italy, and in other European countries (Bianchi, 1988, 1992) - with tenacity, or maybe we should say stubborn persistence! - signs of regional planning. He came across frequent stories of failure and occasionally of success. In fact, all the analyses of the European experience of regional and local planning undertaken in the last twenty years have been unanimous in recognising its theoretical weakness, insufficient technical base and low effectiveness.

In particular, the following shortcomings have been identified:

- considerable differences between the various planning operations in terms of forms and contents (Barras and Broadbent, 1979, Bianchi, 1982, 1988, Williams, ed., 1984);
- a widespread lack of the formal requirements of planning, i.e. quantification of objectives, coherent ranking of priorities, generation of strategies, formal evaluation of alternative strategies, formal and quantified monitoring procedures (Bianchi, Johansson and Snickars, 1984);
- a sort of 'divorce' between decision processes and techniques of decision-making (Bianchi and Magnani, 1985), despite the existence of appropriate methodologies.

It is worth investigating the reasons for this surprising divorce between the domain of 'rational knowledge' and that of planning decisions. A first possible reply could relate (at least in Italy, but also in other parts of Southern Europe) to the fact that the average politician or public decision-maker is reluctant to apply quantitative approaches in the field of planning. In Italy, this is reinforced by the general reticence of most consultants, despite the rarity in Italy of modelling approaches.

It could also be conjectured that the reluctance to use models is due to an unreasonable but understandable fear of the decision-maker of not being able to dominate their complex implications, and hence risking a sort of expropriation. (On this line of reasoning, see Pumain in this volume, p. 354, when she speaks of "apparent lack of control by the user on the results of the models ... The models demand a real effort of confidence from the decision-makers, who firstly have to believe in the specification of the model ... Secondly, it is very difficult for actors to admit that their action very often may not be decisive".) But this is doubly false. Firstly, for the simple reason that models do not decide, they help to decide. Secondly, for an obvious reason: how can one master the complexity of urban phenomena with only common sense and intuition?

In any case, it has for some time no longer been acceptable for the blame to be laid on the techniques. The techniques exist and have proved reliable in experimental tests, they have become more flexible and realistic, the operative costs are being rapidly reduced (due to the falling costs of computing in general, and because the models themselves are generally portable, i.e. they, or, at least their methodological framework, can be transferred from one application to another). Undeniable proof of this is found in the papers collected in this volume and the exhaustive bibliographies accompanying them.

The faults of the tools

Neglected space. It is impossible, nevertheless, to accept the hypothesis of total 'innocence' of the techniques. There are no excuses, for example, for the delay with which economics began to incorporate, and not without reticence, spatial variability in economic processes. Today, it seems a banal statement of common sense to say that the economic and social reality is differentiated in time and space. But, though the time dimension is an integral part of the very concept of economic development (and, as such, was present in the oldest theoretical precepts of economics, if only because of the evident reality of cycles of prosperity and depression), more than half a century has had to pass since *The Wealth of Nations* before awareness of the spatial variability of development was affirmed.

Certainly, von Thünen published his *Isolierte Staat* between the 1820s and 1860s; between the beginning of this century and the 1940s, first Weber, then Christaller and Lösch reproposed, with innovative and penetrating analysis, the theme of location of economic activities, repairing the fracture between space and economic theory. But we had to wait for Isard with his 'regional science', and the middle of this century, before the spatial variability of development was fully integrated in the theoretical domain (although it long remained 'marginalised' with respect to mainstream economic science).

Various explanations (not all equally convincing) have been given for this curious divorce: the intrinsic aspatial nature of classical economic theory (even in the Marxian lessons), the tendency of the elegant neo-classical constructions to see the imperfections of reality as irrelevant, etc. It has even been suggested, perhaps not completely without foundation, that the fact that the basic texts on spatial economic were written and published in German may have prevented their widespread circulation!

However, the reluctance of economic science to tackle spatial phenomena continued even when there were already appropriate conceptual and analytical tools available. Until recently, not even the disturbing dualism between development and underdevelopment, the cause of marked differences between the economies of different countries and their internal structure, served to advance the perception of the spatial multiplicity of socio-economic change. Both classical and neo-classical approaches, united in their faith in the 'single mechanism', glossed over these differences - the latter in the conviction that underdevelopment was 'not yet development' and the former, initially at least, with the idea that development and underdevelopment were mutually essential, one a function of the other.

An example of the dominance of the paradigms of 'standard economics' is given by the development of the Italian economy since the second World War. It took twenty years before the very different forms of development instigated in the so-called 'Third Italy' by the small-enterprise led industrialisation were recognised. Given the dimension and the visibility of the phenomenon, as well as the number of disciplines involved in this failure to recognise the reality, one cannot seriously put forward a hypothesis of a general syndrome of distraction. Directly to blame are the conceptual paradigms and research protocols (see Becattini and Bianchi, 1987, Bianchi, 1994).

Formal elegance, irrelevance of substance

The regional sciences (understood as a regional 'declination' of economics, geography, sociology, etc., rather than the strictly Isardian Regional Science) began to spread rapidly in Italy fifteen to twenty years ago as the result of several very different causes: the setting up of the regions, a cultural international 'hybridisation' and also, perhaps we should add, the impulse of the Italian Regional Science Association (AISRe). It is therefore not difficult, with so much evidence still open to view, to have first hand experience of the embarrassment experienced by the regional sciences, in almost all their specifications (national, disciplinary, methodological), and even in the more consolidated branches on the economics side, such as those relating to the theory of international trade, methods of Operations Research and the analysis of development and underdevelopment processes. This embarrassment depends, in our view, on the fact that the 'concrete-abstract-concrete' cycle (historically at the origin of the regional sciences) is coming unstuck. In other words, the scientific path following the 'problem-solving/theoretical generalisation/new practical application' cycle is being systematically substituted by the more academic 'hypothesis/ theorem/demonstration' procedure.

Already in 1960 Walter Isard warned that "a general theory of location and space-economy is of little direct use in treating concrete problems in reality", if not accompanied "by techniques of regional analysis which are operational - techniques which yield estimates of basic magnitudes for the space-economy and for each region of a system. These magnitudes are requisites for both the proper understanding of social problems and policy formulation" (Isard, 1960, p. VII). Thirty years later, the President of the International Regional Science Association, Rodney L. Jensen (1991), could ask where regional science is going, pointing out that "an increasing number of voices are emerging in recent years, questioning, and sometimes severely criticising, the apparently increasing trend towards mathematical and theoretical sophistication at the expense of relevance, particularly relevance in policy making" (p. 98).

We could ask ourselves whether the delay or difficulty in perceiving the change, as in the case of the 'Third Italy', and the progressive detachment from practical applications are perhaps two manifestations of the same phenomenon, i.e. two faces of a gradual 'slide' of regional sciences not so much from the practical to the theoretical, or from the concrete to the abstract, but simply, and far more dangerously, from the relevant to the irrelevant.

Search for the lost subject

We could go further and ask whether the old and new ambiguities typical of the labelling and identification procedures of urban systems, and of the region as well (sometimes we forget that the subject of study of regional sciences is the 'region'!) are an added difficulty, distinct from the sliding process referred to previously, or perhaps another expression of it, even though not an automatic consequence. We would tend, as the reader has probably already gathered, to opt for this second interpretation. Naturally, this is not the place to demonstrate the grounds, it is sufficient to show that it is not ungrounded.

The definition of the concept of region has given rise to constant debate, especially among geographers, with developments of increasing complexity, though not always of increasing clarity. But, strangely enough, the task of identifying the possible empirical entities fitting into the various definitions of the region has never been seriously attempted. It doesn't seem to have tempted even the brightest analysts, including those who have pioneered the freshest studies on regional differentiation in development, i.e. the studies which, with an *esprit de finesse,* tried to go beyond the usual emphasis on industrialisation and the measurement of development solely in terms of economic performance, to encompass socio-cultural aspects and anthropological profiles. With a notable contrast in qualitative rigour, they accepted 'ready-made' regions, from administrative ones to provinces, without showing the slightest hesitation in superimposing sophisticated interpretations on rough and nebulous territorial units.

In short, development is intrinsically differentiated not only in spatial terms (by development models and levels), but also in temporal terms (by development times and cycles), so it is normal to find regional systems at different stages of development. Despite this, there seems to be little concern about which is the appropriate territorial unit for the study of these space-time differences. Applying zoning *ex-ante,* the relationship between the given and the unknown is completely overturned. Thus, losing the eloquence of real measures of real processes in a fog of random averages, it is not surprising that the spatial differentiation of development (initially considered hastily, and not very shrewdly, as variations in a single process) has taken so long and been so reluctantly recognised. It is, in the end, quite understandable that *post festum* analyses, unable to capture diffrential specificities of development, have ended up by losing practical relevance, to the extent of having to justify, with the alibi of scientific purity, their progressive analytical and formal contortions.

If this really is the situation, as we suspect, there is an irresistible temptation to look for a single matrix uniting the delay in perceiving the multi-regionality of development, the slide form relevance to irrelevance, and the reluctance to tackle the problem of the territorial unit suitable for development analysis. So once again we are questioning the paradigms of standard analysis. In effect, the framework of economic reasoning (aspatial mechanism of capital formation, capital location according to location factors, diffusion of development from centre to periphery according to the product cycle) uses the 'region' as a toponym to provide coordinates for the phenomena. *Where* are the location factors? From where and towards where does capital relocation proceed? The spatial grid is used as a map not to *explain* where, but simply to *know* where.

The framework of regional sciences reasoning, on the other hand, concerned about measuring spatial interactions, identify and measure inter-regional transactions (for example, the people and goods which move from *x* to *y*). The intensity of the interdependencies and the differences of potential which explain the variations in intensity are measured, but again the regions serve only as addresses (where are you from? where are you going?) but are not part of the problem. The multiregionality of the analysis overlies the multiregionality of the object and hides it from view.

To be honest, it seems rather pointless to investigate the space-time differentiation of development without making an attempt to resolve the problem of identifying the appropriate territorial unit. We are reminded by Pumain (in this volume) that the problem exists and is urgent: "The limits of the city itself have become fuzzy ... It is also difficult sometimes to clearly separate a city and a network of cities ... Systems of cities are also difficult to isolate as scientific objects of study" (p. 326).

In order to approach a definition suitable for empirical implementation, there is already something more than just a mere premise: in territorial terms, the 'functional city' (especially in its incarnation as the 'daily urban system'), and in geo-economic terms, the 'local system' (and in particular the 'industrial district', when appropriate requisites - historical-cultural tradition, dimension, specialisation and integration of manufacturing units are given). At this point, it would not be impossible to put together a theoretical background with an eclectic, but coherent (and hence functional) mosaic of references to provide an explanation of the economic efficiency of systems of small firms. This could begin with Adam Smith's model of the division of labour, include the classical theories of location (from von Thünen to Isard), and take on the roles of Marshallian external economies and specialisation (Young, Stigler), especially if we escape (with Hirschman, Myrdal and Perroux) the bigoted worship of equilibrium (especially in relation to urban systems: Batty, 1993, complains that "for a hundred years or more, urban theorists have treated cities as though equilibrium were their natural condition. However, as current events increasingly demonstrate, this is less and less true", p. 14).

The path towards the identification of the territorial unit appropriate for the spatial analysis of development could start with Weber, take us past Von Bertalanffy and, maybe, Miller, to join Berry and the most recent Italian studies on the theme. To select from the family of instruments of regionalisation the algorithm least in conflict with these premises and adapt it to requirements would not seem an operation of enormous difficulty. At the moment, however, our purpose is not so much to find the *solution* to the problem of defining the appropriate territorial unit, as simply to raise the *problem.*

So, while Regional Science in the narrower sense is rapidly becoming an Olympian branch of applied mathematics, the regional sciences - without the capitals! - in a happy cross-fertilisation of economics, geography, urban science, systems analysis, etc. with a fruitful eclecticism of strong theories and weak concepts, are doing their best, despite mixed success and frequent crises of identity, to tackle the problems, large and small, old and new, of our times.

1.5.3 Knowledge is Action

Re-examining the nexus

We now take up once again our discussion on planning (of the city, but also other kinds of planning) and the nexus between knowledge and action. Having redefined the first term as knowability and the second as the capacity to act, we should like to investigate whether there is a real separation between the two. Is it possible (logically in theory and chronologically in practice) to divide the moment of knowledge from that of action, especially in the light of the concept of complexity? The distinction goes back to the classical separation between science and technology which, as we began to see in 1.3, is less and less convincing. It has been upset by the reversibility of the arrow: instead of 'invention/discovery → application' we find with growing frequency 'need/objective → invention/discovery'.

If the city, as we believe, is a system with self-organising capacity, the conoscitive procedure implies a point of view (statement of preference) of the observer who wants to reconstruct the processes of learning, self-elaboration and change.

Symmetrically, every action not only induces reactions from the self-organising capacity of the system which throw new light on its behavioural properties, but also changes the knowledge framework of the actors of the system, in the sense that it influences their preferences and/or interests. In addition, in a participatory (concertative) planning approach, the knowledge (explicit and transmittable) about the structure, its reactive mechanisms, the tendencies and, above all, the possible effects or impacts of possible actions, influences the expectations and, therefore, the behaviour of the actors.

For example, simplifying a great deal, we can assume that the relationship between knowledge and action, with respect to the governance of territorial systems (in its widest sense) could differ according to whether society as a whole is evolving towards a post-industrial scenario (Touraine, 1969, Bell, 1973, Gershuny, 1978) or a post-modern scenario (Lyotard, 1979, Toffler, 1980). The post-industrial scenario is represented by a society evolving without a clear break from the previous pattern of advanced industrialisation: expectations about the outcome of further scientific/technological developments are optimistic, and the 'axial principle' is control and systematic application of scientific knowledge (Bell, 1973). The post-modern scenario, on the other hand, is inspired by a distinct change in the 'system of values', towards one based on pluralism, solidarity, self-realisation, participation, decentralisation and environmental consciousness.

The post industrial scenario

In this scenario, the government functions will grow in two fields:

a. that of economics, in order to stabilise the oscillations, support and orient research, (prevalently technological); the main interlocutors of the government are organised interests (the neo-corporative tendencies of society, and consensus through the vote of political exchange);
b. that of society, in order to manage the social conflicts emerging from marginalisation (handicapped, drug addicts, immigrants, etc.) and deviance (organised and general crime, explosions of gang violence, etc.).

The main task of government is therefore to achieve economic and social stability. But from this assumption it is possible to derive two alternative approaches:

i. the abstentionistic approach: given the costs and shortcomings of public authority planning, policies of deregulation and privatisation are promoted for technical services, such as transport, and social services, such as hospitals;
ii. the interventionist approach: given the frequent failure of the market, which does not automatically guarantee the production of socially necessary or politically strategic goods and services, forms of direct intervention in the economy are promoted, such as control, share holding, even public ownership of companies.

An intermediate approach is that which achieves normative regulation through constraints, controls, incentives and disincentives to the action of market forces and society actors. The action of the government would, in any case, be constrained by the available resources, and the need for consensus, neither of which can be modified, at least in the short term.

Knowledge and its diffusion, for the purpose of consensus, are not a presupposition for action, they *are* action. In effect, if the primary task of the government is to achieve economic/social stabilisation, it will not follow teleological programmes: programming here means forecasting and, if possible, anticipating the emergence of destabilising phenomena. The principal analytical instrument is, therefore, impact analysis, carried out to avoid the need for further actions or the manifestation of exogenous events. Impact analysis requires tools able to analyse the economic, social and spatial structure of the system and its components (evolution, reactivity and diffusion tendencies).

The post-modern scenario

In the post-modern scenario, on the other hand, the government has far fewer and very different functions:

a. from the point of view of the contents, the action concentrates on two issues, promotion of the quality of life and protection of the environment;
b. from the point of view of the form, public involvement is expressed through open and flexible programmes inspired by the need for co-ordination and co-operation between social actors;
c. from the point of view of its role, the government acts essentially as a catalyst of financial, decisional and operative resources, following the principle of decentralisation, self-government, social management and participation.

The government, therefore, has no finalistic autonomy, and acts as a mediator and interpreter of the aspirations society and its communities. The government's role concentrates on:

- the exploration of possible futures, in the search for 'desirable' developments;
- obtaining the widest possible consensus for its programmes, through convincing discussion rather than the suggestions of propaganda;
- encouraging the greatest possible participation in decision-making,

including the adoption of new joint decision methods (concertation).

The tools required are not very different from those above, although there will be different criteria adopted in their use. The exploration of possible futures requires models able to produce forecasts on the structural evolution of the system, i.e. scenarios which are consistent (compatibility between the main economic variables, coherence between demographic and employment forecasts, etc.). Once again, these are tools belonging to the family of dynamic spatialised multi-sectorial models, which can be used to construct scenarios by introducing hypotheses on the spontaneous trends of the system or options concerning its desired evolution. In an application of this kind, in order to ascertain the feasibility of a policy, or identify the actions necessary to reach the desired scenario at time t_n, it is possible to trace the path from scenarios t_{n-1}, t_{n-2}, etc. up to the present.

The search for consensus through conviction presupposes, above all, the widest possible circulation of information and the maximum transparency with respect to the probable consequences for the system of the various hypotheses and options which emerge from society. This implies the need for impact analyses, which can be carried out with the same models, used this time as instruments for *ex-ante* evaluation. The forecasts are made conditional by exogenously introducing the policy options into the model (expressed in terms of modifications of the structural quantities, performance required, changes in the parameters, etc.).

Participation is facilitated by the widespread knowledge of the connection between desired options (causes) and possible results (effects). Two presuppositions indispensable for making the concertation work are the unambiguous specification of the aim of the negotiations and the *ex-ante* definition of the rules of the game.

In all the cases considered, knowledge *is* action. In other words, Einaudi's 'to know in order to deliberate' has now become 'to know is to deliberate'. Hence the increasingly frequent (and fruitful) reference to the self-organising properties of urban systems and the increasingly adopted approach to planning in terms of learning processes. So Asheim (1996) who speaks of *Industrial Districts as 'Learning Regions'* finds an echo in *European Planning Studies* (5, 1, 1997) and the special issue on 'Regional Systems and Learning Economies', while Hassink (1997) argues for 'localised industrial learning'.

1.5.4 Systemic Levels and City Planning

The importance of the meso level

Properties and self-organising activities of urban systems can be found at different levels (Pumain, in this volume): "The city as a scientific object of study should thus be conceived at various levels of spatial organisation. At the very least we need to consider the level of the *individual actors*, that of the *city itself* and the *system of cities* [our italics]. Some intermediate levels, such as the neighbourhoods within cities or some regional subset of cities may also sometimes be of interest" (p. 326).

Four levels have, therefore, been identified:

a. a micro level, which represents the choices of individual actors (individuals, firms, institutions, etc.), on which the reproductive and evolutive (involutive) dynamic of the system depends;
b. a macro level, which represents the whole system, the level at which complexity is more immediately perceived and at which the evolutive trajectory expresses itself;
c. a 'super-macro' level, which represents, if it is systemic, the metropolitan system to which the urban system belongs;
d. some intermediate levels, which constitute parts of the urban or metropolitan system.

If we move from the spatial organisation to the structure of the interdependencies within and between levels, it is useful, ignoring the super-macro and intermediate levels, to introduce a meso level, which represents the system of interdependencies between the choices of the actors at the micro level. This is a network of interdependencies not necessarily connected, nor rigid, but which nevertheless describes the link between the market and organisation *à la* Williamson (1983), i.e. the transactional relations between actors expressed in mutually conditioned choices which begin before and last after the actual moment of exchange.

The convenience of the resulting classification (macro, meso, micro levels), lies in the fact that it permits a link with the three levels of mainstream urban modelling: at the macro level, the models of urban growth and global models of urban structure; at the micro level, the models which apply principles of micro-economics to the behaviour of urban actors and the models which simulate the relations between micro behaviour and macro structures. As Pumain comments (in this volume): "Social and spatial macro-structures may be considered the product of interactions among individuals, each of them following a life-time trajectory, with probability constraints on the transitions from one state to another, through household, professional or migratory 'events'" (p. 349).

What emerges clearly is the lack of knowledge (analytical or empirical) at the meso level (if we exclude some applications of network analysis). A similar situation emerges if we attempt to relate the styles or approaches to planning identified in 1.5.2 to the three levels, combining the *knowability* of the object (the city) and the *capacity to act* of the decision-maker:

a. the rational-comprehensive approach focuses on the macro level, on the knowability and possibility of controlling the urban dynamic acting on the variables of this level;
b. the neo-free market approach trusts in market forces and, therefore, in the choices and creativity of the actors regulating the system at the micro level, denying the opportunity (or even the possibility) of a conscious regulation through planning.

However, it is the meso level which, emphasising the interaction between actors, and the numerous, changing networks of the urban milieu, offers the most promising prospects of analysis and action. This is especially so in a social context where concerted action (between social actors and institutional decision-makers) is proposed as an inevitable response to the demand for participation resulting from the increasing subjectivity of individuals and their relational networks (associations, movements, etc.). It is also one of the ways (not the only way, but maybe the main one) of coping with the complexity of urban government, and avoiding the trend towards the hierarchical simplification (of doubtful practicability in democratic or neo-corporative contexts).

Subsidiarity: a principle with many consequences

It is at this point that the theme of subsidiarity enters the scene. We refer not only to that concerning the relationship between levels of government which European policies and waves of more or less genuine federalism are making felt but, above all, to the subsidiarity relating to the relationship between institutions and social actors, i.e. between state and market, public and private.

Vis-à-vis the relative incapacity of urban and regional governments to manage either the increasing volume of functions assigned from above (an effect of the general tendency to decentralise) or the overload of demands from civil society (other than for reasons of efficiency in the allocation of resources), it is becoming essential to restrict the management sphere of public institutions to fields which cannot be entrusted, for reasons of effectiveness or equity, to civil society or the market. And, it is worth underlining, these constitute motivations which are distinct and quite distant from those of neo-free marketeers which have prevailed until now. In other words, it is necessary to add to the *territorial dimension,* which normally accompanies the principle of subsidiarity, a *functional* dimension, which could represent the most appropriate means for improving the performance of institutions (Grote, 1993, Bianchi, Grote and Pieracci, 1995).

After all, the market is only one of the possible mechanisms for governing economic and social processes (Schmitter, 1985, Thompson, ed., 1991, Streeck, 1992). In the economic sphere, as illustrated in Fig. 2, the existence of a free and non-regulated market (*spontaneous equilibrium mechanism*), as well as forms of hierarchy like the mono-oligopolistic firm (*hierarchy mechanism or external and authoritative control*) are recognised. In an intermediate position between market and hierarchy, are strategic alliances and networks of firms (*self-regulation mechanism*). In the social sphere, we find forms of solidarity between competitors, like those of the 'communitary market' of systems of small firms and industrial districts (*spontaneous equilibrium mechanism*) and the public management of the economy by the political authorities (*hierarchy mechanism*). The intermediate position is occupied by social networks which govern their own sector (*self-regulation mechanism*), almost always in relation with the structure of the spontaneous market and institutions (the case of the role played by social interests organised in sectorial associations).

REGULATION MECHANISMS	ECONOMIC SPHERE	SOCIAL SPHERE
spontaneous equilibrium	free market	common market (industrial district)
self-regulation	strategic alliances networks of firms	social networks (organised social interests)
hierarchy	mono/oligopoly	imperative planning

Fig. 2 Mechanisms of socio-economic regulation

These networks of firms and social networks (in rapid development, due to the proliferation of single issue organisations, from voluntary social work, to environmental protection, cultural activities, etc.) constitute the fabric of the meso level. At this level it is possible to direct new forms of intervention, like those known as 'network policies', i.e. actions aimed at the development of associate forms of enterprise and co-operation (even multiregional and transnational) between groups of firms. These policies have two main objectives: to overcome the isolation of the institution by reinforcing the capacity for dialogue between social actors, and to support organised social interests, freeing them from the practice of lobbying and platonic consultations, promoting them to the role of co-decision-maker, and giving them joint responsibility for the process of planning.

In effect, as stated by Millon-Delsol (1993): "the idea of subsidiarity can survive only in federalist regimes which value autonomy not only of the individual, but also of social groups (p. 4, our translation) "conferring authority and decision-making capacity to groups emerging from civil society and acting independently" (p. 87, our translation).

The decision presupposes a choice, and a choice presupposes criteria. But, today, strong criteria are lacking. The traditional points of reference - religious, moral, ideological - have become relative and no longer represent norms of behaviour. Hence to quote once again Millon-Delsol (1993) (our translation): "in the absence of unassailable ethical criteria, our contemporary entrusts himself to the reassuring arms of science" (p. 30). But a society inspired by the idea of subsidiarity "organises itself on the basis of thousands of daily decisions which owe little or nothing to science, but a great deal to individual judgement and conscience" (p. 89). Even the traditional institutions are eroded by subsidiarity, since the subjective choice of the individual is counting more and more: "public power is no longer the only holder of the real *coup d'oeil,* in the sense in which Max Weber used this expression to characterise politics" (p. 91).

Spontaneity or hierarchy? Regulation...

In the context of the meso level, as we said with reference to the post-modern scenario, the government operates essentially according to principles of self-government and participation. Having reduced its finalistic ambitions, the government plays the role of mediator and interpreter of society's aspirations, guaranteeing the institutional conditions for the dynamics of social actors and their networks. Reducing its aspirations does not mean a reduction in its role, which is simply changed, having to concentrate on the institutional conditions, i.e.

mechanisms of regulation.

And here the corporatist approach *à la* Schmitter and the theories of regulation *à la* Boyer provide useful indications for the analysis and *governance* of the meso level of urban systems with the aim of ensuring their homeostasis. Despite the repeated requiems (Schmitter, 1989), the corporatist approach is still alive and ready to render service (Schmitter and Grote, 1997): "if capitalism requires an effective mechanism to ensure an ordered competition between producers and a mutually acceptable distribution of income between capital and labour ... an active consensus can only be obtained through a systematic dialogue between organisations representing these interests" (p. 6).

The theories of regulation (Boyer, 1987, Regini and Sabel, 1989, Boyer and Saillard, 1995) have shown their continuing vitality, both in terms of theoretical developments and practical applications (Sengerberger and Loveman, 1987, Regini, 1996), even in the models of governance of the industrial district (Bortolotti, ed., 1994), if this can be considered as "a framework of social relations where the role of the market is no longer regulated spontaneously by the interdependencies of the community, but is socially conditioned by the interactions between locally organised social interests, and between these and the institutions of local government" (p. 97, our translation).

1.5.5 Relevance and Pertinence of the Sciences of the City

Pertinent tools for relevant problems

At this point we can once again put *the* question: are *our* sciences (in general, regional sciences; in particular, the sciences of the city) useful? Apart from the not always innocent interrogatives from outside, this question is also being asked within the discipline itself. One sign is the anxiousness with which regional scientists are constantly concerned about the scientific status of their discipline (Isserman, 1995). Certainly, some questions spring from the doubt that the regional sciences are tackling the important issues of today: peace, the environment, migration flows. Books, journals, seminars and professional activities show that this is not true. Other questions arise from the sting of self-critical conscience: yes, we deal with relevant problems, but is it with pertinent tools?

The reply is not easy, but there is a real commitment to improving the tools, found in *few other disciplines*. In fact, we can legitimately argue that between the dissolution into applied mathematics and the fading of research into persuasive discourse, typical of many social sciences, our sciences (the regional and urban sciences) are exercising a sane eclecticism in the effort to identify relevant problems to which to apply pertinent tools.

A substantial proof of this statement is provided by the eight thick volumes in the series 'Modern Classics in Regional Science', edited by Button and Nijkamp (1996). In the introduction to the series, the two editors compose a kind of 'manifesto' (not so much for *Regional Science*, as the for *regional sciences*), in which they draw up the following clear identikit (p. XII):

a. the pervasive transdisciplinary nature: "regional science has, in the second part of the twentieth century, become a major discipline at the edges of regional economics, economic and social geography, environmental science, economic planning, decision theory and political science";
b. research attitudes: "ranging from abstract topological equilibrium analysis to applied and very concrete regional planning issues";
c. themes which have produced significant scientific developments: "location theory, urban economics, transport studies, environmental and resource economics, spatial informatics";
d. common features in the consideration of space, though with disciplinary and thematic differences:
   - "first, space (for example the region or the city) is seen as an important medium for interactions and transaction" which includes "more recently, the transmission of knowledge, data and information";
   - "secondly, space is viewed ... as both an opportunity for economic development (e.g. the modern idea of gateway concepts and teleport strategies) and, by acting as a constraint on the synergy benefits possible through instant interaction, as an impediment to economic development";
e. the community of regional scientists can be regarded as a 'movement' since "it must be recognised that these performances are based on the grassroots work of thousands of supportive regional scientists".

A theme of major current relevance and the centre of attention of regional sciences is - in fact, it couldn't be otherwise - the impact on the city of innovative processes, and in particular those triggered by the new information technologies (NIT), which derive from the combination of telecommunications, informatics and multimedia. In a recent review (Bertuglia, Bianchi and Camagni, 1996), after pointing out that NIT is part of a more general framework of profound transformation (the tertiarisation and globalisation of the economy, the crisis of the welfare state) affecting post-industrial society (see also Amin, ed., 1994, Batten, Casti and Thord, 1995), the authors observe that the effect of NIT on urban dynamics has two distinctive roles:

a. an 'enabling' role, which appears especially at the microscopic level, involving the possibility of activating local factors and/or qualities and individual behaviour;
b. a co-evolutionary role, which relates to the macroscopic level, to the extent that dynamics of change caused by NIT in certain parts of the urban system can in turn aliment or be conditioned (inhibited) by modifications (inertia) in other parts.

The results obtained by the various approaches to the analysis of the impact of NIT on the city (economic approach, urban dynamics approach, spatial interaction approach) have made it possible to conclude that the regional sciences, combining the analysis of technological regimes with the spatial patterns of innovation, have permitted the recognition of the *territorially differentiated nature* of innovation, demonstrating that the tools available, especially the more advanced ones, are substantially valid.

As far as the prospects of regional sciences are concerned, a convincing theoretical repertory is proposed by Camagni, in this volume, in terms of priorities for research, which can be summarised as follows:

1. to reintroduce economics into the theoretical investigation of complex urban systems, concentrating attention in particular on:
   - innovative Schumpeterian type processes, which represent the real evolutive forces and the motor of structural economic change;
   - Ricardian processes linked to the spatial distribution of income as well as rent theory, to include the differentiation between parts of the city, between cities, and between city and country;
   - the Marxian power relations, i.e. the conflicts concerning the distribution of income between the various spatial entities: "If the world of complexity theory remains a nonconflictual world, its ability to interpret real phenomena will remain limited" (p. 375);
2. to reintroduce into the theoretical framework the physical dimension of the city, "easily forgotten in abstract macroeconomic or spaceless reasoning, as for example in much of the literature which sees cities as nodes in trans-territorial, physical or relational networks" (p. 376);

3. to understand better how the different characteristics of the city coexist and interact (how the built-city interacts with the city of culture, the city as a production machine with the community-city, etc.);
4. to give priority to the integration between economic aspects (allocation efficiency), social aspects (social effectiveness) and environmental aspects (environmental equity), i.e. the theme of sustainable urban development.

Still on the theme of urban sustainability (a problem of indisputable importance) an attempt has been made (Bertuglia, Bianchi and Camagni, 1996) to show that the most severe problems of European cities today lie not so much in their decay, backwardness, poverty or conflictuality, as in the irreversible and cumulative processes which find their origin in certain 'vicious circles', like that of *exclusion* (those cities excluded from the major communication networks, physical and nonphysical), *sustainability* (insufficient attention to environmental quality which results in reduced attraction and hence fewer resources available and local well-being), *social segregation* (whereas the historical city integrated differences, the contemporary city risks crystallising the segregation of differences), *loss of identity* (due to the standardisation of construction types and urban morphology).

The question of urban identity is a source of perplexity for our tools, even the more advanced ones. It of course took a semiologist to remind us that urban space has always been significant (Barthes, 1967). The *urbs* can be seen "as writing, a discourse, a text" which speaks to its inhabitants, as these "speak to their own city, from the moment they live in it, walk its streets, decodify that jungle of signs of which it is made" (Mucci, 1990, p. 27, our translation). This is the grammar and lexicon of the city: "the cypresses and marble ... an organism made up of statues, temples, gardens and houses, steps, vases, capitals, and regular open spaces" (Borges, 1984, p. 804, our translation). It is Borges, a blind man, who is able to show us the communicative values of the city.

The fact that there may be limits to the knowability of the city, due to these aspects which are resistant to explicit or model-based knowledge, should not be a cause of embarrassment. It is the artist who penetrates those meanings still impermeable to our tools. William Gibson, master of cyberpunk (*Blade Runner* style) literature, anticipates one of the possible futures, already embedded in our present-day metropolises, describing in *Neuromancer* (Gibson, 1984) the sprawl of BAMA (Boston-Atlanta Metropolitan Axis) as an agglomeration of dreary delinquency and glittering technology, under a grey sky illuminated by the neons of multinationals and obscene holograms of notorious clubs.

The nostalgia for a disappearing past permeates all the works of Daniel Pennac (see, as an example, *Monsieur Malaussène,* 1995). Belleville, the run-down Paris quarter threatened by the axe of modern building, stands as a metaphor for a city which houses a variegated population of immigrants, subproletariate, artists, down and outs, *gauchiste* intellectuals, all united by a robust solidarity.

Nearer home, the illuminist Italo Calvino succeeds in showing us *Le Città Invisibili* (The Invisible Cities) (1972, p. 75), arriving where our models cannot reach. "I have constructed in my mind a model of a city from which all possible cities can be drawn forth ... It contains everything that responds to the norm", says Kublai Khan (the structural/functionalist approach!). And Marco Polo replies: "I too have in mind a model of a city from which all others derive ... It is a city made only of exceptions, preclusions, contradictions, incongruences, and nonsenses" (and here we have the approach based on the conception of complexity and chaos).

In effect, "there is a place in the world whose name is Korogocho. In the language of the Kikuyu, who live on the fertile uplands on the slopes of Mount Kenya, Korogocho means chaos" (Berrini, 1996, p. 7). This is the 'other' city: that not modelled by our tools. Korogocho, the poorest bidonville of the many surrounding Nairobi, with "dwellings of every imaginable material - earth, wood, corrugated iron, sheets of plastic" (Berrini, 1996, p. 9).

The sciences of the city repropose the option of the plan

Having said all this, with the aim of illustrating our awareness of the limits of our approaches (even the most advanced ones) in the face of the dimensions of the city which cannot (yet) be dealt with rigorously, we can nevertheless claim that the conceptual apparatus of the regional sciences for the analysis of development does not seem radically put out by these new phenomena (with the exclusion of global processes, which put even mainstream economics in difficulty).

The five main guiding principles of urban and regional economics (agglomeration, accessibility, spatial interaction, hierarchy and competitiveness) (Camagni, 1993) still account for most of the tools effective in the analysis and interpretation of change and in providing reliable frames of reference for decision-making. Of course, it is essential to beware of the time horizons: the tools provide reasonable indications of best practice for the short term and allow the construction of scenarios as guide for action in the medium term. But the long term remains a question of speculation and does not allow the simple transposition of those tools, analytical findings and normative suggestions which are appropriate in the short and medium term (especially in times of profound structural change).

We therefore have to recognise that hard concepts borrowed from economics (rational expectations, transaction costs, etc.) as well as soft concepts borrowed from geography or sociology (embeddedness, untraded interdependencies, etc.) have been easily incorporated in the analytical 'toolbox' of the regional sciences, making a significant contribution.

There is an increasingly urgent need for a programme of in-depth investigation which, using the labels, metaphors and concepts that have proved so fruitful in going beyond the standard ideas (industrial district, system areas, *milieu innovateur*, etc.) with the appropriate theoretical and analytical tools (input-output, network analysis, sunk costs, regulation models, etc.), proposes new theoretical syntheses on spatial productive systems (Bianchi, 1997, Bramanti and Maggioni, eds., 1997).

To sum up, the regional sciences have tackled and are tackling the critical problems of our time, well beyond the threshold of their traditional fields of enquiry. While some of the theoretical and analytical tools are still doing valuable service, others are clearly obsolete, and it is increasingly evident that there is a need for tools able to deal with the new phenomena. Here there is some work in progress, with promising developments. In any case, the regional sciences lay themselves open, with no particular aversion, to 'contamination' by approaches even distant from their own theoretical origins. So why all these doubts about their scientific status?

If we are to tell the truth, the roots of the discomfort (for those who feel it) come from another source: the (unconfessed) aspiration for the formal status of the discipline, which is academically unrecognised. But in agreement with Isserman (1995), we would observe that though it is "not a science, not a discipline, regional science is a remarkable phenomenon in the sociology of science. It is an international, interdisciplinary association that has produced noteworthy contributions to several disciplines" (p. 274). "Here there has been astounding, spectacular success. The regional science associations, the meetings and the journals are an incredible co-operative effort ... Regional scientists often share more research interests with one another than with other scholars within their own discipline" (p. 273).

If this is true for the American Regional Science Association, imagine our own regional sciences! Hopefully they can find the motivation for renewed commitment. Our studies in fact are, directly or indirectly, both an invitation and a means of reopening the issue of the conscious government of social, economic and territorial processes.

We know that "the collapse of the 'real socialism' is being exploited to dismiss the very concept of planning and to shrink the State's role in the regulation of mixed economies" (Sachs, 1992, p. 3). Despite everything, the free market idea continues to enjoy vast consensus even in Europe. Therefore, to repropose today the option of the plan, especially in Europe, do we not run the risk of becoming (to quote Pope's verse, as do Bateson and Bateson, 1979) the fools who "rush in where angels fear to tread"? More prosaically, we could ask whether to propose planning means falling for the temptation of a utopia which revealed itself a very painful one for Europe between the Oder and the Urals.

But, after the intoxication of the free market ('more market, less State'), the pendulum is beginning to swing the other way, towards the idea of the plan. Europe, through the European Union, is calling on member states (and more than the states, their regions) to experiment with various forms of planning, by proposing partnerships, negotiations, co-decisions and co-management. In this case, applications will not be enough, what is needed are coherent programmes and appropriate tools of implementation, i.e. plans, like those required by the White Paper of Delors, and the plans even more necessary for the regional European policies of the approaching millennium, which seem to be inspired by the style of concerted action and regional pacts, to deal with the social malady of unemployment.

The crisis of welfare in the West, and the historic failure of the planned economic/social systems induce us to proceed with caution as to the possibility of governing or 'planning' the complexity of social, economic and territorial processes. However, when the systems to be governed are territorial systems (regions and urban systems), we know that we are dealing with systems which are complex, due to the large number of interconnections which evolve with non linear dynamics, and which react to actions affecting parts or relations with effects on other parts or relations staggered over time. To the systemic complexity we must add decisional complexity, due to the numerous institutional decision-makers (not necessarily in agreement), as well as the social complexity of the involvement, in the decision process, of the multiple collective preferences supported by the growing subjectivity of individuals and groups in the community.

The *methodological uncertainty* of the analytical procedure, in short, becomes *systemic uncertainty* of the subject to be analysed. The alternative is clear: either to give up the idea of governance, putting a mystic faith in the 'invisible hand' of the market, or to equip ourselves for the conscious governance of complexity (this is the conflict between *strength* and *power*, Ruffolo, 1988) stubbornly setting out once again, *with fewer illusions, but more tools,* on the hard path of planning.

The sciences of the city are here to give us a hand.

1.6 The Papers in this Volume

We now present a summary of the papers collected in this volume, with the purpose of giving a idea of their contents and the overall organisation of the book. They are a series of brief abstracts, which aim simply to provide an outline of the subject matter, without of course being able to do full justice to the arguments developed. We hope nevertheless that this will serve as a stimulus to the reader to examine the individual papers.

Section 1 tackles the theme of complexity in the urban field. This is discussed adopting a vision in which urban systems are interpreted in an abstract way, and considered as dynamic, evolutionary systems, possessing the characteristics typical of complex phenomena. Particular attention is paid to features such as self-organisation and synergy, which are discussed here in a prevalently theoretical light, although references are made to concrete applications.

In the paper which opens this session, *Peter Allen,* aiming at a new creative conception of complex systems not founded on the mechanical paradigm, shows how self-organisation allows a deeper understanding of the origin and evolution of the structure and organisation of urban areas, and how mathematical models can be developed to represent processes of change in an urban system in which the basic taxonomy, as well as the variables, vary over time. Such evolution takes place following the action of fast processes which ensure the maintenance of variety at the microscopic level, and in a context in which there is a systematic examination of the stability of the taxonomy and the existing variables in relation to the changes which they generate over time. In this framework, he also discusses some applications of the concept to models which describe how an urban structure can co-evolve with, for example, the transport system or technological and cultural innovation. The author describes and comments on some self-organising models of cities and other areas, focusing on the question of environmental sustainability and the need for an integrated model of decision support able to explore the global long term effects of socio-economic and environmental policies.

Agostino La Bella examines the potential of the evolutionist approach for expressing the complexity of urban and regional systems and providing an effective paradigm for analysing, forecasting and governing the city and other territorial systems. After some observations on the notion of complexity, he presents an application of the theory of perturbations to multi-regional demographic models, and then discusses some aspects of the fractal approach and catastrophe theory. The author puts forward some criteria for the comparison of theories and models of urban and regional development intending, in the light of these criteria, to stimulate discussion on the evolutionist approach.

Maria Tinacci Mossello highlights the growing difficulty of defining and recognising the city, both due to its changing form, and the problem of distinguishing the city-system from the society-system. She examines the hypothesis of the city as a local system and reflects on its meaning, on the definition of its limits and recognition of the city as a self-organising local system within a framework in which the boundaries of the city are unclear. The author points out the capacity of the modern city to produce information and innovation, stressing that this must not lead us to lose sight of the city as a place, a product of history and dense mixture of 'contents'. She concludes by showing that the complexity of cities, as complex systems in an increasingly complex environment, derives not so much from the composite and multiple nature of the components, as the breadth of the field of opportunities for action.

Dimitrios Dendrinos presents an unusual vision of social dynamics in which he identifies the fundamentals, stressing that usually the representation of social reality is partly a product of the imagination and partly fruit of an effort to replicate and codify reality as closely as possible. He demonstrates the need to appropriately define a series of figures and their corresponding roles: social actors, observers and a supra-observer, as well as an underlying framework or supra-structure within which the processes of social dynamics take place. The supra-structure contains the necessary information to allow a supra-code to act as a propulsor of the social dynamics, and reveal profound and elusive aspects of reality. One interesting result of this attempt is the specification of eight fundamental types of social dynamic.

Federico M. Butera proposes an analogy between learning processes and self-organisation, with the aim of defining a tool which allows the identification of the possible trajectories of a complex system. As learning occurs through communication, self-organisation can be considered a process of communication between the system and its environment. He presents a methodology for reducing the uncertainty of planning, based on an evaluation of the information transmitted in a complex system, within the framework of an approach to urban systems where their development is seen as neither entirely unpredictable, nor entirely deterministic, but a guided self-organisation process.

Valter Cavallaro puts forward an analysis of the centralisation and decentralisation processes inherent in the lifecycle of cities. These dynamics are interpreted as the self-organising phenomena in a complex system, and therefore as a source of information in the context of a circuit of creation of information from disorder, followed by creation of disorder from order (order from noise, noise from order).

Lorenza Lucchi Basili presents a description of self-organisation in complex systems, making particular reference to physical phenomena. She then gives a description of aspects of fractal geometry and an interpretation of the city, especially in the pre-industrialisation phase, in the light of the complexity paradigm.

Finally, *Francesca Bertuglia* identifies and examines, with specific reference to the field of architecture, certain aspects of complexity, such as self-organisation and 'surprise', understood as variety, and the whole set of meanings and interpretations which go to make up architecture and the architectural project. Numerous examples of recent projects are examined in the light of complexity. The author claims that an architectural project should increase complexity, seen as the possibility of choice; complexity is therefore a quality which planning or design should attempt to produce.

In conclusion, the papers presented in this first section allow us to claim that although, as we saw in 1.4, it may be true that the idea of complexity cannot yet aspire to being considered a real paradigm, it has nevertheless already stimulated a totally new way of looking at the city, which is leading to highly significant developments.

The focus of Section 2 is the impact of the conception of complexity on the sciences of the city, i.e. whether and how the conception of complexity has generated, or is likely to generate, changes in the disciplines concerned with the city. In other words, having examined in the previous section the various aspects of complexity and discussed how the phenomenology of complexity - made up of self-organisation, synergy and the unpredictability of the outcome of actions - connotes the abstract concept of the city seen as a dynamic system in evolution, we examine in this section in what way the paradigm of complexity is affecting the sciences of the city.

In the paper which opens the Section 2, *Denise Pumain,* reviews the analytical efforts made in the regional sciences since the 1980s - a period, as we have already mentioned, marked a significant change in approach -

focusing on the use of the new conceptual and operative tools for tackling problems connected with urban complexity. Before examining the new research strategies, the author discusses the main difficulties still being faced in producing a satisfactory definition of the city and in constructing consistent theories about it. She presents a wide-ranging picture, examining the main theoretical questions of urban research, as well as the various methodologies recently proposed for the description of complex urban systems. Pumain underlines the distance that exists between the theoretical features which lead us to consider urban areas as complex systems, and the results of applications of complex system theory to urban research. She claims that either the current state of urban research is not sufficiently advanced to solve the main theoretical difficulties, or it is necessary to review our hypotheses on the nature of urban complexity.

Roberto Camagni identifies in his paper certain unsatisfactory elements which remain, despite the widespread scientific appreciation of current urban modelling, and presses for an in-depth general evaluation of the scientific results of the new theoretical developments. He suggests a number of directions for research, which could favour important new theoretical results deriving from the theory of complex systems. In particular, he draws attention to the need to give greater emphasis to the economic point of view in the examination of urban systems. This should take into account innovative processes (Schumpeter), questions relating to distribution (Ricardo) and conflicts of power (Marx), as well as the physical dimension of the city. Camagni claims that we need to improve our understanding of the interactions between the various aspects of the city (the city as a physical entity, as a machine of production, as culture and as society).

Vittorio Silvestrini, conscious of the link between natural and social science, investigates the modern relationship between science and the city, underlining the clear distinction between theory and the phenomenological model. The latter is only an intermediate step towards theory, but helps to illuminate some aspects of the cause/effect relations which govern the development of the system. The author connects the discussion on the relationship between science and the city to the discussion on the relationship between the industrial factory and the city, stressing the co-existence of the 'city of machines' and the 'city of man' explaining how technological innovation has broken the virtuous Fordist circle, resulting in a series of problems which are due to the unbalancing of the market towards the supply. He suggests that the rediscovery of the use value economy could be a response to the crisis in the 'money-goods-money' chain of consumption and profit.

Alfredo Mela and *Giorgio Preto* tackle the problem of the meaning of the sciences of the city from an epistemological point of view, developing the analysis with separate reference to economics and sociology. They discuss therefore the systemic nature of the city, presenting a framework in which they describe a logical procedure which could be used to identify the conditions in which the city can be represented as a self-organising system.

Dino Martellato proposes a conception of complexity according to which it is only a reflection of the inadequacy of the knowledge of the observer, therefore, what appears complex today, may not be tomorrow. He observes that in economics little consideration has been taken of complexity, and that analysts have resorted to expedients such as the principle of rational expectations or Nash's equilibrium. As soon as we have to deal with the problems of strategic interaction between agents outside these frameworks, or we have to consider economic dynamics, complexity tends to appear, as in the case of interactions in oligopolistic markets for example, in which innovation is important.

Fiorenzo Ferlaino examines, in particular, the role of analytical tools, emphasising the importance of statistical and descriptive models and highlighting the differences between these and formal mathematical models used in the regional sciences (from gravitational models to those dealing with complex phenomena). He proposes a systematic treatment and taxonomy of territorial objects and structures from the objectives and methodologies relating to the various analytical tools. Firstly, a number of definitions of the city are presented, understood as the quantitative result of different analyses and different underlying paradigms. Then he analyses the complexity of the urban sciences through a classification of the dynamic relations and causal links underlying the modelling of complexity.

Carlo Socco puts forward some reflections on an interpretative paradigm of the city relating to the level of the very significance of the city, which precedes the vision of the city as a complex system. The territory is interpreted as a landscape of the life of society, an expression of culture and history of the society which lives there; the landscape is interpreted, therefore, as an aesthetic text and object of a semiotic theory.

Lastly, *Silvana Lombardo* suggests a view of planning as a learning process, which is applied to the continuous control of evolutionary processes of the city and integrated, as a support tool, in the process of public decision-making which underlies the transformations. She demonstrates, in addition, how the weakening of positive rationality in the framework of the complexity resulting from limited knowledge of the system and the consequent non-acceptability of long term forecasts, do not make the planning pointless, as long as we considering it in a different way and adopt appropriate methods to further our knowledge. Assuming complexity to be dependent on the level of knowledge, she proposes a method for measuring complexity based on the theory of information and illustrates this with an application to a problem of location.

In conclusion, the papers presented in Section 2 allow us to outline a framework of the sciences of the city in which complexity emerges as an object of study and a factor for renewal. This approach, although still at an initial stage, appears to have the potential for notable development, which, taking advantage of the renewed exchange between the sciences of nature and the sciences of society, seems able to promote a deeper understanding of urban dynamics and, more in general, complex dynamics.

The central theme of Section 3 is the impact of the concept of complexity on the planning of the city, i.e. whether and how the conception of complexity has generated, or is likely to generate, changes in the way that the planning of the city is conceived or practised.

In the opening paper of this section, *Andreas Faludi* presents a wide-ranging and detailed picture of the history of planning in the Netherlands, from the origins, at the beginning of the century, up to the present period. He places particular emphasis on the various phases in the evolution of thc figure of the urban planner, through the constant controversy between engineers and architects: the former interpreting planning as a technical process, the latter as an intuitive process which is the fruit of individual synthesis. Faludi highlights the influence of certain personalities who have been fundamental to this historical process. Through the review and analysis of these positions, he introduces interesting reflections on the nexus between knowledge and action, as well as the nature of the discipline, presenting a concept of planning doctrine.

Luigi Mazza makes a series of observations on the concepts introduced by Faludi, underlining, in particular, the importance of the concept planning doctrine, the definition of strategic planning and the relationship between the vision of strategic planning and urban design. He also discusses the role of the detailed Town Plan, stressing that we need to arrive at a deeper understanding of such plans, integrating them with more advanced methods and techniques.

Giovanni A. Rabino, adopting a theoretical framework of reference centred on the vision of complexity, tackles the question of the scientific foundation and the nature of urban planning. Having discussed the different kinds of knowledge, he sustains that in planning, as in human actions in general, art and science are not antithetic, they simply refer to two different kinds of knowledge and are integrated in a framework in which the widespread participation requires the transmission of knowledge. He also describes a classification of the conceptions of planning, according to their poietic and cognitive dimensions, underlining the role of creativity and claiming that the approach to urban planning needs to be scientific but not scientistic.

Francesco Indovina discusses the interaction between policies and social practice in the government of urban transformations, suggesting an approach which takes joint advantage of the positive developments in the 'science of the city' and those in the normative aspects of urban planning. He analyses the question of the correct balance between urban government, illustrating the qualities of good government, and the free manifestation of economic, social, technological and cultural dynamism, illustrating in detail some of the current conditions which tend to make the government of urban transformations difficult.

Giuseppe Longhi investigates the nature, myths and paradigms of urban science, after the recent waves of technological innovation, attempting to identify new rules for planning. While aware of the dangers of the tendency to consider the urban project as an autonomous activity and separate from the more general processes of transformation, the author stresses the need for urban planning as an intersection between different forms of knowledge, in the hope that new scientific developments, which are needed urgently, will allow urban space to find its place in the complex network represented by the world city.

Sandro Fabbro examines Science and Technology Parks (STP), with particular reference to the one in Trieste, as examples of policies which aim, among other things, to act on the system of relations existing between the main socio-economic actors of the cities involved. He suggests that STPs can be used as models of certain problems of planning and management of change in the modern city. The author contributes to the debate on strategic planning, highlighting how the pragmatic push 'to do' can trigger a vicious circle if the need to activate local actors capable not only of learning and interacting, but also facing and reflecting critically on change is not recognised.

Mariolina Toniolo discusses the delay in the application of strategic planning in Italy, identifying a number of cultural, academic and institutional reasons. Using the example of the Municipality of Venice, she indicates how the City Plan, if combined with a series of policies which go beyond mere physical planning in the strictest sense, can transform the city also functionally. She claims that new legislative conditions, such as those recently adopted in Italy, can help overcome a limited and sectorial view by modifying the behaviour of administrators and planners, encouraging co-operative behaviour which permits the implementation of projects requiring the interaction between different actors.

Tunney F. Lee traces a detailed picture of the evolution of urbanisation in China from ancient times to the present day, emphasising the profoundly different conceptions of the city in China and Europe in the past, and describing the painful transition represented by the opening to the West at the end of the 18th century, and the advent of socialism in the present century. Comparing the current processes of urbanisation in China with Europe, and observing the lack of control on the rapid and chaotic capitalistic development, he points out that the crisis of traditional values in present-day China is reflected in the serious problems being faced in urban areas, and concludes with the hope that China can learn from some of the planning errors made in the West.

Lastly, *Angela Spence* claims that the use of sophisticated analytical techniques is not enough in itself to ensure the definition of adequate policies for the management of the city. Looking at some contrasts between cities in Britain and Italy, she identifies some of shortcomings of past attempts at 'rationalistic' planning, concluding that it is important to go beyond the false counter-position between the formalised 'scientific' approach and the intuitive approach. She suggests that a fruitful way of dealing with complexity involves integrating the two approaches and hence tapping into wider sources of knowledge. In other words, it is important to learn the art of using the sciences of the city.

In conclusion, the papers presented in Section 3 make it possible, firstly, to delineate the concept of planning which has been recently emerging - a conception in which the evolution of urban systems appears as a process to be guided, rather than forced, and in which urban sustainability is a fundamental condition to achieve - and, secondly, they illustrate how the ideas of complexity help to give a solid foundation to this conception of urban planning.

Section 4 focuses on the impact of the concept of complexity on the methodologies of the urban sciences, i.e. whether and how the conception of complexity has generated, or is likely to generate, changes in the methods and tools used for the analysis of urban phenomena. Various techniques for the collection, management, processing and evaluation of data are presented, from the design, structure and use of geographic information systems (GIS), of which a number of applications are considered, to mathematical simulation models and decision support methods.

In the paper which opens this section, *Marina van Geenhuizen and*

Peter Nijkamp present a systematic framework in which information tools and techniques of analysis are related to the need to manage urban complexity towards the objective sustainability. They discuss the concept of the sustainable city, highlighting the need to possess appropriate information systems and emphasising, in particular, the complexity of urban dynamics. The authors analyse the challenge faced in planning the complex city, and point out the passage from the use of models for forecasting to scenario analysis. Claiming that GIS, together with decision support models, provide a powerful means of improving knowledge and decision-making in urban planning, they suggest an architecture and contents suitable for an urban information system and, finally, describe an information system specifically designed for the coping with the problem of environmental sustainability.

Roberto Tadei and *Marco Dellasette* make a series of observations on the design of an information system for a metropolitan area, describing the specific requirements of a GIS for the planning of mobility and transport. They underline that a GIS should contain mathematical models, performance indicators and multicriteria methods, and can be enriched with processed data alongside the raw survey data. The authors identify the subsystems of the metropolitan system and the interactions between them, and emphasise the need for an integrated approach to the planning of mobility and transport. They focus on the modelling aspects of the information system, the mathematical models needed and their various uses, the evaluation procedures for demand scenarios and supply alternatives, and other aspects concerning more strictly the information and computing.

Luigi Fusco Girard, commenting on the paper by Van Geenhuizen and Nijkamp, pays particular attention to the relationship between communication, decision processes and urban indicators, stressing that the latter can help to improve communication, the quality of decisions, and therefore action. The improvement of communication and decisions is in fact essential, if urban development is to be not only sustainable, but self-sustainable. This depends on the level of urban organisation, i.e. the degree of co-ordination of the system of public, private and social institutions, which the author refers to as 'institutional capital'. To increase this, he claims that evaluation, communication and co-ordination can serve as guidelines, and that some elements of help are provided by information systems, indicators and evaluation methods.

Aura Reggiani considers, first of all, the possible contribution offered by the methodological approaches recently adopted in the economic/spatial sciences, and presents a brief review of these with reference to their potential in the analysis and modelling of urban complexity. She presents an application relating to the identification of structural tendencies in a systemic component, such as the evolution of transport in a complex city, considering the network of Italian cities which will be connected by the future High Speed Rail system - a network which could itself be interpreted as a new and larger complex city.

Ennio Cascetta defines the system of mobility and its relationship with the system of urban activities. Especially in conditions of heavy congestion, these behave as complex systems made up of numerous and very different elements which interact in a direct and also nonlinear way. For their study sophisticated mathematical models are proposed, able to reproduce the principle variables and the relations between them. The author explains the standard paradigm of mathematical models for simulating the most important elements of the system of urban transport. He then mentions some of the most promising areas of research for the modelling of complex sequences of movements in urban areas, the dynamic analysis of transport systems and the dynamic assignment of flows to the network, describing various fields of application of these new types of model.

Sylvie Occelli in response to the recognition of the complexity of urban problems and the need to have adequate tools for the treatment of these problems, discusses some questions relating to the methodologies for the analysis, planning and control of urban systems. These methodologies, as instruments for acquiring knowledge, make it possible to obtain some kind of information gain. On the basis of observations of the connection between analytical tools and planning problems, she identifies the role and function of these tools and examines in some detail the features of a new perspective for their meaning and use, discussing: (i) the rationalisation, operative aspects and spread of tools for producing information, (ii) learning and creative potential, (iii) the relations between analytical tools and images of the plan. She then explains that a conception of analytical tools which is not purely technical makes it possible to evaluate their instrumental role and also recognise a learning role. In this respect, the problem of communication, in its widest sense, constitutes a new propulsive factor for the development of new tools.

Finally, *Frank C. Englmann, Walter Scheuerer, Rainer Carius, Bettina Oppermann, Sabine Köberle* and *Ortwin Renn* present a framework for the discussion of the problem of achieving compatibility between mobility and the environment. They describe the organisation of a study being carried out for the drawing up of a Regional Transport Plan for the Region of Stuttgart, and the structure of a special research project on environmentally compatible mobility, which aims to contribute to the basic research on the relations between the micro, meso and macro levels in an interdisciplinary context. They also discuss the objectives and standards of environmental quality, and methodologies for environmental impact analysis, paying particular attention to techniques of mediation for aware decision-making in a context of public participation.

In conclusion, the papers presented in Section 4 paint a picture which further confirms the idea, already expressed in this introduction, that the complexity of urban systems, in order to be understood and suitably managed, requires a clear awareness of the need for a close link between the gathering of empirical data and the use of methods and models, sometimes highly sophisticated, as well as the opportunity offered by the new information and communication technologies.

References

Albrechts L. (1996) Sul futuro della pianificazione spaziale, *Critica della razionalità urbanistica*, *6*, 84-89.

Allen P.M., Boon F., Deneuburg J.L., de Palma A., Sanglier M. (1978) The Dynamics of Urban Evolution, volume 1: Interurban Evolution, volume 2: Intraurban Evolution, Final Report, US Department of Transportation, Research and Special Programs Administration, Washington D.C.

Amendola G. (1997) *La città postmoderna. Magie e paure della metropoli contemporanea*, Laterza, Bari.

Amin A. (ed.) (1994) *Post-Fordism*, Blackwell, Oxford.

Andersson Å.E., Zhang W.B. (1997) Nonlinearity in Social Dynamics, Order versus Chaos, *Discrete Dynamics in Nature and Society*, *1*, 111-126.

Anselin L., Rey S.J. (1997) Introduction to the Special Issue on Spatial Econometrics, *International Regional Science Review*, *20*, 1-7.

Asheim B.T. (1996) Industrial Districts as 'Learning Regions': A Condition for Prosperity?, *European Planning Studies*, *4*, 379-400.

Bailly A.S., Coffey W.J. (1994) Regional Science in Crisis: A Plea for a more Open and Relevant Approach, *Papers in Regional Science*, *73*, 3-14.

Bairoch P. (1985) *De Jericho à Mexico, ville et économie dans l'histoire*, Gallimard, Paris.

Bara B.G. (1990) *Scienza cognitiva. Un approccio evolutivo alla simulazione della mente*, Bollati Boringhieri, Turin.

Barras R., Broadbent A. (1979) The Analysis in English Structure Planning, *Urban Studies*, *16*, 27-41.

Barthes R. (1967) Semiologia e urbanistica, *Op. cit.*, *10*, 7-27.

Bateson C., Bateson M.C. (1979) *Angels Fear. Towards an Epistemology of the Sacred*, Dutton, New York.

Batten D.F. (1995) Network Cities: Creative Urban Agglomerations for the 21st

Century, *Urban Studies, 32*, 313-327.

Batten D.F., Bertuglia C.S., Martellato D., Occelli S. (eds.) (1998) *Innovation and Urban Evolution*, Springer Verlag, Berlin.

Batten D.F., Casti J., Thord R. (eds.) (1995) *Networks in Action*, Springer Verlag, Berlin.

Batty M. (1993) Cities and Complexity: The Implications for Modeling Sustainaibility, Fourth International Workshop on Technological Change and Urban Form, Berkeley, 14-16 April (mimeo).

Batty M. (1994) A Chronicle of Scientific Planning, *Journal of the American Planning Association, 60*, 7-16.

Becattini G., Bianchi G. (1987) I distretti industriali nel dibattito sull'economia italiana, in Becattini G. (ed.) *Mercato e forze locali: il distretto industriale*, Il Mulino, Bologna, 169-178.

Bell D. (1973) *The Coming of Post-Industrial Society. A Venture in Social Forecasting*, Basic Books, New York.

Berrini A. (1996) *L'anima dei bulldozer. Viaggio nella nuova baraccopoli africana*, Baldini and Castoldi, Milan.

Bertuglia C.S., Bianchi G., Camagni R. (1996) Innovazione, ambiente e sviluppo nella prospettiva delle scienze regionali, paper given at the inaugural meeting of the XVII Conference of the Italian Regional Science Association (Sondrio, 16-18 October) (mimeo).

Bertuglia C.S., Clarke G.P., Wilson A.G. (eds.) (1994) *Modelling the City: Performance, Policy and Planning*, Routledge, London.

Bertuglia C.S., Detragiache A., Rabino G.A. (1993) Elementi per una carta dell'urbanistica e proposta di un centro per le scienze della città, in *Storia, concorso, risultati e poi... la Carta di Megaride 94*, F. Giannini e Figli, Naples, 171-174.

Bertuglia C.S., Fischer M.M., Preto G. (eds.) (1995) *Technological Change, Economic Development and Space*, Springer Verlag, Berlin.

Bertuglia C.S., La Bella A. (1991) Introduzione, in Bertuglia C.S., La Bella A. (eds.) *I sistemi urbani*, Angeli, Milan, 9-54.

Bertuglia C.S., La Bella A. (eds.) (1991) *I sistemi urbani*, Angeli, Milan.

Bertuglia C.S., Leonardi G., Occelli S., Rabino G.A., Tadei R., Wilson A.G. (eds.) (1987) *Urban Systems: Contemporary Approaches to Modelling*, Croom Helm, London.

Bertuglia C.S., Leonardi G., Wilson A.G. (eds.) (1990) *Urban Dynamics: Designing an Integrated Model*, Routledge, London.

Bertuglia C.S., Lombardo L., Nijkamp P. (eds.) (1997) *Innovative Behaviour in Space and Time*, Springer Verlag, Berlin.

Bertuglia C.S., Occelli S. (1995) Gli indicatori territoriali, con particolare riferimento a quelli di performance spaziale: inquadramento storico, presupposti concettuali, problematiche operative, qualche esempio, in Campisi D., La Bella A. (eds.) *Il governo della spesa pubblica e l'efficienza di servizi*, Angeli, Milan, 313-348.

Bertuglia C.S., Rabino G.A. (1990) The Use of Mathematical Models in the Evaluation of Actions in Urban Planning: Conceptual Premises and Operative Problems, *Sistemi urbani*, 12, 121-132.

Bertuglia C.S., Rabino G.A., Tadei R. (1991) La valutazione delle azioni in campo urbano in un contesto caratterizzato dall'impiego dei modelli matematici, in Bielli M., Reggiani A. (eds.) *Sistemi spaziali: approcci e metodologie*, Angeli, Milan,

97-143.
Bertuglia C.S., Rabino G.A., Tadei R. (1992) Review of the Main Conceptual Issues Facing Contemporary Urban Planning, *Sistemi urbani*, 14, 151-171.
Bertuglia C.S., Vaio F. (1997) Introduzione, in Bertuglia C.S., Vaio F. (eds.) *La città e le sue scienze*, vol. 1, *La città come entità altamente complessa*, Angeli, Milan, I-XCIII.
Bianchi G. (1982) L'esperienza di programmazione regionale in Italia: una breve rassegna critica, in Bielli M., La Bella A. (eds.) *Problematiche dei livelli sub-regionali di programmazione*, Milan, Angeli, 1982, 41- 61.
Bianchi G. (1988) *I piani regionali oggi: messaggi politici o atti di governo?*, Le Monnier, Florence.
Bianchi G. (1992) The IMPSs: A Missed Opportunity? An Appraisal of the Design and Implementation of the Integrated Mediterranean Programmes, in Leonardi R. (ed.) *The Regions and the European Community*, Frank Cass, London, 47-70.
Bianchi G. (1993) Programmare oggi, Quaderni di analisi e programmazione dello sviluppo regionale, IRES Toscana, Florence.
Bianchi G. (1994) Requiem per la Terza Italia? Sistemi territoriali di piccola impresa e transizione postindustriale, in Garofoli G., Mazzoni R. (eds.) *Sistemi produttivi locali. Struttura e trasformazione*, Angeli, Milan, 59-90.
Bianchi G. (1997) Local Systems: From Metaphors to Possible Theorisation, paper presented at the ERSA97/37th European Congress of the European Regional Science Association, Tor Vergata University, Rome, 26-30 August (mimeo).
Bianchi G., Grote J.R., Pieracci S. (1995) Dalla coesione economica alla coesione istituzionale. Sussidiarietà funzionale e reti socio-istituzionali nelle politiche regionali, in Gorla G., Vito Colonna O. (eds.) *Regioni e sviluppo: modelli, politiche e riforme*, Angeli, Milan, 305-330.
Bianchi G., Johansson B., Snickars F. (1984) Models as Integral Parts of Regional Information Systems, in Nijkamp P., Rietveld P. (eds.) *Information Systems for Integrated Regional Planning*, North Holland, Amsterdam, 115-136.
Bianchi G., Magnani I. (1985) Discutendo di sviluppo multiregionale, in Bianchi G., Magnani I. (eds.) *Sviluppo multiregionale: teorie, metodi, problemi*, Angeli, Milan, 21-26.
Borges J.L. (1984) Storia del guerriero e della principessa, in Borges J.L., *Tutte le opere*, vol. 1, Mondadori, Milan.
Born M. (1935) *Atomic Physics*, Blackie and Son, London.
Bortolotti F. (ed.) (1994) *Il mosaico e il progetto. Lavoro, imprese e regolazione nei distretti industriali della Toscana*, Angeli, Milan.
Boyer R. (1987) *La théorie de la régulation: une analise critique*, La Découverte, Paris.
Boyer R., Saillard Y. (1995) *Théorie de la regulation. L'état de savoirs*, La Dècouverte, Paris.
Bramanti A., Maggioni M. A. (eds.) (1997) *La dinamica dei sistemi produttivi territoriali: teorie, tecniche, politiche*, Angeli, Milan.
Button K.J., Nijkamp P. (1996) Series Preface, in Thisse J.F., Button K.J., Nijkamp P., *Location Theory*, vol. 1, Series 'Modern Classics in Regional Science', edited by Button K.J., Nijkamp P., An Elgar Reference Collection, Elgar Brookfields, Cheltenham, XII-XIII.
Calvino I. (1972) *Le città invisibili*, Einaudi, Turin.
Camagni R. (1988) Lo spazio della pianificazione, in Gibelli M.C., Magnani I. (eds.) *La pianificazione urbanistica come strumento di politica economica*, Angeli, Milan,

61-71.

Camagni R. (1993) *Principi di economia urbana e territoriale*, La Nuova Italia Scientifica, Rome.

Castells M. (1989) *The Informational City*, Basil Blackwell, Oxford, Cambridge, Massachusetts.

Casti J.L. (1986) On System Complexity: Identification, Measurement, and Management, in Casti J.L. Karlqvist A. (eds.) *Complexity, Language, and Life: Mathematical Approaches*, Springer Verlag, Berlin, 146-173.

Casti J.L. (1994) *Complexification*, Abacus Book, London.

Cini M. (1990) *Trentatré variazioni su un tema. Soggetti dentro e fuori la scienza*, Editori Riuniti, Rome.

Darwin C. (1859) *On the Origin of Species by Means of Natural Selection or the Preservation of Favored Races in the Struggle for Life*, Murray, London.

Dematteis G. (1997) La città come nodi di reti: la transizione urbana in una prospettiva spaziale, in Dematteis G., Bonavero P. (eds.) *Il sistema urbano italiano nello spazio unificato europeo*, Il Mulino, Bologna, 15-35.

Dendrinos D. (1992) *The Dynamics of Cities*, Routledge, London.

Duby G. (ed.) (1980) *Histoire de la France urbaine*, Seuil, Paris.

Edwards R.D., Magee J. (1964) *Technical Analysis of Stock Trends*, John Magee, Boston.

Feigenbaum M.J. (1978) Qualitative Universality for a Class of Nonlinear Transformations, *Journal of Statistical Physics, 19*, 25-52.

Flyvbjerg B. (1990) Aristotle, Foucault and Progressive Phronesis: Outline of an Applied Ethics for Sustainable Development, paper presented at the Conference on "Moral Philosophy in the Public Domain", University of British Columbia, 7-9 June (mimeo).

Friedmann J. (1993) Towards a non-Euclidean Mode of Planning, *Journal of the American Planning Association*, *59*, 482-484.

Fusco Girard L., Nijkamp P. (eds.) (1997) *Le valutazioni per lo sviluppo sostenibile della città e del territorio*, Angeli, Milan.

Gershuny J.I. (1978) *After Industrial Society?*, MacMillan, London.

Gibson W. (1984) *Neuromancer,* Series 'Ace of Special', Carr, New York.

Governa F. (1997) Il milieu urbano. L'identità territoriale nei processi di sviluppo, Angeli, Milan.

Graham S., Marvin S. (1996) *Telecommunications and the City. Electronic Spaces, Urban Places*, Routledge, London.

Grote J.R. (1993) On Functional and Territorial Subsidiarity: Between Legal Discourse and Functional Needs, Working Paper 1993/1 del Schumann Centre, European University Institute, Florence.

Haken H. (1978) *Synergetics. An Introduction*, Springer Verlag, Berlin.

Harvey D. (1990) *The Condition of Postmodernity. An Inquiry into the Origins of Cultural Change*, Basil Blackwell, Oxford, Cambridge, Massachusetts.

Hassink R. (1997) Localized Industrial Learning and Innovation Policies, in Hassink R. (ed.) Globalization, Regional and Local Knowledge Transfer, Special Issue of *European Planning Studies, 5*, 3, 279-282.

Healey P. (1991) The Content of Planning Education Programmes: Some Comments on Recent British Experience, *Environment and Planning B: Planning and Design, 18*, 177-184.

Isard W. (1960) *Methods of Regional Analysis: An Introduction to Regional Science*,

The MIT Press, Cambridge, Massachusetts.

Isserman A.M. (1995) The History, Status, and Future of Regional Science: An American Perspective, *International Regional Science Review, 17*, 249-296.

Jacobs J.M. (1993) The City Unbound: Qualitative Approaches to the City, *Urban Studies, 30*, 827-848.

Jensen R.L. (1991) Quo Vadis Regional Science?, *Papers in Regional Science, 70*, 97-111.

Krugman P. (1991a) *Geography and Trade*. MIT Press, Cambridge, Massachusetts.

Krugman P. (1991b) Increasing Returns and Economic Geography, *Journal of Political Economy, 99*, 483-489.

Krugman P. (1996) Urban Concentration: The Role of Increasing Returns and Transport Costs, *International Regional Science Review, 19*, 5-30.

Laplace P.S. (1814) *Essai philosophique sur les probabilités*, Courcier, Paris.

Le Moigne J.L. (1977) *La théorie du système général. Théorie de la modélisation*, PUF, Paris.

Lepetit B., Pumain D. (eds.) (1993) *Temporalités urbaines*, Anthropos, Paris.

Lotka A.J. (1925) *Elements of Physical Biology*, Williams and Wilkins, Baltimore.

Lowry J. (1964) Model of a Metropolis, RM-4035-RC, Rand Corporation, Santa Monica, California.

Lyotard J.F. (1979) *La condition post-moderne*, Les Editions de Minuit, Paris.

Mandelbrot B. (1997) Dal caso benigno al caso selvaggio, *Quaderni di Le Scienze*, 98, 16-20.

Marcuse P. (1989) 'Dual City': A Muddy Metaphor for a Quartered City, *International Journal of Urban and Regional Research, 14*, 697-708.

May R.M. (1976) Simple Mathematical Models with Complicated Dynamics, *Nature*, 261, 459-467.

McCauley J.L. (1997) The New Science of Complexity, *Discrete Dynamics in Nature and Society, 1*, 17-30.

Mela A. (1990) *Società e spazio: alternative al postmoderno*, Angeli, Milan.

Mela A. (1996) *Sociologia delle città*, NIS, Rome.

Mela A., Preto G. (1990) Alla ricerca della strategia perduta, in Curti F., Diappi L. (eds.) *Gerarchie e reti di città. Tendenze e politiche*, Angeli, Milan, 127-154.

Mela A., Preto G., Rabino G.A. (1987) Principles of Spatial Organization: A Unifying Model for Regional Systems, paper presented at the 5th European Colloquium on Quantitative and Theoretical Geography, Bardonecchia (mimeo).

Millon-Delsol C. (1993) *Le principe de subsidiarité*, Presses Universitaire de France, Paris.

Morin E. (1977) *La méthode. 1. La nature de la nature*, Éditions du Seuil, Paris.

Mucci E. (1990) La città come ambiente significante, in Mucci E. (ed.) *Firenze. Frammenti di memoria*, Ponte alle Grazie, Florence, 19-37.

Nagurney A., Siokos S. (1997) *Financial Networks: Statics and Dynamics*, Springer Verlag, Berlin.

Nicolis G., Prigogine I. (1987) *Exploring Complexity. An Introduction*, Piper, Munich.

Nijkamp P., Rietveld P., Voogd H. (1985) A Survey of Qualitative Multiple Criteria Choice Models, in Nijkamp P., Leitner H., Wrigley N. (eds.) *Measuring the Unmeasurable. Analysis of Qualitative Spatial Data*, Martinus Nijhoff, The Hague, 425-447.

Peattie L. (1994) An Approach to Urban Research in the Nineties, *Planning Theory, 12*, 9-34.

Pennac D. (1995) *Monsieur Malaussène*, Éditions Gallimard, Paris.

Plummer T. (1989) *Forecasting Financial Markets*, Kogan Page, London.

Prigogine I. (1993) *Le leggi del caos*, Laterza, Rome.

Prigogine I., Stengers I (1979) *La nouvelle alliance*, Gallimard, Paris.

Prigogine I., Stengers I. (1984) *Order Out of Chaos*, Bantam Books, New York.

Regini M. (1996) Still Engaging in Corporatism? Some Lessons from the Recent Italian Experience of Concertation, paper presented at the 8th International Conference on Socio-Economics, Geneva, 12-14 July (mimeo).

Regini M., Sabel C. (1989) *Strategie di riaggiustamento industriale*, Il Mulino, Bologna.

Reilly W. (1931) *The Laws of Retail Gravitation*, Pilsbury, New York.

Revelli M. (1997) *La sinistra sociale. Oltre la civiltà del lavoro*, Bollati Boringhieri, Turin.

Romanoff E. (1995) Guest Editorial: Paradigm Shifts of Regional Science, *Papers in Regional Science, 74*, 205-208.

Ruelle D. (1991) *Hasard et chaos*, Odile Jacob, Paris.

Ruffolo G. (1988) *Potenza e potere*, Laterza, Bari.

Sachs I. (1992) What State, What Markets, for What Development, paper presented at the First World Conference on Planning Science, Planning Technologies and Planning Institutions, Palermo, 8-12 September (mimeo).

Salone C. (1997) Le politiche urbane e territoriali nell'Europa comunitaria, in Dematteis G., Bonavero P. (eds.) *Il sistema urbano italiano nello spazio unificato europeo*, Il Mulino, Bologna, 67-117.

Schmitter P.C. (1985) Neo-corporatism and the State, in Grant W. (ed.) *The Political Economy of Corporatism*, MacMillan Publishers, London, 32-63.

Schmitter P.C. (1989) Corporatism is Dead! Long Live Corporatism! Reflections on Andrew Schonfield's "Modern Capitalism, Government and Opposition", *Politics and Society, 24*, 54-73.

Schmitter P.C., Grote J.R. (1997) The Corporatist Sisyphus: Past, Present and Future, European University Institute, Working Paper 1997/4, Florence.

Schmitter P.C., Streeck W. (1991) From National Corporatism to Transnational Pluralism, *Politics and Society, 19*, 133-164.

Schweitzer F. (ed.) (1997) *Self-organization of Complex Structures*, Gordon and Breach, London.

Secciani A. (1997) Ma questo è materialismo storico!, *Borsa e Finanza*, 2, no. 3, 4 January, 30.

Sengerberger W., Loveman G. (1987) Smaller Units of Employment, International Institute of Labor Studies, ILO, Geneva.

Serra R., Zanarini G. (1986) Introduzione alla scienza della complessità, in Serra R., Zanarini G. (eds.) *Tra ordine e caos*, CLUEB, Bologna, 1-20.

Stewart J.Q., Warntz W. (1958) Physics of Population Distribution, *Journal of Regional Science, 1*, 99-123.

Streeck W. (1992) *Social Institutions and Economic Performance. Studies of Industrial Relations in Advanced Capitalistic Economies*, Sage Publications, London.

Thom R. (1972) *Stabilitè structurelle de la morphogénèse. Essai d'une théorie génèrale des modèles*, Inter Editions, Paris.

Thom R. (1974) *Modèles mathématiques de la morphogénèse*, Bourgeois, Paris.

Thompson G. (ed.) (1991) *Markets, Hierarchies and Networks. The Coordination of Social Life*, Sage Publications, London.

Toffler A. (1980) *The Third Wave*, Collins, London.
Touraine A. (1969) *La société post-industrielle*, Denöel, Paris.
Veltz P. (1998) Economia e territori: dal mondiale al locale, in Perulli P. (ed.) *Neoregionalismo. L'economia-arcipelago*, Bollati Boringhieri, Torino, 128-151.
Volterra V. (1926) Variazioni e fluttuazioni del numero di individui in specie animali conviventi, *Memorie dell'Accademia Nazionale dei Lincei*, 2, 31-113.
Wegener (1994) Operational Urban Models: State of the Art, *Journal of the American Planning Association, 60*, 17-29.
Weidlich W. (1991) Physics and Social Science, the Approach of Synergetics, *Physics Reports*, 204, 1-163.
Weidlich W., Haag G. (1983) *Concepts and Models of a Quantitative Sociology. The Dynamics of Interacting Population*, Springer Verlag, Berlin.
Wheatley A. (1971) *The Pivot of the Four Quarters*, Edinburgh University Press, Edinburgh.
Williams R.H. (ed.) (1984) *Planning in Europe. Urban and Regional Planning in the EEC*, Allen and Unwin, London.
Williamson O.E. (1983) *Market and Hierarchies*, The Free Press, New York.
Wilson A.G. (1970) *Entropy in Urban and Regional Modelling*, Pion, London.
Wilson A.G. (1981) *Geography and the Environment: Systems Analytical Methods*, Chichester, Wiley.

SESSION 1: THE CITY AS A HIGHLY COMPLEX ENTITY

2. Cities as Self-Organising Complex Systems

Peter M. Allen

2.1 Introduction

Today, uncertainties and doubts concerning the future of urban society are apparent everywhere. How can the unsustainable lifestyles of the modern world be changed in order to make them sustainable? And, what is sustainability anyway? Is it some static level of production and consumption corresponding to a maximal exploitation of natural resources, or does it concern the capacity to adapt and change, and to develop a diverse and varied abundance of activities, with creativity and innovation as important ingredients? The urban problems with which we are faced result from the success of the traditional scientific view of the world as a mechanical system, capable of ever increasing exploitation. The traditional engineering approach to a problem has always been to specify the exact context and to produce a structure which is optimised according to a particular set of criteria. But the very success and growth of these technological solutions change the context in which they exist: both from the input side - the raw materials and production structures that are required - and the output side, meaning the impacts on society and on the biosphere. This failure to foresee the limits of technology stems from the 'mechanical' philosophy of science, rooted in Newtonian concepts that came from the Enlightenment. In order to create a new science of society, and of the cities and regions that it inhabits, we need a new, creative philosophy which is not based on the mechanical paradigm.

2.2 Evolution and Mechanics

If we examine a region and consider the traces of past populations and the artefacts that litter the landscape, after dating and classifying them, an evolutionary tree of some kind emerges. There are possibly some discontinuities, suggesting disaster and invasion, but nevertheless a changing 'cast of characters' and of behaviours emerges over time. This is represented in Fig. 1.

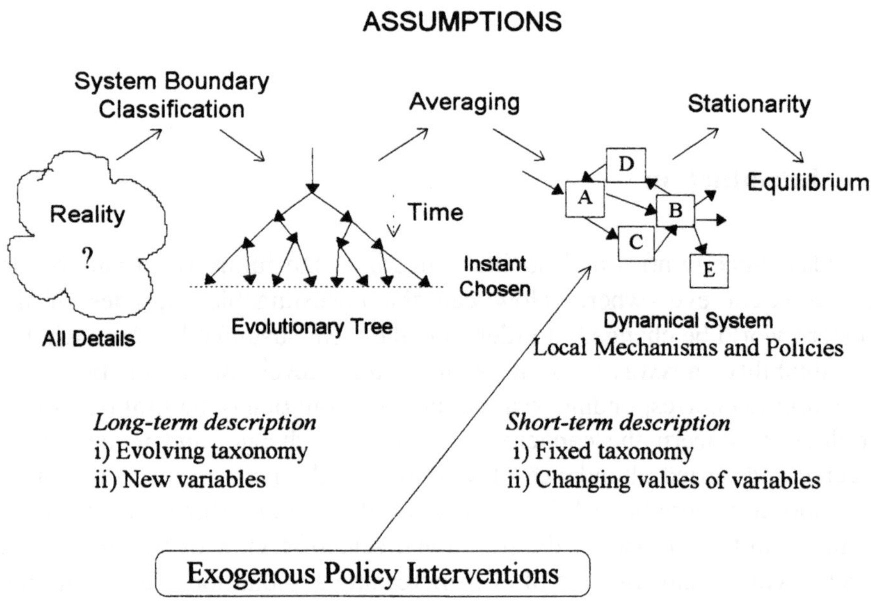

Fig. 1 Data and classification of populations and artefacts leads to the picture of an evolutionary tree of some kind. Mathematical models have concentrated on the causal relations at a given time.

On the left, we have 'reality'. It is drawn as a cloud, since we can say little about it, other than the fact that it includes all details of everything, everywhere, as well as all perceptions and all points of view. However, if we simply list what we see, then it includes a landscape with people of many kinds performing a variety of tasks, businesses, factories, homes, vehicles, and also fossils, disused mines and factories, closed railways, buried cities and evidence of much that has disappeared. By constructing a series of taxonomic rules concerning the differences and similarities of the objects, together with their dates, we can construct an 'evolutionary tree', showing that species, behaviours, forms, and artefacts have emerged and evolved over time.

This process is highly subjective, however, since the differences that we choose to recognise reflect our vision of what is 'coherent' in a social and economic system. The rules of classification that we use result from previous experience about such systems and what matters in them. Are there really socio-economic 'types'? Do firms of the same sector and size behave similarly? What is a sector? Is there as much variation within a group as between groups? Whatever the precise arguments advanced, in order to 'understand' a situation and its possible outcomes, we classify the system into components and attempt to build mathematical models that capture the processes which are affecting these different components.

At any particular moment therefore, we identify the different objects or organisms that are present and attempt to write down some 'population dynamics' describing the increase and decrease of each type. We apply the traditional approach of physics, which is to identify the components of a system and the interactions operating between them, both to and from the outside world and between the different populations of the system. In ecology, this will consist of birth and death processes, where populations give birth at an average rate, if there is enough food, and eat each other according to the average rates of encounter, capture and digestion! In economics, the macroscopic behaviour of the economy is assumed to result from the aggregate effects of producers attempting to maximise their profits and of customers attempting to maximise their utility. This assumes that they know the outcome of what they have not yet tried and also that transactions, production and consumption occur at average rates, changing the GNP, unemployment and other macroscopic indicators. These ideas are all based on the mechanical paradigm of Newtonian physics and assume that all individuals, producers, and consumers of a given type are identical to the average. Such a model expresses the behaviour or functioning of the system at that time as a result of the causal relationships that are present. This gives the illusion that we have a representation of the system which can be run on a computer to give predictions.

However, as we see clearly from our broader picture of the 'evolutionary tree', the predictions that such a model can give will only be correct for as long as the taxonomy of the system remains unchanged. The mechanical model of deterministic equations that we can construct at any given time has no way of producing 'new' types of objects, new variables, and so the 'predictions' that it generates will only be true until some moment, unpredictable within the model, when there is an adaptation or innovation, and new behaviour emerges.

The basis of scientific understanding has traditionally been the mechanical model (Prigogine and Stengers, 1987, Allen, 1988) constructed from the causal relations that exist between the components of the system at a particular time. These are used to construct a pseudo-mechanical representation of the system which can be run forward to provide predictions, and whose component variables reflect the taxonomy of the system. In many cases a further assumption is introduced - the system is supposed to run to equilibrium - so the correspondence between the real object and that represented by the model is made through equilibrium relations between the variables.

In economic geography, as well as transportation and land-use, the models that are used operationally today are still based on equilibrium assumptions. Locations of jobs and residences, land values, traffic flows etc. are all assumed to reach their equilibrium configurations within say, five years, following some policy or planning action. Such an approach fails to take into account the possibility of 'run-away' processes where growth encourages growth, decline leads to further decline and so on, and where actions simply switch the regional system from one evolutionary trajectory to another. Similarly, the equilibrium approach supposes that urban form and hierarchy express some maximised utility for the actors, where consumers minimise distance of travel for goods and services and producers maximise profits. This approach assumes that all the actors know what they want, how to get it, and are doing what they would wish, given the choices open to them. Such ideas give rise in reality to a purely descriptive approach to problems, tracing in a kind of *post hoc* calibration process the changes that have occurred.

System dynamics and simulation were thought to offer a path to the prediction of system behaviour and, because of this, to offer a basis for rational policy and decision-making in complex systems. In fact, those wishing to use equilibrium methods should be obliged to accept the burden of proof, since it is they who make the assumptions. They should prove that the relaxation times of the processes involved in the system are short with respect to the period of interest and, therefore, that the methods are justified. In reality, such evidence has never been presented, but nonequilibrium methods have simply been ignored. The real justification for equilibrium methods was that the computing power was not available to make a proper attempt at understanding such complex systems. But as computers have become more powerful, the need for such strong and unrealistic assumptions has gradually disappeared.

If we look at Fig. 1, we see that however interesting a model of system dynamics might be, it cannot anticipate the changes that may still occur in the evolutionary tree from which the 'moment' studied is taken. The taxonomy of the system will change over time, and therefore the mathematical model of causal relations will be incorrect. It might be good for some time, while the taxonomy is stable and no new classes or types have appeared, but this will only be revealed when the model is shown to be incorrect and in need of re-formulation.

Despite this, however, system dynamic models of problems have been developed, and much attention has rightly been given to the interesting behaviours of nonlinear dynamic systems. Such systems are characterised by equations of the type:

$$dx/dt = G(x, y, z, \ldots) \qquad (1)$$

where G is a function with nonlinear terms, leading to changes in x which are not simply proportional to its size. These functions are also made up of terms which involve the variable x, and parameters expressing the functional dependence on these terms. The parameters reflect two factors fundamental in the working of the system:

- the external factors, which are not modelled as variables in the system. These reflect the 'environment' of the system and may of course be dependent on spatial co-ordinates. Temperature, climate, soils, world prices, interest rates are examples of such factors.
- the values corresponding to the 'performance' of the entities underlying x, due to internal characteristics such as technology, level of knowledge or strategies.

These two entirely different aspects have not been separated out in much of the previous work concerning nonlinear systems, so the whole issue of the evolution of the populations involved in the system has not been addressed clearly. Equations of the type shown above display a rich spectrum of possible behaviours in different regions of both parameter space and initial conditions. They range from a simple approach to a homogeneous steady state characterised by a point attractor, through the sustained oscillation of a cyclic attractor to the well-known chaotic behaviour characteristic of a strange attractor. These can either be homogeneous, or can involve spatial structure as well. This possibility of rich behaviours has proved to be of great significance for many fields of science.

However, in this work, we are concerned with evolution, and that means creating models of systems in which adaptive and structural change can occur. The internal characteristics of the participating actors change endogenously and new variables and mechanisms of interaction can appear spontaneously from within the system itself, leading to a changing taxonomy. The model that we shall describe involves not only the evolution of improved techniques for producing a given good, but also the diversification and growth of markets into new areas.

Let us first consider the assumptions that are made in deriving a system dynamics equation such as (1). In the complex systems that underlie the economy, for instance, there is a fundamental level which involves individuals and discrete events, like making a widget, buying a washing machine, driving to work etc. However, instead of attempting to model all this detail, these are treated in an average way and, as has been shown elsewhere (Allen, 1990), in order to derive deterministic, mechanical equations to describe the dynamics of a system, two assumptions are required:

- Assumption 1: events occur at an average rate;
- Assumption 2: all individuals of a given type x are identical to the average.

The errors introduced by the first assumption can be addressed by using a deeper, probabilistic dynamics called the 'Master Equation', which assumes that all individuals are identical and equal to an average type, but that events of different probabilities can and do occur. So, sequences of events which correspond to successive runs of good or bad 'luck' are included, with their relevant probabilities. As has been shown elsewhere (Allen, 1988) for systems with nonlinear interactions between individuals, what this does is destroy the idea of a trajectory. The evolution is described by a probability distribution for an ensemble of systems that gradually changes its shape, from being sharply peaked and centred on the initial value to spreading and splitting into a multimodal distribution with peaks that correspond to the different attractors of the dynamics. These may be point, cyclic or chaotic attractors, but clearly, if there are different possibilities then the idea of a trajectory for any single system breaks down. The fact is that unpredictable runs of good and bad luck can occur and, as a result, the system can spontaneously change its pattern of behaviour, 'jumping' into some new dynamical attractor, for some time.

However, in thinking about the problem of technological change and evolution, it is more important to discuss the other hypothesis that leads to the mechanical equations, i.e. that all individuals are identical and equal to the average type. Obviously, the first and most important fact about two individuals is that they cannot be at the same place at the same time, and so spatial structure is one of the underlying issues that has been treated inadequately by population dynamics of the usual kind. In the next section we show the result of correcting nonlinear equations in order to take the effects of microscopic diversity into account.

In physics and chemistry the predictive models which work so well rely on the fact that the individual elements that make up the system must obey fixed laws which govern their behaviour. The mechanisms are fixed, and the molecules never learn. But living systems cannot be described by such deterministic laws. To see why, let us imagine a very simple human situation, for example, of traffic moving along a highway or pedestrians milling around a shopping centre. Clearly, movements cannot be predicted using Newton's laws of motion because acceleration, change of direction, braking and stopping occur at the whim of each driver or pedestrian. Newton's laws, the laws of physics, are obeyed at all times by each part of the system but, despite this, they are of no help in predicting what will happen, because the decision to coast, turn, accelerate or brake lies with the human being.

Planets, billiard balls, and point particles are helpless slaves to the force fields in which they move, but people are not! People can switch sources of energy on or off and can respond, react, learn and change according to their individual experience and personality. They can see the potential usefulness for some modification in their timing, technique or tools, and they can tinker and experiment, perhaps finding ways to overcome a problem, or a new way to achieve some desired result. This is where innovation comes from, so the diversity of the experiments performed or ideas tried out will reflect the diversity of the people concerned. The ability of these experiments to be translated into improved and new production and business will reflect the encouragement or discouragement experienced by innovative individuals, and the information flows and scanning that organisations are doing to gather and evaluate such initiatives.

Because of this uncertainty in the longer term, we cannot know what actions are best now. Even if an individual knows exactly what he would like to achieve, as he cannot know with certainty how everyone else will respond, he can never calculate exactly what the outcome will be. He must make his decision and see what happens, being ready to take corrective actions if necessary. Since in business, on the road and in the shopping centre, we are all making these kinds of decision continuously and simultaneously, it is not surprising that occasionally there are accidents, or that such systems run in a 'nonmechanical' way. An important point to remember is of course that human beings have evolved within such a system and therefore that the capacity to live with such permanent uncertainty is quite natural to us. It may even be what characterises living. However, it also implies that much of what we do may be inexplicable in rational terms.

The mechanical approach is softened but not fundamentally changed by statistical models of decision processes where the probability of making a particular choice is proportional to the expected utility derived. This gives rise to probabilistic behaviour for individuals and deterministic behaviour for sufficiently large populations. However, this simple approach ignores the fact that decisions made by individuals are not really independent of each other, and that there is an effect of the communication between individuals. Fashions, styles and risk-minimising strategies affect collective behaviour considerably, and mean that it cannot be derived necessarily as the sum of independent, individual responses.

2.3 Evolutionary Drive

As we have seen above, in deriving kinetic equations in order to model the system that exists at a given time, it has been necessary to derive a reduced description of reality. This is made in terms of typical elements of the system, stereotypes, according to the classification scheme that we have decided to apply. Underneath the model there will always be the greater particularity and diversity of reality.

In the mechanical view, predictions can be made by simply running the equations forward in time and studying where they lead. Is there a unique attractor into which all initial states eventually fall, or are there many possible final end points? Does the system continue in a series of eternal cycles? Or does it display chaotic behaviour, as the trajectory wraps itself around a 'strange attractor'? Despite the interest of these questions, we should remember they are only of any significance if the equations and the fixed mechanisms within them remain a good description of the system and provide an explanation in terms of the internal functioning of the system. But, from the picture of the evolutionary tree in Fig. 1 which we know characterises complex systems, we see that the taxonomy of the system, the variables present and the mechanisms linking them actually change over time. Because of this, the dynamic system that we have adopted as a model of the system will only be a good description for as long as there is no evolutionary change and no new variables or mechanisms appear. In other words, the predictions of the dynamic system model will only be correct for as long as the model itself is a correct description of the system, and this is only for some unpredictable length of time.

Fig. 1 offers us a conceptual framework within which we can understand technological evolution. This has been described elsewhere (Allen, 1994a). In order to describe evolutionary change, we must try to suppress assumption 2 discussed above and put back the effects of innovators. Nelson and Winter (1982) have set out a seminal framework for economics in which internal variabilities and the differential survival of firms are explicitly taken into account as they compete in the production of a particular good. The evolution concerns returns on investment and techniques of production, and has been the basis for many later studies (Goodwin, Kruger and Vercelli, 1984, Anderson, Arrow and Pines, 1988, Silverberg, Dosi and Orsenigo, 1988, Lorents, 1989, Saviotti and Metcalfe, 1991). Clark and Juma (1988) have also set out the essential points concerning the difference between the long and short term view of economic systems, and how this leads to an evolutionary view.

Returning to the general conceptual framework of Fig. 1, we see that in order for us to understand and model a system that can change its taxonomy endogenously we must 'put back' what assumptions 1 and 2 took out in order to get to the deterministic description of nonlinear dynamics. Clearly, the future of any system will be due to two kinds of terms: changes brought about by the deterministic action of the typical behaviour of its average components, and structural qualitative changes brought about by the presence of non-average components and conditions within the system.

We really have a dialogue between the average dynamics of the chosen description (a process that results in what we may call selection) and the exploratory, unpredictable 'non-average' perturbations that result from the inevitable occurrence of non-average events and components. This is a search or exploration process that generates information about the payoffs for other behaviours and leads to the new concept of 'evolutionary drive' (Allen and McGlade, 1988, Allen and Lesser, 1993).

In order to explore the behaviour of systems with endogenously generated innovations and selection, we define a 'possibility space' which represents the range of different techniques and behaviours that could potentially arise. In practice, of course, this is a multidimensional space of which we would only be able to anticipate a few of the principal dimensions. This possibility space will be explored by individuals and groups who investigate the payoffs of new behaviour. In biology, genetic mechanisms ensure that different possibilities are explored, so that the off-spring, and their off-spring in turn, spread out over time from any pure condition. In human systems the imperfections and subjectivity of existence mean that techniques and behaviours are never passed on exactly, and therefore exploration and innovation are always present as a result of the individuality and contextual nature of experience. Local conditions, materials and needs differ and therefore any 'pure' technique or behaviour that migrates into a locality will rapidly diverge in its nature and intent. The diversity of existence itself generates complexity, and hence complexity feeds on itself.

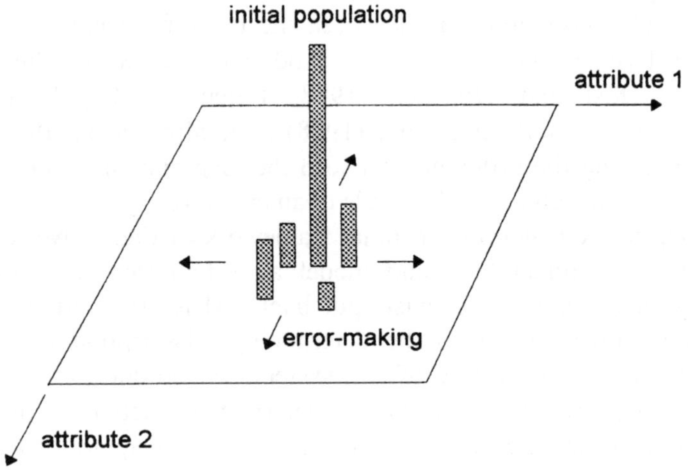

Fig. 2 In 'possibility space', an initially pure behaviour will diffuse outwards as a result of imperfect imitation, learning and features of the local context. Differential success provides 'selection'.

Physical constraints mean that some behaviours do better than others, so imitation and growth lead to the increase of some behaviours and the decline of others. If possibility space is seen as a kind of 'evolutionary landscape', with hills representing behaviours of high performance, then our simulations lead to the amplification of populations which are further up the hill and the suppression of those which are lower down.

By considering dynamic equations in which there is an outward diffusion in character space from any behaviour that is present, we can see how such a system would evolve. If there are types of behaviour with higher and lower payoffs, then the 'up-hill' diffusion is gradually amplified and the 'down-hill' diffusion is suppressed, so the 'average' for the whole population moves higher up the slope. This is the mechanism by which adaptation takes place. It demonstrates the vital part played by exploratory, non-average behaviour and shows that, in the long term, evolution favours populations with the ability to learn, rather than for populations with optimal, but fixed, behaviour.

The self-organising geographic models developed by Allen and Sanglier in the eighties are a simple particular case of these general ideas. Instead of some 'behaviour' space, what we have is real, geographic space. Individuals of any particular type x all differ from one another, being located at different points in space. By using distributions of choice and behaviour around an average, the microscopic diversity of individuals is taken into account, allowing the 'exploration' of seemingly unpopular, irrational and non-average decisions. In this way, changes in the payoffs for novel behaviour can be detected in the system, and innovations can take off. In this case, it concerns 'spatial' innovations, such as the spontaneous emergence of new centres of employment, or of peripheral shopping centres, of industrial satellites and so on. Because of the presence of positive feedback loops, there are many possible final states to which the system can converge, depending on the precise position and timing of non-average events. Information can obviously only come from the paths that were actually taken, not from those that were not taken and, because of this, patterns of change feed upon themselves, resulting in the self-reinforcement of growth and decline. Instead of an objective rationality expressing genuine comparative advantages, the beliefs and the structures co-evolve (Allen and Lesser, 1991).

2.4 The Evolution of 'Communities'

In this section we shall take the evolutionary models a stage further and examine the mutual co-evolution of different populations. Instead of considering the evolution of techniques and behaviours in a fixed landscape expressing higher/lower payoffs, we shall allow for the fact that the payoffs, the adaptive landscapes, are really generated by the interactions of a population with the other populations in the system. In the possibility space, closely similar behaviours are considered to be most in competition with each other, since they require similar resources, and must find a similar niche in the system. However, we assume that in this particular dimension there is some distance in character space, some level of dissimilarity, at which two behaviours do not compete.

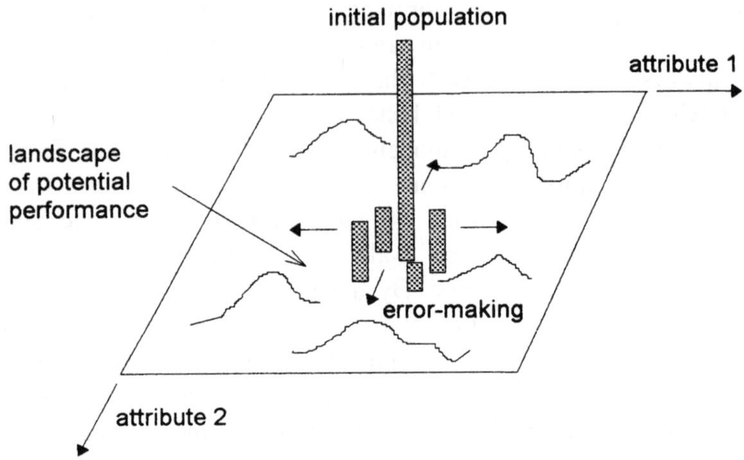

Fig. 3 The effect of 'error-making' in the reproduction of a population produces a diffusion into the surrounding 'character' space. The landscape shows the peaks and valleys of potential performance.

Initially, a single population is placed at the centre of possibility space, and, since it has adequate resources, it grows. However, error-making in the transmission through time of the 'rules' underlying any specific behaviour or technique means that there is a constant tendency for divergent behaviours to appear. This operates just like a diffusion process, except that instead of spreading outwards into geographical space, the diffusion represents the population behaviour spreading outwards to new behaviours and types (see Fig. 3).

During the initial phase of an experiment, we start off with a single population in an empty resource space. Resources are plentiful and the centre of the distribution, the average type, grows better than the eccentrics at the edge. The population forms a sharp spike, with the diffusing eccentrics suppressed by their unsuccessful competition with the average type. However, any single behaviour can only grow until it reaches the limits set by its input requirements or, in the case of an economic activity, by the market limit for any particular product. After this, it is the 'eccentrics' or error-makers that grow more successfully than the average type, and the population identity becomes unstable. The single sharply spiked distribution spreads, and splits into new populations that climb the evolutionary landscape that has been created, leading away from the ancestral type. The new populations move away from each other and grow until in their turn they reach the limits of their new normality, whereupon they also split into new behaviours, gradually filling the resource spectrum as shown in Fig. 4.

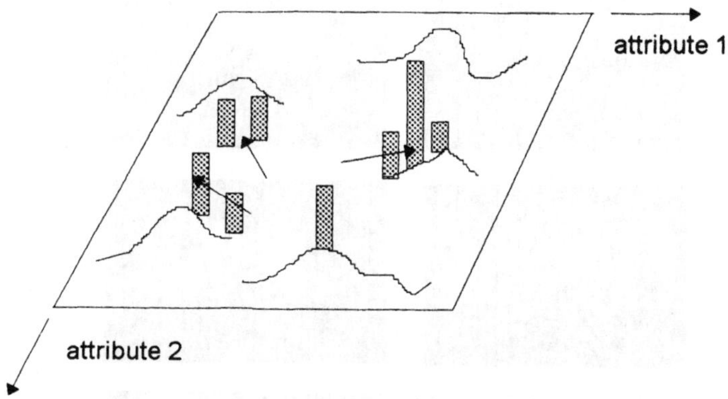

Fig. 4 A single population splits into several different groups, which separate and climb the hills of this fixed landscape. However, in general the landscape will depend on the populations in play.

In Fig. 5 we see the changing qualitative structure of the system over time, in some two-dimensional possibility space. Instead of simply evolving towards the peaks of a fixed evolutionary landscape, through their interactions populations really create the landscape upon which they move; by moving across it they change it. So the different behaviours grow, split off and gradually fill the possibility space with an 'ecology' of activities, each identity and role being formed by the mutual interaction and identities of the others. The limit of such a process would be given by the amount of energy that is available for useful work that can be accessed by the 'technological' possibilities potentially open to the system. This means that evolutionary processes would explore and reinforce mutually consistent technologies and strategies that capture parts of the energy flows through the system, using them to build and maintain their necessary internal structure. The limit would be set by the amount of available energy.

While the error-making and inventive capacity of the system in our simulation is a constant fraction of the activity present at any time, the system evolves in discontinuous steps of instability, separated by periods of taxonomic stability. In other words, there are times when the system structure can suppress the incipient instabilities caused by innovative exploration of its inhabitants, and there are other times when it cannot suppress them and a new population emerges.

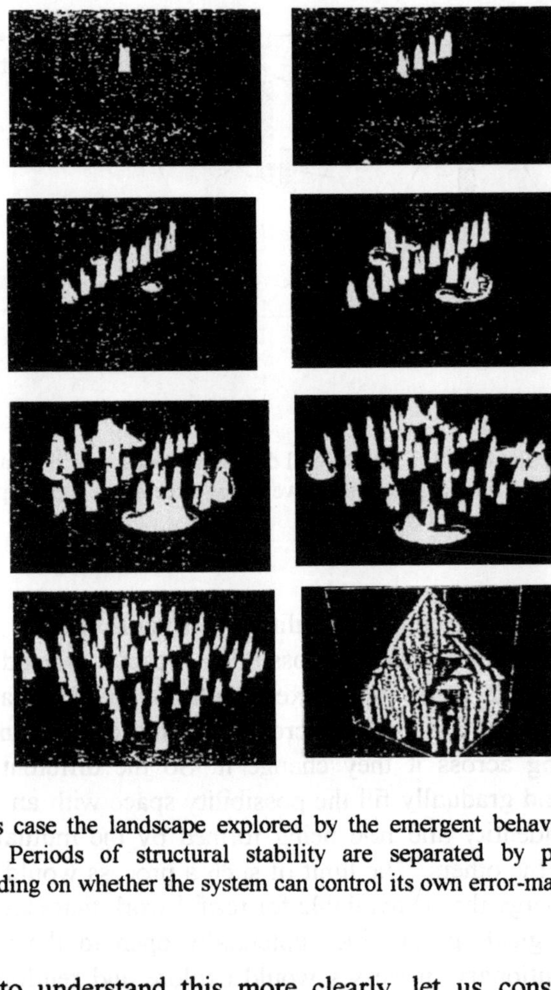

Fig. 5 In this case the landscape explored by the emergent behaviours is shaped by them. Periods of structural stability are separated by periods of change, depending on whether the system can control its own error-making or not.

In order to understand this more clearly, let us consider in detail a simple, one dimensional character space, in which there is not only competition for underlying resources, but also other possible interactions. For example, any two particular populations, i and j, practising their characteristic behaviours, may have an effect on one another. This could be positive, in that the side-effects of the activity of j might provide conditions that help i. Of course, the effect might equally be antagonistic or neutral. Similarly, i may have a positive, negative or neutral effect on j. If, therefore, we initially choose values randomly for all possible interactions between all i and j, then these effects will come into play if the populations concerned are in fact present. If they are not, then obviously there can be no positive or negative effects experienced. To express this more precisely, let us consider 20 points. Between each of them a random number is chosen to represent the value of the potential interaction on i from j:

$$interaction(j,i) = fr \times (2 \times random(j,i) - 1)$$

where $random(j,i)$ is a random number between 0 and 1, and fr is the strength of the interaction.

Each population that is present will experience the net effect resulting from all the other populations present. Similarly, it will affect those populations by its presence:

$$net\text{-}effect(i) = \sum_j x(i) \times interaction(j,i)$$

The sum is over j including i, so we are looking at behaviours that in addition to interacting with each other, also feed back on themselves. There will also always be competition for underlying resources, which we shall represent by:

$$crowding(i) = \sum_j \frac{x(j)}{(1 + \rho \times distance(i,j))}$$

At any time, then, we can draw the landscape of synergy and antagonism that is generated and experienced by the populations present in the system. We can therefore write down the equation for the change in population of each of the $x(j)$. It will contain the positive and negative effects of the influence of the other populations present, as well as the competition for resources and the error-making diffusion through which populations of type i, in small numbers, invade neighbouring behaviours.

$$dx(i)/dt = b(fx(i)+0.5(1-f)x(i-1)+$$
$$+0.5(1-f)x(i+1))(1+0.04net\text{-}effect(i))(crowding(i)/N)-mx(i)$$

where f is the fidelity of reproduction (0.99).

Let us start a simulation with a single population of five individuals placed at 10. In other words, $x(10)=5$. The only population initially present is 10, and therefore the evolutionary landscape in which it sits is in fact that which it creates itself. No other populations are yet present to contribute to the overall landscape.

What matters initially is how population 10 affects itself, since no other populations are present. This may have positive or negative effects depending on the random selection made at the start of the simulation. However, in general, population 10 will grow and begin to diffuse into the types 9 and 11. Gradually, the landscape will reflect the effects that types 9, 10 and 11 have on each other, and the diffusion will continue into the other possible populations. Hills in the landscape will be climbed by the populations, but as they climb, they change their behaviour, changing the landscape for themselves and others. Figs. 6, 7 and 8 show this process taking place over time.

Time = 163 Population = 51 Symbiosis ind. = 8 Competition ind. = 26.1

55 18 -49 -14 7 46 -11 -56 7 13 25 12 19 -37 1 27 52 27 47 0

Population of Each Type

0 0 0 0 0 0 0 0 0 49 2 0 0 0 0 0 0 0 0 0

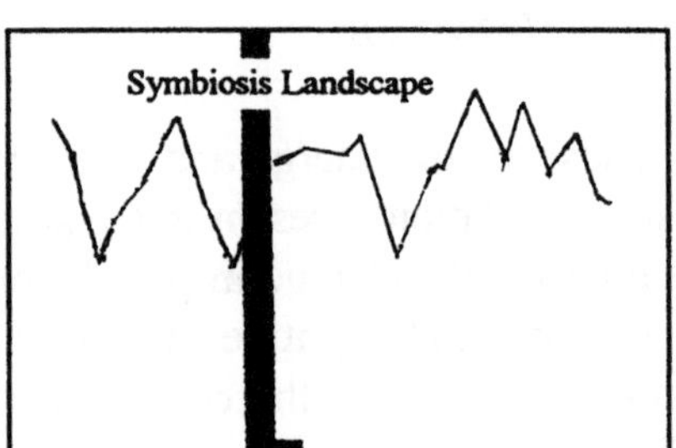

Birth Rate = 0.5
Mortality Rate = 0.1
Competitive Distance = 10.0
Seed = 3

Fig. 6 The initial population and evolutionary landscape of our simulation (t=163).

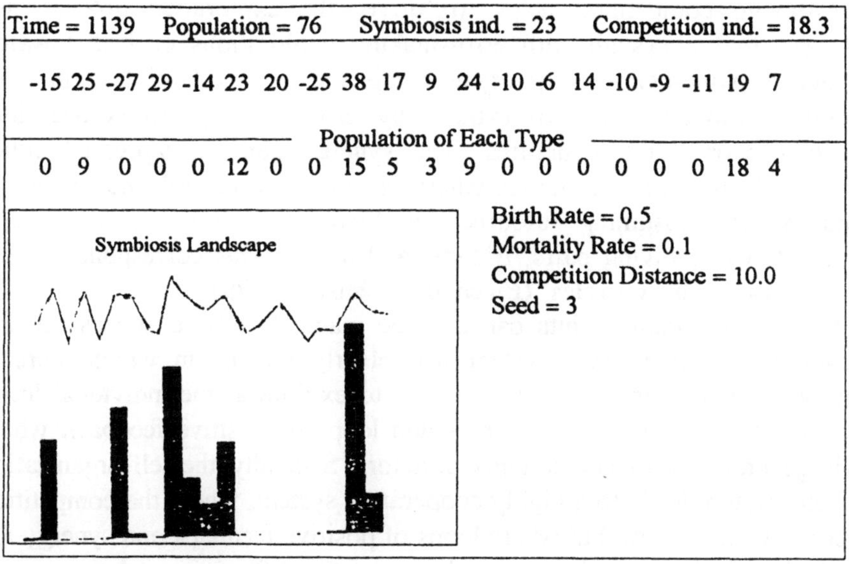

Fig. 7 After 1139 steps several populations have grown and the landscape as a result has changed.

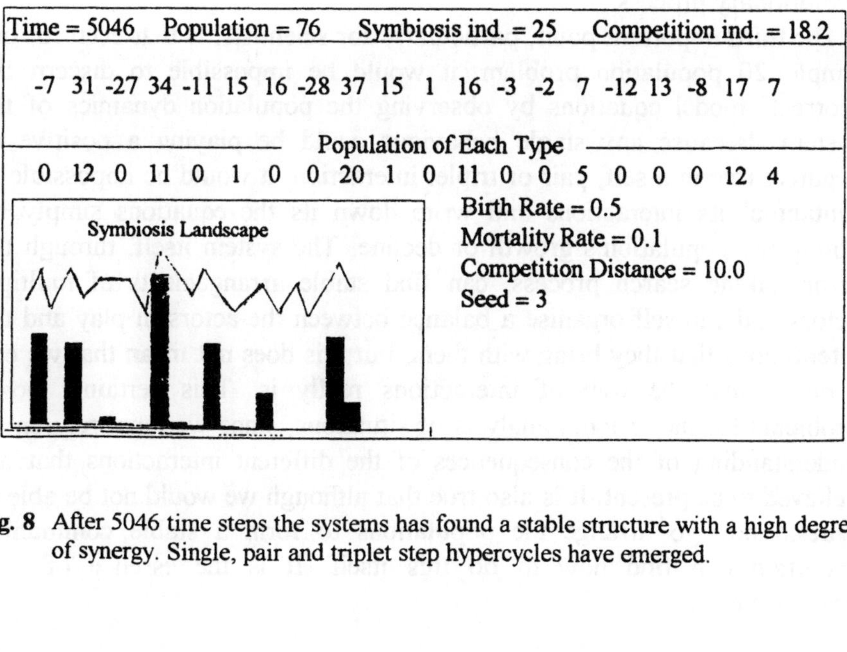

Fig. 8 After 5046 time steps the systems has found a stable structure with a high degree of synergy. Single, pair and triplet step hypercycles have emerged.

Although competition helps to 'drive' the exploration process, we observe that a system with error-making explorations evolves towards structures which express synergetic complementarities. Evolution, in other words, although driven to explore by error-making and competition, evolves cooperative structures. The synergy can be expressed either through 'self-symbiotic' terms, where the consequences of a behaviour, in addition to consuming resources, is favourable to itself, or through interactions involving pairs, triplets, and so on. This corresponds to the emergence of 'hypercycles' (Eigen and Schuster, 1979).

Several important points can now be made. Firstly, a successful and sustainable evolutionary system will clearly be one in which there is freedom for imagination and creativity to explore at the individual level, and to seek out complementarities and loops of positive feedback which will generate a stable community of actors. Secondly, the self-organisation of our system leads to a highly cooperative system, where the competition per individual is low, but where loops of positive feedback and synergy are high. In other words, the free evolution of the different populations, each seeking its own growth, leads to a system which is more cooperative than competitive. The vision of a modern, free market economy leading to, and requiring, a cut-throat society where selfish competitivity dominates, is shown to be false, at least in this simple case. From our example, the discovery of cooperativeness, and the formation of communities of players with a shared interest in each others' success, is the outcome of the evolutionary process.

The third important point, particularly for scientists, is that even for our simple 20 population problem, it would be impossible to discern the 'correct' model equations by observing the population dynamics of the system. Because any single behaviour could be playing a positive or negative role in a self, pair or triplet interaction, it would be impossible to 'untangle' its interactions and write down its the equations simply by noting the population's growth or decline. The system itself, through the error-making search process, can find stable arrangements of multiple actors and can self-organise a balance between the actors in play and the interactions that they bring with them, but this does not mean that we can deduce what the web of interactions really is. This certainly poses problems for the rational analysis of situations, since we must rely on an understanding of the consequences of the different interactions that are believed to be present. It is also true that although we would not be able to 'guess' how to arrange the populations to form a stable community, evolution can find how to do this itself. It is the essence of self-organisation.

Clearly, if we cannot really know how the circles of influence are formed by looking at the data, the only choice, in the case of a human system, would be to ask the actors involved. This in turn would raise the question of whether people really understand the roots of their own situation and the influences of the functional, emotional and historical links that build, maintain and cast down organisations and institutions. The loops of positive feedback that build structure introduce a truly collective aspect to any profound understanding of their nature, and this will be beyond any simple rational analysis used in a goal-seeking local context.

2.5 Self-Organisation of Cities and Regions

In this section, the ideas of 'self-organisation' are applied to the development of cities, with a view to establishing the basis for a decision support framework capable of exploring the longer term consequences of decisions, policies and of technology change. We hope from this to be able to build a model which can predict at least the sort of structure that may evolve under a certain scenario, with the accent on the qualitative features of that structure, rather than on quantitative accuracy.

The first step in the operation is to choose the significant actors, whose decisions, and the interplay of these, will cause the urban system to evolve. In agreement with much previous work, particularly, for example, the philosophy of a Lowry-type model, we first include the basic sector of employment for the city and, in particular, two radically different components: the industrial base, and business and financial employment. Then we consider the demand for goods and services, which will give rise to a local manufacturing and maintenance sector, as well as to tertiary service employment, generated by the population of the city and by the basic sectors. We shall suppose that there are two levels: frequently required, short-range services and a more specialised, long-range set of services. The residents of the city, depending on their type of employment, will exhibit a range of socio-economic behaviour, and for this we have supposed two populations corresponding essentially to 'blue' and 'white' collar workers.

This is our taxonomy of the city. In reality, over long time periods these variables will change, as a blue collar worker ceases to 'be' what he was, and white collar work splits into different types and classes, as new industries and activities appear. Nevertheless, for the model we shall develop, these categories will be considered sufficiently stable in their locational preferences for the time period we wish to consider, for the categories to remain coherent and meaningful during the simulation. Having specified the variables, we now need to define the mechanisms that cause the changes in the value of these variables in each zone. These mechanisms express the average effects of individual events or decisions which lead to the growth or decline of people or jobs of a given type in a zone, or to their in or out migration. In other words they capture the effects of birth, death, and migration of people and jobs.

While birth and death rates are socio-cultural parameters reflecting the religious, social and economic circumstances of individuals, the creation or reduction of jobs in a particular sector reflects in the longer term the profitability of that sector in that zone. If, for example, demand exceeds supply in the retail sector in a given zone, then the excess profits that are possible will lead to investment and job creation. This will increase supply and potentially reduce the excess profits, but in so doing it will have changed the distribution of population as the new jobs created lead to relocation of the employees and to the transfer of their demands for goods and services to the neighbourhood. This is turn will change the pattern of profitability in the other sectors, and will lead to the further creation of jobs and population. So, the linkages between people and jobs, and jobs and people through the spatial expression of intermediate and final demand will lead to a complex cascade of change and re-adjustment as the city grows.

In order to model the mechanisms governing the location and re-location pattern, and the investment pattern, we need to model not just the behaviour that is observed in the data, but the reasons that lie behind it. In other words, we need to represent the locational criteria of the different types of actor, and the changing opportunities that they perceive around them, and from this to generate our 'urban dynamic'. The model is therefore based on the interaction mechanisms of these variables, which in essence require a knowledge of the values and preferences of the different types of actors represented by the variables and, of course, how these values conflict and reinforce each other as the system evolves. In other words, the model is driven by the actions and behaviours of actors who are fulfilling certain roles or tasks corresponding to their job requirements and their cultural 'identity', which dictate how they wish to be viewed and also their pattern of final consumption.

The professional roles that actors adopt concern the successful functioning of their activities, and these therefore reflect their beliefs concerning the functional requirements of jobs in the different sectors. So, heavy industry must ship in large quantities of raw materials, engage in energy and material-intensive transformation processes, get rid of waste, and then ship out the finished products. Its activities lead to a characteristic 'value-added per square metre', which sets limits on the rents which still allow profits and add to the criteria which affect location. Similarly, office headquarters for financial institutions, for example, need to be in centres of communication, in prestigious surroundings, preferably where they can meet with other similar professionals over lunch, so as to keep up with the news and with the latest trends. Obviously, in the commercial and retailing sectors, logistic considerations play a vital role and the spatial organisation, or rather self-organisation of supply chains, can be seen as the underlying dynamic of urban and regional development. Our model consists of equations which express as a set of interacting mechanisms the interdependent location criteria characterising the different urban actors, who need to be near their customers, to have cheap accessible land, etc.

The model is inspired by data relating to Brussels, and so comes much closer to describing reality. The interaction scheme is shown in Fig. 9.

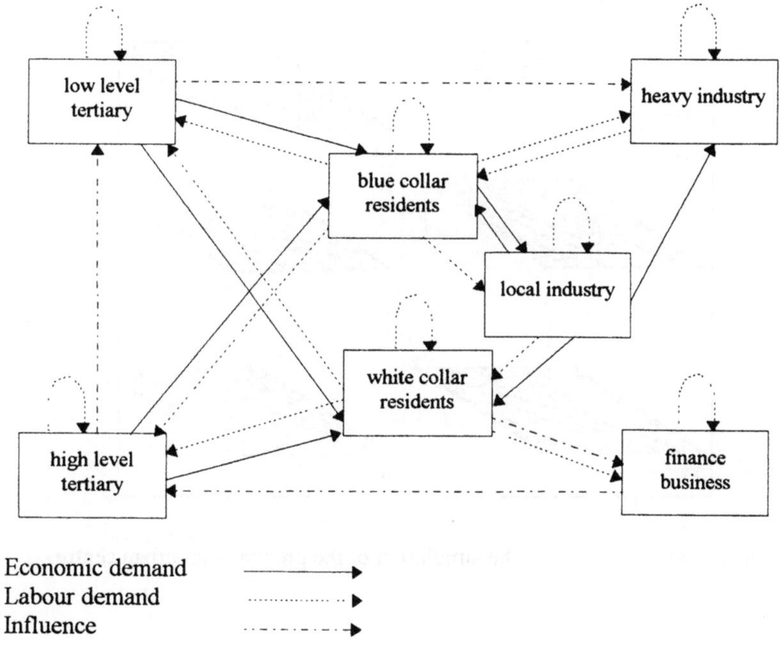

Fig. 9 The interaction scheme for a dynamic, spatially self-organising urban model.

There are five types of employer: industrial, financial, two levels of tertiary activity, and local industry. Each of these has its own locational criteria involving land and infrastructural requirements, as well as differing types of access to road, rail, canal or to air communication. They also have differing labour requirements both in terms of the number of jobs created per square metre and the socio-economic group of employees. Thus, heavy industry requires overwhelmingly blue collar labour, whereas the financial and business firms of the central business district employ predominantly white collar workers. We have therefore chosen to distinguish between these two types of resident, which, together with the five types of employer, form our 'mechanics' of seven mutually interacting variables.

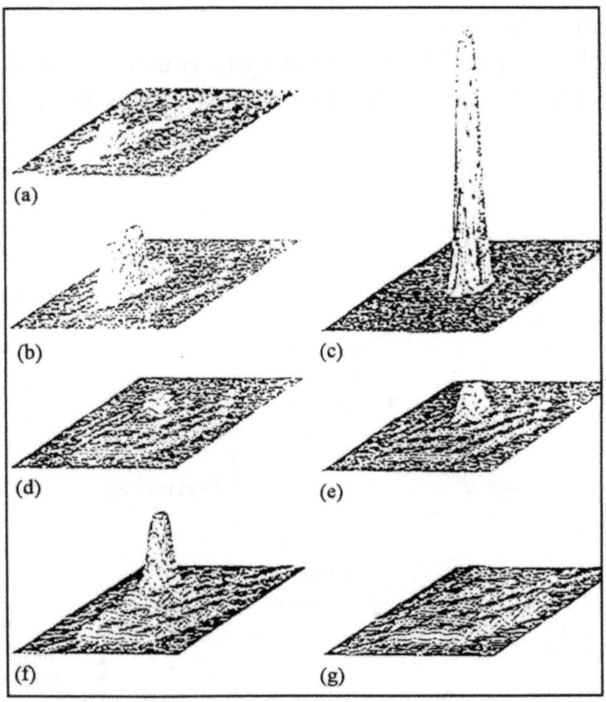

Fig. 10 The initial conditions for the simulation of the growth of an urban centre.

This model was reported elsewhere some years ago (Allen, Engelen and Sanglier, 1983), but recently the whole system has been redeveloped for the PC environment and the models are again a focus of interest. Here we shall simply present a typical evolution and also some results concerning the spatial structure of the tertiary functions, the shopping centres, and their evolution within the urban tissue.

The initial condition of Fig. 10 is the result of an earlier simulation from a smaller centre. Industrial employment (a) lies along a transport axis suitable for heavy goods; employment in local industry (b) is located near heavy industry and also near the residential population; financial and business employment (c) are concentrated in the centre of the city, the central business district (CBD). Low-level tertiary employment (d) simply reflects the population distribution, while high-level tertiary employment (e) is concentrated in the city centre. There is a single shopping district located in the centre. Blue collar residents (f) are mainly concentrated in the centre of the city; white collar residents (g), whose employment is located in the CBD, and also low-level tertiary activities are spread throughout the city.

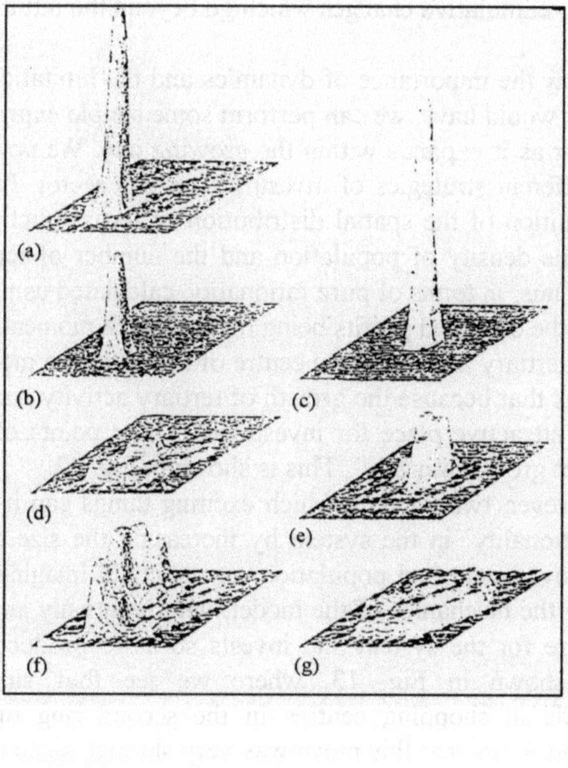

Fig. 11 The simulation at t=900.

In Fig. 11 we see how the growth of the urban centre leads, through spatial instabilities, to the creation of a functional structure where certain activities are concentrated in certain areas, and where residential segregation has also developed. In particular, we find that heavy and local industry (a) (b) concentrates in a single massive complex; finance and business (c) remain concentrated in the central business district (CBD) low-level tertiary employment (d) reflects both the total population distribution and competition for land with that same population, high-level tertiary employment (e) is widespread and blue collar residents (f) are concentrated largely along the transportation axis. Now the highest density is neither in the city centre, nor at the pole of industrial employment, but is in between. White collar residents (g) are more evenly spread through the city with a 'crater' around the industrial area and peak density off-centre.

Clearly, such a model enables us to study, for example, the effects of introducing some new transportation link, or the effects on the urban structure of changing costs of gasoline, as the decisions of employers and residents concerning their location choices interact with each other, causing complex, cumulative changes which go beyond the intuition of any single actor.

In order to show the importance of dynamics and the limitations that an equilibrium view would have, we can perform some simple experiments on the tertiary sector as it expands within the growing city. We now examine the effects of different strategies of investment in this sector. First of all, we have an evolution of the spatial distribution of tertiary activity if the fluctuations in the density of population and the number of tertiary jobs are very small. Thus, in terms of pure rationality, calculated using a simple extrapolation of the observed profits being made at each moment, the place to invest for the tertiary activity is the centre of the city. The mechanics of our system tell us that because the growth of tertiary activity occurs there, this is the most attractive place for investment. Other points of potential investment do not grow 'naturally'. This is shown in Fig. 12.

There are, however, two ways in which exciting things can happen. We can put less 'rationality' in the system by increasing the size of random fluctuations of investment and population, or we could imagine an actor, not described by the mechanics of the model, who intuitively anticipates a different structure for the system and invests so as to produce this new state. This is shown in Fig. 13, where we see four simultaneous investments made in shopping centres in the second ring of the city. Subsequent events show that this move was very shrewd, since the centres attract increasing business, killing off the original shopping centre in the city centre.

Of course, if it had failed, the actor would have been considered 'stupid', and people would have felt a certain satisfaction in seeing that someone behaving abnormally, not to say irrationally, was punished for his sins. However, in our example the investment succeeded, and the actor was therefore 'clever' in guessing the threshold required. In any case, the urban structure is irreversibly marked by this action, which is a small one compared with the scale of its consequences.

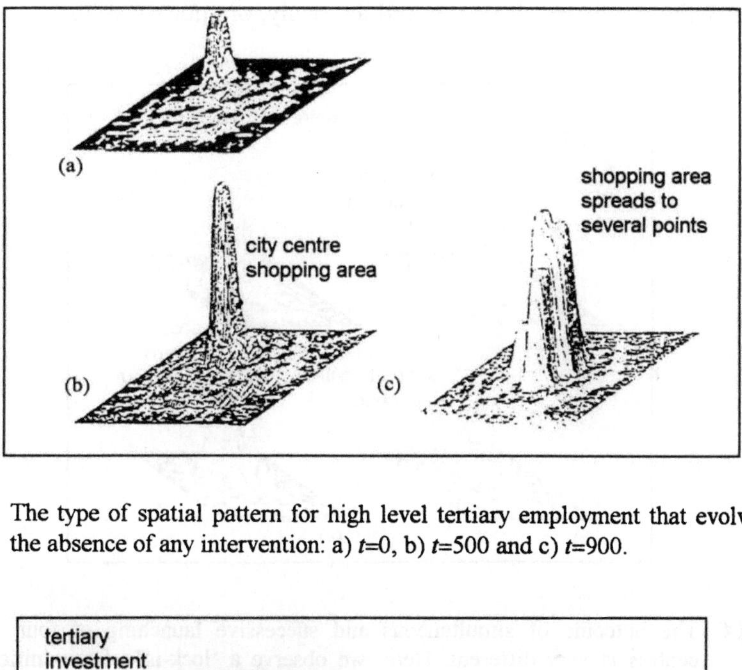

Fig. 12 The type of spatial pattern for high level tertiary employment that evolves in the absence of any intervention: a) t=0, b) t=500 and c) t=900.

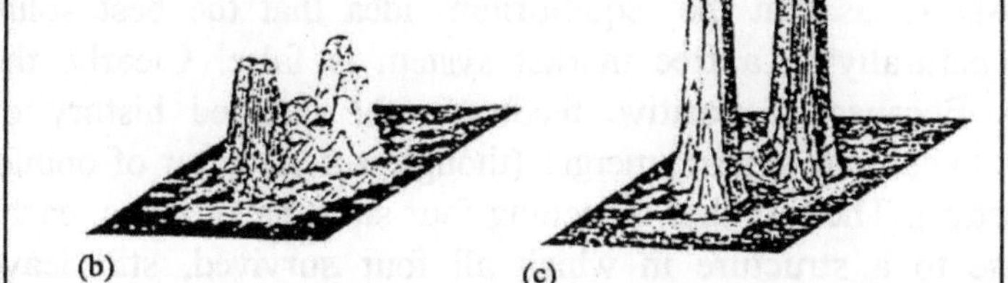

Fig. 13 Intervening in the tertiary sector at t=0 with 4 suburban centres, we find that they successfully grow and establish themselves: a) t=0, b) t=500, c) t=900.

On investigation we find that the emergence of four large shopping centres in the second ring of our city is a relatively stable possibility, providing that the centres are launched simultaneously, as in Fig. 13. This is true even if the action is taken much later, when the central shopping area has already attained a considerable size.

However, if we attempt to introduce the four centres successively, as shown in Fig. 14, we find that after the successful launching of the first suburban centre, the further investments that previously sufficed are unable to survive. The urban evolution is quite different, depending on whether the initiative is taken simultaneously, or successively.

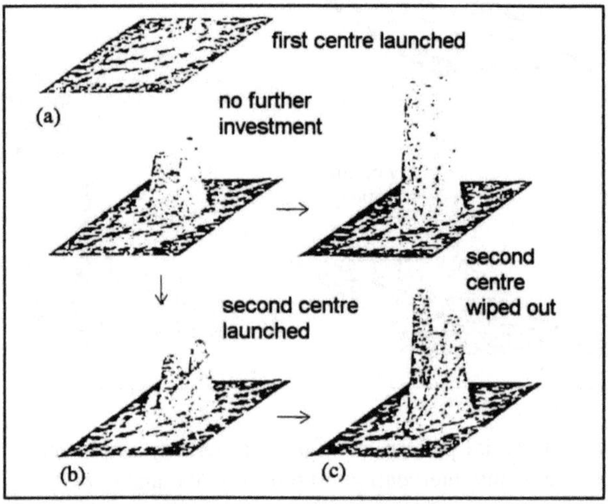

Fig. 14 The outcome of simultaneous and successive launching of four shopping centres is very different. Here, we observe a 'lock-in' of the initial action, showing that at the initial time there is no single equilibrium distribution to which the system must go: a) t=0, b) t=500, c) t=900

This shows us that the 'equilibrium' idea that the best solution will emerge naturally in a free market system is false. Clearly, the timing matters. Because of positive feedback, the detailed history of events changes the structure that emerges (though it is a matter of opinion which is the better). The fact that launching four suburban centres, early or late, gave rise to a structure in which all four survived, still leaves some question as to the 'spontaneous' formation of a hierarchy of shopping centres. To examine the question of whether there is some 'natural separation' of shopping centres, we can perform the following experiment.

Suppose that at the initial moment all twelve locations of the second ring are given a dose of tertiary investment. What will happen? The resulting evolution is shown in Fig. 15 where, as we see, small differences in initial height and accessibility on the transportation network lead to the breakup of the ring of shopping centres into five large centres spaced out at approximately every other point on the second suburban ring. Thus five centres are roughly all that this ring can support, which explains why, when we launched four simultaneously, they all survived. If we tried six or seven, then probably some would be eliminated. Clearly, our model provides us with the basis for software that would design logistics systems and supply chains on the basis of self-organisation. This is subject of current research for commercial applications.

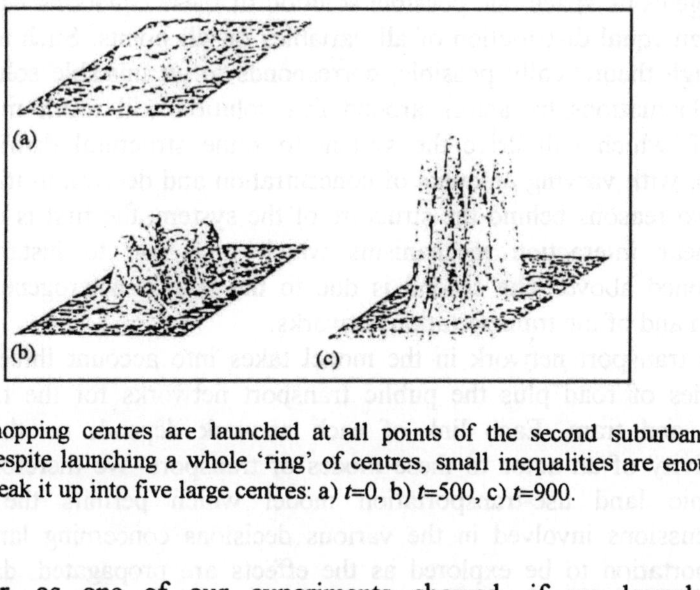

Fig. 15 Shopping centres are launched at all points of the second suburban ring. Despite launching a whole 'ring' of centres, small inequalities are enough to break it up into five large centres: a) t=0, b) t=500, c) t=900.

However, as one of our experiments showed, if we launch our investments successively, an asymmetrical structure can develop. This is stable except in the case of very large interventions. The facility with which a nonlinear system can break symmetries by amplifying fluctuations should warn us against assuming the stability of the present structure, and also of using such assumptions as 'maximum entropy' when dealing with complex systems.

Further improvements can be made to the model described above, which make it more like Brussels. In particular, we can include the various transportation networks that traverse and link the different areas of the city. All the perceived distances and decisions concerning residential location, shopping destinations etc. can be made with respect to the perceived attractiveness of the different possible transport modes and routes available. Can the qualitative evolution of Brussels be generated spontaneously by our model? If it is possible, then this implies that the model contains the 'reasons' for which the structure of Brussels has become what it has and, more importantly, why this might change in the future. It allows an exploration of the possible limits to the stability of this structure, indicating alternative future structures that might evolve under different possible policies, investment decisions and changing scenarios of in and out migration.

Our basic set of urban mechanisms is represented by a set of nonlinear differential equations each of which describes the time evolution of the number of jobs or residents of a particular type at a given point. In a homogeneous space one possible solution of these equations would be to have an equal distribution of all variables on all points. Such a non-city, although theoretically possible, corresponds to an unstable solution, and any fluctuations by actors around this solution will result in a higher payoff, which will drive the system to some structural distribution of actors, with varying amounts of concentration and decentralisation. There are two reasons behind the structure of the system: the first is due to the nonlinear interaction mechanisms which give rise to instabilities as mentioned above. The second is due to the spatial heterogeneity of the terrain and of the transportation networks.

The transport network in the model takes into account three different qualities of road plus the public transport networks for the train, bus, metro and tram. Each link of each network depends on the relative sensitivity of an actor to these modes of transport. We therefore have a dynamic land use-transportation model which permits the multiple repercussions involved in the various decisions concerning land use or transportation to be explored as the effects are propagated, damped or amplified.

The simulations described here have been based broadly on the evolution of a city resembling Brussels. The global characteristics of employment and population are shown in Table 1. The spatial evolution of urban structure is shown in Figs. 16-18 for successive times of the simulation. The simulation times of 0, 10, and 20 are of course somewhat unreal, but they are supposed to describe changes of urban structure which could occur over some 40 to 50 years. The initial condition is once again the result of a previous simulation made without a transportation network.

Variable	$t=10$	$t=20$	$t=30$
Total employment	729.600	669.500	674.300
Total number of active residents	462.670	411.560	414.200
Coefficient of employment	1.58	1.63	1.63
Employment structure			
Industry	25%	22%	22%
Tertiary	75%	78%	78%
Structure of commuter flows from outside to urban centre			
Blue collar	40%	33%	33%
White collar	33%	44%	44%

Table 1 Global figures generated by model simulation for three time periods

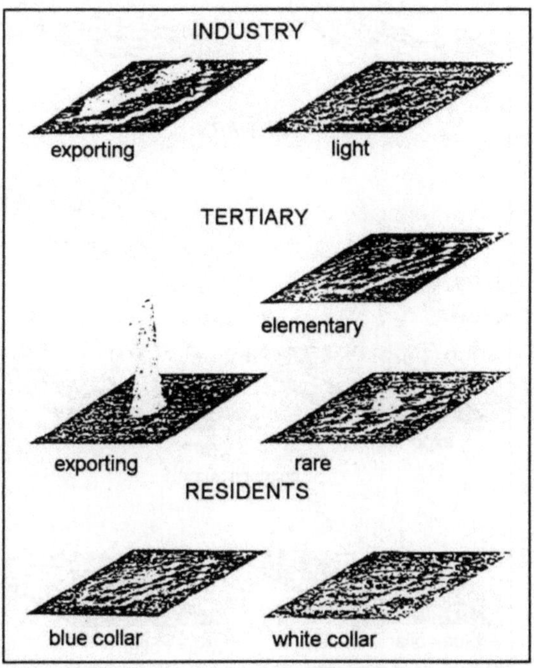

Fig. 16 Simulation at time $t=0$

Industry grows rapidly in the north and the south-west of the city along the transportation axis. Light industry, after some 'spatial indecision' locates at around t=7 near the airport in the north-east of Brussaville. The administrative and tertiary functions, characterised by office employment, after growing intensely in the very centre of the city, face a potential instability at around t=7. If at that moment a seed of investment or planning had been planted in the periphery, then it would have grown rapidly. However, without any such intervention, it is the neighbourhood directly to the east of the centre that attracts these exporting tertiary jobs. This corresponds to the immense concentration of office jobs in insurance, banking, and administration that are in fact packed along the Rue de la Loi, down towards the European Commission building to the east.

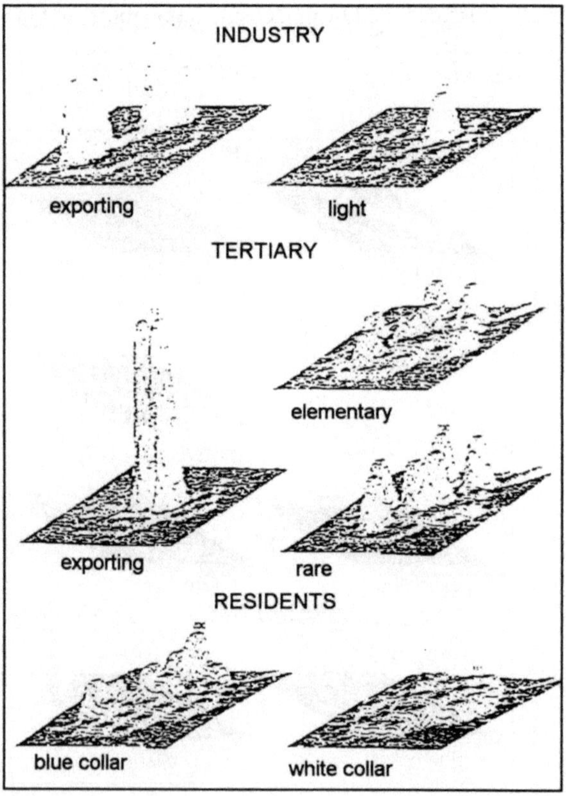

Fig. 17 Simulation at time $t = 10$

Shopping centres and commercial properties grow initially in the centre of the city. However, as land prices soar, a peripheral shopping centre springs up in the north of the city. This is rapidly followed by others, except in the south-east, where the presence of the Bois de la Cambre and the Forêt de Soignes reduce residential densities.

Blue collar residents concentrate in the north, the centre and the south west, while white collar residents live mainly in the east and south. As the time continues through t=20, total industrial employment in Brussaville decreases and is concentrated mainly in the southern industrial pole.

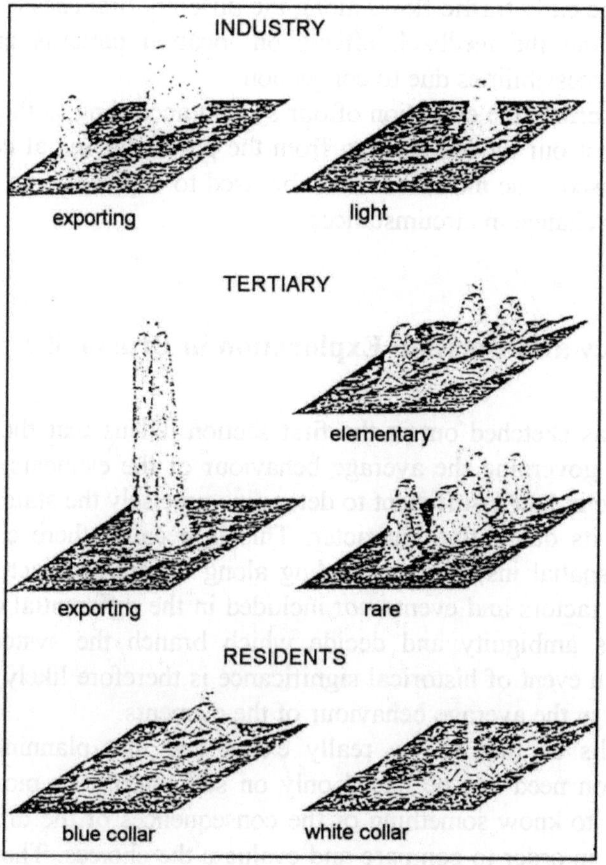

Fig. 18 Simulation at time $t=20$

We see that our urban system evolves to a complex interlocked structure of mutually dependent concentrations. We have two poles of heavy industry, reflected in the distribution of blue collar residents. Financial and business employment in the city centre begins to spread through the urban space at around t=6. Then, at a point adjacent to the centre, it exceeds a threshold and grows dramatically, causing the decentralised locations throughout the city to decrease. The white collar and blue collar residents spread out, many settling outside the system, according to the accessibilities of the networks. A hierarchy of shopping centres appears, serving the suburban population and encouraging further urban sprawl. The model generates not only the locations of employment and residents, but also the daily traffic flows along the different branches of the transport network, and the feedback effects on location patterns caused by the changed accessibilities due to congestion.

This describes the evolution of our system according to the deterministic equations of our model, starting from the particular initial conditions that we have used. The model can now be used to explore some simple policy options or change in circumstances.

2.6 Policy and Decision Exploration in Brussaville

The ideas sketched out in the first section tell us that the deterministic equations governing the average behaviour of the elements of a complex system are in fact insufficient to determine precisely the state of the system and even its qualitative character. This is because there could be many different spatial instabilities, leading along different trajectories. It is the effects of factors and events *not* included in the differential equations that break this ambiguity and decide which branch the system will really follow! An event of historical significance is therefore likely to be one not contained in the average behaviour of the elements.

This tells us that choice really exists and that planning, policy and intervention need not be based only on self interest or pious hope. It is necessary to know something of the consequences of the different options available, in order to compare and evaluate the choices. These evaluations should be made in a broad set of dimensions, corresponding to the different aspects of the 'quality of life' that the various inhabitants of the system may consider important. Clearly, our self-organising models are rather well suited to exploring the question of sustainability, since they examine the longer term implications of decisions and policies, including potential radical re-structuring of urban space. They generate the urban macrostructure (i.e. the overall patterns of flows and activities) from the microstructure within zones, considering the effects of changing the occupation, the pressures and the constraints experienced within them, which in turn feeds on to the macrostructure. Sustainability is about the possibility of finding micro and macro structures which are mutually compatible.

While the model can be used for evaluation, the actual decision concerning which action or policy should be pursued is of course a value judgment which must be made by political decision-makers on behalf of the community. The weighting accorded to different social groups, to the long or the short term, and to the degree of disparity between groups, are matters of social and political judgment. However, in the absence of a valid model, this judgment will be based purely on fictitious future perspectives. Developers will depict the desperate need for some installation, with future demand soaring, job creation, local economic revival and increasing local property values, and all this with apparently no harm, indeed positive good, for the environment! Objectors, on the other hand, will paint a very different image of the same project, pointing out the destruction of the natural beauty, or historical and architectural interest of the area, the threat of ecological collapse and future over-capacity in the area offering only slight short term economic benefits which would certainly not offset the serious reduction in property values that would certainly follow.

The self-organising models proposed here may provide a step towards an improved situation in which the significant consequences of policy can be explored, not just in their narrow context, but also in a wider, systemic, one, since the action may set off a chain of events and repercussions throughout the system. Most disagreements concerning decisions are not about the immediate short-term effects and the narrow context of construction costs, floor area, kilowatts required, immediate traffic changes etc., but concern the long term and wider implications of the decision, which is what our models may be suitable for exploring.

Here we shall briefly illustrate a few different types of urban decisions which can be explored in an evolutionary context of changing spatial organisation and travel patterns.

In the first example, we show in Fig. 19 several possible outcomes of the creation of a new shopping centre in our theoretical city, Brussaville.

Our model shows us that if at time t=10 we launch a new centre of 40 units, involving a total of some 4000 jobs at the location indicated, then it will grow and stabilise the retail structure, preventing similar later initiatives from succeeding. We have assumed that prices are the same as those elsewhere, but if the developer were prepared to accept lower profits during the start up period, then the centre could be launched with a smaller initial size. This question can be explored using our model. However, we see from the second part of Fig. 19 that if the same investment is made in the same place, but at a later time, t=20, then it does not succeed. Clearly, it could be kept going by lowering prices and profits, but intrinsically we see that time t=20 is less propitious than t=10.

In the event, an investment of 50 units at t=20, does in fact succeed, but our model shows us that if only 40 units were available, and normal profits necessary, then it could be successful only if the location of the proposed shopping centre were shifted to the point shown in the fourth part of Fig. 19.

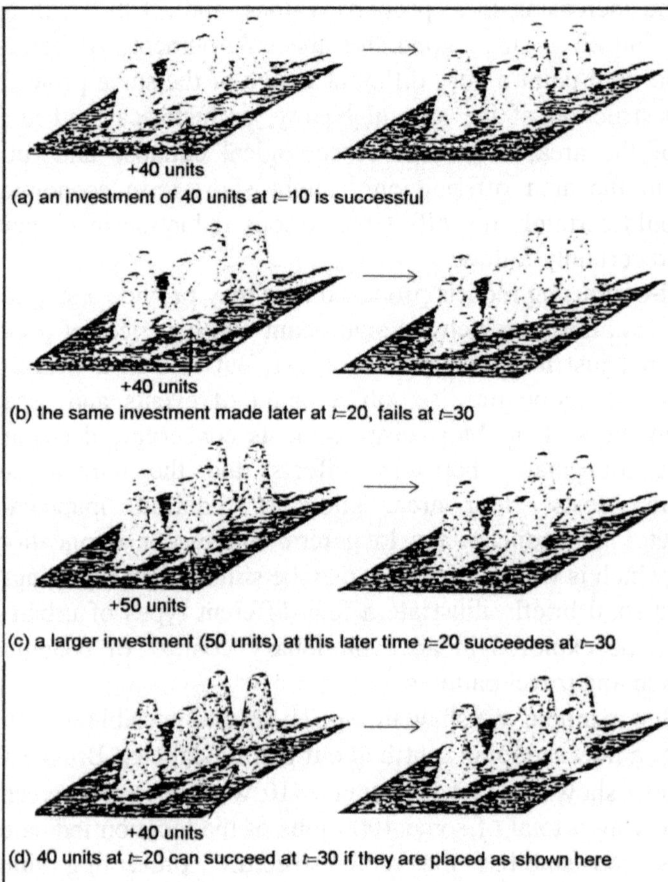

Fig. 19 The introduction of retail investment at different times and places

In the second example, Fig. 20, we show the long term impact of adding a new metro line across the city. Blue collar workers tend to increase in the neighbourhoods at each end of the line, and white collar residents return to the central core indicating some occurrences of gentrification. The model could be used to make a cost-benefit analysis of different possible routes, frequencies and speeds, in order to weigh up the possible long term consequences and arrive at a decision before embarking on the major upheaval that such a project implies.

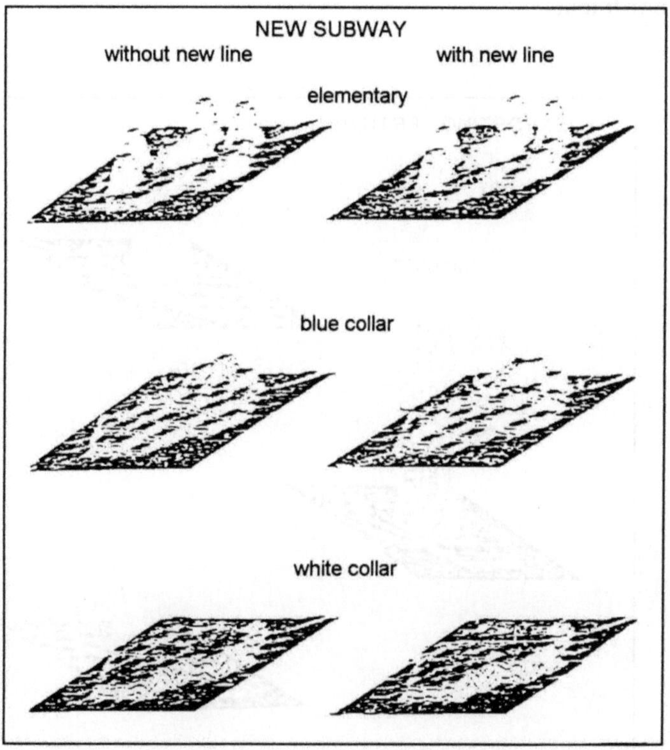

Fig. 20 The effects of introducing a new metro line can be studied using the model. It takes into account the complex chain of effects on land prices, changed accessibility etc.

The next example shows that if the increasing use of computer and telecommunication systems leads to a decrease in the need for business, finance and administration employment to aggregate and form the CBD, then our city could undergo a major structural revolution. In Fig. 21 we see that at a certain critical value the CBD disappears and office jobs are dispersed through and outside the city. Clearly, traffic flow patterns, residences and retail distribution will be vitally affected by this and we see some possible outcomes.

The self-organising model could be used to explore the effect of various types of disaster on the functioning of the city: pieces removed from the transportation system or neighbourhoods devastated by earthquake or floods, and also of course the effects of some man-made catastrophes such as the closure of some large industry on which the city depends, or of war and bombardment.

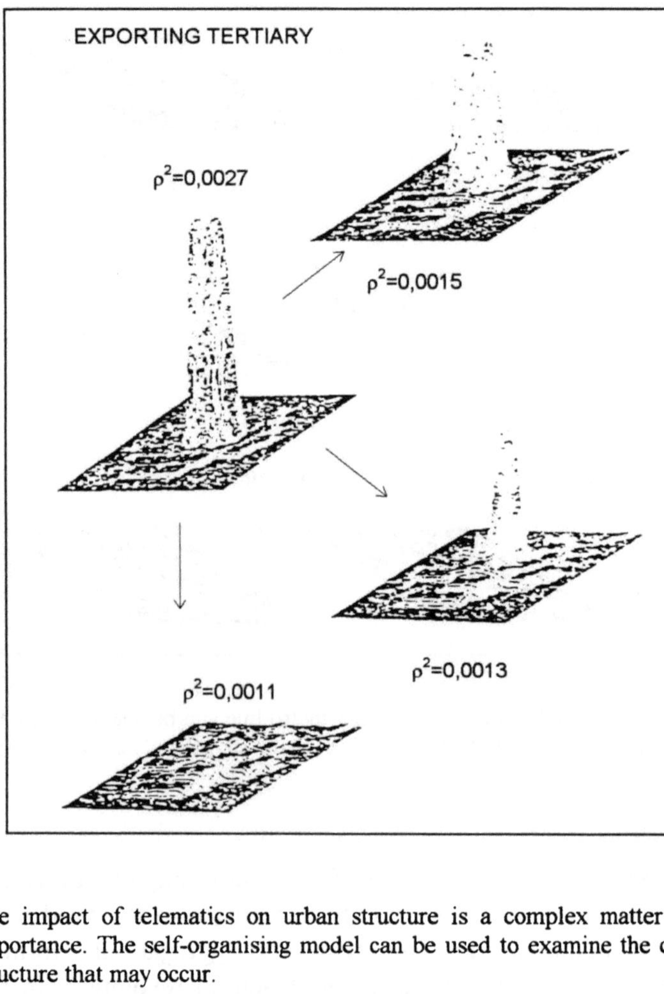

Fig. 21 The impact of telematics on urban structure is a complex matter of great importance. The self-organising model can be used to examine the change in structure that may occur.

One less dramatic example is shown in Fig. 22, where an urban centre grows from exactly the same initial conditions as before, with identical parameter values and locational criteria for the actors, except that instead of the city developing with a canal-river-railway crossing as before, we have in its place a line of hills. The only effect of this on the model is that the accessibility of these points, for the functioning of heavy industry, instead of being privileged is reduced.

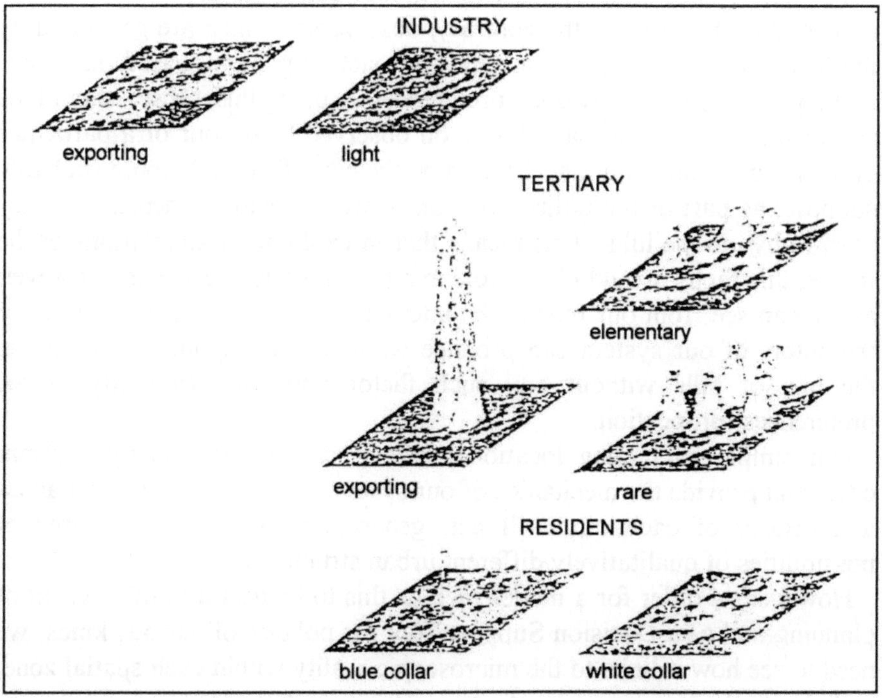

Fig. 22 Slightly altered geomorphology at the site would have given rise to a very different city structure, even for identical parameters for the actors.

We see that the city which evolves is totally different from that of our reference simulation shown above. Industry is dispersed throughout the urban area and with it the blue collar residents and local industry. White collar residents, instead of locating in the south-east, aggregate along the line of hills which have replaced the canal. The distances travelled to work are not the same as before, nor are the costs of shopping trips or the distribution of retail centres. Furthermore, not only are the spatial distributions of the variables different, but also the global quantities of industrial and tertiary activity, and the number of white and blue collar residents are modified. In fact this has been true for all our exploratory simulations, but here it is perhaps more striking because none of the parameters are different, only the terrain has changed. In this case the absence of an axis of good accessibility for industry lowers the attractivity of the whole city for this type of investment. Our model takes into account this change in its global performance.

The above underlines the fact that global quantities are not constraints on evolution but are, on the contrary, observables which are generated by the local events in the system. Our approach is generic, based on simple assumptions about individual preferences and, in this sense, should be contrasted with an approach based on observed behaviour of a particular system. For example, a model based on the city of Fig. 22 would probably suppose, as part of the utility function of white collar residents, that they wish to live in the hills. This means that in exploratory simulations of the future, this factor would play a role in attracting these residents. However, as we can see from our model, the interaction of the locational criteria of the actors of our system can produce white collar neighbourhoods along the line of hills without any such factor appearing explicitly in the preferences for location.

Our simple interacting locational criteria can generate many different cities and provide the mechanics of our system on which the circumstances and history of each city will act, generating an evolutionary tree of possibilities of qualitatively different urban structures.

However, in order for a model such as this to be useful in the context of planning and as a Decision Support Tool for policies of various kinds, we need to see how it links to the microscopic reality within each spatial zone, and also to the larger system of cities within which it sits.

2.7 Nested Complexity

The models above allow us to explore how two particular levels of description are linked by the interactive mechanisms resulting from the decisions and behaviours of their inhabitants. Each zone is characterised by aggregate measures of accessibility and housing and land availability, and is populated by inhabitants with behaviour that is distributed around an average. The 'attractivity' of a zone is given on average, and fails to represent explicitly the possible existence of different sub-localities with perhaps very different characteristics. The model as it stands would therefore fail to capture the real behaviour of firms and residents locating there, who may well find localities within it which are highly attractive and others which are quite unsuitable. In order to improve our representation therefore, we need to examine the lower level of description, and to build up the parameters which characterise a zone as the result of a more detailed calculation carried out at the microscopic level.

This could be done in a variety of ways. One simple way would be to examine the sub-locations in terms of their attractivity, using the criteria developed in the urban model above. Inside each zone therefore, we may trace the patterns of accessibility to the external zones, as well as of the suitability of land, the type of housing and the qualities of the different neighbourhoods. From this, at any given time, the different types of firms and inhabitants could be distributed through the zone according to the suitability of each parcel within the zone.

It is also at this point that these models link up with several other pieces of on-going research. Firstly, it can link to the work of White and Engelen (1993a, 1993b), who have developed cellular automata rules which generate the locational and co-locational features that characterise a particular city locally. It links also to the spatial syntax of Hillier and Hanson (1984) and Hillier and Penn (1992), who look at people's perceptions of the spatial system, and how the pattern of movement and location reflect this, and in turn structure space. Another interesting link is to the work of Batty and Longley (1994), who look at the fractal nature of the spatial patterns both of the boundaries of cities and also the way in which jobs and residences fill the space. Inside our self-organising macromodel of large zones, the constraints and pressures for the growth or decline of different types of inhabitant or employment can be enacted in detail using the micro-model, and from this a more precise and sensitive response will be generated within the localities, giving rise to more accurate representations of the 'average' parameters characterising the zone. This in turn will affect the macrodynamics, and through this the micro-repercussions in the other zones.

In the same way that the 'inner' spaces of the self-organising urban model need to be better represented, so the hierarchical level above also needs to be modelled. In fact, self-organising spatial models were first created for this regional or national level. The first applications (Allen and Sanglier, 1979, 1981) concerned the evolution and emergence of settlement patterns and urban hierarchies. A nonlinear dynamical system of equations underlying the patterns of supply and demand of different goods and services evolved as a result of the random occurrence (parachuting!) of entrepreneurs at different times and points in the system. Consumer demand was assumed to reflect relative prices. The creative interplay of random exploration and rational selection resulted in the gradual emergence of self-consistent market structures and patterns of settlement as economic functions either prospered or declined at their locations, giving rise to a dynamic Central Place Theory. This mixture of 'chance' and 'necessity' characterises all the models of self-organisation, and is clearly present in the work on 'fractal cities' by Batty and Longley, whereby the boundaries and the filling of urban and regional space are shaped by the interaction of probabilistic microevents, and where the probability distribution is shaped by deterministic equations of potential.

In the self-organising models, regional structure emerged as a result of the interplay of positive and negative feedbacks:

- positive due to the urban multiplier, economies of scale and externalities;
- negative through spatial competition both for producers and for residential space.

The emergent structure gradually 'locks in' to a somewhat imperfect, sub-optimal pattern, as a result of its particular history. A very large number of possible stable structures could potentially result, involving different numbers of centres in different locations, offering different levels of cost, utility, or efficiency. They show rather graphically that a 'free market' system does not necessarily run to an 'optimal' solution, but just to one of many possible solutions. The 'invisible hand' of Adam Smith is somewhat shaky, and as a result can lead to a large number of possible structures - each with its own mixture of good and bad qualities. Real choices exist, but we can only successfully make these strategic choices if we can understand, with the use of this kind of model, the qualitative evolution of the system over the long term.

An important point that these models also demonstrate is that the macro-evolution at the regional level is affected by the details of internal structure and spatial configuration of the small localities of which it is made. The hierarchical dynamics are affected by information flows affecting the mental maps of consumers (Gould and White, 1978) and by the perceptions of space by the inhabitants, the focus of the work by Hillier and Penn. The self-organisation of hierarchical structure therefore provides an important link between these different strands of work, and provides a new basis for understanding the strategic evolution of structure as a result of changing technology, transportation, resource availability, and various policies related to the social, environmental and economic dimensions.

Regional and national models have been developed on this basis, showing how the mutual interaction of urban centres has been both influenced by and, in its turn, influenced the flows of investment and of migrants. From these, an evolutionary model of Belgium, described in detail elsewhere (Sanglier and Allen, 1989) has been formulated. The model, which reproduced the spatial pattern of employment and population in the Provinces of Belgium, provided the context within which the model of Brussels has to be placed. The changing investment pattern that affected Brussels was in part understandable in terms of the pattern across the Belgian Provinces. It also reflected, of course, the place of Belgium within the European Union, the very large amount of trade crossing its frontiers and the growth of employment related to the location of European institutions in Brussels, as well as the multiplier effect of financial and business organisations locating in their proximity.

However, the overall 'success' of a city is in general conditioned by two main factors: the costs and benefits that result from its location within the national/international framework, and those arising from its internal structure. So, cities that suffer high levels of congestion, have poor infrastructure, dissatisfied residents with poor educational and training levels, environmental problems and high taxes, for example, will not be characterised by the same parameters of 'functioning' as cities with better internal structures and facilities. Because of this, the particularities of the internal structure, and the success or failure of urban centres with respect to their inhabitants, penetrates upwards to affect their capacity to attract investment and migrants, and hence their long term growth.

Other self-organising models have been developed at the regional level for:

1. the USA, generating inter-state migration (Allen and Engelen, 1985);
2. Senegal, a ten-region model linking population, employment and environmental factors (Allen and Engelen, 1991);
3. North Holland, a self-organising model examining the effects of energy efficiency measures on the local economy (Allen and Engelen, 1987);
4. SIMPOP, a general hierarchical rule-based regional simulation (Bura *et al.*, 1996).

These models have been quite successful in providing a basis for considering economic development, settlement patterns, urban structure, transport and energy. Future research work, however, will be directed at developing a truly hierarchical framework, which will automatically allow at least two levels of description to be modelled simultaneously and simulate the dynamic dialogue between the two levels, as micro and macro structures emerge and interact.

2.8 An Integrated Framework for Socio-Economic, Technological and Environmental Evolution

Now that 'sustainability' has become fashionable, there is a general understanding of the need to consider the long term consequences of our present urban lifestyle. This is a good thing, although it comes somewhat late in the day. The problem is though, that there is no clear view as to the meaning of sustainability, nor the manner in which it can be attained. In the UK, government interest has focused on 'economic' sustainability, which is translated into attempts to encourage commercial ventures using new, cleaner technology to promote energy savings and waste recycling and to charge full economic costs for things. This of course misses the point that it is perhaps the market system itself that threatens sustainability, by forcing high short term economic returns on businesses.

If we think seriously about sustainability, then it should concern the preservation of the options for future productive activities, and should involve a whole range of measures reflecting our 'quality of life'. In other words, in order to evaluate the contribution a policy, technology or action might make to sustainability, we require an integrated framework that could explore its overall, long term effects. For technologies, for example, it would include the implications of the production, use and disposal phases of the products, and also the overall effects of the chain of effects such as spatial re-organisation, which would be involved.

The kinds of model described above clearly offer a possible basis for such an integrated framework but, in the examples given, environmental variables are only taken into account in a very simple manner, and sustainability in environmental terms is not specifically addressed. In some more recent research (the Phaeocystis Report, 1993) the Belgium model was used as the socio-economic part of an integrated model that examined the whole Escaut/Scheldt river basin. The changing pattern of inputs to the river system and the groundwater was generated from the changing pattern of population, employment and land-use of an extended Belgian model which included part of Northern France. These human activities and impacts were then connected to an ecological, biochemical and physical model of the river basin, which allowed the calculation of variables such as the concentrations of oxygen, phosphates, nitrates, phyto and zooplankton, bacterial and organic wastes in each branch of the river as the water descended to the estuary. The integrated model could simulate the water quality in the different branches, the eutrophication of the lower reaches, the output of phosphates and nitrates to the North Sea, and much else. In this way, possible environmental policies and regulations could be tested on the system as a whole, showing their complex consequences.

For example, improved water treatment of urban outflows to the river, led to greater discharges of nitrate and phosphate to the sea, and to eutrophication, because the lower bacteria concentrations in the river were able to de-nitrify less of the nitrates than before. The model also allows an evaluation of the most effective actions/locations for a given investment, and explores the chain of effects that really accompany any particular environmental measure.

Another example of an integrated model that allows an examination of sustainable land-uses, and links environmental and socio-economic variables is that of the Argolid plain of Greece (Archaeomedes Report, 1994). In the Mediterranean, especially in the coastal areas, there is a rapid process of urbanisation, causing increasing population density and also intensification of agriculture. In an attempt to obtain rates of return on capital comparable with those of 'urban' activities, traditional farming practices are being replaced by more modern ones, i.e. with more lucrative crops, requiring increased use of water resources through irrigation.

In the case of the Argolid, the increased exploitation of the coastal aquifers has led to the salinisation of the aquifers and the land. Farming has gradually switched from the production of olives and cereals to the irrigated production of citrus fruits. The efforts of the European Commission's policy to avoid the decline of rural areas has crystallised into price support policies, which encourage the cultivation of citrus fruit.

In recent research a dynamic model has been built which successfully generates the self-destructive process which farmers have engaged upon. It is of interest because it shows how well intentioned policies at one level of the system can have a quite negative effect at another level. The model considers 7 spatial zones and 3 layers: the surface, the subsurface layer and the aquifer. The flows of water through the area of study was then modelled by considering the 3D movements of water onto the surface, through the subsurface layer (when permeable), into the aquifer and out again.

The main human impact has been the decision to grow irrigated crops, resulting in the need to pump water up from wells, boreholes and canals to maintain growth during the hot, dry summer. Before this, for time immemorial, the winter rainfall had fed the aquifers and springs, and given rise to a net positive hydrostatic pressure throughout the zone. Our model considers the chain of effects of pumping of underground water for the increasing area of irrigated crops.

The dynamic model that has been developed assumes that crop choice decisions are made annually, and that this sets out the agricultural requirements for water for the next 12 months. The amount of irrigation that will be required then depends on the profile of crop needs throughout the year, and the rainfall and the evapo-transpiration that actually occurs. The model uses a short time step of 3 days (one tenth of a month) to describe the movement of water and salt over and through the different zones and sections of the model. It simply uses balance equations based on water and salt accounts for each section.

Without discussing the detail of these calculations, we can summarise the model by saying that it allows us to model the farmer's response to his circumstances: market prices and uncertainties, crop choice, and water requirements. This then allows us to model the change in surface and aquifer water, and the salt concentrations in both as sea water is drawn into the aquifers. This in turn produces a pattern of salinisation, the demand from farmers for fresh water to be supplied from elsewhere by canal, and finally the need to increase production and water consumption to make up for falling yields.

The medium which transports the salt around the system is water. Initially, before large scale irrigation occurred, there was a gentle, positive hydraulic head throughout the system, which meant that the aquifers were pure, and that there were some marshy areas of land. There was a steady transfer of the catchment water to the sea. However, as the hectares of irrigated land were increased, the overall water balance of the ground water changed, and in around 1960, it became negative. The coastal zone irrigation rapidly led to the incursion of sea water into the ground water, with a consequent transfer of salt. The continued pumping transferred the salty water from the aquifer onto the surface, where the productivity of the soil was gradually eroded.

In response to this, a small canal was built to bring spring water from the western corner of the Argolid; this water was used for irrigation along the coastal strip. While this allowed intensive fruit tree growing to continue, the farmers further back from the coast continued to expand the area of irrigated crops and hence the amount of water pumped from the aquifer.

The lowered hydraulic pressures led water to feed back from the coastal aquifers, thus transporting sea salt underground some 20 km inland.

Gradually the salt problem has increased, so a large canal project was put in place to deliver spring water over a wide area of the plain. In addition, the lowered water level of the aquifer has meant that farmers on the edge of the Plain have found it increasingly difficult to get water at all, so water needs to be supplied from the large canal both because of salinisation, and increased water demand. The problem is that fresh water is in limited supply, and the sources are growing more salty. Ever greater technological intervention is thus being called upon to maintain the production of citrus fruits in an increasingly artificial landscape, totally reliant on costly infrastructure. When we realise that many of the oranges produced are in fact not consumed but buried to maintain market prices, it seems clear that there is some need to review policy in an integrated fashion!

It is important to realise that the situation in the Argolid is being repeated in many other locations and is, in reality, part of the unsustainable hidden reality of urbanisation. As populations have shifted to the cities, so the decision makers are increasingly divorced from the reality of the natural system that really supports the cities. Cities not only 'self-organise' themselves, but also their own and distant landscapes. The dubious power of economic exchange ensures that cities continue to maintain their supplies, if necessary with more intensive exploitation at greater distances, essentially 'strip-mining' the world's agricultural land.

There is clearly a need for an integrated framework which will allow an appreciation of the net change in real 'wealth', meaning not just the temporary flows of money captured in GNP, but the value of biological potential, the stocks of fertile soil, fresh water and other natural resources, which support the urban as well as the rural population.

2.9 Discussion

The fundamental basis for these models are the decisions of the different types of individual actor, reflecting their values and functional requirements. Although these are represented by very simple rules for each type of actor, when distributed among average and non-average individuals, they give rise to very complex patterns of structure and flow, and to a structural emergence and evolution at the collective level. In turn, the macrostructures that emerge constrain the choices of individuals and fashion their experience, so without the knowledge afforded by such models, there may be little correspondence between the goals of actors and what really happens to them. Each actor is co-evolving with the structures resulting from the behaviour of all the others, and surprise and uncertainty are part of the result. The 'selection' process results from the success or failure of different behaviours and strategies in the competitive and cooperative dynamical game that is running.

The spatial models of urban and regional evolution are examples of this kind of evolution. What emerges are 'ecologies' of populations, clustered into mutually consistent locations and activities, expressing a mixture of competition and symbiosis. This nested hierarchy is the result of evolution, and is not necessarily 'optimal' in any simple way, because there are a multiplicity of subjectivities and intentions, fed by a web of imperfect information. The total pattern emerges as a result of the interaction of imperfect patterns of behaviour for each type of actor, and what this really means is that there is an intrinsic element of unpredictability in the system. Creativity and adaptive response are therefore powered by the degree of heterogeneity of the population, and their microscopic diversity.

The idea that evolution leads to a community of interlocking behaviours is an important result. The history of a successful society within a region, is largely a tale of increasing cooperation and complementarity, not competition. An economy is a 'complex' of different activities that to some extent 'fit together' and need each other. Competition for customers, space, or for natural resources is only one aspect of reality. Others are familiar suppliers and markets, local skill development and specialisation, co-evolution of activities, networks of information flows and solidarities that lead to a collective generation and shaping of exchanges and discourse within the system.

These ideas help us to understand the origins of coherence in human systems. In classical physics, the smooth behaviour of macrovariables such as temperature and pressure, arises from the incoherent, random behaviour of the molecules. It is statistical averaging that leads to smooth behaviour. But, in self-organising systems, the individual elements really are behaving coherently, either as the result of an external parameter that affects each individual separately, or as the result of co-evolution. For example, the day/night cycle affects everyone separately, but leads to the coherent behaviour of people going to work and coming home. However, the interlocking of activities in such a way that food flows from the fields, through various processing and wrapping stages, and arrives on the supermarket shelves in time for the Saturday shopping rush, has demanded an enormous amount of skill and organisation, resulting from a long learning process. The working of a modern economy certainly displays great coherence, and the ideas discussed here attempt to show how coherence, and spatial and hierarchical structures emerge through processes of self-organisation, and that these models can help us anticipate future changes and different types of coherent structure that might emerge. This is really what policy exploration should be about.

The evolution of a society is not about a single type of behaviour 'winning' through its superior performance, since evolution is characterised more by increasing variety and complexity than the opposite. Instead, it is about the emergence of self-consistent 'sets' of activities, with mutually helpful effects. Potential supply and demand are not given independently of one another. People cannot experience what is not made available, but can only be affected by what is produced. Their lifestyle, demands and preferences are shaped by the supply that really occurs, and so a 'learning' dialogue shapes the patterns of consumption that develop in the system. Supply affects demand and vice versa. Cultural structures are formed by the effects of positive and negative feedbacks, imitation, economies of scale, learning by doing, etc. being positive feedbacks, and competition for attention, market and for resources negative ones.

In attempting to model the self-organisation of spatial markets, we must consider the possible effects of speculation in human systems. The important point is that the expected return on an investment is what drives investment, but this must depend on what people believe about the system. What people believe affects what happens, and what happens affects what people believe! This is a positive feedback loop which can be understood on the basis of the kind of models which we are developing. It severely affects the outcome of free markets, as we have seen repeatedly in commodity cycles, land speculation and the prices of almost anything of which there is a limited supply. Instead of free markets necessarily leading to a sensible and effective allocation of investment and resources, we find that prices can be driven by peoples' beliefs, and that these can feed on each other resulting in peaks and troughs, often causing massive misallocations of resources and waste. Clearly, the fact that 'trend creates trend' offers a considerable opportunity for instability and chaos, and this is rendered manageable only by the diversity of perceptions and motivations of human actors. Models can and are being developed to 'learn' robust mutually consistent strategies, and also to encourage diversity in the face of the mass media and instantaneous shared information.

Diversity is absolutely vital to the functioning of the system. Social interaction and mutually advantageous exchange can only occur if two actors are different. If they are both identical and average, then there can be no useful interchange. The evolution of a society and of the urban or regional landscape therefore reflects the specificities of individuals who have different aims, different information and different resources. Imperfect knowledge and plain ignorance all play a role in smoothing the responses of a population to a given situation. Decisions made by some individuals change the conditions and constraints on others, provoking successive responses and adjustments to the evolving circumstances.

The idea that we can solve our problems by simply releasing the forces of the free market is an illusion. The real complexity of the world involves the fact of collective structure, which is not amenable to any simplistic solution, be it central planning or free markets. The goals and strategies, the ethics and the understanding of individuals fashion the collective structure that emerges and give it complex properties which act on each individual uniquely, and which cannot easily be resumed in a few criteria. Similarly, the collective structure that emerges enriches and constrains the experiences and choices that are open to individuals. So one is dealing with the dialogue between individual freedom and beliefs, and the social, cultural, technological and physical realities in which they are embedded, and which they shape.

This discussion of self-organisation in complex systems reveals limits to scientific discourse. The future is inherently uncertain to some degree and, as a result, even the criteria for evaluating possible actions cannot be established with certainty. But knowing that this is so is an important step. If we are to learn from the way that the natural world copes with its inability to predict the future, then we see that parallelism, microdiversity and local freedom are key factors in its ability to deal with whatever happens. We must attempt to find a system that, while evolving enough coherence to function, retains enough individual freedom and microscopic diversity to provide a pool of adaptability and innovation so that it can constantly evolve and restructure in the face of change. In other words, we seek to create a 'learning' society, since this, in effect, is the basis of a sustainable future.

Acknowledgement. The urban and regional models were developed with the collaboration of Michèle Sanglier, Guy Engelen and Françoise Boon. The 3-D visualisation was made by Jack Corliss and M. Lesser at the Goddard Space Flight Center, NASA.

References

Allen P.M. (1988) Evolution: Why the Whole Is Greater than the Sum of its Parts, in Wolfe M.A. *et al.* (eds.) *Ecodynamics; Contribution to Theoretical Ecology*, Springer-Verlag, Berlin.

Allen P.M. (1990) Why the Future Is not What it Was, *Futures*, July/August, 555-570.

Allen P.M. (1994) Evolutionary Complex Systems: Models of Technology Change, in Leydesdorff L., Van den Bessalaar P. (eds.) *Evolutionary Economics and Chaos Theory*, Pinter, London.

Allen P.M., Engelen G. (1985) Modelling in Spatial Evolution of Population and Employment - The Case of the USA, in Ebeling W., Peschel M. (eds.) *Lotka-Volterra Approach to Cooperation and Competition Modelling in Dynamic System*, Mathematical Research, Academie-Verlag, Berlin.

Allen P.M., Engelen G. (1987) *Computer Handled Energy Efficiency Simulation Exploration* (*CHEESE*) Bureau van Energiebespaaring, North Holland, Amsterdam.

Allen P.M., Engelen G. (1991) An Integrated Strategic Planning and Policy Framework for Senegal, European Commission Report, DG VIII, Under Contract Article 8 946/89.

Allen P.M., Engelen G., Sanglier M. (1983) Self-Organising Dynamic Models of Human Systems, in Frehland E. (ed.) *From Microscopic to Macroscopic Order*, Synergetics Series, Springer-Verlag, Berlin, 150-173.

Allen P.M., Lesser M. (1991) Evolution: Travelling in an Imaginary Landscape, in Becker J., Eisele I., Mundemann F. (eds.) *Parallelism, Learning and Evolution*, Lecture Notes in Artificial Intelligence, 565, Springer-Verlag, Berlin, 419.

Allen P.M., Lesser M. (1993) Evolution: Ignorance and Selection, the Evolution of Cognitive Maps. New Paradigms for the Twenty-First Century, *The World Futures General Evolution Studies, 5*, Gordon and Breach, 119-134.

Allen P.M., McGlade J.M. (1987) Evolutionary Drive: The Effect of Microscopic Diversity, Error Making and Noise, *Foundations of Physics, 17*, 7, 723-728.

Allen P.M., Sanglier M. (1979) Dynamic Model of Growth in a Central Place System, *Geographical Analysis, 11*, 3, 256-272.

Allen P.M. Sanglier M. (1981) Urban Evolution, Self-Organisation and Decision-Making, *Environment and Planning A, 13*, 167-183.

Allen P.M., Sanglier M., Engelen G., Boon F. (1985) Towards a New Synthesis in the Modelling of Evolving Complex Systems, *Environment and Planning B, 12*, 65-84.

Anderson P.W., Arrow K.J., Pines D. (eds.) (1988) *The Economy as a Complex Evolving System*, Addison-Wesley, Reading, Massachusetts.

Archaeomedes Report (1995) Land Degradation and Desertification in the Mediterranean, European Commission Report, DG XII, Project EV5V-91-0021, Environment Programme.

Batty M., Longley P. (1994) *Fractal Cities*, Academic Press, London.

Bura S., Guérin-Pace F., Mathian H., Pumain D., Sanders L. (1996) Multi-Agent Systems and the Dynamics of Settlement Systems, *Geographical Analysis, 2*, 161-178.

Clark N., Juma C. (1992) *Long-Run Economics, an Evolutionary Approach to Economic Change*, Pinter Publishers, London.

Eigen M, Schuster P. (1979) *The Hypercycle*, Springer-Verlag, Berlin.

Goodwin R.M., Kruger M., Vercelli A. (1984) *Nonlinear Models of Fluctuating Growth*, Lecture Notes in Economics and Mathematical Systems, Springer-Verlag, Berlin.

Gould P., White R. (1978) *Mental Maps*, Penguin, London.

Hillier W., Hanson J. (1984) *The Social Logic of Space*, Cambridge University Press, Cambridge.

Hillier W., Penn A. (1992) Dense Civilisations: the Shape of Cities in the 21st Century, *Applied Energy, 43*, 41-46.

Lorents H.W. (1989) *Non-Linear Dynamical Economics and Chaotic Motion*, Lecture Notes in Economics and Mathematical Systems, Springer-Verlag, Berlin.

Nelson R.R., Winter S.G. (1982) *An Evolutionary Theory of Economic Change*, The Belknap Press of Harvard University Press, Cambridge, Massachusetts.

Phaeocystis Report (1993) Modelling Phaeocystis Blooms: Their Causes and Consequences, European Commission Report Report, DG XII, Contract CT-0062 (TSTS), STEP Programme.

Prigogine I., Stengers I. (1987) *Order out of Chaos*, Bantam Books, New York.

Sanglier M., Allen P.M. (1989) Evolutionary Models of Urban Systems: An Application to the Belgian Provinces, *Environment and Planning A, 21*, 477-498.

Saviotti P., Metcalfe J. (1991) Present Developments and Trends in Evolutionary Economics, in Saviotti P., Metcalfe S. (eds.) *Evolutionary Theories of Economic and Technological Change*, Harwood, Chur.

Silverberg G., Dosi G., Orsenigo L. (1988) Innovation, Diversity and Diffusion: A Self-Organisation Model, *The Economic Journal, 98*, 1032-1054.

White R.W., Engelen G. (1993a) Cellular Automata and Fractal Urban Form: A Cellular Modelling Approach to the Evolution of Urban Land-Use Pattern, *Environment and Planning A, 25*, 1175-1199.

White R.W., Engelen G. (1993b) Cellular Dynamics and GIS: Modelling Spatial Complexity, *Geographical Systems, 1*, 2.

3. A Short Discussion of Alternative Approaches to Modelling Complex Self-Organising Systems

Agostino La Bella

3.1 Introduction

The search for new theories able to explain and predict irregular phenomena in physical, social and biological systems has been stimulated by the recognition that traditional models, while able to replicate quite well the current behaviour of a system, cannot be reliably used for long-run forecasting because of their inability to take into account sudden structural modifications and their impact on the system. Such modifications are seen to occur frequently in social systems, seriously challenging our ability not only to make predictions, but also to reach a clear understanding of the underlying adjustment processes.

Social systems in fact evolve according to deep and fundamental paradigms that are still largely unknown. Models have until now only been able to deal with the empirical evidence of social phenomena, using it to build theories that have generally been proved wrong as soon as some boundary conditions have changed. This accounts for the delay with which we perceive societal changes, and is the reason why, although we may explain the existing situation, our model fails to forecast subsequent changes.

The evolutionary approach is an attempt to provide a deeper understanding of the mechanisms that determine the long term behaviour of systems in which adaptive and structural changes occur. However, to show that models based on these ideas are better able to cope with the complexities found in social and economic systems and, specifically, with the origins and evolution of human settlements and their organisation, we need to examine whether they respond to a number of requirements relating to the ability to provide a sufficiently stable paradigm to analyse, forecast and govern the growth of our cities and regions.

This work is therefore a preliminary discussion of the evolutionary approach from the above perspective, comparing it with other theories which have been developed in recent years to provide tools able to deal with behavioural patterns not explained by traditional dynamic system models. The starting point is a discussion of the notion of complexity in the field of urban and regional sciences. Then, in Section 3, we present a perturbation theory for demographic models (Campisi and La Bella, 1985) and discuss how it can be used to explain different urban growth patterns and their possible inversion. Section 4 is devoted to the intriguing concept of fractals, discussing how they can be applied in economics, in technology forecasting and also explaining the growth of cities and their ranking. In fact the fractal principle has been extensively used to approximate real world phenomena. Its capacity to reproduce historical series, and even to make predictions, has proved surprisingly good. It seems appropriate to conclude our brief survey with a brief reference to catastrophe theory and a short account of its possible use in the social science field (Section 5). It will be shown how simple models may explain and predict sudden changes of trajectory in state space even when the conditions surrounding the event change slowly and continuously, offering no apparent justification for sudden discontinuity. The main arguments are summarised in Section 6.

It should perhaps be specified at the outset that the author does not claim allegiance to any specific approach to the modelling of self-organising systems. The aim of this work is simply to demonstrate that there exist a number of approaches, many of them brilliant and intriguing, for capturing the essence of the complexities of the real world in mathematical models.

3.2 Complex Cities and Complex Models

It may be useful to review briefly the notion of complexity with specific emphasis on its use in the field of urban and regional sciences. From the fundamentals of system theory, we know that a system can be defined as complex if:

- it has a very large number of components;
- each component is significantly different from the others;
- the network of links between the components is particularly dense;
- each interaction link represents a specific relation between components;
- the single component reacts to signals received through the interaction

links with the other components;
- each relation changes over time, according to stimuli received from other components and from the external environment;
- the pattern of interaction among components is continuously changing.

From this definition it emerges that the study of equilibria, though characterised by a very large number of variables and interrelations, may be complicated (because of the time needed to analyse the system and identify the patterns), but is not necessarily complex. On the other hand, a very simple system, such as a single cell, may not seem complicated, but the nature of the interactions with its environment can mean that in reality it is quite complex.

According to the above definition, cities are certainly highly complex systems. This complexity would appear to manifest itself through a number of elements:

- the very fast dynamics: making it easy to mistake the effects of a phenomenon for its causes;
- an increasing number of interactions with positive feedback: the growth in the number of interactions leads to the development of new channels, i.e. infrastructures, services and, more in general, new or improved technologies. This reinforces the existing interaction pattern and contributes to the creation of new and stronger links, rapidly bringing the system back to a new congestion threshold;
- the information paradox: an excess of information often makes data organisation difficult. Nowadays we have access, as never before, to a huge quantity of information - we appear to know everything, from everywhere, in real time - however, when we need specific data, the right information is almost never available at the right time or in the right place.

Two fundamental concepts emerge when dealing with complex systems: communication and adaptability. Many attempts have been made to capture the features listed above with mathematical models, generally focussing on some specific functions of the city. Some have been only schematic representations, based on theoretical hypotheses and often derived from other scientific fields. Some consist of complicated sets of differential equations aimed at emulating the dynamics of a socio-economic system. Others depict the relationships between variables in equilibrium states (not easily found in reality, but nevertheless often useful for analytical purposes). Only a few models, however, have had a significant impact either on the understanding of the underlying mechanisms of urban development or as tools for supporting policy decisions.

Therefore, before continuing our discussion, it would seem helpful to list a set of 'criteria', derived from the experience of many past efforts in this direction, for evaluating new theories or models relating to the development of cities and/or socio-economic systems in general. These criteria have much to do with what we require from a model, from minimal requisites to very high performance in description, prediction and policy making. Let us therefore review the many purposes that lie at the basis of model-building. Obviously, any model must serve at least one of the following purposes, listed here in increasing order of difficulty:

- the first and most simple requisite of a mathematical model is to provide a meaningful and clearly understandable way of organising data. This function is met, for instance, by many statistical models and the logit models often used to interpolate empirical data. The fit may be very good, but neither description nor explanation of the various phenomena are given;
- the second type of model is 'descriptive'. These models try to capture and depict salient aspects of the phenomena being considered, but do not attempt to provide deep understanding of their working. Like a picture, they can say what is there, but neither what is going to happen in the future, nor why;
- next, we have models attempting to interpret the phenomena. Their aim is to represent the forces that have determined the current state of a system. Concepts of interconnection and communication emerge here, even if the comprehension of causal links may be too weak for forecasting purposes;
- forecasting, which requires models able to represent the learning and adaptive behaviour of a system in response to stimuli from external factors and possibly also internal forces. This is an area where attention should perhaps be focused, since the simple extrapolation of past trends, while useful for producing a possible scenario, may give rise to serious mistakes when used as a basis for policy making;
- finally, there are management models, generally based on optimisation concepts. These aim to identify the best course of action taking into account the functioning of the system, some boundary conditions and a number of constraints. They should be able to anticipate the response of the real world system to alternative intervention policies. Since in the social and economic realm any tampering may cause significant changes in the system's behaviour (even the simple act of observation may

influence its functioning through expectation mechanisms), the building and operation of any such model for a complex system requires a great deal of knowledge about all the above aspects.

The above typology can be used as a basis for the discussion of any modelling attempt. In other words, a model should be judged according to its capacity to satisfy the requisites of one or more of the above classes.

Beginning with the evolutionary approach to modelling socio-economic systems, we find that it does not provide any help in classifying or organising data, it resembles but fails to fit any real evolutionary process and has no forecasting capacity. It would seem that, in these terms, the approach is not yet truly 'mature'. Even the equilibrium models, strongly criticised by Allen (in this volume), have more solid scientific foundations and are probably more useful in practice. It is not always true that equilibrium models are only justified by the lack of computing power to deal with more complex, and realistic, conditions. They are often built as a way of providing better understanding of current trends which, projected at the equilibrium, allow us to identify the underlying forces at work and their impact (which may not be immediately evident). Equilibrium analysis, therefore, often has the same role in the social sciences as the microscope in the natural sciences.

At present stage of development, models based on the evolutionary drive seem mainly to be conceptual schemes based on a biological analogy (the starting point of so many other approaches in the past), but there is still very little empirical evidence of their explanatory power. The idea that the evolution of human settlements, social and economic systems stems from microdiversities due to error-making in the transmission of the 'rules' underlying specific behaviour or techniques is debatable. Certainly, we may use this approach to describe how a population of bacteria grows, but it seems rather difficult at the moment to find more than a pictorial similarity with the evolution of the organisation of geographical space.

It is true, however, that learning processes are important in determining the dynamics of all human activities (Egidi and Marris, eds., 1992). This is not this concept that I want to challenge. I simply argue that the learning processes of human beings, of societies and, more generally, of institutions, even though drawing from past experience, knowledge and errors, are not evolutionary in the Darwinian sense, since they are capable of generating entirely new ideas and structure. Evolution has to do with what is already in existence. This point is better illustrated by the following two quotations from prominent scholars in the field of evolutionary biology:

"We must always bear in mind the crucial fact that evolution is a history-dependent process. Adaptations are not 'designed' *de novo* by nature. Rather, they are jury-rigged, using the material available at the time. Evolution ... is a 'tinkerer' not an engineer!" (Oster and Wilson, 1984). "The jet engine superseded the propeller engine because, for most purposes, it was superior. The designers of the first jet engine started with a clean drawing board. Imagine what they would have produced if they had been constrained to 'evolve' the first jet engine from an existing propeller engine, changing one component at a time, nut by nut, screw by screw, rivet by rivet. A jet engine so assembled would be a weird contraption indeed." (Dawkins, 1982).

My concern, in the remaining parts of this work, will be to show how learning processes can be incorporated into mathematical models which have a wide range of applications and which may be tested against empirical data. The discussion will be limited to a few approaches relevant to the field of urban studies, and to the extent necessary to make the point. For those interested in evolutionary theories in economics, I would like to mention here the well-known works of Nelson and Winter (1982), Williamson and Winter (1991) and Andersen (1994). The brilliant papers of Cohen and Levinthal (1989, 1994) are also useful in showing how the learning process can be incorporated in models of neoclassical derivation.

3.3 The Perturbation Theory Approach

In this section I wish to demonstrate how even very simple models, like the generalised Leslie operator in multiregional demography, may be used to explain sudden changes in the patterns of city growth. The example we take is that of the inversion of the urbanisation process, a phenomenon experienced in many countries. Many authors have discussed the causes (see for instance Blumenfeld, 1979, and Keyfitz, 1980, Keyfitz and Philipov, 1982), citing the diseconomies produced by congestion, the spoiled environment, increasing difficulties in social relations, plus the speed of technological progress which has drastically changed industrial location criteria and improved transportation and telecommunications, making people and economic activities less sensitive to distance.

Obviously, the above factors have produced changes in the fundamental demographic parameters, producing a discontinuity in the response of population dynamics to migration and birth rates. Under specific conditions, which we shall explore here, it may therefore happen that very small changes in those rates produce dramatic changes in the pattern of growth of the population. To understand how a very simple model can capture the essence of this phenomenon, let me first review the fundamental structure of a multiregional population model (for further details see Willekens and Rogers, 1978). In compact form the model can be written as:

$$K(t+1) = G \cdot K(t) \tag{1}$$

where K is the vector of population by zone and age groups, and G is the growth operator of the multiregional demographic system, which is a function of the migration and death rates matrix $M(x)$ and of the age and zone specific fertility rates matrix F through the three matrix functions $P(x)$, $S(x)$ and $B(x)$ given, in the case of five year age groups, by:

$$(x) = \left(1 + \frac{5}{2} M(x)\right)^{-1} \left(1 - \frac{5}{2} M(x)\right)$$

$$S(x) = (I + P(x+5))\, P(x)\, (I + P(x))^{-1}$$

$$(x) = \frac{5}{4}\left[P(0) + I\right]\left[F(x) + F(x+5)S(x)\right]$$

where x identifies the age group.

We know from empirical observation that the non-zero eigenvalues of matrix G are distinct, and therefore the corresponding right and left eigenvectors, say Φ_i and Ψ_j respectively, must satisfy the orthogonality condition:

$$\Psi_j^T \Phi_i = 0 \tag{2}$$

In this case it is natural to normalise the left eigenvectors with respect to those on the right so that:

$$\Psi_i^T \Phi_i = 1 \tag{3}$$

With this normalisation it has been shown (Luenberger, 1979) that the solution of the system (1) can be written as:

$$K(t) = \sum_i \lambda_i^T \Phi_i \Psi_i^T K(0) \qquad (4)$$

where λ_i are the eigenvalues of matrix G.

Since $G > 0$, according to Frobenius-Perron theorem, in the long run we get, for very large values of t:

$$K(t) = \lambda_1^t \Phi_1 \Psi_1^T K(0) \qquad (5)$$

where λ_1 is the dominant eigenvalue of matrix G. Therefore the population trajectory converges along a ray emanating from the origin and defined by Φ_1 and the ultimate growth rate will be determined only by λ_1; Φ_1 is called the stable population distribution, $\Psi_1^T K(0)$ represents the initial population projected on the stable growth ray and is known as the total reproductive value of the initial population.

If small changes in the observed demographic parameters produce relevant changes in the solution of the eigenvalue problem, which gives the rate of growth and distribution of the stable population, the problem is said to be unstable or ill-conditioned. Otherwise, the problem is defined as stable or well-conditioned. From the mathematical point of view there is no precise boundary between stable and unstable problems. In our case, we shall assume a problem to be stable if changes in the solution are no greater than the changes in data. In other words, we want to explore the conditions which make the solution relatively insensitive to a perturbation in the data.

The discussion will be split into two parts. We shall firstly investigate the effect of perturbations applied to matrices M and F on the growth operator G. Secondly, we shall analyze how given changes to matrix G affect the balanced growth properties of system (1).

Now, let $\Delta M(x)$ be a (small) variation of $M(x)$ caused by some exogenous perturbation: in other words, ΔM represents the operator variation Δ applied to $M(x)$ that, acting as a source of perturbation, will propagate its effects to the matrices P and S, giving rise to their variations ΔP and ΔS respectively. These can be estimated as follows.

Let us define the new perturbed matrix of the migration and death rates $M'(x)$:

$$M'(x) = M(x) + \Delta M(x) \qquad (6)$$

Setting:

$$A(x) = M(x) + \frac{5}{2} M(x)$$

we get:

$$P(x) = A^{-1}(x)(2I - A(x)) \tag{7}$$

Let us now assume that the perturbation affects only the (h,k) entry of $M(x)$ by a quantity $\pm\frac{2}{5}\Delta$. In this case we have:

$$\Delta M(x) = \frac{2}{5} uv^T \tag{8}$$

where u is a column vector with elements u_i=0, $i \neq h$ and $u_h = \Delta$; v^T is a row vector with v_j=0, $j \neq k$, v_k=1.

Therefore we obtain:

$$\Delta A(x) = uv^T$$

$$P(x) + \Delta P(x) = (A(x) + uv^T)^{-1} \, (2I - A(x) - uv^T) \tag{9}$$

We now make use of the Sherman-Morrison formula (Wolfe, 1978) to write:

$$(A(x) + uv^T)^{-1} = A^{-1}(x) - \frac{A^{-1}(x)uv^T A^{-1}(x)}{I + v^T A^{-1}(x)u} \tag{10}$$

and, after calculation, from (9) we obtain:

$$\Delta P(x) = \frac{A^{-1}(x)uv^T (A^{-1}(x)uv^T - 1 - P(x))}{I + v^T A^{-1}(x)u} \tag{11}$$

By inspection, we notice that $\Delta P(x)$ takes the form:

$$\Delta P(x) = \pi(x)v^T \tag{12}$$

where $\pi(x)$ is a column vector with elements of order $\Delta + O(\Delta^2)$. With the same reasoning, from (11) and (12) we obtain:

$$\Delta S(x) = \sigma(x)v^T \tag{13}$$

where $\sigma(x)$ is a column vector with elements of order $\Delta + O(\Delta^2)$. We can therefore conclude that a perturbation on element (h,k) of M produces in column k of the submatrix $S(x)$ of G changes no greater than the perturbation applied. It is easy to see that perturbations in the age specific mortality rates produce changes of the same order in the submatrices of G.

We now tackle the analysis of the influence of the above perturbations on the long run growth properties of system (1). Let λ be the dominant eigenvalue of matrix G, and Φ and Ψ the right and left associated eigenvectors, such that:

$$\|\Phi\|_2 = \Phi^T \Psi = 1$$

where:

$$\|\Phi\|_2 = \left(\sum_{i=1}^{N} \Phi_i^2 \right)^2$$

is the Euclidean norm. Let also ΔG be the perturbation applied to G, and:

$$G' = G + \Delta G \tag{14}$$

with $\varepsilon = \Delta \|\Delta G\|_2$ (assumed to be small). It is possible to show (Chatelin, 1983, Kato, 1982) that, if ε is small enough, there exists a simple eigenvalue λ' of G' with an eigenvector Φ normalised by $\Psi^T \Phi' = 1$ such that:

$$\lambda' = \lambda + \Psi^T \Delta G \Phi + O(\varepsilon^2) \tag{15}$$

$$\Phi' = \Phi - R \Delta G \Phi + O(\varepsilon^2) \tag{16}$$

where R is the generalised inverse of $(G-\lambda I)$ relative to the spectral projection $P = \Phi \Psi^T$

$$R(G - \lambda I) = (G - \lambda I)R = I - P$$

Setting $\varepsilon' = \Delta \|\Delta G \Phi\|_2$, since $\left|\Psi^T \Delta G \Phi\right| \le \varepsilon' \|\Psi\|_2$ we get from (15):

$$|\lambda' - \lambda| \le \varepsilon' \|\Psi\|_2 + O(\varepsilon^2)$$

It follows that λ is ill-conditioned if $\|\Psi\|_2$ is large; we can say that $\|\Psi\|_2$ is a condition number for λ when Ψ is normalised by $\Psi^T\Phi = \|\Phi\|_2 = 1$. Hence $\|\Psi\|_2 \geq 1$.

Setting $r = \Delta\|R\|_2$ we get from (16):

$$|\Phi' - \Phi| \leq r\varepsilon'\|\Psi\|_2 + O(\varepsilon^2)$$

It follows that Φ is ill-conditioned if r is large: r is a condition number for Φ. Since it is possible to show that $r \geq \frac{1}{d(\lambda)}$, where $d(\lambda)$ is the distance between the two closest eigenvalues of G:

$$d(\lambda) = \min_{\mu \in \sigma(G) - \{\lambda\}} |\mu - \lambda|$$

where $\sigma(G)$ is the spectrum of matrix G, we can say that if λ is near another eigenvalue of G, then its eigenvector will be ill-conditioned. In general, this is only a sufficient condition, and r may of course be large even if λ is well separated from the rest of the spectrum. Particular cases where this condition is also necessary are discussed in Chatelin (1983), but this does not hold for empirically observed growth operators.

The described techniques are now applied to an example partially based on empirical observations for a region in Central Italy. The region has been disaggregated into four subregions; the demographic characteristics of the subregions are shown in Table 1. For the sake of simplicity, we do not subdivide the population by sex and age.

	In-migration	*Out-migration*	*Fertility*	*Mortality*
Core	High	High	Low	High
Suburbs	Medium high	Medium high	Medium high	Medium high
Rural area	Low	High	High	Low
Rest of region	Low	Low	Low	Medium

Table 1 Demographic features of the subregions

In Table 2 the growth matrix and its eigen-elements are given. Eigenvectors have been normalised such that $\Sigma\Phi_i = 1$ and $\Phi\Psi^T = 1$. It can been observed that the given population has a slightly positive stable growth rate and an even distribution, with the largest percentage of stable population in the urban core. This is due to the large migration flows towards the core which compensate for the higher natural increase in the rural population. We shall now discuss the effects of perturbations applied to the given demographic regime.

The multiregional growth matrix G

0.975098	0.005911	0.005535	0.019556
0.005915	0.980218	0.002917	0.005802
0.005207	0.003507	0.981858	0.010330
0.004238	0.002053	0.007857	0.988851

Eigenvalues

(1.00192, 0.00000)	(0.96962, 0.00000)	(0.97833, 0.00000)	(0.97618, 0.00000)

Right eigenvectors

(0.29165, 0.00000)	(4.17259, 0.00000)	(-0.10354, 0.00000)	(-0.68020, 0.00000)
(0.18601, 0.00000)	(-1.88682, 0.00000)	(-1.19013, 0.00000)	(1.15954, 0.00000)
(0.24891, 0.00000)	(-0.95971, 0.00000)	(0.40443, 0.00000)	(1.27622, 0.00000)
(0.27338, 0.00000)	(-0.32606, 0.00000)	(-0.49123, 0.00000)	(-0.74958, 0.00000)

Left eigenvectors

(0.58250, 0.00000)	(0.19752, 0.00000)	(0.04147, 0.00000)	(165.66563, 0.00000)
(0.48634, 0.00000)	(-0.09180, 0.00000)	(-0.42003, 0.00000)	(686.54667, 0.00000)
(0.95143, 0.00000)	(0.06679, 0.00000)	(0.77242, 0.00000)	(-594.77290, 0.00000)
(1.83921, 0.00000)	(-0.20909, 0.00000)	(-0.37327, 0.00000)	(-102.28108, 0.00000)

Table 2 The parameters of the multiregional model

According to the theory presented above, the condition number for λ is obtained by setting $\|\Phi\|_2 = \Phi^T \Psi = 1$, by $\|\Psi\|_2 = 1{,}116$, implying that perturbations produce changes in λ of the same order. We also observe that the sufficiency condition for the dominant eigenvector to be well-conditioned, and therefore relatively insensitive to perturbations, does not hold because λ is not well separated from the rest of the spectrum. It follows that slight changes in demographic parameters would not significantly affect the resulting stable growth rate; the same changes, however, may completely upset the stable population distribution.

Table 3 presents the results of a sensitivity analysis. An asterisk (*) marks the situations in which the largest stable population switches from the core to one of the other subregions producing a large change in the stable population distribution. An even greater change in the population distribution is obtained in a few cases, where the dominant eigenvalue interchanges with a sub-dominant one when the growth operator is perturbed. Correspondingly, the associated eigenvector (obviously modified in the process) will provide the new balanced growth, i.e. the new stable population distribution. This is a sort of 'catastrophic' behaviour, implying a drastic change in the growth characteristics of the system.

Demographic rates	λ	$\lvert MAX_{\Delta G}\rvert$		$\lvert\Delta\lambda_{MAX}\rvert$		$\lvert MAX_{\Delta\Phi_{MAX}}\rvert$		$\sqrt{\sum(\Delta\Phi_i)^2}$
Fertility, urban core	+10%	5.0	10^{-4}	2.0	10^{-5}	5.0	10^{-5}	7.0 10^{-4}
	−10%	3.9	10^{-3}	6.3	10^{-4}	4.4	10^{-2}	0.05(*)
	−15%	6.0	10^{-3}	1.32	10^{-3}	1.2	10^{-2}	0.15(*)
	−20%	9.9	10^{-5}	1.62	10^{-3}	4.4	10^{-1}	0.18(*)
Fertility, rest of the region	−10%	3.98	10^{-3}	3.2	10^{-4}	4.7	10^{-2}	0.05
Fertility, suburbs	+10%	4.2	10^{-3}	1.17	10^{-4}	6.5	10^{-2}	0.07(*)
	−10%	4.4	10^{-3}	9.1	10^{-4}	5.6	10^{-2}	0.06
Fertility, rural area	+10%	6.3	10^{-3}	3.7	10^{-3}	8.4	10^{-2}	0.10
	−10%	6.3	10^{-3}	3.2	10^{-3}	8.1	10^{-2}	0.07
	−25%	8.0	10^{-3}	9.7	10^{-4}	0.13		0.18(*)
Outmigration, urban core	−50%	2.8	10^{-3}	7.36	10^{-3}	2.7	10^{-1}	0.42(*)
Outmigration, rural area	+10%	2	10^{-3}	1.3	10^{-3}	0.08		0.10
	−10%	2	10^{-3}	1.6	10^{-3}	0.09		0.12

Table 3 Sensitivity analysis

Analyzing Table 3 we can formulate the hypothesis that the inversion of the growth process from centre to periphery may be triggered in some cases by small changes in the fertility rates of a single region. Obviously this depends on the relative importance of fertility and migration rates in determining population dynamics. Further simulations would be necessary to determine the characteristics of an indifference point relating to the sensitivity of the stable population.

3.4 The Fractal Approach

The study of fractals and their applications is a relatively new field, but rapidly advancing on all fronts, from description to explanation. As Mandelbrot points out in his inspiring book (1983), the theory of fractals is the geometric face of the theory of scaling, which was introduced in the social sciences by Zipf (1949). The main purpose of Zipf's brilliant, if somewhat extravagant, work was to establish an empirical law (Zipf's law) which would show that a scaling probability distribution often provided the best possible fit to data in social science statistics. Zipf searched for all sorts of empirical data, ranging from the shape of sexual organs to the size of human settlements. His evidence is always presented in the form of doubly logarithmic graphs. He made a classification of cities according to their population and revealed that urban settlements have a fractal structure which appears to be stable over hundreds of years. Even apparent exceptions to this rule, like for instance Vienna, which is too large to fit a ranking of cities within the Austo-Hungarian empire, falls perfectly into place if ranked after Berlin in the group of German-speaking cities.

The concept of fractals is well suited to analyzing complexity, since it addresses those patterns of nature too irregular to be explained by standard geometry. The family of shapes that can be used to describe these irregularities are called fractals. The most useful fractals involve probabilities, and both their regularities and irregularities are statistical. In addition, the degree of irregularity and/or fragmentation tends to be identical at all scales. But let us introduce the concept in a more formal way.

Definition 1: A fractal is a set for which the Hausdorff Besicovitch dimension D strictly exceeds the topological dimension D_T.

A more intuitive interpretation is provided by Fig. 1, where it is shown how, given an integer b=5, a straight interval of unit length may be divided into N=b subintervals of length r =1/b. Similarly, a unit square can be divided into N=b^2 squares of side r=1/b. In both cases we can define the Hausdorff fractal dimension as $D=\log N/\log(r^{-1})$, which in these examples is 1 and 2, respectively, i.e. the same as D_T, indicating that these two figures are not fractals. We now consider Koch's snowflake which can also be broken down into smaller pieces, with N=4 and r=1/3. As its Hausdorff dimension is given by D=1.2618, greater than D_T=1, we can deduce that Koch's snowflake is a fractal.

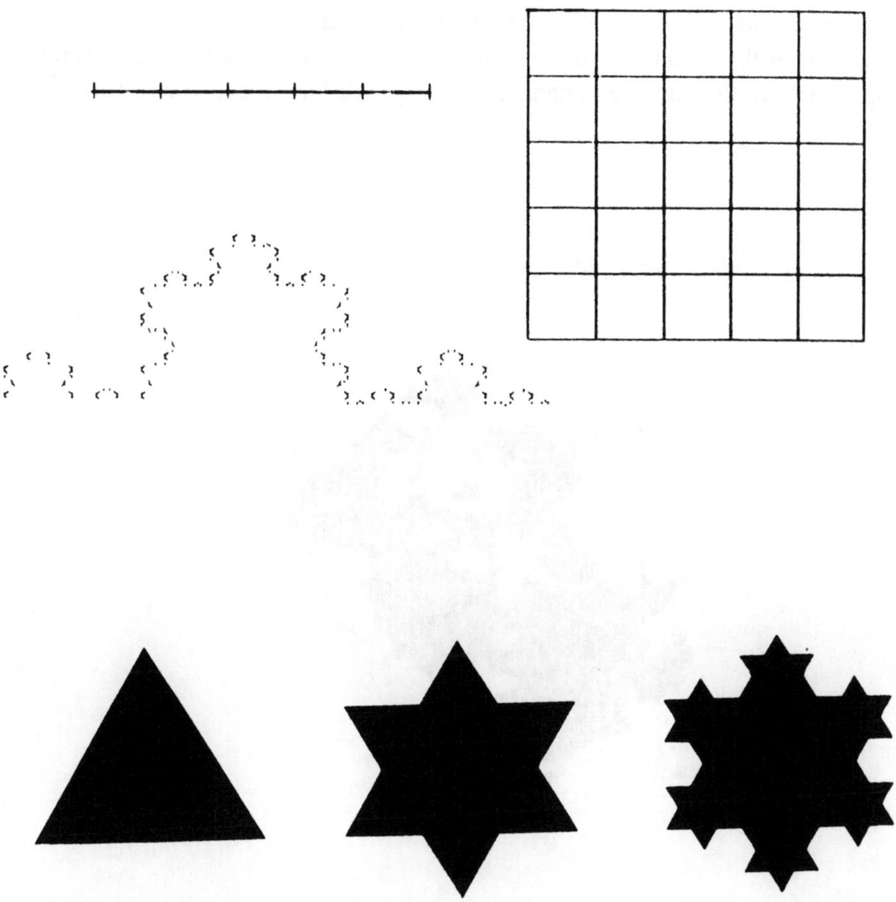

Fig. 1 Self-similarity and fractal self-similarity (Mandelbrot, 1983)

It can be shown that appropriately selected fractals, coupled with random generators, can describe any patterns found in the social sciences. Fig. 2, for example, shows a fictitious fractal Pangea. Its relief was generated on computer by Mandelbrot applying a Brownian function from the points of a sphere (the latitude and the longitude) to scalars (the altitude). 'Sea level' was then adjusted so that three-quarters of the total area was underwater and the resulting coastline projected on a Hammer map. Spatial patterns related to the distribution of people, resources, activities and so forth, can be easily generated with the appropriate selection of the Hausdorff dimension and a stochastic rule (see for instance Fig. 3, where the surface is generated using a different Brownian fractal).

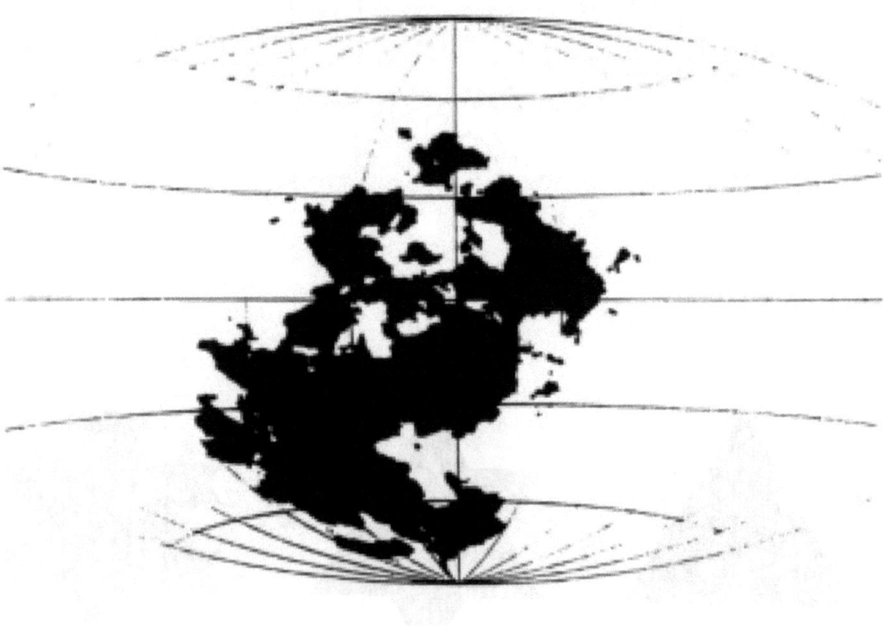

Fig. 2 Brownian Pangea (D=2.5)

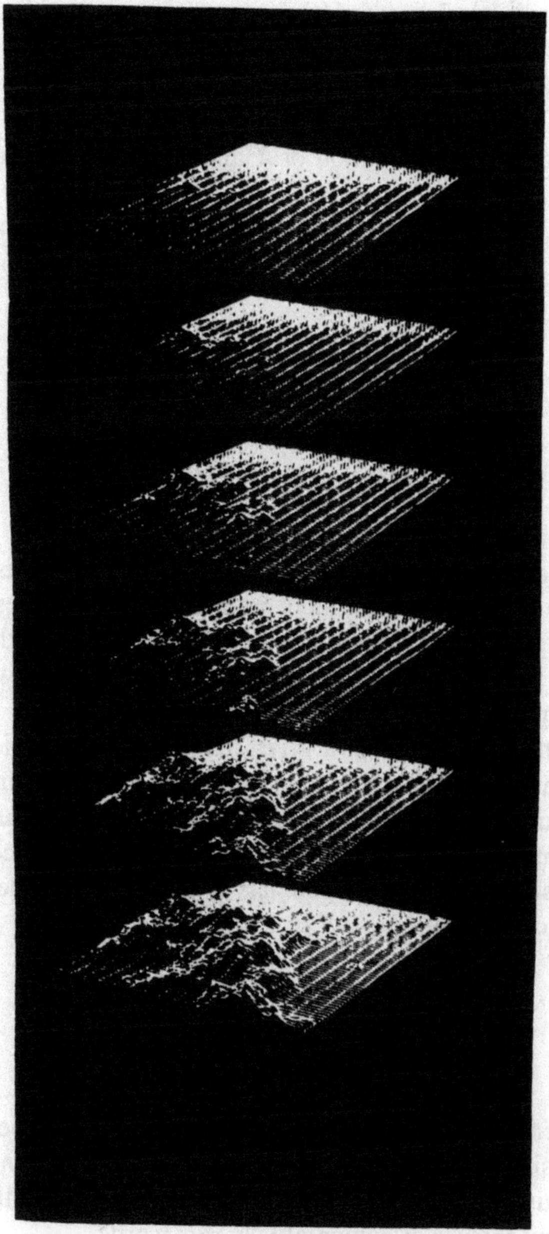

Fig. 3 Brownian fractal areas (D=2.3)

The use of fractals is not limited to the reproduction of geometric shapes. Applications are not yet abundant, but it seems that many empirically based laws in economics or social statistics can be reinterpreted in the light of fractals. To show at least the fundamentals of nongeometrical fractals, let us first provide some definitions.

Definition 2: A random variable X is a scaling variable under the transformation $T(X)$ if the distribution of X and $T(X)$ are identical except for scale.

Definition 3: A random variable X is distributed according to an asymptotically hyperbolic distribution if there exists an exponent $D>0$ such that:

$$\lim_{x \to \infty} \Pr(X<x)x^D$$

$$\lim_{x \to \infty} \Pr(X>x)x^D$$

are definite and finite, and one of the limits is positive.

The first to discover laws of hyperbolic nature in social statistics was probably Vilfredo Pareto, who found impressive regularities in the distribution of personal income (the number of individuals with income exceeding a given large value is hyperbolically distributed). However, we now wish to present a very simple example from Zipf (1949).

Let us suppose that the words in a given text are ranked in decreasing frequency. Designating with R the rank assumed by a word of probability P, one would assume the relationship between P and R to vary wildly according to the language and writer. In fact this is not so. According to Zipf, the relationship is universal and has a simple form, defined by Mandelbrot as $P=F(R+V)^{-1/D}$. Since $\Sigma P=1$, the three parameters D, F and V are related by:

$$F^{-1}=\Sigma(R+V)^{-1/D}$$

which is almost perfectly hyperbolic, and where the main parameter, D, is a similarity dimension (the equivalent of the Hausdorff dimension). Here it serves as a 'measure' of the comparative frequency of words: the bigger D, the more frequent the use of comparatively rare words.

The scaling principle has been applied in economics, in technology forecasting and also, as mentioned before, to explain the growth of cities and their ranking. Fig. 4 shows the ranking of various cities in the world according to their population in 1920. The results of the same exercise for the year 1985 are given in Fig. 5, where the distribution appears to have an elbow around a population level of 5 million. If, however, we introduce the concept of 'urban corridors', i.e. those cities connected by air shuttles or high speed trains, the ranking falls once again into a straight line (Marchetti, 1989). The corridors and their shuttle systems appear to play a unifying role between the cities they connect. In spite of the relatively few people who travel regularly between these cities, the high level such people occupy in the hierarchy of urban functions accounts for this effect.

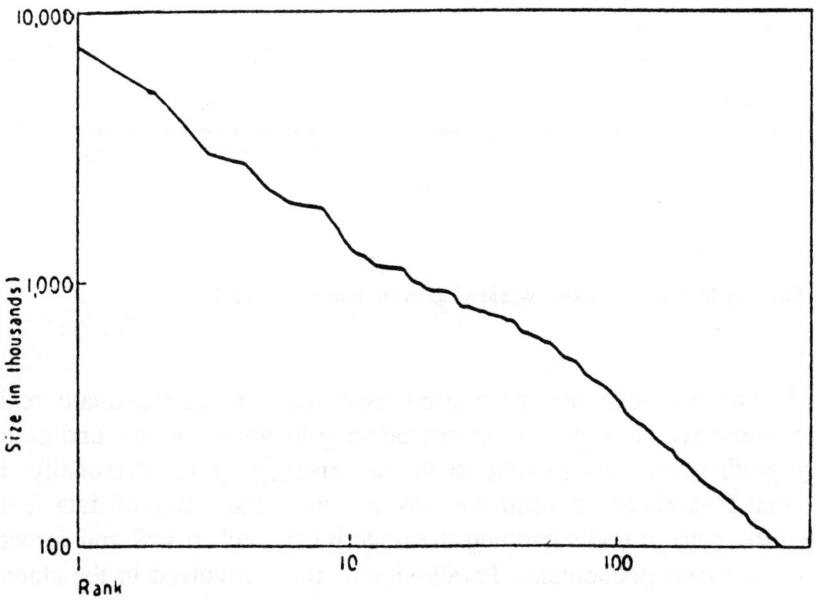

Fig. 4 Rank distribution of cities worldwide by dimension (circa 1920)

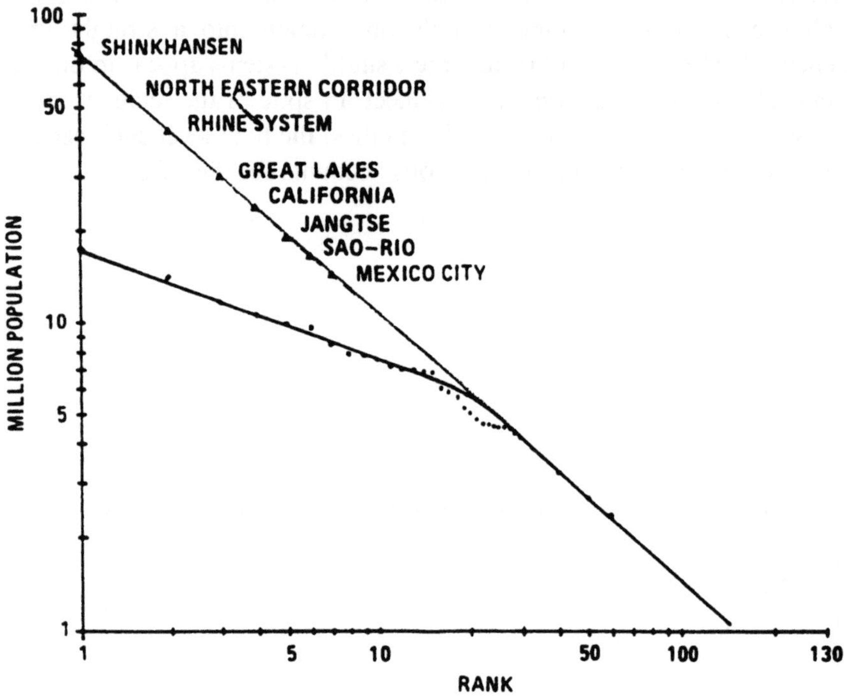

Fig. 5 Rank distribution of cities worldwide by dimension (circa 1985)

The fractal principle has been used extensively to approximate real world phenomena. Its capacity for reproducing historical series, and even making predictions, has proved to be surprisingly good. Personally, I believe that fractals are a wonderful tool for the organisation of data, but of very little help in understanding the underlying mechanisms and forces that govern social phenomena. Paradoxically, those involved in the study of fractals seem to have very little interest in analysing the causes that make the scaling principle so good in interpolating data of any kind. They are apparently satisfied with the good performance of fractals as a descriptive and predictive instrument, and inclined to stress the consequences rather than causes. Nevertheless, the field is certainly fascinating. For our purposes, it shows that even simple models are able to organise data and make good predictions about very complex phenomena and systems. In comparison with the evolutionary type model, it strikes me that, while the latter attempts to go deeply into the mechanisms that trigger changes, but so far have provided little help for organising data and forecasting, the fractal approach deliberately ignores those mechanisms, and concentrates on squeezing statistical properties out of empirical data.

3.5 The Catastrophe Theory Approach

Not more than 30 years ago, a scientist finding strange aperiodic oscillations in, say, a chemical reaction, would have doubtlessly attributed them to some spurious phenomenon or external factors interfering with his experiment. Now, it has become clear that turbulence and chaotic behaviour may be associated with the real essence of a physical or social process. Catastrophe theory was first proposed by René Thom with the aim of providing a general method for dealing with phenomena involving sudden changes in a system subjected to smoothly changing inputs and/or external conditions. The theory had a period of great popularity in the seventies, when hundreds of scientific papers proposed its application to practically everything. After that, many critical contributions based on Thom's and Zeeman's fundamental work, helped to establish catastrophe theory as an intriguing, beautiful field of mathematics with many implications for the study of dynamic systems (Arnold, 1986, Castelnuovo and Hayes, 1993).

The term catastrophe is commonly used to describe an unexpected and disastrous event, like an earthquake, a sudden stock market crash, or a rapid depreciation of currency. It is unexpected because the conditions surrounding the event change slowly and continuously, offering no apparent justification for the sudden discontinuity. We now wish to show that even very simple dynamic models of social phenomena may exhibit a catastrophic behaviour, and that the explanation of sudden changes of trajectory in the state space may sometimes be found in a careful analysis of the structural properties of the system.

Let us consider again migration movements within a demographic system made up of p subpopulations which move among q areas, and define the attraction of region i for an individual of subpopulation k by means of a utility function u_i^k :

$$u_i^k(n) = \delta_i^k + \sum_h z^{kh} n_i^h$$

where $n = \{n_i^k\}$ is the vector of the subpopulations distribution by area and z is a set of suitable weights (agglomeration parameters) which give the relations between the attraction of each region and its 'social' environment in terms of the various population groups residing there: δ is a set of intrinsic preference parameters. We shall assume that the migration probabilities are given by:

$$p_{ji}^k = \nu \exp\{u_j^k - u_i^k\}$$

and therefore:

$$\frac{\partial n_i^k}{\partial t} = \sum_j p_{ji}^k n_j^k - \sum_j p_{ij}^k n_i^k$$

Consider now two interacting subpopulations of size P_1 and P_2, respectively, moving in a system consisting of two areas (e.g. two different zones of a city). Since the size of each group is a constant, we can describe the situation making use of the two dynamic variables $x = (n_1^1 - n_2^1) / P_1$ with $-1 \leq x \leq 1$, and $y = (n_1^2 - n_2^2) / P_2$ with $-1 \leq y \leq 1$. It is easy to verify that the transition probabilities take the general form:

$$p_{ij}^k = \nu \exp\{\Delta u_{ij}^k (x,y)\}, \qquad k = 1,2$$

with

$$\Delta u_{ij}^k = -\Delta u_{ij}^k$$

$$\Delta u^k (x,y) = \pi^k + \phi^k + \sigma^k y$$

where:

π^k is the intrinsic preference differential,
ϕ^k is the relevant agglomeration factor within each subpopulation,
σ^k is a sort of 'affinity' for coexistence with the other population group.

The simple model described above (for a more extensive treatment, see

Weidlich and Haag, 1987) allows us to distinguish three different migration dynamics, relating to different values of the parameters ϕ and σ (for the sake of simplicity, but without loss of generality, we shall assume $\pi^k = 0$ for all k).

Case 1: a homogeneous mix of populations. When parameters ϕ and σ have small values, we obtain the solution depicted in Fig. 6.

Case 2: the separate agglomeration of different population groups (ghettos). When parameters ϕ and σ are large, we get the solution depicted in Fig. 7.

Case 3: dynamic interaction among the zones. When parameters f have large values and $\sigma^1 = -\sigma^2$ (the two population groups have different feelings towards each other, so their behaviour is asymmetric), we get a limit cycle and a stationary quadrimodal probability distribution (see Fig. 8).

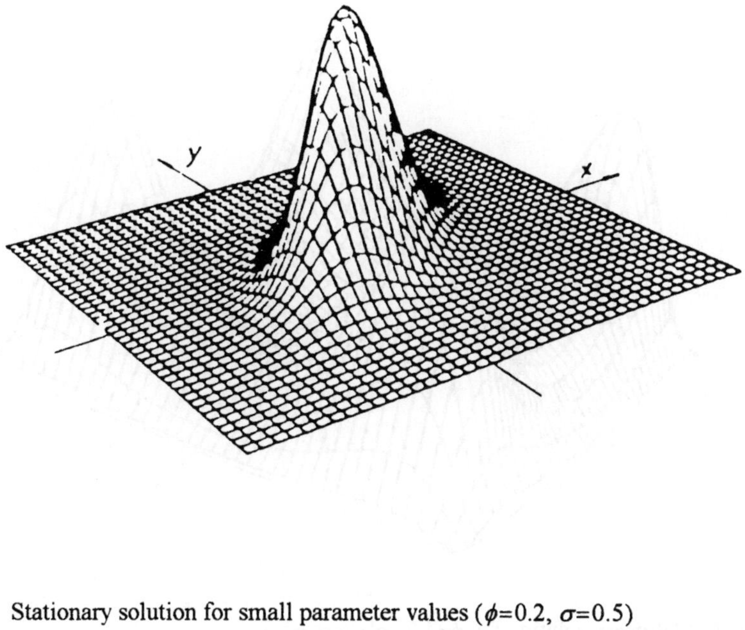

Fig. 6 Stationary solution for small parameter values (ϕ=0.2, σ=0.5)

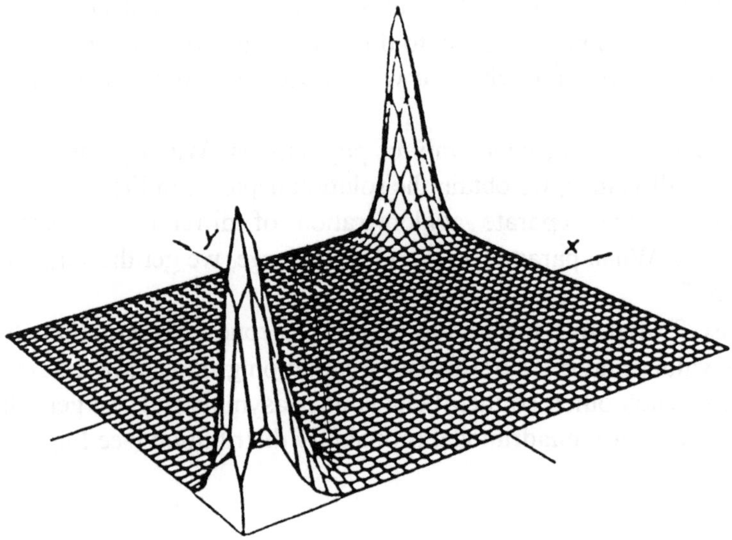

Fig. 7 Stationary solution for large parameter values(ϕ= 0.5, σ= 1.0)

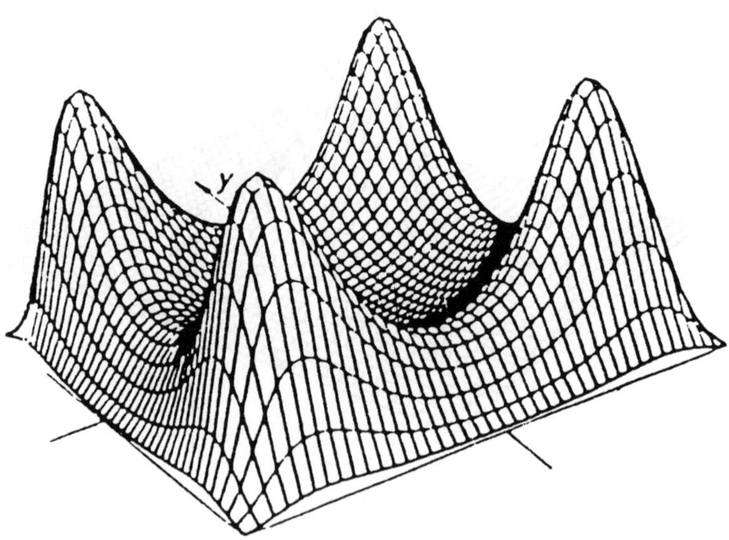

Fig. 8 Stationary solution for ϕ= 1.2, $\sigma^1 = -\sigma^2 = 1.0$

The analysis becomes a little more complicated for a larger number of variables, when strange attractors and therefore chaotic behaviour starts to manifest. Let us consider the case with three population groups moving among three regions, and agglomeration parameters given by:

$$z^{hk} = \begin{bmatrix} 1{,}7 & 1{,}5 & -1{,}5 \\ -1{,}5 & 1{,}7 & 1{,}5 \\ z^{31} & -1{,}5 & 1{,}7 \end{bmatrix}$$

For z^{31}=1.5 we get two limit cycles: for z^{31}=−0.5 the two cycles join into one, for z^{31}=−0.55 we get the limit cycle with multiple period depicted in Fig. 9 with the corresponding Fourier spectrum. Finally, for z^{31}=−1.5 the limit cycle develops into a strange attractor with a continuous Fourier spectrum as shown in Fig. 10.

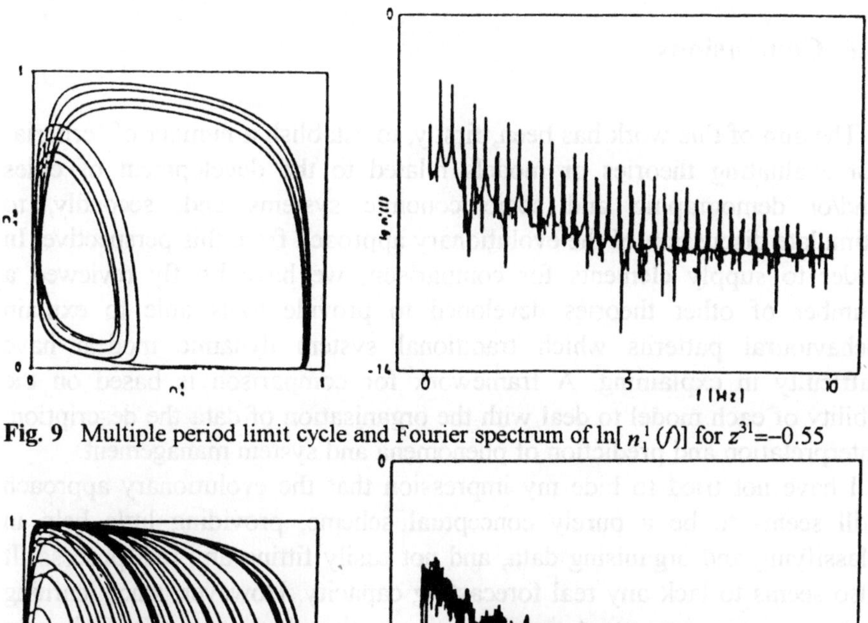

Fig. 9 Multiple period limit cycle and Fourier spectrum of ln[$n_1^1(f)$] for z^{31}=−0.55

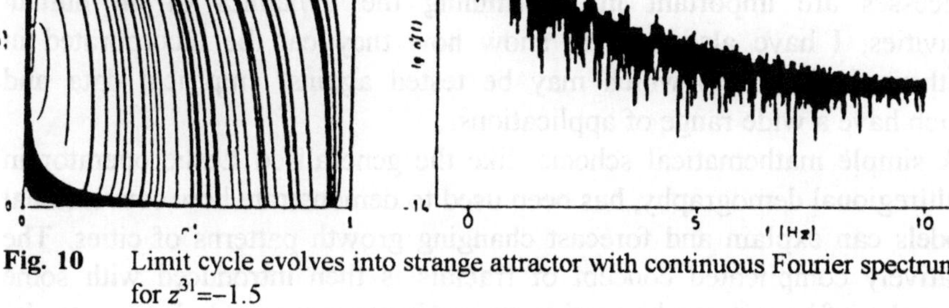

Fig. 10 Limit cycle evolves into strange attractor with continuous Fourier spectrum for z^{31}=−1.5

In more general terms, a numerical analysis reveals that chaotic trajectories appear if:

1. the intra-group agglomeration factors z^{kk} are positive and larger than a critical value;
2. at least two population groups have asymmetrical agglomeration factors;
3. the agglomeration factors matrix is asymmetrical.

It is evident that models of the type illustrated above, when built for the simulation of real situations, are able to accurately represent the behaviour of self-organising systems, including sudden 'phase transitions' triggered by the learning processes of social systems.

3.6 Conclusions

The aim of this work has been, firstly, to establish a number of 'criteria' for evaluating theories or models related to the development of cities and/or demographic and socio-economic systems and, secondly, to stimulate discussion of the evolutionary approach from this perspective. In order to supply elements for comparison, we have briefly reviewed a number of other theories developed to provide tools able to explain behavioural patterns which traditional system dynamic models have difficulty in explaining. A framework for comparison is based on the ability of each model to deal with the organisation of data the description, interpretation and prediction of phenomena and system management.

I have not tried to hide my impression that the evolutionary approach still seems to be a purely conceptual scheme, providing little help in classifying and organising data, and not easily fitting any real process. It also seems to lack any real forecasting capacity. However, since learning processes are important in determining the dynamics of all human activities, I have also tried to show how they can be incorporated in mathematical models which may be tested against empirical data and which have a wide range of applications.

A simple mathematical scheme, like the generalised Leslie operator in multiregional demography, has been used to demonstrate how conventional models can explain and forecast changing growth patterns of cities. The relatively complicated concept of fractals is then introduced with some examples of how it can be used to approximate real world phenomena. Its capacity for reproducing historical series, and even making predictions, has proved to be surprisingly good. The equally surprising fact that the fractal principle, while very good as a descriptive and predictive instrument, completely ignores the real working mechanisms that trigger changes has also been stressed.

The last modelling framework offered for comparison with the evolutionary approach is catastrophe theory, which deals with disastrous events happening in a context of slowly and smoothly changing external conditions. Again, a simple demographic model has been taken as an example of how catastrophic behaviour, like sudden phase transitions triggered by the learning processes of social systems, may be accounted for.

It should also be pointed out that the use of evolutionary concepts in the social sciences is nothing new; it can be traced back to Karl Marx and even to Adam Smith. However, the success of the neoclassical approach completely ousted such concepts. The recent revival has the unquestionable merit of focusing attention on endogenous mechanisms of change, ruling out at least some of the models based on systems theory which, in spite of their tremendous effectiveness in fitting the behaviour of dynamic systems, too often conceal a black box approach - they really investigate input and output variables, completely ignoring internal processes. Moreover, in evolutionary thinking, particular importance is given to the variety and asymmetry found in the real world.

Nevertheless, the many doubts about the practical effectiveness of the evolutionary approach are not to be taken lightly. First of all, there is a very large gap between verbal statements and the quality of the formal models, which do not seem to live up to their conceptual premises. Secondly, evolutionary theories seem better able to explain smooth changes rather than sudden discontinuity or turning points. Therefore, the attempt to combine the evolutionary drive concepts with ideas and methods from other theories better at explaining the dynamics of complex, self-organising learning systems, like the ones outlined in this work, seems very promising.

References

Andersen E.S. (1994) *Evolutionary Economics - Post Schumpeterian Contributions*, Pinter, London.

Arnold V.I. (1986) *Catastrophe Theory*, Springer-Verlag, Berlin.

Blumenfeld H. (1979) *Metropolis and Beyond: Selected Essays*, John Wiley and Sons, New York.

Campisi D., La Bella A. (1985) The Dynamics of Urban Population, *Sistemi urbani, 7*, 221-235.

Castelnuovo D.P.L., Hayes S.A. (1993) *Catastrophe Theory*, Addison-Wesley, Cambridge, Massachusetts.

Chatelin F. (1983) *Spectral Approximation of Linear Operators*, Academic Press, New York.

Cohen W.M., Levinthal D.A. (1989) Innovation and Learning: The Two Faces of R&D, *The Economic Journal, 99*, 569-596.

Cohen W.M., Levinthal D.A. (1994) Fortune Favors the Prepared Firm, *Management Science, 40*, 227-251.

Dawkins R. (1982) *The Extended Phenotype*, Oxford University Press, New York.

Egidi M., Marris R. (eds.) (1992) *Economics, Bounded Rationality and the Cognitive Solution*, Edward Elgar, Brookfield.

Kato T. (1982) *A Short Introduction to Perturbation Theory for Linear Operators*, Springer-Verlag, Berlin.

Keyfitz N. (1980) Do Cities Grow by Natural Increase or Migration?, *Geographical Analysis, 12*, 142-156.

Keyfitz N., Philipov D. (1982) Migration and Natural Increase in the Growth of Cities, RR-82-2, IIASA, Laxenburg.

Luenberger D.G. (1979) *Introduction to Dynamic Systems*, John Wiley and Sons, New York.

Mandelbrot B.B. (1983) *The Fractal Geometry of Nature*, Freeman, New York.

Marchetti C. (1989) Trasporti e città, IIASA, Laxenburg (mimeo).

Nelson R.R., Winter S.G. (1982) *An Evolutionary Theory of Economic Change*, The Belknap Press of Harvard University Press, Cambridge, Massachusetts.

Oster G.F., Wilson E.O. (1984) A Critique of Optimization Theory in Evolutionary Biology, in Sober E. (ed.) *Conceptual Issues in Evolutionary Biology*, MIT Press, Cambridge, Massachusetts.

Rogers A. (1975) *Introduction to Multiregional Mathematical Demography*, John Wiley and Sons, New York.

Weidlich W., Haag G. (eds.) (1987) *Interregional Migrating - Dynamic Theory and Comparative Evaluation*, Springer Series in Synergetics, Springer-Verlag, Berlin.

Willekens F., Rogers A. (1978) Spatial Population Analysis: Methods and Computer Programs, RR-78-18, IIASA, Laxenburg.

Williamson O.E., Winter S.G. (1991) *The Nature of the Firm: Origin, Evolution and Development*, Oxford University Press, Oxford, New York.

Wolfe M.A. (1978) *Numerical Method for Unconstrained Optimization: An Introduction*, Van Nostrand Reinhold Co., New York.

Zipf G.K. (1949) *Human Behavior and the Principle of Least Effort*, Addison-Wesley, Cambridge, Massachusetts.

4. The Possibilities and Limits of Self-Organisation

Maria Tinacci Mossello

4.1 The City as an Object of Study

Geographers have traditionally reserved the term 'urban system' to denote the organisation of sets of cities in space or in a specific environment (Tornqvist, 1981, Herbert and Thomas, 1990). The numerous urban geography studies dealing with single cities have, on the other hand, mainly focused on landscape and on economic function. Berry (1964) was, to my knowledge, the first author to explicitly propose an application to urban analysis of the scientific homomorphism suggested by General Systems Theory (Von Bertalanffy, 1968), inviting readers to think of cities as 'systems within systems of cities'. Bertuglia (1991) clearly refers to this theoretical model though placing special emphasis on single cities, having of course in mind not the study of *a* city, but rather of *the* city.

This position implies the endorsement of a holistic point of view, falling somewhere between the localistic view (long accused of 'exceptionalism' by geographers themselves, Schaefer, 1953) and the systemic view, which is typical of social sciences and basically a-spatial. According to this latter perspective social systems are indeed sources of regional effects, but the role of space, both as a limit and a possibility, is not seen as crucial or even as meaningful for such processes (Tornqvist, 1981). The systemic view seems particularly relevant today in advanced societies, where 'horizontal' relations across space are gaining weight as the territory becomes increasingly accessible and networked. The consequent weakening of the local dimension, both in practice and in terms of perception of place, is leading to the gradual undermining of the 'vertical' relations, i.e. the sense of identity and belonging which individuals (and also objects) establish with their local area. Raffestin (1984) refers to this process as 'de-territorialisation' and has traced its origins back to the introduction of money. The process is clearly being reinforced today by innovations in production and communication technologies. One serious consequence is that decisional mechanisms at the community level become increasingly *complicated* because of the separation between the place where decisions are taken and the place where their effects unfold. Strong and vital local systems can however impose restraints on the globalisation process and its related risks.

In this work we discuss the assumption that the city may be regarded as a local system. We therefore need to fall back upon the basic categories of urban analysis and re-examine their meaning in the light of this hypothesis. We should like, in passing, to point out explicitly that although the paradigm transfer from physical and biological sciences to social science (an increasingly frequent practice in recent times) is theoretically justified by the hypothesis of scientific isomorphism, it is based on the concept of *utility*, the process of evaluation typical of social studies, which implies the non-neutrality of the observer or scholar.

The first issue we wish to examine, because of its particular relevance to contemporary society, is the *recognition* of the city. This seemingly simple question (which, as we shall see, is especially important in the study of the city as a highly complex entity) has traditionally been concerned with the determination of the lower threshold (the minimum number of inhabitants, level of occupational differentiation, level of services, and so on) in order to be able to rank the city within a hierarchical taxonomy of settlements. Today the most difficult task seems to be that of determining the upper threshold, i.e. to establish the limits of the mega-city, the city-region, or the *galactic metropolis* (Knox, 1992). Whereas the problem of setting the lower threshold has a purely taxonomical character, that of the upper threshold involves delineating the limits of the system, and has been a recurrent theme of geographical research. We will return to this point in Section 5.

At a more general level, a further question we wish to investigate is whether it is possible to tell the city-system apart from the society-system. But, first, let us take a step back in order to clarify our terminology and therefore the logic of our argument. We begin with the meaning of the term complexity, which is seen here as the breadth of the 'gap' between the existing reality and the potential for action, i.e. as the range of evolutive possibilities being offered to living experience. The autonomy of the system derives from the wealth of such possibilities, and is never independent of the environment, from which it receives a large amount of information (far greater than the amount of information sent by the system to the environment). It is to this informational stream that the system must apply its selection procedures.

It is an established tenet of the literature (Laszlo, 1985, Luhmann, 1984) that society is a mediator of complexity in that it provides people with more possibilities, but also more constraints, than they would experience living outside any social context. Moreover, society as a whole is a source of its own dynamic processes. Although the hypothesis of the complexity of society seems well founded, we cannot automatically conclude that the city is a social entity to which we can safely attach the systemic qualities of complexity and homeostasis. This question therefore deserves a brief examination.

The city has long appealed to urban geographers, but principally as an interesting *form*. More recently, departing from the tradition which focussed on the spatial distribution of phenomena and on the shape of the landscape, urban geographers have moved towards a multidisciplinary approach, placing more emphasis on the dynamic nature of the city, while being aware that the city is also a social product which is:

a. not easily distinguishable from the non-city;
b. involved in global processes.

Saunders (1981), in agreement with outstanding social scientists, such as Durkheim and Marx, has even claimed that the city in itself is no longer a meaningful object of study. According to some social scientists, the relevance of the study of the city is confined to the particular aspects of scarcity and conflict that characterise the urban market for housing, this being the true nexus between social organisation and spatial structure (Rex and Moore, 1967). Other authors have traced the sense of the city back to its function as reproducer of the workforce (Castells, 1977) and, more generally, to the processes of capitalistic accumulation (Harvey, 1973, 1985).

Whether or not the city is systemic is a central issue for the analysis of post-industrial society, since there are important dynamics which are radically transforming its structure (the existence of structures moved by underlying processes is in fact one of the main conditions for the recognition of a system). At the metropolitan level, these processes tend to take on a global dimension, causing the city to undergo a morphological and functional transformation which is resulting in internal disintegration and extended suburbanisation (Castells, 1989).

But this goes further. There exists, in contemporary society, an anti-urban ideology that emerges at both the ideological-political level and the knowledge-perception level. On the political side, an explicit anti-urbanism has appeared in the planning policy of certain socialist countries: it found a place in the Soviet Union for some time between the two world wars (Coppola, 1986) and in the People's Republic of China after the second World War (Turco, 1988). Although these strategies were never really successful (in fact in the Soviet Union they were not fully implemented), it is probable that they expressed an underlying fear of the complexity and organisational power of the city. A less explicit, but possibly more insidious, anti-urban bias comes from North America, where the city has always been perceived in predominantly negative terms. This has in turn influenced European attitudes (Gravier, 1947, Romei, 1987) and is currently reinforced by the concern with environmental quality and ecological sustainability.

Objections to the assumption that the city is a social system come from complexity theory, and are related to explanations of urban reality provided by the literature on contemporary economic-territorial systems: the city cannot be regarded as a *system* insofar as it is a social *product*. This is *a fortiori* true if society itself is to be regarded as an urban product, as theories of innovation seem to maintain (Camagni, Cappellin and Garofoli, 1984, Del Monte, Imbriani and Viganoni, 1992). It is true that this objection is raised at the functional level and may therefore not apply to the morphological level of urban structure, but reducing complexity to mere morphology would amount to thinking of the city as a self-contained system, thus leading to an overt contradiction not only in functional terms (the city being a market that is open to its environs), but also, in view of the speed of urban expansion, in morphological terms.

This said, there remain at least two good reasons for considering the city as a well-defined entity within society:

a. the city is the most spectacular spatial outcome of social processes, and because of this;
b. it is a fact of great heuristic value for social sciences.

The leading idea of our discussion, namely that of treating the city as a self-organised complex system, is therefore intended as the stepping stone of a research programme (in the sense used by Lakatos, 1974) rather than as a theoretical certainty. Within this framework, the argument proceeds through the analysis of the urban environment, investigated through the traditional categories of the economic base and the urban way of life (Section 2), to a discussion of the subjective knowledge of the city, seen through its history and the perception of its inhabitants (Section 3). Our attention then moves on to more explicit dynamic aspects of urban reality, in order to tackle the question of the 'criticality' of the physical expansion and social integration of the contemporary city (Section 4). We also deal with the central issue of the identification of the city or, more basically, of urban identity within the modern context (Section 5). Finally, we go into the question of urban planning, which although still deemed socially desirable, is somewhat hard to define, both from the point of view of the actual contents - 'urban design' having clearly shown its inadequacy - and of the identification of the decision-makers and reference territory (Section 6).

All of this represents a real and urgent challenge both in scientific terms and also, in our opinion, social terms: we shall try in the course of this chapter to elucidate why.

4.2 The Urban Environment

4.2.1 The New Urban Economic Base

At this point the major question is 'how to interpret city within society' (Herbert and Thomas, 1990). In fact, cities find their meaning only within the social context and by means of it. This is equally true of the urban settlements of pre-industrial society, the large urbanised areas produced by industrialisation, and the sprawling conurbations and metropolises typical of post-industrial society, where the inter-sectoral transition (from the primary sector, through the industrial phase to services) is at the most advanced stage. However, to hypothesise that the city is a complex system means emphasising its continuity and permanence, rather than breaks or disjointedness. Urban evolution therefore needs to be read as a self-governed transformation process, open to the external context, but at the same time endowed with its own, autonomous regulatory code (Mumford, 1961).

This theme is linked to a second one, which is directly connected with the question of the meaning of the city, and therefore the rules of urban self-organisation: *why* has the city appeared? And *under what conditions* does the city survive and develop? The standard explanation, provided by historical urban geography, has a Platonic root and has to do with the division of labour. Lynch (1981) has recently revived this debate concluding that, although cities may initially have been built for symbolic or defence reasons, they have subsequently acquired a general and specific value as places of maximal accessibility to a large number of goods and services. Lynch himself adds that the main resource of an urban area lies in its endowment of transportation and communication facilities, and that its manufacturing activities depend on the supply costs for raw materials, distribution costs for products and supply costs for the workforce. His interpretation therefore shifts the prime meaning of the city from the closed and self-referential 'symbol' model to the 'market' model, whose intrinsic character is open and outward looking.

It is worth commenting here on the traditional distinction between *city-serving* and *city-forming* (or urban economic base) activities. Implied in this distinction is the inherent superiority of the exported base activities, since it is these, according to multiplier theory, which underlie urban growth. If, on the other hand, we consider the city as a self-referential system, we underline the importance of intra-urban relationships. Although extra-urban relations are not specifically included, their existence is implied.

In the urban field, and especially within the context of post-industrial society, the application of complexity theory calls for a deep taxonomic revision. Indeed, the modern city is a place where:

a. nonagricultural activities are diffused outside the urban area;
b. sophisticated business service activities remain within the urban boundaries and employ a large number of commuters living outside the city itself;
c. strategic activities coincide with the production of information/ innovation.

The post-industrial city therefore depends upon its ability to produce information. This is not, however, a sufficient condition for the identification of a complex entity. Although it may be true, as maintained by Allen in this same volume, that the diffusion of information produced elsewhere can also be regarded as an innovation, the city which adopts this strategy will simply assume the role of a *node* within a network, without acquiring self-organisation ability. In this respect, it could even be to some degree inhibited.

The notion of a self-organising city-system is even more problematic when read within the global/local framework. Despite its vagueness from the point of view of formalised knowledge, this framework is so significant in practical terms that it can no longer be ignored in the study of evolution processes. Our knowledge of the evolution of urban culture is more empirical than theoretical. For the last century, industrial growth has been a constant force behind urban evolution, but it has also marked a 'low' in the city's self-regulatory capability (Knight, 1993). Industrialisation, and the turbulent urban growth that it gave rise to - and still gives rise to in the mega-cities of the Third World - is a global process which is beyond the scope of the single city. However, the increasing acknowledgement that globalisation poses serious risks for sustainable development, since it disrupts, or at least subordinates, local factual and cognitive systems, suggests that there remains an important role for the city. According to Knight the possibility of integrating global knowledge and local cultures is the task of the city, provided that it manages to retain enough self-awareness to allow an independent design of its own destiny. Knowledge thus becomes the new *city serving & forming* urban economic base.

4.2.2 The Characteristics and Diffusion of Urban Society

In the post-industrial economy, the heuristic relevance of urban economic activities is becoming far less significant than the social character of cities. Nonagricultural activities, once a characteristic of the city, now tend to spread over the whole territory; the city has ceased to be a privileged niche. This is true not only of industrial activities, but also of services, especially of services to households and lower level business services. Within the tertiary sector, the service activities that keep their urban or, more specifically, metropolitan nature (by metropolis we mean those cities at the top of the functional hierarchy) are the more sophisticated ones which cover vast, discontinuous, and often international areas. As a rule, the city finds its identity as a niche for innovation.

The analysis therefore moves from the urban economic base to the urban way of life and social networks, which are displaying increasing structural complexity. The city has never been, nor it can be, a homogeneous socio-spatial structure, either from a logical point of view, since this would presume the existence of an organised heterogeneous structure, or from an empirical point of view - urban geography studies have always identified growth and diversification as the most common features of city evolution. We could be tempted simply to conclude that the city is *always* a complex system, but such an inference would be mistaken. Complexity, however, is not a consequence of the composite and manifold nature of the elements of a given structure, but of the *intensity of relations* within it and of the breadth of the *field of possibilities* for action (Racine and Reymond, 1973, Bocchi and Ceruti, 1985).

As to the relationship between complexity and the field of possibilities, it is worth noting that action, be it individual or social, is the search for, and instrument of, autonomy (assuming that we can make the distinction between the system itself and an *outside,* which we refer to as the 'environment' of the system). Action reduces environmental complexity and, at the same time, is the source of new complexity within the system. Social action feeds upon and contributes to the intensity of relations which are a necessary condition for the existence of the social system. This raises however an apparent contradiction: in contrast with the rural environment, the social environment of the city is commonly recognised as 'anonymous' (Herbert and Thomas, 1990). This relative scarcity of face-to-face relations in the urban environment would seem to contradict the hypothesis of the role of social contact as a fundamental feature of the city. This can be resolved however by distinguishing between social contact resulting from residential proximity, typical of rural areas, and the *potential* offered by the city for forming wide 'nonlocal' social networks which also carry higher innovative potential.

The hypothesis of contrast between city and country, still so deeply rooted in the social sciences (where it has traditionally been linked with other dichotomies, such as secular/sacred, rational/traditional, innovative/conservative, impersonal/personal), seems to be becoming less and less tenable. The assumption of such distinctions between urban and rural life are countered by important evidence from both advanced and developing countries. In the former, the spread of the 'urban' way of life is now evident even in areas which are not strictly urban, due to the diffusion of services, decreasing transport costs and the overflow of city inhabitants to suburban and nonurban areas. In Third World countries, many authors have observed that there are far stronger similarities between the urban and rural poor than between the poor and wealthy inhabitants of the same city. The huge *gourbivilles*, on the margins of many cities in underdeveloped countries, reproduce the form and style of life of the inhabitants' villages of origin (Lapierre, 1985, Corna Pellegrini, 1989, Herbert and Thomas, 1990). Although it could be claimed that the poor of Third World cities may acquire characteristics more similar to those of other cities of the same type than to the rural population of the same region, the fact that in Western countries the urbanisation process has virtually cancelled any real difference between the way of life in cities and outside cities (Dematteis, 1983) makes it difficult to hypothesise that the city is a social 'entity' (Herbert and Thomas, 1990). This is the crux of the problem: is the modern city the locus of an ever more complex social system without any identifiable spatial centre, or is the city an increasingly complex system in an increasingly complex environment?

4.3 The Value of Historical Memory and of Perception

Space has been a 'container' for possibilities from the very beginning of human experience: the territory is the historical-social expression of the state of things. *Territorialisation*, the gradual building up of a territory, is the outcome of a process of selection between various possibilities offered by space, creating in itself a series of new possibilities. This is particularly true for urban space, where the territorialisation process has been, and still is, particularly intense. Cities tend to be the most 'well-known' places because of the high intensity of human events associated with them. For most of recorded human history, cities have been the centres of political power as well as market, religious and cultural centres.

This does not mean that any large settlement is automatically a complex urban entity (i.e. rich in possibilities), since territorial complexity is not simply the additive result of single territorialising acts. Nevertheless, it is probable that a series of actions coinciding in the same space will generate a 'richer product' than the sum of the same series of actions when spatially independent. One cannot, however, rule out the possibility that a multiplicity of potentially conflicting actions concentrated in the same area can lead to their mutual destruction. In this same volume, Allen identifies this as one of the main factors behind urban decentralisation.

The city therefore represents a well-identified, dense, varied, but also critical territorial phenomenon. It is also, of course, a complex dynamic phenomenon. Allen in fact rejects linear dynamic analyses as nonmeaningful and proposes the use of complex methods more congruent to the analysis of evolutionary dynamics, which are seldom linear and frequently cyclical or even chaotic. It is probably worth thinking of urban dynamics also as *many-faceted*, i.e. made up of a series of elementary dynamics concerning the various urban elements (for instance, institutions, households and firms), their age, location and so on. The observation of such dynamics is suggested by the theory of 'self-organised criticality' (Bak and Chen, 1991), to which we will return in Section 4. This explains the erratic signals (flutterings) noticed in an organised structure as the result of the "superposition of signals of all amplitudes and durations" (p. 24). The statement that "the structure is ... an attractor on the trajectory of the system" in a situation in which "many sources of instability intervene in the evolution of [such] open systems" (Pumain and Haag, 1991, p. 1302) seems to refer to a similar idea.

Every self-organised spatial system possesses by definition a denomination, some concrete elements, and a structure of relationships.

Transferring our argument to the normative aspect, we can say that in order to qualify as a system, the city must first of all have a *name*. This is a necessary condition in order that a *city-sense*, a source of organisation and autonomy, can be associated with the *city-fact*.

Some time ago, Lynch (1960) proposed the study of the images of the city, showing that these may have little, or even nothing, to do with the actual objective forms of urban development, since they are filtered through individual and group evaluations based on taste, culture and experience. Between the space created by economic agents and planners and the space perceived by inhabitants there is a hidden and still poorly known dialectic, which is nevertheless likely to be an important component of the relationships and social conflicts taking place in urban areas. Although perception may be a second level reality with respect to the concrete structure of the city, in dynamic terms, and paraphrasing Pumain and Haag, one can claim that even the perception of the structure may be an attractor of the system's trajectory. We are referring here to the global perception of the city, which is closely associated with the city's name, not simple as a designating word, but as an expression of its *fame* which is, in turn, closely connected with its history.

The nexus between the history and self-organisational ability of the urban system is again not always univocal and unconditionally positive. In some cities, the historical heritage has brought about positive abilities and value-related attitudes; in others, it seems to have been reduced to a disgruntled and sterile contemplation of the past. Making use of a terminology that is closer to the theory of complexity, we can say that, in the former case, history enriches the urban system with tools for self-referential selection in the space of possibilities, whereas in the latter, history acts as a constraint on the system's adaptation within a space of possibilities that is perceived as unsatisfactory. In this respect, it is important to realise that once the charismatic value of its history has been lost, the city will have great difficulty in recovering it (it may at most remain a superficial tourist image or a political-ideological rhetoric glorifying the past).

The problem of historical identity is also relevant from the point of view of the physical planning of the city. The solution is not to be found in 'embalming' the historic city (e.g. through the conservation of its monuments), but in the perpetual reinterpretation of its image. This requires the analysis of lived space (Frémont, 1976). From the point of view of the construction of complexity, the historic nature of a city may be both an advantage and a disadvantage. A strong historically-based urban identity can be the cornerstone for building a satisfactory level of cooperation, which Allen believes to be a fundamental factor for the achievement of coherence in human systems. However, it does seem particularly difficult for the historical city to deal successfully with both the global aspect of its fame and relations with its own hinterland. Florence seems unfortunately to be a case in point. (Although once on a par with major European cities, it is now finding it hard to relate to other towns in its own region which, in economic terms, have overtaken it).

It would seem that a historical pre-existence is not a sufficient condition for the maintenance of an urban system. Speaking of Maurilia, Calvino (1972) tells us that "sometimes different cities succeed one another on the same ground and under the same name, they are born and die without ever having known each other, each incommunicable to all the others" (p. 37). It is very likely, however, that in the absence of political or institutional catastrophes which cause the loss of its *code* of organisation, the historical city will remain and be able to enrich its space with new possibilities. As for less complex local systems, it would seem that the factor permitting the maintenance of territorial identity for the city is its population, which is a source of territorialisation and cultural tradition, as well as innovation (Tinacci Mossello, 1988, 1990).

4.4 The Growth and 'Criticality' of Urban Organisation

4.4.1 Criticality of Urban Expansion

If we examine the urban phenomenon as a whole, ignoring the vicissitudes of single cities, we find that human history has been marked by a steady process of urbanisation. Only recently have some authors begun to speak of counter-urbanisation (Berry, 1976). It was soon realised however that the *clean break* theory was supported only by local analysis: at the regional scale, evidence backed the hypothesis of continued urban expansion, described, for instance, in terms of the *wave theory* proposed by Hall and Hay (1980). Another more elegant, but less convincing proposal, the urban life cycle theory, has been put forward (Van den Berg *et al.*, 1980, Cencini, 1983, Dini, 1988). If examined at the global scale, there is no doubt that the urban phenomenon is becoming more pervasive, due both to the enlargement of existing cities and creation of new ones. Statistical evidence of demographic decline in certain cities, especially central districts, is not sufficient to contradict this statement: neither the overall urban settlement pattern, nor the overall network of urban relationships is showing signs of weakening (Camagni, 1992).

The city is therefore an expanding system and this in itself causes a tension which seems well described by the theory of self-organised criticality (Bak and Chen, 1991). As growth proceeds, not only the quantity, but also the quality and order of possibilities in urban space change. If this transformation does not occur, the city becomes a dysfunctional organism with grave organisational problems. Like Posif's metaphorical horse - an animal which was double the normal size, produced by genetic manipulation, and unable to function because the energy flows were multiplied by a factor greater than two (Paba, 1990) - the city becomes incapable of renewing or even maintaining itself. A variety of subsystems with very different dynamics have to coexist in the city.

Multidynamicity is, according to Bak and Chen, a characteristic of systems in a critically self-organised state, and one can legitimately think of the city as such a system. This amounts to the acceptance of a series of assumptions about the city:

a. it is holistic in character;
b. it exists in a nonequilibrium state;
c. the system is complex and evolves spontaneously "toward a critical state where even a small event suffices to start a chain reaction" (Bak and Chen, 1991, p. 22);
d. the system is subject to a large number of small perturbations and/or to a lesser number of large perturbations; dynamic signals are very erratic in amplitude and duration, and are superimposed according to a pattern that becomes recognisable over a long enough time interval. This effect is known as *fluttering* and indicates that the system is strongly influenced by past events;
e. a small event may bring about a catastrophe.

The theory of self-organised criticality has important links with the theory of fractals, which has recently been proposed as a tool for the study of complex, self-organising human systems, including cities (Bak and Chen, 1991, Batty and Longley, 1987, and Lucchi Basili, in this volume). Although space does not permit further development of this issue, I believe that the hypothesis of a homomorphic relation between the morphological processes described by fractals and urban self-organisation processes is one worth exploring. The self-organised critical state is a situation of *weak* (i.e. foreseeable) chaos which is highly desirable with respect to *strong* chaos, since for the latter it is impossible to make forecasts beyond a certain time interval. This may well be the situation of many urban systems in crisis. The self-organised critical state has in fact already been identified in the case of a simple market network and the behaviour of traffic flows. Indeed, both kinds of chaos can be recognised in the urban system (traffic issues are a concrete example of the critical nature of the organisation of cities).

4.4.2 The Criticality of Urban Integration

It has long been recognised that aggregation levels within cities have a complex structure, in that a large city will have to be read as a world of modules made up of progressively smaller modules, themselves made up of even smaller modules, each however representative of a relatively self-sufficient, hierarchical organism, i.e. an organism endowed with a specific role (Caniggia and Maffei, 1993). Even if we weaken the strong hypothesis of an intra-urban Christaller type hierarchy and adhere to the network view (Dematteis, 1990) (which seems more suited to an urban environment specialised by area as an effect of zoning), it is a fact that the city is an organism that functions on the basis of a multiplicity of relationships most of which are strongly constrained by distance. In other words, the city operates in a space that is criss-crossed by intense flows. This set of flows is precisely what gives rise to and, at the same time, depends on urban traffic.

At least two types of spatial mobility are found in cities: sporadic trips, mostly for shopping and access to services, and regular trips, which are predominantly journeys to work. The latter tends to adopt a stable pattern in urbanised areas, where commuting is a physiological response, on the supply side, to zoning decisions which separate residential and industrial location and, on the demand side, to professional and residential mobility. The spatial diffusion of housing and business areas is reinforced by a similar diffusion of school, health, retail and leisure structures. In this situation, spatial mobility is regarded as natural. Within cities one sees the emergence of clear-cut socio-spatial patterns which are the sum of a large number of individual residential and mobility decisions. The analysis of this phenomenon has been used, especially in the United States, to define urban areas (e.g. the Daily Urban System and the SMSA) which are identified on the basis of commuter catchment areas. This approach is rather limited, however, if compared with the much richer spatial-temporal analysis proposed by Hägerstrand (1970).

The spatial extension and intensity of flows within these networks have tended to grow over time, due to the increasing speed of public and private transport, the expansion of urban areas and increasing land-use specialisation. All these factors lead to the strengthening of the communications network and further increases in the demand for mobility. Haggett (1983) has shown that an improvement in the transportation and communications network initially has the effect of lowering the average time taken for a given journey. Subsequently, however, it provokes an increase in the demand for mobility, which was previously constrained by the shortcomings in the infrastructure. As a consequence average trip times begin to rise once again and a new critical equilibrium is reached at a higher level of overall mobility.

Any infrastructural solution that has an effect on urban morphology must take into account the historical 'memory of place'. The aim should be to minimise disruption, striving to achieve efficiency without bringing about a break with the past. Such discontinuity can be traumatic and lead the urban organism into a situation of crisis, and even chaotic degeneration. We have to steer between two extremes: the problems of congestion and the loss of identity. Haggett (1983) warns us that ill-designed policies run the risk of provoking both; resulting in the loss of identity without resolving congestion problems.

The frequent outcome of all this is *segregation*, i.e. the collapse of relationships between parts of the city, which is all the more likely in periods when the city faces serious financial crisis. This kind of split is found between the inner districts and outer suburbs of many North American cities and in some Third World *mega-cities,* where relationships between the *bidonvilles* and other city districts have become extremely tenuous, reproducing the traditional separation between the 'city of the Europeans' and the 'city of natives' which existed in the colonial period (Corna Pellegrini, 1978, 1989). But another cause of isolation and segregation can be the networks of information flows generated by the new technologies within the *informational city* as a consequence, for example, of teleworking (Castells, 1989).

Segregation often affects the very heart of our cities, but the strength of the centre in our representation of space is, in general, too great to allow the process to go too far. The centre possesses an image which goes beyond its practical function, therefore, as long as the degeneration does not spread to the whole urban organism, sooner or later the municipality will try to promote its recovery. Attempts to regenerate the city centre are more likely when the city has a strong historical image or when the urban system has been affected by post-industrial transformation (Berry, 1978).

This latter process is often accompanied by social turbulence, especially when it concerns the centres of old industrial cities in which there are large abandoned industrial sites, high levels of unemployment and concentrations of marginal social groups. All these elements and disorders characterised many American cities after the late 60s and, more recently, some English cities as well (Herbert and Thomas, 1990). The purposes of regeneration or renewal projects may be social or a reaction to market stimuli, but in either case they are a crucial element of the city's evolutionary trajectory.

4.5 The Inevitability of Urban Diffusion

It is obvious that, within a framework of growth, the alternative to congestion is the spatial expansion of the city, that is to say a more extended urban territorialisation.

In the first chapter of this book, Allen describes this phenomenon, referring to the scarcity of supply of resources and the behaviour of "eccentrics or error-makers that grow more successfully than the average type" (p. 106). Urban development occurs within spatial constraints which, in certain periods, give rise to compact urbanisation, in others to a diffuse pattern of urbanisation. The crucial issue is that of the perception of distance. Technological evolution causes physical distance to take on new values. It is the intensification, speeding up and 'de-materialisation' of transport and communications systems which have led today's city-system to be identified with such vast areas as the megalopolis.

Urban expansion, however, takes very different forms: there are considerable differences in the nature of central areas, the character of suburban areas, and the role of the *edge cities* in different countries and regions. Development in the outer parts of the city sometimes gives rise to relatively autonomous systems, both from the functional and administrative point of view, and may lead to a new *sense of place*. Elsewhere they may be merely dormitory suburbs, representing a housing solution for the lower/middle classes who cannot afford property prices in the centre. Herbert and Thomas (1990) maintain that the former model is more frequent in Great Britain, North America and Australia, whereas the latter is characteristic of European cities. Although this may be generically true, it is likely to be something of an over-simplification. The transformation of the city form may be interpreted in relation to economic transition phases in the following way (Herbert and Thomas, 1990):

pre-industrial stage	–	urban nucleus
industrial stage	–	urbanised area
first post-industrial stage	–	city-region
second post-industrial stage	–	conurbation, metropolitan area, urban field

In the latter stages, the network of relations becomes more intense and complex, and the borders of urban units in the most densely developed areas tend to touch and eventually to merge. This suggests that cities risk losing their identity within a vast and anonymous built-up area. In general, however, central areas, at least, tend to retain their character, but peripheral parts of the city are usually less distinctive and 'fade out' towards the edge of the urban area, making the identification of the boundaries of the urban system problematic, especially from the organisational point of view.

In the modern city system, networks of specialised cities and intense inter-urban functional links are likely to develop (Dematteis and Emanuel, 1992). Friedmann (1978) prefers in fact to speak of 'urban fields' which are extensive areas intensively interconnected with high levels of personal mobility. When the outreaches of these urban fields merge into one another, as in the north-east of the United States - an area including Boston, New York, Philadelphia, Baltimore and Washington - we can speak of the *megalopolis* (Gottmann, 1961). Gottmann (1976) identifies the high potential for international exchange in the fields of commerce, technology, population and culture as the critical factor for the birth of such a system. He also recognised as systems in their own right the distinct and physically separate urban cores located within the *megalopolis*. The high level of overall interaction within this 'system of systems' causes serious problems for urban management and planning. At the worldwide level, Gottmann (1976) identifies five other metropolitan conglomerates, one of which extends across north-western Europe, including Amsterdam, Paris and the Ruhr.

It is clear that the Megalopolis, for which it is difficult to give a precise functional and physical definition, simply represents the most extreme example of the extension of urbanised regions (Doxiadis, 1968, went even further, predicting the appearance of the *Ecumenopolis*).

4.6 The Possibilities and Constraints of Urban Planning

4.6.1 Land-Use Planning

The city is undoubtedly the result of a social project. One can read its evolutionary states through the history of urban morphogenesis whose dominant codes are *efficiency*, which tends to bring buildings together, and *accessibility*, which shapes passages, alignments and non-built areas. Culture will differentiate the forms, but these basic criteria are found wherever there are, or have been, cities (Benevolo, 1975). The subjectivist approach emphasises these variants and the meaning that they take on for the individual-inhabitant, rather than their objectivity. This is not a minor objection, in that it poses the problem of the subjective sense of place: a problem that also concerns Allen, who rightly denounces (in his essay in this same volume) the temptation of the mechanistic and positivistic views of behaviour, which focus on the *average behaviour* of *homo economicus* - a being endowed with perfect knowledge and rational processes of choice, aiming at optimisation.

In the field of economics, the idea of bounded rationality, imperfect knowledge and limited evaluation capacity, has already moved the emphasis towards the concept of satisficing behaviour rather than an optimising choice. Allen goes even further, calling attention to the error-maker, who in his opinion is the real *deus ex machina* of evolution. In the field of urban geography, even the objectivity of space has been called into question, by emphasising its *image dominated* nature (Lynch, 1960, Herbert and Thomas, 1990). The problem is to understand the rules that govern the passage from the level of the phenomenon to that of behaviour. This is made more difficult by the fact that behaviour does not always translate into morphological signs, because of the limited variety of choices that can be made in a predetermined space. Furthermore, the rules of the market and the limited supply of resources may constrain part of the demand for spatial resources into a latent state, eliciting adaptation behaviour or protest, and causing a reduction in the possibilities for action.

We have underlined the two extremes of the vast problem of the organisation of the city: on the one hand, the continued expansion of the city, on the other, the individual nature of decisions. But even without conscious planning intervention, the city is too organised and entropy-negating an entity for it not to be the outcome of a social project. Pumain and Haag (1991) are surely right when they say that the structure is an attractor of urban evolution - though this still leaves much room to chance

- but I believe one must think of the city as a structure sustained by history and translated into a shared image in the minds of its inhabitants, providing an orientation to their choices.

In the face of the city's evident self-organisational capacity, a difficulty arises for both the analyst and the planner. The former has to face the problem of the *duration* of the observation, which must be long enough to allow the identification of the attractor. The latter has to accept the inappropriateness of imposing any geometrical principle to the design of the city; self-organising development is unlikely to follow a regular geometrical form. The poverty of *designed cities*, and the problems faced in evolution subsequent to the implementation of the formal plan, feed the suspicion that rationality of shape is in opposition to the vitality of action and relations. Self-organisation, being a selective process that creates new possibilities for local space, is more likely to translate into some degree of formal disorder (see Lucchi Basili in this same volume) rather than in a clear-cut order. Moreover, at the basis of the many diverse decisions behind the building of the city, there is probably a common and tacit land-use grammar which obeys social and economic rules, which are modified according to both the physical nature of the place and the 'character' acquired in the course of its history.

4.6.2 Suburbanisation and the Crisis of the City Centre

Most problems stem from the extension of the city beyond the optimal limit determined by the system's organisational rules (though for social systems these rules are in fact difficult to establish). This tends to bring the city to the situation of self-organised criticality hinted at above. From the practical point of view, traffic intensifies, the city becomes increasingly congested and polluted and there is a growing separation between its constituent parts.

An important role in the process of expansion-complexification of the city is attributed by many authors to the *edge-cities*, that is to say to the built-up areas which form on the outskirts of the urban area, either through the growth of small existing settlements or through the new spontaneous, or planned, development. Allen hints at this phenomenon, although at a higher level of abstraction, seeing it as stemming from innovative behaviour in physical space; the result of successful *error-making*, or the effect of constraints on the availability of space. The structure of this space, however, will not be the same for central and peripheral areas. It seems in fact that the more successful the marginal areas, the more likely it is that central areas will have problems of development, or even survival, due to processes of decentralisation and segregation. The existence of segregation, already present *in nuce* in the first urban ecology models of the Chicago school (Burgess, 1925), is very evident in recent urban evolution, although the actual form varies considerably from place to place.

Generally speaking, in America city centres have been left to decline from the social-urban point of view in favour of the *local democracy* of suburban areas, usually inhabited by the medium-upper classes (Herbert and Thomas, 1990). In Europe, on the contrary, there has been a serious attempt to thwart the crisis of central areas. Almost everywhere recovery plans have been launched for city centres, trying to avoid handing over the regulation of land use solely to market mechanisms and being stifled by traffic. However, the fiscal crisis of urban municipalities and the strength of the market are too evident for us to be fully confident of avoiding the materialisation of Calvino's *continuous cities* (1972): 'Trude', the faceless city, 'Cecilia' the labyrinth-city, 'Procopia' the crowded city and 'Pentesilea' the endless city. All too often, plans for the renewal of historical centres have been prompted by the lure of high rents and based on speculative development. Their effect has been the weakening, or even the disappearance, of 'social' or public land use in these areas and the reduction of identity to an empty shell. In order to counter these phenomena, it is necessary for the community to allot sufficient financial resources for this purpose but also, and above all, that urban society possesses a strong enough *civic sense* (Putnam, 1993).

4.6.3 The Politico-Social Foundations of Self-Organisation

It should be pointed out that human interaction depends not only on individual preference, but also on policies imposed from above, or arrived at by negotiation with a large (possibly global) social base. At any rate, interactions always occur within an institutional context. In the case of cities, the fundamental institution is the municipality (whose etymology is eloquent: *munia* = duties, burdens; *capere* = to take, to assume), and it is at the municipal level that the organisational rules, above all the rules of urban land-use, are discussed and established.

Any concrete inquiry into the systemic nature of the city needs to take into account the sense of land use of society and community, and the reasons for residential change. This is particularly necessary in the modern city, since it has not only cognitive significance, but great relevance for action as well. Self-organisation means that cities must be able to think and act in their own right. This means that we must first of all redefine the notion of city and citizenship, and that citizens must come to express collective decisions and actions, even if individually evaluated. On the institutional side too, the problem is far from negligible, since *local democracy* is carried out in highly fragmented, suburban spaces, whereas the organisation of large urban areas seems to follow the logic of *local control* (Herbert and Thomas, 1990).

Residential mobility and the increase in nonperiodic trips (i.e. those not for work/school etc.) make it necessary to identify new city limits, since land is needed in the surrounding area for new housing and for recreation. If this is not taken into account, the only feasible way to solve the fiscal crisis of the city is to surrender to renewal policies inspired by market forces, which are likely to weaken the social identity of the centre.

We should at this point tackle the question of the role of the city as a self-organised complex system, bearing in mind that in the social sciences any normative model must have some degree of social usefulness. To capture the full scope of this issue, we must place it within a global-local framework. Indeed, when one reflects that industrial production is today largely located outside urban areas, it is no longer credible to claim that inter-sectoral and territorial division of labour is the reason for urban development. On the other hand, the role of 'information selector' often attributed to the city is still far too vague (Tinacci Mossello, 1992). The local-global paradigm also forces us to consider the role of cities in relation to global processes and the imperative of sustainable development. At the local level, Calvino (1972) poses the issue of urban waste management, describing 'Leonia', the city that remakes itself every day. The problem of waste disposal means that the city still needs a *hinterland*, but in a very different way from the traditional city/country dependence. In relation to this particular environmental problem at least, the city seems to have a distinctly negative role. At the same time, the city plays a central part in the organisation of information society and as a place of technological innovation, as described in Kondratieff's fifth cycle (Tinacci Mossello, 1992).

In positive terms, we can say that, although cities have to a large extent lost their industrial export base, they have until now retained a selective concentration of non-production activities based on knowledge. It is also true that the spread of residential development outside the city is no longer interpreted as a counter-urbanisation process and that nobody really believes that we will live in a world without cities. However, it is necessary to recover a positive sense of the city, which is adequate both at the local and at the global level. Such a sense may, in my opinion, be sought in a development project that aims at the organisation of knowledge for the purposes of development itself. The city seems to be nominated to this role by the post-industrial society, but it is in fact the same role that the city has carried out throughout history (Knight, 1993). What has changed is the context, the globalisation of certain problems and of certain knowledge, which require the city to be the place where a new synthesis between formal global knowledge and informal local knowledge is sought. This creates new possibilities, linked to the increasing complexity of the urban system, and new constraints, linked to the nonexistence of a truly *external* environment (Luhmann, 1984).

4.7 Final Remarks

We can conclude that the city is undoubtedly a complicated but meaningful place in which men, artefacts, functions, symbols and memories gather. It is a place which builds its own sense from a number of different realities: historical, geographical, economic, social and many more. The hypothesis that acts as a common thread to the writings in this volume is the existence of a specific urban complexity, understood as a space of possibilities linked to identity and design. This is a high-profile interpretation which coincides with the hypothesis of the city as system. As we have already remarked, despite the spatial diffusion of the urban way of life and the evident crisis of the city, one of the most important aims of a research programme should be the application of a systemic logic to the study of the city. Complexity and design have always accompanied urban history, which has been marked by the steady spread of urbanisation. This combined characteristic of complexity, design and expansion, makes a systemic hypothesis essential.

The above statements call for some, necessarily brief, terminological qualifications relating to the assumption of the notion of 'system' and the approach to complexity adopted. The system, according to an established meaning that can be traced back, among others, to Von Bertalanffy (1968), Racine and Reymond (1973) and Laszlo (1985), is defined as a set of related elements with a boundary that allows a distinction between the system itself and the outside environment, but towards which it is *open*. It may survive and evolve, retaining its identity, provided that it has enough internal coherence and a minimal degree of self-referentiality. Self-organisation seems to follow from (and to consist of) this, thereby being a typical systemic characteristic, even though Allen (in this same volume) regards it as a specific 'quality' of the system.

The quality of complexity entails a multiplicity of characteristics (Morin, 1985), among which we have emphasised uncertainty, the range of possibilities or choice, the complicatedness of the interactions and the ability to (self-)organise. It is easy to see how these theoretical qualities, although largely developed in other fields of study, fit nicely into a geographical research programme that assumes the city to be a complex entity and, consequently, a system that can self-organise. In fact, the city is a persistent social-geographical entity whose empirical relevance is huge, both at the global and the local level. Using the terminology of Laszlo (1985), we can say that the city shares the characteristics of *generalist systems* despite being a *specialised system* (with respect to its surroundings and the network of cities). Generalist systems are *resilient*, that is to say they have a high potential of adaptation to environmental stimuli, thanks to a sophisticated battery of tools and to an equally sophisticated network of communication flows.

Allen (in this same volume) analyses the characteristics of the city through a series of economic variables. Firstly, he identifies two types of worker, blue collar and white collar, and five types of employer representing the various sectors: industrial, financial, 'lower' tertiary, advanced tertiary and local industry, he then studies the dynamics of the relationships between them. He deals jointly with forecasting mechanisms and decision processes, deriving some important conclusions which include the prevalence of cooperative behaviour with respect to competitive behaviour, and the relevance of innovative behaviour and of deviations from average behaviour, even when due to chance or error (in the model he specifically considers location behaviour). However, a formalised analytical approach such as this, though interesting, has some limitations both with respect to the recognition of lower-level structures - which we might call sub-systemic - and with respect to the time period required (a long enough time interval to allow for the appearance of general environmental change and/or for changes in the state of the system).

As to the first point, it is clear that, if we wish to examine the city as a whole, it is inevitable that some of the subsystems will have to be disregarded, although many are likely to be important for the working of the system as a whole. Within metropolitan areas it is commonly acknowledged that there exist, for example, systems of firms which operate in a Marshallian 'industrial milieu' (Tinacci Mossello, 1982, Scott, 1988), neighbourhood effects (Tinacci Mossello, 1990), segregation/solidarity between ethnic groups (Castells, 1989) and so on.

All of these phenomena are difficult to frame into simple variables. It is therefore difficult to obtain by means of mathematical or statistical language a satisfactory and meaningful representation of urban complexity. It is similarly difficult to capture the essence of complexity *tout court* into formalised, quantifying thought (Morin, 1985).

In the long run, the city tends to persist through transformation, according to the logic of resilient open systems which undergo evolutionary processes (Laszlo, 1985). Indeed, evidence shows that cities have a high probability of survival, passing from the pre-industrial environment, through industrial development to post-industrial transformation. This makes it risky, when attempting a long term analysis, to adopt variables which represent specialisation's of urban systems. Any variable used to describe the state of the system at a given moment, tends subsequently to become permanently enshrined and can become an obstacle to a genuinely dynamic analysis. We could take, for example, the fast and radical change that has taken place in recent decades in the relationship between the location of industry and the city. The study of the dynamics of evolutionary processes teaches us that it is illusory to seek its meaning in the phenomenon itself and that it is much more fruitful to look for the tools and modes of resilience, i.e. the self-maintenance ability of the system (Laszlo, 1985). In the case of the city, permanence has more to do with *sense* than with shape.

A crucial issue, once one assumes the city to be a system, is that of establishing its *limit*. In fact the *city-system*, primarily identified by the intensity of relationships, has never been totally contained within the *city-shape*, either in terms of the classical town/country dichotomy or, more generally, the relationship between the city and its *surroundings*. The limit of the *city-system* therefore lies beyond the city-shape, which makes it hard to meet the requirement of 'operational closure' established by Varela (1985) as a precondition for autopoiesis and therefore for design. This closure relates to the fact that the result of an action by a decision maker should fall within the boundaries of the system itself, and its consequence is that of 'bringing about a world' whose existence is an essential component of organisation.

If the city-system is to be able to self-organise it is necessary that:

a. it has sufficient knowledge of itself and its hinterland to be able to make choices and take decisions which are coherent with a code that is congruent with its own nature and is shared by a large enough number of urban actors;
b. it is also able to decide for a wide enough hinterland.

This would seem extremely difficult to achieve nowadays: not only is the city overflowing into suburban, peri-urban and rural areas, but a large number of processes, sustained by very different spatial logic, are all focussing upon the city. We need to consider, for instance, the phenomena of:

a. residential development, which is occurring in many different forms and locations according, in particular, to the economic characteristics of households. As a consequence, the shape of urban geography is being redrawn, often including noncontiguous areas;
b. industrial location, which is following a logic that is hard to interpret, but is linked to a variety of location factors on the demand side and to environmental economies on the supply side. The field of analysis (and of representation) of alternatives range from the local scale (comparing cities with other places or, simply, comparing the city core with its outer areas) to the world-scale;
c. the deep transformation of the urban labour market, which, due to the transition from industry to services, is revealing significant changes within the dominant tertiary sector: a declining demand for medium-level jobs with employee status (including public employment), and an expansion of nonpermanent jobs at the two extremes, low unskilled jobs and highly qualified professional jobs (Castells, 1989, traces the genesis of the *dual city* back to these complex processes of growth and decline taking place in metropolitan areas);
d. the networking of space, which is less and less satisfactorily explained in terms of hierarchical models *à la* Christaller and more appropriately described through nondeterministic models driven by complementarity phenomena (Dematteis, 1990), up/down discontinuity of ordering (Borlenghi, 1990) and so on. At the infrastructural level, networks are increasingly complex (due to multiple modes, the growth of inter-modal interconnections, and the coexistence and superposition of various orderings, etc.) and lead to congestion in urban nodes. Such congestion is increasingly difficult to manage as it derives from poorly integrated logic and policies which are often independent of the interests of the cities involved.

The city, which, as a result of these processes, risks no longer being able to recognise its boundaries (not only in the physical sense, but also in terms of its interrelations) could lose its sense of identity and, *a fortiori*, the capacity to govern itself. Where then would the political-social governance of space be carried out? It is hard today to imagine a world without cities, even if there have been examples of the practical application of anti-urban ideologies.

It is in this respect that the hypothesis of the city-system, able to organise its own complexity, seems socially so desirable. Spatial complexity would therefore become an essential tool for a socially conscious development, with cities representing the places characterised by the highest intensity of information exchange and social communication. It is in principle possible to imagine a complexity *without* cities, and it is equally possible to imagine a *post-urban* spatial organisation. This latter however, within a world of rapid urbanisation, would appear a highly negative concept. To take it as a hypothesis would imply accepting that the fabric of human settlements is losing its global meaning - a conclusion which, apart from being politically discouraging, raises a logical contradiction with the hypothesis of increasing of complexity.

We must, however, reflect upon the apparent contradiction between the tendency towards non-territorialised space (free from historical-cultural and institutional constraints), which is the support of global relationships and mass communication, and the city, laden with historical and cultural *sense*, physically structured in relatively stable, change resisting forms, nevertheless *framed* in a vast, multidimensional networked exogenous environment. Turco (1988) argues that cities may be affected by an 'excess of territorialisation', reversing the normal relationship between the social and the geographical system, where the former usually dictates the latter. If this is the case, the city, far from playing the crucial role that theories of perception, of innovation, the economic base, etc. have traditionally attributed to it, could become a deadly machine which, while carrying out its self-referential strategies, consumes an *excessive* share of resources.

There is no doubt that the city is in crisis, both from the point of view of its functionality and the quality of urban life. The crisis of the contemporary city has been read in various ways (Dematteis, 1983, Knight, 1993), but today we are obliged to make a new interpretation in the light of the local/global context and the need to achieve a sustainable way of living: that is to say the city *must* be seen in terms of change, rather than of decline. As already pointed out, here and elsewhere, the economic base of the city is no longer to be found so much in material production, as in the production of knowledge. But we must not make the mistake of thinking that this is a novelty of the so-called post-industrial stage of development: the elected role of the city has *always* been that of the production of knowledge (Knight, 1993). Its association with an economic base focussed on material exports derives from only the fact that the city, over the last two centuries, has been so closely linked with industrial location.

In relation to the production of knowledge, the notion of a system is particularly appropriate, since it evokes the existence of an autonomous behavioural code, as well as the ability to select information coming from the environment. In concrete terms, this requires the city to offer a good quality of life, so that it can attract and develop the skills and abilities which make it possible to discriminate, incorporate and expand global knowledge and innovation. In this respect, Knight (1993) recognises the specific advantages and role of European cities, which more than any others have performed an important historical function as producers of culture and development, even beyond their own borders.

The uncontrolled growth of cities and their hinterlands, as well as creating anonymous urban agglomerations, such as the Third World's mega-cities, is threatening the identity of many European cities which have a long outstanding tradition of self-governance and well-planned development (Dematteis, 1983). The congested and fragmented space of those cities facing functional, economic and administrative crisis due to demographic explosion or processes of residential and industrial decentralisation, can perhaps find new meaning within a network model, finding their place as a *node* within a *network* of relations rather than fixed locations in physical space (Dematteis, 1990). The horizons of urban physicality, the built city with its social, and sometimes segregated, space, therefore loses its meaning. The knowledge produced by the city and in the city would primarily be related to the network with its features of complementarity and 'relationality' rather than of hierarchy.

We should perhaps at this point very briefly examine the meaning and content of knowledge which, for our purposes, can usefully be distinguished into two categories (Knight, 1993):

a. 'global' knowledge, which is formalised and transmissible, and whose nature is scientific, quantifiable and universal;
b. 'local' knowledge, which is tacit, informal and socially shared, and whose nature is cultural and ecological.

The former type is sustained and amplified by reticular space and allows us to look at the city as a node, abstracting from its physical substance, although this carries the risks of social polarisation, dualism and segregation, described by Castells (1989) in his *informational city.* The latter type of knowledge, on the other hand, requires all the systemic qualities of the city to be safeguarded and expanded, from the fabric of relationships to the ability to self-govern. Concluding his journey through the invisible cities, when the 'infernal city' was already appearing to Kublai Khan and himself, Marco Polo said: "the first [way] is for many the easiest: to accept inferno and become part of it, until one does not see it anymore. The second is more risky, calling for constant attention and practice: to seek and learn to recognise who and what, in the very middle of inferno, is not inferno, to make it last and to give it room" (Calvino, 1972, p. 170).

There is no doubt that the imperative of sustainable development means that global knowledge must be integrated with local knowledge, and it seems that the most suitable place for this integration in today's organisation of space, is the city. Even though, at first sight, the contrary may appear to be true, collective values find their place at the local level, whereas the global level is the realm of elitarian, meta-social choices and production-oriented values and behaviour (linked to the economics of multinationals and advanced technology).

Sustainable development will not occur by chance. It forces us to think in terms of long-run horizons and of the interrelationships of different scales of space. In this context, the role of the city must be sought in a multiplicity of projects and actions, with a system of governance that transcends the urban scale (moving towards the regional scale). It will require greater understanding of the economic and social transformation of the *inner city* and of relationships with the national and global networks of cities and information; the rationalisation and restructuring of intra-urban relationships in order to turn them from random and congested networks into rich fabrics of social communication; a new design for the enhancement of the quality of urban life and a more rational use of resources; a rediscovery of history and culture; and an expansion of research facilities, with an eye to the integration of local and global knowledge.

It is difficult to say to what extent this new knowledge can be defined in terms of 'exchange' and hence as part of the urban economic base. Maybe the problem of quantification no longer has any meaning. In a global context, the milieu remains, but distance loses its importance. This leaves us however with a vital challenge. What is at stake is the preservation of the character of spatial systems and the meaning of place. If we fail to meet this challenge, we risk turning the city into a labyrinthine node, lost within a global network of relations. The costs would be borne not only by cities themselves, but by the whole of society.

References

Bak P., Chen K. (1991) La criticità organizzata, *Le Scienze*, 271, 22-30.
Batty M., Longley P.A. (1987) Urban Shapes as Fractals, *Area*, 19, 215-221.
Benevolo L. (1975) *Storia della città*, Laterza, Bari.
Berry B.J.L. (1964) Cities as Systems within Systems of Cities, *Papers and Proceedings of the RSA*, 13, 157-176.
Berry B.J.L. (ed.) (1976) *Urbanization and Counterurbanization*, Sage, London.
Berry B.J.L. (1978) Comparative Urbanisation Strategies, in Bourne L.S., Simmons J.W. (eds.) *Systems of Cities: Readings on Structure, Growth and Policy*, Oxford University Press, New York, 502-510.
Bertuglia C.S. (1991) La città come sistema, in Bertuglia C.S., La Bella A. (eds.) *I sistemi urbani*, Angeli, Milan, 301-390.
Bocchi G., Ceruti M. (eds.) (1985) *La sfida della complessità*, Feltrinelli, Milan.
Borlenghi E. (1990) L'industria innovativa e la sua città, in Borlenghi E. (ed.) *Città e industria verso gli anni Novanta*, 3-29.
Burgess E.W. (1925) Urban Areas, in Smith T.V., White L.D. (eds.) *Chicago: An Experiment in Social Science Research*, Chicago, University Press, 113-138.
Calvino I. (1972) *Le città invisibili*, Einaudi, Turin.
Camagni R. (1992) Nuovo paradigma tecnologico e mutamento nei modelli organizzativi, in Del Monte A., Imbriani C., Viganoni L. (eds.) *Sviluppo regionale e attività innovative: esperienze a confronto*, Angeli, Milan, 159-178.
Camagni R., Cappellin R., Garofoli G. (eds.) (1984) *Cambiamento tecnologico e diffusione territoriale*, Angeli, Milan.
Caniggia G., Maffei G.L. (1993) *Lettura dell'edilizia di base*, Marsilio, Venice.
Castells M. (1977) *The Urban Question*, Arnold, London.
Castells M. (1989) *The Informational City*, Blackwell, Oxford.
Cencini C. (1983) Individuazione delle aree marginali in corso di rivalorizzazione attraverso un indicatore demografico: metodologia della ricerca, in Cencini C., Dematteis G., Menegatti B. (eds.) *L'Italia emergente*, Angeli, Milan, 85-104.
Coppola P. (1986) *Una introduzione alla geografia umana*, Liguori, Naples.
Corna Pellegrini G. (1978) *Periferie urbane nel terzo mondo*, Vita e Pensiero, Milan.
Corna Pellegrini G. (1989) *Esplorando polis*, Unicopli, Milan.
Del Monte A., Imbriani C., Viganoni L. (eds.) (1992) *Sviluppo regionale e attività innovative: esperienze a confronto*, Angeli, Milan.
Dematteis G. (1983) Deconcentrazione metropolitana, crescita periferica e ripopolamento di aree marginali: il caso dell'Italia, in Cencini C., Dematteis G., Menegatti B. (eds.) *L'Italia emergente*, Angeli, Milan, 105-142.
Dematteis G. (1990) Modelli urbani a rete. Considerazioni preliminari, in Curti F., Diappi L. (eds.) *Gerarchie e reti di città: tendenze e politiche*, Angeli, Milan, 27-48.
Dematteis G., Emanuel C. (1992) La diffusione urbana: interpretazioni e valutazioni, in Dematteis G. (ed.) *Il fenomeno urbano in Italia: interpretazioni, prospettive, politiche*, Angeli, Milan.
Dini F. (1988) Controurbanizzazione nei paesi occidentali: riscontri empirici e assunti di valore nella letteratura internazionale, *Rivista Geografica Italiana, 93*, 331-342.
Doxiadis C.A. (1968) *Ekistics. An Introduction to the Science of Human Settlement*, Hutchinson, London.

Frémont A. (1976) *La région, espace vécu*, Presses Universitaires de France, Paris.
Friedmann J.R. (1978) The Urban Field as a Human Habitat, in Bourne L.S., Simmons J.W. (eds.) *Systems of Cities: Readings on Structure, Growth and Policy*, Oxford University Press, New York, 42-52.
Gottmann J. (1961) *Megalopolis, the Urbanized North-Eastern Seaboard of the United States*, The Twentieth Century Fund, New York.
Gottmann J. (1976) Megalopolitan Systems around the World, *Ekistics*, 243, 109-113.
Gravier J.F. (1947) *Paris et le désert français*, Le Portulan, Paris.
Hägerstrand T. (1970) What about People in Regional Science?, *Papers and Proceedings of the RSA, 24*, 7-21.
Haggett P. (1983) *Geography: A Modern Synthesis*, Harper and Row, New York.
Hall P., Hay D. (1980) *Growth Centres in the European Urban System*, Heinemann, London.
Harvey D. (1973) *Social Justice and the City*, Arnold, London.
Harvey D. (1985) *The Urbanization of Capital*, Basil Blackwell, Oxford.
Herbert D.T., Thomas C.J. (1990) *Cities in Space, City as Place*, David Fulton Publishers, London.
Knight R.V. (1993) Città globali e locali, in Perulli P. (eds.) *Globale/locale. Il contributo delle scienze sociali*, Angeli, Milan, 107-137.
Knox P. (1992) Facing up to Urban Change, *Environment and Planning A*, *24*, 1217-1220.
Lakatos I. (1974) Falsification and the Methodology of Scientific Research Programmes, in Lakatos I., Musgrave A. (eds.) *Criticism and the Growth of Knowledge*, Cambridge University Press, Cambridge, 91-196.
Lapierre D. (1985) *La cité de la joie*, Pressinter, Paris.
Laszlo E. (1985) L'evoluzione della complessità e l'ordine mondiale contemporaneo, in Bocchi G., Ceruti M. (eds.) *La sfida della complessità*, Feltrinelli, Milan, 362-400.
Luhmann N. (1984) *Sozial Systeme. Grundgriss einer allgemeinen Theorie*, Suhrkamp Verlag, Frankfurt.
Lynch K. (1960) *The Image of the City*, MIT Press, Cambridge, Massachusetts.
Lynch K. (1981) *A Theory of Good City Form*, MIT Press, Cambridge, Massachusetts.
Morin E. (1985) Le vie della complessità, in Bocchi G., Ceruti M. (eds.) *La sfida della complessità*, Feltrinelli, Milan, 49-60.
Mumford L. (1961) *The City in History*, Harcourt and Brace, London.
Paba G. (1990) Limiti e confini della città: un'introduzione, in Paba G. (eds.) *La città e il limite*, La Casa Usher, Florence, 8-21.
Pumain D., Haag G. (1991) Urban and Regional Dynamics, Towards an Integrated Approach, *Environment and Planning A*, *23*, 1301-1313.
Putnam R.D. (1993) *Making Democracy Work*, Princeton University Press, Princeton, New Jersey.
Racine J.B., Reymond H. (1973) *L'analyse quantitative en géographie*, Presses Universitaires de France, Paris.
Raffestin C. (1984) Territorializzazione, deterritorializzazione, riterritorializzazione e informazione, in Turco A. (ed.) *Regione e regionalizzazione*, Angeli, Milan.
Rex J.A., Moore R. (1967) *Race, Community and Conflict*, Oxford University Press, London.
Romei P. (1987) Ambiente e migrazioni in un'area della Toscana: la provincia di Pistoia, CNR, Progetto Finalizzato 'Economia Italiana', Quaderno no. 10.
Saunders P. (1981) *Social Theory and the Urban Question*, Hutchinson, London.

Schaefer F.K. (1953) Exceptionalism in Geography: A Methodological Examination, *Annals of the Association of American Geographers*, *43*, 226-249.
Scott A.J. (1988) *Metropolis. From the Division of Labor to Urban Form*, University of California Press, Berkeley-Los Angeles-London.
Tinacci Mossello M. (1982) Economia e geografia. Dall'analisi delle economie di agglomerazione alla teoria dello sviluppo regionale, *Rivista Geografica Italiana*, *89*, 303-331.
Tinacci Mossello M. (1988) Labour and Resources in Regional Organisation, in Cori B., Fondi M., Zunica M. (eds.) *Italian Geography in the Eighties*, Giardini, Pisa.
Tinacci Mossello M. (1990) *Geografia economica*, Il Mulino, Bologna.
Tinacci Mossello M. (1992) Capacità innovative dei distretti industriali. Formulazione di ipotesi e verifica nel caso di Prato, in Del Monte A., Imbriani C., Viganoni L. (eds.) *Sviluppo regionale e attività innovative: esperienze a confronto*, Angeli, Milan, 377-400.
Tornqvist G. (1981) On Arenas and Systems, Space and Time in Geography, Lund Studies no. 48, Gleerup, Lund, 109-120.
Turco A. (1988) *Verso una teoria geografica della complessità*, Unicopli, Milan.
Van den Berg L., Drewett R., Klaassen L., Rossi A., Vijverberg C. (1980) *Urban Europe. A Study of Growth and Decline*, Pergamon, Oxford.
Varela F. (1985) Complessità del cervello e autonomia del vivente, in Bocchi G., Ceruti M. (eds.) *La sfida della complessità*, Feltrinelli, Milan, 141-157.
Von Bertalanffy L. (1968) *General Systems Theory*, Braziller, New York.

5. On the Foundations of Social Dynamics: An Efficient Mathematical Statement of a General Framework Underlying a Complex Nonlinear Social Determinism, Incorporating a Supra-Observer and a Suprastructure

Dimitrios S. Dendrinos

5.1 Introduction

What social scientists, as detached observers, write about society's dynamics carries in it an inevitable contradiction: on the one hand, it is always a product of the imagination and thus it is fiction; on the other hand, it is an effort to replicate social reality as closely as possible and code it in a transmittable manner, while minimising the always present and requisite variety for reasonable interpretation and extension. It is the requirements for precise replication and transmittable coding of social reality which distinguishes social scientists from novelists.

Thus they produce models or abstractions of social reality. Their products are society's replicas, virtual social realities, modern 'high-tech' parables, mathematically-stated myths, in that they contain grains of what one might consider 'social truth'. As abstractions of social reality, they are not close-ups which propel social actors to action. These models do not and cannot convey the same details and complexity as reality itself.

As observers of social behaviours, social scientists attempt to model the actions of society's actors and their effects. The actors, through their actions, aim to alter social reality in accordance with their desires and expectations about the future state of the social system. So, to some extent, they alter what social scientists attempt to model as observers. Their actions also draw from imagination. Since the objectives, constraints and stakes are different in these two cases, the resulting models (and novels) of reality are not the same. Nonetheless, description of social reality is the participants' inescapable, permanent and consequential task; it always results in social action. Meanwhile, it is the observer's quixotic pursuit.

In this essay, one objective is to elaborate on this contradiction (i.e. societal prose being composed of both fiction and reality). Within this elaboration there are certain propositions that broadly apply, coming quite close to what might be thought of as 'social laws'. They seemingly define the dynamics of this contradiction and, at the same time, they identify a classification scheme to label all social dynamics. Another objective is to take a closer look at the associated interaction between perception of social reality, expectations and purposeful social action. This interaction is fundamental, as it results not only in updating social reality, but also in changing its workings and thus the manner in which it is perceived.

The mechanics of this complex interaction perhaps constitute the basis for stating a general classification scheme of social dynamics, and possibly the only laws one may depict from social reality. This is the essence of the distinction between the subject matter of social and natural sciences. In the latter, perceptions regarding nature do not interfere with or alter natural laws; they may result, however, in different ways of stating these laws.

These efforts inevitably lead to a broad logical framework within which the roles and functions of observers and actors in social reality can be conceptualised. This logical framework is shown to contain a *suprastructure* (a supracode) of theoretical social behaviour, identifying the point of view of a *supra-observer*.

The suprastructure contains the necessary and sufficient information of a social supracode which sets off social dynamics. Put differently, it contains the instructions of social dynamics. A supra-observer is an entity which realises the presence of such a supracode, although this entity is not omnipresent, omniscient or omnipotent, i.e. a supreme being; rather, the supra-observer's properties are of minimum intrusion and knowledge, merely aiming at deriving a theory of social science theories. As the existence of a supra-observer who decodes the supracode can be referenced by another supra-observer operating at a higher level of observation and subject to an even higher level supra-supracode, an infinite regress sets in. This is where issues of symbolic logic creep in, and the insurmountable obstacles in addressing them become of the essence in social dynamics. They will be revisited a little later.

Most of the ideas expressed so far were first stated on a preliminary basis in qualitative terms in a book by the author (Dendrinos, 1992). Here, an effort is made to mathematically state and further elaborate on these fundamental propositions.

By identifying the presence of a supra-observer, the effort is not simply to define another layer of social observation. That would certainly not be enough, no matter how important it may be to the achievement of a more complete understanding of the underlying process giving rise to social dynamics. More basically, in identifying the existence of a suprastructure, subtle aspects of social reality are revealed and some qualitative perceptions of social action (particularly self-fulfilling/defeating aspects of it) are fully incorporated into the statement of social dynamics.

It should also be noted that the intention here is not to derive yet another observer or participant model of social dynamics, thus adding to the already abundant supply of such models. Instead, it is shown that an efficient statement of a general framework within which a theory of social dynamics can be constructed, brings into focus the need for identifying the origin, necessity and role of social actors and society's observers responsible for any theory of social behaviour. Further, the analysis demonstrates that the various entities responsible for producing models of social behaviour originate from elementary propositions in symbolic logic and universal computing machines. References to the field of 'artificial life' may also be warranted, although not pursued here.

5.2 Reality and its Agents

5.2.1 The Primitives

At the start, there are three primitive notions in the definition and description of any social system: an *observer*, either in the form of 'pure' or 'quasi-pure'; a recorded *reality* which is perceived by an observer; and social *actors* (quasi-observers), whose actions change (update and modify) reality.

While actors and reality are necessary and sufficient for each other's existence, i.e. one cannot be defined without resorting to referencing the other, the presence of a pure observer at the outset can be seen as a luxury. The necessity for a supra-observer is even less apparent at first glance! However, a more careful review of the fundamental statements behind a social science theory reveals the necessity for the existence of those two entities, the pure observer and the supra-observer.

5.2.2 Reality

Reality consists of two components: first, there is a set of *recorded time series observations* of one or more variables, to be referred to as the *state variables, x*; they contain a repetition or a reproduction process. Second, there are one or more *models* of its *inner workings* (what was referred to in Section 1 as 'fiction' or 'novels'). These models are attempts to break reality's code, by decoding the underlying replicating process of social dynamics.

Reality usually contains a vast menu of variables, assumed to be available in the form of a bounded time series (contained within initial and terminal time periods), forming a *memory string M*, representing a *recorded reality stream* (RRS)[1]. Each variable's memory string (by itself, or in combination with other variables' memory strings *N*) is used differently, however, by actors and pure observers in deriving models of the RRS.

Variables can be referenced along a time dimension, or along any other applicable dimension, such as space. Since social dynamics are the exclusive focus of this essay, only the time dimension will be used as the relevant reference dimension.

In formulating any of the numerous possible social science models of an RRS, only a very limited number of state (endogenous, relatively fast moving) variables are included together with a vector of *parameters*, *P*. In the social sciences, these parameters are nonobservable entities, which are there to depict the effects that all other (exogenous, relatively slow moving) variables have on the recorded time series. None of these other variables is included in the specific model, as they constitute (as a bundle depicted by the model's parameters) the *environment* (*E*) within which the state variables operate according to the model in question.

There could be multiple recordings of the same variables over the same time horizon. Multiple recordings could either correspond to different observations obtained over the same time horizon by different social actors through independent measurements, or they could be different observations among which various social actors' recordings oscillate. Thus, at any point in time, there may be a number of possibly coexisting past realities in recorded form, as well as in the minds of observers and actors in a social system. Some features of this proposition have been explored by the author using the mathematical formalism of quantum mechanics

---

[1] Throughout the paper, $\bar{x}$ indicates past observed data, $\hat{x}$ indicates future expected data, $\bar{\bar{x}}$ indicates desired (or targeted) levels of a variable.

(Dendrinos, 1991). This issue will not be addressed here, however.

A model of the RRS is not simply a mathematical statement of the inner workings of the state variable(s) x, but also a story (interpretation) of that reality. Good models are those which contain good (interesting) stories, parables or myths; they are notable efforts in breaking reality's code. These stories are simply what social scientists might define as descriptions of social *events* or, more broadly, of social *phenomena*.

5.2.3 The Pure Observer and the Supra-Observer

In any social system there is usually one or more detached or pure observer(s). The pure observer allocates all resources available at any point in time simply to reflect on the past, contemplate the present and speculate on the future course of the social system without intervening in its functions. Further, the observer does not express preferences over these states.

A pure observer has no other purpose than to describe the RRS, which contains the combined, aggregate and end-effect of all actors' actions. The pure observer also has expectations regarding the *future reality streams* (FRS), but not desired end-states for these streams. The observer is not a goal-seeking agent within social reality, although not entirely goal-free: the observer's goal is to *effectively* describe social dynamics.

The outcome of a pure observer's thoughts is of no use to actors in deciding their action: were it of any use, it would mean that the observer is an agent for action (or the instrument of an agent) and would thus cease to be a pure observer. A pure observer's function is *inconsequential* to the dynamics proper of the social system, or to the event under analysis. Pure observers and actors are *mutually exclusive* entities. Thus, a pure observer describes the RRS through a detached model; that model is not part of (and does not affect) the set of elements involved in these dynamics.

A supra-observer, on the other hand, is an entity which describes the behaviour of all entities (except itself) in the social system, including that of the pure observer(s). Without the existence of a supra-observer, the pure observer would not exist, since the observer draws from, but does not in any way affect, reality. And conversely, a supra-observer cannot exist without a pure observer to be referenced. A supra-observer does not explicitly derive any of the social system's models, but supplies the suprastructure of social reality. (It should be noted that the term 'suprastructure' used here has very different connotations from the Marxist 'superstructure'.)

A supra-observer's statement of a general framework, within which any theory of social dynamics (model of RRS) can be represented (including those theories derived by pure observers and actors), is not a part of the observers' or actors' models and does not affect them. Moreover, the suprastructure is neither affected by nor does it affect social actions or a pure observer's model of the RRS. Assuming differently would create logical inconsistencies along the lines of B. Russell's well-known paradox regarding sets and their definition (is the set of all sets a part of the set?).

5.2.4 The Actors

In contrast to the detached (pure) observer(s), social actors are entities very close to, and indeed attached to, social reality. Actors, again contrary to pure observers, are complex agents performing a number of tasks: they are firstly quasi-observers of reality, and secondly they act upon the social system. They allocate their limited resources between observing (i.e. acquiring and processing information) and acting (i.e. producing information).

Their purposeful efforts aim to reach the *objectives* they have at any point in time, while they form *expectations* regarding the future state of the social system and its relevance to them as constituent elements. In so doing, they are the agents for change in the social system and the cause for its dynamics. Their perceptions of reality, their expectations and desires, and through them their actions, have *consequences* for the social system.

There are two types of consequences of social actions: one is to update the social dynamics by generating new records for the state variable(s) x, i.e. the strings. The other consequences is more fundamental and results in changes in the *inner working* of the social system, i.e. the very instructions for social dynamics.

The inner working of a social system will be defined here as containing two components: first, the specific mathematical statement of the system's dynamics, an element to be referred to as the 'inner form' of the system (in other words, the system's model); and, second, the forces (or functions) which give rise to this form, referred to as the 'inner structure' of the system. In Dendrinos (1992), these forces were defined as 'ecological-economic' forces. Nothing will be added to this here.

Usually, there are numerous actors in any social system, but they basically fall into two types of agent for change: one type is the simple *participant* actor, while the other type of agent for social science is a *controlling* agent.

There are also numerous participant actors, again subdivided into two major categories: those whose perceptions and actions are *individually consequential* to the dynamics of the system and those who belong to a *mass* with perceptions and actions *collectively consequential* to the system's inner form and ensuing dynamics, but not individually so.

A controlling agent or agency is the manager or governing entity in and of the social system of whom there are also a great many.

5.3 The Efficient and General Statement of a Suprastructure

5.3.1 The Pure Observer's Function; Direct and Indirect Parts of Reality

Let us assume some RRS as a unique (for simplicity) memory string $M_{\bar{x}}$ consisting of observations on a single state variable, $x(t)$, within bounded time limits $0 \le t \le T_p$; here 0 represents the initial time period corresponding to some arbitrarily chosen initial state of the social system, and T_p is some present time period. Instead of a single state variable, one could incorporate a vector of state variables, but this complication would have no significant impact on the analysis that follows. More on this issue will be supplied later.

The observations are recorded in *discrete* time intervals or periods, which are *equal in length but heterogeneous*. The length of time periods and the nature of such heterogeneity is not of interest here.

$M_{\bar{x}}$ is available to the actor(s) and to a single pure observer (referred to henceforth, for the sake of simplicity, as 'the observer' and considered to be the sole observer of the social system). In this reality stream, the end-effect of all participants' and controllers' actions is depicted by and recorded in the string. The string M has some interesting properties, the central one being that it triggers the social code's dynamics.

Taking a detached and *aggregate* view of reality, on the basis of $M_{\bar{x}}$ and by assuming some regularity through periodic (or nonperiodic) *repetition* of social action over time, the observer formulates a *deterministic* model of the inner form of social dynamics (in other words, of the event):

$$\hat{x}_o(t+n) = F_o[\hat{x}_o(t)] \qquad 0 \le t \le T_p \qquad (1)$$

where x_o represents *estimated* values simulating the $M_{\bar{x}}$ of the RRS, n is a time lag and F_o is a model *form* obtained through a *calibration* procedure involving the model parameters, P. This model is a product of the observer's imagination. If such a model exists (the observer's quixotic attempt to simulate, reconstruct, interpret and thus extract meaning out of the reality stream), some temporally *replicating* deterministic process is identified. Put differently, through the *back-casting* procedure implied in (1), there is a story to be told about the social event: this is the observer's novel. The illusion of having broken the social code sets in.

A note is warranted here: if an observer considers that the social code has been broken, then this might switch the observer to being a participant actor; thus the elusive breaking of society's code might act as a bridge or transformer from a state of observation to a state of action. Alas, by that transformation, society would be modified and the breaking of the code prove to be only an ephemeral illusion.

On the other hand, if no replication process can be derived, i.e. if there is no regularity in social action, then obviously, for the observer, the RRS' $M_{\bar{x}}$ is a *random process*. Indeed, reality might be the only source of a random process, since all mathematical definitions of a random process are inconsistent or incomplete. In this case, the observer's *raison d'être* ceases to exist: there is no story there. Thus the presence of a model is another necessary condition for the existence of an observer. Furthermore, if a model form F_o cannot be found, there is no need for a supra-observer or a suprastructure to exist.

The exact nature of the model form F_o, as well as the nature proper and number of parameters in the vector P, are of no interest here. The observer's model may be efficient (when the number of parameters does not exceed that of the state variables). A supra-observer's suprastructure does not imply an efficient observer's model.

The forces shaping the form of F_o in (1), i.e. the inner structure of the RRS in the eyes of the observer, are of no interest to the supra-observer. Obviously, the observer's perceptions of the RRS and also how F_o models are formed at any particular point in space-time are elements of this structure.

Possible environmental, probabilistic or stochastic perturbation effects in (1) are of no interest in the present analysis either. Whether or not there is a menu of F_o functions, a vector F_o, from which an observer selects one, is also of no direct importance here. More broadly, neither the composition of this menu nor the changes of F_o functions over time are at present of concern.

The RRS' $M_{\bar{x}}$ contains the *direct* part of the reality observed, whereas the F_o (its inner form) is the *indirect* part of reality and only an observer's conjecture.

One could look at this model specification in two ways: either there is simply one state variable to be recorded, or this state variable is the subject of a *selection* process on the part of an observer; the observer has *condensed* the behaviour of the social system's reality into the behaviour of one variable (or a limited set of state variables). In such a case, the direct part of reality can be seen as the effect, recorded along some variable x, of a highly complex process involving x.

The complete book of the disaggregate (micro)reality is obviously not available to the observer, the observer *being neither omnipresent nor omniscient*. An observer cannot have the memory required to store all the information that all participant agents in social reality possess at any point in time. Neither does the observer have access to the computing instructions which every agent of social reality implements in its actions. If a choice is made by the observer to selectively pick or devise substitute data and instructions, then automatically that choice shifts the detached observer from a non-participant to a participant agent of social reality.

It should be noted that there is no need to make any assumptions regarding the knowledge base (source of information, classification of information under some hierarchical structure, etc.) of a supra-observer. However, as already mentioned, the fact that there could be multiple layers of supra-observers under the construct outlined here is of central concern. Further, the indirect part of reality, the model F_o, is an attempt by the observer to simplify, encode and transmit this reality in a manner which one observer perceives as efficient and effective, although another observer might not find it so.

Indeed, whether condensing, coding and transmission are at all feasible is largely a matter of opinion: it depends on the criteria of simplicity and efficiency set by the observer and the associated requirements for coding and transmission, i.e. the ideology behind the model (which is possibly connected with reality and the model's inner structure). There is an interesting interaction between requirements for condensing, coding and transmission, on the one hand, and the inner workings of reality and/or its model(s), on the other, but we shall not go into this any further at present.

As already stated, the model formulation, containing the indirect part of reality, is calibrated (the model's parameters are estimated) on the basis of the available recordings during the time period $0 \le t \le T_p$. In back-casting, the observer makes use of the total memory string $M_{\bar{x},t}$.

At each t, the difference:

$$\sigma_o(t) = [\bar{x}(t) - \hat{x}_o(t)]^2$$

is the observer's model *current deviation factor* (CDF), while the quantity:

$$\sigma_{o,\hat{x}} = [1/(T_p + 1)] \ \Sigma_t[\bar{x}(t) - \hat{x}_o(t)]^2$$

is the model's *average deviation factor* (ADF). The observer obviously considers the model to be valid if these deviations are within admissible limits for the observer domains.

Issues associated with more sophisticated statistical techniques dealing with model calibration, model specification and validation are of no interest to this analysis. Obviously, theory *in* social science is the quest for formulation of such F_o functions, whereas theory *of* social science is an attempt by supra-observers to set a suprastructure along the lines given here.

Assuming a valid model formulation, the observer constructs future *expectations* of reality stream, E_x, defined over a time horizon $T_p<t \leq T_F$, T_F being a terminal future time period over which the observer wishes to *speculate:*

$$\hat{x}_o(t + n') = F_o[\hat{x}(t)] \qquad T_p<t \leq T_F \qquad (2)$$

where n' is a time lag in the future horizon, not necessarily the same as n. This is the product of social science fiction, too. In (2), the model form F_o is the same as in (1), including the parameters P. The observer contemplates simply the state of reality and its course. There is no need at present to assume that the observer resets and recalibrates this model every time an additional set of recordings is made available, thus shifting the indirect reality model through time. Such an activity would not be difficult to incorporate, but the conclusions arrived at here do not critically depend on such extensions.

Conditions (1) and (2) are the only components generated by the observer in this suprastructure. Given that there could be different observers with different expectations about the future, a supra-observer's suprastructure must be capable of accommodating numerous coexisting futures at any time period t among observers. This will not be pursued here, however.

5.3.2 *The Participant's Functions*

It is assumed here for simplicity that there is only one social actor-participant, and thus the actor is by necessity individually influential. Some extensions along the lines of a number of interacting, individually influential actors in a backdrop of mass actions can be incorporated into the analysis. References to game-theoretic issues in such a context are warranted, but this is left to the interested reader to pursue.

The actor-participant, being a quasi-observer, also has access to the $x(t)$, $0 \leq t \leq T_p$ and, in addition, to a string of controls imposed by an actor-controlled $c(t)$, $0 \leq t \leq T_p$, upon the actor-participant. Numerous controls may be imposed by the (possibly many) controlling agents upon actors-participants in a social system but, due to limited resources, a specific actor-participant focuses exclusively on those controls which (directly or indirectly) significantly affect his actions and perceptions. As the actor-participant is not omnipresent or omniscient, errors in perception may be made.

The multiplicity of controls and the selective, and possibly erroneous, perception of their impacts on the actors are among the key reasons why an observer cannot possibly account for all controls affecting all actors' perceptions and actions. Thus, controls and actors are not explicit in an observer's model, i.e. they are absent in the statement of aggregate and detached reality found in conditions (1) and (2).

If an observer attempts to explicitly incorporate controls and actions by specific actors into (1) and (2) and expand this model formulation in a manner which becomes of use for action, then from the supra-observer's standpoint, the observer has switched from being a pure observer to a quasi-observer and thus to an actor. Since it has been assumed for simplicity that there is only one actor, one controller and one control instrument, the full menu of controls would be reduced to a single entity. Thus, the problems of selection by individual actors among items on the menu and actors to interact with would be eliminated.

Utilising a close-up of reality through a *limited* memory string $M_{a,\bar{x},\bar{c},t}$, extended over a portion of the past horizon t_a', $0<t_a'<T_p$, the participant actor formulates perceptions about the past state of the social system through an actor's model; the actor calibrates it using a subset of the recordings $x(t)$, $0<t<T_p$. In contrast to the pure observer, such a model *always exists no matter how imperfect it might be*, since the actor always acts. It is not only an indirect reality model, but also a *partial* model of reality. The model reveals the individual participant actor's *bias* in

perceiving reality.

In the case of numerous participant actors, we cannot determine whether such models are independently formulated or not. This could be a very interesting angle from which to extend this analysis. It would necessitate the institution of 'leaders' and 'followers' in model building.

The model (again, a product of the imagination) is a set of two interdependent replication functions on the state variable x and the control instrument c:

$$\hat{x}_a(t+h) = F_{a,x}[\hat{x}_a, \hat{c}_a] \tag{3}$$

$$\hat{c}_a(t+k) = F_{a,c}[\hat{x}_a, \hat{c}_a] \tag{4}$$

where h, k are possible time lags. Conditions (3) and (4) represent perceptions by the actor-participant about the past through back-casting.

Like the observer, the actor speculates about the future by deriving *expectations* based on this model found in conditions (3) (4), $\hat{x}_a(t'')$ and $\hat{c}_a(t'')$, given by:

$$\hat{x}_a(t+h') = F_{a,x}[\hat{x}_a, \hat{c}_a] \tag{5}$$

$$\hat{c}_a(t+k') = F_{a,c}[\hat{x}_a, \hat{c}_a] \tag{6}$$

The expectations concern the state of the system within an immediate future time horizon $T_p<t_a''<T_{F'}$, where $T_{F'}<T_F$; in (5) and (6) h', k' are time lags not necessarily equal to h, k. The time horizon $T_{F'}$ is dictated by the necessity to act.

From the supra-observer's viewpoint, there are possible multiple coexisting futures perceived by observers; there are certainly many such futures in the expectations of the many social agents-participants. In back-casting, there are similarly numerous coexisting pasts.

Due to the limited recordings used, the quality of the actor's model $F_{x,a}$ may not be equal to the observer's. On the other hand, the model specification by the actor might be superior to that of the observer. The observer has no access to the models $F_{a,x}$ and $F_{a,c}$ and neither does the supra-observer. In back-casting, the actor experiences a current and an average deviation factor, just as the observer does by employing a model of past reality. Errors and improper model specifications may be the rule in these $F_{a,x}$ and $F_{a,c}$ functions.

In contrast to the observer, and in addition to expectations about the future of the system, the actor-participant also has desires, goals, or *targets* concerning the state variable at some time in the future. These targets (either in the form of a desired end-state $\bar{\bar{x}}_a(T_a)$ or a desired future path $\bar{\bar{x}}_a(t)$, $T_p<t\leq T_{F'}$) are present at current time period t and apply for a limited time horizon into the future $T_{F'}$: $\bar{\bar{x}}_a(t'')$. These targets are not available to the observer (and certainly not needed in the work of the supra-observer).

Individual preferences, as well as numerous objectives and constraints underlie the actions of a participant actor in a social system. An individual's book of behaviour and action is not known to the observer or needed for a supra-observer; only the end-result of the actor's action becomes available, through the recordings $\bar{x}$.

Actual *updating* of the social dynamics $\bar{x}(t)$ takes place through a process which may have the form:

$$\bar{x}(t) = \phi_a[F_{a,x}, F_{a,c}, F_c, \bar{\bar{x}}_a] \tag{7}$$

where F_c is a function of the controller's action, to be specified later. Conditions (7) is a model unavailable to the observer (not being omniscient), but used by the (single) actor as a basis for action. The impact of the action is directly related to the current *motivation* of the actor to act; in turn this driving force is directly related to the difference between the desired level of $\bar{\bar{x}}_a(t'')$ and the expected level of $\hat{x}_a(t'')$.

It should be underlined here that the model ϕ_a is different from model $F_{x,a}$; moreover, due to the iterative form of both ϕ_a and $F_{x,a}$, and the interdependence between $F_{x,a}$, $\hat{x}(t)$ and ϕ_a, the form of $F_{x,a}$ is likely to be *highly nonlinear* in the state variable $\bar{x}(t)$. It is also stressed that the key element which makes the social system function is the desire by the actor to achieve some goals, subject to expectations about the future.

A particular form of the updating function ϕ_a could be:

$$\bar{x}(t) = F_{a,x}(t) + E(t) \tag{8}$$

where $E(t)$ is the function depicting current urgency, motivation and impact of the following differential:

$$\delta_{\hat{x}.\bar{\bar{x}}}(t) = [\hat{x}_a(t) - \bar{\bar{x}}_a(t)]$$

For the actor-participant, the difference between the actual recording of $\bar{x}(t)$ and the expectation $\hat{x}_a(t)$ depicts the *current miss factor* (CMF), with the expression:

$$\sigma_{a,\hat{x}} = [1/(T_{F'} + 1)] \sum_t [\hat{x}_a(t) - \bar{x}(t)]^2$$

being the *average miss factor* (AMF). Further, the difference

$$\delta_{\bar{\bar{x}},\bar{x}}(t) = [\bar{\bar{x}}_a(t) - \bar{x}(t)]$$

represents the actor's *current over/underachievement factor*, CO/UF (depending on whether the difference is positive or negative and similarly evaluated by the actor), with the expression:

$$\delta_{a,\bar{\bar{x}}} = [1/(T_{F'} + 1)] \sum_t [\bar{\bar{x}}_a(t) - \bar{x}(t)]^2$$

being the average level of achievement factor over the time horizon $T_{F'}$. Put differently, this difference may pick up the level of *utility/disutility* for the actor, depending on whether its overall effect is positive or negative.

A social actor's actions are in response to the actions of other actors (participants and controlling agents). As such, they represent a choice among a menu of functions ϕ_a at each time period t. These choices, made simultaneously by numerous actors, go far beyond merely updating societal dynamics; they change also the inner workings of the social system partly through changes in the $F_{a,*}$ functions. These changes are due to selection switches in menus for such functions. Put differently, the actor's actions besides updating the recordings of x, have the more profound impact of altering the $F_{a,*}$'s. This is the second major issue this work was set to address.

The incorporation of a number of agents and choice of functions $F_{a,*}$ as a result of strategies adopted in a game-theoretic context where interactions among these agents are modelled, provides a framework which accounts for such a proposition from the point of view of a supra-observer. These choices link perceptions of reality to actions and in turn to new perceptions in a highly nonlinear and complex context. These propositions can only be stated by a supra-observer.

Thus, the participant is described by four functions from (3) to (6), while the end-effect of the actor-participant's action(s) is depicted by condition (7), possibly specified as (8), but unavailable to the observer. Only the updating stream $\overline{x}(t)$ is recordable and available to the observer at each current time period t.

5.3.3 *The Controller*

Similar to the actor-participant, the controller-participant of the system utilises part of the reality stream of the past, $\overline{x}(t)$, together with part of the history of controls, $\overline{c}(t)$, to derive a controller's indirect and partial model of reality. Using a limited memory string $M_{c,\overline{x},\overline{c},t''}$ the controller obtains through back-casting:

$$\hat{x}(t'+k) = F_{c,x}[\hat{x}(t'),\overline{c}(t')] \tag{9}$$

where $0<t_c' \leq T_p$ and k is a time lag. *The controller is not omniscient.* Assuming that such a model can be constructed and accepted by the controller, the estimated values (obtained after calibration) minus the actual past observations supply the controller with a degree of accuracy and acceptability given by the controller's imposed criteria.

Further, on the basis of such a model, the controller extracts expectations on the variable x about the future:

$$\hat{x}(t''+k') = F_{c,x}[\hat{x}_c(t''),\overline{\overline{c}}(t'')] \tag{10}$$

where $T_p<t''<T_{F''}$, with $T_{F''}$ designating the controller's time horizon into the future, and k' which is a possible new time lag.

The central function of the controller is to manage the system by guiding it either to some end-state $\overline{\overline{x}}_c(T_c)$, desired by the controller, or through a desired future path $\overline{\overline{x}}_c(t)$, $T_P<t\leq T_{F''}$. Thus, the controller *wishes* to impose upon the system an optimum control function for the instrument $c(t)$:

$$\overline{\overline{c}}(t+m) = F_c[\hat{x}_c(t+1),\overline{\overline{x}}_c(t+1),\overline{x}(t),\overline{\overline{c}}(t)] \tag{11}$$

where m is a time lag; condition (11) is a replicative discrete iterative dynamic function over the controller's time horizon $T_{F''}$. The observer and the actor-participant have no access to the (possibly numerous) functions F_c. The actor-participant only becomes aware of the current (updating) actual value of the control $\overline{c}(t)$ given by the abstract expression:

$$\bar{c}(t) = \phi_c[F_{c,x}, F_c, F_{a',x}, c(t)] \tag{12}$$

There could be multiple ϕ_c functions; the one chosen from the menu in each time period being a selection under a game-theoretic condition involving strategies and payoffs attached to the controller.

The controller is not omnipotent. By forming the difference:

$$\sigma_c(t) = [\bar{\bar{x}}_c(t) - \bar{x}(t)]^2$$

the controller obtains the *current controller's failure factor*, with the expression:

$$\sigma_{c,\bar{\bar{x}}} = [1/(T_c + 1)] \; \sum_t \; [\bar{\bar{x}}_c(t) - \bar{x}(t)]$$

identifying the average failure over the time horizon. Similar to the observer's and actor's miss factors, the controller obtains equivalent miss factors, which constitute the motivation and urgency for the controller to act. As such, one could assume a specification for the controller function, similar to (8) or the actor-participant.

In summary, the controller is described by the conditions (9), (10) and (11), while the updating of the control takes place through the process (12), unavailable to the observer at any time and to the participant ahead of time.

5.3.4 Discussion

If there is partial or complete equality over the expectations $\hat{x}_0(t)$, $\hat{x}_a(t)$ and $\hat{x}_c(t)$, $T_p<t<T_F$, for either one or more time periods, then there is partial or complete *consensus*, i.e. disagreement or agreement, among society's agents regarding that future time horizon. Equivalent statements can be derived about a consensus (or lack thereof) among society's actors and the observer(s) regarding the past. Functions $F_{a,x}$, $F_{a,c}$, $F_{c,x}$, F_c could contain a considerable degree of disagreement in their form and underlying structure. That is the underlying premise of this work.

A limited game-theoretic setting could be incorporated into this formalism by the supra-observer: the actor's action $\bar{x}(t)$ and the controller's control $\bar{c}(t)$ at current time period t can be seen as simultaneous choices, selected among alternative functions ϕ_a and ϕ_c, available to these actors at t. Again, various strategies and expected payoffs could provide the selection mechanism among these functions.

To the pure observer, the *only* and eternal question remains: is it possible to currently derive a model F_o, given a set of past recording of some reality stream, without knowing all factors known to all actors involved, their limitations, expectations and aspirations?

To the supra-observer, the *only* question is: is there a more efficient general formulation of societal dynamics?

But beyond this, the specification under lack of consensus regarding society's past and future among its agents produces a suprastructure utilising eleven conditions, from (1) to (7) and from (9) to (12), and thirteen variables [$\hat{x}_0$ (past), $\hat{x}_0$ (future), $\bar{x}$ (current), $\bar{c}$ (current), x_a (past), $\hat{x}_a$ (future), $\hat{c}_a$ (past), $\hat{c}_a$ (future), $\bar{\bar{x}}_a$ (future), $\hat{x}_c$ (past), $\hat{x}_c$ (future), $\bar{\bar{c}}$ (future), $\bar{\bar{x}}_c$ (future)]. Of these thirteen variables, two are unspecified, the desired future levels $\bar{\bar{x}}_a$ and $\bar{\bar{x}}_c$. Are indeed social dynamics basically formulated in eleven dynamic equations and no more? Are only thirteen variables needed to state them in their simplest possible form? Why? These questions are to be addressed in future work.

Can this formulation of societal dynamics account effectively for a particular subject which falls *exclusively* under the prerogative of a supra-observer's statement of social dynamics, namely that of *self-fulfillment/defeat* of social action? It takes nine equations, from (3) to (7) and from (9) to (12), and eleven variables (thirteen minus the two associated with the observer) to address it. Furthermore, these conditions lead to the statement of a general classification scheme of social dynamics and to propositions closely resembling 'laws' in social dynamics.

5.4 Self-Fulfillment/Defeat

A central characteristic of any social action, as seen by a supra-observer under this suprastructure, is this: the expectations of a society's actors at any time period and the actual end-effect of the social actions they take are linked through self-fulfillment or self-defeat in time. This is the result of a basic feedback mechanism inherently built into social reality and manifested through an ensuing agreement or difference between expectations and actual events. It also links the achievement or non-achievement of desired outcomes, given expectations, through actions. It is this fundamental interaction which constitutes the statement of a proposition resembling that of a 'social law'.

Any general framework of social dynamics (efficient or not) must be able to identify such a feedback mechanism and its effects. Further, it is only through a supra-observer's statement of a suprastructure that such interaction can be incorporated. No model by either type of actor can do so by itself and, at the same time, comply with the requirements for internal logical consistency. This fundamental interaction can only be checked by an outside agent, not an actor. The observer, by the definition given here, is not the appropriate agent to do so; the observer's task is to produce a detached model of social dynamics utilising quantities which are recorded. This leaves only the supra-observer's suprastructure as a candidate to incorporate and qualitatively address the self-fulfilling/defeating aspects of social reality. But it turns out that self-fulfillment/defeating processes are not the only ones a supra-observer can identify in social dynamics. Indeed, there is a general classification scheme which includes eight classes of possible social dynamics.

First, let us define the class of social dynamics most widely discussed, namely the self-fulfillment of expectations (referred to at times in popular literature as 'self-fulfilling prophesies'). This is the process by which an actor, having expectations about the future of the social system and an expressed *love* towards this particular future, takes actions which bring about such future, although such expectations were false according to an observer. The yardstick to determine whether such expectations were off the mark at the outset, is the detached observer's model, assuming that such a model exists and is valid. In this case, the observer's expectations were wrong and the actor's expectations proved right; thus the characterisation of this interaction as 'self-fulfilling' is merely a defensive one attempting to hide *observer failure*. The presence of 'self-fulfilling' processes shows once again the existence of a pure observer as defined within this suprastructure.

On the other hand, self-defeating expectations (often referred to as 'Catch-22' phenomena) is the process by which an actor, having expectations about the future of the social system and an expressed *aversion* towards this particular future, takes actions to avoid it; these actions however are ineffective and, in fact, make it even more likely to occur. Again, the yardstick to determine whether there was *actor failure* is the observer's detached model of reality.

Thus, at the core of this self-fulfilling/defeating process are false expectations. It is only when the expectations are correctly perceived that the self-fulfilling/defeating process is not present. But this is highly unlikely in social reality. Assuming that expectations are defined as a continuous variable, falseness or correctness can be set in terms of a neighbourhood around the observer's expectations. The difference between the manner in which expectations are used in this context and the way 'rational expectations' are used in neoclassical economic theory should be noted.

i. The general classification scheme of social dynamics

A table can be drawn up (see Table 1) containing an actor's expectations (and whether they are true or false) and the actor's attitude towards them (acting by seeking or avoiding them). By classifying these two attributes regarding actions in conjunction with an observer's expectations, a complete list of circumstances becomes apparent to a supra-observer, including self-fulfilling/defeating conditions. In all, there are eight possible basic types of dynamics.

Actor's expectations at t about T_a

		True	False
at t:	seeking outcome	T (1a) / F (1b)	T (2a) / F (2b)
	avoiding outcome	T (3a) / F (3b)	T (4a) / F (4b)

Table 1 Interplay of actor's expectations and purposeful action

Notes:
(a) the observer's expectations at t prove to be true (T) at T_a;
(b) the observer's expectations at t prove to be false (F) at T_a;
(1a), (3a): both observer and actor are right;
(2b), (4b): both observer and actor are wrong;
(1b), (3b): actor is right, observer is wrong;
(2a), (4a): actor is wrong, observer is right;
(1b): the 'self-fulfilling' case;
(4a): the 'self-defeating' case.

ii. A special case

Define the difference $\delta(t) = [\hat{x}_o(T) - \bar{x}(t)]$ for the same arbitrary time period T into the future, $T<T_F$. The sign of $\partial\delta(t)/\partial t$ would indicate whether the social dynamics move away from or toward the expected value of the state variable $\hat{x}_o(T)$, i.e. whether there is a positive (sign>0) or negative (sign<0) feedback between actions and actors'/observers' expectations under lack of consensus, i.e. $\hat{x}_o(T) \neq \hat{x}_a(T)$.

Further, define the difference $\Delta(t) = [\hat{x}_a(T) - \bar{x}(t)]$ for the same time period T in the future. Then if $\partial\Delta(t)/\partial t > 0$ and the actor is an $\hat{x}_a$-lover this implies that there is a self-defeating process under way. The observer's expectation is the actual future state, either under positive or negative feedback; the cumulative sum of these discrete differences identifies the amount of self-defeat built into the social dynamic process at hand. Equivalently, one can measure the extent of self-fulfillment in the process, if $\partial\Delta(t)/\partial t < 0$. The general classification scheme, however, is not impacted by these (fast) dynamics.

iii. Other cases

Case (1a) from Table 1 represents a situation whereby social dynamics are characterised by 'harmony'; whereas, case (4b) is that of social turmoil. The characterisation of the rest is left to the interested reader.

5.5 Conclusions

This article provides a statement of social dynamics which calls for a suprastructure and supra-observer, along with observers of social reality as well as actors and controllers of this reality. Further, a general classification scheme of social dynamics was indicated with eight major categories of social dynamics including self-fulfillment/defeat. The framework is broad enough to accommodate any social science theory and any social action.

In stating the general classification scheme, the controller was not included. To incorporate the actions and expectations of a controlling agent in the social dynamics, in conjunction with observers and actors' expectations and actions, would significantly complicate the classification scheme and considerably increase the number of social dynamics. It is left, thus, to the interested reader to explore the possible consequences.

Acknowledgement. The author wishes to thank Franco Vaio for significant comments and suggestions and for editing the text originally written in 1994.

References

Dendrinos D.S. (1991) Methods in Quantum Mechanics and the Socio-Spatial World, *Socio-Spatial Dynamics, 3*, 81-110.

Dendrinos D.S. (1992) *The Dynamics of Cities: Ecological Determinism, Dualism and Chaos*, Routledge, London.

6. Urban Development as a Guided Self-Organisation Process

Federico M. Butera

6.1 Introduction

The analysis of complex systems, viewed as thermodynamic systems far from equilibrium, shows that any spontaneous evolutive process (self-organisation process) can be described as a succession of unexpected jumps from a given structure into a new, unpredictable one. It is also recognised that, when dealing with the evolution of complex systems, prediction is possible so long as the structure of the system remains unchanged: any structural change creates a new system to which predictions built up for the old one are no longer valid. But does this apply to the complex city system? In principle, yes, but in practice the evolution of a city follows a different pattern since the process is subject to continuous pressure consisting of actions aiming to drive the evolution towards a pre-defined model. It is rare that something really new and unpredictable changes the structure of a city. What usually takes place is an imitative process with some minor variations. In general, the development of the city is an adaptive process: models already working elsewhere are adapted to a specific urban environment. Sometimes, but rarely, new subsystem models are imagined by some long-sighted and creative politician or decision-maker.

The evolution of a city, therefore, can seldom be considered a spontaneous self-organisation process, since it is artificially driven towards pre-conceived models, implying given components and structures. It is, however, also true that this kind of guided evolution rarely goes as planned; the actual development pattern, as a consequence of unpredictable events and behaviour, is often quite different from that predicted.

The intention of this work is to examine this particular kind of guided self-organisation process, where some new components are artificially introduced, where new relationships appear among components and where unpredictable events tend to divert the system from the pre-defined target model.

6.2 The City as a Neg-Entropy Processor

Any physical system, exchanging energy or matter with the surrounding environment can be studied as a thermodynamic open system[1]. While the second principle of thermodynamics states that, for an isolated real system, entropy (i.e. disorder, disorganisation) cannot decrease, for closed and open system we have the equation:

$$\Delta S + \Delta S_o = \Delta S_i + \Delta S_e$$

where ΔS is the system's entropy variation per unit of time, ΔS_i is the system's production of entropy (which is always positive because the maintenance of internal organisation involves an unavoidable degradation of energy and matter) and ΔS_e represents the flow of entropy coming from the external environment. This is always negative (which is why it is called neg-entropy[2]) and represents the flow of organisation balancing the production of entropy ΔS_i; ΔS_o represents the possible neg-entropy flow coming from the system. In order to maintain its organisation, the system must keep ΔS=0; if ΔS>0 the system's organisation is decaying; if ΔS<0 an evolutive process is occurring (the organisation of the system is improving). According to this approach, the most synthetic representation of the relationship between an urban system and its environment is the one depicted in Fig. 1, where the city is described as a neg-entropy processor, i.e. an open system whose neg-entropy input from the environment consists of matter, energy and information (ΔS_e) and whose outputs are waste (ΔS_i), a neg-entropy flow (ΔS_o) made up of matter and energy (of higher quality and smaller quantity) and information.

---

[1] In thermodynamics, systems are defined as isolated when they do not exchange energy or matter with the environment, closed when they exchange only energy, and open when they exchange both energy and matter.

[2] Neg-entropy can be imagined as an external organisation flow which can maintain or improve the system's organisation.

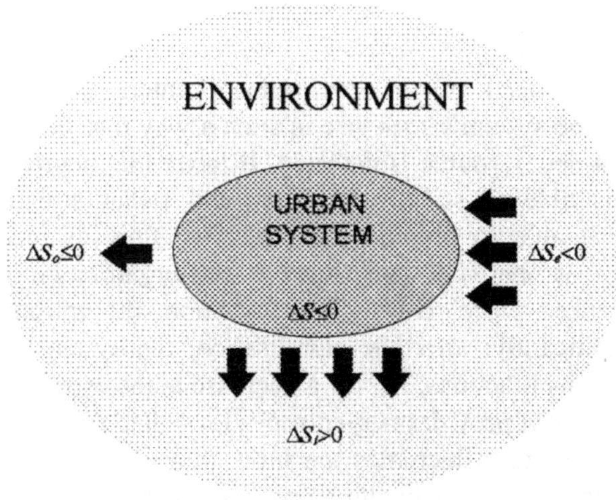

Fig. 1 The city as an open thermodynamic system

The nature and the amount of the neg-entropy input flow crossing the boundary of the city depends on the ability of the system to recognise and metabolise it. Two different systems interacting with the same environment will recognise different sets of resources or neg-entropy sources - just as a carnivore would not recognise a vegetable store as a resource, nor a herbivore a meat store. In urban terms, a city without cars, for example, would not recognise a gasoline pipeline as a resource. In other words, an environment does not have any resource 'per se', the richness in resources is that perceived by the system.

The evolution of the city is characterised by structural changes that make it capable of recognising and metabolising new neg-entropy resources[1]. This means that the environment changes with the system or, more exactly, what changes is the way the environment is perceived by the system. A structural change occurs when some apparently meaningless

---

[1] Any complex system in dynamic equilibrium is always fighting with the excess of matter, energy and information that it is unable to metabolise. The neg-entropy flow that feeds a system is rarely pure, since it contains unmetabolizable elements (i.e. not recognised by the system). Thus the system has to continuously separate out what it recognises as a neg-entropy flow from what it does not recognise as such. The evolution is the progressive recognition of new environmental resources, i.e. the improvement of the system's ability to sift out resources (neg-entropy) from what was previously considered as pure noise. In a detective story, the solution of the investigation is a self-organising event: a set of unrelated data (noise) is integrated into a rational system able to interpret it, thus turning it into information.

fluctuation (or noise) of the environment enters into phase with a fluctuation of the system's structure in such a way that the noise becomes a signal and a new resource is detected. If, after this unpredictable event, the system is able to stabilise its new and richer structure, and the environment is able to provide the necessary flow of recognised neg-entropy, then it can be said that a self-organisation process has successfully taken place. During the transition, the entropy production increases significantly; after the stabilisation, the system strives for an optimisation of its functions, i.e. for a decrease in the entropy production.

If the environment provides neg-entropy flows that are 'poor' (with little noise), or the structure fluctuates too little, it is very unlikely that a self-organisation process, i.e. an evolutive process, will be triggered[1].

To plan city development means to trigger a self-organisation process in the direction of a given model by creating or reinforcing structures and connections among components and by promoting internal fluctuations, i.e. new ideas and activities.

6.3 Evolution by Balancing Novelty and Confirmation

The environment is rich in potential neg-entropy flows or packages: sometimes they are present but unrecognised, mostly they are carried together with other flows and packages of recognised neg-entropy. The system usually filters out the recognised part of the flow and metabolises it, discarding the rest.

The same applies to information. We use the term 'information' here with a different meaning from that used in the theory founded by Shannon and Weaver (1949). They state that the amount of information emitted by a source is given and can only decrease in transmission due to an unavoidable noise effect. As in equilibrium thermodynamics, order can only decrease. Moreover, the information referred to in Shannon's theory is not related to any specific meaning: it attains only to the syntactic level, not the semantic one. The information we consider here, on the other hand, is what Jantsch (1980) calls *pragmatic information*, defined as information that generates information potential, i.e. that changes the receiver. In other words, pragmatic information is newly discovered neg-entropy, recognised and metabolised because of a structural change in the

[1] This is, for example, why cities where entrepreneurs show little innovative capacity are condemned to stagnation.

system. In the previous state, it was only noise.

A parallel can be drawn between the evolution of a system and the learning process. Pragmatic information is made up of two complementary components, novelty and confirmation. Novelty is that part of an information (or neg-entropy) package not recognised by the receiver, whereas confirmation is that part of the package which is recognised and metabolised. Learning is the process through which noise (novelty) is transformed into information (confirmation). A system fed by a flow of matter, energy and information which is perceived as pure novelty, i.e. entirely unrecognised, would die. It is a system in a totally new, unsuitable environment: like a fish out of water, a man on the moon or a person warned of an imminent explosion by a telephone call in an unknown language. On the other hand, a flow perceived as pure confirmation, i.e. not containing any new information at all, stands for stagnation and death. Since both pure novelty and pure confirmation cause the system to die, somewhere in the middle there must be a finite maximum, depending on the information exchanged and on the structure of the sender and the receiver.

Pragmatic information, according to Jantsch, reaches a maximum when both components are balanced. Complex systems transform novelty into confirmation and may evolve through states characterised by maximum novelty (instability thresholds) to new states characterised by a balance between novelty and confirmation (Fig. 2). In this transition, entropy production reaches a maximum (area A), whereas in a state of balanced novelty-confirmation it tends towards a minimum (area B).

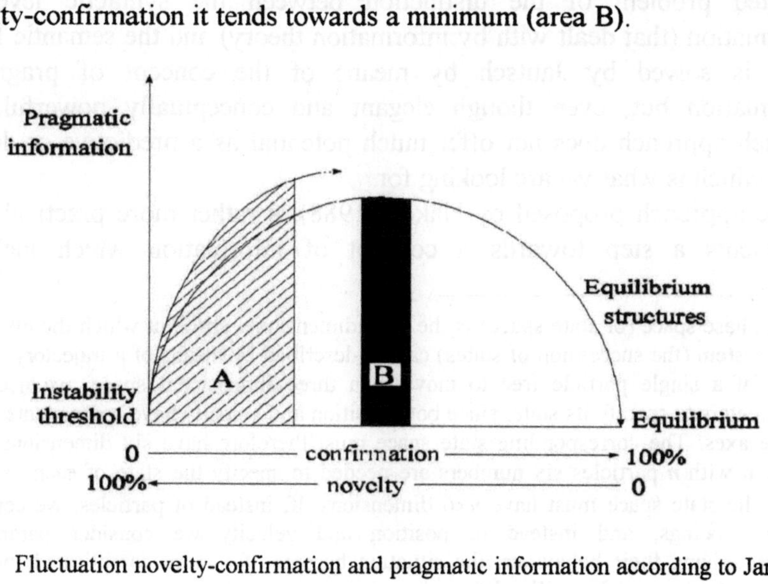

Fig. 2 Fluctuation novelty-confirmation and pragmatic information according to Jantsch

6.4 Information, Communication and Evolution

Nicolis and Prigogine (1987) explored the interaction between systems far from equilibrium, information and chaotic trajectories. Despite using a different approach, these authors are clearly on the 'same wavelength' as Jantsch, connecting physical complexity with algorithmic complexity. The latter is defined as the shortest possible description of a given finite sequence. An entirely random sequence - pure noise - has the maximum possible algorithmic complexity, since the algorithm necessary to describe it is as long as the sequence itself. A sequence such as AAA... has the minimum possible algorithmic complexity, since a single instruction is sufficient to reproduce the entire sequence. In the former case the information reaches its maximum, in the latter it is zero.

According to Nicolis and Prigogine, the complexity of natural systems lies somewhere between these two extremes - the same extremes that Jantsch calls pure novelty and pure confirmation - because, besides randomness, complexity also implies some large scale regularity. They also assert that the self-organised states of matter permitted by the physics of nonequilibrium give us just this kind of complexity. Among these states, the most important for our purpose is chaos dynamics, since the instability associated with chaos allows the system to continuously explore its phase space[1], thus creating information and complexity.

Whenever the concept of information is introduced, it raises the long debated problem of the distinction between the syntactic level of information (that dealt with by information theory) and the semantic level. This is solved by Jantsch by means of the concept of pragmatic information but, even though elegant and conceptually powerful, the Jantsch approach does not offer much potential as a predictive or design tool, which is what we are looking for.

The approach proposed by Haken (1988) is rather more practical, and represents a step towards a concept of information which includes

---

[1] The phase space (or state space) is the multidimensional space in which the evolution of a system (the succession of states) can be described by means of a trajectory. In the case of a single particle free to move in a three-dimensional space, we need six parameters to specify its state, since both position and velocity have components in the three axes. The corresponding state space must therefore have six dimensions. In a system with n particles six numbers are needed to specify the state of each particle, thus the state space must have $n \times 6$ dimensions. If, instead of particles, we consider human beings, and instead of position and velocity we consider parameters characterising their behaviour, the situation becomes far more complicated, but the representation model is still valid.

semantics. His basic proposition is that we can attribute meaning to a message only if the response of the receiver is taken into account. On this basis, it is possible to introduce the concept of the 'relative importance' of a message, i.e. its information content (using the word information with its current meaning). Haken considers a set of messages as a string of numbers and he models the receiver (a gas, a biological cell, a city) as a dynamic system whose states can be characterised at the microscopic, mesoscopic or macroscopic level by a set of quantities, q_l, that may change with time.

We may bring together all the q_l's into a state vector $q(t)$ and describe the time evolution of q, i.e. the dynamics of the system with the following differential equation:

$$\frac{dq}{dt} = N(q,\alpha) + F(t)$$

where N is the deterministic part and F represents fluctuating, random forces. If there are no fluctuating forces ($F(t)=0$), once we have established the value of q at the initial time and the control parameter α, the future values of q are univocally defined. In the course of time q will approach an attractor: this may be a simple point attractor, a limit cycle or a chaotic attractor[1].

The simplest way to visualise the process is to consider the model of a landscape with hills and valleys being swept by a random wind (Fig. 3). By fixing α, we define a specific landscape in which a ping-pong ball, for example, rolls along the sides of the valleys. By fixing the value of q at the initial time, we define the position of the ball on the valley side. From there

---

[1] A dynamic system can be examined in relation to the type of attractor that represents its behaviour. Three principal types of attractors can be defined:

a. periodical attractor: the system periodically oscillates between two or more positions This is the case of an ideal ping-pong ball rolling, without friction, in the bottom of a bowl;

b. point attractor: the system tends to a position of stable equilibrium. This is the case of a real ball, with friction, which comes to rest after a while in the bottom of the bowl;

c. chaotic (or strange) attractor: the system does not tend towards any equilibrium position, neither does it stabilise on a specific trajectory.

The chaotic behaviour of a system is extremely complex and apparently entirely random. A certain recurrence, however, does exist and the trajectory occupies a well defined volume in the state space. The system never retraces the same path, but its behaviour remains within a confined space. Although it is impossible to predict its evolution in detail, i.e. the system is locally unpredictable, it may be globally stable.

the ball rolls down until captured by the point attractor representing the bottom of the valley. If we let $F(t)$ work, i.e. leave the ball subject to random gusts of wind, it may jump from one attractor to another.

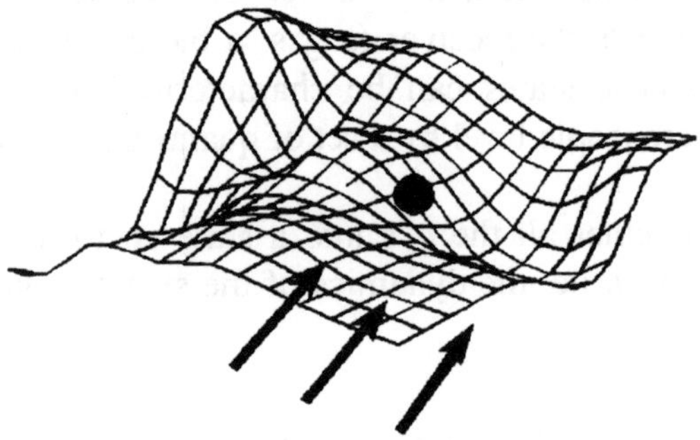

Fig. 3 Landscape with hills and valleys: the bottom of a valley represents an attractor; arrows indicate the instantaneous wind direction

In order to attribute meaning to a message, Haken assumes that a message received by the system sets the parameter α and the initial value q_o. This implies that there is an action on α and q_o only if the message is received, i.e. recognised by the system (otherwise it is only unintelligible noise) and that this reception is accompanied by a change in the system's structure. Let us suppose that, before the message arrives, the system is in an initial attractor q_o. After the arrival of the message there are two possibilities: either the message has been received/recognised and the parameters α and the initial value q are newly set, or the message has left the system in the q_o state. The latter occurs when the message has not been recognised and, therefore, is considered meaningless.

In the case where the message has been recognised, the system goes into a new attractor. As shown in Fig. 4, there are three possibilities:

1. the attractor is determined uniquely by that specific message;
2. more than one message may give rise to the same attractor (redundancy of the messages);
3. one message gives rise to several attractors (ambiguity of the message).

This is the first step of an endless procedure. The reception of a message causes the system to enter into the new state. Even the arrival of the same message a second time will affect the new system in a different way, since the information already received has changed the receiver.

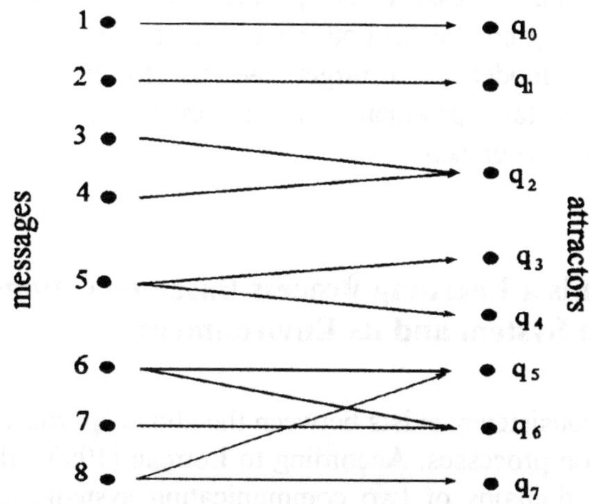

Fig. 4 Messages-attractor interaction according to Jantsch

Not all messages, of course, have the same importance, nor all attractors. It is possible attribute a relative importance p_j to the attractors, in such a way that $\Sigma_j p_j=1$. The value assigned to p_j depends on the task that the system has to perform. In a similar way we can define the relative importance p_i of the messages emitted by the source and reaching the system. For p_i too, the condition $\Sigma_i p_i=1$ holds. Of course, the relative importance of the message p_j depends not only on the system, but also on the tasks activated.

In order to evaluate whether a system annihilates, conserves or generates information, it is possible, as suggested by Haken, to use the tools of the information theory. Instead of considering the relative frequency of symbols, we can examine the relative importance p_i within a set of messages and the probability p_k that an attractor k is reached as a consequence of the message i. With this approach it is possible to introduce the quantities (entropy):

$$S^{(0)} = -\sum_i p_i \ln p_i \qquad\qquad S^{(1)} = -\sum_k p_k \ln p_k$$

and to evaluate whether the message has produced annihilation ($S^{(1)}<S^{(0)}$),

conservation ($S^{(1)}=S^{(0)}$) or generation ($S^{(1)}>S^{(0)}$) of information. Generation occurs when the system has been upgraded through a self-organisation leap, i.e. a learning process. To design the development of a city is like designing a teaching course. In both cases we are dealing with a learning process. With a model as a target, we try to create the necessary prerequisites (mental, physical and organisational) to permit the information to be recognised.

6.5 Evolution is a Learning Process Based on Communication between a System and its Environment

A surprising consistency exists between the above approach and analysis of communication processes. According to Lotman (1993), if A and B are the information domains of two communicating systems, three possible conditions may be fulfilled (Fig. 5):

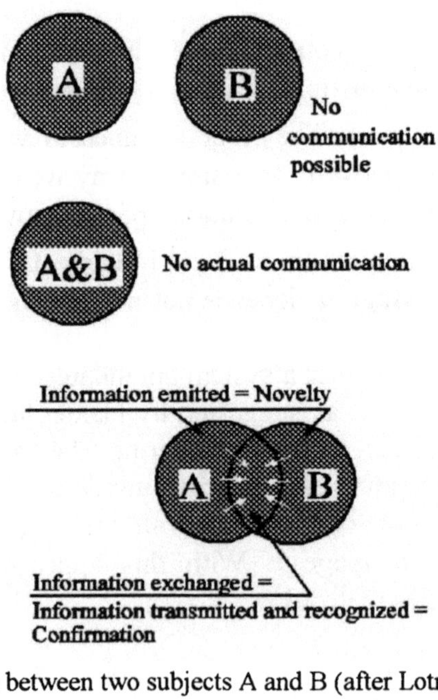

Fig. 5 Communication between two subjects A and B (after Lotman, 1993)

1. *No intersection*. No communication is possible since any message sent is unintelligible to the receiver. Jantsch would say that the information

flow is made up of pure novelty;
2. *Complete overlap*. No communication is possible because the information exchanged is without content as the two information domains coincide. In Jantsch's terms, the information flow consists of pure confirmation;
3. *Partial overlap*. In this case communication occurs; this is measured not by the area of overlap (where no information exchange takes place), but the transformation rate of novelty into confirmation. As pointed out by Lotman, the intersection area pulsates as a consequence of *two contrasting tendencies: the aspiration to facilitate the comprehension, which constantly leads to attempts to enlarge the intersection area, and the aspiration to enlarge the value of the message, linked to the tendency to maximise the difference between A and B.*

Adopting this concept, we can make a simplified application of the approach developed by Haken. We begin with a simple definition of communication, i.e. the process though which the behaviour of one individual modifies the probability of behavioural acts in other individuals. Wilson (1975) indicates how to measure Lotman's overlapping area, i.e. the information transmitted between two subjects (one considered the source of signals, the other as the receiver). The problem is that, as previously noted (see Fig. 4), analysing a large number of couples of communicating subjects, we find that a signal does not receive a single response, but a spread of responses. This, however, can be considered in terms of the 'disturbed channels' of information theory. It is possible in this way not only to quantify the transmitted information, but also the amount of information emitted and not received (referred to as equivocation) and the amount of information induced (ambiguity), as shown in Fig. 6.

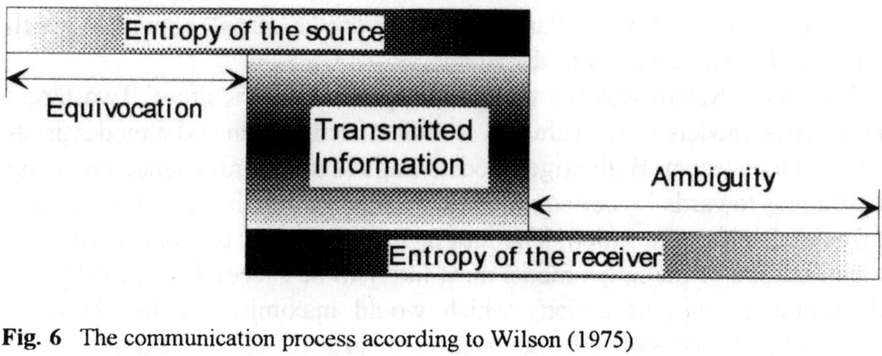

Fig. 6 The communication process according to Wilson (1975)

There is no reason, in principle, why this approach should not be extended to the communication between a system and its environment - in our case between a city and the rest of the world, or between a social actor and the city. It is possible to identify the signals of the system and the response of the environment, or vice-versa.

Going back to our main objective, we can say that if we accept that the development of most cities is a guided self-organisation process, i.e. a sort of teaching course in which we try to shape the changing system (the pupil) as closely as possible to our model, then we can make pragmatic use of the latter technique.

It should be added that this approach seems to fit well with the current developments in computational methods based on neural networks. Also in neural networks, the learning process is based on the progressive change of the relative importance (weight) of the interactions between the components of the system in which *information gradually emerges from noise by learning* (Parisi, 1989). In our case, the system we wish to analyse by means of the neural network techniques is the communication system, which is made up of signals and responses interacting with different probabilities of connection.

6.6 From Theory to Practice

The approach described above has not yet been applied to the interaction between a city and its environment, but some tests on simpler systems have been made. One example is a study of the Mediterranean island of Pantelleria (Barbera and Butera, 1992), in which an analysis was made of the agricultural production system of the island, whose economic activity is largely based on specialised agricultural products. This system was considered to communicate with an environment made up of all those activities carried out on Pantelleria and any external activities which interacted in some way with the island.

The basic system was the traditional system of agriculture. Two target production models were evaluated: a modernised system and a modernised-sustainable system. Both target models implied structural change involving upgrading towards higher complexity, in other words, it was a typical case of guided self-organisation. The aim of the study was twofold: firstly, the identification of the target model most likely to be successful, secondly, the definition of support actions which would maximise the likelihood of triggering a process of self-organisation towards the chosen structure.

The first step was to identify the components of the environment as perceived by the system. In practice, this was achieved by identifying all the social and institutional actors involved, directly or indirectly, with the production process, and establishing their perceptions of each step (action) of the process. A detailed analysis was therefore made of the production system (*action* by action), of the *actors* (individual, social, institutional) and the *factors* (climate, transportation facilities, energy availability, etc.) potentially affecting each action. Two input-output tables were drawn up, one showing actions and actors, the other actions and factors.

The second step concerned the quantitative evaluation of the perceived relative importance of the interactions. A ranking was obtained by means of a series of interviews (Table 1).

	Actors (environment's response)				
ctions (system's essages)	A	B	i	Z	
1	$S_{A,1}$	$S_{B,1}$	$S_{i,1}$	$S_{Z,1}$	$\Sigma_i S_{i,1}$
2	$S_{A,2}$	$S_{B,2}$	$S_{i,2}$	$S_{Z,2}$	$\Sigma_i S_{i,2}$
j	$S_{A,j}$	$S_{B,j}$	$S_{i,j}$	$S_{Z,j}$	$\Sigma_i S_{i,j}$
n	$S_{A,n}$	$S_{B,n}$	$S_{i,n}$	$S_{Z,n}$	$\Sigma_i S_{i,n}$
	$\Sigma_j S_{A,j}$	$\Sigma_j S_{B,j}$	$\Sigma_j S_{i,j}$	$\Sigma_j S_{Z,j}$	

Table 1 Relative importance $S_{i,j}$ assigned by the actors to each action of the production process

For each action, actor and factor, the values of the information transmitted, the existence of equivocation and ambiguity, as well as the information contained in the message and response were then calculated. These are shown in Table 2, where $p(X_i, Y_j)$ is the probability that the action (message X_i) is perceived with a value scored S_{ij} by actor Y_j and is given by the ratio (corresponding to the relative importance in the Haken's model):

$$p(X_i, Y_j) = S_{ij} \Big/ \Sigma_{ij} S_{ij}$$

		RESPONSE								
		Y_1	Y_2	…	Y_j	…	Y_n	$p_X(i)=p(X_i, Y)$	$H(X_i)$
MESSAGES	X_1	$p(X_i, Y_1)$	$p(X_1, Y_2)$	…	$p(X_1, Y_j)$	…	$p(X_1, Y_n)$	$\Sigma_j p(X_1, Y_j)$	$-p_X(1)\log_2 p_X(1)$
	X_2	$p(X_1, Y_1)$	$p(X_2, Y_2)$	…	$p(X_2, Y_j)$	…	$p(X_2, Y_n)$	$\Sigma_j p(X_2, Y_j)$	$-p_X(2)\log_2 p_X(2)$
	…	…	…	…	…	…	…	…	…
	X_i	$p(X_i, Y_j)$	$p(X_i, Y_2)$	…	$p(X_i, Y_j)$	…	$p(X_i, Y_n)$	$\Sigma_j p(X_i, Y_j)$	$-p_X(i)\log_2 p_X(i)$
	…	…	…	…	…	…	…	…	…
	X_n	$p(X_n, Y_1)$	$p(X_n, Y_2)$	…	$p(X_n, Y_j)$	…	$p(X_n, Y_n)$	$\Sigma_j p(X_n, Y_j)$	$-p_X(n)\log_2 p_X(n)$
	$p_Y(j)=p(Y_j, X)$	$\Sigma_i p(X_i, Y_1)$	$\Sigma_i p(X_i, Y_2)$	…	$\Sigma_i p(X_i, Y_j)$	…	$\Sigma_i p(X_i, Y_n)$	$\Sigma_i p_X(i)=\Sigma_j p_Y(j)$	
	$H(Y_j)$	$-p_Y(1)\log_2 p_Y(1)$	$-p_Y(2)\log_2 p_Y(2)$	…	$-p_Y(j)\log_2 p_Y(j)$	…	$-p_Y(n)\log_2 p_Y(n)$		

Table 2 Calculation of the quantities characterising the communication between a source of messages $X_1, X_2, .. X_n$ and the environment, whose corresponding responses are $Y_1, Y_2, ... Y_n$

$H(X) = \sum_i H(X_i)$ Information of the source

$H_{Yj}(i) = -\sum_i p(X_i, Y_i) \log_2 p(X_i, Y_i)$ Information of the message

$H(Y) = \sum_j H(Y_j)$ Information of the receiver

$H_{Xi}(j) = -\sum_j p(X_i, Y_i) \log_2 p(X_i, Y_i)$ Information of the response

$H_X(Y) = \sum_j p_X(i) H_i(j)$ Ambiguity

$H_Y(X) = \sum_i p_Y(j) H_j(i)$ Equivocation

$T(X, Y) = H(X) - H_Y(X) = H(Y) - H_X(Y)$ Information transmitted

$R = 1 - H_Y(X) / H(X)$ Redundancy of the source

The information values of the messages are indices of complexity. If their calculated values are plotted in a bar diagram, as in Fig. 7, we obtain a detailed picture of the system's structure, which is like the spectrum of a light or sound source. This sort of 'spectral analysis' of the complexity of a productive system allows us to identify the actions, actors or factors that are perceived as being most critical for the proper working of the process (even though the interviewed subject is often unaware of this interpretation). Comparing the spectrum of the present process with the spectrum obtained by analysing an improved, innovative process, we can identify the subsystems (actions, actors, factors) with the highest perceived 'novelty' content. These are the subsystems on which we have to work hardest if the innovated process is to be successful. When many subsystems show a high degree of novelty, it may be wise to reduce the scope of the innovation, since the probability of success would be very low.

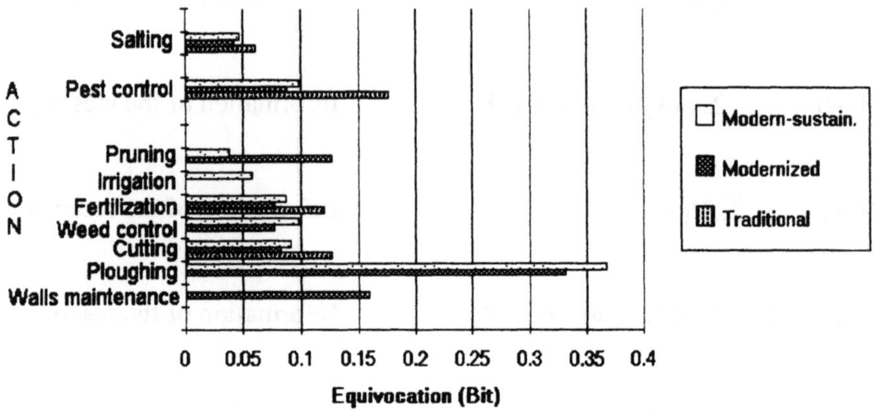

Fig. 7 Equivocation of each action of the productive process: the higher the value of the equivocation, the more novelty must be transmitted and recognised by the environment (the actors), and thus the more unlikely its success. For example, modernisation implies a simplification for salting, pest control, fertilisation and cutting. Actions implying much novelty (in the sense that the environment is not ready to provide the necessary neg-entropy flux), such as irrigation, mechanical ploughing and the adoption of pre-fabricated walls are more difficult to put into practice.

A further evaluation can be made in order to identify the most critical actions, actors or factors, by analysing the sensitivity to perturbation of each subsystem. This is done by perturbing the subsystem, i.e. modifying randomly the relative importance of different elements and analysing the effect on the spectrum of the whole system.

If the interviews are repeated at time intervals and a series of spectra produced, these can be used as photograms, giving a dynamic description of the development process and representing a way of monitoring the self-organisation process.

One question we may ask is: are there optimum values for transmitted information, equivocation and ambiguity? According to Jantsch, an optimised, autopoietic, system should fluctuate around 50% novelty and 50% confirmation. Studies on the English language show that the amount of confirmation, measured as redundancy (see table 1), reaches values around 70% (Brillouin, 1962). From other studies on DNA (Gatlin, 1972) and on ecological systems (Odum, 1972) it can be derived that:

a. the more a system moves upwards in the evolution scale, the lower the redundancy of the communication process;

b. the richer the environment, the lower the redundancy of the ecosystem;
c. an absolute optimum of redundancy does not exist: the level depends upon several factors, among which the stability of the environment plays a strong role.

In other words, the evolution/learning process leads to the perception of a richer and more stable environment because of the increase in the number of system-environment connections and the relative importance of interactions moving towards more evenly distributed values.

6.7 Final Remarks

The semi-predictive tool proposed here is conceptually positioned somewhere between those treating self-organisation as an entirely unpredictable event (pure novelty) and those based on traditional mechanical models of deterministic equations (pure confirmation). It seems to be particularly suitable for application to complex systems, such as cities, regions or diffused production systems, where the concept of guided self-organisation processes apply.

As a tool, it has the advantage of giving the planner a synthetic index of the transmitted information which makes it possible to evaluate and monitor the relative quality not only of the system as a whole but also each of its components. (By quality, we mean here the distance between the target model and the present system.) There is, of course, a need for further research, mainly for reducing the uncertainties deriving from the choice of actors and factors and from the evaluation of the relative importance of the interactions.

References

Barbera G., Butera F.M. (1992) Diffusion of Innovative Agricultural Production Systems for Sustainable Development of Small Islands: A Methodological Approach Based on the Science of Complexity, *Environmental Management, 5*, 667-679.

Brillouin L. (1962) *Science and Information Theory*, Academic-Press, New York.

Gatlin L.L. (1972) *Information Theory and the Living System*, Columbia University Press, New York.

Haken H. (1988) *Information and Self-Organisation*, Springer-Verlag, Berlin.

Lotman J.M. (1993) *La cultura e l'esplosione*, Feltrinelli, Milan.

Jantsch E. (1980) *The Self-Organising Universe*, Pergamon Press, Oxford.

Nicolis G., Prigogine I. (1987) *Exploring Complexity: An Introduction*, R. Piper, Munich.

Odum H.T. (1972) *Environment, Power and Society*, Wiley Interscience, New York.

Parisi D. (1989) *Intervista sulle reti neurali*, Il Mulino, Bologna.

Shannon C.E., Weaver N. (1949) *The Mathematical Theory of Communication*, University of Illinois Press, Urbana, Illinois.

Wilson E.O. (1979) *Sociobiology, The New Synthesis*, Harvard University Press, Cambridge, Massachusetts.

7. Strange Loops, Tangled Hierarchies and Urban Self-Regulation

Valter Cavallaro

7.1 Introduction

Dynamic models of self-organisation deriving from the physical and mathematical sciences have now become part of the 'tool kit' used by many analysts of territorial systems. Growth and decline dynamics, and the evolutionary trajectories of urban systems, are often studied through models that use differential equations. In this way it is possible to underline nonlinear and nonequilibrium processes that describe the shift of patterns in territorial systems. However, the use of these models implies the need to look into their theoretical validity when applied to systems in which the behaviour of components, unlike that of quanta or gas particles, is self-conscious.

In accordance with Jayet (1993), it is possible to assert that "the current generation of self-organising urban models seems to be at a half-way stage. The introduction of randomness and of nonpredictive determinism is done from the outside. It has as yet modified neither the fundamental statute of the predictive systems, nor the associated verification and measure procedures" (p. 78). It is quite clear that complex systems with random behaviour cannot be assimilated with complex systems with self-conscious and synergetic behaviour. They need to be separated and dealt with in different ways.

Self-organisation models based on the introduction of probabilities handle the components of the system as if they were equal to the average (Allen, in this volume), without taking into account the behaviour of single components. Besides depending on a self-conscious choice, they become a source of information for the other components, which therefore find themselves in a new situation. In the strategic choices which territorial actors define and carry out, such a process is fundamental, therefore the information deriving from it cannot be fitted to the system and modelled by an external observer as a series of variables. On the contrary, they constitute the internal characteristics of the system, i.e. they distinguish one component from the others. Hence, "the future of any system will be due to two kinds of terms: changes brought about by the deterministic action of the typical behaviour of its average components, and structural qualitative changes brought about by the presence of non-average components and conditions within the system." (Allen, in this volume, p. 11).

The importance of non-average components is not proportional to their number, but relates to the communicative value that their action acquires. The decision to behave differently from the average represents a communicative act affecting the state of the system and the possibility of describing this state. The dynamic analysis of the evolution of such systems must therefore introduce the communicative and self-conscious value of the behaviour of non-average components. In other words, we need to change from the application of mechanistic models (those that introduce random variables to deal with the behaviour of components) to communicative models. These have to analyse the average and non-average behaviour of components by studying the communicative aspects of self-observation and self-description. In this way the meaning of the term 'complexity' becomes clearer and closer to the interpretation of Turco (1988), who claims that complexity is the number of possible choices.

There is no absolute definition of complexity (the world itself is complex), but only a relative definition: a is more complex than b. We can say, for example, that the situation in which an individual must choose the way to take at a crossroads is more complex than when there is a single road. The subject will have to choose without ever knowing the consequences of his choice, even if he knows what lies at the end of the road. In this example it is possible to understand the projectual value of the complexity. I think the following simple definition can clarify how complexity should be distinguished from the concept of complication. The case in which following the road, even if it is the only one, involves climbing over a wall, going through bushes and crossing a ditch could be called *complicated* as it requires various tools and equipment. The direction is nevertheless clear, i.e. the information reaching our walker is simple and understandable. On the contrary, in a complex case, the problem shifts to the analysis of ambiguous information (or noise, as it will be defined), involving choices with very strong projectual value.

Complexity is not therefore a property of a single system, but refers to a situation in which the systems of the observer and the observed are so intricately interconnected that they become inseparable. We are concerned not so much with the 'complicated' situation in which the difficulty in acting is that of knowing an object, but the situation in which it is terribly difficult to act knowingly or to know how one is acting. This should make it clear that complexity can refer to the more recent mathematical instrumentation (fractal, theory of chaos and catastrophes, fuzzy sets), because it is complicated, but the opposite is not true. Very often these models are used in a predictive way, i.e. they are used to make *ex-ante* evaluation of a possible and well-known set of alternatives. But the change in paradigm should be separated from the use of new tools (with regard to the need to distinguish tools and paradigms, see the chapter by Ferlaino, in this book), even though these changes occur in parallel.

These are the preliminary statements on which the present work is based. The work is divided into two parts: the first (Section 2) analyses some central concepts concerning territorial systems as complex and self-organising systems. I take appropriate steps with the help of self-organisation theory (Atlan, 1979, Von Foerster, 1981, Dupuy, 1982, Morin, 1977), and pay particular attention to the production of order from noise and production of noise from order. In the second part (Section 3) I advance the hypothesis that the theory of cycles in city evolution can be interpreted as the cycle 'noise → order → noise'. I attempt to demonstrate that this hypothesis allows us to interpret the theory of cycles as a one which can take into account local factors, i.e. the ability of local systems to react to an increase or decrease in demographic dynamics according to their specific characteristics. The contribution ends with a diagrammatic representation of this interpretation.

7.2 Self-Organising Principles of Systems and Self-Regulation

7.2.1 Some Definitions

An interesting approach to the study of self-organising systems and their evolution, concerning the behaviour and choices of its components, is described by Von Foerster (1981). This approach is based on two principal assumptions relating to the dynamics arising between components of the system and between the system and its environment.

The first assumption specifies that any event represents a source of information for the system, i.e. it is a stimulus that the system will interpret according to its own internal properties and capacities (which are different from those of any other system, even in the same class[1]). These stimuli, which bring about structural changes, may originate from two different sources:

1. the outside environment, i.e. events that do not depend on the organisation or components of the system;
2. the system itself or, more correctly, its components and the interactions that occur between them.

The latter can be interpreted as a second order source, as it is possible to assume that the internal components can give rise to an information flux which is the consequence of decisions and actions carried out as a reaction to external stimuli, i.e. a first order source.

The second assumption specifies that the actions, reactions and compensations which the system carries out and which cause changes in its structure and organisation, are due to the will of the system to keep its own identity. The term *will* is used intentionally: it implies that the system is not a mechanical one, it has a self-conscience and recognises its own identity. This identity is formed through self-observation and materialises in self-descriptions (an urban plan can be read as a self-descriptive document, but many other texts or actions have the same value).

As we have said already, the structural changes that the system activates in order to keep its own identity, can be:

1. *conservative* changes - changes in which the type of system component remains the same, while the type of interaction between those components is modified;
2. *innovative* changes - changes that modify the type of component and, as a result, the type of interaction.

In the first case, there is a substantial variation in the network filters regulating the information flow, while in the second it is possible to observe a variation of interfaces. We therefore have to distinguish changes that produce structural modification, i.e. *structural innovative* changes, from those that modify only the types of interaction, but not the

[1] When we speak about systems of the same class, we mean that such systems can be compared, as they belong to the same group and same level.

components or their qualities. The best way to explain this is to give some examples.

A variation in tax rates is a conservative change. The aim is to guarantee the level of service offered by the State, but the relationship between the public organism and the citizen remains as defined in the constitution. It may sometimes be true that a modification of filters at a higher level can produce innovative changes at the local level (an increase in taxes could lead to a reduction in the competitiveness of the local system, causing some components to leave the system for others of the same class), but a simple variation in dimension cannot be defined as an innovative change. A variation of local city taxes that produces demographic shrinkage or increase can be interpreted as a conservative change, as it does not modify the structure of the local urban system (the component systems and their relationships). The structure would be modified if all the factories, i.e. the base economy of the city, were to leave the system. On the other hand, a variation in electoral law or constitutional rules can be considered an innovative change of system organisation, since these rules can be interpreted as interfaces that permit the participation of citizens in political decisions. Their modification causes a change in the relations between component-systems. Finally, a revolution or a *coup d'état* represent an innovative structural change. In this case, the role played by the component-system of citizens and the component-system of political institution would be abandoned in favour of new ones.

As we have seen, the concept of innovation proposed here differs considerably from that normally used in economic and geographic sciences, or rather, it acquires a different meaning. In fact, it is possible to speak of innovation only when there is innovation in the organisation of the system. Organisational innovation is often closely linked with technological innovation, for example in the industrial restructuration which follows the introduction of automation. But this assertion has to be verified and cannot be assumed *a priori*. From the link between technological innovation and organisational innovation it is possible, however, to build an example of the self-organisation of a system.

Technological innovation has the effect of decreasing the production costs (in terms of money and time) of a consumer good. This can produce various effects in the organisation of the system: it can cause a decrease of manpower involved in production, or it can cause an increase of production, resulting in an increase in the standard of living. More simply, the greater revenue due to the decrease in production costs may be distributed though an increase in wages or concentrated in an increase of profit. These are different evolutive trajectories that depend on the power relationship between workers and management, or between the population component system and enterprise component system. Changes in these represent changes in the organisation of the system. The system follows one trajectory or the other according to its organisation, i.e. the relationship between the two component systems, in this case between trade unions and the employers. When the trade unions are strong, the workers will earn much more, in the opposite case, profits will increase. The specific properties of the organisation of a system will cause different responses (greater concentration or greater distribution of wealth) to the same stimulus (the introduction of innovation).

The system therefore filters the stimuli and follows a given trajectory by selecting stimuli suited to its organisation. It reduces complexity, selecting evolution trajectories based on its own specific distinguishing properties; it is these properties which define its identity. In this case, the reduction of complexity can be defined as the production of *order from noise.*

To put this more simply, the changing dynamics of the system's organisation and structure depend on the reactions caused by the self-organising system with respect to certain stimuli. These stimuli can be classified as:

a. organised impulses,
b. noise.

Organised impulses are those stimuli that, from the analysis of the allopoietic properties of the system, are deemed capable of leading to known and desired transformations. Usually, according to General Systems Theory (Von Bertalanffy, 1968), they are defined as inputs. These inputs contain the future organisation of system, but in this case we cannot speak of self-organising dynamics. This does not mean that the reacting system is not a system with autopoietic and self-organising properties, it only means that operating stimuli use allopoietic properties as a fulcrum. In other words, the system behaves like a simple machine, whose evolution is predictable and expected.

Noise, on the other hand, can be defined as a number of aleatory perturbations without any causal relationship with the organisation of the future state of the system. If the system, under the effect of these random perturbations, instead of being destroyed or disorganised, reacts with an increase of complexity and continues to function, we can affirm that it is a self-organised system (Atlan, 1979).

The planning of self-organised systems has always involved the first type of stimulus, considering the system to be regulated by causal actions. It is clear however that if we consider the self-organising properties, the effects offered by the second type of stimulus must be evaluated very carefully.

7.2.2 Order From Noise

The principle of 'order from noise' can be considered a process through which a self-organising system activates its reaction to stimuli arising from its relationship with the outside environment. This requires some clarification of terminology. First of all, it is necessary to explain the meaning of 'order' and of 'noise'.

The essential references are the Biological Computer Laboratory's studies on the cybernetic approach. To measure the concept of order, Von Foerster follows the concept of redundancy (Shannon and Weaver, 1949). To explain the concept of noise, he uses a very simple, but significant example.

If we put some polarised cubes (each face has an opposite charge) in a box, then close it and shake it, when we open the box the cubes will be arranged in a random and orderly, but also stable structure.

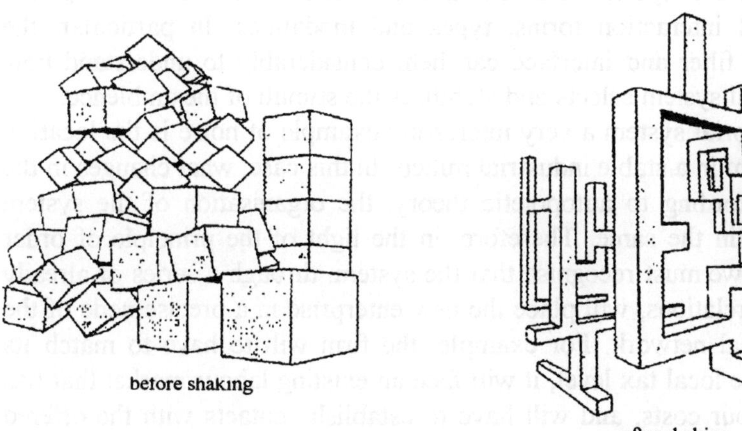

Fig. 1 Order from noise: before and after shaking

The system has absorbed low cost energy (the shaking) and has produced an orderly structure. The energy can be called noise because it does not have a defined direction, i.e. its aim is not to produce the specific order that comes out at the end of the process, when the box is opened.

Noise therefore involves an energy flow or, more in general, an information flow without a defined direction, that does not fit into a defined information channel (in the relations and hence in the system's organisation). The system, thanks to its organisation and structure, selects the stimuli deriving from the environment to produce a reduction in the level of internal entropy.

Morin (1977) summarises the necessary conditions for the described process to occur as follows:

1. determination of the characteristic constraints of the material elements we are dealing with (here: cubic form, metallic constitution and magnetic difference) which constitute the order principles;
2. the possibility of selective interactions that can connect these elements in determined situations and possibilities (magnetic interactions):
3. availability of nondirectional energy (disorderly shaking);
4. the production, due to this energy, of many encounters, some of which establish in a selective way the stable interactions that have an organisational role.

In a territorial system, the four above conditions can be reinterpreted with typical interaction forms, types and modalities. In particular, the concepts of filter and interface can help considerably to understand how the territorial system selects and stabilises the stimuli of the ambience.

In a territorial system a very interesting example of noise is the location of a new firm in a stable industrial milieu. In this case, what changes in the milieu? According to autopoietic theory, the organisation of the system should remain the same. Therefore, in the light of the principle of order from noise, we must recognise that the system, through a series of already established relations, will place the new enterprise in a precise node of the organisational network. For example, the firm will to have to match its budget to the local tax level, it will face an existing labour market that has its own labour costs, and will have to establish contacts with the offered services. The system offers many interfaces and establishes some filters (basically economic filters), so the new entry does not modify the system's organisation. It is possible that noise would have this intention, but autopoietic theory explains that the system reacts in a conservative way in order to preserve its identity.

Although the organisation tends to remain stable, the structure of system will be modified. This change of structure produces a very important effect on the cognitive domain of the system. In fact, the new component introduces new interactions that the system can integrate without losing its identity. These interactions are those that the new component establishes with the whole system and those the system had with other systems, which now become part of the new component's environment.

In theoretical terms we can affirm that if the new entrance increases the complexity (by increasing the number of possible internal interactions and hence the level of entropy and chaos), the reaction of the system will be to reduce the complexity by assigning a defined role to the new element.

This kind of evolution trajectory can be deduced from the analysis of a differential equation of redundancy. If the redundancy is measurable with:

$$R = 1 - \frac{H}{H_m} \tag{1}$$

where $\frac{H}{H_m}$ is the ratio of entropy H of a information source[1] and the highest value H_m that the entropy could have in a self-organising system increases as time goes by, then we will have:

$$\frac{\partial R}{\partial t} > 0 \tag{2}$$

If we replace in the disequation (2) the value of redundancy R as measured in (1), we have:

$$\frac{\partial R}{\partial t} = -\frac{H_m({\partial H}/{\partial t}) - H({\partial H_m}/{\partial t})}{H_m^2} > 0$$

But, as it is always true that $H_m^2 > 0$ (it is impossible to imagine a perfect, orderly system in which the maximum entropy would be equal to zero) then we obtain:

$$H\frac{\partial H_m}{\partial t} > H_m\frac{\partial H}{\partial t} \tag{3}$$

---

[1] The value H of entropy is measured as the number n of elements multiplied by its logarithm: $H = n\ln(n)$.

To understand the meaning of this difference, it is useful to discuss two particular cases: those in which the two terms, the maximum entropy and the entropy of the system, remain constant in the long run.

Case a: H_m=constant

In this case the derivative with respect to the time of maximum entropy is equal to zero, therefore the first term of disequation (3) is equal to zero. We obtain:

$$\frac{\partial H_m}{\partial t} = 0 \qquad \text{and} \qquad \frac{\partial H}{\partial t} < 0$$

which means that the entropy of the system has a tendency to decrease as time goes by. This is very common and means that if the maximum entropy is constant, i.e. if new components do not enter the system and vary its structure, and if the system is self-organising, then the relations among its components will be kept stable. This stability can be achieved by the construction of filters and interfaces that regulate the relations. In the case of social and territorial systems, the introduction of norms or the stabilisation of prices in an isolated system can be interpreted as agents that push the system in this direction.

Case b: H=constant

In this case, the derivative of the system's entropy with respect to time is equal to zero, therefore the second term of disequation (3) is equal to zero, and we have:

$$\frac{\partial H}{\partial t} = 0 \qquad \text{and} \qquad \frac{\partial H_m}{\partial t} > 0$$

This means that the maximum entropy of the system, and hence its maximum disorder, increases. But as the system is self-organising, this conclusion sounds absurd. Instead, it means that when new components enter the system, this reacts by producing order (the entropy remains constant) and the new components become part of the system's organisation. In this way the principle of identity conservation is respected. If we go back to the example of the magnet, this is what happens when we have shaken the box, opened it and put in a new cube without paying attention to its position. After closing the box again and shaking it a second time, on re-opening it we will find the cube fitted into the previous architectural organisation. Probably, unless we have marked the cube, it would be hard to recognise it. Although the maximum entropy is increased (there is an additional element), the system has guaranteed the maintenance of the organisation, stabilising the system's entropy. This phenomenon is particularly interesting because it permits us to explain self-organisation as a continuous tension between the increase of complexity (variety) and the decrease of complexity (redundancy). The first is caused by an increase in number of structural or organisational components and hence an increase in the number of internal relations, the second is due to the system's ability to introduce normalisation elements. The system's ability to increase the cognitive domain by increasing the number of relational components introduces a very important new property of self-organising systems: in parallel with the concept of Von Foerster we call it *noise from order*.

7.2.3 Noise from Order

In 2.2 our intention was to make clear the tendency of a self-organising system to order the stimuli of the environment according to its rules. We have followed step by step the reasoning of Von Foerster (1981) and Morin (1977), but it does not emerge clearly how a self-organising system can also produce noise from order.

The production of noise from order can have two main forms. The first corresponds to the second principle of thermodynamics. Ilya Prigogine has deeply studied the meaning of entropy conservation from the point of view of the analysis of a system. These studies have led to the definition of dissipative systems. A dissipative system is one which produces entropy (Prigogine and Stengers, 1988). A system that produces order, i.e. a system that increases the redundancy level (Von Foerster, 1981), can be seen as a dissipative system, as it produces entropy that flows into the environment. Any production of order is inevitably linked with the production of entropy that flows into the environment. The production of entropy is demonstrated, for example, by the impossibility of avoiding any energy production system also becoming a source of pollution.

In a social-territorial system, the production of environmental chaos takes on different forms which coincide with the types of order (or redundancy) that the system, as a self-organising system, produces. If we use the network paradigm, these different forms can be interpreted with the concept of isolation[1].

The modification of the relations between certain components of a system implies changes in the relations with other components. Essentially the sum is a zero sum. A very obvious example of the mechanism of isolation is disused industrial premises. The decision to adopt new production technologies frequently causes obsolescence of industrial buildings. These buildings are then abandoned and isolated by the system. In reality, they no longer belong to the industrial system, and the relationship with the whole territorial system has a lower intensity than when they housed a productive activity. They could be interpreted as noise, and seen as representing a stimulus for the production of order by the system. This is a circular relationship, as there may be a constant modification of the system's boundaries, as a result of the isolation/reintegration of components.

The production of noise from order can take on a second form. It can be linked to the willingness of the system or of components to produce noise. It is possible to recognise this phenomenon in the emission of information by some of the system's components, i.e. information which is not precisely directed, but launched into the environment awaiting a reaction.

This is the case of policies that tend to activate networks, but which initially do not have a defined partner. More generally, it applies to all activities that contribute to the creation of a milieu. In a territorial system the components of a milieu constitute a set of information (noise) that becomes a specific property of the system and that can be received from other systems (or their components) located in a given geographical environment. These systems will in turn produce order from this non-directed noise. Using a sligtly different terminology, we can say that the system receives complexity from the environment, reduces it, but through this reduction produces a new complexity. This is what happens with all policies, actions and plans that originate either from the desire to carry out the reduction of complexity necessary to implement a project, or the desire to send real messages into the 'ether' of the global system (e.g. urban marketing policies).

---

[1] The word isolation here implies expulsion from the system. This can happen in a complete way, when all relations such a component established with the others of the system are severed, or partially when only some relations are cut. We change the usual sociological meaning of the word in order to give it a more general sense that will explain the behaviour of systems.

This analysis can be applied to the whole set of historical, social and cultural factors which become established in a community or in cultural and political institutions (Garofoli, 1991). These factors, whether materialised or not, are the elements that constitute the particular milieu of a given territorial system. They permit the system to differ from others and to build its own identity by which it is recognised from outside.

In the case of isolation, the potential nodes of network (elements that were previously connected) are known, but it is not possible to identify the network. Any identity (or meaning) belongs to the isolated elements inside the system. In this case the situation is complementary: there is a meaning (or identity) that the system ascribes to itself or to some of its parts, but some, or all, of the terminals are absent. The terminals are those elements that establish a connection with the system after having recognised its identity. In both cases, however, we can interpret the presence of noise as a stimulus to the production of order and innovation and, eventually, the production of self-organisation.

If we try to draw a simplified representation of the dynamics of a system that produces order from noise, subsequently producing internal and external noise, we arrive at the loop proposed by Morin (1977) (Fig. 2).

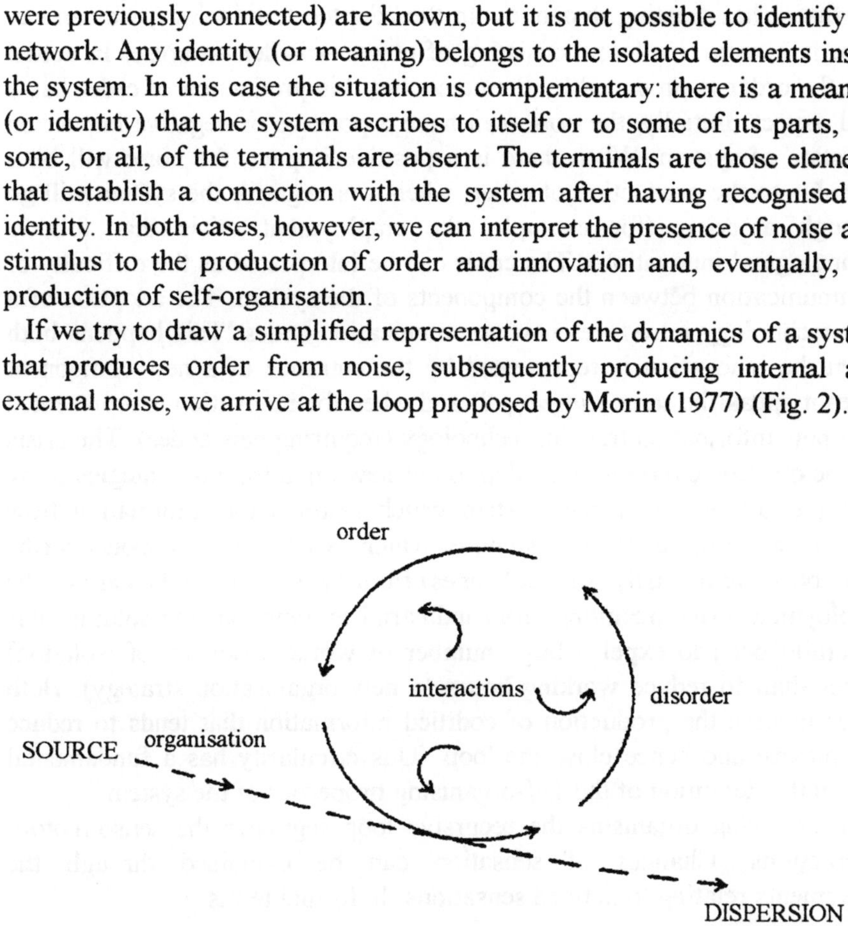

Fig. 2 The self-organisation cycle (after Morin, 1977)

The graph shows that the loop is a spiral and is therefore not a vicious circle, i.e. when the loop closes on itself it does not come back to the starting point. The sequence noise → order → noise involves irreversible transformations.

7.2.4 *The Noise → Order → Noise Loop*

If we follow the reasoning presented in 2.2 and 2.3, we can conclude that the history of the system can be described as a continuous alternation of choices which reduce complexity (producing order from informative noise) and choices which increase complexity (producing informative noise).

We can say that the evolutionary trajectory of the system is not linear, but that it has floating dynamics. In the adopted terminology, we can say there is a *recursive* loop consisting of noise → order → noise. In actual fact the order is produced by the noise, but this production of order has a dual aspect. Initially, the noise can have a positive or negative impact on the state of system. When there is a positive impact, the noise will be a stimulus to the production of order; when it is negative the system will go through a crisis (for example, the employment crisis that follows technological innovation). The crisis can be interpreted as the difficulty of communication between the components of the system. The adoption of a new technology introduces new communication codes. This happens both when the innovation is represented by the entrance of a new component (does it speak the same *language*?) and when the innovation is represented by a new information transfer technology (requiring new codes). The crisis can be overcome through the adoption of new organisation strategies or by coercive action, i.e. a new system which restores the information flow between all subjects of the system, or which expels some components (the more obsolete or easily isolatable ones) from the system. In the case of the employment crisis resulting from industrial automation, the solution has generally been to expel a large number of workers (an act of isolation) rather than to reduce working hours (a new organisation strategy). Both cases involve the production of codified information that tends to reduce complexity and hence close the loop. This circularity has a fundamental role in the definition of the self-organising properties of the system.

In the living organisms the recursive loop regulates the senso-motory interactions. Changes of sensation can be explained through the movements relating to defined sensations. In formal terms:

$$s = S(m)$$

and its movements through the reactions to given sensations. In formal terms:

$$m = M(s)$$

If the two explanations are related a recursive loop takes form. Formally we can write:

$$s_i = S(M(s_j)) = SM(s_j)$$
$$m_k = M(S(m_l)) = MS(m_l)$$

where s is the sensation, m is the movement, S and M are the senso-motory and motor-sensory operators.

This type of formulation can be adopted for the relations between order and noise in a territorial system. As we have said before, it is possible to state that the construction of order depends on noise and hence:

$$R_a(rum_k) \rightarrow ord_i$$

where R is an operator of a system element a that produces order. We can also claim that noise production depends on order and hence:

$$P_a(ord_i) \rightarrow rum_k$$

where P_a is an operator of a system element a that produces noise.

If we read the two applications contemporarily we can recognise a recursive loop:

$$P_a(R_a(rum_k)) \rightarrow rum_l$$
$$R_a(P_a(ord_i)) \rightarrow ord_j$$

If we take the first application and define the operator $P_a(R_a)$ as OP, we can write:

$$OP(rum_k) \rightarrow rum_l$$

Running along the loop, the new noise causes new order operations that produce in turn further noise. Therefore:

$$OP(OP(rum_k)) \rightarrow OP(rum_l) \rightarrow rum_n$$

If n approaches infinity and if, to make things easier, we omit the intermediate terms, we can write:

$$OP(OP(OP(OP(OP \ldots etc.) \rightarrow rum_{\infty}$$

It is obvious that if we apply the operator to an infinitesimal term, we obtain the same infinitesimal term, i.e.

$$OP(rum_{\infty}) \rightarrow rum_{\infty}$$

This formulation is very similar to the one used by Von Foerster (1981) and that used by Piaget (1975) to explain children's capacity for balance. It seems very promising for explaining that the loop is a creative loop, and not a vicious loop. The most important conclusion we want to stress here is that stability can be achieved through a recursive loop of subjective operations[1] . Here, we can introduce the concept of a tangled hierarchy.

According to Simon (1969), hierarchy is the topological form that guarantees stability. Through the application of recursive operations, we have demonstrated that stability is also guaranteed by an interaction process. This process is called a strange loop (Dupuy, 1982) or *tangled hierarchy* (Hofstadter, 1979).

The term 'hierarchy' means that at instant *t* the informative flow has a defined direction. Therefore it is possible to recognise a stimulus and a reaction to this stimulus. In a classical hierarchy the direction is constant

---

[1] To understand the meaning of stability, it is necessary to give the example from Von Foerster (1981): one should take as a recursive operator the function costg deriving from the combination of cosine and tangent, and as the starting point of the chain of events an arbitrary value, m_0 = 0,5:

m_1 = costg (0.500) = 1.204
m_2 = costg (1.204) = 0.375
m_3 = costg (0.375) = 1.342
m_4 = costg (1.342) = 0.231
m_5 = costg (0.231) = 1.470
m_6 = costg (1.470) = 0.101
m_7 = costg (0.101) = 1.540
m_8 = costg (1.540) = 0.031
m_9 = costg (0.031) = 1.556
m_{10} = costg (1.556) = 0.015
m_{11} = costg (0.015) = 1.557
m_{12} = costg (1.557) = 0,014
m_{13} = costg (0.014) = 1.557
m_{14} = costg (1.557) = 0.014
....
$m_{\alpha-1}$ = costg (1.557) = 0.014
$m_{\alpha-2}$ = costg (0.014) = 1.557

Here the fascinating phenomenon of bi-stability emerges: any value implies the other.

and immutable unless a disturbance drives the system to the end. In a tangled hierarchy, on the other hand, at instant $t+1$ the direction changes and the reaction becomes a stimulus, causing a new reaction in the system or in the component which has launched the first stimulus at time t. The functioning of tangled hierarchies guarantees the self-organisation of the system. The identification of tangled hierarchies and of the noise → order → noise loop represents a fundamental step in the analysis of the evolution of self-organising territorial systems.

A highly explicative formal theoretical approach to the evolution of a self-organising territorial system which changes itself through circular interaction is Eigen's hypercycle (Cramer, 1988). Eigen uses the hypercycle to describe the self-organising mechanism of biochemical molecular reactions. These reactions are catalytic, i.e. they are activated in the presence of special triggering actions in enzymes. Many enzymes can become associated with catalytic cycles, which take the form shown in Fig. 3.

Every catalytic reaction triggers off the next reaction. In this process we have the production of catalysts E_n A. catalytic hypercycle (Fig. 4) consists of a circuit of catalytic cycles which have a double function: to reproduce themselves according the form foreseen by the catalytic cycle and to sustain the subsequent cycle. There are two information vectors **I** for every reaction: the first is trained to self-reproduce, the second is trained to trigger the reactions of the next catalytic cycle. The architectural form of Eigen's hypercycle is particularly suited to the description of the production of external and internal noise in the ordering of component systems and comprehensive territorial systems.

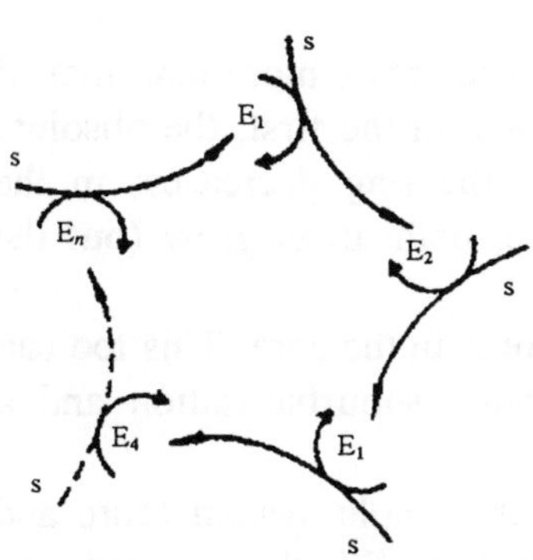

Fig. 3 Catalytic cycle

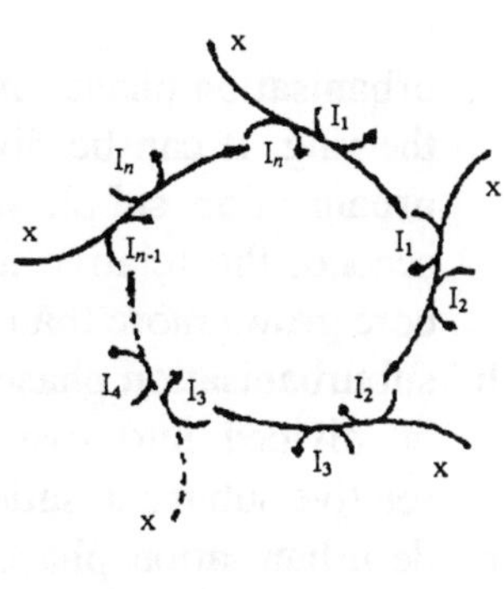

Fig. 4 Catalytic hypercycle

The Eigenian hypercycle can describe the process through which the territorial system produces a share of goods for export and a share for the domestic market The analysis becomes more interesting if we attempt to describe the activities that produce the stable components which represent the substratum of local communities. This analysis could be a way of recognising the relationship between the role of stable components in the local network and their role in the global economic competition network. Coming back to systemic terminology, the hypercycle, the tangled hierarchy and the strange loop can provide a way of analysing the organisation of the system: a system in which we can establish a circularity between the activation of relations (the movement described by Piaget) and the production of the domain of the system (the sensation). Now the problem is to explain the evolution of the territorial system as a self-organising system.

7.3 The Decline/Growth Loop as a Principle of Self-Organisation

7.3.1 *The Spatial Cycle of Urbanisation*

In this last section we shall try to verify whether the noise → order → noise cycle can explain the cycle of urban decline and growth according the urbanisation cycle hypothesis (Berry, 1976, Van den Berg *et al.*, 1982, Van den Berg, 1987). This is a descriptive model that shows an empirically verified statistical regularity (Camagni, 1992). According to the model, the population of an urban system, divided into core and ring, changes in a cyclical pattern that can be described as follows:

a. urbanisation phase: the population of the core grows more than that of the ring. It can be divided in two subphases: in the first, the absolute urbanisation subphase, the population of the ring decreases; in the second, the relative urbanisation subphase, both areas grow (but the core grows more than the ring);
b. suburbanisation phase: the ring grows more than the core. This too can be divided into two subphases: an absolute suburbanisation and a relative suburbanisation;
c. de-urbanisation phase: the population of the whole system (core and ring) declines: this can result from the decline of both core and ring (absolute de-urbanisation) or occur because the decline of the core is

greater than the growth of the ring (relative de-urbanisation);

d. re-urbanisation phase: this can be defined as absolute re-urbanisation if the core grows, and relative re-urbanisation if the decline in the core is less than the growth in the ring.

The four phases can be summarised in the following diagram:

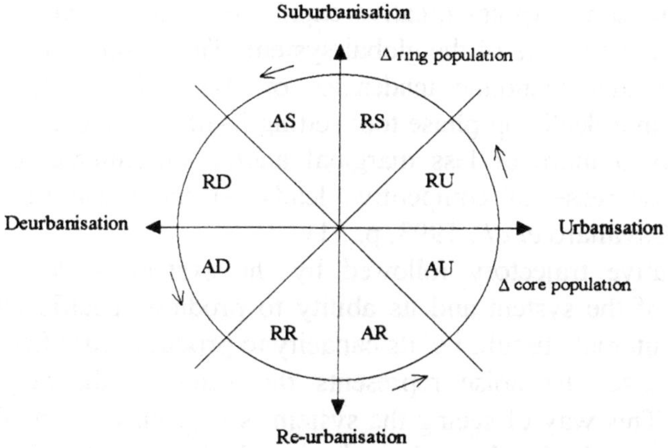

AU	absolute urbanisation	RD	relative de-urbanisation
RU	relative urbanisation	AD	absolute de-urbanisation
RS	relative suburbanisation	RR	relative re-urbanisation
AS	absolute suburbanisation	AR	absolute re-urbanisation

Fig. 5 Urbanisation spatial cycle (source: Camagni, 1992)

The aim of the urbanisation spatial cycle model is descriptive, there is therefore no attempt to give a theoretical explanation of the reversals in population dynamics. Is it possible then, and how, to find the upper limit of growth? Is there a critical dimension which it is not possible to exceed? What are the elements that trigger off the growth process after a period of decline? To answer these questions and others relating to demographic dynamics, researchers have appealed to various interpretations, particularly those deriving from the application of Schumpeterian economic cycles and from ecological models, such as the prey-predator model (Camagni, 1992). One of the merits of these interpretations is that they do not identify a static upper limit of growth, beyond which a process of decline begins (as some researchers thought possible by applying theories of marginal benefit). The empirical researches effectively show the possibility of demographic decline occurring at any urban dimension

(Camagni, 1992), hence the impossibility of finding a synthetic indicator of economies of scale and diseconomies that are a function of the absolute dimension of the city.

Here we propose a third possible interpretation: the concentration and dispersion of population are a function of the self-organising capacity of the urban system, and of its adaptability to stimuli from the environment. This assertion is the generalisation of the idea according to which the urban system can "express creative capability through adapting itself to the new strategic waves of the global system. This happens by 'tuning in' with the world's economic tendencies or 'lying down' on the wave, following it in a declining phase too; setting itself, in a more or less stable way, to play a more or less marginal and/or subordinate role, with a substantial decrease of competitive leads on the scene of worldwide relations" (Cavallaro *et al.*, 1993, p. 51).

The evolutive trajectory followed by the system is defined by the uniqueness of the system and its ability to produce specific reactions to external or internal stimuli, i.e. its capacity to produce order from noise. In the above case, the noise represents the value of the agglomeration economies. This way of seeing the system is very close to the theories of urban *milieu* or the incubator hypothesis, which claim that the *ambience* of the city has to attract and create new activities, leading to growth in the dynamics of the system. It is certain that the willingness to go to Silicon Valley could override many considerations, such as the high cost of transportation which would result. The powerful attraction of such an urban system cannot be measured with dimensional indicators: it is based on the *ambience quality* of the system.

The indicators which permit us to verify the system's ability to produce self-representations which can be projected to the outside, to achieve synergies among the subjects within the system and to benefit from the application of new technology, will therefore relate to the socio-environmental web, which is the most important factor in defining the attraction of the system (Governa, 1995). In brief, these are the *human resource* indicators, which have not only a qualitative meaning. From this point of view it seems right to evaluate the system's growth not only in terms of *extensive growth*, in the sense used by Gille (1992), but in terms of *intensive growth*, i.e. the growth that depends on the number and the strength of the interconnecting subjects in the network[1] .

---

[1] In the extreme, we can say that a city with 12 million inhabitants barricaded in their homes certainly does not represent a point of attraction.

7.3.2 *The Concentration → Dispersion → Concentration Cycle as a Noise → Order → Noise Cycle*

If the concentration/dispersion/concentration dynamics and growth/decline cycles can be interpreted as information-producing phenomena, i.e. stimuli to change in a conservative or innovative way, then the noise→order→noise cycle can represent an effective analytical model.

Fig. 6 attempts to show how the cycle, represented by the equations:

$$R_a(rum_k) \rightarrow ord_i$$
$$P_a(ord_i) \rightarrow rum_k$$

could be used for the analysis of the growth/decline dynamics of territorial systems.

The aim of this diagram is to emphasise how it is possible to analyse the decline/growth dynamics according to our starting hypothesis. In this model, growth and decline are not described as in opposition to each other, but as the result of the same self-organising process. The possibility of including points of bifurcation in the noise → order → noise cycle is particularly interesting. These points allow us to describe disequilibrium states of system that drive it to evolve towards continuous growth or inevitable decline. The 'choice in the dilemma' depends on the unique state of the system, and its ability to trigger a development process.

In this perspective, the cycle is no longer a descriptive cycle of evolutive dynamics, but it acquires an interpretative evaluation of the system's ability to react and produce self-representations or projects that will allow it to modify its trajectory. Moreover, it is fundamental to consider that such representations are not necessarily carried out by the system as a whole, or by controlling or regulating systems that give overall and unanimously valid representations of the system and of its components. In other words, it is not possible to affirm that the most effective representation would be the average component's representation (see Section 1). It is possible that the representations will be produced by the *lower part,* introducing innovations that are able to modify the system's organisation and then to influence the choice of the evolutive trajectory.

Exceptional cases can be interpreted in this way: e.g. the case of cities in developing countries (situations which would be considered unendurable in Europe and would certainly result in demographic decline), or the case of the average Italian city, in particular the difference between the evolution of the cities in Central Italy or in the South where the virtuous cycle of local development has not caused levels of economic growth comparable to the other Italian regions, and in some cases (e.g. Puglia) has shown signs of regression (RUR, 1992).

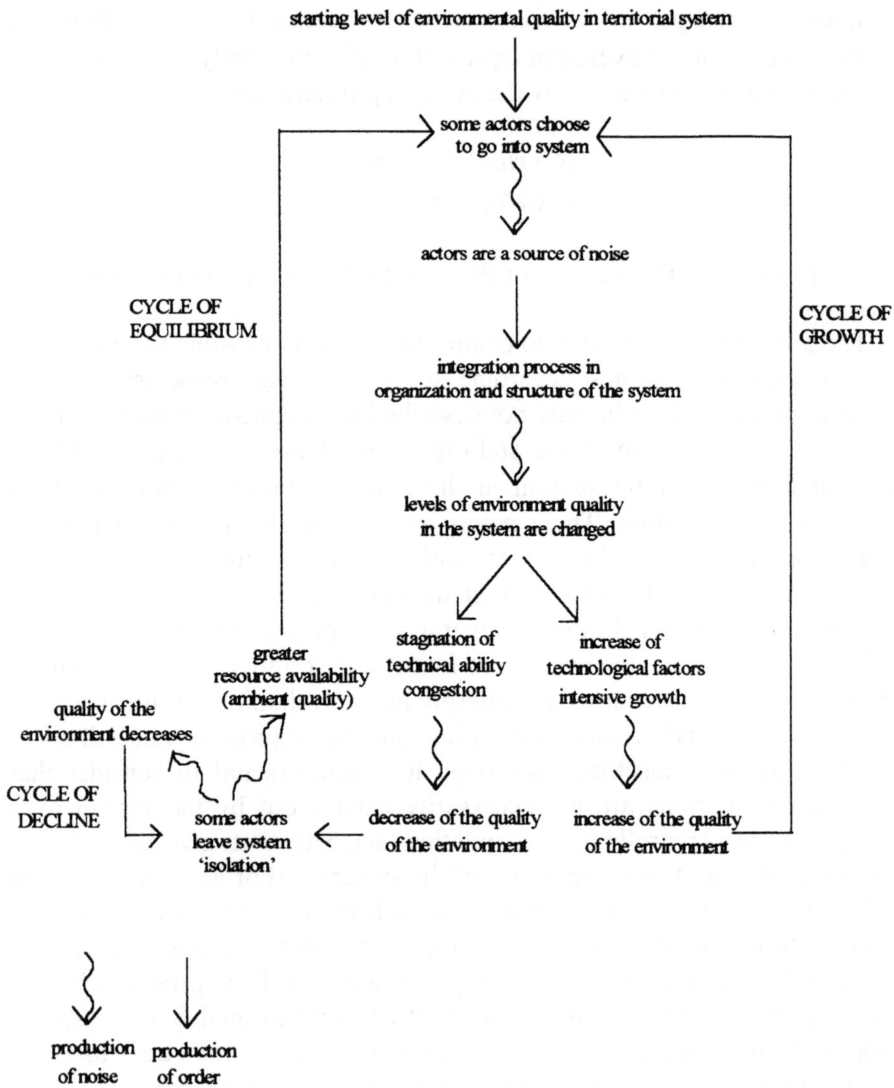

Fig. 6 The urbanisation spatial cycle as a self-organised loop

It should be taken into consideration that to analyse the growth/decline dynamics through dimensional variables (the number of inhabitants) can lead to macroscopic mistakes. Levels of congestion or the availability of resources are a function of both the number of potential users and their ability to exploit the resources or enjoy benefits deriving from their use. It is clear that diseconomies can be caused by high levels of exploitation. In this case re-equilibrium policies can aim for dimensional control, but also for control of the level of exploitation. It is also possible that growth may be of a kind that cannot be recorded through the variables that measure the dimension of system.

7.4 Conclusions

In conclusion we can affirm that the urbanisation spatial cycle theory can be enriched by adopting a perspective that introduces consideration of the complexity of the system, since growth and decline concepts assume a more complex meaning. As urbanisation, de-urbanisation and re-urbanisation must refer not only to the resident population, but also to the use of resources and benefits that derive from the position in the system, it is necessary in empirical research on urban system evolution to use indicators that are able to represent the unique features of the system and its ability to trigger off reactions to external stimuli. The growth and decline (economic or demographic) needs to be investigated in a perspective that permits us to understand the level of exploitation of resources achieved by the components of the territorial system. The indicators will represent qualitative phenomena such as intensive growth, and not only quantitative phenomena such as extensive growth (the dimension of an urban system).

The second conclusion relates to the importance of moving along two parallel paths: on the one hand, to consider the possible contribution of the theories of self-organisation and of systemic complexity and, on the other, to take up the classic theories of urban systems so as to re-interpret them in the light of more recent systemic assumptions. In this way we can produce models which can be submitted to empirical verification.

References

Atlan H. (1979) *Entre le cristal et la fumée. Essai sur l'organisation du vivant*, Seuil, Paris.

Berry B.J.L. (ed.) (1976) *Urbanization and Counterurbanization*, Sage, New York.

Camagni R. (1992) *Economia urbana. Principi e modelli teorici*, NIS, Rome.

Cavallaro V., Ferlaino F., Mela A., Preto G. (1993) Per una teoria dei sistemi metropolitani, in Lombardo S., Preto G. (eds.) *Innovazione e trasformazioni della città. Teorie, metodi e programmi per il mutamento*, Angeli, Milan, 38-66.

Cramer F. (1988) *Chaos und Ordnung. Die komplexe Struktur des Lebendigen*, Deutsche Verlags-Anstalt, Stuttgart.

Dupuy J.P. (1982) *Ordres et désordres. Enquête sur un noveau paradigme*, Seuil, Paris.

Garofoli G. (1991) *Modelli locali di sviluppo*, Angeli, Milan.

Gille L. (1992) Interconnexion des réseaux, in Curien N. (ed.) *Économie et management des entreprises de réseau*, Economica, Paris, 31-52.

Governa F. (1995) L'analisi del locale nelle dinamiche territoriali: una prospettiva in termini di reti e milieu, Doctoral thesis, DIITE, Architecture Faculty, Turin.

Hofstadter D.R. (1979) *Gödel, Escher, Bach: An Eternal Golden Braid*, Basic Books, London.

Jayet H. (1993) Les modèles d'auto-organisation urbaine à mi-chemin?, in Lepetit B., Pumain D. (ed.) *Temporalités urbaines*, Anthropos Economica, Paris.

Morin E. (1977) *La méthode*, Seuil, Paris.

Piaget J. (1975) *L'equilibration des structures cognitives*, Presses Universitaires de France, Paris.

Prigogine I., Stengers I. (1988) *Entre le temps et l'éternité*, Fayard, Paris.

RUR (1992) *Città e impresa. Una regola per i progetti urbani*, Edizioni RUR, Rome.

Shannon C.E., Weaver N. (1949) *The Mathematical Theory of Communication*, University of Illinois Press, Urbana, Illinois.

Simon H.A. (1969) *The Sciences of the Artificial*, MIT Press, Cambridge, Massachusetts.

Turco A. (1988) *Verso una teoria geografica della complessità*, Unicopli, Milan.

Van den Berg L. (1987) *Urban Systems in a Dynamic Society*, Gower, Aldershot.

Van den Berg L., Drewett R., Klaassen L., Rossi A., Vijverberg C. (1982) *Urban Europe. A Study of Growth and Decline*, Pergamon Press, Oxford.

Von Bertalanffy L. (1968) *General Systems Theory: Foundations, Development, Applications*, Braziller, New York.

Von Foerster H. (1981) *Observing Systems*, Intersystems Publications, Seaside, California.

8. The Fractal Geometry of Urban Organisation: Beyond the Crisis of 'Spontaneous Order'

Lorenza Lucchi Basili

8.1 Introduction: the Paradigm of Complexity

Modern science has been marked, from its birth to our days, by an *analytical* approach to the complexity of real world phenomena: complexity has been decomposed into simple objects and 'basic laws' deriving from observation at this fundamental level have been proposed[1] (see e.g. Oldroyd, 1986, for an extensive and detailed discussion). One could say that this approach tries to understand the world in the same way that children do when they discover the internal structure of a toy by taking it apart. In recent years the scientific approach to the study of complexity has undergone a deep transformation that can be interpreted as an overall change of paradigm[2] (see e.g. Ceruti, 1992).

The changing attitude toward complexity originates from the idea that the information built into complex systems is richer than that coming from single constituents. One cannot, therefore, explain the complex through the

---

[1] The ambition of perfect predictive ability, nurtured by the reductionist-deterministic approach, has been thwarted by the increasing awareness of the inescapability of measurement errors highlighted by ***Heisenberg's principle of indetermination***. From this premise, dynamic systems theory has shown that the unavoidable approximation in the determination of the initial position of the system tends to be exponentially ***amplified*** as time unfolds, soon reaching macroscopic mass, leaving no scope for reliable prediction (see e.g. Ruelle, 1991, Vulpiani, 1994).

[2] "By paradigm we mean the whole body of the universally recognized scientific accomplishments, which have, for some time, provided a model of the set of problems and solutions that are acceptable to those who work within a given research field"; this is the original definition by Kuhn, cited in Oldroyd (1986, Italian ed. p. 419). See Vallega (1995) for an up-to-date discussion of the notion of paradigm within the context of geographical sciences.

relatively simple without taking into account that, when arranged into a whole, constituent parts gain additional meanings and 'purpose'. As a consequence, the approach turns from analytical to synthetic and, as is common in synthetic thinking, this brings to the fore increasingly global and complex levels of organisation at the expense of an accurate understanding of details (see e.g. Arecchi and Arecchi, 1990).

Chaos theory and, more generally, the theory of nonlinear dynamic systems[1] are the new mathematical tools that allow us to achieve a comprehensive understanding of complex systems of all kinds. These are the conceptual foundations supporting the epistemological shift that is at the origin of the new paradigm, known as the *paradigm of complexity*. This is a strongly interdisciplinary paradigm that has been applied in fields as diverse as physics, biology and sociology[2]. By increasing the scale of scientific observation, it becomes evident that among apparently heterogeneous fields there are important analogies and a common feature: the existence of systems made up of a large number of interacting elementary units. The single units that make up a complex system establish reciprocal relationships through the exchange of information, in other words they 'organise' themselves. This is the source of extra informational value not perceivable when observing single constituents in isolation. The organisation process basically involves choosing one possibility among the many available, privileging the one which allows a harmonious, nonconflictual interaction among the constituents of the system. Organisation is therefore the means through which the nonlinear system controls and governs its own complexity.

As an example, we take the case of a fluid heated from below. The particles in closest contact with the source of heat increase their mobility and begin to rise. This movement clashes however with the colder particles which tend to move downwards. When a cold particle reaches the bottom of the container, its temperature increases and it begins to move upwards, whereas the opposite happens to the formerly hot particles, which gradually cool as they rise through the fluid. This phenomenon creates a convection current that gradually spreads over the whole fluid. It has been noted, however, that convection does not occur in a disorderly fashion, but follows a clear and somewhat surprising logic: a pattern of ascending and descending columns of fluid appears. Such columns channel the motion of particles, thereby smoothing it and, so to speak, organising it. This is

---

1 For an introduction to nonlinear dynamical systems theory see Vulpiani (1994).

2 Most recent developments suggest in fact a further widening of the scope of application toward fields like psychology. See, for example, Serra and Zanarini (1986).

clearly the most 'economical' and efficient way to design convection, in that it prevents the particles moving in one direction from being disturbed by those moving in the opposite direction.

The mechanism underlying the above phenomenon should be clear: once some particles find a route offering low resistance along which they can ascend, an ever increasing number of particles will be induced to follow the same route. This becomes what we might call a route of 'minimal resistance'. The gathering of hot particles along such 'pathways' allows cold particles in turn to find a way along nonoccupied pathways, until neatly perceivable *convection cells* emerge. This result does not hinge upon a conscious search for the optimal organisational solution, but is the outcome of the joint action of a large number of particles whose only apparent aim is that of choosing the most favorable direction, that of minimal resistance.

Analogous instances of collective behaviour may be found, under suitable environmental conditions, in other physical phenomena such as magnetism. Single magnets, for example, exchange information about their magnetic orientation, eventually reaching a common orientation causing the emergence of magnetisation at the macroscopic level. Also in social phenomena, such as fashion, there are cases of behavioural conformism where the main propagation channel is the mass media.

Organisation is induced by particular external conditions (in the case of fluid convection, the temperature of the heat source, for instance) but, as we have seen, order does not come from an act of will: it is the system itself that finds its own order principle, choosing a particular configuration among the many available as a response to external stimuli. It is for this reason that we speak of *self-organisation*[1]. This is a *dynamic order principle*, i.e. one that does not consolidate into geometrically regular structures, like those of crystals, within which single atoms can only oscillate around predetermined positions. To make a clarifying though somewhat unorthodox analogy, this type of order is similar to that created by a juggler whirling a set of balls through the air, constructing a sort of flying wheel. The important difference is that in the case of a complex system the juggler is absent (Zanarini, 1994). Many natural systems

[1] Self-organisation is a reflex of living and non-living systems' ability to spontaneously find dynamically ordered configurations. Matter can therefore self-organise, generating previously inexistent dynamical structures that cannot be traced back to, and are not characteristic of, any single system component. Examples of self-organising non-living systems are convective currents, 'complex' chemical reactions like the Belusov-Zapotinski's 'wave' reaction and laser light beams; see e.g. Haken (1981), Carrà (1989), Arecchi and Arecchi (1990).

possess a *dynamically ordered structure*: they are made up of a myriad elementary objects, such as atoms or molecules, which organise into moving structures, e.g. clouds or vortexes. The remarkable feature is the absence of a juggler, or external 'supervisor' who induces the organisational principle by acting on single system components following a purposeful strategy. The idea of self-organisation therefore expresses the possibility of highly organised collective behaviour, despite the absence of a recognisable 'design intention'.

Just as natural selection tends to reward those organisms that manage to find the best adaptation to changing environmental conditions, nonlinear dynamic systems perpetuate themselves by opposing destructive actions coming from outside through a modification of their internal organisation (adaptation) that preserves its overall integrity (see e.g. Cini, 1994). This reactive ability is typical of such systems and originates from their very complexity, i.e. from the behavioural richness, which always carries the risk of becoming unmanageable and descending into chaos but, for this very same reason, proves to be extremely flexible and ready to react to stimuli. 'Self-organised criticality' denotes that type of equilibrium at the edge of chaos which is reached spontaneously by some nonlinear systems, allowing them to switch easily between dynamic regimes as external conditions change (see Bak and Chen, 1991). In other words, in the presence of self-organised criticality the system finds its way to a *critical* configuration of parameter values, i.e. to a set of parameter values lying on the frontier between regions that are characterised by different dynamic behaviours and properties, that is to say to a situation of maximum adaptive flexibility.

In recent years we have witnessed an increasing accumulation of evidence, from both the natural and social sciences, supporting the idea that nonlinear systems are capable of complex adaptations. The new paradigm of complexity seems to lead to a relatively nonintuitive conclusion: that through disorder and chaos a complex system undergoes fluctuations and apparently random modifications whose aim is to let the system explore a number of configurations from which the one best fitted to the environmental stimuli will be selected[1] .

For a better appreciation of this point, let us consider again the example of heat convection. As we have seen, when the fluid is heated, particles within it begin moving from the bottom to the top of the container and vice versa (convection currents). The innumerable microscopic movements become evident at macroscopic level as temperature gets close to a critical

---

[1] Some interesting related considerations are provided by Cavallaro (in this volume).

point and gives rise to a patterned rippling of the fluid surface. Under suitable experimental conditions (e.g. specific fluid characteristics, the shape and size of the container, the thermal gradient between the top and the bottom of the container), the convective currents arrange themselves into regular macrostructures known as *Bénard's rollers*[1]. This phenomenon, which is unfortunately observable only under laboratory conditions, signals the onset of a high and unexpected *degree of cooperation* between particles. It is all the more surprising if one considers that physicists have long believed that heating an otherwise closed system induces an increase in disorder and consequently a loss in the level of organisation. This kind of organisation becomes meaningful and recognisable only at a spatial-temporal scale quite different from that of the particles themselves. Bénard's rollers are billions of times larger than the molecules or the typical interaction distances between them, and last much longer than the average time lapse between two subsequent molecular bumps. Order therefore appears at a macroscopic scale.

Likewise, if we consider self-organised complex systems of a 'social' nature, such as anthills, we notice that single components (ants) neither heed explicit organisational prescriptions, nor apparently have access to general information on the organisational design ruling the social life of the community. They seem to move along random trajectories and to change their behaviour on the basis of local interactions, i.e. on the basis of information exchanges with other ants they happen to meet. A global picture of the system is available only to outside observers with access to a different (and in particular larger) spatial-temporal scale.

To sum up, a complex system: (i) is made up of many interacting elements, (ii) is not isolated, but exchanges matter and energy with the environment, (iii) displays organisational characteristics which are neither deducible at the level of the single constituents nor the outcome of external prescriptions or regulation.

The paradigm of complexity is an expression of the spreading awareness that the world is not entirely predictable. It defeats our understanding in that it is far more complex than our mental schemes: it is self-organised and chaotic and, at the same time, partially ordered. The idea of universal determinism and unconditional predictability is now untenable, since complex systems cannot be broken down into simple constituents and thus the very premises of the reductionist approach have been undermined. The

---

[1] When the container is uncovered and the fluid presents a high *surface tension*, rollers tend to interact with each other and to self-organise spatially: the outcome is the characteristic 'beehive' surface structure of *Bénard's cells*; see e.g. Zanarini (1994).

science of complexity and chaos is not helpful if the purpose, as it was for classical physics, is to 'simplify' the world. On the contrary, it invites us to develop ways of thinking which are as complex as the world around us. Complexity is oriented towards the understanding of systems that have been only partially explained by classical physics, often having been put aside as 'uninteresting'. Looking through these new glasses at real world systems (and even seemingly familiar and extensively studied ones, such as the solar system[1]), one finds many apparent irregularities that challenge our intuition. These can be 'retranslated' and understood through the new language of complexity.

8.2 The City as a Complex System

The natural interdisciplinary character of the paradigm of complexity, that is to say the apparent universality of self-organisation phenomena, seems solid enough to deserve credit as the basis of a new theoretical reading of spatial configurations as highly structured and articulated as urban settlements. Is it plausible to see the city as a nonlinear dynamic system which self-organises and then adapts to the evolution of the housing, production, social and cultural demands of its inhabitants?

There is no doubt that the city meets the requirements that characterise a dynamic system: it is a system of interacting parts which evolves through time. Moreover, it is clear to anyone who has been involved with the problems of the management of traffic or other city networks that the city is a nonlinear system. In such systems, the complexity emerges from the fact that even seemingly negligible causes tend to be *amplified,* bringing about significant effects on the existing order. For example, simply by establishing a 'one way restriction' within the urban road network, the allocation of traffic may be modified over the whole network and effects felt even on distant routes apparently independent of the ones being regulated.

As to the self-organisational ability of the city, it is unfortunately all too evident that our contemporary urban conglomerations are unable to

[1] Saturn's rings, for example, are chaotic systems: an infinity of particles, whose individual behavior is erratic and unpredictable, are distributed along bands whose structure is clearly fractal (see Section 4 below). When observing them through a telescope, we realize that each ring and each band is itself made up of other smaller rings and bands.

manage themselves without extensive 'support' from outside. In some cases, one could even claim that cities are right in the middle of chaos rather than on the edge of it! Things have not always been that way, however. The management of urban settlements is a specific problem of industrial and post-industrial societies where the pace of change has undergone unprecedented acceleration and where the growing complexity of land use patterns has led to the emergence of more complex functional hierarchies within the urban organisational tree (see e.g. Mumford, 1961). For this reason, it is useful to examine the historical processes that have led to the birth and growth of urban settlements. It is particularly instructive to consider the extensive case history of small settlements which underwent '*non-designed*' development (Secchi, 1984) and, going back even further, the genesis of ancient cities.

The founding of the ancient cities coincided with the beginnings of human socialisation and organisation. As mankind realised the usefulness of distributing among the population the everyday tasks instrumental for survival, the appearance and diffusion of well-structured settlements soon followed. This was the first instance of urban 'organisation', or rather 'self-organisation', since it occurred in the absence of any explicit design. We have seen above that, for a nonlinear system to self-organise, it is crucial that specific *external conditions* prevail.

In the case of early urban settlements, the 'critical' condition for organisation was the *division of labour* (see e.g. Liverani, 1988). The need to aggregate into small communities, within which a system of social relationships could develop, emerged when single households discovered the rewards of giving up the autarchic mode of production to specialise in the production of exchangeable goods. *Sociability* became a channel of *communication* that helped to disseminate a form of organisation throughout the whole system. In other words, social contact allows the circulation of the information necessary for the emergence of global order which, in this context, also entails the formation of characteristic spatial patterns. In order to facilitate exchange, some households are induced to orient the entrance of their houses toward a common space which soon acquires the status of a public space. Gradually this conditions the orientation of initially noncompliant houses to arrive at a global (i.e. widely shared) logic concerning the meaning of private and public space[1] (Hillier, 1988).

This simple example concerning early forms of urbanisation forcefully

[1] The orientation function may in some cases be performed by a socially meaningful feature such as the temple or the chieftain's house.

underlines once again the two basic conditions for self-organisation: (i) the existence of specific external conditions that spur the system to find an organisation; (ii) the existence of a level of communication between constituent parts that is high enough to *amplify* the information that is produced locally. Here, as elsewhere, the self-organisation process is not the outcome of a 'design intention'. The urban organism does not control its evolution: it is the interaction of locational choices for housing and industry that brings about a global structure. Its features cannot be traced back to the intentions or aims of single individuals[1]. The settlement's global organisation lies far beyond the immediate purposes of any specific individual, acquiring a logic and meaning of its own. More precisely, people are not motivated by a holistic principle of spatial order when making their location decisions. Their only purpose is to accommodate their current needs with the minimum feasible cost.

If we think of urban systems as the result of the locational choices of a large number of individuals, this leads us to the conclusion that individuals will opt for those solutions that grant them the greatest benefit, i.e. 'minimal resistance' solutions. This is the very same logic that led fluid particles to look for free routes and organise into convective currents. In most cases, due to the pressure of immediate everyday needs, 'minimal resistance' solutions are those which entail the minimal adjustments with respect to the status quo.

An example of an application of the principle of minimal resistance to an urban context is the progressive formation of a built block. As shown by Caniggia and Maffei (1993), the urbanisation process tends initially to occur along a linear path linking places which are especially valued by the community. Caniggia and Maffei call this the *matrix path*. As development along the matrix path grows beyond certain limits, communication problems arise between the extreme poles of the settlement. This state of things is clearly a source of practical problems and discourages further linear extension along the path. The logic of minimal resistance then suggests breaking through the main frontage to enable further expansion along secondary paths, called *implantation paths*, which are generally perpendicular to the matrix path. Communication between implantation paths turns out to be impractical, however, because it requires using the matrix path. This leads to the development of secondary *connection paths* to link the implantation paths, thus closing off the block (Fig. 1a). The

---

[1] Remember that the amount of information associated with the global organisation of the system is greater than the sum of the information associated with each of the constituents.

need for communication between parts of the settlement is therefore accommodated through *gradual adjustments* of the existing structure, responding not so much to issues of global coherence, but simply current demands.

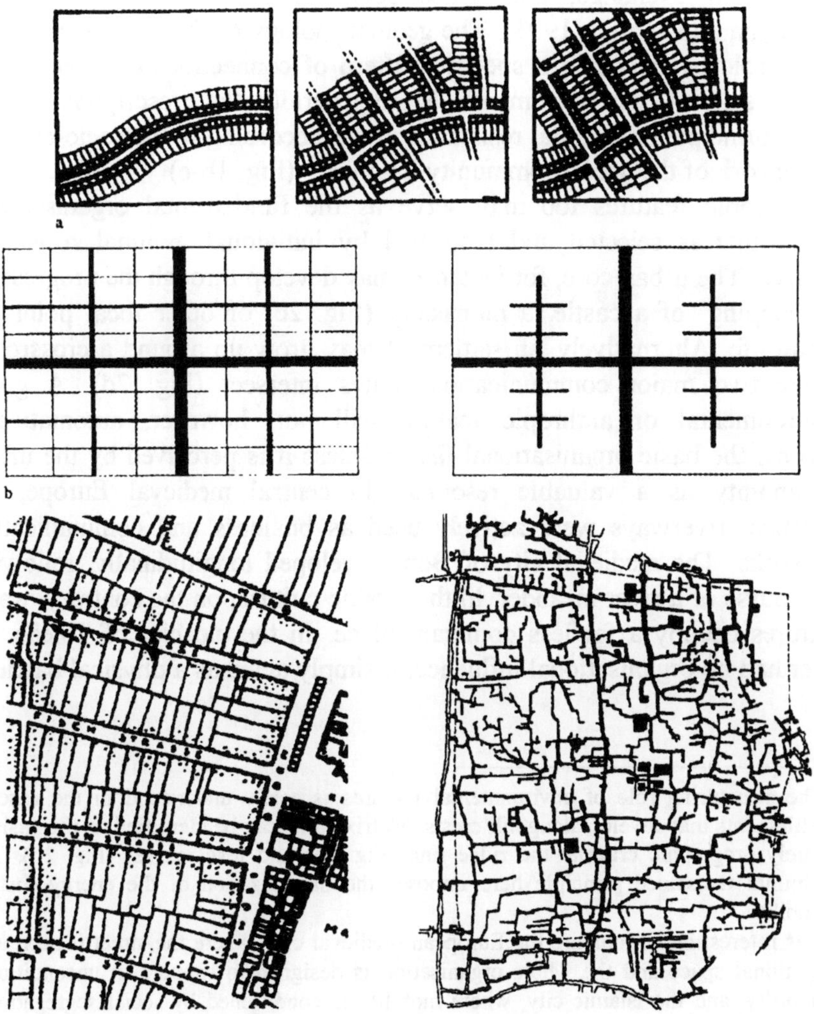

Fig. 1 (a) Progressive formation of closed blocks by breaking across the matrix path; (b) comparison of two different structures: network hierarchy with easy access and tree hierarchy with limited access; (c) two historical examples: Lubeck, medieval urban network (network hierarchy); Cairo, early 19th century road network (tree hierarchy)

To sum up, non-designed urban order is the outcome of the superimposition of a myriad 'minimal resistance' choices taken in response to a precise set of housing, economic or social demands. Housing location choices in medieval European settlements, for instance, were a response to the need to choose a place that guaranteed easy access to the urban core as well as to the activities of other inhabitants (in order to maximise information exchange, and hence social and business opportunities) (Caniggia and Maffei, 1993). The geomorphology of the site often plays a major role by imposing a specific pattern of connections between places for the sake of easy movement[1]. More generally, the prescriptive content of the principle of minimal resistance varies according to the socio-cultural framework of the urban community involved[2] (Fig. 1b-c).

Anthropic features too may serve as the fundamental organisational factor that is selected and amplified by individual minimal resistance choices. The urban core, for instance, may develop through the progressive 'enveloping' of a castle, a monastery (Fig. 2c) or other focal points of urban life. Alternatively, an settlement may grow up around a crossroads where two major communication routes intersect (Fig. 2d). A given environmental or anthropic feature will not, however, automatically become the basic organisational factor unless it is perceived by the urban community as a valuable resource. In central medieval Europe, for instance, riverways were actively used as business and communication networks. The medieval city of Bern developed longitudinally, following the course of a river meander. In the Arabic culture, on the contrary, water features occupy a far less dominant place. In the old city of Toledo the river had no organisational influence, it simply acted as a physical barrier.

[1] The structuring role of environmental features is easily understood in the case of settlements that develop along hill crests. Matrix paths can be identified on the basis of purely orographic criteria: the ridge line (Fig. 2a) and level curves (Fig. 2b). The minimal resistance principle here imposes the minimization of the energy cost of mobility.

[2] It is interesting to compare the European medieval city, where the street is seen as a relational space and the whole organisation is designed to support an unconditional mobility, and the Islamic city, where mobility is conditioned by social segregational constraints for family groups. Here, the street, although keeping its connotation as a public place, does not acquire the status of a relational space and favors the isolation of the (family-controlled) alley-courtyards. The typical street networks, consequently, have different patterns: in the former, there is a 'network structure' which makes the urban fabric fully permeable to pedestrians, whereas in the latter there is a 'tree structure', which does not facilitate pedestrian circulation (see Schumacher, 1982, Guidoni, 1992).

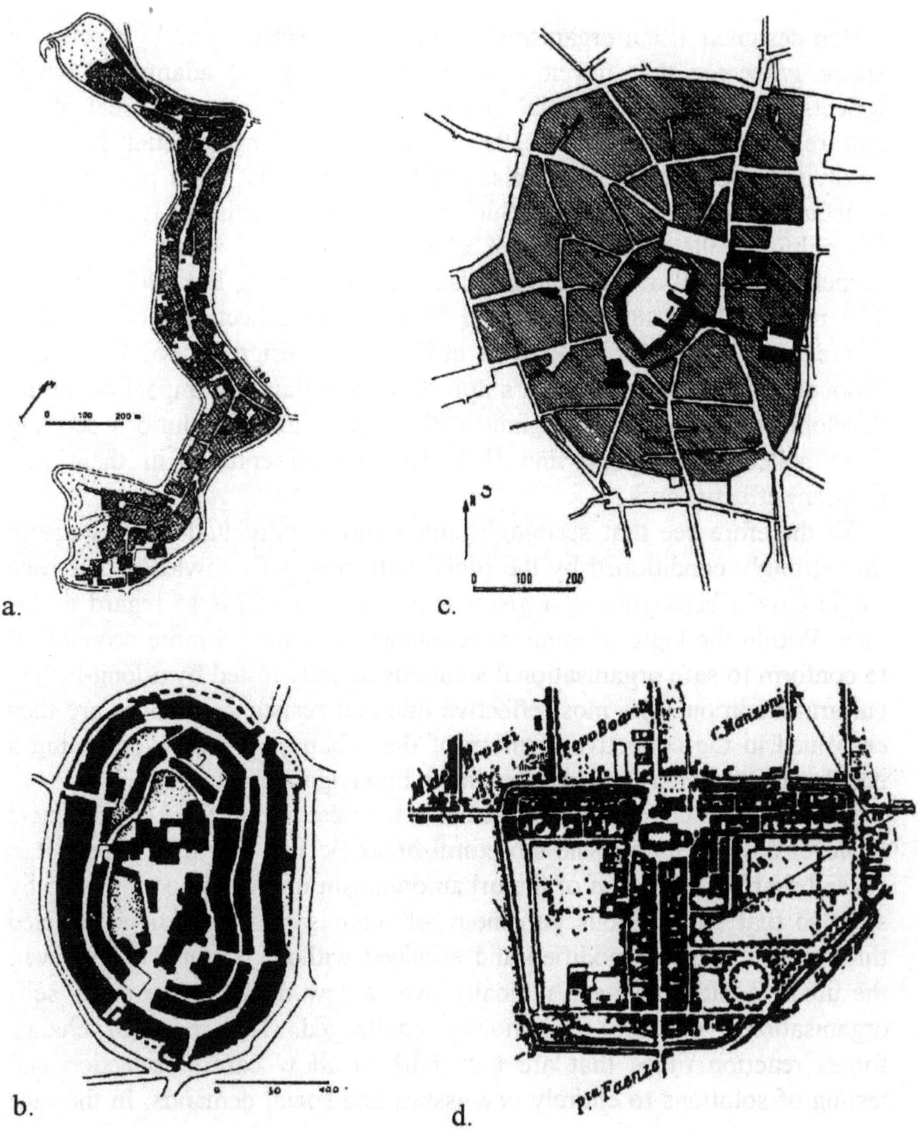

Fig. 2 Environmental features as organisational factors: (a) Recanati: settlement built along a crest; (b) Lucignano: settlement following contours; (c) Vigevano: progressive 'enveloping' of a castle; (d) Lugo: development around a crossroads, showing how growth favours the directions leading to major towns (Bologna, Ravenna, Faenza)

Non-designed urban organisms are therefore characterised by a coherent *space grammar* that develops through a process of adaptation which reflects the logic of the social and economic relationships typical of the culture that generates it. Order is reached after a long selection process[1] involving a repertoire of organisational solutions that have proved to be satisfactory and reproducible. Such a repertoire constitutes the historical and cultural collective *memory* of an urban community which, although in perpetual renewal and evolution, also retains a strong link with the past. The role of the designer is played here by the collective memory. But, whereas the designer/planner will evaluate his design proposals through rational analysis of the project's appropriateness and its impact on future development of the urban organism, the organisational solutions deriving from collective memory find their legitimation entirely in their past (proven) effectiveness.

We therefore see that seemingly autonomous individual choices are in fact strongly conditioned by the social patrimony of knowledge that each single citizen belonging to a given social context learns to regard as his own. Within the logic of minimal resistance, it is indeed more economical to conform to safe organisational solutions already tested by a long-lasting cultural tradition. The most effective minimal resistance choices are then enshrined in the collective memory of the urban community, becoming a permanent patrimony that can be drawn upon again and again.

Solutions that have proved effective in the past, retain their effectiveness however only as long as no structural break occurs. For this reason, it is crucial that the evolution of the urban organism unfolds smoothly, step by step, so that the precious patrimony of organisational solutions selected through time can be modified and enriched without trauma. If, however, the urban context changes radically over a few decades, the city's self-organisational potential is seriously challenged. The speed of change forces reaction times that are too short to allow careful selection and testing of solutions to entirely new issues and social demands. In the face

---

[1] To establish an analogy between biological processes and urban transformations, one can see city development as the result of the action of a selection mechanism based on adaptation to site conditions. Adaptation takes place in time through a series of trials during which the urban organism explores some of the feasible settlement patterns. If a given morphological solution fits social demands, that is to say it allows the development and the consolidation of a viable system of urban relationships, this solution tends to 'fix' and to persist through time. As experience is accumulated, the most effective solutions are encoded into urban 'types', which not only reproduce through time but also across space, as similar environmental and social conditions emerge elsewhere.

of the need for complete restructuring of the urban fabric, a logic of adaptation based on minimal resistance solutions proves inadequate because of its 'shortsightedness', i.e. its inability to keep the global dynamics of change under control.

It thus becomes clear why (self-)management of large, modern urban settlements proves so hard. This is even more evident when we realise that much of the repertoire of solutions offered by the collective memory has developed from a pre-industrial context. On the threshold of the fourth information-based industrial revolution, these solutions are clearly inappropriate to the vast array of unprecedented organisational and adaptational problems currently being faced.

In conclusion, the idea that the city may be studied as a nonlinear dynamical system deserves some credit. Following an approach that emphasises the city's dynamic potential for adaptation to its socio-environmental context, we can see how the post-industrial city has lost its ability to self-organise successfully, in that it has not managed to keep up with the growth and transformation induced by the accelerating pace of economic development[1]. It has thereby evolved into a congested and fragmented[2] conglomerate of ill-connected parts, which are unable to exchange those informational flows indispensable for coherent global development.

8.3 Managing Complexity: Issues and Policies

Any organism grows and changes through time. The resulting transformation, however, does not merely involve a change of scale or linear expansion of the original form: it requires a deep restructuring of the organism. But, there is a limit to an organism's rebalancing and self-repairing ability. As pointed out by Paba (1990), with reference to the contemporary city, once this limit is exceeded, there is the risk of ending up like Posif's horse. According to the anecdote attributed to the psychologist Gregory Bateson, Posif, a scientist working on genetic

[1] See e.g. Mumford (1961) and, with reference to urban management issues, the remarks of Avarello and Cuzzer (1982).

[2] The emergence of organisational difficulty is revealed through the alternating sequence of fragmentation and congestion phases: a bad spatial distribution of urban functions causes *congestion* in communication (mobility: long-range movements needed to accommodate basic demands) which eventually leads to *fragmentation* (isolation: difficulty in long-range movements to access more specialized functions).

manipulation, managed to modify the DNA of a common draught horse to obtain a laboratory animal twice as large. Posif's horse was however unable to stand as it was too heavy to support its own weight, its heat regulation system did not work as it had to deal with a body volume eight times larger than usual, the horse was continually out of breath because its windpipe was too narrow, and so on. The organisational crisis of the contemporary city is traceable to a similar type of functional dissonance. It is therefore a necessary to resort to external *design control* to 'repair' the many self-regulatory breakdowns arising from the abrupt jump of scale.

The design experiments of past decades have often been unable to provide satisfactory solutions to the problems raised by the modern city. One of their major shortcomings has been the lack of a working compromise between the 'global design' approach (the attempt to design the whole urban structure with the same level of detail typical of the architectural scale) and the pragmatic 'urban management' approach (the provision of general guidelines and certain land-use constraints at city-wide scale). The former approach is too rigid to be able to keep up with the pace of urban change, tending to force it into over-constrictive organisational solutions. The latter, on the contrary, tends to focus on 'negatively phrased' constraints on urban development (which are not always respected), rather than proposing positive organisational solutions (Campos Venuti, 1994). There is, in addition, the difficulty of dealing with the unintentional consequences of design choices and unforeseen contingencies that arise during the implementation of the project. These difficulties are accentuated by the persistence of a *sequential* logic in the planning process which still too often imposes a sharp distinction between the design phase and the implementation phase, not allowing for the reworking or rethinking of the project as fresh information comes in. Finally, 'designed' urban spaces often suffer from a fundamental shortcoming: the inability to reproduce the complexity and the richness of the traditional urban fabric.

Although we recognise that non-designed urban areas have an intrinsically high spatial quality, it is equally true that its mechanical or nostalgic imitation is a bad design policy. When the design, however, bears no relationship to the logic and nature of the organisational processes reflecting the city inhabitants' demands for efficiency and accessibility, the risk of 'rejection' becomes substantial. The imposed design will gradually be dismantled by the 'urban organism' which reacts to the designer's 'attack' through an inexorable process of re-appropriation and re-colonisation of urban space by the traditional organisational model.

One should not underestimate the importance of change. As we have already pointed out, the city is a dynamic system. Architectural and urban designs, on the other hand, tend to be conceived (possibly as a consequence of the authoritative and rationalistic approach) as definitive, self-contained endeavours. For this very reason they often bring about a violent discontinuity with the existing urban environment, aiming to 'tame' it and force it into the designer's logic. Such a methodology is clearly too schematic in its abstract 'functionalism' and is incapable of reproducing the rich perceptual and cultural stimulation that is a characteristic of the complexity of traditional settlements.

Space has a symbolic code of its own. Its 'meaning' is the result of a slow historical process and its value lies in the ability to provide a clear 'common orientation' to all users (see e.g. Lynch, 1981, Cortesi, 1979). Even in large chaotic contemporary cities, it is the traditional, usually central, districts which still provide the deepest, long-lasting 'collective image' of the city. New suburban and peripheral settlements may have become autonomous realities with an important functional role, but it is the historical centre which remains the 'true' city of collective memory and serves as the yardstick against which the quality of urban relationships will be evaluated.

The town planner must therefore learn to *cooperate* with the city's natural organisational processes, formulating a design policy that is able to guide these processes, while leaving sufficient room for spontaneity and compensating for the communication breakdown caused by the system's sudden, excessive growth. As noted by Tinacci Mossello (in this volume), the underlying rationale of urban design should be to transcend the city's organisational constraints resulting from the continued expansion of urban areas (requiring global control), taking into account the many individual decisions whose aggregation generates urban dynamics. The designer is therefore under pressure to avoid overt conflict with individual incentives. A credible policy must produce a design, or plan, which takes these constraints into account and guides urban transformation in the direction that best meets social demands.

The implication of the above discussion is that self-organisation is no longer a viable route to global coherence for the contemporary city. A better option is what we could call *designed hetero-organisation* (Arecchi and Arecchi, 1990) where the order principle is not simply the outcome of a series of autonomous choices taken from within the 'city-system', but is also subject to external rational supervision. This guidance will prove 'acceptable' as long as it respects and solicits positive interaction with the system's natural adaptation mechanisms.

Such a goal is certainly ambitious. To have a reasonable chance of being achieved, it is essential to decipher the linguistic code of the urban organisation, in particular, the spatial patterns transmitted by the historical memory of the urban community. Our basic assumption, which will be discussed in Section 4, is that such a linguistic code may be reconstructed by means of fractal geometry. This approach is being increasingly used by researchers in diverse fields to describe the 'deep organisational level' of complex phenomena.

8.4 Fractal Geometry: towards a 'Grammar of Urban Morphogenesis'?

To avoid our interpretation of the city as a nonlinear dynamical system remaining a vague analogy, we must now investigate further the organisational characteristics of spontaneous order as they emerge from nonlinear systems theory. The most powerful language available to date for the study of the organisation of complex systems is *fractal geometry* (Mandelbrot, 1975).

Fractals are geometrical objects with a highly complicated 'indented' appearance. Their descriptive effectiveness emerges at different scales. In some cases, it is the very morphology of the organised system that displays a fractal structure (objects like sponges or snowflakes; see Briggs, 1992), whereas in others the fractal structure becomes apparent only through a suitable graphical display of certain properties of the dynamic system, e.g. the convergence of a pendulum towards a magnet (for a rigorous analysis see Peitgen and Richter, 1986).

In general, fractal geometry is better suited than Euclidean geometry to the description of natural forms: from the indentations of a mountain ridge to vegetal structures and body organs. One could say that the world has been created by means of a fractal *language*. But we can push our argument even further: as noted by La Bella (in this volume), the use of suitably chosen fractal structures coupled with random generators allows us to reproduce virtually any data pattern observable in the social sciences. On the other hand, he also remarks that however successful the use of fractal geometry as a *descriptive* tool, it leaves entirely open the problem of *understanding* the forces which have generated the observed data. A stimulating challenge would be to establish to what extent fractals can prove useful in explaining and interpreting social phenomena. More specifically, one may legitimately ask whether it is in fact possible to translate into fractal algorithms the rules of adaptation to environmental conditions that govern social processes and, in particular, the process of urban morphogenesis.

In this section, after a brief presentation of the basic tenets of the fractal modelling approach, we shall focus on a specific class of fractal algorithms that look promising for building a credible model of the adapting city: the so-called L-Systems. As fractals allow us to find regularities within complexity, they open up a new approach to the understanding of what, up to now, has seemed chaotic. We no longer need to break complexity into pieces to be studied separately without ever having a 'full-blown' picture of complexity. We are now realising that chaos too has rules of its own and today, thanks to this new tool that elucidates the morphological structure of complexity, it is beginning to look somewhat less 'chaotic'.

To classify an object as a fractal, two requirements must be met. Firstly, it must be 'morphologically rich' at whatever dimensional scale it is observed. In other words, it must have *complexity at all scales*. In other words, as any detail is exploded, further detail that was not perceivable at less fine-grained scales must be revealed. To make a visual analogy, one should imagine taking photographs of a terrain from different heights. The only difference is that with fractals the zoom-in process may go on indefinitely.

A fractal is therefore an unbounded store of morphological information that is hierarchically organised as a function of the scale of observation. As observation proceeds, it becomes apparent that fractal objects are characterised by another important property which is linked to the first, i.e. *self-similarity*. The landscapes that are revealed from sequential zooming are all 'similar' to one another. Various degrees of self-similarity are possible, each representing a distinctive fractal family.

Firstly, there is the *linear family* that includes the *strictly self-similar* fractals, for which any exploded detail produces a form identical to the original one. A strictly self-similar fractal is therefore made up of an infinite number of reduced copies of itself (Fig. 3). These are the least 'interesting' instances of fractals but, being generated by linear transformations, the geometrical construction on which they are based is easily identified. Fractals belonging to the *nonlinear family* are more complex. Sequential zooming yields details which resemble the original, but are not identical. The best known example of a non-strictly self-similar fractal is the Mandelbrot set, whose morphological articulation is truly exceptional. When exploring the Mandelbrot set, one sometimes comes across reduced copies, but most are not perfect and some, in fact, bear only a vague resemblance (Fig. 4).

The above fractal families are *deterministic*. One can however also construct *stochastic* fractals for which the generating procedure admits random operations. For this reason, as observed by La Bella (in this volume), the latter class of fractals is well suited to the modelling of real-world phenomena in which random factors always play an important role. Such fractals can serve as valuable and credible models of natural objects, such as coastlines or mountain reliefs, where the enlargement of a detail, although in fact different from the whole, is in practice 'indistinguishable', as it is generated by an identical morphological matrix.

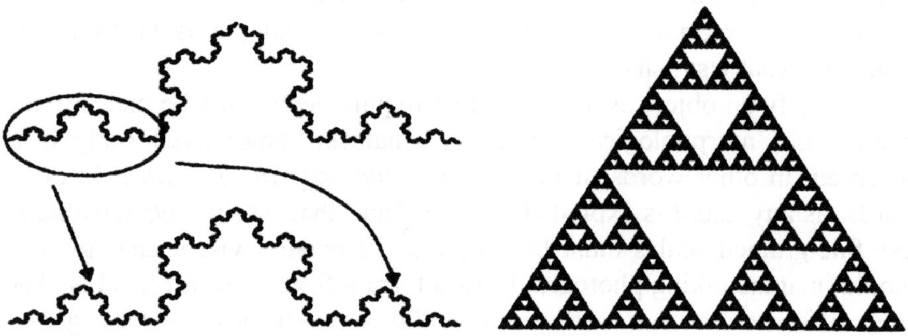

Fig. 3 Two examples of strictly self-similar fractals: Koch's curve and Sierpinski triangle

The second requirement for the acknowledgement of the fractal nature of an object is the so-called *fractal dimension*. Unlike Euclidean objects, fractals are characterised by a fractional dimension. As already observed, a fractal is a storehouse of compressed morphological information. A Euclidean object, on the contrary, always reveals its structure completely: at whatever the scale it is observed its appearance is unchanged. This basic difference is at the root of the notion of fractal dimension.

If, for example, we consider an Euclidean segment such as a straight line, it is easily confirmed that its length will not vary even if measured with rulers of different lengths. If, on the other hand, we consider a fractal object, such as Koch's curve, which is full of 'kinks', it is immediately evident that, however short the ruler used to measure it, there will always be an infinity of details not recorded. Therefore, the shorter the ruler employed, the greater the total length of the curve, since more and more details are captured.

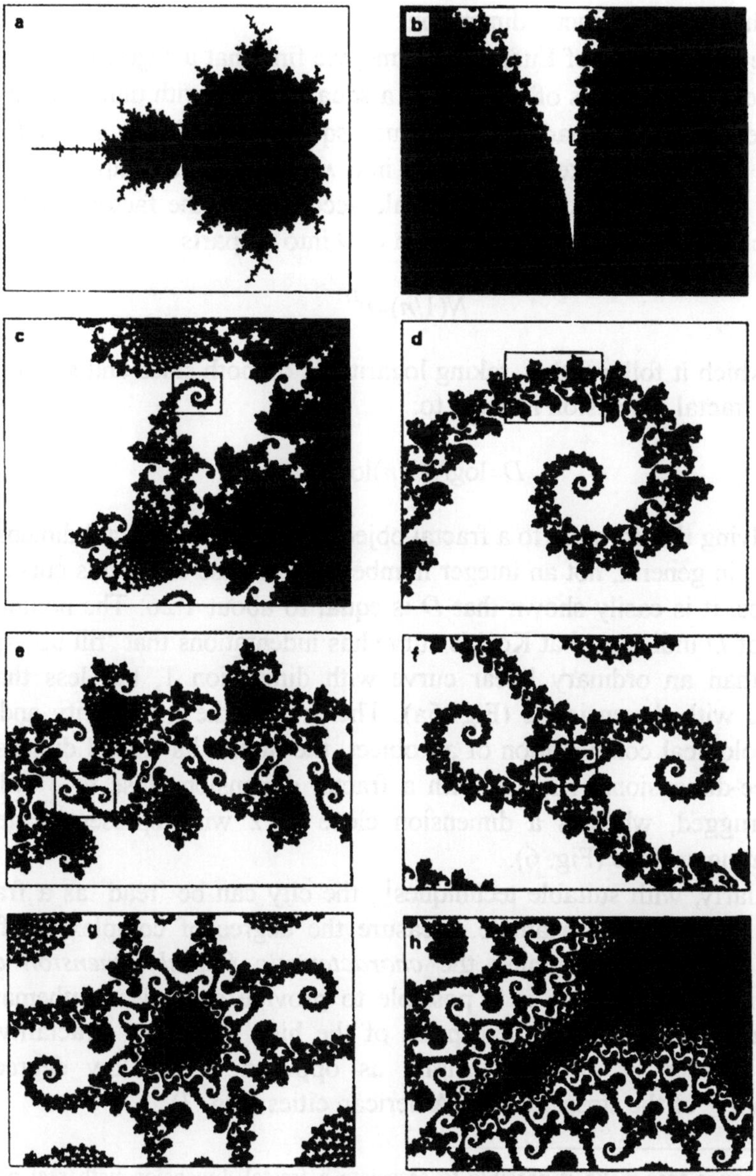

Fig. 4 Non-strictly self-similar fractals: sequential 'zooming in' on a Mandelbrot set

As any detail can always be further exploded with an informational gain, this process is virtually unlimited, so we can conclude that the curve has infinite length. For objects such as fractal curves, the notion of 'length' therefore has no meaning; a meaningful geometrical description must rely upon the notion of fractal dimension.

If we take a series of Euclidean forms, we find that a segment (D=1) can be covered by n rulers of length $1/n$, a square (D=2) with unit side divided into n equal parts is made up of n^2 small squares, and finally a cube (D=3) whose unit side is likewise divided into n parts is made up of n^3 small cubes. In general, a reduction of scale according to the factor $1/n$ breaks an Euclidean object whose dimension is D into n^D parts:

$$N(1/n)=n^D$$

from which it follows that, taking logarithms on both sides and solving for D, the fractal dimension is equal to:

$$D=\log N(1/n)\log^{-1} n$$

Applying this formula to a fractal object, we obtain its fractal dimension. This is, in general, not an integer number. In the case of Koch's curve, for instance, it is easily shown that D is equal to about 1.26. The noninteger value of D indicates that Koch's curve has indentations that 'fill up' space more than an ordinary linear curve with dimension 1, but less than a surface with dimension 2 (Fig. 5a). The greater the irregularity and the morphological complication of an object, the higher its fractal dimension. A three-dimensional surface with a fractal dimension close to 3 will be quite rugged, whereas a dimension close to 2 will represent a gently undulating surface (Fig. 6).

Similarly, with suitable techniques[1], the city can be 'read' as a fractal object. We can, for example, measure the degree of complexity of the street network, determining the *characteristic fractal dimension* of a settlement. In this way, it is possible to provide a precise mathematical translation of the visual perception of the high degree of 'fractality' of urban fabrics of medieval origin, as opposed to the low degree of 'fractality' of the street grids of American cities (Fig. 7).

---

[1] Techniques that make it possible to associate a fractal dimension with real objects whose morphology is less regular than that of self-similar fractals are now available. The most widespread method is the so-called '***box dimension***'. The object is covered with a set of square 'boxes' of side $1/n$. The total number $N(1/n)$ of boxes needed to fully cover the object is determined as a function of box side $1/n$. As the latter is gradually decreased, if N increases at an approximately exponential rate and, in particular, proportionally to n^D, one can conclude that D is the approximate fractal dimension of the object (Fig. 5b), see Batty and Longley (1994). The box dimension technique may be applied to a street network, for instance, to determine the characteristic fractal dimension that best fits its spatial articulation.

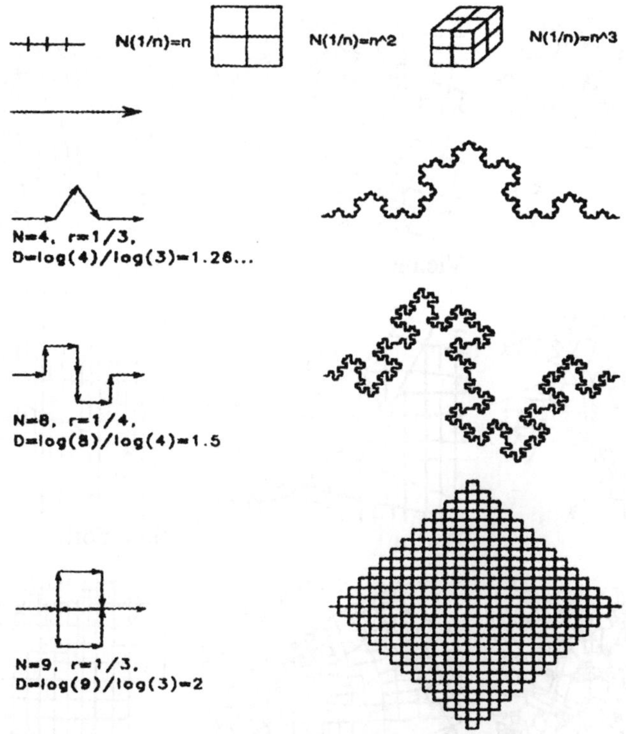

Fig. 5a Euclidean and fractal forms: calculation of fractal dimensions

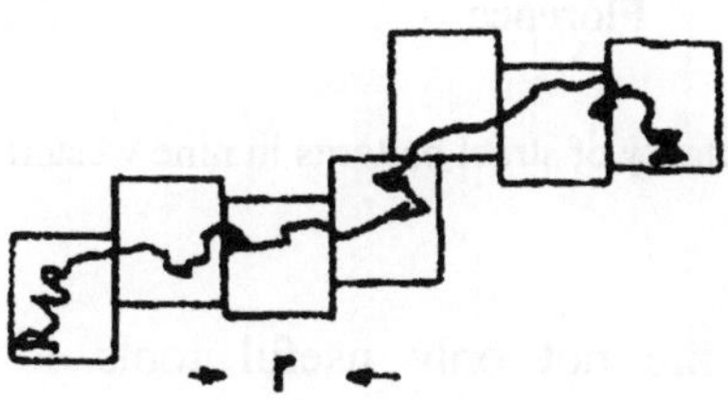

Fig. 5b Calculation of fractal dimension with the 'box dimension' technique

Fig. 6 Fractal simulation of a mountain relief: the ruggedness becomes more accentuated

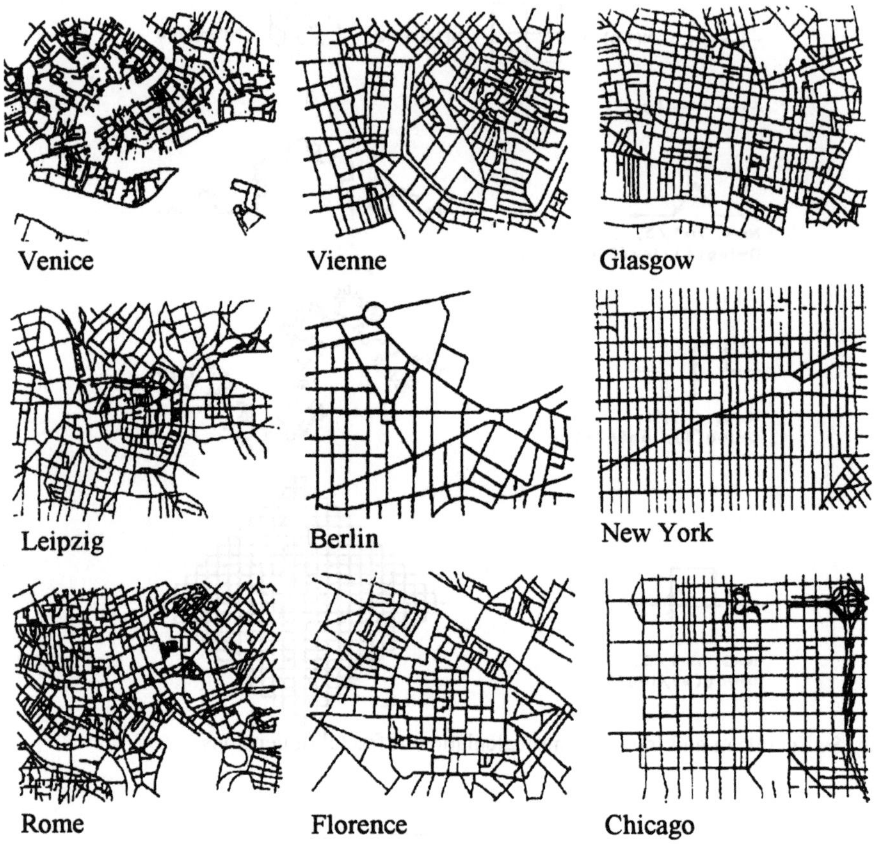

Fig. 7 Comparison of fractality of street patterns in nine western cities

Fractals, however, are not only useful tools for describing highly complex morphologies. Self-similarity allows fractals to represent a wide variety of morphologies through the infinite variety of juxtapositions of a relatively limited repertoire of forms. It is therefore an extremely 'economical' way to generate complexity, as it is based on the repetition of similar morphologies. In mathematical terms, this implies the infinite repetition of similar operations. In spite of their complex nature, fractals are thus the result of fairly simple mathematical laws which produce complexity only insofar as they are infinitely *iterated*[1] . To generate a

[1] 'To iterate' means repeating several times the same instruction or series of instructions according to a predetermined sequence.

fractal object, we need only a simple *algorithm*[1] consisting of a few *instructions*. From the very first iterations, the algorithm already contains all the information needed to produce the whole object. As the iterations unfold, the information that was potentially present in compressed form is 'exploded', revealing its full richness.

This procedure is similar to that used by nature to *encode* its *morphogenetic processes* (it is indeed characteristic of nature to economise on resources). Just as DNA encodes the tantalising complexity of living beings into a simple *chain of aminoacids*, the fractal algorithm concentrates within itself the manifold morphology of the object to which it corresponds. For certain classes of fractals, such as L-Systems, which are able to reproduce with impressive accuracy the morphology of living structures like herbaceous plants, this analogy proves to be extremely close. In such cases one can legitimately speak of a 'genetic code of form' that contains within itself the whole morphogenetic process (Prusinkiewicz and Lindenmayer, 1990).

It is interesting at this point to conjecture upon the existence of a genetic code of form able to synthesise within itself a generic morphogenetic process, such as the birth and development of a small urban settlement. If nature 'makes use' of DNA to reduce evolution to the variation of a few fundamental traits of genetic heritage, the collective memory of an urban community could well boil down to an informational heritage that is transmitted, more or less consciously, from one individual to another. This is subject to a continuous process of trial and error, making possible the development, evolution and diffusion of spatial organisational models by testing their qualities in a variety of different situations. The possibility of encoding urban morphogenesis by means of a fractal algorithm provides us with a theoretical representation of such process and a simulation of the most likely scenarios of future development.

L-Systems appear especially suited to this purpose since they have been conceived not only as a descriptive model, but as a true algorithmic characterisation of the morphogenetic potential of vegetal organisms. The rules that generate form are activated in different ways depending on the type of impulse coming from the environmental context. L-Systems thus provide a method for translating the morphogenetic rules defined by the collective memory of the urban community, into simple 'conditional production rules' of the form 'if x happens under conditions y, then do z'. Morphogenesis hence becomes the outcome of the interaction among the

[1] It is assumed here that an algorithm is a list of instructions that may be interpreted and carried out by a computer.

various production rules that are activated by changing environmental conditions, spelling out the adaptation strategy of the urban organism. A preliminary simulation study of the birth and development of an urban street network 'encoded' by means of a fractal algorithm of the L-Systems type, is presented in Donato and Lucchi Basili (1996).

To reach a deeper understanding of the applicability of fractal geometry to the study of urban processes, it is necessary to come back to our characterisation of the city as a self-organising complex system. The principle of minimal resistance ensures that, as the urban organism grows, urban functions will redistribute through space according to a Central Place-type logic[1]. In other words, those functions and primary activities that require frequent access will tend to have a capillary distribution, whereas more specialised functions will be strategically located to ensure accessibility for all potential users. The higher the hierarchical level of a given function, the wider its catchment area of potential users and hence the more important its physical connections with other parts of the city. The efficiency of hierarchical network structures as organisational models of urban activities will depend upon the nature of the adaptive self-organisation mechanisms.

Arlinghouse (1985) has shown that all Central Place geometries can be obtained as variants of the so-called generalised Koch's method, which is one of the simplest procedures for the generation of fractal objects (Fig. 8). An important and deep connection is thus established between the *hierarchical modularity* of the fractal structure, which originates from basically morphological considerations, and the *hierarchy of polarities* of Central Place geometry, whose rationale lies in the interaction between individual location decisions. It follows that the organisational efficiency of the Central Place hierarchy can be traced back to the organisational efficiency of fractal structures, and thus to self-similarity. The spatial distribution of urban functions induced by the logic of minimal resistance will therefore have a fractal and tendentially self-similar nature.

Urban growth processes, however, do not usually give rise to regular and immediately recognisable geometrical patterns, like those generated by *deterministic fractals,* since the spatial distribution of the activities and functions is influenced by a large number of random and heterogeneous factors. The deterministic self-similar pattern therefore has in this case only a qualitative and abstract value. One must take into account the fact that each group of functions has a Central Place geometry of its own, and

[1] See, for example, the analysis of the growth of late-medieval Florence carried out by Caniggia and Maffei (1993).

that these geometries may all coalesce into a coherent global organisational pattern as in Lösch's model (Tinacci Mossello, 1995). A detailed description of the structural/geometric properties of urban organisation will therefore need to be expressed through with a multifractal system (Schroeder, 1991) of interwoven self-similar hierarchies, each corresponding to a distinctive group of urban functions.

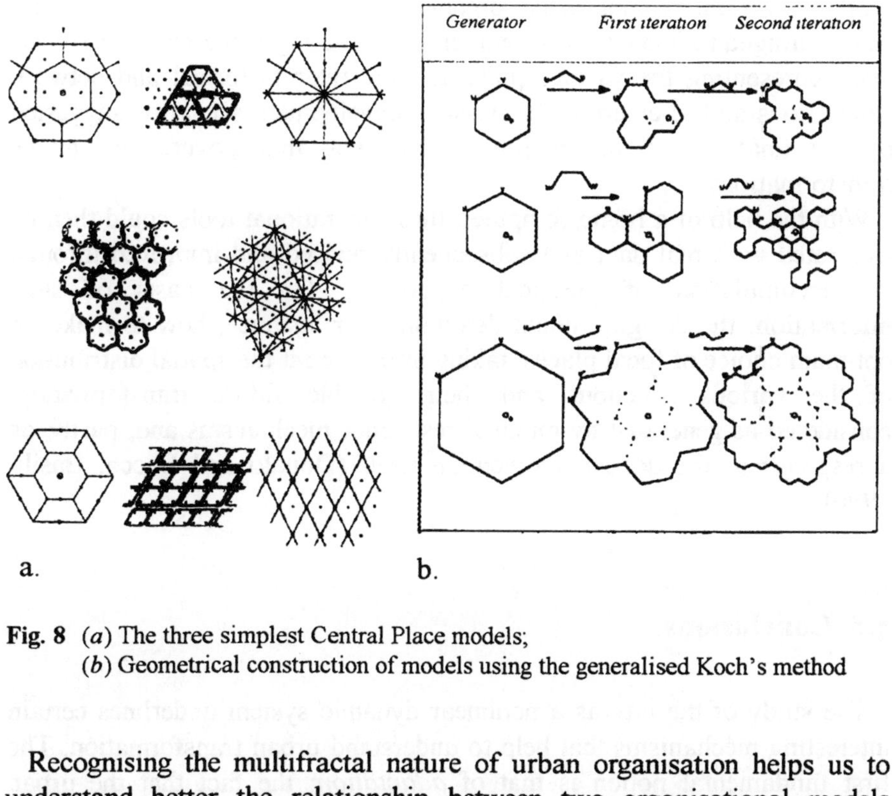

Fig. 8 (*a*) The three simplest Central Place models;
(*b*) Geometrical construction of models using the generalised Koch's method

Recognising the multifractal nature of urban organisation helps us to understand better the relationship between two organisational models which are often presented as antithetic: the hierarchy and the network. If we realise that hierarchical organisation is differently structured for different functions (each of which is characterised by a typical 'fractal architecture') and that the polarities for the various functions need not coincide spatially, we can deduce that the hierarchy and the network can in fact play complementary roles (Tinacci Mossello, in this volume). The distinctive role of the network is in establishing communication between the various central place hierarchies, connecting the poles relating to different ranks of the hierarchy. Translating this into a precise formal analysis is however not an easy task. The formulation of a geometrical network model which lies within a multifractal structure is a research task still to be explored.

The preliminary nature of the research presented here means that it is rather early to advocate ready-to-use methodologies and prescriptions. Possible future developments will require a series of clear-cut operational goals. It would be useful to apply theoretical multifractal models as design support and, more specifically, as tools for evaluating the impact of design proposals on the existing urban organisation. The designer/planner would be encouraged to imagine this intervention as taking place within a *fractal map* representing the various polarities and their historical and network connections and to examine the possible development scenarios suggested by the analysis of the nonlinear dynamics that govern the map's transformations.

With the help of a home computer, these operational tools could then be calibrated with real data and subsequently revised and improved through the accumulation of practical experience. On the basis of such information, the designer could determine, for instance, how to make an optimum choice of focal places, taking into account the spatial distribution of the various functions and their possible future transformation, considered as generated by minimal resistance mechanisms and, partly, as a response to the design intervention itself (Donato and Lucchi Basili, 1996).

8.5 Conclusions

The study of the city as a nonlinear dynamic system underlines certain interesting mechanisms that help to understand urban transformation. The first fundamental notion is that of *adaptation*: the fact that the urban organism modifies itself as a response to stimuli from the environment outside the system. A more detailed examination of the adaptation mechanism brings us to the second basic notion: the *principle of minimal resistance*. According to this principle, non-designed urban change (i.e. the aggregation of a large number of individual choices) tends to privilege those possibilities that entail the greatest immediate benefit and the least resource outlay for individuals. Under suitable and relatively stable environmental conditions, the joint effect of social adaptation and individual demands may give rise to organised collective behaviours. At a larger scale than that of individual choices, these can be recognised as stable models of the spatial organisation of activities, supported by intense communication between system constituents. This is what we refer to as *self-organisation.*

In this section we have underlined the problems that self-organisation entails. On the one hand, it requires the active involvement of all system components into a rich and complex process of interaction and informational exchange. When this occurs, the spatial organisation itself is shaped in such a way that it ensures the perpetuation of these relationships. This is the so-called 'spontaneous relationality' of non-designed urban fabrics. On the other hand, relational richness requires the existence of a widely shared symbolic code through which individuals assign meaning to the various elements of urban organisation. People thus arrive at a collective image of the city that orientates their choices, making them mutually compatible.

Such processes are however challenged by the sudden 'jump of scale' characteristic of post-industrial societies. Firstly, the sheer size of the contemporary city tends to preclude the possibility of a harmonious communication between constituent parts. Congestion of the communication channels causes the fragmentation of the city into 'urban islands'. It thus becomes more and more difficult to establish relationships within the overall organisational scheme. Such a breakdown of communications undermines the unity of the urban culture, leading to the birth of many different local identities, often in conflict with each other.

In order to regain coherence in the urban fabric, it is necessary that individual and possibly conflicting local viewpoints, even though sometimes shortsighted and parochial, should be valued as expressions of adaptational talent, and supported by an external rational supervision that superimposes a global orientation. This *cooperation between spontaneous adaptation and rationality of design* could generate a new organisational model which we have called *designed hetero-organisation*. In order to make this model operational, we must arrive at an understanding of the deep organisational level of urban dynamics and the linguistic code of spatial organisation that is ingrained in a specific urban culture.

We maintain here that a tool which can enable us to reach this level of knowledge is fractal geometry, since it gives us a compact and tractable description of complex spatial patterns, increasing not only our level of information, but also our understanding of systemic interactions, which is the basic premise for the formulation of any responsible urban intervention project. Building methodologies and design tools that make an active and creative use of the chain reactions and modifications that the designer's intervention causes on urban organisation is an ambitious but stimulating goal for future research.

Acknowledgments. I wish to thank Prof. C.S. Bertuglia for his precious and detailed observations on the previous version of this paper and Prof. Maria Tinacci Mossello for some useful conversations. The responsibility for what is written remains however that of the author.

References

Arecchi F.T., Arecchi I. (1990) *I simboli e la realtà. Temi e metodi della scienza*, Jaca Book, Milan.

Arlinghaus S.L. (1985) Fractals Take a Central Place, *Geografiska Annaler, 67B*, 83-88.

Avarello P., Cuzzer A. (1982) *Urbanistica e mercato edilizio*, Sansoni, Florence.

Bak P., Chen K. (1991) La criticità auto-organizzata, *Le Scienze,* 271, 22-30.

Batty M., Longley P. (1994) *Fractal Cities*, Academic Press, London.

Briggs J. (1992) *Fractals. The Patterns of Chaos*, Simon and Schuster, New York.

Campos Venuti G. (1994) *La terza generazione dell'urbanistica*, Angeli, Milan.

Caniggia G., Maffei G.L. (1993) *Lettura dell'edilizia di base*, Marsilio, Venice.

Carrà S. (1989) *La formazione delle strutture*, Bollati Boringhieri, Turin.

Ceruti M. (1992) *Il vincolo e la possibilità*, Feltrinelli, Milan.

Cini M. (1994) *Un paradiso perduto. Dall'universo delle leggi naturali al mondo dei processi evolutivi*, Feltrinelli, Milan.

Cortesi A. (1979) Cultura urbana, progetto, disegno urbano, in Paoli P., Cortesi A. (eds.) *Disegno urbano. Una proposta per la città come sistema*, Pitagora, Bologna, 95-124.

Donato F., Lucchi Basili L. (1996) *L'ordine nascosto dell'organizzazione urbana. Un'applicazione della geometria frattale e della teoria dei sistemi auto-organizzati alla dimensione spaziale degli insediamenti*, Angeli, Milan.

Guidoni E. (1992) *L'arte di progettare le città. Italia e mediterraneo dal medioevo al settecento*, Kappa, Rome.

Haken H. (1981) *Erfolgsgeheimnisse der Natur*, Deutsche-Verlags, Anstalt.

Hillier B. (1988) La morfologia urbana e le leggi dell'oggetto, in Zanella P. (ed.) *Morfologia dello spazio urbano. Questioni di analisi e di progetto*, Angeli, Milan, 29-64.

Liverani M. (1988) *L'origine della città. Le prime comunità urbane del vicino oriente*, Editori Riuniti, Rome.

Lynch K. (1981) *A Theory of Good City Form*, MIT Press, Cambridge, Massachusetts.

Mandelbrot B. (1975) *Les objets fractals*, Flammarion, Paris.

Mumford L. (1961) *The City in History*, Harcourt, Brace and Jovanovic, New York.

Oldroyd D. (1986) *The Arch of Knowledge. An Introductory Study of the History of the Philosophy and Methodology of Science*, Methuen, New York.

Paba G. (1990) Limiti e confini della città: un'introduzione, in Paba G. (ed.) *La città e il limite*, La Casa Usher, Florence, 8-21.

Peitgen H.O., Richter P.H. (1986) *The Beauty of Fractals. Images of Complex Dynamical Systems*, Springer-Verlag, Berlin.

Prusinkiewicz P., Lindenmayer A. (1990) *The Algorithmic Beauty of Plants*, Springer-Verlag, Berlin.

Ruelle D. (1991) Determinismo e predicibilità, in Casati G. (ed.) *Il caos. Le leggi del disordine*, Le Scienze, Milan, 13-21.

Schroeder M. (1991) *Fractals, Chaos, Power Laws. Minutes From an Infinite Paradise*, Freeman, New York.

Schumacher T. (1978) Buildings and Streets: Notes On Configuration and Use, in Anderson S. (ed.) *On Streets*, MIT Press, Cambridge, Massachusetts.

Secchi B. (1984) Piccoli centri, *Casabella, 504*, 14-15.

Serra R., Zanarini G. (1986) Auto-organizzazione in universi artificiali, in Serra R., Zanarini G. (eds.) *Tra ordine e caos. Auto-organizzazione e imprevedibilità nei sistemi complessi*, Clueb, Bologna, 115-151.

Tinacci Mossello M. (1990) *Geografia economica*, Il Mulino, Bologna.

Vallega A. (1995) *La regione, sistema territoriale sostenibile. Compendio di geografia regionale sistematica*, Mursia, Milan.

Vulpiani A. (1994) *Determinismo e caos*, Nuova Italia Scientifica, Rome.

Zanarini G. (1994) *Finestre sulla complessità. Ordine e caos nella natura*, Editoriale Scienza, Trieste.

9. The Complexity Paradigm in Architecture

Francesca Bertuglia

9.1 Complexity and the Architectural Product

The complexity paradigm has successfully penetrated many fields of study. Among those to which there have not yet been significant applications is that of architecture. For this reason, the present section, which examines the problem of how to interpret the architectural product in terms of the complexity paradigm, is simply intended to be an introductory investigation, representing a first approach to the problem.

Studies of complexity (see, for example, Waldrop, 1992) have always underlined the plurality of meanings contained within the concept of complexity. In the architectural context, the meanings which would seem particularly relevant are the following:

a. *self-organisation*, understood as a 'bottom up' organisation process involving a large number of people who in some way demonstrate respect for a set of rules, producing nevertheless a non-predetermined outcome characterised by a high degree of variety;
b. *surprise*, caused by the unexpectedness of the results;
c. a *varied and not easily exhausted set of meanings and levels of interpretation*, which overlay each other and intermingle.

With reference to self-organisation, we should note that, before the industrial revolution, the city (or aggregations of buildings in general) behaved as self-organised systems in which individual actors, pushed by specific needs, modified existing constructions or built new ones (see the chapter by Lucchi Basili, in this volume). This was done with or without the assistance of 'experts', respecting a set of rules written not so much in manuals as in the collective memory of the community to which they belonged. From this process, in which respect for the rules was interwoven with the initiative of many actors, grew the towns and cities of the pre-industrial era, producing settlements in which fantasy did not lead to unruliness and order did not mean repetition. Some splendid examples have remained to the present day.

This kind of self-organisation process was possible because the slow speed of change meant that the 'rules' were based on experience. With the industrial revolution came a rapid acceleration in the rate of change and, as a result, the self-organisation mechanism lost the basis which made it operable. For the city as a whole, or the construction of any group of buildings, it was necessary to resort to some form of external planning or design. It is in this context that, since the industrial revolution, the activity of the planner or designer should be placed.

We can deduce that the role of the city planner, whether designer of a group of buildings or a single building, is not to suffocate but 'cooperate' with the spontaneous processes which occur in the city. This must be done in such a way as to leave room for these processes to unfold, but at the same time making it possible to guide them and compensate for the lack of communication resulting when the speed with which the systems grow or change becomes too great. In other words, the planner and designer have to attempt to remove any hindrance to the self-organisation process, which we could refer to as a 'guided self-organisation process' (see the chapter by Butera, in this volume).

In relation to points b. and c. above, we can observe that a work of art surprises by its unpredictability and ability to give rise to many different interpretations. The task of the planner and designer is to add complexity to the whole city, not only the part directly affected by their project, and to increase the possibility of choice open to the users. In other words (see the chapter by Lombardo), complexity becomes a quality which the design or plan should produce or increase.

At this point, we focus our attention on the activity of the designer. We do this by examining certain architectural projects and evaluating the extent to which they:

a. stimulate and guide self-organisation processes;
b. introduce surprise, increasing the number of meanings and interpretations, i.e. the complexity of the object designed (and the possibility of choice of the users).

In order to do this, in Section 2 we shall attempt to define what we mean by complexity in an architectural 'product', then in Section 3 use these criteria to analyse a number of architectural projects. Finally, in Section 4 we draw some brief conclusions.

9.2 A 'Complex' Reading of the Architectonic Product

The identification of the elements which permit us to add complexity to an object (or design) is a difficult task. In fact, from what has already been said, it is evident that it is not the 'simple' or easily identifiable elements, such as the size of the project or the type of urban function involved, which determine the degree of complexity. As we shall see from the analysis of specific architectural projects, the complexity of an object may be due to characteristics or aspects of the project itself, sought by the designer, or it may be due to aspects of the context of which the project is part and with which it is in interaction (the complexity is often in fact a result of this interaction).

Most of the examples chosen concern urban renewal schemes. This should not be surprising, since it is in projects situated in a structured urban fabric that we find: (a) developments which have accumulated and been superimposed over time, (b) a combination of different functions, (c) a multiplicity of social actors, and (d) networks of relations and flows, which already possess the capacity for self-organisation, surprise and a variety of meanings and interpretations. In other words, these are situations in which there already exists a degree of complexity which a well designed project can augment.

Even though at first glance an architectural project might appear to have a simplifying effect, it may well in reality add complexity to the context, especially when it respects the history of the community or users involved and their interactions. This statement perhaps requires some explanation. While we recognise that classical science was inspired by simplicity, and hence the principle of reduction, it is nevertheless also true that, despite the discovery of complexity, modern scientific analysis still tends to simplify since this is often unavoidable. But in relation to management (and architectural design can be understood as a form of management in its widest sense), even when the project 'takes away' something and gives the impression of simplifying, it often in fact adds complexity (for a general and abstract discussion, see Casti, 1986). For example, when the project respects what already exists, though it may remove superfluous elements accumulated over time or eliminate certain functions, this is done in order to clarify the historical and social relations considered to be particularly significant, i.e. to emphasise the underlying complexity. In our analysis, we have therefore focussed on architectural projects which clarify complexity and/or add some new form of complexity.

The review has been divided into four categories: (a) new towns, (b) city centre renewal, (c) renewal in the periphery, and (d) projects involving elements of cityscape, which - we should add - have not only the function of making complexity more comprehensible, but also presenting in a more orderly fashion elements which otherwise would have very little homogeneity. We therefore consider schemes at completely different scales, from renewal projects involving vast areas (including the extreme case of 'new towns') where there is emphasis on the overall design, to the insertion in existing urban areas of single objects or buildings (elements which complete the city) with strong semantic value in a highly structured context.

The redesign of an urban area (cases b and c), whether monofunctional or multifunction, raises the problem of the chain of effects which may be provoked, not only in the area itself, but also in the rest of the town or city. The fact that we are operating in the context of a self-organising system means that the identification of these effects must be considered as a learning process for those involved, with the inevitable errors and correcting action. This is particularly crucial in the case of multifunctional areas because of the presence of many different kinds of operator and user, each with different objectives, different types of interaction, and demand for different types of 'containers' or 'channels' for the activities and flows, requiring different architectonic solutions. Most of the projects described are in fact located in multifunctional urban areas. As we have said, even totally new urban projects can be included (case a), but these are more rare and also less interesting from the point of view of complexity. In these cases it is probable that, at least in the initial stages, self-organisation processes are missing. If we reflect on the criticisms of new towns, we discover that almost all can be traced back to the lack of complexity, that quality possessed by older cities as the result of a long and tormented historical process, i.e. self-organisation.

9.3 Schemes and Projects Analysed

9.3.1 New Towns

The example we adopt is not a new town in the traditional sense but, inspired by a highly perceptive observation by Sudjic, a rather specific case: the expo. In the modern era, Sudjic (1993) pointed out that the expo is for the city what fast food is for the restaurant. In fact, within it we find entertainment and work, shows and social activities, as well as many other functions which are also an integral part of city life. Although artificially created and concentrated (hence the validity of the parallel of the fast food restaurant), it is worth noting that exhibitions or large fairs have sometimes been catalysts of urban development. In other words, they represent an artificial and condensed form of city, but can give rise to urban development which is neither artificial nor condensed.

Exhibitions, fairs or fun parks concentrate many different types of urban space: piazzas, walks, theatres and so on, within a single huge area which, with time, becomes integrated into the city itself and, occasionally, acts as a generator of urban growth. From the first exhibitions which were seen as sporadic events, we have come to the idea that the site of a fun park or a large fair may be planned in such a way as to have an nonephemeral impact on the urban structure which hosts it. A significant example was the World Exhibition of Barcelona in 1888, with which the Catalan capital finally succeeded in expanding beyond the Medieval walls, rapidly to become a modern city.

Almost a century later, we find the appearance of a mixed form combining the traditional expo and the world of cartoons. Walt Disney supplied his product to many large exhibitors (Ford and Pepsi, for example), but he also nursed the dream of reconstructing the city as it should have been. Disney in fact was closer to the architectural culture of the period than it may seem at first sight. In his hands, the fun park was transformed from a spontaneous popular phenomenon into a cultural event involving famous architects and artists. To give one example, Disney World, which opened in 1971 near the city of Orlando, now comes near to being a real city with 23,000 employees and 20 million visitors a year, almost as many as attracted by London.

From these observations it should be clear why we have chosen as an example Frank O. Gehry's *Entertainment Center*, a project located within Eurodisneyland (Cohen, 1993). *Eurodisneyland Resort* extends over an area of 1943 hectares, a third of the area of Paris, and is sited on the edge of the new town of Marne La Vallée. The whole scheme (which in the future will cover an area far larger than that made available for development in 1919 after the demolition of Thiers' military defence wall around Paris), is an example of how it is possible, taking a fun park as a pretext, to create an entire city from nothing, even if for the moment it is only a tourist city. The site chosen for the Eurodisneyland resort, with development planned up to the year 2017, can be reached in a few hours from most of Western Europe using an advanced integrated transport system (road, rail and air) entirely developed by the French State (Bédarida, 1992).

The internal organisation of this vast area possesses many elements already familiar in similar schemes in the United States and Japan. The attraction park and the reception area follow a radial plan which includes both a railway line and a circular boulevard; the spatial distribution of the activities is based on sectoral zoning and a division into subareas without any particular order (Fig. 1). The constructions in the fun park, the hotels and the retail centre have intentionally been designed without an overall harmony. In fact, each theme is developed according to its own particular style, but to avoid any clash caused by the close proximity of such different elements, the planning of the external spaces and the pedestrian links has been undertaken by landscape architects who have operated according to the 'cinematographic' idea of a series of sequences with discontinuity between themes (Miller, 1990).

The Entertainment Center designed by Frank O. Gehry is a kind of shopping centre which has the role of linking various elements: the fun park, another park - yet to be built - dedicated to the cinema, and the hotel area. This link serves as a walkway (Fig. 2) connecting the railway station with an artificial lake overlooked by the hotels, as well as a shopping centre consisting of a large covered area along an internal backbone. The complex covers an area of 18,000 square meters and includes shops, cafeterias, a discothèque overlooking the lake, a series of restaurants, an arena theatre and an exhibition space (Brandolini, 1992). The architectural style intentionally 'breaks the rules', mixing numerous materials: concrete, zinc, painted steel, wood on a ground surface of asphalt and coloured cement. This part is used mainly in the late afternoon and evening, not only by the hotel guests, but also the inhabitants of Marne La Vallée, offering the former a view of urban life and creating an exchange between the real city and the simulated city.

Fig. 1 View of Eurodisneyland from the hotel district (Bédarida, 1992, p. 29)

Fig. 2 Views of the central road in the Entertainment Center of Eurodisneyland (Cohen, 1993, p. 87)

As we have already stated, this is an extreme case: a shopping complex sited within a structure which is - at least in its initial stage - basically monofunctional and part of an overall project of which practically all details are predefined. Interestingly, it is located close to another urban structure, the new town of Marne La Vallée, which also has a limited number of functions. The Entertainment Center represents a situation in which it is difficult for self-organising processes to occur. The consequent lack of complexity is recognised in the design and an attempt has been made to eliminate, or at least to disguise, some of the visible effects by using a built form: (a) without overall harmony or any particular sense of order, (b) a transgressive form of architecture and (c) a mixture of many different materials. In other words, an attempt is made to generate surprise and allow different interpretations of the environment. As already mentioned, a further idea is to create some form of interaction between the local inhabitants and those staying in the Entertainment Center.

Although possibly eliminating the impression of lack of complexity, these strategies do not create real complexity and, above all, do not increase the possibility of choice of users. At most, the ruse may succeed in hiding the lack of complexity for the long period until a real process of self-organisation begins, i.e. when interaction occurs between many different types of person (residents, tourists, tourist operators etc.). We could add that the observations relating to this example of a 'simulated' city are also valid, though in a lesser degree, to many new towns.

9.3.2 City Centre Renewal

Manuel De Solà-Morales, in an interview in 1988, stated that in the contemporary city the monument and the piazza no longer play the role of central places; the real poles of attraction are now places like shopping centres, stadia and railway stations (Menditto and Rossi, 1995). In our investigation of city centre renewal, we examine some very large scale projects which have permitted: (a) the creation of new public spaces, not necessarily in the open air, but with buildings dedicated to communications, services and culture, (b) the refilling of gaps in the urban fabric and (c) the reconstruction of parts of the city.

De Solà-Morales himself, together with Moneo, was responsible for the design of a large building on the *Avenida Diagonal* in Barcelona, conceived as a mixed development including offices, a residence and an elaborate shopping centre (Parcérida, 1994). It is the presence of different urban functions which distinguishes this building, designed in 1986 but completed only recently. The huge complex occupies an entire block and has been called the 'city building': an expression which re-echoes the basic aim of the project, to reproduce a piece of city.

In the old city, it is rare for a single building to occupy a whole block; in general there are a number of buildings with different functions, constructed independently. The idea of rebuilding a block as a single construction, even though not in a homogeneous style, may seem artificial, but was a significant choice. The parts of the building visible from Avenida Diagonal are of different height and have different setbacks from the street. Each part has its own character, emphasised by the different composition of the windowed facade (Fig. 3).

Referring back to the parameters by which we can recognise complexity of a project, the distinguishing feature of this example is certainly the concept of creating many different levels of interpretation. To explain in more detail, when seen in longitudinal section, the building looks like an enormous liner moored in Avenida Diagonal, anchored to the ramps, linked to the pavement by vertical access to the upper floors and galleries crossing them and held in tension by the diagonals which link them to the nearby streets; the car parks are located in the ship's 'hold'. Inside, the space seems larger than one would imagine from the outside. The shopping gallery is a combination of different elements (a piazza, a grand patio and a 19th century gallery) imitating the canons adopted in the large American-type shopping centres: with extremes as poles of attraction, the idea of the microcosm, the impression of the juxtaposition of large and small, the attention paid to detail and the fragmentation of space to the point where the layout of the shops seems temporary, as if the individual premises had been acquired at different times and on the basis of separate initiatives.

Fig. 3 Façade of De Solà-Morales' building looking onto Avenida Diagonal in Barcelona (from Monastiroli, 1994, p. 12)

The next two examples are very large schemes which have in common the fact that they are located near major railway stations. Both are representative of a phenomenon typical of many big cities, the transformation of the station or disused railway land by means of a large-scale construction project.

The *Broadgate* complex, in the financial centre of London, includes a number of multifunctional blocks arranged around three central spaces. The project, closely connected with the renewal of Liverpool Street station, uses disused railway land to transform this part of the city into an important node of exchange between work, transport, residential and retail development. Broadgate provides 300,000 square meters of new office space, with vast multifunctional zones in the basement suitable for computer cabling, and offers a level of flexibility that allows the introduction of substantial modifications in only 24 hours (Hardingham, 1994). The offices have a grandiose appearance, with vast atriums, open-plan work areas furnished to make them more attractive for employees working under conditions of stress, restaurants, cafés, gymnasia and shops. The operation, made possible by joint public and private finance, includes 13 buildings and three piazzas built in 14 successive phases, each involving different architects. The first four, completed over a period of six years (1984-1991), were designed and built by Arup Associates following a plan based on a detailed study of the flows of commuters from the nearby Liverpool Street station. The speed of construction was made possible by the use of prefabricated elements and units which were inserted in a steel skeleton designed to house different types of 'container', from an auditorium to retail stands. More than half of the building is supported by a base constructed above the station platforms.

Liverpool Street station is a splendid example of the kind of late Victorian railway architecture still referred to today as 'cathedrals to steam'. It is currently also an example of the significant transformation occurring from the traditional station to a new type of exchange point which brings together services for travellers and other services for the local inhabitants, uniting image and technology, and underlining the growing importance of mobility within the city. We focus here not so much on the restoration project (which can perhaps be criticised for its over-faithful reproduction of certain elements), as on the overall transformation made possible by the demolition of Broad Street station in the 1970s. The main objectives of the scheme (1985-1991), undertaken by the Architecture and Design Group, were to build a direct link with the Underground and rationalise access to the platforms, as well as to modernise the station. This is an interesting and highly successful example of functional reorganisation originating from the interaction and integration of a large number of private and public bodies. Liverpool Street station has since become one of the most lively in the London area.

Another significant example of renewal in the city centre is the *Charing Cross* station project by Terry Farrell. This station was already a characteristic landmark on the London skyline in 1864, located alongside the Thames between New Scotland Yard to the west and Somerset House to the east. After the collapse of the roof during repair works, the station now has not only a new roof and totally new appearance, but nine floors of offices built over the platforms (Davies, 1991). Farrell resolved the problem of combining these functions with a design in which the different parts - railway station and office block - are easily distinguished (Fig. 4). The building has the right dimension, the right appearance and fits well into the surroundings, completing the skyline.

The scheme has led to the improvement of the surrounding area: Villiers Street has been redesigned, the spaces under the Thames bridge near the Underground station have been revitalised (Gough, 1993). The skill of the integration emerges at different scales, from the distant view to small details. The grandiose outer facade of the building is the most visible part of a highly complex design intended to improve a part of the city which was congested and in decay. It gives the appearance of three separate buildings, each with a different treatment corresponding to its function (Davies, 1991). In *Embankment Place* it is easy to recognise architectural elements borrowed from other nearby buildings, as well as a clear influence of Ledoux in the Villiers Street facade and some elements copied from one of Moscow's underground stations. Inside the building, vertical and horizontal circulation, public and private spaces, and vehicle and pedestrian traffic are interconnected. In addition, under the platforms there is a car park which at the weekend is transformed into a market.

In the last example we are not concerned so much with the project itself as the design method, proposed by Mangurian and Ray (1993), which is based on the principle of nonuniformity and stratification, and has led to the plan for the *Grand Arts and Entertainment Center* of St. Louis. The aim of the project is to respect and emphasise urban complexity. It originated from a creative process conducted on seven levels which will eventually all be superimposed and make it possible to increase the richness and variety of this part of the city dedicated to the theatre and entertainment. Each level leaves a strong imprint on at least one of the architectural aspects. The 'Acropolis' level, for example, develops the topography of the district; the 'Cluster' level is concerned with the definition and characterisation of the focal points; the 'Overlaid Patterns' level involves the design of the small scale elements and the materials making up the environmental structure; the 'Patchwork Quilt' level introduces a series of modifications to the landscape and the design of the ground surfaces to make it easier to cross the district on foot; the 'Grand: On-Stage Off-Stage' level divides the district into different urban scenes dedicated to theatres and large restaurants, including a retro-scene with shops, small cafés and workshops; the 'Street and Green' level studies a system of street lighting as a 'natural' system, and the green spaces as an artificial expression of nature in an urban environment. Finally, the 'Discrete Elements' level pays attention to the nonhomogeneity of the buildings - one of the distinctive characteristics of the Grand Center - treating this as a feature to preserve and emphasise.

These first four examples indicate how city centre renewal projects, even those on a large scale, can manage not to hinder processes of *self-organisation*, but may even stimulate and guide them by increasing the possibility of choice of users and, in a general sense, adding complexity. This always (or, at least, often) happens when a large number of users is involved, since with many different types of people, it is easier for interaction to occur. This mechanism spreads to the new buildings and their users, with a reciprocal effect. It can occur even when, as in the case of Liverpool Street station, the quality of the architecture is debatable and hence not stimulating (i.e. not helping to create interaction). Lack of architectural quality, even though an obstacle, in general is not insurmountable, due to the strength of the self-organisation mechanism. The mechanism is reinforced when the building design is specifically oriented towards its creation, as in Broadgate where the aim of the project was to remove blocks to flows of users, and the Embankment Place project, which triggered visible improvement in the surrounding areas. One objection could be that the success of these projects, Broadgate, Liverpool Street and Charing Cross, can be explained by the fact that all three are associated with stations. These, like any structure connected with mobility, necessarily generate interaction and are natural stimulators of self-organisation processes. However, since the presence of structures connected with mobility is common in city centres, it would seem that these examples remain valid. It is worth making a final observation about the last case. As explained above, the St. Louis plan followed a methodology purposely conceived to emphasise the variety of the district and, through this variety, increase the levels of interpretation, the users' choices and, therefore, the complexity. In this sense, it was innovative, since the methodology was specifically oriented towards creating complexity as a quality of the project.

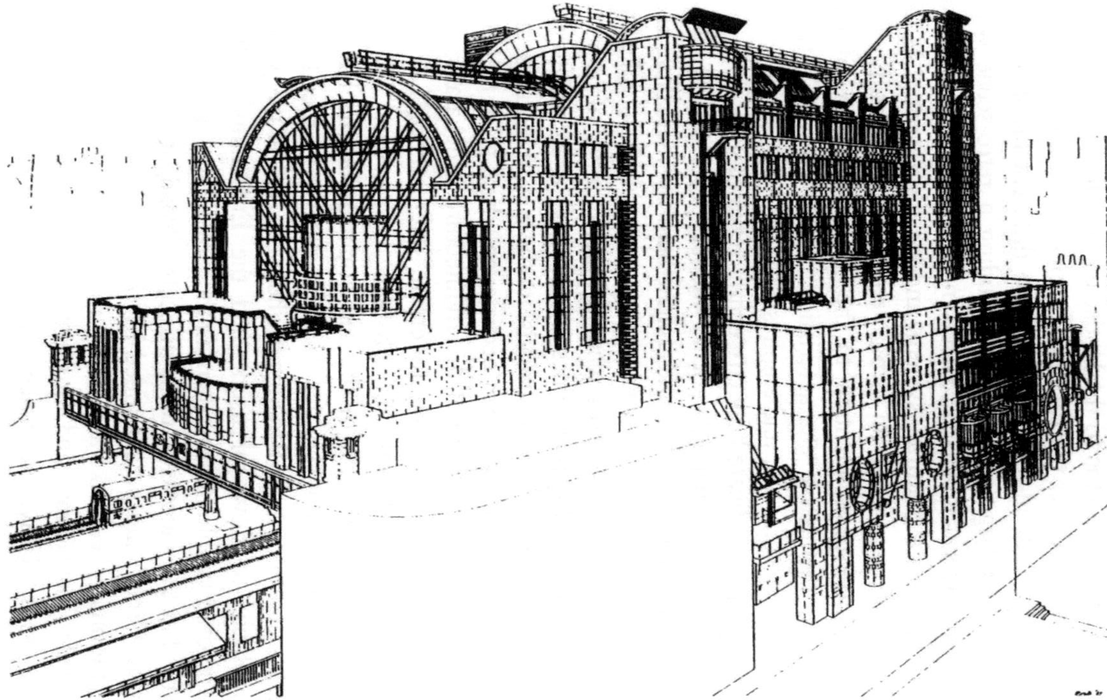

Fig. 4 Perspective of Charing Cross Station building and office block in London (Gough, 1993, p. 55)

9.3.3 Renewal in the Periphery

In those metropolises which have experienced rapid growth, it frequently happens that the greatest potential for development and change tends to be located towards the peripheral parts of the urban area rather than the centre. In Paris, for example, the warehouses at Bercy, the site for the future French Library, and the areas left empty by the closure of the Citroën factories in the west have provided opportunities for large-scale urban projects (Bédarida, 1995). After many years of discussion on how to integrate these previously isolated areas into the metropolitan system, the idea of developing parks and gardens emerged as a central component. The result was the creation of two new green areas: the Citroën-Cervennes park and the Bercy park.

The idea for the project goes back to the early 1970s, at the time when the Plan for the south-east Seine was being drawn up. This scheme foresaw the building of a large park, public buildings and a new residential area with services and retail activities as well as other new infrastructure. One of the distinctive characteristics of Bercy, which had been isolated from the rest of the city for more than a century, was the number of roughly paved tracks crossing the area and the abundant vegetation, including over 500 trees. The authorities decided to emphasise this unique feature by creating a new urban park covering 14 hectares. The group which won the project competition concentrated on defining an overall strategy for the multitude of scattered fragments. A general plan, coordinated by Jean-Pierre Buffi, established the volume, the design of the facade, the choice of materials, etc. for each subproject, including the way in which each would be divided up and attributed to the various participants.

To the north of this park and east of the residential area is the new Bercy Arts Centre. This structure, designed by Frank O. Gehry, serves as a meeting place for artistic and intellectual activities in the capital and responds to a variety of needs. Here we consider specifically the building housing the *American Center* which is interesting for two reasons: firstly, it is part of a vast and varied urban project, and secondly, it is particularly complex, both in terms of the number of functions catered for and its architectural image. The public areas, consisting of exhibition spaces, a library, a restaurant and a 400-seat theatre, are located on the ground floor (Fig. 5). More specialised functions include a theatre gallery, a language school, administrative offices, a library and an art gallery overlooking a 'terrace sculpture'; other spaces dedicated to the visual arts are located on the upper floors.

The *Edgemar Development* by the same architect, designed in 1988, has been described as a 'complex in city form' (a kind of development not uncommon in Los Angeles, which has many 'mini-cities'). This new retail and cultural centre situated on Main Street, parallel to the ocean and linking Santa Monica and Venice, opens towards the city becoming part of it (Fig. 6). The project involves the construction of new factories and the re-use of an existing one (Zardini, 1992). The *Edgemar Development,* with shops, offices, car park, art museum and restaurant, constitutes an extension to the city. The new development does not substitute the original with something completely different, nor is it cut off from the surrounding urban area, but proposes re-use, transformation and integration in an urban context of strong character. The buildings conserve the scale of the existing facade of Main Street with the introduction of a courtyard extending the public space within the block. The facade of an old shop has been skillfully preserved using a combination of restoration, reconstruction and reinvention. The designer has succeeded in reinterpreting Los Angeles using some existing elements: the two-storey buildings, an old factory, the car park and Main Street.

Renewal schemes in the periphery raise problems similar to those faced in central areas, with the added difficulty that in such areas self-organisation mechanisms are often less vigorous. Nevertheless, if the new projects do not create barriers, cutting themselves off from the surrounding context, they can be rapidly integrated and even serve as a stimulus or provide a new orientation to self-organisation processes. The project must favour this by not creating obstacles which, for the reasons given above, especially in peripheral areas could delay the onset of self-organisation processes, and entail considerable social costs.

The first example is typical of a recent phenomenon, the creation of gaps in the urban fabric caused by the disuse or closure of factories, military areas, prisons etc. These often leave large empty sites on the edge of the central area, and represent an opportunity for reconstruction of part of the city, reintegrating it with the whole and encouraging lively new activities which can trigger self-organisation processes. As such projects are usually on a smaller scale and also more scattered than those in central areas, the impact of the architecture itself can easily be diluted, or even lost completely, unless the design is of particularly high quality.

Fig. 5 The American Center, view towards the Seine and the Grande Bibliothèque de France (from Bédarida, 1995, p. 85)

Fig. 6 View of Main Street in the Edgemar Development, between Santa Monica and Venice (from Zardini, 1992, p. 100)

9.3.4 City Infill Projects

The schemes described in this last section are on a much smaller scale than the preceding ones. Their interest and significance is therefore strongly conditioned by the context in which they are sited. Due to their scale they can easily be integrated into the self-organisation processes operating around them, but must be able to make a sufficiently strong impact to avoid being overwhelmed by the surroundings.

The *Fishdance Restaurant*, designed by Frank O. Gehry, which is located in a garden on the seashore of Kobe and surrounded by a jumble of features with no particular character, has become the major architectural element of the area (Gehry, 1993). It consists of three distinct volumes (Fig. 7): a spiral tower, a fish-shaped sculpture, and a rectangular base with a sloping roof of blue metal. These house a bar, a dining room, kitchens, a kushi-katsu counter and grill, which together create an informal and intentionally varied series of eating places.

A similar concept has been applied in Yokohama, where the architect Ito has used an existing concrete tower (a water tank and ventilation channel for the shopping centre below) to make a strong architectural impact. With the addition of a new elliptical outer covering, it has been transformed into the *Wind Tower*, which shines in the daytime and is illuminated at night (Ito, 1988, 1993). This redesign of an existing structure, purely for the sake of its visual impact, shows particular sensitivity to the urban landscape (Fig. 8). It is a clear example of an approach to urban maintenance attentive to the historical character of the city, while demonstrating the capacity to add innovative elements which complement and blend with the existing architectural structures.

In the heart of the city of Toronto, the area between Young Street, Wellington Street East, Bay Street and Front West Street, has recently been transformed into a multi-use social and cultural complex: BCE Place (Calatrava, 1993). The design of the gallery which links the various sections is one of many examples of schemes involving public space. The architect, Santiago Calatrava, has connected and covered the space between Heritage Square, the atrium of the Canada Trust Tower, the old Clarkson Gordon building and Garden Court on Bay Street with a 27 metre high 'backbone' (Fig. 9). From the BCE Place Galleria there is access to an underground system of pedestrian passages as well as to the lower floors of the complex. The gallery serves as a corridor of light: the supporting pillars are like trunks of trees whose interlacing branches form the roof. This is a particularly good example of how an indeterminate space can be made into a distinctive feature.

Fig. 7 View of the Fishdance Restaurant in the city of Kobe (Gehry, 1993, p. 97)

Fig. 8 The Wind Tower in Yokohama, before redesign and after, with illumination at dawn (Ito, 1988, p. 40)

The last example, the *Galleria Paseo del Caminante*, is a small two-storey shopping centre in the Argentinian city of Cordoba. The centre, designed by Miguel Angel Roca, echoes the structural elements of the urban fabric: the street, the corner, the pergola and the neogothic tower (Manger, 1993). The new building complements the adjacent buildings despite its geometric design and modern facades. The architecture of the gallery combines geometrical rigidity with pink coloured walling, similar to that of the local churches and houses. This is a successful example of the insertion of a fragment of city in the context of Cordoba through the union of popular and religious culture, public and private use.

As we have seen, these projects involve single elements with one, or very few, strictly limited functions whose modest task is to complete a part of the city. From the observations made above in relation to renewal projects in the periphery, we can deduce that here too self-organisation processes spread relatively easily. The main problem is to avoid the corrosion of the architectural statement, which must therefore be incisive and of an extremely high quality, even when, as in the extreme case of the Wind Tower in Yokohama, it is limited to the refurbishing of an existing structure.

Fig. 9 Internal view of BCE Place Galleria in Toronto (Calatrava, 1993, p. 107)

9.4 Conclusions

As observed in Section 1, the field of architecture, unlike many other areas of study, and in particular the closely related area of urban analysis, has not until now been subjected to systematic analysis in terms of the complexity paradigm. There seems to be no reason why this area should continue to be neglected.

As we have attempted to show in Section 2, complexity provides a particularly stimulating point of view for the interpretation of architecture and, in particular, the analysis of relations between an architectural construction and the surrounding urban context. As indicated in Section 3, the complexity approach can provide useful guidelines for the designer, providing awareness of the risk of building 'cathedrals in a desert', dominant but isolated structures in the middle of large urbanised areas. More importantly perhaps, it may help to indicate how this can be avoided. It also provides a warning of situations when, on the other hand, the architectural statement risks being diminished, or even completely overwhelmed by its surroundings. Complexity analysis can provide a way of distinguishing between these two opposite dangers.

Similarly, the complexity paradigm helps us to understand an important aspect of the architectural quality of a project, especially in cases where it is considered debatable or the intention is in some way unclear. Naturally, we present these ideas simply as an indication of the usefulness of applying the complexity paradigm in architecture. The analysis could of course be carried much further. We hope that what has been demonstrated in this study may serve as a stimulus to continue in this direction.

References

Bédarida M. (1992) Le due città, *Lotus*, 71, 24-35.

Bédarida M. (1995) La memoria in gioco, *Lotus*, 84, 66-85.

Brandolini S. (1992) Entertainment Center di Eurodisneyland a Marne La Vallée, *Casabella, 61*, 596, 54-63.

Calatrava S. (1993) Padiglioni neogotici, *Lotus*, 78, 96-113.

Casti J.L. (1986) On System Complexity: Identification, Measurement and Management, in Casti J.L., Karlqvist A. (eds.) *Complexity, Language and Life: Mathematical Approaches,* Springer-Verlag, Berlin, 146-173.

Cohen J.L. (1993) Main Street Blues: il Festival Disney di Frank O. Gehry, *Lotus*, 77, 80-91.

Davies C. (1991) Underneath the Arches, *Architects' Journal*, May, 30-39.

Gehry F.O. (1993) Kobe, Fishdance Restaurant, *Lotus*, 76, 96-103.

Gough P. (1993) Three Urban Projects: Alban Gate, Embankment Place, Vauxhall Cross, *Blueprint Extra*, 9.

Hardingham S. (1994) *London*, Artemis, London, 210-219.

Ito T. (1988) La Torre dei Venti, Yokohama, *Domus*, 691, 40-45.

Ito T. (1993) La Porta di Okawabata e la Torre dei Venti di Yokohama, *Lotus*, 75, 54-59.

Mangurian R., Ray M.A. (1993) Sette livelli, *Lotus*, 75, 78-103.

Manger P. (1993) Centri commerciali, *Tecniche Nuove*, Milan, 24-27.

Menditto S., Rossi R. (1995) Il ritorno alla progettazione dei luoghi, *Costruire in laterizio, 8*, 45, 206-209.

Miller R. (1990) Eurodisneyland and the Image of America, *Progressive Architecture, 10*, 92-95.

Monastiroli A. (1994) L'idea dell'isolato Diagonal, *Lotus*, 82, 6-29.

Parcérida J. (1994) Uno shopping mall anonimo, *Lotus*, 82, 30-35.

Sudjic D. (1992) *The 100 Mile City*, Flamingo, London, 213-231.

Waldrop M.M. (1992) *Complexity. The Emerging Science at the Edge of Order and Chaos*, Simon and Schuster, New York.

Zardini M. (1992) Los Angeles come contesto, *Lotus*, 74, 98-108.

SESSION 2: THE SCIENCES OF THE CITY

10. Urban Research and Complexity

Denise Pumain

10.1 Introduction

During the last two decades, a number of interesting innovations have appeared in the field of urban research. New paradigms such as the dynamics of open systems, self-organisation, synergetics, chaos and evolution theory, have been recognised as conveying fruitful analogies for urban theory. New types of modelling, such as sets of nonlinear differential equations for spatial systems, cellular automata, multi-agents models, fractal growth, neural networks and evolutionary models have been investigated. By reviewing the orientations of contemporary urban research, it is possible to recognise among these new urban models several types of strategy which are useful for dealing with the complexity of urban systems.

It is not our purpose here to provide an impossibly exhaustive review of new trends in urban research. Several specialised reviews have already been published about operational intra-urban models (Wegener, 1994), computer-oriented urban modelling (Batty, 1992), or about models of systems of cities (Mulligan, 1984, Pumain, 1991), and urban models in general (Bertuglia and La Bella, 1991). We should therefore like to focus on how these new conceptual or operational tools may be used for solving some of the problems linked with urban complexity.

Before analysing some of these research strategies, it is necessary to recall the main sources of difficulty which urban research has to face in defining its area of study and building consistent theories about the city. The main theoretical issues in urban research will then be examined through the various methodologies recently proposed, whose oxymoronic task is that of making complex urban systems simple.

It would also seem valid to ask how much the new ideas have added to urban theory and general knowledge about cities, bearing in mind that a large number of the models built have never been applied, nor have they been rigorously compared with other models. Some far more classical types of approach without any connections with complexity studies, on the other hand, have led to a number of new results and interesting reflections. Closures and delays in communication are frequently observed in science but, if prolonged, they may be detrimental to urban research and planning.

10.2 Various Sources of Complexity in Urban Systems

10.2.1 Introduction

Ever since urban systems have been considered significant objects for research, a large part of the theoretical thinking has been devoted to the scientific appraisal of the urban realm. All authors recognise that cities are too complex and too diverse to be approached as a unified conception. The difficulty about building a single concept of a city has its roots in the uncertain origins of urbanism, in the manifold aspects of urban life and is reflected in the variety of measures which are used for defining urban areas.

10.2.2 Definitional Complexity

Historians agree upon the universality of the urbanisation process: cities appeared independently on various parts of the earth's surface at about the same time. However, there is controversy about the nature and significance of urbanisation. A few authors insist on the economic dimension: two or three thousand years after the invention of agriculture, towns began to emerge in order to commercialise the agricultural surplus by trading manufactured goods: "not only has agriculture been an absolute prerequisite to the emergence of true urban systems, but there exists also an inverse relation: agriculture leads almost ineluctably to the town" (Bairoch, 1985, p. 631). Other authors maintain that the main function of cities belongs to the symbolic order, it is part of the exercise of power: "the city is characterised neither by its number of inhabitants nor by their activities, but by particular features of legal status, of sociability and culture... those features are consequences of the main role of the urban organism. That role is not economic, it is political" (Duby, 1980, p. 13).

For others, it is the collective relationship to the feeling of the sacred (Racine, 1993) which has created cities. In any of the seven regions of the first urban generation, if one traces the characteristics of the emerging urban form, one arrives not at a settlement dominated by commercial trade, or centred on a citadel, an archetypal fortress, but more likely at a ceremonial complex (Wheatley, 1971).

This diversity in the theoretical interpretations of the origins of towns and cities is further increased by the variety of theoretical definitions and points of view which may be adopted, whether anthropological, cultural, economical, geographical, political, religious or social.

According to a *demographic* definition, a city is a permanent grouping of a resident population in a given area. The size and density of the aggregated settlement are linked within a *morphological* definition which also recognises the persistency over time and spatial continuity of building, and the urbanistic rules in its organisation. Such a conception implies that the city, in the limited area that it occupies, is unable to produce all the food that it needs. Its survival therefore depends on trading between the manufactured goods and services that it produces and the agricultural goods which are necessary to its population. This provides us with a *functional*, or a socio-economic definition: the city groups nonagricultural activities and innovates by developing a relatively complex social division of labour. This diversified economic base forms the carrying capacity of the town.

The commercial trade between towns and the countryside is always unequal, as manufactured goods are over-valued when compared to foodstuffs (Aydalot, 1985, Camagni, 1992). The consequent economic advantage of the town is inseparable from its *legal status* as a place of power. Even if this power has a religious or political origin, it is linked with an economic and territorial privilege and gives the town dominance over the neighbouring localities (Racine, 1993).

This large variety of definitions is one aspect of the complexity which characterises towns and cities. In order to get a full understanding of such a manifold object, it is indeed impossible to reduce it to a single dimension, even in the realm of a well-established discipline. A theoretical economic approach to the city cannot ignore its political and symbolic aspects, a pure social theory of the city would miss out the spatial and morphological features which are essential to its characterisation. The study of cities cannot therefore be confined to one specific field of investigation, as it is sometimes possible and fruitful to do for other objects of study in social science. On the contrary, a meaningful insight into urban dynamics can only be obtained by cross-fertilising the urban disciplines: the explanation of economic success is to be sought for in social and symbolic valuation processes, social interaction cannot be understood without knowing the material conditions for its existence in urbanism, and so on. As a consequence, the theoretical concepts elaborated within the framework of scientific research lose their explanatory power when applied to cities, as they are generally inadequate for dealing with urban phenomena.

10.2.3 Spatial Complexity

It has long been noticed that towns and cities develop links with the surrounding area: a settlement specialised in nonagricultural activities carries out central functions for its complementary region (Reynaud, 1841, Christaller, 1933). Unlike places (e.g. mining settlements) which exploit resources at their *site*, or in their close neighbourhood, towns and cities make a living from the wealth created by their *situation*. They exploit their position in trade networks, whose spatial range depends upon the size and specialisation of the city (Reymond, 1981a, 1981b). For this reason, a city cannot be conceived of as an isolated entity, it always belongs to a network of cities, it is 'a system within systems of cities' (Berry, 1964). Systems of cities have their own regular properties which are not only the sum of the individual cities they contain, and hence constitute a meaningful level for the analysis of urbanisation (Pred, 1977, Pumain, 1992).

The city as a scientific object of study should thus be conceived at various levels of spatial organisation. At the very least we need to consider the level of the individual actors, that of the city itself and of the system of cities. Some intermediate levels, such as the neighbourhoods within cities or some regional subset of cities may also sometimes be of interest.

However clear the identification of those levels of organisation from a conceptual point of view, it is often difficult to recognise them in reality. The neighbourhoods may have clear-cut edges, but several different partitions of an urban space may be considered relevant. The limits of the city itself have become fuzzy, due to suburbanisation and long distance commuting which has broken the spatial continuity of daily urban systems. It is also difficult sometimes to clearly separate a city and a network of cities, as in the case of large conurbations or very densely connected regions like Randstad Holland, the Ruhr, the rivieras or the Megalopolises. Systems of cities are also difficult to isolate as scientific objects of study. They are generally defined within the limits of a single country, since international borders greatly reduce the intensity of spatial interactions. However, it is obvious that interurban interactions are not negligible despite international borders. Moreover, the degree of openness of the system of cities varies according to the situation of each city within the urban system. The largest cities, or the ones which are specialised in international activities, are more open than others to external interaction. It is therefore very difficult in practice to define precisely the effective limits of a system of cities.

Because they are structured at various organisational levels, urban systems belong to the category of complex systems. However, when compared to physical or biological systems, the handling of that complexity is made more difficult by the fact that the separation between the levels is not always easy to determine in practice.

10.2.4 Temporal Variations

Another source of complexity which makes the description and modelling of urban systems difficult is the multiplicity of time scales operating comtemporarily in the same city. One only has to consider how the timing of daily life adjusts (for instance by commuting) to the more stable patterns created by the location of jobs and housing facilities in the city. The life of buildings is generally longer than the duration of stay of their users or inhabitants, or even of a whole generation (Whitehead, 1977). This leads to the recognised pattern of moves from central locations to the periphery and back to the centre, linked with successive stages in the life cycle of individuals. But other time-scales, which may have a decisive and sometimes catastrophic effect on its inhabitants, intervene in the life of a city. The duration of a cycle of economic specialisation (adoption of a large set of innovations) may last from a few decades to one century or more, inducing alternate periods of rapid growth, stability and slow decay. Even if cities succeed in adapting to several successive waves of innovation, the speed of change of the economic and social functions is generally faster than the speed of transformation in the town plan and infrastructure. All these differences in the time scales of the components of an urban entity cause severe dysfunctioning in the everyday life of cities, and also make the description of the dynamics of the urban systems very difficult. Not only do the relevant time stages have to be identified, but also mechanisms articulating various temporalities have to be considered.

Without being too deterministic, or playing with ideas about finalism, we could suggest that it is the existence of such different time scales, an inherent aspect of urban complexity, which explains the survival of cities over very long periods of time. The same structures are able to respond to very different social needs and economic activities. It is this adaptability and flexibility which ensures the sustainability of the whole urban system.

Momentary disadaptation of form to function, congestion phenomena, time lags in adjustments to change, mismatch in facilities and infrastructure, discrepancies between real needs and the objectives of policies may result from the inequalities in the intrinsic duration of the life-cycle of each of the components in an urban system. The analytical handling of such a large variety of time-scales is very difficult and is a major source of the problems faced by urban theory and modelling when we try to conceptualise urban change.

This challenge is reinforced by another consequence of the temporal variability of urban systems, which is their historical, or rather, evolutionary character.

10.2.5 Urban Systems are Evolutionary

Urban systems are also complex because they cannot be wrapped up in a stable taxonomic description. The social and economic content of cities and its significance changes over time. Different 'generations' of towns and cities have to be characterised differently, according to their morphological aspect as well as to their urban functions. These changes lead to the disappearance of some cities (e.g. abandoned citadels or the 'ghost towns' of the gold rush), but generally the transformations and adaptations occur within a permanent urban entity. Cities are very often considered as the main vectors of social transformation and technological innovation. This *adaptation* process induces more or less pronounced specialisations of towns and cities into functional types. A series of successful adaptations, or failures to transform, result in marked inequalities in city size. Meanwhile, the significance of size thresholds is also evolving over time. The qualitative content referred to by concepts such as a 'small' town, a 'large' city, an 'industrial' or an 'administrative' one, has to be constantly revised. As it concerns innovation and creation, this particular process of urban change is very difficult to include in models which attempt to make predictions.

This process is a further source of complexity in the sense that the integration of social interactions over time produces networks of relations which may persist through migrations of people and even the passage of generations. Towns and cities become indivisible objects and cannot be analysed as the juxtaposition of their constituent parts. It is highly significant, for instance, that it will take a conurbation resulting from the merger of two or three independent cities a very long time (often decades or centuries) to reach the same functional level as a single monocentric city of the same size.

At another scale of study, systems of cities also have an evolutionary character. Archaeology supports the idea that, since the very beginning of urban history, towns have appeared not in isolation, but in groups with linkages which make them function in networks (Fletscher, 1986, Bairoch, 1985). The two main theories of the genesis of city systems, bottom-up and top-down, are probably complementary. Corresponding to the idea that towns and cities derive from the surrounding rural region which feeds them, is the theoretical approach which sees city systems as occupying and controlling a territory to ensure its defence and meet the needs of its inhabitants (Botero, 1588, Vauban, 1707, Le Maître, 1682, Reynaud 1841, Christaller, 1933). Relating to the observation that cities have always been organised in networks, is the view of cities as nodes within systems of relations - in this case interdependencies are analysed without giving particular importance to the immediate environment (Reclus, 1895, Bird, 1977, Batten, 1992, Dematteis, 1985, 1990, Camagni, 1990).

In both cases, an intrinsic property of city systems is considered to be the fact that their growth is linked to increases not only in population, but also in the productivity of the agricultural, industrial or tertiary activities supporting them. The space-time contraction caused by the increasing speed of communications is another major historical trend in the evolution of geographical space, and strongly influences the development of urban systems (Chevalier, 1832, Reclus, 1895, Juilliard, 1972). The spacing and hierarchical structure of systems of cities is closely connected to the speed of intra-urban and inter-urban communication (Pumain, 1993). Systems of cities cannot therefore be modelled as systems in equilibrium, as proposed for instance by central place theory, since any explanatory theory has to take into account the fact that they are produced by the historical process of development of settlement systems, i.e. *urban transition*. This is another important feature of urban complexity.

Until now we have used the concept of complexity without defining it. The concept has become very fashionable, because it raises echoes in various fields of knowledge, especially the biological and social sciences. However, contrary to the views of some specialists in 'complexology', for instance those developed at the Santa Fe Institute by S. Kauffman (1993), we are sceptical about the possibility of finding universal 'laws of complexity', applicable to all complex systems, which is why we began by specifying how the concept of complexity could be understood when dealing with urban theory. This specification helps us to understand which questions can be addressed to more general theories of systems dynamics and which notions or models could be usefully borrowed from other fields of research. One strategy for dealing with all four main aspects of urban complexity has been to make use of ideas and techniques from theories developed in the field of mathematics, physical or biological sciences.

10.3 New Paradigms for the Modelling of Complexity

10.3.1 Introduction

One aspect of urban complexity is the fact that the same mechanisms, rules or planning principles may lead to different structures, according to the situation of the city and the phase of its evolution. The same action (external perturbation or internal change) may have no effect if the city is on a stable dynamic trajectory, but it can deeply alter the structure if it intervenes at a moment of instability when bifurcations are possible. The ability of sets of mathematical nonlinear equations to reproduce this type of qualitative behaviour has been exploited in several types of urban model. Most of these models have been transferred from physics and rely upon analogies between urban systems and self-organising physical systems.

10.3.2 Principles of Self-Organisation

The aim of dynamic models of urban structure is to understand better and predict the evolution of socio-spatial systems. The new dynamic models draw mainly on analogies from physics, for instance chemical kinetics (Allen, 1978) or laser theory (Haag and Weidlich, 1984). These systems possess a very large number of elements and are described on at least two levels: system-wide (macroscopic variables) and elementary (microscopic variables). A third level of description is sometimes added, with subsystems comprising unfixed numbers of elements. Differential equations describe the variation of a macroscopic variable over time. These are obtained from definitions of interactions at the microscopic level, between the elements of the system. The passage from the microscopic to the macroscopic description of the system is the most arduous problem in such model building. It may lead to stochastic formulations when interactions between the elements do not have a deterministic form, but their results are expressed only as probability distributions.

According to the theory of dissipative structures or synergetic theory (Haken, 1977), self-organisation and bifurcation phenomena may occur in open systems when these are maintained under an influx of energy. The systems may organise themselves into structures which are created or destroyed during their evolution. This evolution is at the same time deterministic (following a trajectory which can be predicted using a model describing interdependencies between variables) and random or undetermined (during periods of instability when a change in structure, or a phase transition, can occur). The model may therefore admit several solutions or multiple dynamic equilibria. The various trajectories correspond to qualitatively different structures, and the system may be driven towards one branch or another by the amplification of a small fluctuation (Prigogine and Stengers, 1979, Allen and Sanglier, 1981).

Many sources of instability intervene in the evolution of such open systems when they are far from equilibrium. On the one hand, they continually undergo internal fluctuations, variations in the level of their characteristic variables (which may result from changes in the micro-states of the elements of the system). On the other hand, they are also subject to external perturbations from their environment. An open system is thus continually adjusting the number of variables and subsystems. The structure remains relatively permanent only when, under given conditions, this constitutes a stable state, i.e. a state towards which the system returns after having distanced itself a little. In this case, the structure is viewed as an attractor on the system's trajectory. Dynamic instability may induce a jump from one trajectory to another, or from one structure to another, i.e. from one qualitative behaviour of the system to another, passing through a bifurcation point.

This paradigm provides a new framework for urban theory and urban modelling. According to Batty (1993), the change in paradigm can be related to a change in the city itself, "for a hundred years or more, urban theorists have treated cities as though equilibrium were their natural condition. However, as current events increasingly demonstrate, this is less and less true. As the local and the global intersect in unanticipated ways, and as the role of historical happenstance is seen to be more influential than ever, as the evolution of ever more complexity seems to be the norm, we sorely need new theories of how cities form, how they evolve, and how we might control or at least influence their development" (p. 14).

10.3.3 Transferring Theory from Physics to Urban Research

To what extent is it useful to develop an analogy between physical and urban systems?

When formalising an urban entity as a dynamic spatial system, one has to define it as a set of elements. These can be either located elements or geographical zones. Location (at least relative location) is a basic property of the elements of the system; the interactions between them are in part spatial interactions, linked to expressions of the absolute location (site) or relative location (situation) of the elements. The state of the system is defined as the configuration of its characteristic variables. A geographical structure is therefore given by a particular relative size and evolution for the state variables and/or located subsystems which are used for the description of the urban system.

At the most disaggregate level, an urban system can be formalised as a set of localised and interacting actors (individuals or groups as persons, households, firms, associations, etc.) who are using and continuously re-creating geographical differences and spatial configurations. These interactions are varied in nature and form competition for space, propensity to agglomerate, segregation tendencies, imitation and so on. According to the case, they generate the homogeneity or the diversity of contents, an increase or decrease in gradients, concentration or dispersal in distributions. One has to hypothesise that the dynamics of these interactions is creating the spatial structures, for which they are both an expression and a condition for their existence.

Such a normalisation has many appealing features:

- it allows stress to be placed on the linkages between the macroscopic descriptors of the system, the individual behaviour of the actors and interactions defined at the microscopic level. Much research has been conducted and empirical regularities established at the micro and macro scale separately, but a clear connection between the two levels of observation is not always established. Here, self-organisation phenomena, meso or macro-scale structures, are described as consequences of an interaction game between individuals, each animated by their own objectives. These consequences are not always intuitive to the observer, they are often not concerted and generally not perceived by the actors.
- this approach can also provide an interpretative framework for observed regularities in urban systems. Due to the instability of the elements being considered, it is only because bifurcations occur in the evolution of

dynamic systems and because their trajectories 'jump' from one attractor to another, that identifiable and separable categories, or large-scale regularities may be observed. Temporal series therefore appear as 'possible' sequences of complex dynamics. Very often the problem for social sciences, where experimentation is impossible, consists of identifying the dynamics which produced a particular temporal series of observable structures, i.e. a specific trajectory (Prigogine and Stengers, 1979).

- another interesting feature of this approach is that the historical dimension of social systems is taken into account via the concept of irreversibility. On the one hand, the explanation of the state of a system at a given date integrates its previous trajectory (i.e. its history) as the contemporary structure includes the 'memory' of previous bifurcations. On the other hand, the characteristic fluctuations of dynamic systems imply that it is impossible to prepare initial conditions which would lead to identical futures. The impossibility of exact prediction is then given as a theoretical 'a priori'. However, analysis of the dynamic behaviour of the systems and their sensitivity to changes in parameter values allows the exploration of a limited number of possible futures, according to the assumptions made about the evolution of the parameter (Allen and Sanglier, 1981).

These ideas may appear seductive, but it is not because of their novelty - all of them were originally expressed a considerable time ago. It is because they allow the relaxation of some of the more oppressive restrictions imposed by the previous methods of model building and because new experimental tools related to these old ideas are therefore available. The new models can from now on be in agreement with these old ideas, reformulated as a 'new' urban theory.

Since systems theory has been applied to urban systems, a noticeable evolution has occurred in the scientific domains from which analogies are drawn. The concept of entropy, for instance, was referred firstly to statistical mechanics and then to information theory. System dynamics had their first inspiration in hydraulics (Forrester, 1969). The dynamics of cities as open systems far from equilibrium was first conceived with reference to 'dissipative structures' or 'synergetics', conveying analogies with physical systems. Another way of thinking is increasingly guiding the work of urban theoreticians and modellers. They still use the mathematical tools which were first proposed by physicists (or mathematicians) for building and testing their models, but they refer more often to theoretical analogies with living systems. A shift has occurred from the concept of pure 'dynamics' to the concept of 'evolution' (Allen, 1991, Rabino, 1993). Although urban systems can be described as largely self-organised open systems, whose structure and evolution depends upon internal fluctuations as well as external fluctuations, they also have a historical and evolutive behaviour, an adaptive hierarchical structure and a power of creativity which invites us to develop more analogies with living systems.

The field of ecology is by no means a new source of inspiration for urban scientists; we can go back to the work of the famous Chicago school. But besides some tentative transfers of ecological concepts (Dendrinos and Mullally, 1985) and some interesting use of Atlan's work by urban modellers (Camagni, Diappi and Leonardi, 1986, Diappi and Ottana, 1994), there has been very little further serious research into the possibility of introducing the concepts of evolution theory into urban theory, despite the increasing use of the word itself. The rather bad reputation of social Darwinism has perhaps been too dissuasive. On the other hand, even though urban systems seem to have many features in common with living systems, they also have distinctive social properties, like the intentionality of their organisations as well as their short-term learning capacities. This could lead to search for other analogies, for example, in the direction of computational automata or artificial intelligence (Huberman, 1988). Devices like cellular automata or neural networks are now being explored for applications in urban modelling. These could be connected to evolutionist theories and incorporate more classical dynamic tools. If well related to empirical research, a new and more specific theory of urban evolution could emerge from these experiments.

A short review of some applications of dynamic modelling to intra-urban as well as interurban structures demonstrates the diversity of inspiration which has served urban modellers. Probably not all these experiments can become part of urban theory or operational tools, even though they were all inspired by the idea of reducing or explaining better the complexity of urban systems.

10.4 Quantitative Models of Qualitative Structures

10.4.1 Introduction

When using nonlinear equations and/or interdependencies between variables, even very simple mathematical models can produce very complicated dynamics (May, 1976).

The property of differential nonlinear equations in generating complex behaviour for the state variables of a system has been used in several ways. When realistic projections are needed, one safe approach is the search for analytical solutions which lead to simpler formulations of urban models. Another trend is to develop more realistic and complete representations of cities, but only with an exploratory use of the models. In both cases, the mathematical tools are used to provide a variety of possible qualitative structures from the same set of equations, i.e. the same kind of urban mechanisms, by changing the value of only some parameters. Applications are always made for aggregate variables, either at a city level or for a system of cities.

10.4.2 Analytical Models of Urban Growth

The formulations used for these models are more or less directly drawn from findings in other fields: catastrophe theory in mathematics, Volterra-Lotka models of interacting species in biology, and master equation techniques from physics. All these models are conceived for giving analytical solutions, with the implication that one knows how to identify the equilibrium states of the system, how many they are and whether they are stable or not. One is then able to situate the observed trajectory with respect to these equilibria and to make predictions about the stability of the structure according to hypothetical variations in the values of parameters.

a. Catastrophe Theory

R. Thom's (1974) catastrophe theory provides very good analytical tools and theorems about the number and precise qualitative shape of the functions linking the state variables to the elementary catastrophes. However, it allows too few variables and parameters (respectively a maximum of two and four) to describe the system, so it cannot be of great help in most urban problems. Lung (1985) underlines the difficulties in constructing models which satisfy the basic assumptions of the theory: how to define a potential function, how to incorporate competition for space, and how to find relevant state variables which may undergo sudden changes, yet produce another qualitative structure for smooth variations in the values of parameters.

Reviews of these applications are given by Wilson (1981) and Rosser (1991). Usually the spatial structure is summarised by a global measure of size, or density, or intensity of spatial interaction, whose variations are related to a few control parameters. For instance, Amson (1975) considers urban density as a state variable dependent upon two parameters: it is proportional to rents and inversely proportional to 'opulence', which is a measure of the benefit from urban interaction. He then shows that urban density follows variations according to the fold model if a saturation effect is introduced, with rental becoming a logistic function of density. If a minimal threshold of available space per person is added, the model for variations in the density becomes a cusp. The condition for smooth and continuous behaviour of the density is that one of the parameters, rental or opulence, will have high values.

Catastrophe theory is used by many authors to describe discontinuities in urban growth. For example, Mees (1975) gives an interpretation of the revival of cities in Europe between the 11th and 13th century, following a long period of urban decline, as a discontinuity in the relationship between the urbanisation rate and the possibility of external trade. This discontinuity appears at a critical value in decreasing transportation prices. A more comprehensive model involving a butterfly catastrophe takes as a state variable the urban population working in manufacturing. This is a function of four control variables: density of regional population, average productivity, urban-rural productivity differential and difficulty of transport. Papageorgiou (1980) tries to explain the cases of sudden urban growth observed in the 19th and 20th century by existence of scale economies in the utility function of cities for individuals. Even when utility increases in a continuous manner, technological changes may at a given level introduce a discontinuity in the urbanisation rate, which then jumps from one equilibrium point to another in accordance with a fold catastrophe. A simpler test has been provided more recently by Casetti (1991) who describes a transition between two equilibrium states in the urbanisation rates of the world in the 19th and the 20th century as a cusp.

A third approach consists of explaining the existence of thresholds and hierarchical levels in urban systems by the jumps and hysteresis phenomena of catastrophe theory. In this fashion Casti and Swain (1975) suggest that the functional level of a central place may be seen as a cusp function of two control variables: population and per capita income, whereas Wilson (1981) relates the size of various urban facilities and urban centres to the benefits of facility size and to the disbenefits of travel. If the first control variable is a logistic function of size, and the second a linear one, the size may vary with the slope of this last function according to a kind of fold catastrophe.

The most interesting feature of this family of models is that they seem to provide good theoretical insights into the discontinuous behaviour of some aggregate variables, but they are still of little help in modelling the spatial structure or spatial interactions in urban systems.

b. The Volterra-Lotka Model

This formulation depicts the evolution of the number of individuals belonging to two interacting biological species. Each species is also characterized by a birth-death process of the logistic type. Translations of this formulation to geographical analysis have been attempted for instance by Dendrinos (1980, 1984) and Dendrinos and Mullally (1981, 1985), who use the city's population and its mean per capita income as variables in the place of species. Their model simulates the evolution of the share of an SMSA's population in the total national population x and of its relative level of per capita income y. Equations of the model are of the following form:

$$dx/dt = \alpha\,(x\,(y-1) - \beta x)$$
$$dy/dt = y\,\gamma\,(X - x)$$

where X is a carrying capacity of the SMSA, α and γ are speeds of adjustment and β a coefficient of 'urban friction'. This model was chosen because it allows replication of one of the most frequent behaviours observed among the 90 SMSAs of the United States between 1940 and 1977. Such behaviour is the sink-spiral type, with oscillations around successive equilibria. However, the scarcity of available temporal data does not allow strong empirical support to this model.

Volterra-Lotka formalisations are appealing because of the diversity of dynamics that they can generate from relatively simple mathematical equations, and because the trajectories and the stable states can be analytically computed. They have been used for instance for reproducing cyclical behaviour in the relative development of an urban centre and its periphery (Orishimo, 1987, Zhang, 1990) according to the hypothetical model proposed by Klaassen (pre-urbanisation, urbanisation, de-urbanisation and re-urbanisation). Volterra-Lotka's equations are also used in various models of competition for city space between two types of land use or between two categories of urban citizens. There is still much work to be done in testing such models, selecting the more often observed cases among the six theoretical equilibrium states according to an ecological analogy (symbiosis, competition, commensality, amensality, isolation and predation). For several of the models which have been proposed, the exact interpretation and measurement of the model parameters is not entirely clear.

Spatial interaction is not easily included in this kind of model structure. In most cases, it appears only in a weak and implicit manner, since the size of one element in the system is expressed by a relative measure, as a share of the total sizes of the elements of the system.

c. The Master Equation Approach

The master equation method relies upon theoretical principles used in the field of synergetics. It is potentially of great interest to urban research, because it explicitly links the state transition probabilities of individuals at a micro-level and the evolution of some variables describing a macroscopic structure. The master equation gives the variation in time of the probability of given configurations in the space of the state variables. The probability of transition from one configuration to another depends on the assumptions made about the number and nature of parameters affecting the individual transition probabilities. This stochastic formulation is then used to derive a deterministic equation for the evolution of the mean values which, in turn, allows the estimation of the parameters.

This procedure has been used for deriving a two-population/two zone model of the Volterra-Lotka type (Haag, 1984). The migration rate of each population group from one zone to another depends on the preference for a given part of the city, the propensity to cluster ('internal sympathy'), to join members of the other group ('external sympathy') and a general mobility level (flexibility). The authors explore analytically all possible types of spatial configuration and levels of urban segregation according to the sets of values taken by these four parameters. In this simple version, such a model is a good pedagogical presentation of elementary spatial dynamics.

The master equation approach has also been used for reformulating an intra-urban model of residential rent and density interactions (Haag and Dendrinos, 1983) and has been applied to twelve SMSAs in the United States (Dendrinos and Haag, 1984). The SMSAs are divided into two parts: the city centre and the suburbs. Utility functions depending on rent and density levels through land availability are used to describe the individual behaviour of typical land-buyers (who move) and sellers (who transfer rental value from one zone to another). Aggregation of these individual behaviours produces a stochastic master equation, whose deterministic mean value describes the evolution of the population share and of the relative rent price in the central part of the urban unit. From simulations with empirical data, the authors predict a probable reversal in the suburbanisation trend for the 1990s, if the parameter values which represent the general conditions of the SMSA's environment do not change.

A migration model for a whole set of regions has been constructed by Haag and Weidlich (1984) using the same approach. Individual transition probabilities from one spatial subdivision to another are defined as functions of the difference in the attractivity of regions. They are aggregated to define an equation of movement of the probability of states of the system (e.g. all possible configurations of population repartition among regions), whose mean value equation is used for the estimation of parameters. The dynamics of the spatial system is then related to trend parameters whose values may be compared from one system to another, or over time (a global mobility rate, a degree of 'cooperation', an agglomeration effect, and a saturation effect), and to a set of preferences for each region. Socio-economic variables which could explain these preferences are not included in the dynamic model. However, when fitted with a regression model to a temporal series of preferences established from the model (using, for instance, annual migration tables), they allow the prediction of future migration patterns and the evolution of the spatial configuration of population. The current spatial configuration of population can also be compared with the stable state which would emerge if the observed pattern of migration flows were prolonged. The difference between the two gives an idea of the extent of the ongoing structural reorganisation in the system.

Sanders (1992) applied this model to a large set of French cities observed between 1954 and 1982. She relates the shift of migratory preferences, from the cities of the north-east to the south, to the emergence of business services as a major source of differentiation in the contemporary dynamics of the system of cities. She also provides various scenarios of possible further evolution, which can be obtained from different hypothesis about the future migratory trends (Haag *et al.*, 1992).

Because of their computational ease, we can classify models using a master equation approach with the models allowing analytical solutions of dynamical structures, even though the solution in some cases can only be obtained by iteration. These models are also analytical in the sense that they give very detailed and precise information about the dynamics of a very small number of state variables (one or two). Like catastrophe theory or Volterra-Lotka's model, they mostly deal with state variables at the aggregate level of an entire city or of a system of cities. This could seem to be a misunderstanding of the method, as the master equation theoretically provides a clear relationship between the spatial behaviour of individuals and the global dynamics of the population as measured by aggregated variables. However, despite this theoretical possibility, the handling of the master equation itself would be too difficult in practice and one has to make simplifying assumptions to define a mean value equation (Haag, 1984). What is defined in the migration model as a 'utility' at the individual level is actually an aggregate measure of the 'attractivity' of each zone (Sanders and Pumain, 1992). In reality, the migration model functions and gives results for the aggregate level of the state variables of the system.

10.4.3 Global Models of Urban Structures

A second group of modelling approaches uses nonlinear differential equations to simulate changes in the spatial structure of an urban area and give less attention to the identification and computation of precise analytical solutions for stable equilibria. They emphasise the production of a variety of possible spatial structures which are studied in an exploratory way with the help of simulations. The equations of the systems are more complex since a large number of state variables and interactions between them are used for the description of the system. These models are therefore more realistic representations of a city.

a. The Dynamics of Urban Spatial Structures

By using continuous differential equations instead of difference equations and by drawing analogies from chemistry or physics instead of hydraulics, these models are more flexible than the dynamic model proposed by Forrester (1969). They also allow a description of the spatial structure of cities in several zones, whereas the attempts to spatialise Forrester's model turned out to be very cumbersome (Fournier, 1990). The models explore how a variety of urban forms can be generated from deterministic nonlinear equations and mainly refer to intra-urban structures. They describe the evolution of the location of populations and economic activities within an urban area. This is usually done by counting the number of jobs and residents in various categories for each neighbourhood and computing interaction flows between them. These models can be considered dynamic extensions of the Lowry model (1964), as they integrate the same general principles well-established by urban empirical research (economic base theory, agglomeration economies, logistic growth and spatial interaction functions) within the framework of an atomistic and market-oriented urban economy. These traditional elements of urban theory were for the first time brought together in a single dynamic context by Harris and Wilson (1978) and Allen's intra-urban models (Allen, 1978, Allen *et al.*, 1981).

Another very interesting improvement is that these models take into account the effect of external relationships on a city's development, whereas Forrester located the main source of a city's dynamics within the city itself, in its internal structure. In most models of the Leeds school (Wilson, 1981), as well as Allen's intra-urban model, there is an external demand which plays a major role as a driving force in making a city grow or decline. Some authors have argued that these urban models were only kinetic and not really dynamic since they did not integrate endogenously the causes of the evolution of the urban system. This interpretation would seem rather severe: the hypothesis of an external driving force is interesting from a theoretical perspective, since it is an acknowledgement of the multilevel character of urban systems and of their dynamics. Observations on the autonomy of a city as a 'system' - in Berry's sense (1964) - show that whereas the city's 'settlement system', including the housing and all the associated infrastructure and services, has some degree of autonomy (its evolution depends mainly on local decisions), the city's economic system has no real local autonomy, as its evolution is strongly determined by external linkages and decisions located outside the city - sometimes very far away, and even in other parts of the world (Reymond, 1981a).

b. Identifying Bifurcations

The models of urban spatial structures developed by Wilson and Allen have indeed thrown light on urban spatial dynamics by establishing explicit connections between some critical parameters values and the shape of spatial urban patterns. The models developed at Leeds under Wilson's direction are more closely linked to the analytical approach mentioned above since they can be subdivided into submodels generating journey-to-work or shopping trips, allowing some analytical descriptions of the morphogenesis of the location of places of residence or shopping centres for instance. Bifurcations and equilibrium points can be studied analytically for variations of one parameter, the other being held constant. The determining effect of the sensitivity of people to travel costs in shaping the pattern of residences and the strong impact of a parameter describing the sensitivity of consumers to scale economies in generating a concentrated or dispersed pattern of shopping centres, illustrate the concept of bifurcation. Above a critical value of the parameter, the city evolves towards a concentrated pattern, whereas below this value, the pattern will become fully dispersed. More complex formulations of the models, linking the supply-side and the demand-side of urban dynamics, and disaggregating the variables into various income groups, types of housing or kinds of economic activities, can be used to study the global dynamics of the system by means of simulation (Beaumont, Clarke and Wilson, 1981a, 1981b, Clarke and Wilson, 1983a, 1983b).

In Allen's work, the whole set of interactions is integrated into a single model, so analytical solutions for some subsystems cannot be computed. Spatial interactions are modelled mainly by means of attractivity functions, which characterise the advantages of one location in comparison with all other possible locations for each state variable in the system. The mathematical expression of such an attractivity function is very complex and the dynamic behaviour of the variables can be studied only by means of simulation. However, it is clear from the applications of the model to real cities (Allen, Engelen and Sanglier, 1984a, 1984b, Pumain, Sanders and Saint-Julien, 1989) that bifurcations may occur and totally transform the spatial structure of the city, even for a very small variation of some parameter (for instance the parameter measuring scale economies or the propensity of economic activities to agglomerate).

So, in both cases, the same mechanisms of spatial and economic interactions occurring within a city may give rise to a variety of spatial structures. Interpreting the structure of a given city according to this theoretical framework allows us to conceive that particular shape as one of the possible results of more general urban dynamics, which could have generated many other shapes as well. Because of the irreversibility of the evolution, successive bifurcations determine the uniqueness of the actual pattern of a particular city. Although each city is unique, a general theory of the rules governing the functioning and evolution of urban structures no longer seems an unattainable aim for urban research.

In practice, the identification of the timing and nature of bifurcations remains a difficulty of this approach. Even though they may be well defined in theory, they are not easy to recognise from the observations about a real city or when calibrating a model, because changes in social systems occur most of the time with slow transitions, and because the limits between what may be considered as two different structures are very fuzzy. Over a relatively short time period, a change in a trajectory may be interpreted as a shift towards a distinct new trajectory through a bifurcation point. However, a longer period of observation may reveal that this apparent move was only a momentary fluctuation in a general evolution which actually maintained the same overall direction. For the same reasons, it is also very difficult to distinguish between two possible interpretations of a change in urban structure, i.e. whether it is a jump to another trajectory remaining within the same type of dynamics, or a change which implies another definition of the urban structure and hence a change in the dynamic model itself. Usually, the slowness of transitions and the inertia of the structures are not well reproduced by dynamic models. The applications of the models have shown that they usually produce bifurcations more easily and more frequently than can be observed in reality (Wegener, 1983, Sanders, 1992).

c. *The Question of Temporal Scales*

The bifurcations which can be observed in the spatial structure of a city are, in reality, very rare events associated with nonpredictable external perturbations (such as seismic activity or war), or happen only if very long time periods are considered. Bifurcations may occur more frequently at smaller scales, for instance in the decisions taken in the everyday life of individuals, but their effects on the urban environment are in most cases negligible. Bifurcations can be noticed only over longer time scales at higher spatial levels, for instance in the reconstruction of blocks of buildings, or the change of function of a neighbourhood. Such local changes can be considered bifurcations since they alter the structure and significance of a neighbourhood, but may nevertheless not cause any perceptible effect on the structure of the city as a whole. It is likely that it would take much longer for the whole structure of the city to be affected by the changes in the interactions among urban actors. The problem is therefore to find a proper adjustment between the temporal and spatial scales of the models and of their applications.

Wilson's model supports the hypothesis that there are at least one or two situations where a city is in equilibrium. In the short-term, the commuting pattern satisfies the maximum entropy principle, and there may also be an equilibrium in the size and spatial distribution of shopping centres according to the preferences of consumers. Analytical solutions may be computed in both cases from the model and permit short-term predictions about the urban structure. Allen's model, on the contrary, is not concerned with these short-term dynamics and considers only the longer term dynamics of the urban structures generated from an external demand by the economic, social and spatial interactions within the city. The right use of this model is in exploring the various possible urban structures which may emerge over medium or long-term periods due to local changes, such as massive investment in the infrastructure, for instance, the building of a new shopping centre, or the value of some global parameter like transportation costs or preferences of households concerning living conditions.

From these observations it is obvious that the performance of different models is very difficult to compare, even when they are apparently formulated in very similar ways (Lombardo *et al.*, 1988). The integration of different speeds of evolution into the urban models, in connection with the spatial and time scales of impact and with the degree of reversibility of such changes, is still a major challenge for the modelling of urban structures. This means that better connections should be developed between urban modelling and urban analysis at lower scales of observation.

d. Models of City Systems

Similar formulations have been developed for the simulation of the evolution of systems of towns and cities. Central place theory remains the major conceptual framework in this field. The dynamic aspects of this theory, although mentioned in early work (Christaller, 1933), have received only sporadic attention from urban theoreticians and are still not well formalised, despite a recent surge of interest in the question. Difference equations, including spatial interaction, are used by White to describe the relationships between retail activities and consumers through cost equations. He links the relatively concentrated or dispersed spatial pattern of centres to the interaction parameter, and tests the compatibility of the simulated urban hierarchy with a rank-size distribution for one sector (White, 1977) and two sectors (White, 1978). The model however was not applied to observed systems. Allen and Sanglier (1978, 1979a, 1979b) simulate the development of a system of central places. A system of differential equations is used to simulate the growth and differentiation process in a set towns and cities in competition for the attraction and development of urban functions which may be allocated to certain locations either hierarchically or randomly. However, although they obtain final distributions of city size which satisfy the rank-size rule, the development of the spatial patterns of centres over time are more typical of a market progressively colonised by entrepreneurs than the genesis of a realistic central place system, since the model starts with only two central locations which share the whole region as their market areas (Allen, 1978).

Other insights into the dynamics of city systems have been gained by giving better specifications to the old 'over-identified' rank-size rule model and connecting it with dynamic processes. The statistical shape of the distribution of city sizes is not a sufficient description of the hierarchical structure of urban systems. A considerable improvement in measuring the degree of universality of Zipf's law can be obtained if reliable and comparable information is used (Moriconi-Ebrard, 1993). It has been shown that the law is connected with an almost stochastic spatial distribution of urban growth (Gibrat, 1931, Pumain, 1982, Roehner and Wiese, 1982, Winiwarter, 1984). However, the purely random process has to be combined with the historical process of space-time contraction in order to become a proper description of the demographic evolution of an urban system (Pumain, 1982, Guérin-Pace, 1992). This has been related in a dynamic model to a competitive migration process, where the city size distribution can be conceived of as a stable attractor (Haag, 1994).

More recent work examines the effects of spatial competition between cities. This field of research has long been neglected because of the relative lack of data about flows of any kind, except population, between cities. Interesting empirical studies have demonstrated the strong interdependency of towns and cities belonging to the same urban system, when they are highly connected by information flows (Pred and Tornqvist, 1973). Studies have demonstrated the propagation of short-term fluctuations (Marchand, 1981), as well as similarities in medium-term socio-economic transformations (Pumain and Saint-Julien, 1978) or the long-term diffusion of innovations in systems of cities (Pred, 1966, Pred and Tornqvist, 1973, Rozenblat, 1992). The spatial representation and modelling of interurban connections are not easy. For instance Cauvin and Reymond (1980, 1985), after Dacey (1974), have tried to model the spacing of cities, while Muller (1983) attempted to represent their differences in accessibility by using non-Euclidean spaces. Fik and Mulligan (1990) modelled internodal flows within the traditional framework of central place theory, and Kremenec and Esparza (1993) have developed a model of flows including exchanges of spatial goods (for which demand decreases with distance) as well as nonspatial goods (which can be exchanged between two centres of the same level). More research still has to be done in order to integrate the hierarchical concept of central place theory and the process of urban specialisation into a dynamic theory of systems of cities.

The models proposed until now seem more suited to simulating the competition between already established urban centres than simulating the genesis and adaptation of a settlement system. One may ask, nevertheless, whether the methods available are capable of dealing with the full complexity of urban systems. Modelling with differential equations has some advantages - the model is written fairly concisely and it can be tested under certain conditions with a guarantee of repeatability - but such models treat geographical space as an isotropic function of distance, using various synthetic spatial interaction formulations. It is difficult for them to take into account a large variety in the range and scope of spatial interactions and to consider more than two different geographical scales at the same time. More flexible simulation tools may give better results in this respect (see Section 5).

10.5 Analytical or Micro-Level Approaches

10.5.1 Introduction

Cartesian philosophy still has some adepts who do not hesitate to deal with urban complexity by developing analytical investigations. If urban macro-structures are thought of as products of interactions between actors at the microlevel, it is logical to start with a good understanding of the behaviour of those actors. Some authors like Boudon (1984) deny any usefulness and real theoretical value to models from the social sciences which do not give a meaningful interpretation from the point of view of the individual. 'Methodological individualism' may be used as a basis for analytical urban research and it can be illustrated either by highly theoretical or by empirical studies.

10.5.2 New Urban Economics

Typical of this approach is the set of works gathered under the generic name of 'New Urban Economics'. These apply in a systematic way the concepts and methods of micro-economics to the analysis of urban form. William Alonso, in his thesis "*Location and Land Use*", first published in the sixties (Alonso, 1964) would appear to be the initiator of this field of research which has since proliferated, being reported mainly in the journals of Regional Science. Reviews were first given for instance by Richardson (1977) and then later on by Fujita (1989).

The interesting contribution of these works is that they integrate the principles of economic theory with a strict definition of concepts and a rigorous use of mathematical tools. The formalism starts with hypotheses about individual behaviour and ends with deductions of urban spatial patterns at the macrolevel. The main issues considered are the spatial distribution of population and employment densities, the market valuation of accessibility inside cities, the effects of nuisances and urban amenities on the location of households, the search for agglomeration economies by firms, and the genesis of polycentric agglomerations through the appearance of secondary job centres in the periphery (Gannon, 1993).

Since the first applications of Von Thünen's ideas about the game of bid-rents and the determining effect of the distance to the centre in explaining the spatial pattern of urban land use and the location choices of households (arbitrating between the cost of land and transportation costs in Alonso's model), several improvements have been brought into the models in order to relax the over simplistic hypothesis of 'homo economicus' behaviour: various forms of individual utilities have been integrated in order to take into account the diversity of urban actors (White, 1988), the mechanisms of social spatial segregation have been considered (Rose-Ackerman, 1975), the competition between traffic and other types of land use has been investigated (Solow, 1972, Kanemoto, 1975), more complicated metrics have been proposed for the description of urban space (Huriot and Perreur, 1990), and the effect of uncertainties and incomplete information on spatial forms have also been considered (Andrulis, 1982, Beaumont, 1990).

Two main difficulties are still preventing this highly theoretical approach from totally dominating urban research by its explanatory power and the precision of its analytical tools. The first, less applicable in British, American or Japanese cities, but highly relevant in other countries, concerns the respective role of market mechanisms and public decisions in the shaping of cities. Weighting the two types of effects within models still remains a delicate exercise. The second criticism is also linked with urban complexity. Despite some uncertainties or imprecisions, the models of the New Urban Economics are still inspired by concepts of relative optimisation in the determination of the urban spatial structure and of a trend towards an equilibrium in the explanation of urban change. In this respect, this approach is highly questionable for the theoreticians who hold cities to be evolutionary, complex systems.

10.5.2 Individual Behaviour in Urban Space

Another direction of research follows the same theoretical line with different methods of investigation. It is related both to empirical surveys of the spatial behaviour of urban actors and to their modelling in the framework of discrete choice theory, resting upon the principles of random utility theory (Macmillan, 1993).

Empirical research on spatial behaviour has shed new light on the underlying motivations and determinants of moves in urban space. Sample surveys are treated by using methods of categorical data analysis and of longitudinal data analysis. Reviews of this field of research are given by Timmermann and Borgers (1985), Wrigley and Longley (1984). The main applications have improved knowledge about residential mobility, identifying several types of residential strategy (depending on the characteristics and life story of individuals) and have completed previous investigations about the effects of the characteristics of the supply-side in the process of housing choice (Clark, Deurloo and Dielman, 1984, Van Wissen and Rima, 1988). The same type of approach has been applied to shopping trips, evaluating, for instance, the relative importance of multipurpose trips in the shopping behaviour of consumers. Timmermann (1980) contradicts one of the main hypotheses of central place theory - the assumption that the consumer chooses the closest place of supply for every level of service is demonstrated to be false. Consequences for the geometry of central places are derived, both inside cities and in regional networks of cities (Toninato, 1979). The rationales for the location of stores within cities have been investigated by the same type of survey and discrete choice modelling (Wrigley, 1988).

By trying to link in a very explicit manner the observations made about individuals and the aggregate structures, a method such as micro-simulation seems promising for meeting one of the hardest challenges imposed by the complex character of urban systems, which is the integration of these levels in a holistic way. Urban research still has a great deal to accomplish before improved knowledge about the behaviour of individuals, their aspirations and mental representations can be successfully integrated into aggregate models of cities. Empirical investigations and specific surveys are too rare. They should be related in a systematic way to urban modelling in order to provide relevant variables and parameters for the description of urban evolution. This need for relevant descriptors has become even more urgent since flexible and powerful tools, such as cellular automata, multi-agent systems, or fractal models, have been developed and used for urban simulations.

10.6 Simulating the Relation between Micro Behaviour and Macro Structures

10.6.1 Introduction

The growing demand for models which generate various urban spatial forms has encouraged the development of a class of methods which derive city maps from hypotheses made about the behaviour of urban actors under specific assumptions or constraints. Contrasting with the models quoted above (Section 4), these do not try to make exact predictions but are mostly exploratory. They are mainly simulation models, not making any hypotheses about the optimisation of a given constraint, nor supposing any trend towards equilibria.

This field of research also has a long history. Already in the fifties T. Hägerstrand was perhaps the first to develop simulation of spatial patterns over a grid representing a geographical space and the evolution of a spatial pattern over time according to precise hypotheses about imitation behaviour leading to spatial diffusion effects. The uncertainty about the exact location of potential imitators was simulated by a Montecarlo method, introducing stochasticity into a process guided by the spatial distribution of information around each decisor (mean information field). The method was used to reproduce the spatial extension of black ghettos in some North American cities (Morrill, 1965a) and for a theoretical simulation of the development of an urban system under a process of migration of population (Morrill, 1965b).

Various types of application have improved those first ideas and methodologies for developing simulation models of urban spatial patterns.

10.6.2 Urban Micro-Simulation

Social and spatial macro-structures may be considered the product of interactions among individuals, each of them following a life-time trajectory, with probability constraints on the transitions from one individual state to another, through household, professional or migratory 'events'. This idea has been applied to the reconstruction of global social or spatial structures from very large samples of simulated life stories (Holm and Oberg, 1984, 1989). Despite the large number of individuals surveyed (about 100,000), and despite giving a reasonable result at the aggregate level of the whole of Swedish society, the method still does not allow us to generate spatial patterns which are detailed enough for a transposition to the simulation of urban space.

Researchers at Leeds (Clarke, 1990), use an inverse approach to derive representative individual profiles from urban information available at the level of urban blocks. The probabilities of occurrence of a set of characteristics are computed from census block records, and sets of fictitious individuals are generated from a random game simulating them. This method allows further studies of the interactions between individual characteristics (for instance, unemployment and nationality). It also makes it possible to integrate various levels of analysis and to make predictions about needs in transportation or services. However, the method is rather cumbersome for a large number of city blocks and it is difficult to generalise the observations to other cities.

At much more detailed scales of observations and extending Hägerstrands' ideas about a 'time geography' (Hägerstrand, 1970), Janelle, Goodchild and Klinkenberg (1988) and Goodchild, Klinkenberg and Janelle (1993) discuss the question of spatial and temporal aggregation of data describing the everyday space-time behaviour of individuals within a city. More investigations in other urban contexts need to be made before the results can be generalised. Microanalytical urban modelling techniques will probably improve by integrating geographic information systems and microsimulation models (Wegener, 1986).

10.6.3 Cellular Automata

A cellular automaton consists of an array of cells which may be in any one of several qualitative states. At each iteration, the state of each cell may remain the same or change to another state according to the state of the neighbouring cells. Some models, like the famous 'game of life', although very simple (cells have only two possible states, alive or dead) may give rise to a full variety of spatial configurations. More complex cellular automata are useful in urban research if they allow for several possible states and sophisticated definitions of the neighbourhood. Tobler (1979) first mentioned cellular automata as the 'geographical type' of models. Couclelis (1985) drew attention to their use for modelling micro-macro relationships in spatial dynamic models and for deriving complex dynamics from simple rules (Couclelis, 1988). White (1991) applied this formalism to simulate the evolution of land use patterns within urban areas, according to the probability of change from one type of use to another. White and Engelen (1993) checked the fractality of the resulting simulated urban spatial structures. Portugali, Benenson and Omer (1994) applied the method for generating spatial distributions of communities inside a city according to their preferences for a given type of social neighbourhood. A model of urban growth using a two-state cellular automata to represent the long term process of infill of space in peri-urban fringes and also in an entire urban area has been developed by Batty and Xie (1994). This last paper, instead of exploring only the theoretical possibilities of the model, also tries to calibrate it with observed urban evolution.

10.6.4 Multi-Agent Systems

The methodology of multi-agent systems is a part of Distributed Artificial Intelligence. These systems are conceived as societies of autonomous agents who are able to act both on themselves and on their environment. The agents can communicate with other agents. Their behaviour is the result of their observations, knowledge and interactions with these agents. The determinants of an agent's behaviour have a local character; there is no global constraint on the system's dynamics. The general behaviour of the system is produced by the combination of actions of the agents. Such multi-agent systems can in some way be considered as experimental devices materialising the hypotheses of self-organisation theory.

One could imagine several applications of multi-agent systems in the urban realm, simulating for instance interactions between urban actors, applying principles of game theory or replicating a game like SIMCITY. However, the first application of this methodology has been given for another level of organisation of the urban systems. The method has been transposed in order to simulate the emergence of a system of towns and cities from a rural settlement system. Through a competition process for accumulating the surplus of agriculture in their vicinity, drawing benefits from long distance trade and developing new sources of profit from innovations, some settlements become towns and cities. Each settlement can be considered an agent whose behaviour is defined by the properties of the aggregate entity that it represents. Their interactions and transformations, which are simulated by the multi-agent system, produce a system of cities whose structure and evolution can be compared with empirical evolution of historical urban systems. The SIMPOP model is able to simulate several types of settlement systems, with more or less specialised and hierarchised patterns (Bura *et al.*, 1996).

The multi-agent system method of modelling has the same ability as cellular automata to simulate a large variety of spatial configurations. Compared to classical cellular automata, it also allows a greater variety of spatial interaction, including variable extension of the spatial range of interactions, which can be defined by the connectivity of a network as well as by contiguity, according to the characteristics of each agent. Moreover, instead of allowing only a few possible qualitative states for each cell, the method is able to integrate any qualitative or quantitative description of an agent, whose behaviour may be very complicated. Compared to the simulations by sets of differential equations, the method is very flexible. It allows for a much more detailed representation of spatial interactions and of some local properties and also makes it possible to introduce new agents or new rules into the model without changing the other parts. The risk in applications could be a loss in generality of the model, since the rules may easily be adapted to specific or local cases. It is also not easy to express them in the usual mathematical formulations, since the model mixes quantitative and qualitative rules.

As in the case of cellular automata, and more generally with all simulation models, the use of this method is more deductive than experimental. Well-known rules can be introduced and parameters evaluated at the level of individual agents in order to reproduce some observed behaviour at the global level of the system. However, it cannot be inferred from a satisfying correspondence between simulation and reality that this set of rules is the only one leading to such a result. The validation of the model remains a difficult problem.

10.6.5 Fractals

The conception of urban entities as fractal objects is another way of connecting them to the field of study of complex systems. Evidence of the fractality of urban forms has been given by several authors, by looking at city boundaries, by exploring the relationship between the area and perimeter of cities, or by computing fractal dimensions for the built-up areas of cities (Batty and Longley, 1994, Frankhauser, 1994). The same authors have also used simulation models for reproducing fractal structures in a process of urban growth. Batty and Longley mainly used a 'diffusion limited aggregation' approach, whereas Frankhauser was able to connect several more concrete rules of urban growth into a simulation model of fractal growth. Connection between building and network development, inclusion of old villages in an expanding urban area, and the preservation of green spaces around already built-up structures were the main processes leading to fractal urban forms. Other investigations have been made on the fractal character of a hierarchy of central places (Arlinghaus, 1985, 1989, Frankhauser, 1994), but these remain for the moment theoretical.

The application of fractal ideas to urban structures not only enables us to improve the morphological description of cities, it may also be useful in changing the references which are normally used for understanding their genesis. For instance, the concept of density applied to urban residential population or land use is not perhaps all that relevant to a real understanding of aggregate urban population or built-up area, whereas the concept of fractality (or density gradient) might well relate more usefully to a theory of urban development. The same transposition may also be fruitful for a direct interpretation of the hierarchical structure of systems of cities as a fractal process of occupation of space, instead of considering it as the abnormal product of some externalities (Pumain, 1997).

10.7 The Question of Chaos

In discussing the dynamics of complex systems, we cannot ignore the question of chaos. It seems generally however to be formulated at levels which are too abstract to be of relevance for urban research. It is true that some of the mathematical models used for simulating urban evolution, such as migration models, or models of change in land use, as well as May's famous logistic equation, may exhibit chaotic behaviour. But, in all such cases, the range of parameters which can give rise to chaos are outside the realm of variation of real world parameters. May's equation starts being chaotic at growth rates of over two hundred per cent per unit of time. The migration model derived from a master equation approach could exhibit chaotic behaviour if the same migratory preferences were maintained for a thousand years (Haag, 1994). The cellular automata simulating changes in land use would give chaotic results if a store changed its activity twice in the same day (White, 1983)! Should we think in more philosophical terms about the consequences of such highly hypothetical possibilities when applying those models to real urban objects, or can we neglect this behaviour as beyond the normal range of observations? After all, a regression line is a mathematical model with two infinite ends, but this vertiginous possibility never prevented anybody from making use of the restricted portion of the model applicable to real data.

Instead of asking: '*Is our world chaotic*?' perhaps the reverse question would be more interesting: '*Why is our urban world not chaotic?*'. What are the social regulations which prevent the rise of chaotic behaviour of urban systems and ensure relative stability to the trajectories of the objects that we are studying?

10.8 Conclusions

There is an obvious distance between the theoretical features which lead us to think about urban areas as complex systems and the results which are now available from applications of complex system theory to urban research. Either we have made wrong hypotheses about the nature of urban complexity, or the present state of urban research is too little advanced to solve the main theoretical difficulties.

What are the reasons for this relative failure? We can make a few hypotheses. Firstly, the fascination of pure mathematics is undeniable. The beauty of a cusp catastrophe curve, the attractiveness of trajectories spiralling towards a sink and the appealing aesthetics of strange attractors are more seductive than the ordinary poorly fitted discontinuous curves of empirical observations. It must also admitted that the mathematics of complexity are difficult to understand and explain, and maybe we should not expect at such an early stage to achieve both initiation to the theory and successful experimentation in the urban realm. Another possible explanation lies in the unease aroused by the apparent lack of control by the user on the results of the models. The normal output is only exploratory, revealing just a few of the possibilities which could happen in reality. The models demand a real effort of confidence from the decision-makers, who firstly have to believe in the specification of the model - this is more difficult with simulation models than with calculation ones, due to the difficulties of calibration when bifurcations are possible. Secondly, it is very difficult for actors to admit that their action very often may not be decisive, and that it may result in a fluctuation of little importance in the more general urban dynamics within which it is embedded.

So it is perhaps not surprising that the modelling of urban complexity has not until now received much in the way or feedback or interest from urban planners. It is more worrying that theoreticians have not been sufficiently concerned by the need for links between the new approaches and the empirical evidence required for the development of a comprehensive theory of urban change. Despite this gap in theoretical research, the present partial review shows that a number of new possibilities are now available for exploring the dynamics of urban structures. Perhaps the main challenge is that of developing better connections between the mathematical and computational tools and their applications to observed urban evolution and planning operations. The question of the transfer of the concepts should be investigated more deeply than it has been until now.

References

Alonso W. (1964) *Location and Land Use*, Harvard University Press, Cambridge Massachusetts.

Allen P.M. (1978) Dynamique des centres urbains, *Sciences et techniques, 50*, 15-19.

Allen P.M. (1991) Spatial Models of Evolutionary Systems: Subjectivity, Learning and Ignorance, in Pumain D. (ed.) *Spatial Analysis and Population Dynamics*, Congresses and Colloquia, 6, John Libbey, INED, London, Montrouge, 147-160.

Allen P.M., Boon F., Deneuburg J.L., de Palma A., Sanglier M. (1981) *Models of Urban Settlement and Structure as Dynamic Self-Organizing Systems*, US Department of Transportation, Washington DC.

Allen P.M., Sanglier M. (1978) Dynamic Models of Urban Growth, *Journal of Social and Biological Structures, 1*, 265-280.

Allen P.M., Sanglier M. (1979a) Dynamic Models of Urban Growth, *Journal of Social and Biological Structures, 2*, 269-298.

Allen P.M., Sanglier M. (1979b) A Dynamic Model of Growth in a Central Place System, *Geographical Analysis, 11*, 256-272.

Allen P.M., Sanglier M. (1981) Urban Revolution, Self-Organization and Decision Making, *Environment and Planning A, 13*, 167-183.

Allen P.M., Engelen G., Sanglier M. (1984a) Computer Handled Efficiency Stimuli Exploration, Final Report, Contract for Provinciaal Bureau Energiebesparing, North Holland.

Allen P.M., Engelen G., Sanglier M. (1984b) Self-Organizing Systems and the Laws of Socio-Economic Geography, *Brussels Working Papers on Spatial Analysis, serie A, 4*.

Amson J.C. (1975) Catastrophe Theory: A Contribution to the Study of Urban Systems, *Environment and Planning, 2*, 175-221.

Andrulis J. (1982) Intra-Urban Workplace and Residential Mobility under Uncertainty, *Journal of Urban Economics, 11*, 1, 85-97.

Arlinghaus S.L. (1985) Fractals Take a Central Place, *Geografiska Annaler, 67B*, 83-88.

Arlinghaus S.L. (1989) The Fractal Theory of Central Place Geometry, *Geographical Analysis, 21*, 2, 104-121.

Aydalot P. (1985) *Economie régionale et urbaine*, Economica, Paris.

Bairoch P. (1985) *De Jericho à Mexico, ville et économie dans l'histoire*, Gallimard, Paris.

Batten D. (1992) Network Cities, Infrastructure and Variable Return to Scale, Congress of the Regional Science Association, Palma de Mallorca.

Batty M. (1992) Urban Modelling in Computer Graphic and Geographic Information Systems Environments, *Environment and Planning B, 19*, 663-688.

Batty M. (1993) Cities and Complexity: The Implications for Modeling Sustainability, 4th International Workshop on Technological Change and Urban Form, 14-16 April, Berkeley.

Batty M., Longley P. (1994) *Fractal Cities, a Geometry of Form and Function*, Academic Press, London e San Diego.

Batty M., Xie Y. (1994) From Cells to Cities, *Environment and Planning B, 21*, 31-48.

Beaumont C. (1990) *Contribution à l'analyse des espaces urbains multicentriques*, thèse du 3me cycle, Universté de Dijon.

Beaumont J.R., Clarke M., Wilson A.G. (1981a) The Dynamics of Urban Spatial Structure: Some Exploratory Results Using Difference Equations and Bifurcation Theory, *Environment and Planning A, 13*, 1473- 1483.

Beaumont J.R., Clarke M., Wilson A.G. (1981b) Changing Energy Parameters and the Evolution of Urban Spatial Structure, *Regional Science and Urban Economics, 11*, 287-315.

Berry B.J.L. (1964) Cities as Systems within Systems of Cities, *Papers of the Regional Science Association, 13*, 147-163.

Bertuglia C.S., La Bella A. (eds.) (1991) *I sistemi urbani*, Angeli, Milan.

Bird J. (1977) *Centralities and Cities*, Routledge and Kegan, London.

Botero G. (1588) *Della ragion di stato: delle cause della grandezza delle città*, Luigi Firpo, Turin.

Boudon R. (1984) *La place du désordre, critique des théories du changement social*, Presses Universitaires de France, Paris.

Bura S., Guérin-Pace F., Mathian H., Pumain D., Sanders L. (1996) Multi-Agent Systems and the Dynamics of Settlement Systems, *Geographical Analysis, 2*, 161-178.

Camagni R. (1990) Strutture urbane gerarchiche e reticolari: verso una teorizzazione, in Curti F., Diappi L. (eds.) *Gerarchie e reti di città: tendenze e politiche*, Angeli, Milan.

Camagni R. (1992) *Economia urbana*, Angeli, Milan.

Camagni R., Diappi L., Leonardi G. (1986) Urban Growth and Decline in Hierarchical Systems: A Supply-Oriented Dynamic Approach, *Regional Science and Urban Economics, 16*, 145-160.

Casti J., Swain H. (1975) *Catastrophe Theory and Urban Processes*, RM-75-14, I.I.A.S.A., Laxenburg.

Casetti E. (1991) Testing Catastrophe Hypotheses, *Socio-Spatial Dynamics, 2*, 2, 65-80.

Cauvin C., Reymond H. (1985) *L'espacement des villes*, CNRS, Mémoires et documents de géographie, Paris.

Cauvin C., Reymond H., Schaub R. (1989) Accessibilté, temps de séjour et hiérarchie urbaine, *Sistemi urbani, 3*, 297-324.

Chevalier M. (1832) Exposition du sistème de la Mediterranée, *Le globe*, July 12.

Christaller W. (1933) *Die zentralen Orte in Süddeutchland*, Fischer, Iena.

Clark W.A.V., Deurloo M.C., Dielman F.M. (1984) Housing Consumption and Residential Mobility, *Annals of the Association of American Geographers, 74*, 29-43.

Clarke M. (1990) Regional Science in Industry and Commerce: from Consultancy to Technology Transfer, *Environment and Planning B, 17*, 257-268.

Clarke M., Wilson A.G. (1983a) The Dynamics of Urban Spatial Structure: Progress and

Problems, *Journal of Regional Science, 23,* 1-18.

Clarke M., Wilson A.G. (1983b) Exploring the Dynamics of Urban Housing Structures: A 56 Parameter Residential Location and Housing Model, 23th Congress of the Regional Science Association, Poitiers.

Couclelis H. (1985) Cellular Worlds: a Framework for Modeling Micro-Macro Dynamics, *Environment and Planning A, 17*, 585-596.

Couclelis H. (1988) Of Mice and Men: What Rodent Population Can Tell us about Complex Spatial Dynamics, *Environment and Planning A, 20*, 99-109.

Dacey M.F. (1974) One Dimensional Central Place Theory, *Studies in Geography, 21*, Northwestern University.

Dematteis G. (1985) Verso strutture urbane reticolari, in Bianchi G., Magnani I. (eds.) *Sviluppo regionale: teorie, metodi, problemi*, Angeli, Milan.

Dematteis G. (1990) Modelli urbani a rete, in Curti F., Diappi L. (eds.) *Gerarchie e reti di città: tendenze e politiche,* Angeli, Milan.

Dendrinos D.S. (1980) A Basic Model of Urban Dynamics Expressed as a Set of Volterra-Lotka Equations, Catastrophe Theory in Urban and Transport Analysis, Department of Transportation, Washington D.C.

Dendrinos D.S. (1984) The Structural Stability of the US Regions: Evidence and Theoretical Underpinnings, *Environment and Planning A, 16*, 1433-1443.

Dendrinos D.S., Haag G. (1984) Toward a Stochastic Dynamical Theory of Location: Empirical Evidence, *Geographical Analysis, 16*, 287-300.

Dendrinos D.S., Mullally H. (1981) Evolutionary Patterns of Urban Populations, *Geographical Analysis, 13*, 328-344.

Dendrinos D.S., Mullally H. (1985) *Urban Evolution: Studies in the Mathematical Ecology of Cities*, Oxford University Press, Oxford.

Diappi L., Ottana M. (1994) City Network System: The Neuronal Approach, Milan Polytechnic, unpublished paper.

Duby G. (ed.) (1980) *Histoire de la France urbaine*, vol. 1, Seuil, Paris.

Fik T.J., Mulligan G.F. (1990) Spatial Flows and Competing Central Places: Toward a General Theory of Hierarchical Interaction, *Environment and Planning A, 22*, 527-549.

Fletscher R. (1986) Settlement in Achaeology: Worldwide Comparison, *World Archaeology, 18, 1*, 59-83.

Forrester J.W. (1969) *Urban Dynamics*, MIT Press, Cambridge, Massachusetts.

Fournier R. (1990) Le modèle CARPE, thèse du 3me cycle., Université Paris X, Nanterre.

Frankhauser P. (1994) *La fractalité des structures urbaines*, Anthropos, Paris.

Fujita M. (1989) *Urban Economic Theory, Land Use and City Size*, Cambridge University Press, Cambridge.

Gannon F. (1993) Modèles de la ville et politiques urbaines optimales, thèse du 3me cycle, Université Paris X, Nanterre.

Gibrat R. (1931) *Les inégalités économiques*, Sirey, Paris.

Goodchild M.F., Klinkenberg B., Janelle D.G. (1993) A Factorial Model of Aggregate Spatio-Temporal Behaviour: Application to the Diurnal Cycle, *Geographical Analysis, 25,* 277-294.

Guérin-Pace F. (1992) *Deux siècles de croissance urbaine*, Anthropos, Paris.

Haag G. (1984) A Dynamic Model for the Migration of Human Populations, in Griffith D.A., Lea A.C. (eds.) *Evolving Geographical Structures*, Nato Advanced Institute Series, Martinus Nijhoff, The Hague.

Haag G. (1994) The Rank-Size Distribution of Settlements as a Dynamic Multifractal Phenomenon, *Chaos, Solitons and Fractals, 4*, 4, 519-286.

Haag G., Dendrinos D.S. (1983) Toward a Stochastic Dynamical Theory of Location: a Non-Linear Migration Process, *Geographical Analysis, 15*, 269-286.

Haag G., Weidlich W. (1984) A Stochastic Theory of Interregional Migration, *Geographical Analysis, 16*, 331-357.

Haag G., Munz M., Pumain D., Sanders L., Saint-Julien T. (1992) Interurban Migration and the Dynamics of a System of Cities, *Environment and Planning A, 24*, 181-198.

Hägerstrand T. (1970) What about People in Regional Science?, *Papers of the Regional Science Association, 24*, 7-21.

Haken H. (1977) *Synergetics, an Introduction*, Springer-Verlag, Berlin.

Harris B., Wilson A.G. (1978) Equilibrium Values and Dynamics of Attractiveness Terms in Production-constrained Spatial Interaction Models, *Environment and Planning A, 10*, 371-388.

Holm E., Oberg S. (1984) Migration in Micro and Macro Perspectives, *Scandinavian Population Studies, 1*, 61-84.

Huberman B.A. (1988) *The Ecology of Computation*, Elsevier, Amsterdam.

Huriot J.M., Perreur J. (1990) Distances, métriques et représentations de l'espace, *Revue d'économie régionale et urbaine, 2*, 197-237.

Janelle D.J., Goodchild M.F., Klinkenberg B. (1988) Space-Time Diaries and Travel Characteristics for Different Levels of Respondent Aggregation, *Environment and Planning A, 20*, 891-906.

Juilliard E. (1972) Espace et temps dans l'évolution des cadres régionaux, *Études de géographie tropicale offertes à P. Gourou*, 29-43, Mouton, Paris.

Kanemoto Y. (1975) Congestion and Cost-benefit Analysis in Cities, *Journal of Urban Economics, 2*, 246-264.

Kauffman S.A. (1993) *The Origins of Order*, Oxford University Press, New York.

Kremenec A.J., Esparza A. (1993) Modeling Interaction in a System of Markets, *Geographical Analysis, 4*, 354-368.

Le Maître A. (1682) *La métropolitée ou de l'établissement des capitales, de leur utilité passive et active, de l'union de leurs parties, de leur anatomie, de leur commerce*, B. Boekholt, Amsterdam.

Lombardo S., Pumain D., Rabino G., Saint-Julien T., Sanders L. (1988) Comparing Urban Dynamics Models: The Unexpected Differences in Two Similar Models, *Sistemi urbani, 2*, 213-228.

Lung Y. (1985) A la recherche de nouvelles techniques ou d'un nouveau paradigme: à propos d'approches récentes de l'espace économique, *Cahier d'économétrie appliquée, 1*, 69-77.

Macmillan W.D. (1993) Urban and Regional Modelling: Getting it Done and Doing it Right, *Environment and Planning A, 25*, Anniversary Issue, 56-68.

Marchand C. (1981) Maximum Entropy Spectra and the Spatial and Temporal Dimensions of Economic Fluctuations, *Geographical Analysis, 13*, 95-116.

May R.M. (1976) Simple Mathematical Models with Very Complicated Dynamics, *Nature*, 261, 459-467.

Mees A.I. (1975) The Revival of Cities in Medieval Europe. An Application of Catastrophe Theory, *Regional Science and Urban Economics, 5*, 403-425.

Moriconi-Ebrard F. (1993) *L'Urbanisation du monde depuis 1950*, Anthropos, Paris.

Morrill R.L. (1965a) Migration and the Spread and Growth of Urban Settlement, *Lund*

Studies in Geography, B, 26, Gleerup, Lund.
Morrill R.L. (1965b) The Negro Ghetto: Problems and Alternatives, *Geographical Review*, 339-361.
Muller J.C. (1983) La cartographie des espaces fonctionnels, *L'Espace Géographique, 2,* 142-152.
Mulligan G.F. (1984) Agglomeration and Central Place Theory: A Review of the Litterature, *International Regional Science Review, 9*, 1-42.
Orishimo I. (1987) An Approach to Urban Dynamics, *Geographical Analysis, 3*, 200-210.
Papageorgiou Y.Y. (1980) On Sudden Urban Growth, *Environment and Planning A, 12*, 1035-1050.
Portugali J., Benenson I., Omer I. (1994) Socio-Spatial Residential Dynamics Stability and Instability with a Self-Organizing City, *Geographical Analysis, 26*, 321-340.
Pred A. (1966) *The Spatial Dynamics of US Industrial Growth 1800-1914*, MIT Press, Cambridge.
Pred A. (1977) *City Systems in Advanced Societies*, Hutchison, London.
Pred A., Tornqvist G. (1973) Systems of Cities and Information Flows, *Lund Studies in Geography, B,* 38, Gleerup, Lund.
Prigogine I., Stengers I. (1979) *La nouvelle alliance*, Gallimard, Paris.
Pumain D. (1982) *La dynamique des villes*, Economica, Paris.
Pumain D. (1991) City Size Dynamics in Urban Systems, Symposium on Dynamic Modeling and Human Systems, Department of Archaeology, Cambridge.
Pumain D. (1992) Les systèmes de villes, in Bailly A., Pumain D., Ferras R. (ed.) *Encyclopédie de Géographie*, Economica, Paris.
Pumain D. (1993) L'espace, le temps et la matérialité des villes, in Lepetit B., Pumain D. (ed.) *Temporalités urbaines*, Anthropos, Paris, 135-157.
Pumain D. (1997) Pour une théorie évolutive des villes, *L'Espace Géographique, 2*.
Pumain D., Saint-Julien T. (1978) *Les dimensions du changement urbain*, CNRS, Paris.
Pumain D., Sanders L., Saint-Julien T. (1989) *Villes et auto-organisation*, Economica, Paris.
Rabino G. (1993) The Evolution Theory and Urban Modeling, 8th Colloquium in Theoretical and Quantitative Geography, Budapest.
Racine J.B. (1993) *La ville entre Dieu et les hommes*, Presses bibliques et universitaires, Genève, Anthropos, Paris.
Reclus E. (1895) The Evolution of Cities, *The Contemporary Review, 67*, 2, 246-264.
Reymond H. (1981a) L'ouverture informatique en géographie urbaine: de l'analyse multivariée socioéconomique à la simulation organique des systèmes urbains, *Informatique et Sciences Humaines, 50*, 9-20.
Reymond H. (1981b) Une problématique théorique de la géographie: plaidoyer pour une chorotaxie expérimentale, in Isnard H., Racine J., Reymond H. (eds.) *Problématiques de la géographie*, Presse Universitaires de France, Paris.
Reynaud J. (1841) Villes, in Encyclopédie Nouvelle, in Robic M.C., Cent ans avant Chistaller, une théorie des lieux centraux (1982) *L'espace géographique*.
Richardson H.W. (1977) *The New Urban Economics and Alternatives*, Pion, London.
Roehner B., Wiese K. (1982) A Dynamic Generalization of Zipf's Rank-Size Rule, *Environment and Planning A, 14*, 1449-1467.
Rose-Ackerman S. (1975) Racism and Urban Structure, *Journal of Urban Economics, 2*, 85-103.
Rosser J.B. (1991) *From Catastrophe to Chaos: A General Theory of Economic*

Discontinuities, Kluwer, Boston.

Rozenblat C. (1992) Le réseau des entreprises multinationales dans le réseau des villes européennes, thèse du 3me cycle, Université Paris I.

Sanders L. (1992) *Système de villes et synergétique*, Anthropos, Paris.

Sanders L., Pumain D. (1992) La formalisation du changement dans trois modèles de dynamique urbaine: une étude comparative, *Revue d'économie régionale et urbaine, 5*, 773-794.

Solow R.M. (1972) Congestion, Density and the Use of Land in Transport, *Swedish Journal of Economics, 1*, 161-173.

Thom R. (1974) *Modèles mathématiques de la morphogénèse*, Christian Bourgeois, Paris.

Timmermann H. (1980) Central Place Theories and Spatial Shopping Behaviour, Ph.D. thesis, University of Nijmegen.

Timmermann H., Borgers A. (1985) Spatial Choice Models: Fundamentals Trends and Prospects, University of Technology, Eindhoven.

Tobler W. (1979) Cellular Geography, in Gale S., Olsson G. (eds.) *Philosophy of Geography*, Reidel, Dordrecht, 379-386.

Toninato G. (1979) Une nouvelle approche de la théorie des places centrales, le concept de centralité imparfaite, thèse du 3me cycle, Université de Strasbourg.

Van Wissen L., Rima A. (1988) *Modelling Urban Housing Market Dynamics*, North Holland, Amsterdam.

Vauban S. (1707) Projet d'une dîme royale, in Pirou S., Simiand F. (eds.) (1933) *Collection des principaux économistes*, Félix Alcan, Paris.

Wegener M. (1983) A Simulation Study of Movement in the Dortmund Housing Market, *Tijdschrift voor Eonomische en Sociale Geografie, 73*, 267-281.

Wegener M. (1986) Integrated Forecasting Models of Urban and Regional System, *London Papers in Regional Science, 15*, 9-24.

Wegener M. (1994) Operational Urban Models State of the Art, *Journal of the American Planning Association, 60*, 1, winter.

Wheatley P. (1971) *The Pivot of the Four Quarters*, University Press, Edinburgh.

White M.J. (1988) Location Choice and Commuting Behaviour in Cities with Decentralized Employment, *Journal of Urban Economics, 24*, 129-152.

White R.W. (1977) Dynamical Central Place Theory, *Geographical Analysis, 9*, 226-243.

White R.W. (1978) The Simulation of Central Place Dynamics: Two Sector Systems and the Rank-Size Rule, *Geographical Analysis, 10*, 201-208.

White R.W. (1983) Chaotic Behaviour and the Self-Organisation of a Retail System, *Brussels Working Papers on Spatial Analysis, A, 3*.

White R.W. (1991) A Cellular Automata Approach to the Evolution of Urban land Use Patterns, 7th Colloquium of Theoretical and Quantitative Geography, Stockholm.

White R.W., Engelen G. (1993) Cellular Automata and Fractal Urban Form: a Cellular Modelling Approach to the Evolution of Urban Land-Use Pattern, *Environment and Planning A, 25*, 1175-1199.

Whitehead J.W.R. (1987) *The Changing Face of Cities*, Blackwell, Oxford.

Wilson A.G. (1981) *Catastrophe Theory and Bifurcation: Application to Urban and Regional System*, Croom Helm, London.

Winiwarter P. (1984) Iso-Dynamics of Population-Size Distribution in Hierarchical Systems, rapporto presentato al meeting annuale della Society for General Systems Research, Los Angeles.

Wrigley N. (1988) *Store Choice, Store Location and Market Analysis*, Routledge, London.

Wrigley N., Longley P.A. (1984) Discrete Choice Modeling in Urban Analysis, in Herbert D.T., Johnston R.J. (eds.) *Geography and the Urban Environment*, 45-94, John Wiley and Sons, Chichester.

Zhang W.B. (1990) Stability versus Instability in Urban Pattern Formation, *Socio-Spatial Dynamics, 1*, 41-56.

11. Beyond Complexity in Urban Development Studies

Roberto Camagni

11.1 Introductory Remarks

Denise Pumain, in this volume, presents us with a fascinating fresco of fifteen years of theoretical elaborations on the dynamics and evolution of urban systems. The picture is wide-ranging: a detailed inspection is made of the specific features of different approaches, their innovative characteristics, limits, and logical position with respect to each other. It would seem difficult to add further suggestions or integrations to this picture, given its completeness, the general consistency, and the careful selection operated among the myriads of contributions to a burgeoning literature.

Her contribution gives us an idea of the importance of the analytical effort which took place during the 1980s and which permitted a visible methodological and theoretical jump, representing a qualitative discontinuity with respect to previous research. New paradigms are now being applied to the analysis of urban phenomenon (the paradigm of complexity and self-organisation), new analytical methods and approaches are being utilised, borrowing from other sciences (catastrophe theory, bifurcation theory, synergetics, the theory of fractals) and interesting results have been attained in developing dynamic versions of older economic and spatial models and in the utilisation of nonlinear specifications. There seem to be new possibilities of achieving unexpected and innovative results, from multiple solutions to periodic and chaotic ones resulting from minor shifts in the control (or 'slow') variables of the systems. I refer here in particular to the dynamic version of the spatial interaction model by Harris and Wilson (1978) and the macroeconomic models of urban growth by Miyao (1981).

A complex and multifaceted evolutionary theory of urban development is emerging, parallel to the developments in economics over the same period, which gave rise to an evolutionary theory of the firm (Simon, 1972, Nelson and Winter, 1982, Dosi *et al.*, 1988). In addition, a new concept of time has emerged, a time which is very different from the mechanistic and chronological traditional concept (a time which is reversible, running in either direction from past to future and from future to past). The new concept of time is based on creative and morphogenetic processes which we could define as the rhythm of innovative processes in space[1]. Time, when conceived in this way, is made up of synergy and feedback effects, due to the random and cumulative nature of combinatorial processes.

The central feature of this conceptualisation is irreversibility, which is able to describe and explain two apparently contradictory phenomena: on the one hand, the cumulativeness and sequentiality of (spatial) innovation processes, which always have their roots in specific (local) pre-conditions and, on the other, the discontinuity that they determine. This is visible in the unexpected breaks in the static pattern of spatial and production conditions, characterised by perfect information and predictability[2]. Irreversibility means path-dependency, i.e. the impossibility of leaving a path once it is started or of starting the same path twice, even in apparently similar conditions. Urban 'dynamics' therefore becomes urban 'evolution': a different scientific research programme concerned with the interpretation of structural change, territorial innovation and chance.

In spite of these theoretical achievements, it is worth giving some further thought to the present state of that part of the regional science literature coming under the label of 'complexity approaches to urban research'. In Pumain's contribution to this book there are a number of critical statements that in my opinion deserve greater attention. She says that "a large number of the models built have never been applied, nor have they been rigorously compared with other models" (p. 324), and that "There is an obvious distance between the theoretical features which lead us to think about urban areas as complex systems and the results which are now available from applications of complex system theory to urban research"

---

[1] This concept of time finds its premises in the reflections of Bergson and Heidegger. According to Bergson (1989), "time is invention or it is nothing at all" (p. 341); according to Heidegger and Deleuze, time 'plays' in space, generating a 'space-of-play' or of 'representation' ('*Zeit-Spiel-Raum*') through the infinite potentialities of combinatorial processes. See Camagni, 1992, and Camagni, 1995a.

[2] Aydalot's model of 'break and continuity' ('rupture/filiation'), proposed within GREMI's reflection on territorial innovation processes is based on the same apparent contradiction between sudden disruption and cumulativeness. See Aydalot, 1988.

(p. 354), leading us to throw some doubts either on the rightness of our hypotheses or the current state of the theoretical research. These critical observations should, in my opinion, be taken seriously. Although we are still at an early stage, it is time for a general tentative assessment of the results achieved by this theoretical trajectory.

The present contribution therefore focuses on this issue with a view to indicating some possible 'cross fertilisation' and research directions that could allow new theoretical achievements from complex system theory applied to urban phenomena.

11.2 From Complexity to Simplification

11.2.1 Introduction

The initial sensation is that after the astonishing acceleration of the 1980s, pure and applied research on urban dynamics and evolution has been, for a number of years, in a position of stalemate. My feeling, like that expressed by Pumain, is that the enormous methodological effort of the last decade has proved relatively sterile in terms of the interpretation of urban phenomena and provision of policy suggestions.

This stalemate could provide the opportunity for a critical assessment of the methodological achievements of the last decade (and be advantageous for textbook writers, like myself, since there is less need for continuous updating!), but is also intriguing and to a certain extent worrying. Why should it be so? What is wrong with the complexity approach, which was welcomed by the scientific community with such enthusiasm?

11.2.2 Towards a New Field of Specialisation?

My first tentative answer to these questions, which is subject to further reflection and to many possible exceptions, is that in the regional sciences the complexity approach, though originally developed in a multidisciplinary framework, has remained mainly the field of inquiry of a single discipline. Unlike previous breakthroughs in regional science thinking, self-organisation and complexity-related models have mostly been elaborated by quantitative geographers (in the Anglo-Saxon and French tradition). When outsiders entered the same field - namely economists, spatial theorists and planners - they were so fascinated by the new approach that they remained locked into the same logic and methodology.

Previous breakthroughs in regional science were generally the result of a real interdisciplinary effort. If we take for example the 1960s, which were characterised by the emergence and development of the spatial interaction and land-use models: economists like Alonso, Stevens and Mills, land-use scientists like Perloff and Muth, model builders like Lowry, geographers with a physics background like Wilson, and planners like Britton Harris collectively built up a coherent and comprehensive new approach to urban spatial theory with strong normative and policy fall-outs.

By the same token, the 1970s saw the convergence of some 'mainstream' economists like Solow and Mirrlees towards urban problems. An easy integration of their ideas by the regional science community gave rise to interesting synergies and that subsystem of regional economic thinking that was (self-)labelled 'new urban economics'. Once again, the normative outcomes of the new general approach were highly relevant and innovative, especially in the sphere of local public finance.

The complexity approach seems to have followed a different trajectory. The language of the pioneers and the outstanding representatives of this approach - like Papageorgiou, Dendrinos, Wilson, Leonardi, Allen, Pumain, Bertuglia - is the language of quantitative geography and mathematical ecology. Very rarely have economic concepts and methods been integrated into this methodology, with the notable exception of the proof of asymptotic equivalence between the spatial interaction entropy model and the microeconomic models of discrete choice based on random utility theory, due to Giorgio Leonardi, 1985 (who, with his creative intelligence, achieved an interdisciplinary view within his own culture)[1].

11.2.3 Towards an Over-Simplification?

Within the complexity approach, the analogies with other sciences - from biology and the chemistry/physics of dissipative processes - have generally been proposed in a mechanistic way, paying more attention to the originality of the methodological approach than to the real meaning of these analogies for urban analysis. The result is that the application of

---

[1] Interestingly, and not by chance, the sub-trajectory of logit and probit models, which derive from random utility theory, appears one of the most fruitful in terms of theoretical advances and practical application to transport policy, as shown in the recent works of Nijkamp and Reggiani (1988, 1991). We remain nevertheless mainly in the field of applied quantitative geography.

such models to the urban sphere seems to have been made with little effort to show which urban development problems or policy issues could be tackled through their application.

In aggregate ecological models, for example, the city is observed through only one indicator, its size. Simulations show us complex potential dynamics, which could easily refer to any other system subject to complex evolutionary paths. In biology, such a process of over-simplification of the variables and the dimensions of the problem may result from the acknowledgment of the 'complexity' of causal and interdependence relationships underlying the observed phenomena and an explicit acceptance of the limits of our understanding. In this field in fact some empirically defined macro-behaviours, whose structural relationships remain widely unknown, are studied. But the explicitly stated aim of research programmes in other sciences, such as physics, economics and also urban economics, is that of examining these structural relationships, isolating the relevant variables, and analysing system behaviour in particular conditions of nonequilibrium. It does not seem fully acceptable that the lack of a commonly agreed theory or conjecture among scholars should be used, as by Dendrinos and Mullally (1985), to justify the absence of any effort to explain the simulated time path of a city's size.

A symptom of this contradiction is the uncritical transfer of the concept of carrying capacity - which is highly relevant in the economics and ecology of natural environments - to the field of urban ecology, where its relevance is doubtful (Camagni, 1995b). An effect of the insufficiently critical appraisal of external concepts and methods is the limited exploitation, on the other hand, of the potential of the prey-predator model, beyond the study of Malthusian or Darwinian scarcity of physical resources[1].

In simulation procedures, description frequently prevails over interpretation. I acknowledge that the description and reproduction of dynamic processes, and in particular of complex ones, is quite different from the traditional description of static, structural elements of the urban environment, and that a quantum jump in scientific interest has been achieved. Nevertheless, it often happens that the new interesting 'facts' emerging from advanced model simulation - like a change in the nature of the solutions, a sudden bifurcation or catastrophe, a sudden structural change - only result from the mathematical properties of the models, and

[1] Interesting exceptions to this critique are some recent applications of the model to transport economics and planning, where congestion is considered as the predator and mobility demand as the prey.

are coupled *ex-post* with real 'facts'. The values attached to the points of structural change (and these points themselves) often have no real significance or interpretive meaning, beyond the fact that they indicate a discontinuity in the behaviour of the system.

Finally, we could point to the non-infrequent cases where complexity is used in theoretical terms not so much with a view to achieving a deeper understanding of the real world but, paradoxically, to escape from this understanding. Complexity in fact is sometimes used more as an alibi than as a challenge, ending up with a convenient listing of the multiple interactions it would be 'interesting to analyze', but that nobody will ever actually analyze. To simply say that 'everything holds together' - which is true, but probably trivial - or to underline the intrinsic unpredictability of events may be useful if we wish to criticise methodological determinism or some econometric simplifications but, if detached from the search for new theoretical tools, may easily turn into regressive scientific propositions and a disincentive to intellectual curiosity.

11.2.4 Some Open Questions

It seems to me that the best theoretical exercises, i.e. those that utilise the complexity approach in the most innovative and methodologically sound way, succeed in merging all the theoretical premises of the new paradigm - multiplicity of interactions, nonlinearity of relationships and irreversibility - and that optimistically explore its heuristic potentialities, leaving aside the comfortable niches provided by the 'weak' interpretations of the 'weak thinking' approach. These do not interpret the crisis in traditional spatial theory as the logical impossibility of formulating laws and conjecturing causal relationships, but as the need to consider the possibility of an exponential multiplication of possible outcomes, resulting from the co-presence of a multiplicity of interrelationships and feedbacks of a micro nature in particular conditions of instability[1] . Prigogine has said that 'far from equilibrium, matter sees': complex systems, both natural and social, self-organise in unpredictable forms.

The big advantage of the new paradigm, in relation to previous and more traditional ones, lies precisely in the possibility it opens to explore 'possible worlds' through computer simulation. However, this poses a new and important methodological question: what validation and assessment

---

[1] A reflection on the effects of the full acceptance in urban planning theory of the complexity approach and of 'procedural rationality' à la Simon (1972) is presented in Camagni, 1996.

criteria can we use in order to distinguish between good and bad models, or to know whether to accept or refuse their results? It is clear that the traditional econometric validation criteria are useless in this case, as the new paradigm tries to capture the emergence of new structural relationships and explicitly refuses the extrapolation of past relationships, however complex.

In some ways, the models and procedures utilised nowadays bear strong conceptual similarities to the Forrester family of dynamic models of the late 1960s, though in a more modern, conscious way and with a sounder mathematical basis, e.g. in their use of nonlinear relationships and stochastic variables. The Forrester models (1969) were in fact also used in the simulation of urban dynamics and were based on the interactions between a large number of variables whose relationships remained largely unexplained. These models had a double theoretical advantage that we can perhaps appreciate more fully today: firstly, they underlined the relevance of endogenous variables and decision-making processes in the determination of the development path of the city and, secondly, they allowed a 'general equilibrium' simulation of the evolution of the urban system, focussing on the feedback effects that keep such a system together, in spite of the possibility of diverging or explosive behaviour on the part of single variables. This type of model has, however, as a consequence of the theoretical weakness and lack of sound validation criteria, been completely abandoned, except by a limited number of fans.

Today, it is possible to say that evolutionary theories stemming from the complexity approach are facing a similar challenge. In fact they need to:

a. supply sound validation criteria as regards the model structure and results;
b. demonstrate, in their internal logic, the reason for the stability of territorial systems, despite the fact that synergy elements, increasing returns and positive feedback effects multiply the possibility of explosive trends in some variables or specific sub-systems (Pumain rightly points out how two hundred years of urban development and three technological revolutions have failed to alter the structure of the urban hierarchy in France);
c. strengthen the theoretical and interpretative base of the existing models.

This is by no means an easy task. We gladly leave the first to methodology experts and philosophers of science! In our modest view, what model builders can be expected to provide are some likelihood criteria that could lead to the choice of likely parameters for simulation procedures, and the consistency of results not with a single empirical pattern but with some more general, macro-territorial, relationships which have proved to be valid in different space-time conditions[1].

The second aspect is relevant also in a wider perspective, as it concerns the consistency between micro and macro behaviours, a logical relationship that in economics is usually handled by going from the micro to the macro level (the micro-foundations of macro-economics). What is probably needed in our field is the reverse path: to indicate how, why and which macro-constraints guaranteeing the stability and coherence of the entire system may become constraints to micro-behaviour, i.e. the macro-foundations of micro-economics. As Pumain rightly says, the question we should be asking is: 'why is our world *not* chaotic'?

In spatial economics, we can pose a further interesting question: at which territorial level do macro-constraints work? Let us examine a common constraint in macro-economics, the constraint of the balance of payments or trade balance. Is there an equilibrium to be maintained between imports and exports in a given territory? Which feedbacks are triggered when this does not occur in the short or medium run? What is the relevant territorial level - the city, the region, or the nation? Moreover, in a condition of monetary union, when the accounting of a trade balance becomes impossible, can we still confirm that such an equilibrium is relevant?

When the model simulates an explosive or implosive trend for a given territory, are we sure that in the long run there will not be any feedback effects capable of correcting or reversing these trends? At the level of the nation state, if development trends imply a surplus or a deficit in the trade balance, a condition of imbalance cannot be maintained without generating adjustments in the exchange rate or the internal income level - two effects that are intrinsically rebalancing, even in the absence of explicit policies. At the regional level, a permanent deficit may be maintained (or generated) by external income transfers, external investments or sales of internal assets such as firms or buildings. In the absence of these flows, population has to emigrate. May this be considered as an equilibrium? Are there longer run forces that can re-establish internal competitiveness?

Let us now take up the urban level: here there is greater mobility of people and capital, and therefore fewer constraints exist with respect to trade surplus (competitiveness) or deficit (decline). In this case probably the strongest feedbacks come from the environment (congestion, quality of

[1] One of these macro relationships or statistical uniformities, regarding urban systems, is the rank-size rule.

life) or socio-psychological aspects (the collective reaction to a condition of urban decline). Once again, the proper countervailing forces have to be identified, specified in terms of their appropriate development time and included in the simulation models. Without doing this, it is difficult to interpret or understand why, notwithstanding some catastrophic trends in single cities, the urban structure as a whole in the long run is relatively stable.

The last aspect, closely linked to the previous one, regards the strengthening of the theoretical underpinnings of the simulation models. This is in my opinion the most important and urgent task, rather than, as some have suggested, the improvement of the goodness of fit of the models, for example through the multiplication of exogenous variables or constraints. When attempts have been made to apply models to empirical realities, as in the case of some self-organisation models utilised in the planning process, the results have been poor, and have risked trivialising the entire theoretical approach - which, we should remember, is based on the morphogenetic role of *endogenous* relationships.

11.3 New Theories for New Problems

Perhaps the most worrying aspect of the new wave in urban theoretical and methodological approaches is their inability to detect empirical problems worthy of new theorisation.

Highly advanced methodologies are frequently applied, for example, to 'old' migration problems which have already been thoroughly explored with more traditional techniques and are probably of little current relevance due to a slow down of population mobility. Similarly, it is astonishing to observe the continued interest in the problem of supermarket location in cities, hardly a fascinating subject from the theoretical point of view, and already the trigger of Reilly's methodological elaborations sixty years ago, of Lakshmanan's location model thirty years ago and of Harris and Wilson's most valuable contribution on dynamic interaction models fifteen years ago. Why this wide gap between the ambitious methodological targets and the low profile of interpretive targets?

Perhaps this is the source and the most relevant symptom of the maturity disease not only in urban dynamics but in the entire Regional Science field, discussed autoritatively in recent issues of such journals as *Papers in Regional Science* and the *International Regional Science Review.* Rightly and convincingly, Denise Pumain shows us the wide spectrum of open theoretical problems which concern the nature, role, performance and behaviour of the city, rarely taken up directly by the complexity approach.

11.4 A Changing Concept of Time

In assessing the real theoretical advance achieved by the complexity approach, a further danger concerns the treatment of time.

As already stated, the complexity approach potentially provides us with a totally new conceptualisation of time as the rhythm of structural change and morphogenesis. Sometimes though, especially when models are used to interpret an empirical reality, there is a risk that simulation exercises trivialise this conceptualisation of time, due to the lack of explanation of the relationship between model-time and real time.

The meaning of an iteration within the model is often unclear: either because many different kinds of reactions and feedbacks take place, implying different paces and rhythms, or because the time needed for each reaction to be fully developed is not clear in theoretical terms. In these conditions, the problem is often overlooked or evaded.

The complexity approach, having freed the modern conception of time from the limits of mechanistic and chronological interpretations intrinsic to traditional dynamic modelling, risks losing much of its advantage as a result of the lack of a convincing theory of change. In fact, it often ends up with a return to a nominalistic and conventional interpretation of time, i.e. time as the unit measure of the iterations of the model.

We now understand that change is an unavoidable feature of complex systems; we know approximately how it happens and are able to model it in abstract terms. What we still do not completely understand is *why* it happens, through what processes it balances cumulativeness and randomness, rule and invention, positive and negative feedbacks, unpredictable accelerations and constrained dynamics, breaks and continuity.

Building a true theory of urban (and regional) change will be in my opinion the main challenge of future theoretical thinking in Regional Science. Not having achieved this is not of course the fault of complexity approaches. On the contrary, they have helped us to understand that this is the main theoretical problem today, and have provided us with the right analytical and methodological tools.

As we have already said, a much stronger interdisciplinary effort is needed - an idea also expressed in a recent excellent collective book edited by

Lepetit and Pumain (1993). As a modest contribution to this wide and ambitious research programme, I suggest in the next section a series of research priorities.

11.5 Some Priorities for the Researcher

11.5.1 To Re-Introduce Economics into Urban Complexity Theory

Economics needs to be re-introduced into the theoretical reflections on complex urban systems. This should not however be in the sense pursued by the 'new urban economics', which represents the city as a campsite where caravans of different size locate and move freely - the monocentric city approach is probably not suitable for the new tasks, as it concentrates on marginal adaptations around an equilibrium point and is too tied to a single principle of urban organisation: the accessibility principle[1] . Nor should it be an attempt to create dynamic versions of neoclassical or neo-Keynesian macro-economic models of urban growth, which have probably achieved their ultimate results[2] , being constrained by assumptions of perfect competition and by an approximate, mainly exogenous, treatment of technological change.

Other elements of economic reasoning ought to be integrated within a more general reflection on urban evolution. My own research programme has for some years attempted to include the following elements:

A. *Schumpeterian innovation processes,* which are the true evolutionary forces and triggers of structural change in the economic sphere. This element was successfully integrated into a recent family of dynamic urban self-organisation models developed in the eighties at Milan Polytechnic by the author with a number of colleagues (Camagni,

---

[1] In order to fully take into account the complexity of urban systems, other principles have to be included and integrated: agglomeration, interaction, hierarchy and competition principles; see Camagni, 1992.

[2] More specifically, they have overcome the superficial criticisms of the neoclassical model, demonstrating that it is perfectly able to allow for cumulative growth paths, sudden bifurcations and even catastrophic outcomes, when non-linear relationships are taken into account (see Miyao, 1981, Camagni, 1992). Similar results may be achieved by introducing learning processes, as in the new wave of endogenous growth models represented by Lucas, 1988, Romer, 1990. See the interesting debate presented in the *Journal of Economic Perspectives, 1*, 1994, with a critical comment by Solow, 1994.

Diappi and Leonardi, 1985).
The eclectic nature of these models, called SOUDY - supply oriented urban dynamics - far from bringing a patchwork solution, proved to be highly beneficial, allowing the close integration of economic and spatial aspects. In fact, these models include:

- elements of urban self-organisation, triggered by innovative processes (the birth or the attraction of new functions in the city),
- dynamic urban interaction elements, based on an urban attractiveness function defined in terms of location profits,
- dynamic elements of the hierarchisation of centres, tied to the relative profitability of innovations occurring in the single centres, and to spatial demand constraints.

Through a dynamic simulation process, some relevant common wisdom hypotheses have been demonstrated. First of all, that a unique 'optimal' city size does not exist - there is a series of efficient urban dimensions which depend on the rank of the highest functions that the city has developed; secondly, that for the formation of a complete urban hierarchy one needs increasing returns to urban scale (reached in the model through the successive attraction or development of higher-order functions by the single city) (Camagni, 1992);

B. *distributional aspects of a mainly Ricardian origin.* This is a broad field of economic research that has been to a large extent neglected in spatial economic theory, with the exception of rent theory. It is not sufficient to consider only the remuneration of the production factor represented by urban land and its variation in urban and rural space. We also need to include in our reasoning on spatial heterogeneity all the other income aspects that differentiate neighbourhoods within the city, cities of different sizes and, more in general, the 'city' and the 'countryside'.

An interesting example of the integration of distributional aspects into a model of spatial dynamics *à la* Volterra-Lotka was presented recently (Camagni, 1992). The model simulated the interaction between urban profits (determining growth) and urban rents (the predator), cyclically reducing the growth of the city. The economic logic of the model was intrinsically classical and Ricardian.
Also in this case, the results of an eclectic approach look fruitful. The superimposition of a sound economic fabric, provided by

Schumpeterian business cycles theory and Ricardian rent/profit conflict theory, over the skeleton of the ecological mathematical model would allow, in my opinion, the achievement of two important results: firstly, to overcome the evident limits in economic relevance and interpretive capability of the early applications of these models in the urban sphere and, secondly, to supply the urban life-cycle hypothesis with a sound and relevant economic interpretation. In this last respect in fact, it was increasingly clear that the model was mainly a descriptive tool, based on a simple physical interaction between the urban core and the urban ring;

C. *Marxian power relationships and conflicts,* as reflected in income distribution between different territorial units. If the world of complexity theory remains a nonconflictual world, its ability to interpret real phenomena will remain limited.

The model of classical economics, based on the 'contradiction' between city and countryside, which has been used to interpret the evolution of territorial relationships for centuries, may still be an interesting analytical tool even today and deserves to be 'revisited'. The countryside is increasingly the *locus* of diffused industrial activities - the Common Agricultural Policy currently provides incentives to cease or reduce agricultural production in favour of providing land of high environmental value for the advantage of city dwellers (Camagni, 1994). The city, having lost most of its manufacturing activities, is increasingly becoming the *locus* of tertiary activities and company headquarters, the natural habitat for the economic and power élites.

What will be the effects of this profound structural change? Does the previous 'contradiction' still characterise the city and the countryside, embedded into the relationships between tertiary-quaternary activities and diffused industrial activities? Or is it shifting to other geographical spaces, e.g. to the relationships between the Norths and the Souths of the planet?

Some early tentative answers were suggested by Aydalot and Camagni (1986), utilising Baumol's model of urban crisis (Baumol, 1967) in the context of a city-countryside dichotomy. The most interesting theoretical result related to the relevance of aspects of distribution, determined by the change in relative prices between the two spaces. In fact, it was shown that, even in the absence of monopolistic practices by the city, if the two territories are characterised by a specific productive 'vocation' (industry in the dispersed countryside and services in the city) and if services are crucial for industrial productivity increases, then the city may postpone the crisis, predicted by Baumol, through a rise in the relative price of its service production with respect to industrial production.

Even if we adopt Baumol's hypothesis of zero productivity growth in service activities and equal growth in wages in the two sectors, the risk of 'urban crisis' is perfectly counterbalanced by a distributive mechanism that classical economists would have labelled 'exploitation' of the countryside by the city. Only a process of tertiarisation of the countryside, with a parallel freezing of the present specialisation of the city - and consequently the disappearance of the distinction between city and countryside - would avoid the perpetuation of the old 'contradiction'.

The relevance of the above examples in the present discussion resides precisely in the fact that they identify fields in which deeper reflection is urgently needed, as well as the advantages of an eclectic, interdisciplinary approach, rather than specific, though interesting, theoretical results.

11.5.2 To Take the Urban Physical Dimension into Full Account

The second aspect we need to re-introduce into the theoretical framework is the physical dimension of the city. This dimension is easily forgotten in abstract, macroeconomic or spaceless reasoning, as for example in much of the literature which sees cities as nodes in trans-territorial, physical or relational networks.

It is important to recognise that physical space matters, and that the city is not, and cannot, even in abstract terms, be represented as only a spaceless point, a railway station or an airport. In the city the two dimensions of *place* and *node* co-exist, in forms that still have to be fully understood. The city is above all a place and an agglomeration of diverse activities in direct proximity: an agglomeration that derives its social relevance, locational advantage and economic value from the difficult equilibrium between proximity advantages and agglomeration diseconomies, between the reduction of mobility costs and increase in congestion costs, between the density of settlement and lack of open space - in brief, from elements and variables of a mainly physical and territorial nature.

Certainly, the city is also the geographical operator that allows the worldwide networking of economic activities. On the one hand, the network city is less and less able to live with its own physicality, which means congestion, diseconomies of scale and social conflict, which hamper the de-materialisation which is implicit in information flows. On the other hand, it is the territorial nature of the city, its economic activities and social overhead capital, the density of its relationships, which provide the critical mass and potential demand for both internal information and global networking, as well as the content of the interrelations and information exchanged with the external world, deeply linked to its culture and identity.

But the logical relationship between the two natures of the city, place and node, does not operate only in one direction, from the former to the latter. The external networks, linking the city with external markets of goods, factors and information, have strong feedback effects on the character of the place, its competitiveness and external visibility, and consequently on its competitive advantage.

We have to treat jointly the two dimensions, place (*lieu*) and node, in their multiple interactions, as shown in the ancient Egyptian hieroglyphic representing the city: a cross inserted in a circle (the cross does not refer to the internal streets, but to external physical networks[1]). Unfortunately the methodological instruments we use at present are able generally to handle only one dimension at a time.

A research programme aiming to investigate directly the interaction between the two dimensions would need to include urban morphology and planning among the relevant aspects to analyse.

11.5.3 To Interpret Urban Complexity

This brings us to the third consideration, and the third aspect we have to introduce into our scientific programme. We need to arrive at a better understanding of how the different characters of the city coexist and interact, giving rise to the complex and fascinating unity. We have to understand how the city-artifact relates to the city-culture, the city-production machine, the city-community, and how the combination of these elements relates to the natural environment. In a dynamic setting, we have to fully understand how the *décalage* between the rhythms of evolution of the different dimensions of the city - the artifact, the machine, the social and cultural institutions - gives rise to multiple potential combinations of use, re-use, transformation, rejuvenation and relaunching of the city. These are combinations that may be sometimes regressive, but always innovative.

This constitutes a major intellectual challenge. The difficulty of such a research programme is undeniable, given the distance and even incompatibility between the analytical instruments, the disciplinary

[1] I owe this suggestion to Jean-Marc Offner.

languages, the models and tools of representation with which the single aspects have been traditionally handled.

11.5.4 Urban Sustainability

Growing interest is rightly devoted nowadays to the problem of the sustainability of urban development, an issue located at the crossroads between economic aspects (allocative efficiency), social aspects (social or distributional efficiency) and environmental aspects (environmental equity). I devote the following section to this issue, as it represents a clear example of the heuristic potentialities of a nonconventional approach inspired by complexity and by the evolutionary theory of economics (Nelson and Winter, 1982, Dosi *et al*., 1988) and territorial systems.

11.6 An Evolutionary Approach to Urban Sustainable Development

The approaches to the problem of urban sustainable development recently proposed in the international debate suffer generally, in my opinion, from a basic misconception: they use and mechanically transfer tools and definitions from the economics of natural resources or of global sustainability to the urban environment. They forget that the city is by definition a nonnatural environment, an artifact created by man, perhaps his most valuable creation.

The historical rise of cities through separation and independence from the surrounding countryside implies a clear-cut division between activities and professions - those which exploit natural resources and those which do not. Cities represent the emergence of social interactions enhanced by proximity, unthinkable in a model of sparse settlements; the development of activities linked to control, culture, art, social and technological innovation; and the development of values of individual freedom as opposed to the 'ethical life' of peasant communities (Camagni, 1992).

The existence of cities therefore implies a fundamental choice: the abandoning of a model of life and social organisation wholly based on the integration of man and nature in favour of one wholly based on the integration of man and man; abandoning production functions based on the factors of land and work in favour of functions based on fixed social overhead capital, information and energy.

In other words, we should be examining not so much cities themselves - phenomena found in all civilisations, whose right to existence can only be refused by a superficial 'ecologistic' romanticism taking to extremes some persistent anti-urban elements of urban planning culture - as much as some highly important trends involving cities and which risk jeopardising their primary role as points of social interaction, creativity and collective wellbeing. I refer here to the processes of disordered and limitless growth which cities often undergo during periods of economic take-off and rapid industrialisation, involving widespread urbanisation, variously labelled as 'sprawl', 'metropolitisation', 'peri-urbanisation', 'urban diffusion', '*ville éclatée*' or 'edge-city development' (Camagni, 1994). These processes have rendered empirically ambiguous the conceptual distinction between city and country, leading towards a non-city and a non-countryside. They are processes which have, above all, exacerbated the problem of mobility and energy consumption because they result in a settlement model wholly dependent on the private car (Boscacci and Camagni, 1994).

I also refer to the processes of 'ghetto' creation which are increasing in large cities, due partly to global social transformations and partly to the difficulty (and delay) with which public policies have dealt with the problem. These processes too, which have a more directly social or planning nature, must be borne in mind because they mean that some sectors of the population have insufficient access to the benefits of the urban environment, and influence the entire functionality and attractiveness of cities.

In conclusion, research on urban sustainability must have as its model of reference not an earthly paradise of eco-biological equilibria, nor cities designed according to ideals or utopias. On the contrary, its model should be an (albeit simplified) multidimensional archetype, in which the various functions of cities are recognisable: the economies of agglomeration and proximity, accessibility and social interaction, and links with the outside world. It should be a place in which the maximum of collective wellbeing emerges from positive dynamic integration (co-evolution) of the natural environment, the built and cultural heritage, the economy (and thus employment) and society.

It is very difficult to tackle this complex research programme with traditional approaches, even in their most advanced forms. The urban system moves on the basis of (and due to) phenomena of feedback, synergy, cumulativeness, network externality, increasing returns and indivisibility, i.e. nonlinearities which generate all kinds of possible outcome - explosive development, sudden catastrophic leaps and chaos - and, above all, irreversibility. These phenomena are intrinsic to urban evolutionary processes and imply a conception of time which encompasses morphogenesis, innovation and irreversibility - a concept at odds with the deterministic reversible view of the models of classic mechanics and neoclassical economics (Camagni, 1995a).

Consequently, the three elements on which sustainability policies may be based - technology, urban form, and habits - evolve along trajectories which are cumulatively reinforced and which, once set in motion, can be made to change their direction only with difficulty. Examples are individual mobility patterns which become consolidated through residential choice, transport technologies which lead to the development of a myriad of complementary assets, and individual residential choices which give rise to integrated, stable, territorial systems.

At this point, the distinction between short term and long term becomes crucial and must be clarified. In the short term, interventions which have a strong influence on the relative prices of resources and products can certainly shift consumer preferences (e.g. towards products with lower environmental impact), mobility patterns (towards public transport, if it exists) and production techniques (in the direction of energy saving). All this takes place, however, in a limited interval around the point in which a historical condition occurs or, in economic language, around a point of local equilibrium (the configuration of the settlement pattern, technologies used and stock of private and public fixed capital all being equal). In this case, the neoclassical model can be fruitfully utilised (Fig. 1).

Nevertheless, in the long term, if learning and feedback processes consolidate the existing territorial or technological pattern, and if large sunk costs are involved in the passage to other models, historical equilibria may remain 'locally' stable, even in the presence of theoretically more efficient equilibria (Erdmann, 1993). The new equilibrium point, indicated as the most efficient one, may therefore never be reached.

In a more specifically territorial framework, these phenomena of irreversibility are even more evident. For example, in the early 1980s, when the question of the possible reduction of energy consumption arose (not for reasons of sustainability, but due to problems of scarcity, price rises, and possible interruptions in supply and also because, in the United States especially, urban sprawl was being severely criticised), the problem of irreversibility reared its ugly head and, finally, more people became aware of it. It was stated that "the United States has invested the bulk of its capital development since World War II in an increasingly centrifugal fashion. We cannot declare this obsolete without bankrupting the country" (Sternlieb and Hughes, 1982).

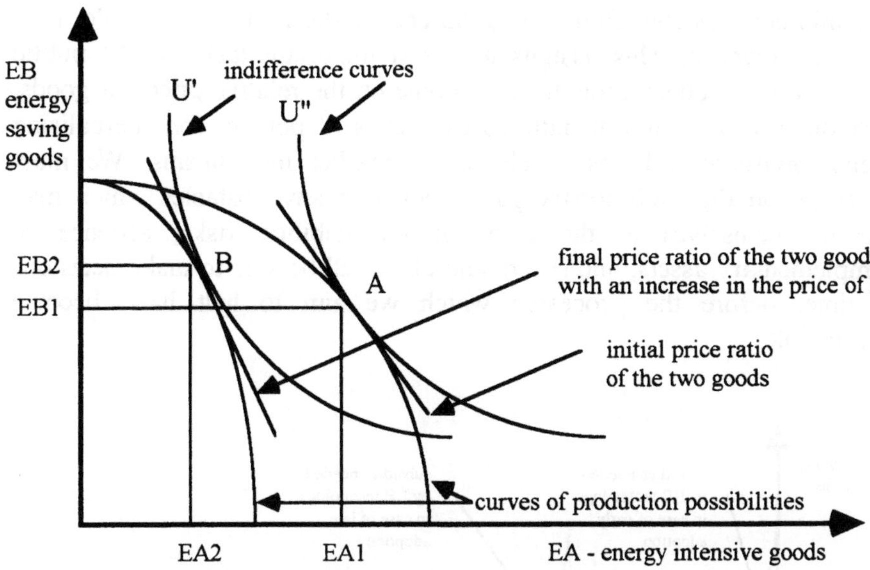

Fig. 1 Effects on two types of commodity of a rise in energy price. A rise in energy price (e.g. because of a carbon tax) generates an inward turn in the curve of production possibilities, shifting the price of the energy intensive good (EA) upwards and reducing its demand.

In these processes, time becomes a strategic variable for territorial or industrial policies due to the strong effects of learning and positive feedback which strengthen those technologies or territorial processes which are the first to appear. As Fig. 2 shows, if an environment-benign (EB) technology (or territorial model) starts its development trajectory at time 0, the subsidies or incentives required may incur extra costs with respect to a traditional, environment-adverse (EA) technology (or territorial model). On the other hand, if the EB is introduced at time 3, whereas EA starts at time 0, the disadvantage in terms of private costs increases, as does the cost of possible future public incentives.

In conclusion, approaches based on typical neoclassical assumptions - continuous and differentiable production functions, continual substitutability between production inputs or between the arguments of utility functions, and the absence of nonconvexities (all of which deny the existence of evolutionary phenomena such as those quoted above) - may lead to erroneous or misleading conclusions as soon as we enter a truly dynamic environment. Points of apparent equilibrium, for example, may never be reached. This assigns new and important tasks to the public sector. It is no longer enough to intervene on the relative prices of goods, technologies or territorial patterns by means of policies for internalising social environmental costs, relying on market mechanisms. We must intervene on the evolutionary path, removing any obstacles which may present themselves in the form of indivisibility, risks, absence of complementary assets, and so on, and above all be able to make decisions in time, before the processes which we aim to halt have become irreversible.

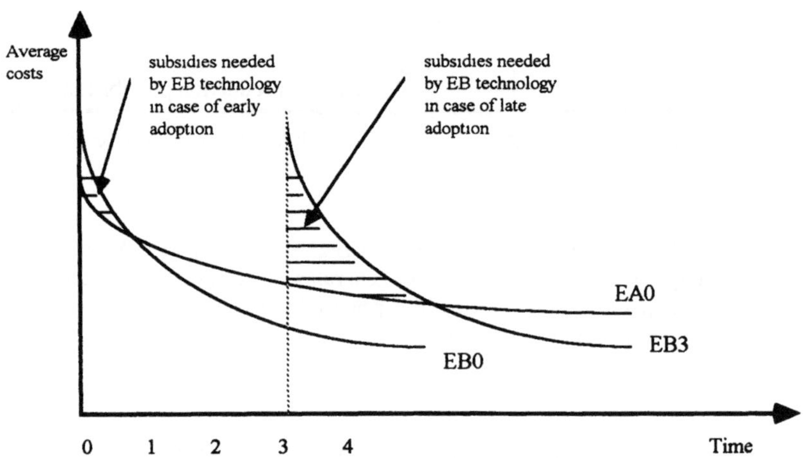

Fig. 2 Evolution of the relative advantage of alternative technologies over time
EB0 = environment-benign technology adopted at time 0
EB3 = environment-benign technology adopted at time 3
EB0 = environment-adverse technology adopted at time 0

11.7 Conclusions

The complexity approach - and its language - has many merits in spatial economics. The first resides in pointing out that our world, and the city as its specific and most advanced expression, is the outcome of the interplay of complex forces and of largely unforeseable combinations. It is the result of evolutionary, creative paths in which randomness and necessity are deeply interwoven, and in which the same individual decision may determine completely different outcomes as a result of small variations in the starting conditions. Time is marked by structural change and itself becomes the rhythm of change. The second merit resides in its ability to open our view to more global visions and make us suspicious of models and interpretations which are deterministic, mono-causal and over simplified.

In terms of disciplinary methodologies, the complexity approach has introduced sounder and more advanced modelling devices, more suited to the new theoretical framework, where self-organisation processes, sudden catastrophic jumps in state variables and shifts in the nature of solutions are intrinsically allowed. But along with these undeniable benefits, mainly achieved in the last decade, some points of weakness are emerging. In particular, the specification of the conditions for the acceptance of interpretive models and their results is unclear. Many theorisations have an excessive degree of abstraction, often ending up in highly generic statements. There is a mainly *ex post*, sometimes artificial comparison of the results of simulation procedures or analytical findings with respect to empirical reality.

Perhaps its major drawbacks lie, however, in the limited interdisciplinary effort and the lack of a deeper understanding of the structural relationships specific to urban phenomena. Analogies are proposed with other scientific approaches, and the models and their laws prevail over theoretical and interpretive issues. This is particularly evident when new, sophisticated methodological tools are utilised for the investigation of old problems for which perfectly good interpretations have already been provided by more traditional, even though seemingly 'uninteresting', tools.

New synergies among territorial disciplines are needed, and in particular a more wide-ranging effort by spatial economists in the exploitation of the rich potentialities of the new approach. In this work, we have indicated some fields in which the cross-fertilisation of different disciplines looks more fruitful, and where some initial 'eclectic' reflections already exist. The complexity approach deserves application in new, interesting and relevant fields. Among these, the urban sustainability issue - which is multidimensional and highly complex - appears one of the most promising, due to the advantage of nonconventional approaches when confronted with evolutionary, innovative and irreversible processes.

References

Aydalot P. (1988) Technological Trajectories and Regional Innovation in Europe, in Aydalot P., Keeble D. (eds.) *High Technology Industry and Innovative Environments: the European Experience,* Routledge, London, 22-47.

Aydalot P., Camagni R. (1986) Tertiarisation et développement des metropoles: un modèle de simulation du développement régional, *Revue d'économie régionale et urbaine, 25,* 171-186.

Baumol W. (1967) Macroeconomics of Unbalanced Growth: the Anatomy of Urban Crisis, *American Economic Review, 57,* 415-426.

Boscacci F., Camagni R. (eds.) (1994) *Fra città e campagna: periurbanizzazione e politiche territoriali,* Il Mulino, Bologna.

Camagni R. (1992) *Economia urbana: principi e modelli teorici,* La Nuova Italia Scientifica, Rome.

Camagni R. (1994) Processi di utilizzazione e difesa dei suoli nelle fasce periurbane: dal conflitto alla cooperazione fra città e campagna, in Boscacci F., Camagni R. (eds.) *Fra città e campagna: periurbanizzazione e politiche territoriali,* Il Mulino, Bologna, 13-85.

Camagni R. (1995a) Global Network and Local Milieu: Towards a Theory of Economic Space, in Conti S., Malecki E., Oinas P. (eds.) *The Industrial Enterprise and its Environment: Spatial Perspectives,* Aldershot, Avebury, 195-214.

Camagni R. (1995b) Lo sviluppo urbano sostenibile: le ragioni e i fondamenti di un programma di ricerca, Working Paper on The Economics and Planning of Sustainable Urban Development, Dipartimento di Economia e Produzione Politecnico di Milano, no. 95-01, in Camagni R. (eds.) *Economia e pianificazione della città sostenibile,* Il Mulino, Bologna.

Camagni R. (1996) La città come impresa, l'impresa come piano, il piano come rete: tre metafore per intendere il significato del piano in condizioni di incertezza, in Curti F., Gibelli M.C. (eds.) *Pianificazione strategica e gestione dello sviluppo urbano,* Alinea, Florence.

Camagni R., Diappi L., Leonardi G. (1986) Urban Growth and Decline in a Hierarchical System: a Supply Oriented Dynamic Approach, *Regional Science and Urban Economics, 16,* 145-60.

Dendrinos D., Mullally H. (1985) *Urban Evolution: Studies in the Mathematical Ecology of Cities,* Oxford University Press, Oxford.

Dosi G., Freeman C., Nelson R., Silverberg G., Soete L. (eds.) (1988). *Technical Change and Economic Theory,* Pinter, London.

Erdmann G. (1993) Evolutionary Economics as an Approach to Environmental Problems, in Giersch H. (ed.) *Economic Progress and Environmental Concern,* Springer-Verlag, Berlin, 65-96.

Forrester J.W. (1969) *Urban Dynamics,* MIT Press, Cambridge, Massachusetts.

Harris B., Wilson A.G. (1978) Equilibrium Values and Dynamics of Attractiveness Terms in Production-Constrained Spatial-Interaction Models, *Environment and Planning A, 10,* 371-388.

Leonardi G. (1985) Equivalenza asintotica fra la teoria delle utilità casuali e la massimizzazione dell'entropia, in Reggiani A. (ed.) *Territorio e trasporti: modelli matematici per l'analisi e la pianificazione*, Angeli, Milan, 29-66.

Lepetit B., Pumain D. (eds.) (1993) *Temporalités urbaines,* Anthropos, Paris.

Lucas R.E. (1988) On the Mechanics of Economic Development, *Journal of Monetary Economics, 22,* 3-42.

Miyao T. (1981) *Dynamic Analysis of the Urban Economy,* Academic Press, New York.

Nelson R., Winter S. (1982) *An Evolutionary Theory of Economic Change,* Harvard University Press, Cambridge, Massachusetts.

Nijkamp P., Reggiani A. (1988) Entropy, Spatial Interaction Models and Dicrete Choice Analysis: Static and Dynamic Analogies, *European Journal of Operational Research, 36,* 186-196.

Nijkamp P., Reggiani A. (1991) Processi spazio-temporali nei modelli logit dinamici, in Bielli M., Reggiani A. (eds.) *Sistemi spaziali: approcci e metodologie,* Angeli, Milan, 33-55.

Romer P. (1990) Endogenous Technical Change, *Journal of Political Economy, 98,* 71-102.

Simon H. (1972) From Substantive to Procedural Rationality, in McGuire C.B., Radner R. (eds.) *Decision and Organisation,* North Holland, Amsterdam.

Solow R. (1994) Perspectives on Growth Theory, *Journal of Economic Perspectives, 1,* winter, 45-54.

Sternlieb G., Hughes J. W. (1982) Energy Constraints and Development Patterns in the 1980's, in Burchell R., Listokin D. (eds.) *Energy and Land-use,* Center for Urban Policy Research, Rutgers University, 127-144.

12. Model, Plan and Process for the Control of Urban Systems

Vittorio Silvestrini

12.1 Introduction

The approach chosen by Denise Pumain, in this book, to deal with the sciences of the city has been to analyse the results which can be obtained from the application of mathematical models derived from physics to the study of urban systems. She has presented a wide-ranging review of the models recently developed to describe the evolution of the city or the city system. As I am not an expert on models, my own contribution adopts a different approach. But to begin with, I should like to make some observations on the intrinsic limits of mathematical simulation models, in particular when these are applied to urban systems.

Physics, and natural sciences in general, make ample use of models for the simulation of natural systems. It is important however to make a clear-cut distinction between a *phenomenological model* and a *theory*. A phenomenological model is described by an algorithm of varying degrees of complexity obtained in strictly empirical terms from measurements taken on the system. It is able to simulate the general behaviour of the system when this finds itself in the same conditions as when the measurements were taken. A theory is much more. It is an organic set of laws which, probing into the merits of cause/effect relationships, and starting from prescribed initial conditions, explains the behaviour of systems belonging to large categories. The transfer, based on analogies, of mathematical algorithms from one field of application to another (in our case from physics to city planning) falls into the category of simulation models.

Science, which often resorts to simulation, considers phenomenological models as an intermediate instrumental phase. They are used either to obtain partial forecasts of the evolution of a system, before developing a theory, or as an intermediate step which helps to arrive at a theory, i.e. as a tool to shed light on some aspects of the cause/effect relationships governing the development of the system.

In the case of complex systems, the forecasting capacity of models is usually limited and can be expressed only in terms of probability. The models are therefore generally used for other objectives, such as the identification and classification of instability or bifurcation points, with the aim of preventing the system from going through such states and adopting catastrophic patterns of development. They are also used to establish a sort of hierarchy in the causes determining the evolution of the system, with a view to the development of a theory.

Great importance has been attached to Prigogine's model of the evolution of open catalytic systems (a famous example of great conceptual importance), because it proves the possibility that a structural and functional organisation can be spontaneously generated within a system far from equilibrium without the intervention of a 'creator' external to the system and without single constituents of the system being subject to ordering codes. This conclusion opens new horizons to research on the origins of life on our planet (Prigogine and Stengers, 1979). The attempt to transfer Prigogine's model from physical to social systems[1] involves determining which features of social evolution depend on the ethical codes of single individuals, and which depend on 'macro' (political/institutional) relationships, or are even independent of any law of internal correlation, depending only on boundary conditions imposed from outside.

First of all, we should ask ourselves what basic objectives we can expect to achieve by the application to urban systems of mathematical models developed for the simulation of the behaviour of complex physical phenomena. This is a crucial question, raised more or less explicitly in many of the contributions to this volume. Perhaps we should begin by asking what is meant in physics by a 'complex' model. In everyday language, and in that used by city planners, this adjective has multiple and often contradictory meanings. But only if the city is in fact complex, in terms of the definition given by physics, is it valid to apply those models developed and validated for complex systems.

A system is considered complex when it consists of a very large number of (more or less simple) constituent elements interacting with one another in such a way that its spatial arrangement (structure) and its development

[1] The first proposal along these guidelines was made by Prigogine himself (see Prigogine and Stengers, 1981).

over time (evolution) cannot be ascribed to properties of its single constituents, but depend also (and mainly) on collective phenomena. According to this definition, there is no doubt that the city - considered not as an empty container, but as a whole with an organised structure and a number of functions and activities - is a complex system. In fact, due to the inward and outward flows of energy and matter brought about by the human community living and working in the city, the modern city can be defined an 'open' system and 'far from equilibrium'. Therefore, according to Prigogine's theory, the city is also, to a certain extent, susceptible to self-organisation.

Given that the city corresponds to the definition of a complex system, it is legitimate and reasonable to make reference to models normally used to imitate the behaviour of such systems. Of course we should remember that we are dealing with models and not with theories, and that there is no universal model which can be applied to all complex systems or to all behavioural aspects of each single system. In addition, we should note that there are some basic cognitive limitations which, independently of the model used to formalise the description, are due to the nature of complexity. In fact, as already mentioned above, the behaviour of a complex system on the local scale can be described and forecast only in terms of probability.

A description in merely statistical terms is nevertheless useful and also predictive (and can be experimentally validated) in relation to large sets of systems, since it tells us what proportion of the total will behave in which way. From the point of view of control (or, if we prefer, in terms of planning), these systems are manageable and optimisable, operating in what decision theory calls 'conditions of uncertainty'. Just as the forecasts of these systems are statistical, their planning too is of a statistical nature. The system needs to be planned in such a way that it maximises the probability that, as it evolves, it will approach the configuration which is considered optimal. This is like saying that it maximises the number of systems approximating the optimal configuration.

Every city however is a unique and unrepeatable system, thus an approach which achieves a merely statistical optimisation cannot be considered satisfactory. But there is more to this. Because of certain peculiarities of the city, the planner usually has in reality to operate in those conditions which Collingridge (1982) calls 'conditions of ignorance'. His or her predictive capacity relating to the evolution of the system (with respect to its objectives) can only expressed in terms of probabilities, and in reality is practically nil.

Does this mean that we should give up the whole idea of planning? Not at all. It has been proved (and we shall briefly discuss this in Section 5) that even in conditions of uncertainty and ignorance a system can be effectively planned, provided suitable learning and operating tools are used, according to the rationale that I have elsewhere called *aware planning* (Silvestrini, 1986).

From the cognitive point of view, the attempt to construct predictive scenarios is replaced by a *sensitivity analysis*, which characterises the way in which the system reacts to external stimuli (the manipulation of its 'control parameters'). From the operative point of view one aims towards a system which is as flexible and controllable as possible. I return to this point in more detail later.

We now come back to the question of simulation models. Considering their weak predictive power, what is their usefulness? Within the context of 'aware planning', they can play an important role both on the cognitive and the operative sides. In fact, while the local behaviour of a complex system can generally be described only in stochastic terms, the interactions between the system and its surroundings are usually characterised by the average value of suitable collective parameters. With the use of a spatial and temporal averaging operation, these parameters generally lose their stochastic and uncertain character and acquire a substantially deterministic behaviour. This generally holds true both for the system as a whole and for its parts or subsets (which are generally complex systems themselves, whether spatial subsets, such as districts, or functional subsets, such as the transportation or residential systems). Simulation models can thus play the fundamental role of determining and characterising such collective parameters and examining the way in which the behaviour and evolution of the system reacts to changes in the values of the parameters.

As these parameters are the means through which the system can be 'read' from the outside, the result of this kind of analysis (basically a sensitivity analysis) has great cognitive importance. Moreover, considering that it is through changes imposed on these collective parameters from outside that the behaviour of the system can be modified, the analysis is also the necessary premise to any aware planning intervention (i.e. control of the system). These general methodological considerations will hopefully be better explained in the Section 5, where we investigate their application to a particular type of urban planning problem. I should first however like to discuss some more general issues concerning the relationship between science and the city.

Human civilisation, and in particular that of the industrialised and rich countries, has over the past decades become strongly permeated with science which, either directly through its cultural canons or through its technological derivations, has profoundly modified our systems of production, our ways of trading and living. This has produced deep changes in social and civil organisation and, as a consequence, in the functions and roles of the city. Science has, on the one hand, burdened the city with new tasks and problems, but on the other hand has also provided powerful new potential and management tools.

These issues could give rise to many more considerations, but I intend to deal with just two. First of all, in Section 2, I shall be examining the large infrastructural networks which serve cities in the industrialised countries (the transport and communications networks, energy supply systems etc.). Although not always fully appreciated, these networks have a significant effect on urban development. The conclusions we reach about the features of these systems help us in defining the problem of planning them, discussed in Section 5. Secondly, I deal with the crucial issue of the relationship between the industrial factory and the city. In the past century the factory has gained enormous social importance. Its manufacturing role has in many ways been secondary and instrumental to many other more important functions. It has been not only a producer of consumer goods and profit, but has also produced jobs, wages and the whole 'work culture', i.e. the culture of rights/duties and cooperation. The city has in fact been able to give up many of its own functions and delegate them to this industrial structure.

Over recent decades however, due to the automation of production processes and the progressive expulsion of manpower (machines replacing workers), the factory has gradually become merely a place for the production of goods and profit. It is now necessary for the city to take care once again of the social functions previously adopted by the factory, but this is creating serious political and urban planning problems as well as methodological difficulties. In Section 3 I analyse this issue in more detail, and in Section 4 discuss a series of related proposals. From the technical point of view, in the rationale of the planning the city system, these considerations are useful for defining the 'objective function' of the optimisation process.

12.2 The City in the Technological Era

12.2.1 The Machine People

Let us begin by examining some of the characteristics of the current technological civilisation, which is affecting urban systems, creating radical changes in the age-old tradition of cities.

Technological man has enormously increased his own working capacity by developing and putting to his own use a large number of machines powered by artificial and nonrenewable sources of energy. At present the annual world consumption of energy is around 8 billion Toes (tons oil equivalent)[1] .This is equivalent to the energy produced by 100 billion workers (Silvestrini, 1990b). Subdividing this by sector, we find that a little under half of these 'mechanical slaves' work in factories (the industrial sector), about 25% in transportation and 25% in our homes (Silvestrini, 1980).

Approximately 80% of total energy is consumed in the industrialised world, where a little more than one billion people live. So, we can say that every human being (man, woman and child) living in an industrialised country has at his/her service the equivalent of about eighty mechanical slaves. If we look at the distribution of energy consumption, we see that a large proportion of our 'mechanical slaves' operate in large cities or metropolitan areas.

The expression 'mechanical slaves' is particularly apt because these technological beings produce not only work, but also need to be fed, and produce waste which must be disposed of (Stoppini, 1988, Zorzoli, 1988). The problem of the supply, transportation, transformation and storage of the 'foodstuffs' for these machines are comparable - in terms of complexity and size - to those of the management of the human food supply system. The difficulties of the disposal of waste produced by the machines is far more serious and, from many points of view, more complex.

Thus, a large modern industrial metropolis with a population of, let us say, three million people, actually contains the equivalent of many scores of billions of inhabitants. These are not included in the census returns, but they are physically present: they work, feed, occupy space, move, pollute the soil, water and air, and generate problems which did not exist in the

---

[1] A systematic analysis of global energy needs has been carried out by IIASA (see Marchetti, 1988).

city of the past. On the other hand, they also provide the means of improving the quality of life and of solving the problems of the city (including those they produce themselves). Using an effective comparison, we may say that the city of the technological era consists of two parallel cities: the 'city of man' and the 'city of machines' (Marchetti and Nakicenovic, 1979).

The planning of the great infrastructures (the technological networks which enable the mechanical slave people to function) is one of the main problems of city planning today. This was not so much a problem in the classical city, but there is nevertheless an interesting analogy with the Greek water supply system of Naples, an underground and 'invisible' network matching the geometry of the city above, or the remote-controlled heating plant in ancient Ostia. The planning of technological systems on an urban scale is a constraint which affects the evolution of the human city, but is also a highly effective tool for its management, as it makes possible solutions which in the past would have been technically impossible. Many of the problems of the modern city of man result from the scarce attention paid to the planning of the city of the machines. It is for this reason that this issue is examined in detail in this work.

There are two important features relating to large technological systems which we should like to underline:

i. the fact that their successful operation has become an essential condition for the functioning (often even the survival) of the city of man. Mobility, the production, transformation, storage and distribution of foodstuffs, the habitability of buildings, the functioning of the health assistance system, security and safety systems, and so forth, are all primary city functions which would be critically affected by an energy black-out, for example (Silvestrini, 1986);
ii. the fact that they are characterised by different development and control times from those of the human communities (this point is discussed further in Section 2.3).

12.2.2 Telematic Functions

In the previous subparagraph I drew attention to the large technological systems involved in production and in the supply of energy (industries, transportation, heat supply, etc.). Over the past few decades, a further category of technological devices has been rapidly developing. This too is organised within large systems, but is dedicated to the performance of nonmaterial functions (such as information management) rather than the production and management of material goods. I refer to remote-controlled and computerised systems and devices.

Through both direct and indirect mechanisms, these systems are already conditioning urban planning and development, and in the future seem likely to do so to an even greater extent than the 'classical' technological systems (Beguinot and Cardarelli, 1992).

i. *Urban functions*. Every urban function performed and used in the 'classical' city requires the physical co-presence of two subjects: *the operator* producing the function and *the user*. This produces *mobility* needs (of people and also objects). Using remote control devices, these functions can now be requested and supplied from a distance, offering urban planning new, unforeseen degrees of freedom.
ii. *The organisation of work*. Due to telematics and the automation of production and services, the organisation of work in industrialised countries is undergoing a radical evolution. This is leading to changes in the social and economic structure, as well as urban organisation. The whole of Section 3 is dedicated to the analysis of this important matter, and Section 4 to related proposals.
iii. *The remote control of wealth* (Petrella, 1995). Remote-control technologies allow the management of information and organisation of work from a distance, and also the transfer in real time of value added and wealth flows from one side of the world to the other. This new potential, which closely affects the two issues above, has in addition a strong impact on the geographical distribution of markets and of human activities in general. The consequences will be discussed later in this chapter.
iv. *The remote control of technological systems* (Silvestrini, 1990b). The new technologies mean that not only can each machine acquire a 'brain' and therefore replace man in the performance of an increasingly number of complex and intelligent functions, but it can also undertake the remote control in real time of technological systems (see Section 2.2). The planning of the 'city of machines' can benefit from a series of powerful new control instruments, as we shall see in more detail in Section 5.
v. *The management of interactions between the two cities* (Silvestrini, 1986). Computerised and remote-controlled technologies affect the organisation of the city of man and provide new instruments for the control of the city of machines, but they also offer new possibilities for management of interactions between the two. For example, by means of

transmission in real time of information on the distribution of urban traffic, it is possible to divert vehicles away from the most congested routes. This opens up new ways of planning large infrastructures.

12.2.3 The Time Factor

The techniques and ways of controlling complex systems depend on their reaction times. In spite of the enormous progress made by mankind in terms of construction techniques, it still takes a very long time to build or modify the large structures required by modern technological systems. It takes decades, for example, to build or reconvert the transportation infrastructure of a large city (motorway crossings, overpasses, tunnels, railroads, etc.) or to construct large electric powerplants and all the infrastructures necessary to connect and supply them (power lines, oil and gas pipelines, oil tankers and harbour facilities, etc.) - no less than the time needed to build large, ancient structures such as Hadrian's Wall or the Egyptian pyramids! If we take into account the political and bureaucratic/administrative procedures required to implement the decisions, these times are even longer.

The peripheral technological devices which represent the interface with the user (vehicles, machine tools, electrical appliances, etc.) can be modified in terms of performance, power input, number, etc. at a much faster pace (the usual time is around a few years), mainly due to the phenomenon of *product innovation*, which we shall come back to in the next section. This difference means that by the time large technological systems have been completed, they have to interface with a reality which may already be very different from that which existed at the planning stage, causing problems of adaptation. We can further observe that the computerised systems which permit the control of the increasingly capillary network of technological systems, including their peripheral user systems and the interface between the two are generally characterised by practically negligible time delays (fractions of a second) (Silvestrini, 1986). This can be exploited in the planning/designing process, as we shall see in Section 5.

12.3 Market, Technological Innovation and Work Organisation

Over the past few decades the organisation of work has been undergoing radical changes as a consequence of technical/scientific evolution and changing economic/political conditions (Petrella, 1995). The failure of the socialist regimes has caused the abandonment of planned economy models, and left a single model: the market economy. This does not mean of course that government-planned economic measures have been completely abandoned of, but that they tend mainly to be used to support the market, assuming that economic stability (and the welfare of the whole community) can be guaranteed by the self-stabilisation capacity typical of the free interaction between demand and supply.

Until about ten years ago this assumption was theoretically supported by Ford's so-called virtuous circle, which states that the only real economic incentive should be the profitability of industrial investment. In mathematical terms, profitability (profit) is attained when the market (sales volume) expands at a constant rate. This condition is automatically achieved in Ford's model thanks to a positive reaction mechanism within the system. In fact, the same investments which determine an increase in production (and hence the supply of products), also lead to an increase in the number of workers and their total wages, which produces an expansion in the market, creating demand for the products. In Ford's virtuous circle the worker in the industrial sector plays a triple role: he/she fosters the expansion of product supply through production work, his/her wage increases consumption, and his/her savings (which, through the credit system go into the pockets of the investor/owner) support industrial investment. A well-balanced fiscal system allows an enlightened, free-trading state to distribute the benefits produced by the industrial system among the whole community.

This mechanism not only guarantees dynamic equilibrium between demand and supply, which is necessary to guarantee market expansion and therefore the health of the free-trade economy, but also contains an automatic mechanism which means that a flourishing market (i.e. expanding system of goods) produces benefits for the whole 'system of mankind'. The enlargement of the market implies capturing a larger and larger number of people within the virtuous circle. This assumption led to the utopian idea that all people would be gradually drawn within the sphere of industrialisation and welfare. The only limit to this process could be a global boundary condition, i.e. the finite supply of the nonrenewable resources needed for production and/or the inability to metabolise the waste of an exponentially expanding volume of consumption (the global environmental problem) (Meadows D.L. and Meadows D., 1972).

However, even before this limit was reached, a new parameter intervened to upset Ford's circle and draw attention to more urgent needs: the so-called 'technological revolution'. New information technologies began to be applied in the manufacturing (process automation) making it possible to increase production (in both quantitative and qualitative terms) and also productivity (per worker and per unit of time). The result was that the total number of workers, rather than increase, in fact decreased.

At the same time, alongside the traditional goods market, a new *functions market* has developed, consisting in particular of computerised services. These functions have penetrated the service market in a capillary way, making it possible to achieve unforeseen improvements to both the quality and quantity of services, extending however also to this sector competition between man and machine. The quality and number of products - devices, functions and services - has rapidly increased, however this expansion no longer corresponds to an overall expansion in prosperity as it did in the past, but to a progressive narrowing of the distribution of wealth, dooming new nations, regions and social classes to poverty and underdevelopment.

As a result of the new possibilities offered by computer networks, the industrial system has expanded worldwide and undergone a huge restructuring process (Petrella, 1995). Mass production, semiprocessing and component production now tend to be located in the periphery of the urban system, and often in countries outside Europe and North America where manpower costs can be minimised. Planning, organisational and commercial management, and the control of value added and investment flows are now concentrated around the most structured and strongest poles. Marginal areas have undergone a galloping de-industrialisation over the past few decades. The gap between rich and poor countries, regions and classes has been increasing, causing an outbreak of social, political and military conflict, widespread unemployment and other ills.

This phenomenon has had serious repercussions on the organisation of cities, and in particular of large metropolises. Large cities in the underdeveloped countries, which previously seemed attractive as potential bridges for transition to the wealthy world, have not been able to keep their promise and are overcrowded with degraded, hungry people willing to trade all their resources (even their bodies and organs) to survive. The industrialised boundary areas between the Northern and Southern hemispheres, mainly producing semifinished and finished products, have suffered a surge of de-industrialisation and hence unemployment. This has released a huge mass of manpower and has also freed precious urban land which has now become available for new uses, but is vulnerable nevertheless to urban and social degradation. Even in the wealthy strongholds, however, only a small sector of the population is really well off. Many are abandoned to poverty, while the 'brains' of the new industrial economy are constantly looking out for new ways of becoming or staying competitive in a shrinking market.

The automation of production, and hence the possibility of producing more and more at lower and lower costs, in theory promises growing benefits for all, but in fact only for a lucky minority has this become a reality. The breaking of Ford's virtuous circle, as the increase in production is accompanied by a reduction in the number of workers - tends to offset the market towards supply. Larger quantities are produced, but the number of potential buyers is decreasing. Whenever possible, large companies buy up competitors and then close down their factories, as they are interested in acquiring their markets rather than their production capacity.

To compensate for the lack of demand, several countermoves have been made by the free-trade economy. First of all, an expansion of the market is sought not through territorial expansion, but an acceleration in the replacement times of products. This is pursued through constant *product innovation* which tends to stimulate new needs and new responses to old ones through the *consumption rationale* (Barcellona, 1987*)* promoted by mass-media. The media, in so doing, have become instruments of wealth and power production - creating new status symbols and new values (or rather, antivalues).

Government measures are taken to support the market, using economic resources gathered both through fiscal drag and through the public debt. These measures can be direct, such as the ordering of large public works (e.g. for the military) or indirect, including various mechanisms for the redistribution of wealth. The mechanism adopted in Italy as part of the government's 'special measures', supposedly to correct regional and social imbalance, in fact aims to transform citizens' savings into consumption. Neither category of intervention has however avoided the scourge of unemployment nor chronic overproduction and is in fact causing very evident damage. The negative effects include the widening of the gap between rich and the poor (nations, regions and social classes), the worsening of the environmental crisis (peaks of local environmental distress are added to the global problem), lack of protection of fundamental human rights (due to the gradual dissolution of the social state) and corruption, as well as crisis in democratic institutions.

In Section 4, I suggest a more far-sighted solution to the overproduction crisis caused by automation, and investigate the role of the city within this alternative project.

12.4 The City as a 'Factory' of Values

The crisis described above has its origin in the competition between man and machine on the work market. It has been aggravated by the response, which is to increase levels of consumption within the wealthy sector of the population. The consequences are the hypertrophy, or over-expansion, of the goods system and its progressive separation from the 'people system', hence a shrinking share of the population has access to an accelerated flow of goods. Technically speaking, a consumption and exchange value economy is prevailing, while the parallel 'use value economy' is being increasingly asphyxiated and abandoned.

In other words, we can say that production and the economy are becoming characterised by a *money-goods-money* chain, whose only purpose is profit. The goods are merely an instrumental link in the chain. No attention is paid to whether the goods satisfy real needs or are 'virtual goods' invented by worshippers of the status symbol. As a consequence, profit is concentrated in few pockets, and the growing production of goods is accompanied by the development of larger and larger areas of unmet needs. In the past, on the contrary, it was the market economy which led the money-goods-money chain. Final goods were functional and met the needs of the individual (*homo sapiens*, *homo faber*) and of society as a whole.

The rediscovery of the use value economy, based on the conservation and enhancement of territorial resources to satisfy the real needs of man[1], could be an alternative and strategic solution to the crisis characterising the end of this millennium. It is an economy which cannot be expropriated, unlike the consumption economy which is indifferent to the features and history of the territory, as well as to the culture and needs of those living there.

The respect of use values requires more personalised products, tools and functions, tailored to the particular situation (the local requirements,

[1] By 'real consumption' we mean that which is not induced merely by the goods market, aiming only at profit.

tastes, needs and culture of the user), in contrast with the 'complete indifference' of consumer goods. This means that such an economy requires a different and deeper relationship between science and society. In order to fully use its potential, it must acquire scientific know-how and that of the technologies produced by science. The bridge between science and its applications should not result simply in the supply of more and more sophisticated products, as in the case of the consumption economy, it is necessary to graft the new knowledge and technologies onto the traditional culture, in such a way that it thoroughly permeates all human activities (from craftsmanship, to small and medium firms, schools and professions). Starting from the bottom (the education of users, so that the demand is tailored to the needs of man), this knowledge should become an ingredient of welfare, rather than just an instrument of profit. So, what we really need is not so much transfer channels which allow us to connect the temples of science to the premises of industrial production (from research to the supply of products), but open places where science can discuss with society, talking and listening at the same time, submitting itself to criticism, mixing with other knowledge, rediscovering the past while looking towards the future.

During the past century, large factories (and in particular the industrialised areas on the edge of cities in which many factories with synergetic relations were concentrated) have played a multiple and vital role. They were not only the place of production, they also became the core of the economy by fostering and expanding the flow of Ford's virtuous circle. They sustained supply and demand, and caused trickle-down effects in the tertiary sector of the surrounding territory (especially the nearby city). The role of the factory was not however only an economic one. It played an important social and cultural function since it stimulated an awareness of rights and duties. It was a meeting place, a place of conflicts, mediations and reconciliations, serving as the school and university of workers. It was therefore not only the location for the production of goods, but a place for cultural exchange and confrontation of diversities. It was through the factory that knowledge of the new achievements of man became actions and entered the life of the community. The factory was the source of development and growth in the economy, general awareness and democracy.

As long as the city possessed factories, which served as reference points for the whole metropolitan area, it no longer needed to fulfill certain historical roles. Its citizens no longer used the city squares to air their debate, and the city could separate its functions (thanks to modern systems of transportation each district could have a different specialisation). But as processes become automated, machines have replaced people, and the factory is gradually losing all its other functions too to become more and more specialised as a producer of consumer goods. It no longer guarantees work from generation to generation, as it did in the past, and is no longer a place for the exchange of ideas and cooperation because of the separation and remote-controlled management of functions. It is no longer the place of expansion and conquest, but one where each worker defends his/her privilege to work. It is no longer the place of collective growth, but of isolated alienation (though it should be noted that the Japanese case deserves a completely separate analysis, Erto, 1995).

All the functions which the city delegated to the factory are now being abandoned. But since they are absolutely necessary, not only for the economy but to guarantee the quality of life and civil society, it is necessary for the city to take them on once again. Otherwise, the city will become an asphalt jungle, a place of isolation and conflict, of underground economies and of crime. The city should become what the factory used to be: not a place which only produces consumer goods for many and profit for a few, but the distributor of certainties (wages, social status, dignity as citizens). It must be seen as the place of civil growth and of the culture of rights and duties.

So, first of all, the city must regain its *productive function*. It should provide activities and jobs generated not only by consumption, but by the need to preserve and improve the quality of the territory and the life of those living there. This requires maintenance and restoration as well as durable goods. In order to encourage these activities, city administrations and national government need to provide incentives, such as special loans for the maintenance of real estate, investment in environmental protection and in the artistic and cultural heritage. This should be a primary concern of city planning.

Secondly the city should be the source of culture for society: a culture deriving from a diversity of peoples enriching each other's experience, promoting tolerance and solidarity. To achieve this, it is necessary for the 'piazza' to become once again a meeting place and centre for communication and exchange. We also need to regain complexity by putting back together the various urban functions, so that each district, periphery, and section of the territory is complete in itself, with all the urban functions present. The city must, in addition, be the place where the community has access to new knowledge and achievements, becoming a place of learning, where man manages technologies rather than being subjected to them. In conclusion, the city must become the instrument of response to social needs.

It is necessary therefore for technologies to stop sustaining consumption and start being consumed. They must help man expand and free his capacities and potential, instead of debasing him and depriving him of power. Transport and communication systems need to enhance the territory rather than destroy and devalue it, providing multiple forms of mobility without forcing man to always be a commuter. Virtual meeting points must be enriched by personal exchange and not just offer a dreary, unidirectional flow of prepackaged images and messages. But, for all this to happen, it is necessary that new approaches to the planning of technological systems and new instruments for their social control be developed and spread. This is the topic I shall discuss in the next section.

12.5 The Planning of Complex Systems

The profitability of industrial investment, which is the main incentive for the free-trade economy, requires an exponential growth of the market and thus of the volume of goods produced. Since technological systems also have to foster and serve the production system, the result is a natural tendency to over-expand, leading to a strongly invasive impact on the territory and on natural resources. The use value economy, as it is motivated not by profit but by the need for a better quality of life, does not have an inbuilt tendency to continuous expansion of the supply of goods. This would suggest that to achieve the transition from a consumption economy to a use value economy, it is enough to prevent the over-expansion of technological systems. But, due to the criteria usually adopted in the planning of large systems, which I have elsewhere defined as 'presumptuous planning' (Silvestrini, 1986), it is not so simple.

The starting point of our reasoning has already been indicated at the beginning of Section 2.1 and in the issues discussed in Section 2.3 It proceeds as follows:

i. the good functioning of large technological systems is vital to a large modern metropolis;
ii. the implementation of structural measures on such systems requires very long time periods, whereas the reality in which they operate changes at a much faster pace, due to the effect of multiple and unpredictable causes.

As a consequence of the first of these points, the planner is obliged not to make mistakes, i.e. is expected to be infallible, but due to the second, he or she operates in reality in what we have previously called conditions of ignorance. Let us take as an example the energy system. It is absolutely necessary for the planner today to come up with all the operative decisions which permit the energy system to produce the quantity and quality of energy required to meet the future needs of the city (and of society in general). This is conceptually possible, since the energy system - although heavy and invasive - is not complex in the sense defined in the introduction. Its evolution can therefore be deterministically planned. Nevertheless, as the time required for plan implementation is very long, in the meantime society (its culture, tastes and consumption) is likely to have radically changed. Since society is a complex system, it is not possible to predict its future state, and hence its energy demand, with any reasonable reliability. The construction of predictive scenarios is still vague and ineffective, which is why the planner operates in conditions of uncertainty.

The way out of this apparently insoluble dilemma (need for infallibility despite ignorance) adopted by the presumptuous planner, who is looking for perfection by taking decisions which produce self-justifying conditions, involves entering what Collingridge calls the ‘caution spiral’. The energy system is therefore prudently planned much larger than necessary, stimulating as a result an induced growth of the user system. This triggers a positive feedback process. It is evident (confirmed both by mathematics and experience) that the net result is the tendency to hypertrophy of the whole system (technology plus users), i.e. to explosive exponential growth that is independent of the real benefits (or damage) produced for mankind, as well as for the environmental system.

The alternative, proposed elsewhere, is the adoption of ‘aware planning’ (Silvestrini, 1986) which is based on a supposition which is opposite to that of presumptuous planning. It implies accepting the condition of fallibility, i.e. to live with mistakes. For this to be compatible with the need for good functioning of the planned system, it is essential that the errors have nondestructive effects, i.e. can be minimised to the ‘non-upsetting’ level. This is achieved by adopting two criteria: a cognitive one (a sensitivity analysis) and an operative one (flexibility). Fig. 1 shows the structure of such a process.

The sensitivity analysis goes through the following phases:

i. collection of a data bank on the system and preparation of a general descriptive map;
ii. analysis of the interactions between subsets and subsystems, and of the interactions between the whole system and its environment; the

determination and characterisation of the collective parameters describing such interactions;

iii. the construction of a simulation model imitating the reactions of the system when varying the values of the collective parameters;

iv. simulation of the behaviour of the system and its reactions to upsetting interventions on its collective parameters.

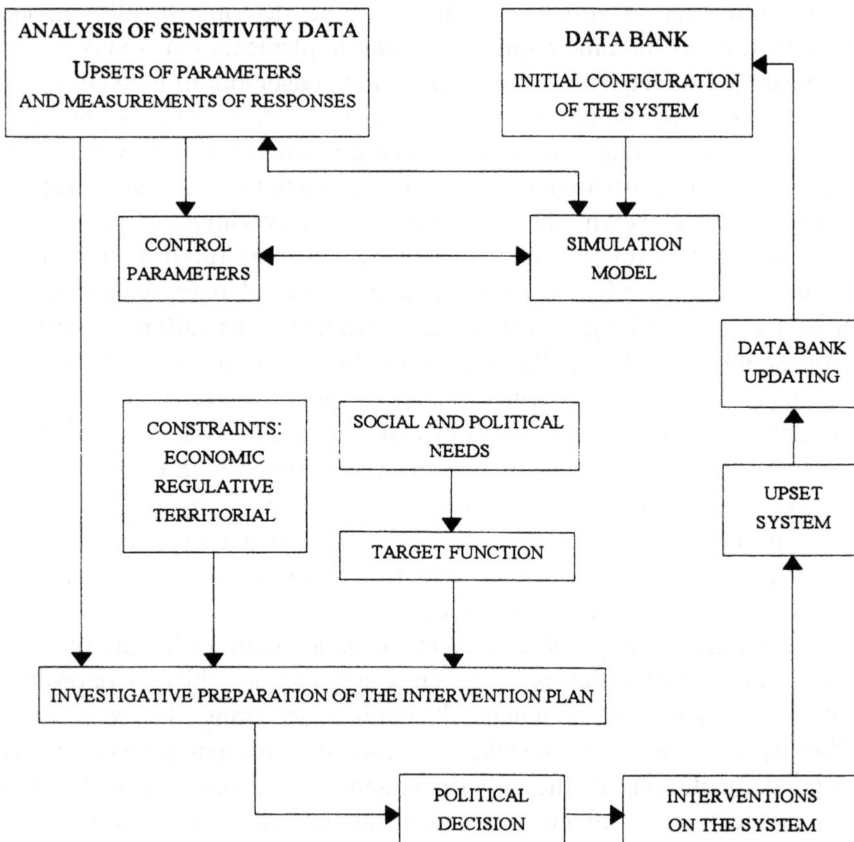

Fig. 1 Diagram of a plan-process with inclusion of simulation model and sensitivity analysis.

It should be noted that the activities taking place in the city and the interactions with the environment (most of all the anthropic ones) are generally more complex than the structures making up the city. The models used for the sensitivity analysis are therefore applied to the functions, while the structures (which will also be affected by interventions in the planning phase) are adopted as boundary conditions. Bearing this in mind, the sensitivity analysis should be validated in the field. By introducing ad-hoc upsets in a given function (for example, the closing or opening a main road, or modifying the system of traffic lights, if we are analysing road traffic), it is possible to experimentally test the reactions of the system.

From the cognitive point of view, the sensitivity analysis is less ambitious than a predictive model. To make a forecast it is necessary to know not only how the system will react to external stimuli, but also how these stimuli will evolve over time. Nevertheless, the sensitivity analysis (which plays a determinant role in the aware approach to planning) is a very sophisticated tool. As far as I know, it has never been applied to a city as a whole, but only to functional subsets (e.g. the domestic energy supply system or transportation system). The possible operative structure of the planning process for a regional energy system is shown in Fig. 2.

The sensitivity analysis, as we have seen, is merely a cognitive phase, but is absolutely essential to the successive phases of intervention. Planning affects the structural and functional features of the system, with the aim of optimising its behaviour in relation to the objectives set. For this to be possible (considering that one usually operates in conditions of ignorance), it is necessary to act on the system so as to make it flexible. First of all, we must try to make it as insensitive as possible to the most uncertain parameters, i.e. not susceptible to control by the planner, and secondly, we must find tools which will allow us to determine the gaps between the real state of the system and the one which is considered optimal (the early diagnosis of errors). Early diagnosis is usually performed by monitoring the state of the system in real time. This can be effectively done in many cases by means of computerised and remote-controlled technologies. Together with sensitivity analysis (which establishes which is the greatest error or upset the system can absorb without prejudice to its good functioning), early diagnosis allows the planner to intervene and correct, on condition the system can be controlled.

Controllability is achieved through procedures of various kinds (Silvestrini, 1990b). First of all, the system's reaction times are shortened as much as possible, and it is equipped with networks of control parameters which allow it to be automatically controlled (the computerised control network for this purpose can be the same as that which gathers data for the early diagnosis).

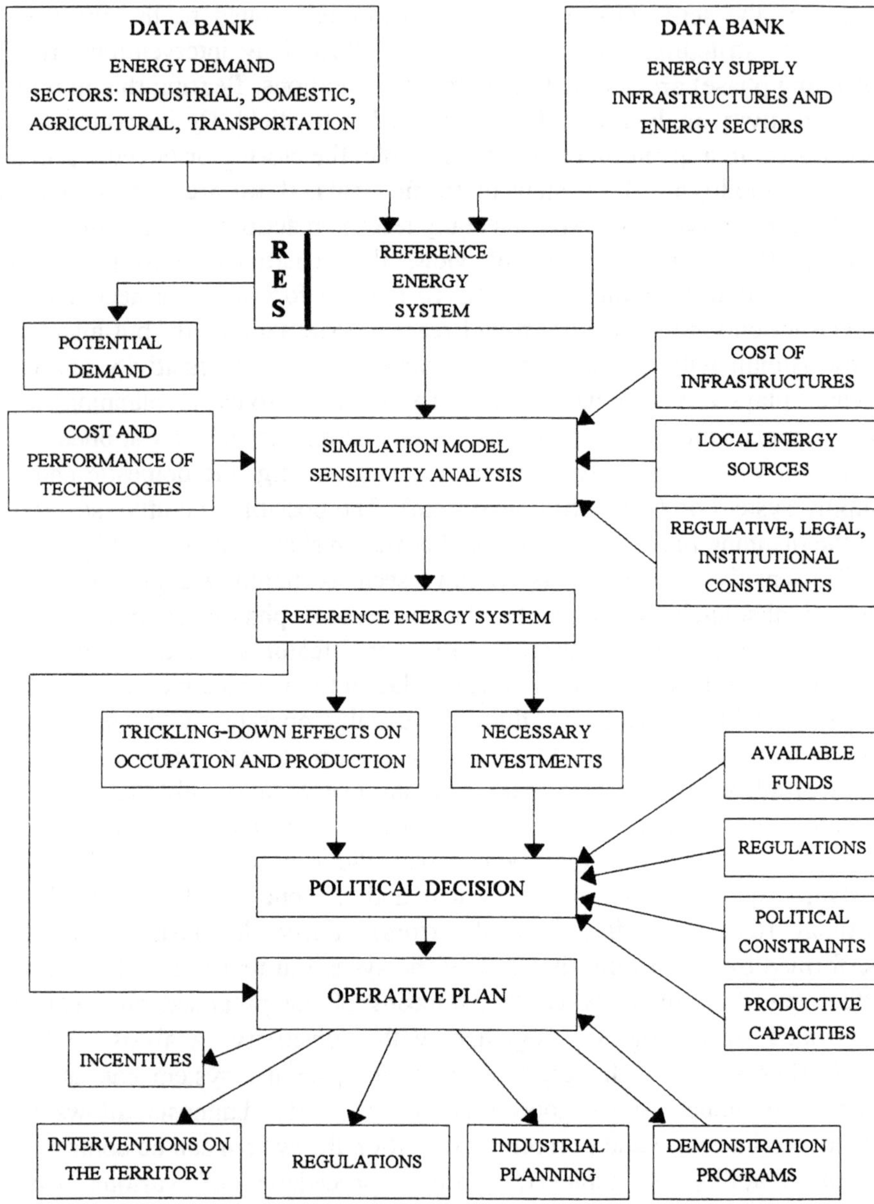

Fig. 2 Possible logical and operative structure of the plan-process for an energy system at the regional scale

The system should also be provided with as many 'alternative paths' as possible, enabling it to respond to each function in different ways. A 'soft interface' is created between the technological (supply) system and the user (demand) system, which absorbs fluctuations or errors on both the supply side and the demand side. This interface informs and influences the user system as much as possible, so that it can cope with any upsets which may appear on the demand side.

In the aware approach, the plan is replaced by a flexible instrument which we could call the 'plan-process'. This is an evolutive instrument which continuously adjusts to changes in the state of the supply system and in the needs of the demand system (Silvestrini, 1986). Some examples of the application of this approach to technological systems have been described elsewhere (Silvestrini, 1990a, 1990b, Collingridge, 1983). In general, the maximum flexibility appears to be achieved in medium-sized systems, rather than very small scattered ones or very large-scale ones. In the case of urban systems, the most satisfactory scale seems to be that of the district. This conclusion, which was reached theoretically, is also supported by the results of field experience in projects aiming to achieve the compatible city, seen as a mosaic of 'aware pieces' which are organically assembled to create the metropolis.

References

Barcellona P. (1987) *L'individualismo proprietario*, Bollati Boringhieri, Turin.

Beguinot C., Cardarelli U. (eds.) (1992) *Città cablata e nuova architettura*, Giannini, Naples.

Collingridge D. (1982) *The Social Control of Technology*, Frances Pinter, London.

Collingridge D. (1983) *Il controllo sociale della tecnologia*, Editori Riuniti, Rome.

Erto P. (1995) *Qualità totale, in cui credo*, Cuen, Naples.

Marchetti C. (1988) Storia e prospettive delle energie primarie, *Proceedings of the meeting "Energia, Sviluppo, Ambiente"*, Editrice Compositori, Bologna, 21-34.

Marchetti C., Nakicenovic N. (1979) The Dynamics of the Energy Systems and the Logistic Substitutional Model, RR 79/13, IIASA, Laxenburg.

Meadows D.L., Meadows D. (1972) *The Limits to Growth*, Universe Books, New York.

Petrella R. (1995) *I limiti della competitività*, Manifestolibri, Rome.

Prigogine I., Stengers I. (1979) *La Nouvelle Alliance*, Gallimard, Paris.

Silvestrini V. (1980) *Uso dell'energia solare*, Editori Riuniti, Rome.

Silvestrini V. (1986) *Come si prende una decisione*, Editori Riuniti, Rome.

Silvestrini V. (1990a) Energy Planning and Flexibility of the Energy System, *Energy Systems and Policy, 14*.

Silvestrini V. (1990b) *Ristrutturazione ecologica della civiltà*, Cuen, Naples.

Stoppini G. (1988) Energia da fonti fossili convenzionali: tecnologia, sviluppi, impatto ambientale, in Proceedings of the meeting 'Energia, Sviluppo, Ambiente', Editrice Compositori, Bologna.

Zorzoli G.B. (1988) *Il pianeta in bilico*, Garzanti, Milan.

13. Self-Referential Processes of Spatial Integration

Alfredo Mela, Giorgio Preto

13.1 Introduction

The question we explore in this essay is one which has been in the background of the discussion of this whole book: to what extent can we justifiably speak of 'city sciences'? Though expressed here in simple form, this question in reality raises two fundamental problems, as well as the relationship between the two.

Firstly, it throws doubt on the scientific character of the sciences concerned with the city, i.e. the disciplines which come together in occasion of debate such as this. At the very least, it questions their scientific nature when they are used to conduct a coherent investigation, open to empirical verification, about the city. This doubt relates in particular to those disciplines involved in the analysis of aspects concerning the behaviour of social actors in an urban context - the disciplines which belong prevalently to the human sciences, such as economics, sociology and human geography. In what sense are they able to construct a scientific dialogue about the city?

In this book, doubts have been expressed by others too in this connection. Silvestrini, for example, suggests that the models used to analyse the dynamic behaviour of urban systems are only of a phenomenological nature and do not represent the formalised expression of a theory as such. Other contributors, such as La Bella, have proposed a classification of analytical models which admits a more detailed typology. So, to put the question in more explicit terms, we can reformulate it in the following way: are the human sciences applied to the city able to offer interpretations with theoretical significance and not solely sets of empirical observations?

The second problem raised by our initial question concerns the scientific nature of the object of our investigation: the city. Certainly, the common understanding of the term leads us to accept that there exists an empirical object which can be defined as a 'city', even though, as observed by Tinacci Mossello in this volume, it is not easy to establish the criteria which distinguish this from 'non-city' settlement forms (e.g. the rural town) or very large urban agglomerations (the urban region or megalopolis). However, the main problem is to establish whether this empirical object has independent status, i.e. can stand as an object of study in its own right.

In many of the contributions to this volume it has been claimed (explicitly or implicitly) that the city's scientific status derives from the fact that it can be considered a system with self-organising capacity. Others question this assumption or, at least, deny that it can be automatically accepted in a context such as this, increasingly characterised by the process of internationalisation of the economy and of many social activities. In either case, even if we do not fully share this doubt, the problem remains of establishing what properties serve as a basis for deciding whether or not the city can be considered an entity with systemic qualities, and what conditions are necessary to accept the existence of self-organising activities at the urban scale.

These aspects will be examined in the present study and serve as a basis for two distinct areas of reflection. In Sections 2, 3 and 4 we pose the question of the epistemological status of the sciences of the city, looking specifically and in greater depth at urban economics and sociology, then bring together some brief general observations in Section 5. In Sections 6 and 7 we move on to a discussion of the question of the systemic nature of the city. In this connection we present a schema which describes a logical approach to the determination of the conditions under which the city can usefully be represented as a self-organising system. It is obvious that the problems raised are of great complexity, and we make no claim to having resolved them exhaustively with the present reflection.

13.2 The Sciences of the City and their Epistemological Status

The city is a highly complex phenomenon whose study, like that of other objects, can be undertaken in a systematic way, involving the definition of specific theories, appropriate methodologies and analytical techniques, i.e. it can take on the aspect of a science endowed with its own specific instruments. So why, therefore, do we speak of the 'sciences' rather than the 'science' of the city? We feel that this is for the same reason that we speak of the sciences and not the science of nature, referring in a general way to all the various disciplines - physics, chemistry, biology and so on - which investigate natural phenomena.

The reason lies in the complexity of the subject, which obliges us to select subsets of phenomena, which are then analysed according to different criteria, making use of different analytical tools or conceptual apparatus and techniques. This selection leads however to a separation and, therefore, loss of some connections which are of strategic importance to our understanding. The attempt to recompose the whole has resulted in the need for recognition of the intersecting areas between these various fields: chemical-physics, biochemistry, biophysics etc. Perhaps, in reality, the problem has been stated incorrectly. There is no such thing as the 'science' of an object, there is only an approach, i.e. an intention with which we observe an object, and hence a series of points of view or intentions. These are the various aspects of the search for knowledge that constitute the different 'sciences'. This general observation applies equally to the 'city sciences' - urban economics, urban sociology, human geography and its various subdivisions, (economic, urban, regional etc.), cultural anthropology and so on.

The human sciences appeared far more recently than the physical sciences: modern economics was born at the end of the 18th century, sociology around the mid 19th century, and the application of ethology to human behaviour is even more recent. The relatively late formation of the human sciences is not, it seems to us, due to a lesser urgency of understanding (physics and politics, since Aristotle, have been the two mainstays of philosophical reflection), but to the greater difficulty of dealing with their complexity. It is far harder to achieve the separation of phenomena into different points of view (the possibility of αναλυειν). The Machiavellian distinction between ethics and politics is only the first of many steps. The fact that the development of the human sciences was a response to urgent economic, socio-political and cultural demands has continued to influence the scope of various areas of study, i.e. encourage the development of separate fields whose definition requires the use of different conceptual analytical tools and which have rarely been given unified treatment (some of the few examples are Marx's "*Manifesto*" and "*Capital*").

The demands have been inspired by different motivations and aims, and have therefore generated projects with very different characteristics. Even though there are large overlapping areas, they lack a general unifying paradigm or general agreement on method. But this separation, as we have said, is a necessary first step towards the definition of specific scientific status (and coherent conceptual and analytical frameworks), i.e. towards coherent points of view. The process of recomposition, the search for unification, belongs to a successive phase in which a comparison is made between clearly defined disciplinary areas.

13.3 Economics, Space and the City

It is not by chance that economics was the first behavioural science to claim special scientific status, since it is far easier to separate from other social phenomena those connected with the exchange and production of goods. This claim is supported by the existence of specific theories, methodologies and formalised analytical techniques, as well as by the fact that their field of application, i.e. the economic aspect of human affairs, is clearly defined. (It is a definition based, in other words, not on established criteria but analytically, since it relates to a particular dimension, a particular 'point of view'). The field can be rigorously defined in accordance with the simultaneous observance of the four conditions proposed by Robbins: the multiplicity of aims, the possibility of classifying their relative importance, the limitation of means and the limitation of their alternative uses (Robbins, 1932). It is possible on this basis to demonstrate the formal identity of the theories which refer to the various manifestations of economic action (Samuelson, 1947).

Economics is essentially a deductive science. Its epistemological suppositions may be questioned, as may its initial postulates (like, for example, that of utility), but the rigorous definition of its field of application and its neutrality with respect to value judgement, exclude arbitrary interpretations. (In the same way, the specification of the field of application of the 'universal' law of gravity means that it cannot be invalidated, even when subatomic analysis demonstrates its incapacity to explain certain phenomena.)

This still holds when economic analysis, going outside the limits of a conventional single-pointed conception, encompasses the spatial dimension as an economically significant variable in terms of costs and benefits. In fact, the reference to space does not introduce reductive specifications at the level of abstraction of 'economic laws', i.e. limiting the degree of generality of the theories or their applicability to empirically observable phenomena. Unlike classical sociology (as we shall see in Section 4), the introduction of the spatial dimension in economics does not signify the passage from a 'pure' science to an 'applied' science, valid only in contexts characterised by special features (which take on the connotation of exogenous phenomena with respect to the theory).

The age-old use of a symbolic good, money, as the perfectly operable intermediary in the exchange of all other goods, has provided the basis for the quantification of human impulses, such as desire, and therefore for the calculation of utility and disutility, i.e. benefits and costs. If the consideration of the spatial dimension of economic phenomena can be achieved through the transformation of physical measurements of distance and density through monetary measures (in such a way that the effects of space can be evaluated as monetisable costs and benefits), the degree of generality of a theory - defining for example the optimal conditions for the configuration of a system - remains unchanged, or even increased. Its general applicability is also unchanged, or even extended. The abstraction of Walras' general equilibrium is reflected, and its explicatory power increased, in Lösch's *räumlische Ordnung*.

The question becomes more complicated when, in the interests of achieving closer adhesion to the actual geographical manifestations of the phenomena we wish to analyse, the Euclidean spatial convention is replaced by a topological conception - unidirectional relations are substituted by the consideration of multidirectional flows, and discontinuous non-homogeneity substitutes continuous homogeneity. The need to refer to more complex spatial organisations, which is particularly felt in field of geography, poses serious problems. To solve them we can proceed by analogy, seeking ideas from more mature disciplines, such as physics, with its systematic treatment of space, making use of physical paradigms and transposing concepts of mass, impedance and energy to the spatial analysis of human activities.

We are referring here to a series of models which have had wide application in territorial analysis. Although certain reserve has been expressed about their use, it seems to us that to proceed by means of analogy is not unusual for science in its nonlinear path to knowledge: from Anaximander's fire to the hypothetical Big Bang (which has led the way to so many deductions). The central question however is not the validity of analogy, but the distinction between a phenomenological model and theory, i.e. between empirically defined representations of system components and laws which explain the behaviour of the system by "probing into the merits of cause/effect relationships" (Silvestrini, in this volume, p. 387). The fundamental discriminating factor is whether or not we are able to specify the cause/effect relationships, in other words, whether we wish to (or, in the case of complex systems, are able to) arrive at the definition of the causal relations determining the behaviour of the phenomena being studied.

If the system is complex, there can be points of instability or bifurcations at which the behaviour of the system may even become chaotic. To ensure that the system does not descend into chaos means not only being able to prevent the system from passing through such points (and therefore possessing suitable technical instruments), but also being able to define the validity of the deterministic relations which govern its behaviour. We can take as an example the model proposed by Allen (Allen and Sanglier, 1979) in which the time period between the initial instant and the saturation point of the 'carrying capacity' is an interval during which an explanation is provided of the evolution of the system 'entering into the merits of cause and effect relationships', on the basis of the convergence of the theory of firms, the market, the urban economic base and spatial interaction.

With reference to spatial interaction, we return for a moment to the question of adopting analogies from physics. The so-called gravity models are in our view incorrectly defined as theory, since the description/ simulation of phenomena is carried out by means of a phenomenological simulation model which, on the basis of analogy, reproduces the formalism of another model (a real theory) used to explain a completely different phenomenon (universal gravity). Nevertheless, the probing empirical tests to which the gravity model has been subjected have stimulated reflection on the causal relations which generate spatial phenomena (Camagni, 1993). Thus, from the construction by analogy of a simulation tool which serves to make "partial forecasts of the evolution of a system before developing a theory", we have moved on to its consideration as an "intermediate step which helps to arrive at a theory" (Silvestrini, in this volume, p. 388). These two stages need not necessarily be in opposition if we are not content with partial predictions and make an effort to discover the specific theoretical implications.

Following a similar nonlinear path, and through another analogical procedure, some useful results have recently been achieved in connection with the concept of entropy. Entropy models are founded firstly on an appropriate theoretical basis, making reference to the theory of random utility (McFadden, 1974) and the maximisation of entropy (Coelho and Wilson, 1976) and secondly, the demonstration of equivalence with the entropic approach (Leonardi, 1985). Here, the use of the term 'theory' is no longer inappropriate.

In relation to these convergences and the role played by space, Leonardi and Rabino (1984) have stated that "around concepts such as space, it is possible to construct a theoretical unity of human sciences (and in particular regional sciences, which study the role of space in human settlements and industrial activity)" (p. 9). We share this conviction, while recognising that the task is far from easy and that the paradigmatic use of principles and methods from other disciplines often constitutes a necessary step in the construction of this unity.

Our observations in relation to spatial economics (and economic geography) are obviously of general application, independent of the nature of the economic phenomena considered and the spatial scale of their manifestation. It is clear, however, that by reducing the field of observation to a very limited area characterised by a high density and variety of phenomena (due to the multiple forms of interaction between the phenomena themselves), the difficulty of the analysis increases. If, on the other hand, the field of observation is an urban system, an analysis of specific economic phenomena involves the problem of the complexity caused by the ecological relations (Dendrinos and Mullally, 1985), i.e. between different economic functions and also between economic, socio-political and cultural functions. At this scale, space no longer has the connotations of an abstractly definable space, manoeuvrable as an independent variable, but takes on the characteristics of a physical subsystem with a relative basic autonomy, but strong interactions with the other subsystems.

It is therefore from the depths of the tradition of spatial economics (and economic geography) that urban economics (and urban geography) have arisen and acquired their independent status.

13.4 Sociology, Space and the City

In relation to the questions touched upon so far, sociology presents rather different problems which, from certain points of view, are a mirror image. Whereas economics owes its success to the relative clarity with which it can define its field of interest, proposing a specific and rigorous view of society, sociology arises from the attempt to construct a science of society as a global entity, deriving from the interaction between its parts or heterogeneous constituent elements. It therefore adopts, at least implicitly, a plurality of views in relation to its subject.

Economics succeeds, though not without epistemological problems (which emerge in the construction of more refined theories) in attributing to economic actors a rationality definable in more or less univocal terms and, on this basis, is able to propose verifiable relations between the micro-level (at which individual choices are situated) and the macro-level (where choices come together, giving rise to system dynamics).

Sociology, on the other hand, has been forced to recognise the plurality of forms of rationality present in various parts of the same social system. Important attempts have been made to classify these forms, but it has not been possible, except for single sectors or in the form of theories of limited range of application (Boudon, 1984) to resolve the problem of achieving coherence between the micro and macro level. This does not mean that we are obliged to choose between these two levels, but that the ways in which sociology constructs its representations vary considerably depending on the analytical levels considered (the social action of the individual, phenomena relating to parts of society, the dynamics of large social systems or the whole system) and that the resulting representations are not always easy to recompose into a single coherent construction.

This situation also explains why, in sociology, the activity of description and interpretation in a causal sense, takes on such different forms depending on the objects concerned and the analytical level. It ranges in fact from the attempt to verify hypotheses of small/medium range through traditional tools of empirical analysis and the statistical representation of the results, to attempts to represent the dynamics of societal macro-systems on the basis of an original interpretation of the systems paradigm (of the kind undertaken, for example, by Luhmann). More rarely, use is made of models based on mathematical formalisms or simulation algorithms. It is significant, however, that these cases seem to involve very specific problem areas, situated either very close to the micro extreme of analysis, like Gallino's model of the ego (Gallino, 1992a), or in equally close proximity to the macro extreme (e.g. the analysis of international relations).

These general observations, which hold for the whole framework of sociology, are even more valid when referred to the sociological analysis of the city. In this case, to the difficulties deriving from the multi-perspectives of sociology and the extreme complexity of the city itself, we have to add the difficulty of defining the society/space relationship. Once again, the comparison with economics, even though undertaken in a schematic fashion, can be useful. In economics, the introduction of the spatial dimension involves the integration of certain fundamental principles (such as the principles of agglomeration or distance impedance) into the general body of theory. Although this integration is not without difficulties, it constitutes a recognised problem area with a relatively consolidated tradition and unity. In the history of sociology, on the other hand, the introduction of the spatial dimension comes from a series of dissimilar theoretical strategies and, as a result, the relations between the disciplines concerned specifically with spatialised objects and general theory have taken on different forms.

For the sake of simplicity, we classify the theoretical strategies through which the problem of space has been tackled in sociology into three types. The first is what we could define the 'deductive' strategy and is the most widespread point of view in classical sociology. It considers sociological theories to be endowed with general validity, at least within limits (defined implicitly or explicitly), and as such does not refer to any specific spatial context. In this sense, therefore, 'pure theory' is nonspatial. Reference to space appears when we pass to a less abstract level. Here, the theory is no longer pure, but applied to a particular context, the specific features of which appear as exogenous factors of a spatial nature. If the context in question is a particular city (or even a group of cities belonging to a given societal model), this is interpreted as the place in which examples are found - within limits determined by the context - of general processes which the pure theory describes and interprets. In any case, the passage from the theoretical nonspatial context to one involving reference to a specific place, unlike the situation in economics, is understood as a loss of generality, or even a throwback to a premodern approach now for the most part superseded (Agnew, 1989).

A second strategy which, although not such an organic or influential paradigm as the first, has also given rise to classical studies, is the 'empirical/comparative' approach. We refer here to the many studies which focus on specific spatial contexts, the so-called community studies, often carried out on medium-sized cities. Beyond the apparent ideographical nature of this kind of research, there is often an underlying attempt to identify, through case studies, the presence of phenomena of general validity, without however resorting to theorisation which would complicate the link between the global and the local. In any case, an essential aspect of this strategy is the implicit conviction that general laws of social action do not manifest themselves without the mediation of the specific characteristics of the context, within the globality of their determination in space and time.

More recently, alongside these two long-standing strategies, a third more explicitly 'spatialist' strategy (Ledrut, 1987) has been taking form. This is based on the assumption that all social phenomena, whether at the micro, meso or macro level, possess intrinsic space/time characteristics. As stated by Pred (1990) "Society is always and everywhere constituted by the time-space specific projects of actual people, by the execution of time-space specific projects that are the medium and outcome of social relations" (p. 11). In addition, social space always presents itself as a local combination of the presence, or absence, of agents, resources, human artefacts and elements of the natural world which favour or hinder social activity. From this point of view, therefore, the dualism between the nonspatial and spatial levels of sociology theory no longer has any meaning: space and time take their place in the heart of general theory and, at this point, it is the abstraction from these two dimensions which needs to be justified in function of specific heuristic requirements.

The three theoretical strategies considered here have implications in relation to the sociological approach to the city, as they result in three distinctly different ways of representing this particular social phenomenon.

In the deductive approach, the city is seen essentially as the field in which confirmation can be found of general laws. At the most, a theory of the city can be understood as a study of the way in which these laws react to spatial constraints (often understood as a form of conditioning which depends on a level of pre-social human interaction), and give rise to morphological effects which can be studied with special concepts and methods. This view, typical of the urban sociology mainstream, opens the way to the description of urban space and also to attempts at generalisation. Such attempts have, however, revealed themselves to be of limited theoretical validity, since the consideration of the spatial dimension is situated transversally, so to speak, to the development of more general sociological theory.

The empirical/comparative approach on the other hand gives importance to case studies, providing the sciences of the city with empirical analyses of great value. Sociological analysis, however, by not tackling the problem of theorisation, implicitly leaves this task to other disciplines. For one reason or another, therefore, both strategies risk leaving the representation of the city without the expression of the sociological point of view, of which the strong point, as we have already said, should be the globality of interpretation. It is not surprising that, in these circumstances, regional and urban sciences have generally taken more account of the economic point of view, giving the impression that the role of social behaviour in the dynamics of urban processes is held in low consideration (Bailly and Coffey, 1994).

The 'spatialist' approach is still in its early stages and it is difficult to predict to what extent it will be able to fill the gaps mentioned above. We can certainly say that it is in harmony with our previous suggestion that space and time are dimensions around which the joint interests of the social sciences could come together. Obviously, bringing these dimensions to the centre of sociology theory would not be sufficient in itself to provide an antidote to the tendency to fragmentation which has characterised the development of the sciences of the city until now. If it is to become a systematic programme of research, it would be useful if it favoured a reformulation of the problem area, which in turn could stimulate a partial reintegration of different disciplinary points of view.

13.5 The City, Its Problems and Related Tasks

In the two previous sections we came to the conclusion that both urban economics and urban sociology can be considered 'city sciences', inasmuch as the city is considered not only the background scenario to economic and social actions, but is in itself the object of observation. Put in another way, what we are interested in observing is the urban space generated by those actions.

The above specification seems to us necessary since, if every socio-economic discipline applicable to urban issues could be considered a city science, the 'sciences of the city' would lose any specific quality, being identified with the entire complex of sciences of society. In same way, the solution of urban problems would be identified with the solution of social problems in general; urban planning would be identified with politics, losing its specific status.

Nevertheless, to say that urban economics and urban sociology are city sciences does not mean that they possess a unified body of theory on which it is possible to base an independent discipline. It simply implies that they are potentially able to merge their theoretical apparatus and methodological tools, and reach results which could be incorporated in the framework of acquisitions accepted by each. This applies however not only to these two disciplines, but extends to a vast range of approaches which, while retaining their own special characteristics and not aspiring to a definitive unification, have in common the city as object of observation in the sense given above.

We therefore need to insist on the use of the plural when referring to the 'sciences' of the city. This also implies that advances in understanding of the urban phenomenon now depend, and as far as we can see will continue to depend, at least in the near future, on advances achieved in the fields of research of a multiplicity of disciplines, including economics, sociology, geography, anthropology, psychology, semiology, and also technology (in view of the fact that the physical aspects are not pure accidents or passive manifestations of the action of the reference system, but made up of elements of the structure of the system and therefore capable of conditioning its actions). Obviously, these disciplines, although they are not watertight compartments, use different languages, which express different ways of approaching their subject.

To these observations, we would add some comments relating to the task faced by town planners, i.e. the need to examine, and in some way to resolve, the city's problems, many of which are serious and urgent. It is their urgency, and not only the conceptual and methodological difficulties involved, that forces us to adopt a nonlinear path in deriving methodologies, techniques and operative applications from the theory (proceeding from the 'scientific' definition of the problems to the 'engineering' of their solution). We should perhaps not be concerned by the fact that the urgency of the situation often inverts this order, meaning that the connections are frequently recognised only in retrospect. (The theoretical reflection of the laws of thermodynamics came after and not before the work of engineers on the steam engine - the machines nevertheless worked and the industrial revolution went ahead!)

So, to prevent certain types of development from becoming catastrophic, to understand which features of social evolution depend only on 'macro' political/institutional relations or "boundary conditions imposed from outside" (Silvestrini, in this volume, p. 388), or to reflect on the "general issues concerning the relationship between science and the city" (Silvestrini, in this volume, p. 390), are all aspects which, independently of the need to provide a systematic scientific character to the process, constitute integral parts of this task.

This does not mean that we can proceed in a disorderly or random way, chasing desperately after the problems as they emerge, in the hope that success lies in developing strategies determined by organising the problems into coherent categories and classifying appropriate tools for their solution. An example of this is the classification of models proposed by La Bella (in this volume), defined in order of increasing difficulty of the objectives achievable by their application: from statistical models to descriptive, predictive, then optimisation models. This represents a taxonomy for the engineering of systems considered at different levels of complexity (Le Moigne, 1977), a detailed classification of tools applicable to different objects depending on their complexity (or the level of complexity we wish to take into consideration). This kind of systematic approach (especially when it involves classification) carries certain risks. Even though we are obliged to proceed in this way when our intention is to gain knowledge (otherwise, how is it to be gained?), the presumption of total understanding (despite being only an intellectual understanding and, at the present state of development of the sciences of the city, still very uncertain, incomplete and fragmentary) can generate the illusion of complete dominion over the system and the way it functions.

This presumption can lead, and in fact has already led, to negative results for planning policy: totalising approaches (deriving the specific from the general) inspired by rational conceptions of a rigidly mechanistic type have run into unexpected complexities, exposing its 'weak flank' to attacks which have forced town planning to a position of defence. This has occurred as a result of the underevaluation of certain aspects of complexity due, essentially, to the nonhierarchical (or non-elementarily hierarchical) interdependence between spatial subsystems which have strong self-organisation properties.

The adoption of a socio-economic point of view therefore obliges us to deal with the problem of identifying these subsystems and their generative processes, which are highly varied, both in type and extent.

13.6 Systems and Integration

The framework which we now present aims to outline a logical sequence for reflection on the city, seen as a system with self-organising potential (Preto, 1995) and resulting from processes of spatial integration (though we limit our attention here to processes of a socio-economic nature). This approach attempts to reconcile the needs of analysis with those of synthesis. In other words, it seeks to achieve a breakdown of the processes determining the formation of the city-system, in order to define a number of subsets corresponding to the fields of interest of the various sciences of the city. At the same time, it takes into account the overall picture in which these processes occur, identifying the lines of a possible reunification. For this reason, it is inspired by a 'spatialist' conception of the sciences of the city in the sense described in Section 3 and 4, i.e. where space is not simply the scene in which the action takes place, but is contemporarily a condition and a product of this action.

The main steps of our reasoning are as follows.

a. There exist a multiplicity of actions carried out by social actors, which define a given space-time context. In fact, each of these has an intrinsic space-time valency, or involves specific ways of occupying physical space and interacting with material and symbolic elements existing in space. It is characterised by rhythms which mark the succession of specific actions.

b. These actions can be classified with reference to theories from the field of human sciences and divided into sets (which are necessarily fuzzy) made up of relatively homogeneous classes of actions.
c. As we are dealing with socio-economic actions, each actor will interact with other actors. Given that each action has an intrinsic space-time dimension, the interactions too have an intrinsically space-time character.
d. The regular repetition and institutionalisation of these interdependences (whatever the basis that determines them) can give rise to the formation of systems with self-reproducing capacity, themselves endowed with specific space-time characteristics. These, in turn, and always by means of a theory, can be classified into categories according to the prevalent type of interaction. For example, it is possible to reformulate Parsons' theory of social subsystems as suggested by Gallino (1992b), distinguishing between four types of socio-spatial system: economic systems, political systems and systems aimed at biopsychic or socio-cultural reproduction.

We now concentrate on this last aspect, i.e. the formation of systems capable of stabilising and reproducing interaction. As time is not directly implicit in the determination of spatial contexts, to simplify the discussion we do not specifically take into account the dynamic aspects of the systems considered. The analysis is therefore limited to the spatial dimension.

As we have said, such systems necessarily have a spatial dimension and must therefore be seen, *ipso facto,* as systems of spatial interdependence. They can be defined according to the ways in which the potential relations of interdependence take place and the time dimension within which they occur, generating a specific space. As spatial systems, they develop at different scales, but the scale is not only a dimensional indication (measured along a continuum according to the size of the area in which the interdependencies take place), it is also a qualitative indication, since different scales presuppose different means of communication (for example, a prevalence of face-to-face interactions, or various forms of telematic communications), which in turn imply different time dimensions.

We should like to add some further observations to highlight two important features of complexity. One derives from the fact that actors of different kinds establish different kinds of interdependence (horizontal complementarity relations, hierarchical dependence, or commensalistic or symbiotic relations). These translate into actions which are part of the field of actions specific to each actor, although interconnected with each other in the establishment of the relationship. These relations, and hence the specific actions connected with them, generate spaces - which we define 'interspecific relational spaces' - and are the result of the intersection of several specific spaces.

To clarify, we give a simple example involving two systems: the industrial system and the political/administrative system, represented respectively by two actors, an entrepreneur and an officer of the local planning department. A request for planning permission for the construction of an industrial building is an action specific to the entrepreneur (contributing to the definition of the specific space of the industrial system). In the same way, the decision to grant or refuse planning permission is an action specific to the local planning officer (contributing to the definition of the space of the public administration system). The two actions are interconnected, and the two spaces generate an interspecific relations space.

Although these spaces can be defined in an abstract way, a factor of complexity is introduced when we consider that the relations actually take place in a specific geographical space. In the above example, this can be identified as the public desk of the planning office, i.e. the physical infrastructure which serves as interface between the two systems (see Cavallaro, in this volume). Such spaces in fact often overlap, generating physically determinable, multispecific spaces which cannot be given a single definition because of the presence of a number of different kinds of action. This also happens when components of different systems coexist in the same geographical space (i.e. when the spaces they occupy are superimposed).

Urban space is *par excellence* a space of this type - a road, for example, can serve contemporarily as a place for shopping, for walking, for the transport of goods, for journeys to work, and also as the channel for utility supplies and cable networks. The combined presence of so many functions is made possible by the technical characteristics of the road network, the compatibility of the activities, or simply by their complementary timing (some activities being carried out during the day and others, such as refuse collection, at night time).

In order to introduce such complexity factors, we have made use of examples relating to the urban scale, but the problem of differentiating the levels at which territorial manifestations of systemic relations occur can be resolved - for the sake of simplicity - with the consideration of two scales:

the 'local' and the 'supralocal'. The former corresponds to the urban dimension, the latter to a larger area, i.e. the macro-regional, national or international scale.

Fig. 1 presents, in a highly simplified form, the effects of the structuring processes of various kinds of spatial system at different scales. However, the sciences concerned with space, and with the city as a spatial system, are concerned not so much with the overall picture, as the nature of the processes occurring and the extent to which they succeed in integrating the actions of the multiplicity of actors, hence attributing self-organising capacity to the aggregate system, in such a way that it recognises itself and is perceived as an indivisible entity.

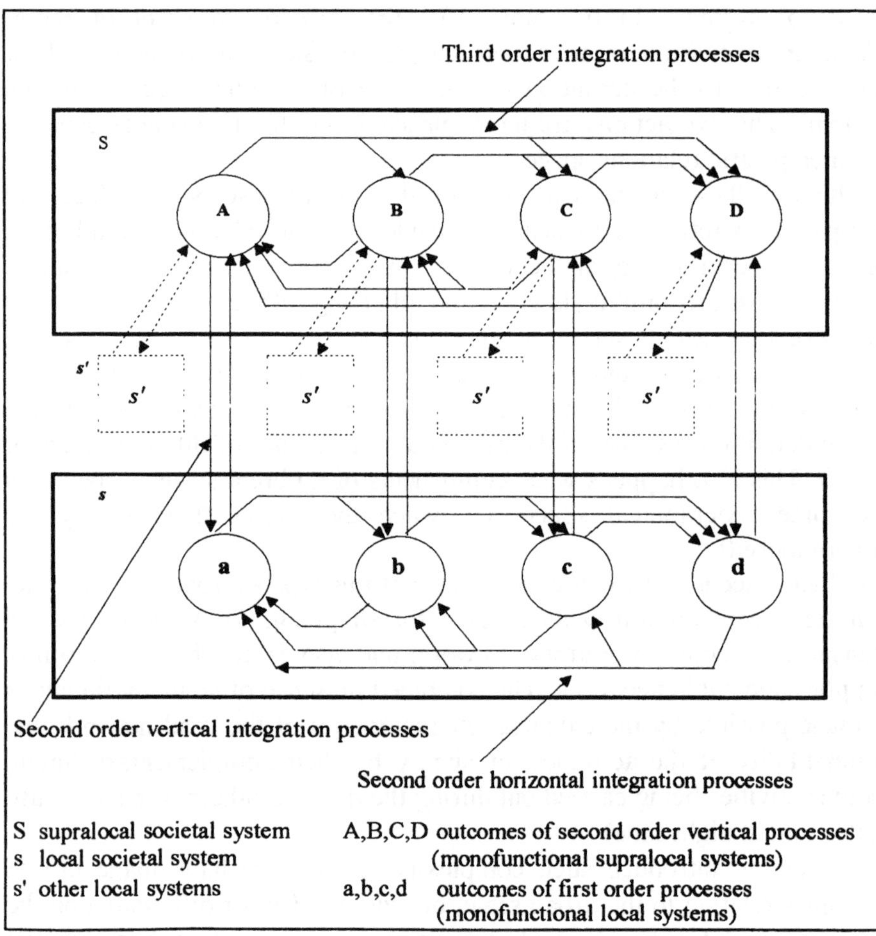

Fig. 1 Levels and orders of spatial integration processes

How can these processes of integration be classified analytically? Once again, for the sake of simplicity, we limit ourselves here to describing a few fundamental types, listing them in an order which goes from the local to the supralocal, and from the specific (sectorial) to the synthetic (multisectorial). Hence we shall speak of processes of different orders of vertical and horizontal integration (Bagnasco and Negri, 1994).

The processes we define as 'first order' integration processes bring together a range of actions of a similar nature (e.g. economic actions) which take place at a local scale (e.g. the urban scale). The outcome of such processes is the genesis of specific systems at a reduced territorial scale. These systems can be of different kinds, depending on the degree of necessity and intensity with which the integration processes occur. They extend from a lower limit, where single actions simply coexist in a given space (the actions, though of a similar type, are weakly integrated, e.g. with simple 'commensal type' relations), to an upper limit where the actions are closely interconnected (relations of complementarity). The more potentialities that exist, and the more successful the integration, the more likely we are to obtain an economic system capable of reproducing itself and giving rise to relatively stable urban development, at least over a certain time period. We can consider integration to have been successful when the principle of complementarity between different economic functions is put to effective use (in ecological terms we could speak of cases of 'symbiosis', or of 'commensalism' in cases where there is joint exploitation of externalities; there can also be a combination of these two principles). In different contexts and phases of development, growth poles and Marshallian districts represent positive examples of economic integration at the local scale.

There are two kinds of 'second order' integration process:

i. Horizontal. This is the kind of integration process which, in the same local context, relates one kind of specific local system with another (e.g. economic systems with political systems, those of biopsychic or socio-cultural reproduction). If the integration is successful, it generates 'interspecific' local systems. When the integration concerns the totality of local specific systems, we can speak of 'local societal systems' or 'local societies'. In cases of this kind of integration, a city can, and in fact must, be considered a local society. Cities of very different types can represent cases of second order 'horizontal' integration. One is the 'world city' which, in addition to a highly diversified economic system, possesses an efficient and active political system, and a cosmopolitan, culturally innovative society. Another may be the city with a specialised economic base, which can constitute an integrated local society, so long as there is sufficient coherence and integration between specific systems.

ii. *Vertical.* This kind of integration process links specific local systems with similar systems located elsewhere, i.e. not belonging to the same local system. If the integration is successful, it generates large homogeneous systems (specific supralocal systems) in which each specific local system is networked with the others (e.g. an urban economic system connected with other local economic systems to form a regional or national economic system). Second order vertical integration processes can arise in very different ways. In a general way, we can distinguish 'top down' and 'bottom up' types of integration. In the former, the integration depends on the initiative of the supralocal system, as in the case of a large multi-branch firm which decides to locate in a given town, activating the integration of the economic system of that town within a wider economy. In the latter, the initiative is taken by local economies which, establishing reciprocal links, give rise to synergies located at a wider geographical scale.

Finally, where there are many local societies composed of specific systems, which are in turn connected within supralocal networks, we can speak of 'third order' integration processes. These describe processes which relate various kinds of supralocal network to determine the formation of large scale interspecific systems, or 'supralocal societies' (societal systems at regional, national and international scale). As stated above in relation to second order vertical integration processes, here too we can distinguish top down and bottom up processes. Here, the role of the political system is decisive. Centralised political formations tend to favour a model of societal integration which emphasises hierarchical relations and the subordination of local societies - we could call this an 'imperial' model. Non-centralised political formations, on the other hand, tend to favour a 'federative' model in which local societies maintain strong independence and the relations between them tend to be on a par. Similarly, the market plays an important role. We can have centralised vertically hierarchical urban structures, acting like polarised energy fields, in which the 'urbanity' and 'completeness' of the functions tends to fade out towards the periphery, or else specialised structures which generate horizontal exchange, i.e. complementary networks. More frequently, in reality, we find a combination of the two.

When we speak of integration processes of any order, we are alluding to a multiplicity of subprocesses which favour relatively stable synergies between actors. In this sense, therefore, it would be mistaken to suppose that the outcome can be evaluated only in binary terms, i.e. that it either produces or does not produce integration. It would be more correct to speak of degrees of integration achieved by the systems. Going more deeply into the evaluation, we could distinguish various degrees of integration, referring to specific sectors. A city could for example have a well-integrated economic system, but a weak and fragmentary political system.

To sum up, a local (urban) system is a societal system which, spatially, can be defined as a *Daily Urban System* (the 'ecological space' of that societal system). It is a system consisting of the intersection of the action areas (interspecific and multispecific spaces) of a range of specific systems whose relations are not all contained within the local system. (If they were all perfectly contained within the local societal system, we would have a completely self-contained 'Isolierte Staat'.)

A local societal system (an urban system) therefore:

- as a system, has an organisation of which the level of connectivity depends on the outcome of first order and second order horizontal integration processes;
- as a society, has a structure with different degrees of completeness depending on the variety of the specific systems of which it is composed;
- as a local phenomenon, is inserted in a network of relations consisting of second order vertical and third order integration processes of the specific component systems which appear at a higher scale and can therefore have different degrees of openness, depending on the density of the mesh of the supralocal network.

Taking this into account, it is suggested that the diagram could serve to generate a taxonomy of urban systems. We can, for example, classify cities on the basis of the degree of completeness and coherence of the single specific systems which make it up, the degree of horizontal integration within the local society, or the degree of connection with the supralocal systems.

If this taxonomy is used with reference to specific theories it will also be possible to attribute the role of an *a priori* classification, open to empirical verification through the construction of indicators. This could be done for example through theories (e.g. of Marxist, or at least, conflictualist derivation) based on the idea of a spatial division of labour. From this we could derive criteria for establishing what characteristics we might expect in an urban context in function of the role it plays in wider geographical systems. Such criteria would allow us to distinguish between cities with a 'central' and those with a 'peripheral' role, permitting us in both cases to make suppositions *a priori* on the degree of internal coherence between the component systems, the level of integration of the urban economy and so on.

The proposed diagram is of course a highly simplified representation into which additional, more complex factors could easily be introduced, without changing its basic nature. To give just one example, it would be possible to eliminate the constraint (implicit in the representation in Fig. 1, 2, and 3) of independent operation of the vertical and horizontal integration processes. In reality, the degree of internal integration of an urban system certainly influences the way it integrates in supralocal systems and vice versa. More generally, we could say that complex processes of feedback occur between all the integration processes mentioned above. This would however introduce into the analysis the explicit consideration of the formative and evolution processes of the system and, therefore, the time dimension (which has been excluded from the present treatment).

The very large number and the variety of integration processes are some of the fundamental factors of complexity in urban systems. We need to know how to tackle this problem and whether the variety can be reduced to a number of categories based not only on the number of relations, but also on their typical manifestation in the formation of integration processes.

13.7 Self-Referential Processes and Principles of Integration

The taxonomy introduced in Section 5 is not intended as a classification permitting the identification of types according to the requisites of the object being studied (i.e. a pre-scientific classifying typology of the kind proposed by Linneus). It is suggested above all as an analytical tool making it possible to identify typical forms of functioning and evolution of urban systems. While the consideration of the number of relations may allow us to define different levels of spatial integration, other features of the relations are important for understanding the morphogenesis of the systems. These are the features which govern the 'piloting' (Le Moigne, 1977) of the systems, i.e. the hetero-directionality and/or self-referencing abilities. We are concerned with the organisation of the systems, the consideration of their self-organisation capacity and, therefore, the nodes (margins and interstitial areas) which permit their control and strategic planning (Gibelli, 1993).

In order to construct a taxonomy for this purpose, it is necessary first of all for there to be an analytical base which makes it possible to trace the integration processes back to causalities relating to some fundamental principle. As we are concerned here with socio-economic spatial integration processes, this principle must inevitably be economically or sociologically oriented and refer specifically to the field where both aspects converge, i.e. spatial organisation. Economics and sociology identify respectively the market and the organisation as the areas in which fundamental principles regulating integration processes can be defined, inasmuch all relations between actors of the social system (and therefore the territorial system) can, in a general sense, be traced back to actions of exchange or production.

The market principle and the organisation principle therefore have a major role in orienting the analysis. To these, in a subordinate role, we could add the principle of networking, which allows the analysis of the more aleatory aspects connected with processes based on 'optional' choices relating to the activation of potential virtualities (Dupuy, 1991). While the first two are principles founded on firm theoretical bases which are sufficiently structured to allow the use of generalisable analytical methodologies, the third is presented in interestingly paradigmatic terms and is useful for analysing phenomena connected with integration processes (not directly related to the market or the organisation, but nevertheless important for the overall configuration of the kind of spatial systems being studied).

Although reference is made to two basic principles, one of economic derivation and the other sociological, this does not mean that the former is applied exclusively to the interpretation of economic phenomena and the latter to noneconomic phenomena; they are both applied to specific aspects of either kind of phenomena.

The market principle is adopted with reference to conditions abstractly definable as conditions of perfect competition, with the necessary exceptions due to the consideration of its spatial dimensions. In this way, it is particularly suited to the interpretation of integration processes which constitute a field of interacting forces, where the actors converge in order to define agreements and conventions, shared but always renegotiable. The field is a place which is neutral and non-appropriable. Its dimension can be determined in consideration of the intensity of relations and also the breadth of their convergence. This refers not only to the goods market, but to all those forms of exchange which take place in the forms indicated above and therefore include all negotiations relating, for example, to the exchange of information, the formation of electoral consensus, and so on.

The implementation of processes of integration on the basis of the market principle presupposes the following conditions:

- there must be a large number of actors interacting individually according to a utilitarian logic, the relations between actors being regulated by the laws of demand and supply;
- the outcome of the interaction must depend on impersonal mechanisms, i.e. independently of the objectives of individual actors. This implies that no single actor can influence the outcome.

For these conditions, there are two levels of observation:

- the micro level, based on the analysis of deterministic, and often directly quantifiable, relations between actors;
- the macro level, based on the analysis of the aggregated effects of relations between actors (which can also be quantified, though in a less direct way) and the observation of variations between given time intervals describing the characteristics of the resulting field.

For example, we would locate at the micro scale the analysis of the technological conditions which underlie (and make possible) the interdependencies between actors of the exchange system (i.e. the effects of a technology on the scope of such interdependencies) and at the macro scale the analysis of the outcome of these interrelations and hence the formation of areas of exchange.

We now examine the ways in which the conditions relating to the principle of organisation come into being. This principle is particularly suited to the interpretation of integration processes which go to set up long lasting systems with specific objectives (often in conflict with other similar systems with competing aims). The processes are based on the control of the individual behaviour of those actors belonging to them. The organisation principle applies to all situations involving the coordination of the efforts of a number of actors for the purpose of producing some kind of output, which can be a good or service for the market or, in the noneconomic context, simply an objective.

The integration process based on the principle of organisation presupposes the following conditions:

- there must be a number of actors who behave according to logic defined in a superindividual way (by a project requiring the coordination of

actions and the planned use of resources);
- the outcome of the cooperative process must depend on the effectiveness of the coordination, i.e. the control over actors and resources, in view of a specific purpose, taking into account the presence of organisations with conflicting positions.

There are essentially three levels of observation for these conditions (see Fig. 2):

- the micro level, based on the analysis of the ways in which individual actors belonging to the organisation interact (in a controllable and partly predictable way);
- the meso level, based on the analysis of the internal structural properties of the organisation and the observation of the variations over time of these properties in function of both endogenous and exogenous factors;
- the macro level, based on the static and dynamic analysis of the effects produced by the conflict/competition or cooperation between organisations which, in their relational space and in the absence of institutionalised coordination, take the form of a self-organising type of super-organisation (Collins, 1988).

In addition, each level of observation lends itself to analysis at different levels of disaggregation through which we can examine the organisational structure of the whole spatial system, as well as the analysis of specific organisation types (Perrow, 1967):

- the micro level, for example, lends itself to the analysis of organisations which are small and self-sufficient (in the determination of their objectives as well as their form), whose output is usually one-off production, and where there are strong interpersonal relations (e.g. a research institute or theatre company);
- the meso level is that of organisations whose output is characterised by associated production where several parallel structures are centrally co-ordinated (e.g. a school or an insurance company);
- the macro level corresponds to an organisation which produces a wide range of products (including intermediate goods), characterised by mass production and assembly, or processes associated with a massive bureaucracy with a strongly hierarchical structure (e.g. a holding company, the Catholic Church) (see Fig. 3).

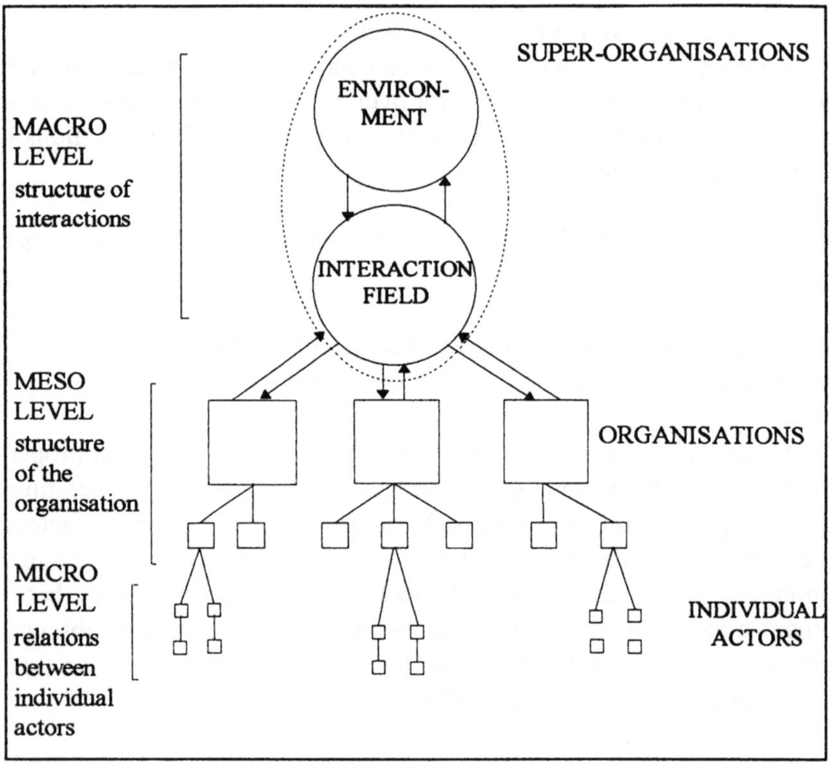

Fig. 2 Levels of observation for integration processes

On introducing the principles of the market and the organisation, we stated that their potential for classification of interdependencies which form a spatial system of relations was exhaustive. It was also specified, however, that there exist relational fields which cannot be univocally or clearly defined. This is the case of relations which, though informal, are not infrequent nor unimportant to the structuring of the spatial system. In this category we could consider for example relations within a neighbourhood, or those generated in a literary salon, i.e. any relations based on unplanned exchange. For the analysis of this kind of relation, the network paradigm could possibly be a suitable conceptual instrument.

On the basis of the current level of development of the network paradigm, it is difficult to establish whether in fact it represents a third principle independent of the first two, or simply a different way of representing phenomena which can be explained through the market or the organisation. In any case, it seems particularly suited to the interpretation of integration phenomena which give rise to intermediate outcomes with respect to those of the first two types. The presence of networks, in fact, (like the market principle) generates a field in which agreements and conventions are of a kind which, like the principle of organisation, give rise to long-lasting forms collective projects, even though founded only on a weak control of individual behaviour (Mela, 1995).

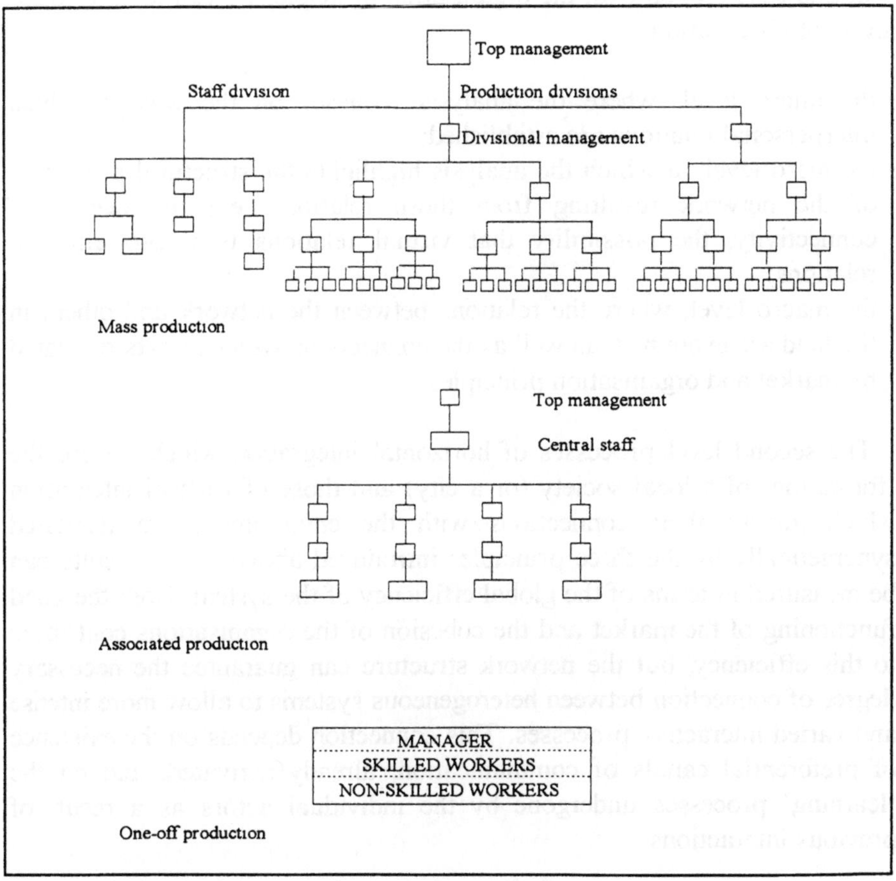

Fig. 3 Types of organisation

The integration processes on the basis of the network principle presuppose the following conditions:

- there must be a large number of actors who, in principle, act individually

but establish cooperative relations which hold for a sufficient period to make possible the development of logical inter-individual actions;
- the outcome of the process must depend in part on impersonal mechanisms, and in part on the effectiveness with which the network coordinates individual action. In any case, the outcome is the object of a learning process which modifies the structure of the network.

In this case too, as with the organisation principle, we can identify three levels of observation,:

- the micro level, where the analysis focusses on the way in which interpersonal relations are established;
- the meso level, at which the analysis highlights the structural properties of the network resulting from those relations, e.g. its degree of connectivity, the possibility that virtual relations turn into effective relations;
- the macro level, where the relations between the network and others in the field are examined, as well as the connections with contexts regulated by market and organisation principles.

The second level processes of horizontal integration which govern the 'formation' of a local society (or a city) and those of vertical integration which govern their connections with the environment are regulated synergetically by the three principles introduced above. Their results can be measured in terms of the global efficiency of the system. Both the good functioning of the market and the cohesion of the organisations contribute to this efficiency, but the network structure can guarantee the necessary degree of connection between heterogeneous systems to allow more intense and varied interactive processes. This connection depends on the existence of preferential canals of communication, already activated, and on the 'learning' processes undergone by the individual actors as a result of previous interactions.

13.8 Some Final Observations

The aim of this work has essentially been to indicate a way of defining the organic connections between urban economic analysis (the well explored field of market relations) and urban sociology (the less explored field, at least as far as the urban implications are concerned, of social organisation and networks). A systemic approach has been adopted in an attempt to capture the fundamental regulatory principles determining the nature of the relations which have contributed to the structure of the urban system and its organisational conformation.

These principles are considered fundamental because they underlie the self-referential behaviour of the system and hence its self-organisation and autopoiesis (see Cavallaro, in this volume). For this reason, we retain that it is the joint reference to consolidated theoretical contributions from sociology and economics (even though here, in fact, they have been dealt with only in parallel, they could in the future be interrelated) which make it possible to verify empirically the structural characteristics of a systemic construction not only postulated in abstract form.

The reproposal of a systemic approach to the analysis of urban phenomena is dictated not only for cognitive purposes, but also practical reasons, i.e. the need for an orientation for the construction of a methodology for urban planning. In fact, there is an increasingly felt need to develop planning procedures which take into account the ways in which spatial organisation processes evolve, their location behaviour and, therefore, the relations between locations i.e. the virtual network of spatial relations (Dupuy, 1991). These processes are evolving extremely rapidly, generating significant modifications in land use in urban areas even over short periods. If the relationship between demand and supply of land is regulated by the statutory plan, the plan must be an instrument capable of controlling these dynamic processes. It must, in other words, be able to constantly adapt to them and orientate them in order to reconstitute patterns allowing the pursual of the objectives of strategic urban planning policy as well as their coherent redefinition (Mazza, 1993).

The focus on the dynamics of the processes and their evolution means that the spatial analysis must refer fundamentally to conceptual schemes of support which, by emphasising the relations, allows the understanding of the pattern of evolution of territorial dynamics (in this work we have limited ourselves to the consideration of the structural aspects). The most recent contributions to systems theory seem well suited to this purpose. It should be specified, however, that the adoption of this point of view for the observation of urban phenomena does not mean that we should automatically assume the city is a complex system and, as such, possesses intrinsic self-organisation properties. This latter assumption underlies approaches of the systemic/objectivist type, inspired by the mathematical and biological theories of complexity, which attempt *a priori* to define general 'laws' of behaviour and of the evolution of an entity, as well as to demonstrate that refined conceptual instruments are available today for its aggregate modelling.

The rejection of this assumption does not mean of course that we refuse to employ these instruments, if and when they prove useful, but that we reject the idea that the systemic nature of an object is an objective immanent property. It implies that a correct systemic prospective derives from the point of view of an observer and that the systemic nature results from the interaction between the observer and observed. We would agree with the affirmation of Tinacci Mossello, in this volume, that the idea of treating the city as an organised complex system constitutes a research programme and not a theoretical certainty. This does not weaken the research by making it less sure of itself. The same observation in fact holds for any subject of study (even considering living organisms as self-organising systems represents a research programme; the idea seems interesting to us today, not because its constitutes a theoretical certainty, but because it has more explanatory power than, for example, the Cartesian approach which considers the organism as a machine).

Nevertheless, even though we can justifiably state that the attribution of systemic qualities to the organisation also depends on the 'subjective' intention of the observer, this does not imply closing the discussion in a relativistic or post-modern vein (as if no intention were worth more than another, or as if the choice of one research programme rather than another depended on the rhetorical capacity of the group of researchers and had only moral or aesthetic implications). In fact, the 'deconstructive' and anti-systemic position of the post-modernist thesis denies the possibility of this programme using two types of reasoning: firstly, because it criticises the application of systems theory to social phenomena on epistemological or ideological grounds, secondly, because it seeks to demonstrate the non-applicability of systemic paradigms to the urban scale, given the global, and practically nonspatial, nature of social relations today.

We maintain here, in contrast to the above position, that research programmes can be more or less fruitful, possess more or less intrinsic coherence, and be more or less able to explain an object or, at least, integrate the (falsifiable) affirmations made about it in the framework of the scientific knowledge of a given period. (The programme which interpreted the organism as a machine was able to account for certain regularities of physiological functioning and stimulus/response sequences. It was integrated into the framework of available scientific knowledge in the 17th century and served until it proved insufficient, when the way was opened to new programmes.)

The programme based on the hypothesis of the city as a self-organising system can be fruitful (by this we mean coherent, integrated with other branches of science and able to resolve problems), as long as it does not proceed with excessive simplification or 'forcing'. Its strong point is not the *a priori* assertion of the systemic and autopoietic nature of the city, which needs to be corroborated with the use of refined tools imported from other disciplines, but it will prove effective if it is able to indicate the processes which lead a set of economic and social activities, carried out in the context of a densely populated local society, to give rise to a degree of coherence and to synergies with mechanisms able to provide self-organising capacity.

To identify these processes means defining criteria which allow us to establish whether, and to what degree, that which appears a simple spatial aggregate of activities and social practices can be considered a self-organising system. These criteria would in effect need to be defined *a priori,* but their application to actual spatial aggregations means that we can only judge *a posteriori* whether or not these are in fact systemic, by comparing them with the criteria and the results of empirical observation. It is clear that the outcome could be positive or negative. It could confirm that while certain cities have systemic properties, others (many large Third World cities, for example) are only an accumulation of nonrelated parts, or nodes relating to separate, non-interconnected networks. If this were the case, the consequences would be not only of academic interest, but also relevant at the practical level, since it would imply the need for completely different forms of policy intervention.

References

Agnew J.A. (1989) The Devaluation of Place in Social Science, in Agnew J.A., Duncan J.S. (eds.) *The Power of Place. Bringing together Geographical and Sociological Imagination*, Unwin Hyman, Boston, 3-25.

Allen P.M., Sanglier M. (1979) A Dynamic Model of Growth in a Central Place System, *Geographical Analysis*, *11*, 256-272.

Bagnasco A., Negri N. (1994) *Classi, ceti, persone. Esercizi di analisi sociale localizzata*, Liguori, Naples.

Bailly S., Coffey W.G. (1994) Regional Science in Crisis: A Plea for a More Open and Relevant Approach, *Papers in Regional Science*, *73*, 3-14.

Boudon R. (1984) *La place du désordre. Critique des théories du changement social*, Presses Universitaires de France, Paris.

Camagni R. (1993) *Economia urbana*, NIS, Rome.

Cavallaro V. (1995) Costruire il presente, doctoral thesis, Turin, Venice, Milan (unpublished).

Coelho J.D., Wilson A.G. (1976) The Optimum Location and Size of Shopping Centres,

Regional Studies, 10, 413-421.
Collins R. (1988) *Theoretical Sociology*, Harcourt Brace Jovanovich, Inc., Orlando.
Dendrinos D., Mullally H. (1985) *Urban Evolution: Studies in the Mathematical Ecology of Cities*, Oxford University Press, Oxford.
Dupuy G. (1991) *L'urbanisme des réseaux. Théories et méthodes*, Armand Colin, Paris.
Gallino L. (ed.) (1992a) *L'incerta alleanza. Modelli di relazioni tra scienze umane e scienze naturali*, Einaudi, Turin.
Gallino L. (ed.) (1992b) *Teoria dell'attore e processi decisionali. Modelli intelligenti per la valutazione dell'impatto socio-ambientale*, Angeli, Milan.
Gibelli M.C. (1993) La crisi del piano fra logica sinottica e logica incrementalista: il contributo dello strategic planning, in Lombardo S., Preto G. (eds.) *Innovazione e trasformazioni della città*, Angeli, Milan, 207-239.
Ledrut R. (1987) L'espace et la dialectique de l'action, *Espace et Société*, 48-49, 131-150.
Le Moigne J.L. (1977) *La théorie du système général. Théorie de la modélisation*, Presses Universitaires de France, Paris.
Leonardi G. (1985) Equivalenza asintotica fra la teoria delle utilità casuali e la massimizzazione dell'entropia, in Reggiani A. (ed.) *Territorio e trasporti: modelli matematici per l'analisi e la pianificazione*, Angeli, Milan, 29-66.
Leonardi G., Rabino G.A. (1984) Prefazione, in Leonardi G., Rabino G.A. (eds.) *L'analisi degli insediamenti umani e produttivi*, Angeli, Milan, 7-12.
Mazza L. (1993) Descrizione e previsione, in Lombardo S., Preto G. (eds.) *Innovazione e trasformazioni della città*, Angeli, Milan, 181-196.
McFadden D. (1974) Conditional Logit Analysis of Quantitative Choice Behaviour, in Zarembka P. (ed.) *Frontiers in Econometrics*, Academic Press, New York.
Mela A. (1995) Innovation, Communication Networks and Urban Milieus: A Sociological Approach, in Bertuglia C.S., Fischer M.M., Preto G. (eds.) *Technological Change, Economic Development and Space*, Springer-Verlag, Berlin, 75-91.
Perrow C. (1967) A Framework for the Comparative Analysis of Organisations, *American Sociological Review, 32*, 194-208.
Pred A. (1990) *Making Histories and Constructing Human Geographies*, Westview Press, Boulder, Oxford.
Preto G. (1995) The Region as an Evolutive System, in Bertuglia C.S., Fischer M.M., Preto G. (eds.) *Technological Change, Economic Development and Space*, Springer-Verlag, Berlin, 257-275.
Robbins L. (1932) *Essay on the Nature and Significance of Economic Science*, Macmillan, London.
Samuelson P.A. (1947) *Foundations of Economic Analysis*, Cambridge, Massachusetts.

14. Innovation, Agglomeration and Complexity in Urban Systems

Dino Martellato

14.1 The Difficulty of Defining Complexity

Many of the contributions to this volume embrace the idea that the sciences of the city belong to the sciences of complexity, and that the city is inextricably tied up with the concept of complexity.

This is not altogether surprising, since the social sciences, almost by definition, deal with complex phenomena. The brain is complex, human behaviour, learning and evolutionary processes are complex. Why then should the multilevel interaction of thousands of individuals be simple? It would be difficult to take such a claim seriously. Nevertheless, it would be nice to find some easy-to-manage tools able to give us clear descriptions and reasonably correct forecasts of the dynamics of social systems, and especially urban systems. But perhaps it is too much to hope that a model could provide us with a faithful reproduction of the complexity of an urban system without itself being so incomprehensible as to be useless. One major problem to overcome is that of the distance between the object and the observer - an object inevitably seems small to the distant observer.

Everyone seems to have their own notion of complexity. Complexologists in particular, it would seem, often have more than one. In a recent article, Horgan (1995) provided a list of more than thirty known definitions of complexity! It is unlikely however that all of these notions lend themselves to the quantification of the degree of complexity of problems. In many cases this is extremely tricky because of the difficulty of translating a qualitative concept into a quantitative index.

Complexity is certainly a qualitative and elusive concept which emanates from the relationship between the observer and the observed. But what seems complex today may well be considered straightforward tomorrow and simple or even trivial the day after. I am convinced that the degree of complexity that an observer attributes to a phenomenon depends largely on the state of his or her knowledge. Silvana Lombardo, in her contribution to this book, reminds us that we should define as complex not simply that which we do not know, but that which can never be known. I completely share this relativist approach to the definition of complexity. The observed system seems complex to the observer when there is insufficient information, when the observations are affected by a strong 'noise' component, when the analysis requires a calculation beyond the capacity of the observer, or simply, when there is only partial understanding of the phenomenon observed.

Of the two alternatives: (i) that which we do not know is complex and, (ii) that which can never be known is complex, I believe we must opt for the former, if only because the latter would lead to a paradox. As no-one knows the limits of what is knowable, i.e. the point to which knowledge can arrive, we are unable to say what is really complex. But, in reality, complexity is an overworked concept, we have too many rather than too few definitions. In economics alone, there is already confusion in the use of these terms. The definition of complexity as a synonym of partial understanding on the part of the observer is certainly very common and reductive, but maybe for this reason more useful than the others, in view of the impossibility of formulating a unifying theory of complexity (Horgan, 1995).

14.2 Complexity in Economics

In the study of economics in general, and in urban economics especially, it has always been held that to possess a simplified, if possible mathematical, description is indispensable. Firstly, it serves to overcome the cognitive limits of the observer-analyst and, secondly, to compensate for his inability to develop models which can reproduce in manageable form the complexity of evolutionary systems. Economists have long been using this approach - already adopted by Von Thünen (1826) - along with the hypothesis that people always behave in a perfectly rational way and according to the principle of marginal benefit. Only much later was an attempt made to relax the axiom that economic agents always seek to maximise something (profit or utility, depending on the case).

Alchian (1950), for example, assumed that there could be uncertainty, incomplete information and more than one motivation, and suggested the adoption of a criterion of choice other than maximisation of benefit. More recently, Akerlof and Dickens (1982) have applied the psychological theory of dissonance to construct a model which drops the classical assumption of perfect rationality of choice.

Interaction between individuals, each of whom takes strategic decisions, i.e. decisions requiring prediction of the future decisions of others, creates uncertainty because these forecasts are necessarily uncertain. Whereas the hypothesis of rational expectations, so common in economics, supposes that each individual possesses 'perfect infinite foresight', here we are at the opposite extreme because, in normal conditions, the individual possesses forecasts which are, to say the very least, limited and imperfect.

There is no doubt that interaction in the market is the oldest, and possibly the most important, vehicle of complexity in economics. Already in 1962 Herbert Simon stated that, in order to be a useful tool for the study of systems with a large number of intricately interacting parts, a model must have special characteristics and cannot be based on rationality. Imperfect competition in its various forms provides numerous examples of structures whose dynamics are dominated by the interaction of individuals taking strategic decisions. In all of these cases, it is the functioning of the market which creates uncertainty in its own future configuration. The principle of strategic interaction is general and, even though in a simplified form, underlies both the theory of rational expectations and the concept of Nash equilibrium. The idea of rational expectations has two components: (i) the rational behaviour of individuals, who are assumed to be able to maximise the value of some objective function with respect to perceived constraints, (ii) the mutual consistency of their perceptions (Sargent, 1993). Here, strategic interaction occurs between individuals who, so to speak, 'see in the same way', even though they may not necessarily realise it.

The Nash equilibrium presupposes that participants in the game are able to identify in advance an equilibrium position, i.e. the state at which no player/agent, even when not wishing to collaborate with the others, has any interest in modifying his or her strategy since a change would not increase expected utility (Simonsen, 1988). This means that to establish the Nash equilibrium it is necessary that: a) there is a 'common knowledge' about the nature of the situation, b) everyone knows *a priori* everyone else's strategy, c) there is a single equilibrium, d) everyone is capable of calculating the solution and of translating it into strategic action. Interaction therefore occurs between agents who have full knowledge before acting and know that they cannot choose any solution other than that which satisfies everyone. The logic of models based on rational expectations, or those referring to the Nash equilibrium, implies firstly that there always exists an equilibrium and secondly that all agents share and use that representation or model of the decision-making process which produces the equilibrium.

Looked at from this point of view, the two paradigms (rational expectations and Nash equilibrium) are simply expedients to 'keep at a distance' the complexity which arises from strategic interaction in a world dominated by uncertainty. If there is no 'common knowledge' and not everyone uses the representation of strategic interaction which produces equilibrium, in a system in which there are many thousand players with only partial information, strategic interaction ends up inevitably by making it difficult for there to be one equilibrium. As a result, the hypothesis of uniqueness implied by rational expectations is invalidated. This means that there must be a large number of solutions or possible equilibria and, therefore, complexity. The system does not converge to a state of equilibrium since this requires a common perception of the state of the system and mutual consistency between the strategies.

A further example of the problems which can arise from strategic interaction is the possibility that the market produces 'sunspot' type equilibria, i.e. flares in activity. Especially in financial markets, strategic interaction between agents can give rise to temporary equilibria so completely out of line with the market fundamentals that external events (which can be likened to random variables) are able to influence the equilibrium simply because the agents think that they will (Azariadis, 1981).

Recognising that even relatively simple dynamic models can have such complicated solutions that it becomes impossible to forecast the future behaviour of the system, economists have had to face up to the problem of complexity (Day, 1982, Baumol and Benhabib, 1989, Cass and Shell, 1989, Scheinkman, 1990).

If we move from the question of time to that of space, we run into problems which are rather different, but no less interesting. We intend here to consider just a few studies concerned with this aspect. Perhaps the most important are those which examine the interaction between innovation, market structure and urban agglomeration. We should state immediately that these studies do not adopt typical models of complexity, although they deal with problems similar to those dealt tackled by evolutionary theory and the development of economic webs (Kauffman, 1988), both tools of the science of complexity. The next section is dedicated to these studies and Section 4 examines further examples of complexity in a spatial context.

14.3 Endogenous Innovation, Market Structure and Agglomeration

Technological innovation is influenced by the economic environment and, therefore, the market structure and spatial configuration of the economy for the simple reason that firms, in reacting to the environment and attempting to adapt it to their purposes, produce innovations and make use of practices such as publicity, rent-seeking and internal restructuring. A firm, however, like any other evolutionary organism, must respect a principle of vital importance, i.e. it must continually adapt and evolve in order to maximise its capacity for understanding and controlling the environment. Failure to do so is likely to put its survival at risk.

In the economic field, the environment is made up of a large number of firms, individuals and official bodies which are highly interdependent. They interact, above all, through the exchange of products, factors and information. But there are other forms of interaction which are no less important, though often less evident. These concern the various types of externality which, when positive, produce economies of agglomeration, and in fact provide essential conditions for the existence of the city. They represent an influence which interconnects firms and also consumers. As illustrated by Von Thünen as long ago as 1826, accessibility is just one of the factors of association. Agglomeration economies occur in different forms: from the point of view of production, they can exist within the firm (economies of scale and scope at the level of the firm), within the industry (economies of location, which act as economies of scale at the city level) or within the urban area (economies of urbanisation, which act as economies of scope of the city).

There is no lack of techniques for the analysis of these mechanisms, but they are not always very satisfactory. The two main approaches look at the process from opposite points of view. Seen in terms of the production function, agglomeration economies are assumed to result in higher levels of productivity; seen however in terms of the cost function, agglomeration economies imply lower production costs. From the consumption point of view, on the other hand, agglomeration economies arise from the fact that certain public goods and services cannot be provided unless a minimum scale is reached. It has been shown consumers appreciate, and are prepared to pay more for, the availability of a large variety of goods and services.

The existence of economies of agglomeration justifies the formation and existence of cities as much as the question of accessibility. Agglomeration economies are not present, however, in all types of city. As we have already said, innovation is influenced by the structure of the market and the spatial structure of the economy, i.e. by agglomeration, but the relationship is reciprocal - the nature of the market is itself influenced by innovation. In certain cases, innovation induces deconcentration, in other cases concentration, both vertical and horizontal. At the spatial level, the weakening of agglomeration economies implies dispersion, that is the weakening of the vertical or horizontal links between firms. The complexity of the relationship between technical progress, the market and urban structure is of a circular nature. The firm realises that in the market and in space there is a given degree of concentration and level of differentiation of products. These obviously have an effect on the firm's profits and, as a result, on its decision to spend money on investment and R&D. Dasgupta and Stiglitz (1089), among others, have demonstrated the micro basis of a theory which claims that market structure depends on expenditure on R&D and process innovation. Their Schumpeter inspired model explains a very commonly observed fact, that R&D expenditure increases when the fragmentation of the oligopolistic market increases. However, if we assume free entry, it can be shown that, in the long term, the degree of concentration and R&D expenditure can be determined sumultaneously.

As far as spatial concentration is concerned, the theory is less developed. Eaton and Lipsey (1978), for example, have proposed a model of product innovation in which the market area, and therefore spatial concentration, plays a major role. The firm makes its investment in fixed capital and R&D, does marketing and spends on publicity; in other words, like all firms, it does what it can to improve its position. In doing so, it constantly seeks to turn product differentiation, the market and spatial structure to its advantage. This is the opposite situation from that studied in the previous cases. Davies and Lyons (1982) in fact demonstrated that the degree of concentration is partially explained by technology and partially by a stochastic component, or a probability of entry which depends on certain specific characteristics of the environment in which the firm operates. These are just a few examples of theories providing a joint explanation of innovation, externalities and market structure.

We cannot claim that the state of development is altogether satisfactory, as we do not have anything like a complete or coherent description of the determinants of the degree of concentration, or dispersion, observed in the market or geographically. We are therefore not in a position to understand the determinants of the passage from vertical to horizontal concentration and vice versa. Nevertheless, complexity arises not only from the circular nature of these relations, nonlinear dynamics, the dimension of the problem or other aspects of urban phenomena, it is also the result of our inability to describe many aspects of evolutionary social systems (Anderson, Arrow and Pines, 1988). Considering the situation simply from the economic point of view, we still do not possess an adequate description of the production process at the level of the firm, the sector or the city, neither are we in a position to determine whether the long-term success of the strategy of imperfect competition is based on the Darwinian principle of the selection of the fittest or on the principle of coexistence (in connection with the analysis of the automobile market, see Fairn, 1996).

14.4 Examples of Complexity in the Spatial Dominion

If we turn our attention to studies of complexity in the spatial area, or look for studies in spatial economics in which traces of complexity emerge, we realise that there are in fact very few. We also find that they tend to be mainly concerned with methodology and to be exemplificative. The brief review that follows is intended to indicate the principal trends emerging and has no claim to being exhaustive.

First of all, although it may seem rather scholastic, it is perhaps useful to make the distinction between stock and flows. An urban system consists of people, who carry out their activities connected with work, travel, recreation, shopping, study, and so on, and of a series of fixed physical entities (houses, factories, shops, infrastructure, etc.) which make these activities possible. While the individuals and goods produce flows and have a fixed point distribution, the entities correspond only to stocks.

Some studies emphasise the physical element of the city, neglecting the human aspect to focus on the stocks of housing, jobs and infrastructure. The result is that even very simple hypotheses about the behaviour of the stocks tend to lead to complex urban structures or 'landscapes'. One of the first examples of this kind of approach is the study by Clarke and Wilson (1981), a more recent example is the work by Krugman (1994). These two contributions, although very different, have in common a kind of 'minimalism' in their hypotheses and a 'maximalism' in their dynamics - a characteristic which was also, if one looks carefully, a distinctive feature of the work by May (1976).

Clarke and Wilson (1981) examined the evolution of the structure or spatial distribution of retail outlets. Their hypothesis was that location choice is rational and dictated by the profit principle, applied in a very simple - one could even say simplistic - way. They assumed that the retail supply in a given area varies immediately according to the difference between income received and the expenses incurred. This principle, which was also the backbone of the Harris and Wilson model (1978), assumes that the income depends on two components: the first, deterministic, element is the demand potential, and the second, which is stochastic, is an additional element introducing a purely cyclical component into the evolutionary model. In an application to the city of Leeds - with a total of 900 retail outlets and 30 points of concentration of demand in the corresponding census areas - the interaction between the bifurcations and randomness in the demand gave the spatial structure some particularly interesting features. Urban structure becomes complex since there may sometimes be a bifurcation, or even chaos.

Krugman (1994) begins with different assumptions, but arrives at results which are not dissimilar, at least from the methodological point of view. He points out that well-known economic principles, such as increasing returns and comparative advantage, can produce complex spatial structures or landscapes, i.e. structures in which there are several attractors and hence multiple equilibria. The decisive factor in the determination of a trajectory is the initial state of the system, with the consequence that the 'path-dependence' is produced by Marshallian type dynamics. It is assumed that there are increasing returns, costs of transport and mobility of factors. In this respect, the approach derives from 'new trade theory' which, unlike more traditional theory, gives little weight to comparative statics or the supply of resources (intended in the widest sense), but assigns great importance to increasing returns and dynamic advantages. The result is an urban system with concentrations of economic activities at points of highest accessibility. This self-organisation process has features similar to those of a Lösch or gravity type system. The system evolves until it finds its equilibrium - which is only one of many possible equilibria - compatible with the initial conditions and the principles of optimality. Krugman considers the system to be complex, not so much because there are bifurcations and chaotic oscillations, but because of the nature of the spatial structure, i.e. the fact that, even when it begins with a disorderly pattern, the system organises itself into an ordered spatial structure which can be interpreted in terms of simple principles, such as the rank-size rule. The structure, as a hierarchical organisation, is also independent of historical contingency.

Although in the Clarke and Wilson study the system has a complex (and in certain cases unpredictable) evolution, in the Krugman study the complexity is synonymous with its self-organising capacity. Leaving aside these differences of interpretation, however, the two models produce similar hierarchical urban structures.

Other studies associated with flows of people and goods are perhaps even more abstract, but serve to complete the picture. Nowak and May (1992), for example, present a spatialised version of the 'prisoner's dilemma'. This allows them to examine how cooperative behaviour can evolve in a space formed of n^2 cells (in the geographical sense). A well-known feature of this game is that, when played only once, the preferred strategy is collaboration. If the game is repeated however, the best strategy is to betray.

In a two-dimensional context, each player or group of players corresponds to a cell which has four diagonal neighbours and four nondiagonal neighbours. A deterministic mechanism, such as the prisoner's dilemma, governs the spatial structure in a population distributed over a lattice: what happens in one cell depends on what happens in the eight neighbouring cells (and, possibly, in their neighbouring cells too). The result obtained with a large number of simulations is a pattern which, at each simulation, is qualitatively different from the previous one. There are very evident elements of complexity (understood as unpredictability), as well as regularity (order), which depend exclusively on the parameter representing the advantage gained by the player who betrays over those who collaborate, i.e. the score that each makes in relation to his neighbours. As there is no memory, the past has no effect and each simulation therefore depends on synchronic and not diachronic comparison.

If we take a territorial system in which there is no movement and therefore no cost of transport, as the state of each cell depends only on that of its neighbouring cells, the above experiment indicates that interaction alone is able to determine the state of the system. This conclusion, which could have important implications for the dynamics of spatially organised physical or proto-biological systems, does not seem to have particular significance for the analysis of the spatial organisation of social systems, since in the latter the interaction is far more complex than the game described above or any other evolutionary type of game.

The central problem is that of finding a model able to describe interactive behaviour. This problem is tackled, but not completely resolved, by Arthur (1994), who excludes at the outset the hypothesis that a functioning model can assume rationality. The insurmountable difficulty of applying a model based on perfect rationality and complete information obliges us to substitute the principle of perfect rationality with something else. But what? As the modelling of bounded rationality is only in its early days, Arthur suggests substituting the optimisation model with a process of learning based on trial and error, i.e. the formation, validation, and possibly rejection of alternative hypotheses. This process of reasoning is inductive, but different from the genetic algorithm proposed by Holland, involving a society made up of artificial 'individuals' who evolve and learn through the constant appearance, and disappearance, of new individuals.

Arthur observes that, to act in a complex or ill-defined context, it is impossible for the individual to use perfectly rational frames of reference for two reasons. In the first place, it would assume that man possesses an unbounded rationality and logical capacity, whereas in fact above a certain threshold of complexity, absolute rationality is limited or fails completely (we could say, in this regard, that complexity is the antonym of rationality). In the second place, when interacting in highly intricate situations, we know that others no longer act rationally, so we are obliged to constantly reformulate our hypotheses about their behaviour. In this situation, we have to abandon the illusion that purely deductive behaviour is sufficient and make use of an inductive form of reasoning.

Arthur supposes that the individual makes use of provisional hypotheses on the behaviour of others, as well as the recognition of behavioural patterns. These hypotheses are the fruit of a process of learning which is constantly evolving, although the 'right' model or pattern is never discovered. The individual nevertheless bases his or her forecasts and decisions on these hypotheses, which are abandoned as soon as they are shown to be unrealistic and, above all, not useful in arriving at a good decision. Since every agent reacts in this way, the system becomes co-evolutionary because everyone continually modifies their view in relation to the feedback they receive from the others. What is lacking, in other words, is one of the fundamental characteristics of the paradigm of rational expectations: consistence. In these cases, none knows what model any of the others uses to predict and therefore act - this is the opposite of what happened in the case of the prisoner's dilemma. The individual does not possess, in fact cannot possess, a predictive model on which to base his decisions rationally. He can therefore only proceed inductively, making and discarding hypotheses or models on the basis of their predictive ability and therefore their performance.

This, essentially, is the methodological premise underlying Arthur's work. It is of great importance because it represents a complete reversal of the view which gained popularity in economics with the famous article of Lucas (1975). The theory of rational expectations (Muth, 1961) sprang from the idea that the dynamic models used until then in economics had not assumed sufficient rationality. Arthur, on the other hand, and before him Simon (1956), claim that as knowledge of all the alternatives and exogenous events is not possible, and there is sufficient calculation capacity, rationality is necessarily bounded. Therefore, while the new classicists are inspired by the ideas of Lucas, assuming that decisions, and therefore events, are based on predictions deduced rationally from a model, Arthur maintains that behaviour is determined by a learning process, and proposes an approach which, at the time of Simon, was called 'information processing psychology'.

Unfortunately, Arthur does not deal specifically with the structure of urban systems, but he does consider a problem of congestion which, by definition, involves decisions affecting the equilibrium between space (as a resource) and demand. He imagines that N individuals must decide whether or not to go to a bar. The decision depends on the expected level of congestion (a person will decide to go if there are fewer than 60 people). As there is no possibility of knowing in advance how many people will decide to go, there can be no universal model and therefore no rational expectations. Each person decides simply on the basis of inductive reasoning, knowing the past levels of crowding. No-one knows how the others will decide, so there is no rational or correct forecast. If everyone decides to go to the bar because they think there will be few people, they will create congestion and vice versa. The correct behaviour would be to decide to go when the others decide (wrongly) not to go, thinking it will be crowded.

Using a computer experiment which assumes that each person decides on the basis of a number of prediction models or hypotheses (e.g. the average of the last four occasions, or the difference from 100 of the last time), Arthur generates a series representing the number of people in the bar. This series oscillates or, to use a more appropriate term, self-organises around 60%, in the sense that the oscillations remain, but the average is always 60/100. The battery of hypothesis-models is continuously modified on the basis of their performance. The 100 agents have around sixty models which tell them to go (a set which changes in composition, but not in number) and about forty which tell them to stay at home. Individuals do not have a single model, or even the same set of models, nor is there a genetic algorithm which adapts and improves with the passing of time. Nevertheless, the changing aggregate or ecology of a hundred 'stupid' prediction models, which each time appear to the individuals as the best, generate an average forecast which coincides with the threshold of 60/100.

The representation of behaviour which emerges from this example and other studies of Arthur seems fairly realistic, since no individual is considered capable of using of a genetic algorithm, nor of being inductively rational in the classical sense. A particular model is used only if it performed well the previous time, otherwise another one will be chosen. It derives from this that the set of operative models adapt over time (it would not be correct to say that they improve). The result is not precise because the exact number of people in the bar is impossible to predict, but the system self-organises in the sense that 40% of the models say that more than 60 people will be present, advising the person not to go, and 60% say that there will be fewer than 60, therefore advising the person to go.

In this work there is a methodological element of general interest, going beyond the example relating to crowding in a given location. As there is no reason to believe that decisions of a spatial type should be taken in a way that is structurally different from those concerning consumers, financial speculators, business managers or others, the principle of inductive reasoning can be an instrument for representing the complexity even in models of spatial flows.

14.5 Conclusions

It is impossible to draw conclusions on the complexity of urban systems for the simple reason that we are still at the beginning of a process of fertilisation which is likely to be long. Even adopting a purely economic approach, things do not improve greatly in the sense that there are no firm points or conclusions. There is however the sensation of breaking into problems which were once confined to the margins, and often completely ignored.

One difficulty is that of choosing from among the many concepts of complexity that which is most useful in economics. In the opening section we proposed a relativistic concept according to which complexity is only a reflection of the inadequacy of the knowledge of the observer/analyst. This is a concept which includes many other definitions of complexity. In the studies considered, we find ourselves facing phenomena which are difficult to dominate and often, in part, inexplicable. In the few works on complexity in the spatial domain, use is made of various other concepts relating to complexity, such as unpredictability and the self-organising capacity of the urban system.

In economics, or at least in mainstream economics, various expedients have been employed to avoid the problem of complexity - rational expectations and Nash equilibrium are examples of hypotheses which remove the possibility of strategic interaction leading to complexity. But as soon as we consider problems posed by strategic interaction between agents, complexity appears in various forms, as discussed in Section 2. The most symptomatic case is that of the interactions which occur in a oligopolistic market in which innovation is important, as we saw in Section 3. There are however few studies of complexity in which space, or the urban system, have an explicit role. The four examples considered in Section 4 not only use very different behavioural criteria, but refer to very different concepts of complexity.

References

Alchian A.A. (1950) Uncertainty, Evolution and Economic Theory, *Journal of Political Economy, LVIII*, 211-221.

Akerlof G.A., Dickens W.T. (1982) The Economic Consequences of Cognitive Dissonance, *The American Economic Review, LXXII, 3*, June, 307-319.

Anderson P.W., Arrow K.J., Pines D. (1988) *The Economy as an Evolving System*, Santa Fe Institute, Addison-Wesley, Redwood City, California.

Arthur B. (1994) Inductive Reasoning and Bounded Rationality, *The American Economic Review, Papers and Proceedings*, May, 406-411.

Azariadis C. (1981) Self-Fulfilling Prophesies, *Journal of Economic Theory, 25*, 380-396.

Baumol W.J., Benhabib J. (1989) Chaos: Significance, Mechanism, and Economic Applications, *Journal of Economic Perspectives, 3*, 77-105.

Clarke M., Wilson A.G. (1981) The Analysis of Bifurcation Phenomena Associated with the Evolution of Urban Spatial Structure, in Hazewinkel M., Jenkovich R., Paelinck J.H.P. (eds.) *Bifurcation Analysis - Principles, Applications and synthesis*, Reidel, Dordrecht.

Cass D., Shell K. (1989) Sunspot Equilibrium in Overlapping Generations Economy with an Idealized Contingent-Commodities Market, in Barnett W.A., Geweke J., Shell K., *Economic Complexity Chaos, Sunspots, Bubbles, and Nonlinearity*, Cambridge University Press, Cambridge, 2-19.

Day R.H. (1982) Irregular Growth Cycles *The American Economic Review, 72*, 3, 406-414.

Dasgupta P., Stiglitz J. (1980) Industrial Structure and the Nature of Innovative Activity, *The Economic Journal, 90*, 266-292.

Davies S.W., Lyons B.R. (1982) Seller Concentration: the Technology Explanation and Demand Uncertainty, *The Economic Journal, 92*, 903-919.

Eaton B.C., Lipsey R.C. (1978) Freedom of Entry and the Existence of Pure Profit, *The Economic Journal, 85*, 455-469.

Fairn V. (1996) A Replicator Theory Model of Competition through Imitation in the Automobile Market, *Journal of Economic Behavior and Organisation, 29*, 141-157.

Harris B., Wilson A.G. (1978) Equilibrium Values and Dynamics of Attractiveness

Terms in Production Constrained Spatial Interaction Models, *Environment and Planning A, 10*, 371-388.

Horgan P. (1995) Dalla complessità alla perplessità, *Le Scienze*, 324, 80-85.

Kauffman S.A. (1988) The Evolution of Economic Webs, in Anderson P.W., Arrow K.J., Pines D. (eds.) *The Economy as an Evolving Complex System*, Santa Fe Institute, Addison-Wesley, Redwood City, California, 125-146.

Krugman P. (1994) Complex Landscapes in Economic Geography, *The American Economic Review Papers and Proceedings, 84*, 412-416.

Lucas R.E.Jr. (1975) An Equilibrium Model of the Business Cycle, *Journal of the Political Economy, 83*, 1113-1144.

May R.M. (1976) Simple Mathematical Models with Very Complicated Dynamics, *Nature*, 161, 459-467.

May R.M. (1978) The Evolution of Ecologic Systems, *The Scientific American*, 239, 3, 161-175.

Muth R. (1961) Rational Expectations and the Theory of Price Movements, *Econometrica, 29*, 315-353.

Nowak M.A., May R.M. (1992) Evolutionary Games and Spatial Chaos, *Nature*, 359, 826-829.

Sargent T.J. (1993) *Bounded Rationality in Macroeconomics*, Clarendon, Oxford.

Scheinkman J. (1990) Nonlinearities in Economic Dynamics, *The Economic Journal, 100*, 33-48.

Simon H.A. (1962) The Architecture of Complexity, *Proceedings of the American Philosophical Society, 106*, 467-482.

Simon H.A. (1979) Rational Decision Making in Business Organizations, *The American Economic Review, 69*, 4, 493-513.

Simonsen M.H. (1988) Rational Expectations, Game Theory and Inflationary Inertia, in Anderson P.W., Arrow K.J., Pines D. (eds.) *The Economy as an Evolving System*, Santa Fe Institute, Addison-Wesley, Redwood City, California, 205-241.

Von Thünen J.H. (1826) *Der isolierte Staat in Beziehung auf Landwirtschaft und Nationalökonomie*, Pythes, Hamburg, Leopold, Rostok.

15. The Sciences of the City: Analytical Tools and Complexity

Fiorenzo Ferlaino

15.1 Introduction

The city can be defined in various ways, depending upon the characteristics we wish to emphasise. We can carry out anthropological, sociological, economic, geographical analyses, and many others, which each highlight one or more specific aspects of this complex entity. The concept of complexity can be referred therefore to the ways in which the city is seen when observed from different points of view. Over the last ten years, many new analytical tools and models have been developed. These have formalised particular complex aspects of the urban system which cannot be expressed through systems of normal equations or dimensions of Euclidean space. We refer to catastrophe theory, principles of self-organisation - the city as a 'dissipative structure', as a 'synergetic' or an 'autopoietic structure' - also to neural simulations, multi-agent systems, fractals morphology and chaotic dynamics.

Nevertheless, like gravitational or entropic models of the first or second generation, none of these models is able alone to explain the concept or the reality of the city. So, alongside the development of these theories and methods for investigating complexity, efforts have continued through statistical/descriptive approaches (Pumain *et al.*, 1992) to provide more precise explanations and to trace clearer boundaries to urban space. There seems to exist, however, at least in the academic world, a form of *schizophrenia:* despite recognition of the usefulness of formal models for representing aspects of complexity, when it comes to providing a quantitative description or definition of the city, use is made of models of statistical derivation. We wish in this work to discuss the contrast between these two ways of approaching the theme of the city, since it represents a fundamental problem which is far from being resolved.

What relationship is there between 'hard' modelling using complex mathematics and the 'soft' approach using statistical methods and models? Is there any relationship between the different spheres of interest which analyse the city through different perspectives?

A second aspect concerns the concept of *complexity*. A number of speakers have stated that the city is a complex entity, nevertheless, although attempts have been made to examine the concept of complexity in epistemological terms, it has not yet been given a precise and systematic definition. Here, we propose a heuristic model which investigates a generic causal relationship and attempt, using this model, to highlight the underlying structure and different dynamics of complex systems. What should we understand by complexity? What mechanisms govern it? Is complexity an innate quality of the 'observed' or is it in the 'eye of the observer'? We shall attempt to answer these questions, proposing a new classification of the 'sciences of the city'.

15.2 Interpretative Paradigms and Analytical Tools

When we ask what is understood by the city, the replies are likely to make reference to very different concepts of the territory and its dynamics, many of which would appear to be in contrast with each other, making it very difficult to find any single uniting synthesis. It would therefore seem useful to attempt to examine systematically some of the main concepts and the categories used for investigating the territory.

The city can be examined through:

1. descriptive and theoretical analysis;
2. statistical analysis and models;
3. mathematical models of various degrees of complexity.

This last category has already been described and evaluated by Denise Pumain. I should like however to reformulate the analytical system which was provided, giving more space to statistical tools and indicators in order to arrive at a different, and I hope more systematic synthesis of the analytical approaches to the city.

I shall begin by stating that, in general, to each kind of tool corresponds a particular analytical methodology. Surveys and statistical analyses are primarily based on the *inductive* method, placing emphasis on the city as a specific object which can be defined by measurable indicators. Each generalisation derives from a survey, and the theory consists of the synthesis of these generalisations. The *deductive* method is rather different. In this case, the city is seen as a set of relations which are complex but nevertheless definable and coherent. It is assumed that there exists some kind of causal relation between structure and urban functions which can be 'read' and hence modelled, or deduced. Mathematical models are therefore based on a deductive methodology, and presuppose the possibility of representing in formal language the complex causal links existing in the city or region. To formulate a theory, a process of *abduction* is used, making reference both to concrete data and to logical derivations which suggest a hypothesis. To be more precise, abduction provides the theory which induction verifies.

Firstly, we may observe that there is, in general, a close link between the method and the tool used. Secondly, in order to clear away any possible misunderstanding, it should be specified that the various tools used, though involving different methods of investigation, do not necessarily belong to different fields of study. In other words, the same paradigm, or the same view of the city can be used as a basis for analyses using very different tools and methodologies. Gravity theory, for example, which was first applied in the sixties in models which referred to 'urban mass' as the indicator of attraction, is now used rather differently; the concept of local attraction being expressed through a series of factors such as the presence of organisational headquarters, research institutes, innovative milieus etc. These may be difficult to translate into mathematical terms, but nevertheless still represent the idea that one region or city may be more attractive or more central than another (see, for example, the well-known DATAR and GIP-Reclus study, 1989). Similarly, the organic paradigm, derived from Spencer's theories, was adopted by the Chicago school and continues to be used today in complex models of evolution and competition. These make reference to catastrophe theory rather than the theory of dynamic self-organisation adopted by the Brussels school, or the Lotka-Volterra models which were developed in particular by Dendrinos (Dendrinos and Mullally, 1985).

In many cases the tools and methods used to analyse the city may coincide, but they are certainly not based on the same paradigms. The city has been seen in many different ways. We now look at some of the most significant.

a. The dichotomous approach in which 'town' is opposed to 'country'. This dichotomy originated with the physiocrats who considered the role of the country as driver of the economy, and that of the city the control and distribution of the income produced. The distinction remained in 18th century thought. It was referred to by Ricardo and then Marx, who reversed the roles: the industrial city was seen as the propulsive force behind the economy and the main place of production, while the country played a subordinate role. The concept of dichotomy remains today, appearing for example in the environmentalist view which contrasts the urban lifestyle with the 'natural' rural lifestyle.

b. The gravitational view, already referred to above, which considers the city as an 'attractor'. There are many possible factors of attraction, each relating to a different theory or model: the attraction is sometimes considered to be the weight of population, or a leading firm capable of generating economic growth, or it may consist of specific functions relating to internationalisation, the innovative milieu, or the use of advanced technology.
c. The structuralist paradigm in which the city is seen as a system organised on different levels. This may highlight the physical elements (the roads, housing, green spaces, industrial areas etc.) which make up the city, or its functional structure (the central business district, residential areas, industrial zones etc.), investigating the organisation links and interconnections between them, or can consider the social structure of an urban area, examining the spatial organisation of different social groups and changes over time in their internal structure and relative location.
d. The organic view considers the city to be a complex organism with its own collective identity, although made up of many different roles and functions. This organism is 'born', grows through a number of development stages, and finally reaches a phase of decline which leads it to modify its collective identity. A development of this approach, as already mentioned, was adopted by the Chicago School and can be recognised today in the evolutionist approach.
e. The systems approach sees the city as a complex but open system consisting of a set of functional subsystems. The system may be in a state of equilibrium or nonequilibrium, may be evolving towards more or less stable forms, or maintaining its functional organisation in a far from equilibrium state. In this last case, the city becomes a dissipative structure, transforming and dissipating energy to maintain its self-organisation and its socio-functional complexity.
f. Another view of the city is the approach which considers it, above all, as a place of interaction. The interactions can be of various types. They may be seen in terms of exchange, understood both in an economic

sense and in a more general social or cultural sense. If we focus on the interaction between economic subjects, the city becomes an enormous and complex market. If we consider social interaction between individual or collective subjects, or between decisional networks, then the city appears as the carrier of a network of relations and interests of different social groups. Finally, there are also symbolic interactions which are represented by nonmaterial forms of communication between places and people: the city now becomes a complex text expressing its own language and contents perceived, for example, in terms of identity or belonging.

The interaction paradigm is undoubtedly the least structured of the various approaches and requires a more precise classification. The common element is the importance given to communication - carried out in relatively explicit or symbolic form - between decision-makers and between those with particular interests.

g. There also exists a normative view where the city is seen primarily as a space controlled by regulations which organise the social, property and symbolic values of the city. This is the 'planning' approach to urban space.
h. Finally, there is the historical approach which sees the city in terms of its evolution through historical and cultural dynamics and attempts to provide a classification of the different phases, emphasising their most meaningful aspects.

In relation to these different approaches to the city, it is important to make some general observations. The first is that they are not mutually exclusive categories. There is in fact considerable overlap between the different views: they are simply different ways of interpreting urban space. The 'catastrophic' model of the butterfly used, for example, by Mees to describe the transformation and renaissance of the Medieval city in Europe (Mees, 1975), can also be seen in systemic terms, in dichotomous terms or related to a historical classification.

The second observation is that the paradigms are necessarily nonexhaustive. As theoretical models, they are subject to the geographer's paradox: the most precise map, at 1:1 scale, is obviously unusable, and similarly, the most precise classification with *n* classes for *n* texts, is equally impracticable.

Thirdly, it is interesting to observe that the approaches evolve over time, surmounting their limitations and finding new ways to express the changing nature of the city. In this way, as explained above, the dichotomous approach has evolved into the environmentalist view, the gravitational approach is now expressed in terms of different attractors, the organic view has become the evolutionist view, and the systems approach appears to be continuously subject to new fashions!

A further observation which needs to be stressed is that the analytical tools are independent of the approaches (Fig. 1). The town/country dichotomy, for example, is adopted in classical models (Von Thünen, 1826) and in ecological approaches as well as ISTAT classifications (1986) where it is used to define and distinguish urban and rural areas. It also appears, as already mentioned, in Mees' model (1975) which contrasts *farmers* and *city dwellers* in the populations of regions.

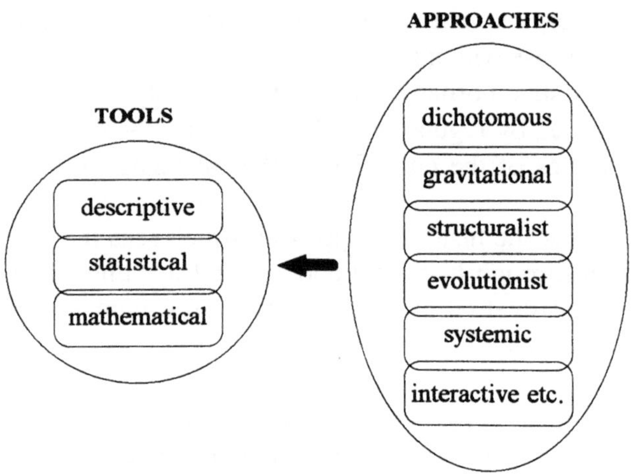

Fig. 1 Relationships between interpretative approaches and tools

There is a third way in which we can construct a systematic interpretation of the sciences of the city based on the set $\{e,k,F,R\}$. This set defines any formal model or, if one prefers, any axiomatic language. To have a complete model of the city, or of the whole territory, we must firstly define, as in any process of axiomatisation, the linguistic objects with which we are operating. In other words we need to know the *elements* $\{e\}$ of which it consists, define the *constants* $\{k\}$, and hence the *functions* $\{F\}$ which link and structure the elements and constants and, lastly, the *relations* $\{R\}$ and their dynamics.

15.3 The Nomological Level

The first type of study involves the identification of elements and establishment of boundaries or thresholds through which we can define the concept of the city. By establishing empirically significant criteria, we are able to explore the structure of an area. The aim is to give a precise and measurable definition to the various *partitions* of the territory. By partition we mean a complete and homogeneous subdivision of the territory.

Given a region R and areas $A_1, \ldots, A_n \subset A \subseteq R$ then A is a partition of R if, and only if:

$$A_1 \cup A_2 \cup \ldots \cup A_n = R$$
$$A_1 \cap A_2 \cap \ldots \cap A_n = 0$$

The most obvious partitions are administrative districts, but they may also be commuter catchment areas, provinces, regions, river systems and so on. There are procedures and methods for defining and identifying the elements of each kind of partition. In Italy, there are more than one hundred forms of partition associated with different laws and for the various sectors of the public and private administration. There exists in fact a discipline, administrative geography, whose purpose it is to define the basic elements and organise them in a consistent way, according to specific parameters and objective scientific categories.

In urban analysis, probably one of the most common partitions used for the definition of the urban area is the commuter catchment area. A series of studies carried out in Britain in the seventies was based on the statistical concept of *labour areas* (Hall *et al.*, 1973). These consisted of the worker catchment or 'self-containment' areas around towns employing over 20,000 people. These areas, also referred to as a Metropolitan Economic Labour Areas (MELA), were made up of an urban nucleus, the SMLA (Standard Metropolitan Labour Area) and those outer suburbs with the most intensive journey-to-work flows. The nucleus itself consisted of a core including the city centre, inner suburbs with a density of over 1235 residents per square kilometre, plus a ring of districts in which more than 15% of the working population had jobs in the centre. Smart's self-containment index (1974) was used to define nonmetropolitan areas: aggregations of districts with an overall self-containment index of over 75%.

In Britain in the late seventies and early eighties, there was an important debate around the concept of the *daily urban system* (Coombes *et al.*,

1978a). Methodologies and algorithms were specified to define both *'functional regions'* (Coombes *et al.*, 1978b) and *'travel-to-work' areas* (Coombes, Green and Openshaw, 1985). The echo of this debate reached Italy (IRES, 1986) and was reflected in the work of ISTAT-IRPET (1989, 1994) whose aim was to subdivide the entire national territory into *sistemi locali di lavoro* and *regioni funzionali*. A prior partition into *sistemi urbani* (Hall, 1980) did not respond to the objective criteria established for defining the travel-to-work areas, so use was made of the Carta Commerciale of Italy drawn up by the UICCIAA and published in the Atlas of Italian Commercial Areas by Tagliacarne (1973). On the basis of these commercial areas, Hall arrived at a partition of Italy into 84 metropolitan regions and 95 nonmetropolitan regions. In 1981 the ISTAT-IRPET studies grouped Italy's 8086 communes into 955 local labour areas and 177 functional regions. It is interesting that the same criteria applied to the 1991 census results produced only 784 local labour areas. The most significant change was in the northern industrial areas where the intra-regional migration growth rates were the highest.

It should be noted that this latest partition belies, at least in Italy, the hypothesis that the largest cities are affected by a counter-urbanisation process with a diffusion of the population into rural areas. The twelve major metropolitan areas, with the exception of Palermo, all in fact grew, and the number of districts they included increased from 290 to 466. In 1981 they represented 26.2% of the total population. Today 28.5% of Italians live in the major cities (a growth of over 2% compared with the national rate of 0.3%). This data confirms the 'ring' effect, consisting of growth around the main metropolitan areas accompanied by depopulation of the central core. (There was nevertheless an overall increase, not decline, in the attraction of Italy's metropolitan areas between 1981 and 1991.)

Similar analyses have been carried out in many other countries including France with its 365 *'zones d'emploi'* established by INSEE. Studies using methodologies more similar to Hall's have been applied in Ireland, the Netherlands and Germany.

At the next level of study we attempt to define more complex objects. Once a partition has been established, it can be divided, with the use of indicators and statistical thresholds, into subsystems. In other words, in a given spatial partition, with statistical methods we can define *constants* which determine *subsets*. The concept of city is relevant here, appearing as a constant of particular elements defining the morphology of space. Pumain (in this volume) puts forward a *demographic* definition according to which the city consists of a permanent concentration of population resident in a confined area, a *morphology* given by the urban structure and form of the built-up area, and a *functional* definition relating to its social and economic nature. These definitions reflect, to a large extent, those studies undertaken by statistical institutes and urban researchers who have attempted to define the city through a series of indicators.

From the academic point of view, an important *socio-demographic* definition of the city was that introduced in 1950 by the Bureau of Census in the United States - the Standard Metropolitan Area (SMA). The problem faced by the Bureau was to define the main urban agglomerations, using an existing partition, the county, as a starting point. The SMAs constituted an aggregation of counties forming a central city with a minimum population of 50,000 and an outer ring of counties, which were included if they had a density of at least 150 inhabitants per square mile, a minimum threshold of 10,000 workers, or 75% of the working population in nonagricultural occupations. A minimum level of commuting was also established to indicate a level of economic and social integration between the central city and other cities (25% and 15% respectively) (Berry, Goheen and Goldstein, 1968). In the sixties, these areas were renamed Standard Metropolitan Statistical Areas (SMSAs) and finally, in the seventies, to take account of further aggregation the *Standard Consolidated Statistical Area* (SCSA) was defined.

Reformulating this model in Italy (Cafiero and Busca, 1970), SVIMEZ identified metropolitan areas as aggregations of adjacent districts with a minimum threshold of 100,000 inhabitants, 35,000 nonagricultural workers and a density of at least 100 inhabitants per square kilometre. In the eighties, a further development of the model (Cafiero and Cecchini, 1990), which introduced new criteria relating to the relationship between place residence and place of work, identified 39 urban areas. Three of these, with over 3 million inhabitants (Milan, Rome and Naples), were defined as metropolitan areas, 11 were defined large urban areas, and 25 minor urban areas. Altogether, these extended cities covered 11% of the national territory, included 1449 administrative districts (communes) and accounted for 31.7 million people (55% of the total population), approximately half of these in the Milan metropolitan area.

From these analyses the city appears as a territorial subset which can be determined by a series of measurable indicators with empirically defined thresholds. The process of classification, which in this case identifies a single subsystem, distinguishing the city from the rest of the territory (non-cities), can be made more or less complex and can define different constants or clusters, which may be relatively open or fuzzy.

It is interesting to note that there has been a statistical 'crystallisation' in the classification of various urban areas. In Italy, ISTAT (1986) divided all administrative districts (communes) into four groups using one main factor extracted from a principal component analysis carried out on a dozen variables. These included density, % of population of working age, % population in nonagricultural employment, % with school diplomas, average family size, % working outside the district of residence, % population in employment, % owner-occupied properties, type of housing and number of telephones per inhabitant. The first factor, defining the degree of urbanisation, resulted in 862 communes (10.7% of the total and 51% of total population) being defined urban districts, 2815 were defined semi-urban districts (34.8% communes and 23.7% of population), 2259 semi-rural districts (27.9% of communes and 6.8% of the population) and 2150 rural districts (26.6% communes and 18.5% of the population).

Similar classifications exist in France, where INSEE since 1962 has distinguished *communes-centres* (isolated towns or communes which give their name to an urban agglomeration), *communes de banlieue, communes rurales* belonging to *zones de peuplement industriel et urbain* and stagnant or declining rural districts which are defined as *communes rurales profonds.* In Spain and Greece, the territory is grouped into urban zones, towns with over 10,000 inhabitants, semi-urban, those with 2,000-10,000 inhabitants, and rural, those with under 2,000. In the Netherlands, districts are classified as rural, urbanised rural, dormitory or urban. In Denmark the classification is rather more complex, grouping districts into three types and twelve classes according to the dimension of the principal urban nucleus.

Both the choice of partition and the definition of its constants belong to the sphere of administrative geography. This is an autonomous area of scientific activity consisting of a nomology concerning both the *nomós*, i.e. the province, district, region or other divisions of the territory and the *nómos*, i.e. the norms, laws, thresholds and indicators which define them. In general terms nomological study is concerned with identifying the basic elements constituting the territory. Studies of this kind use statistical instruments and in general an inductive methodology, making use of empirical data to formulate theories and criteria of definition.

In this sphere of analysis, complex mathematical models would appear to have little to offer. This is evident from Batty and Longley (1994), who provided a list of the relative fractal dimensions of a number of European and American cities. Though interesting from the point of view of urban morphogenesis, this provided no additional information to the concept of density or for defining indicators of urban density. It would be equally unhelpful to attempt to define the city on the basis of its gravitational attraction, rather than a simple index of migratory flows or the demographic weight.

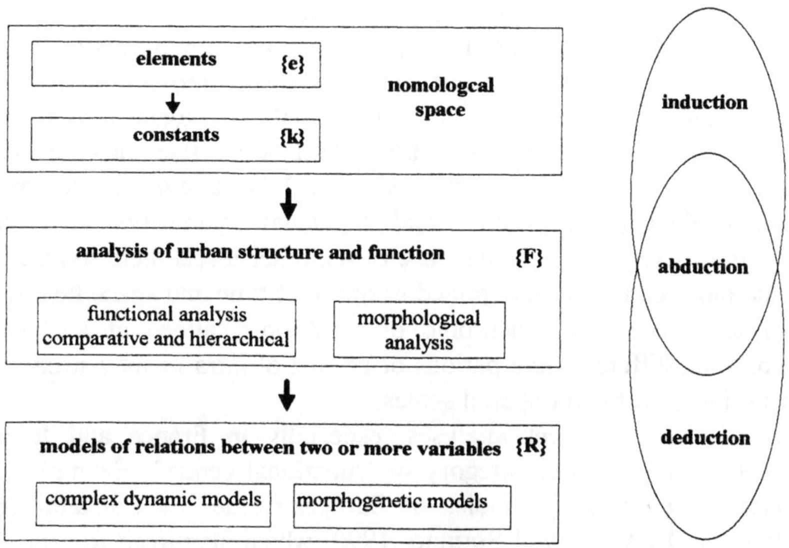

Fig. 2 Classification of the sciences of the city

15.4 Functional and Morphological Analysis

There is a third level of analysis in urban and economic geography which is concerned with defining the organisational and functional structure of the city. In order to provide a more complex picture of reality, analyses of this kind take the elements and constants of the urban area and attempt to highlight their structure. The use of description, of statistical indicators and modelling is interwoven. The tools range from the classical Christaller type models describing the functional hierarchisation of space, cluster analysis and factorial or multicriteria models, to models inspired by Thom (1988) relating to catastrophe theory and aiming to identify the relational and functional structure of the city and its equilibrium profiles. The models are used here however not so much to underline the dynamic relations, but to explain the structure, functions and forms of organisation inherent in the city.

In a general sense, there is a connection between this analytical level and the relational level. The static interpretation of the functional structure of a society or a city can be seen as an equilibrium profile of a dynamic system or a cross section of a topological map, a creodo or an epigenetic landscape at a given time. The interest is in formulating theories concerning the different aspects which make up urban space. In practice, a process of abduction is used: from the description and statistical analysis, the functional structure is 'read' and, once formalised, appears like a static photograph or a specific 'section' of the relational model.

The common feature of this third level is the fact that the analytical approach is based on what has been defined the 'circumstantial paradigm', used to identify the structure and functional organisation underlying the city and its region. As in the case of the other levels, here too this can be made more complex and applied to different time and space horizons. The approaches can be synchronic or diachronic, aimed at analysing and comparing different time periods or consist of intra or inter urban studies, focussing on different spatial scales.

A number of recent analyses, especially in France and Italy, have extended the classical category of 'functional centre'. Examples are the series of comparative studies of European cities by Camagni and Pio, 1988, and by Conti and Spriano, 1990, which attempted to rank them in hierarchical order. Other analyses have aimed to link the subset of urban functions with the concept of 'milieu' - the studies by Camagni, 1989, and Crevoisier and Maillat, 1991, carried out under the auspices of GREMI (Groupe de Recherche Européen sur les Milieux Innovateurs) founded by Aydalot in 1984. Their definition of the city depends on two major categories: (i) a set of indicators relating to metropolitan functions and (ii) a group of 'milieu components', i.e. a set of environmental conditions which are assumed to support the interactions and development processes typical of the metropolis (IRES, 1991). The bringing together of these two measures provides a way of comparing the structure of different cities. It is a methodology with several strong points. It can determine the internal organisation of a city through the interpretation of certain statistical indicators and, by referring to both functional and 'milieu' components, it provides a synthesis of the characteristics of the city. In addition, it makes it possible to define a hierarchical and/or comparative order between the constants and to interpret the diachronic evolution of the city.

These advantages are shared by other methods at this third level of analysis, which is the level of 'theorisation'. As they make use of an abductive methodology, analyses of this kind tend to occupy a dominant space in the literature. In general, an attempt is made to define theories which are sufficiently comprehensive to recognise the complex functional structure of the city without worrying about the coherence of the formal tools used. In fact, use is made of both inductive and deductive methods in order to provide support for the hypotheses put forward. The descriptions therefore have the role of connecting the various methods and contributing to the formulation of the theory.

15.5 The Analysis of Dynamic Relations

Finally, there is a fourth level of analysis which takes as given the urban features and their functions (expressed generally in terms of variables) and attempts to discover the dynamic relations which exist between them. The focus is on the interactions and processes taking place in time and space. Unlike the previous levels of analysis, we are concerned with constructing a systemic and dynamic model, rather than systematic theories. Descriptive and statistical tools are inappropriate here because they are incapable of providing coherent and synthetic analyses.

There are two main areas of interpretation: morphogenic analysis, which aims to explain the modifications over time in urban form, and relational or dynamic analysis, which attempts to highlight the dynamic causality between functions or different variables of the urban system. In the former kind of analysis, fractal theory and, above all, neural networks and cellular automata are used, as well as catastrophe theory, for the recognition of forms and codes of linguistic and morphological communication. Dynamic analysis makes use of a number of different kinds of model, including the family of *gravity* models (Lowry, 1964), the *entropic* models (Wilson, 1970) and *dynamic-simulative* models (Forrester, 1968), as well as the *complex* dynamic models which give rise to the categories of autopoiesis and self-organisation, synergy, catastrophe, bifurcations and so on, analysed by Pumain in this volume.

It is a vast field which lacks a synthesis and systematic organisation. The studies at this level are more concerned with internal coherence than the creation of a systematic and comprehensive theory of the city. To formulate such a theory, it is necessary to know which dynamics are fundamental to the city. Are they finite or infinite? Which are the most meaningful? Is it possible to classify them? Much progress has been made in classifying these archetypal dynamics. Two studies which would seem to be particularly significant are: the classification of ecological relations defined by Dendrinos and Mullally (1985) and the spatial and temporal interpretation of universal trends derived from catastrophe theory (Thom,

1980), used in various regional models (Wilson, 1981).

The first is an extremely valuable classification, constituting a kind of methodological reference, even for analyses more closely connected with theoretical interpretations, i.e. at the third level. Many of the processes and functions associated with the city can be included in this classification. In formal terms, it classifies the relations for dynamic systems of the kind:

$$dx/dt = x\,(a + bx + cy)$$
$$dy/dt = y\,(d + fy + gx)$$

In this case, it is the sign of the parameters c and g which explains the type of relation existing between the two variables x and y. The parameters a and d are usually < 0 as they define the rates of growth of the populations x and y. In advanced society and in many post-industrial systems this has become negative. The parameters b and f are positive if we assume there is a symbiotic relationship in the population, and negative when we assume a competitive relationship. In geographical systems these parameters are in general positive and capture the polarising effect of the city and the local systems. Parameters c and g change easily in geographical systems and have relational dynamics defined by:

$c > 0$ and $g > 0$	*symbiotic* relationship between populations x and y. The two populations depend on each other; the competition creates a homologous rather than oppositional synergetic relationship.
$c = 0$ and $g > 0$ $g = 0$ and $c > 0$	x is *commensal* of y, y is *commensal* of x. This occurs when the growth of y does not depend on x, while the greater presence of y in the same ecological niche facilitates the increase of x.
$c < 0$ and $g > 0$ $c > 0$ and $g < 0$	x is the *prey* of y, x is the *predator* of y. There exists a complementary and opposing relationship between the two populations. The existence of one depends on the other in a synergetic not a complementary way.
$c = 0$ and $g < 0$ $c < 0$ and $g = 0$	*non-commensalistic* preying of y on x, *non-commensalistic* preying of x on y. This is symmetrical to the commensalistic relationship. The growth does not depend on x, but the greater presence of y in the same ecological niche is in opposition to the development of x.

$c = 0$ and $g = 0$	*isolation* of x with respect to y. This occurs when there is no interaction between the two populations.
$c < 0$ and $g < 0$	*competitive* relationship between the two populations. This occurs when there is no complementarity between x and y and the survival or domination of one implies the disappearance or total submission of the other.

The richness and synthesis of the description of the relationships between the two variables achieved with the above classification is in fact far superior to any statistical description (which in general highlights only the *differences* between factors and their *complementarity*).

The second classification uses catastrophe theory and attempts to identify certain archetypal relations. Thom (1980) proposes the following semantics:

Elementary catastrophes	Function V	dynamics and semantics
Fold	$V = x^3+ax$	finish, begin
Cusp	$V = x^4+ax^2+cx$	become, capture, emit
Swallow-tail	$V = x^5+ax^3+cx^2+dx$	link, fail
Butterfly	$V = x^6+ax^4+cx^3+dx^2+fx$	give, receive
Hyperbolic umbilicus	$V = x^3+y^3+dxy-ax-fy$	recover
Elliptic umbilicus	$V = x^3-y^3+d(x^2+y^2)-ax-cy$	negate, block
Parabolic umbilicus	$V = x^3y+y^4+dx^2+fy^2-ax-cy$	connect, close

The question we need to ask is: which urban facts or events underlie the archetypal dynamics listed above? Until the sciences of the city have defined in an unequivocal and meaningful way the situations which correspond to these formations, systematic urban modelling will remain a complex but insufficient method, incapable of being an independent science.

15.6 Complexity

We have underlined above the limitations of formal modelling and should now like to look at some of its strengths by investigating one of the most controversial concepts affecting urban analysis: complexity.

What is complexity? To reply to this question, we shall use a heuristic model and a dynamic lagged system known as May's model (1976). This model can provide a classification of the complexity of a causal system. The problem relating to the definition of complexity therefore becomes: how many typologies and what kind of deterministic causal relations can exist between two or more variables, two or more territorial systems or between their functions?

If we consider the system $\{C,F\}$, where C indicates a family of causes and F a family of effects, we can say that:

a. the set of effects F is generated by the causes C;
b. the lasting effects F become causes C after time t when the previous generation of causes is removed from the model;
c. each generation of C and F is regulated by unitary time.

The model proposed is a heuristic model of action and is based, as we shall see, on simplifications that nevertheless do not prevent it from providing sufficient information to make an initial classification of complex relations. We define:

- C_t the causes at time t;
- F_t the number of effects at time t generated by causes assumed as initial conditions;
- p the effects which after time t+1 become causes;
- m the average number of effects which are generated by a cause in one unit of time.

The model equation will therefore be:

$$C_t = p\,F_{t-1} \quad (1)$$
$$F_t = m\,C_t \quad (2)$$

with t=1,2,3, ... , n.

In practice the model tells us that the number of causes at time t are generated by the lasting effects in the preceding time period, while the effects at time t are generated by the causes existing at the same time. p is an index of complexity which measures, in a heuristic and quanto-qualitative way, the degree of transformation of the effects into causes in a given time period and m measures the generation of effects produced by a single cause. This is a relatively complex model compared with the methodological interpretations of causal relations founded on a bi-univocal and constant relationship between cause and effect.

Let us assume, for the moment, that the values p and m are given:

$$C_t = p_o F_{t\text{-}1} \tag{3}$$
$$F_t = m_o C_t \tag{4}$$

In these conditions, substituting the value C_t we obtain the recursive formula:

$$F_t = m_o p_o F_{t\text{-}1} \tag{5}$$

which tells us that today's effects depend on the control of effects generated by yesterday's actions. This formula in reality expresses infinite formulae of the type:

$$\begin{aligned} F_1 &= R_o F_0 \\ F_2 &= R_o F_1 = R_o (R_o F_0) = R_o^2 F_0 \\ &\ldots \\ &\ldots \\ F_t &= R_o F_{t\text{-}1} = R_o^t F_0 \end{aligned} \tag{6}$$

where R_o=$m_o p_o$.

If we assume that p_o is constant, or that the number of effects which, after unit time, are transformed into causes is constant, then the growth of the effects will depend exclusively on the behaviour of m, or the average number of effects generated by a cause in unit time. In this heuristic model, the relationship between cause and effect which define the context in which we insert the action is exponential. The relative graph is shown in Fig. 3. R_o is the existing rate of implementation between cause and effect. The model therefore gives us the following obvious and reasonable result:

a. if the rate of implementation is positive and greater than one, the relationship between the cause and effects is cumulative and exponential. We can state that the complexity of the system is increasing over time;
b. if R_o equals one, implementation does not exist and the complexity of the system remains constant;
c. if R_o is less than one, the implementation is negative and therefore the causes and effects are decreasing over time, simplifying the causal relationship and hence the intrinsic complexity of the system.

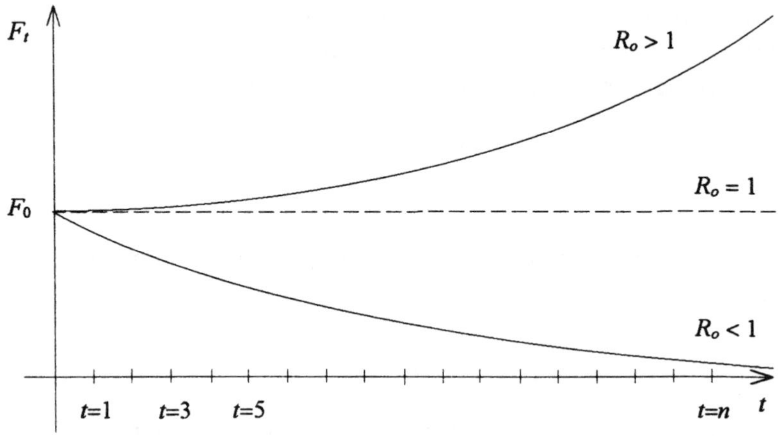

Fig. 3 Growth or decline of a simple system of causal action

These conclusions are based on a series of simplifying assumptions. In general, the rate of implementation is not independent of the causal mechanisms deriving from actions within the system, but depends on the set of causes affecting a system of action $C \Rightarrow F$. This is the same as saying that the average number of effects m will depend on the set of causes C and cannot be considered a parameter, but a functional variable of the system. In other words, m is a function of C:

$$m = \phi(C)$$

and as C_t is a function of F_{t-1}, $C_t = \Psi(F_{t-1})$, we obtain the equation:

$$F_t = \Psi^2(F_{t-1}) \tag{7}$$

This is a nonlinear difference equation and there is no standard procedure which can provide an exact solution.

It is more realistic if we assume that the rate of implementation between cause and effect will tend to decrease when the quantity of generating causes increases, i.e. that there is a homeostatic internal regulation which protects it from an exponential explosion in the causal relations. For the sake of simplicity, we assume that this regulation is linear:

$$m = m_o(1 - C_t/H) \tag{8}$$

where m_o and H are positive and constant.

m_o expresses, as usual, the original number of effects generated by a cause in unit time; H is a quantitative factor defining the social and economic milieu affected by the causes C. In practice, it represents the *resource potential* available in the system.

Substituting in (4) we have:

$$F_t = m_o\,(1 - C_t/H)\,C_t \tag{9}$$

and as $C_t = p_o\,F_{t\text{-}1}$ (3), we obtain:

$$F_{t\text{-}1} = C_t/p_o$$

We obtain a system of equations:

$$F_t = m_oC_t - m_oC_t^2/H \tag{10}$$
$$F_{t\text{-}1} = 1/p_o\,C_t \tag{11}$$

where (10) and (11) are respectively a descending parabola passing through the origin with roots $C_t = 0$ and $C_t = H$, and a straight line with slope $\text{tg}\alpha = 1/p_o$, passing through the origin, because the known term is missing (Fig. 4).

With variations in time t, the indices C and F will change and form a cobweb graph which defines the relationship between cause and effect. There can be different situations depending on the rate of implementation R_o.

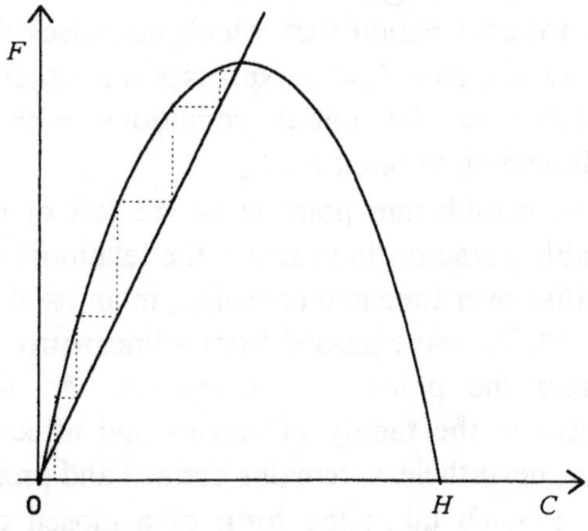

Fig. 4 Converging dynamics of the point of equilibrium

1. For $1<R_o<2$, the point of equilibrium (intersection) is at the top left of the parabola. The cobweb is contained within the area between the parabola and the straight line and converges at the point of equilibrium. It has a stable, nonoscillating monotonic (decreasing) behaviour. This means that the interaction between cause and effect will tend towards stabilisation on an equilibrium value F_e. The number of effects produced by *n*-causes will always be the same and hence the system is predictable and controllable over time. In other words, the system of causal interaction has a teleological finality. The actions and successive steps converge on a fixed point which can be considered, in a global sense, to be the ultimate purpose of the dynamics and therefore the structure or equilibrium profile of the system. We can state that a causal action system will have a teleological finality within certain fixed thresholds of the rate of implementation R_o. Social and economic systems of this type have often been examined in the literature and interpreted in terms of general equilibrium.
2. For $R_o=2$, the point of equilibrium is the top of the parabola. In a general sense, we can consider this point as the optimum rational purpose of the relationship *C-F*. The causal action is always predictable and the local effects do not affect the ultimate purpose, which is therefore controllable. A form of generalised induction principle also applies; each action produces a predictable result. In other words, the result does not depend on the initial conditions but on parameters which structure the dynamics; despite different conditions, it always tends towards the same point. An extension of this principle underlines the relationship between actions of different degrees of rationality. The equilibrium point may be reached within few or many steps, over a shorter or longer time period, but nevertheless is reached. If we define a rational action that which optimises the time taken to reach the final result, then $F_e=C_e$ expresses the situation of maximum rationality relative to the initial conditions with many possible rationalities, all tending to point $F_e=C_e$.
3. For $2<R_o<3$, the equilibrium point is on the left of the parabola and represents a stable attractor. In practice, the relationship between *C* and *F* tends to stabilise over time and converge, in an oscillating fashion, on this point (Fig. 5). We have passed from a linear process to oscillatory dynamics around the point of convergence. As we can see, the relationship between the family of causes and effects becomes more complicated but, nevertheless, remains defined and predictable.
4. For $R_o=3$, the cobweb takes the form of a closed cycle around the

equilibrium point. This can be rectangular (in the case of non-perpendicularity) or quadratic (in the case of perpendicularity) and illustrates a state of double neutral stability around the point of intersection. Here too, the relationship between cause and effect is predictable and defined either by stable cyclicity (R_o=3) or by decreasing cycles around an equilibrium point given by $2<R_o<3$. Many social and economic phenomena seem to give rise to this kind of dynamic when the underlying structure remains stable (e.g. economic cycles). Also in this case, given the oscillating equilibrium around a point, the system has a teleological finality.

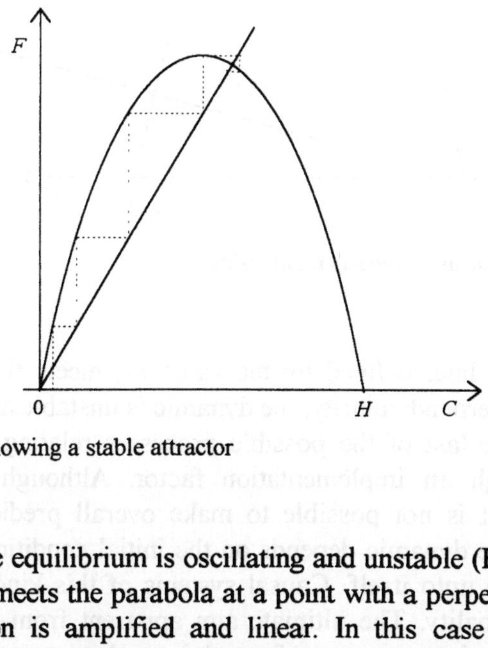

Fig. 5 Dynamic showing a stable attractor

5. For $R_o>3$, the equilibrium is oscillating and unstable (Fig. 6). When the straight line meets the parabola at a point with a perpendicular tangent, the oscillation is amplified and linear. In this case the cause/effect relationship is no longer controllable, but the behaviour of the system is nevertheless predictable. It acts in fact within a *symmetrical schismogenesis* relationship which gradually amplifies the scope of the phenomenon and the response to it. An example in the socio-political field is given by the arms race - each action of the strategic actors is totally predictable, consisting of an amplified response to the actions of one of the other actors. Many dynamics of this kind can be found in the fields of sociology, regional science, economics and psychology. They are extremely dangerous as they evolve in a circular and cumulative fashion. Systems like this have pseudo-teleological finalities in the

sense that, although there is no intentional orientation of the system towards a given objective or point of equilibrium, they in fact proceed along a predictable path, by virtue of the responses of one actor to the actions of the other(s).

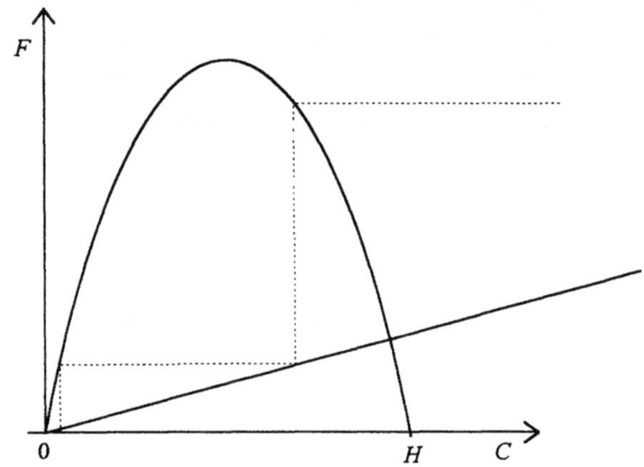

Fig. 6 Symmetrical schismogenesis relationship

6. If the straight line, defined by the value p_o, meets the parabola at a point of non-perpendicularity, the dynamic is unstable and random (Fig. 7). This is the last of the possible processes relating the causes and effects through an implementation factor. Although it is formally controllable, it is not possible to make overall predictions about its behaviour. The dynamic depends on the initial conditions, hence every path is a case unto itself. Causal systems of this kind do not have a teleological finality. The ultimate aim apparent from its stability and reproducibility becomes transformed into chaos: predictable at new local points, but unpredictable at the global level.

The study of stability tells us therefore that as the rate of implementation grows, the system $C \Rightarrow F$ tends to adopt a chaotic and noncontrollable form *a priori* in that it has no teleological finality. The relationship between cause and effect has complex dynamics which cannot be reduced to simple linear correlations. The same teleological finality can be defined by various dynamic configurations of the system of relations: it can converge on a point, have a cyclical nature or act as an attractor. When we speak of complexity and of social actions which do not have only causal relations, it is useful to take into account the explanation provided by this simple heuristic model. It needs to be remembered that the city moves within this complexity.

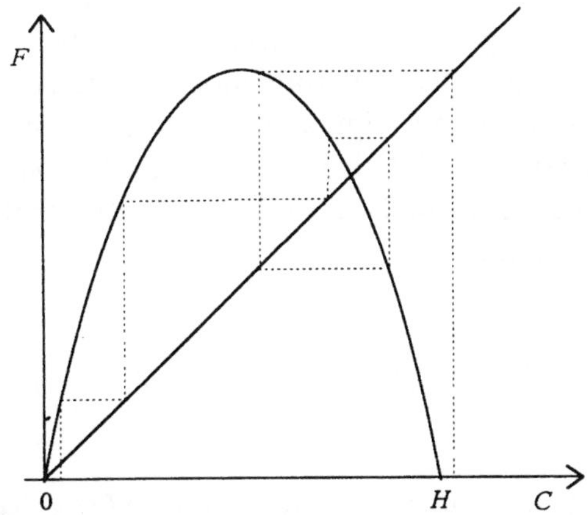

Fig. 7 Chaotic dynamics

15.7 Conclusions

As we have seen, for certain thresholds of complexity, May's model leads to chaos. As many recent studies have demonstrated, however, the progression towards chaos does not necessarily occur suddenly. There may be regularities which are more or less periodical, giving rise to attractors with cyclical or other 'strange' forms. In a global sense, prediction is not possible in that the system no longer has a single kind of behaviour. Nevertheless, being subject to an attractor, it moves towards a situation in which the definable surroundings in the phase space are not equiprobable. This means that even chaos may be structured and possess partial forms of order. Between order and disorder there are nonempty sets defined by attractors with their own morphology.

Do socio-economic and urban systems belong to this class? We can certainly say that they are complex systems in which chaotic behaviour is possible. Does this mean that formal systems of analysis are useless? We are bound to answer no, but must at the same time admit their incompleteness for representing a reality which is complex and, within certain limits, unpredictable. It is nevertheless through a formal model that we have been able to rationalise the cause/effect relationship and the teleological finality of a system. It is here that we see both the strength and paradox of mathematical formalism.

Normalisation remains the clearest instrument with which to explain its own underlying relativity. This is an important paradox which underlines the limits of formalism, while at the same time demonstrating its aesthetic and explanatory strength. This paradox should be accepted, without attempting to concede the interpretation of reality exclusively to mathematical tools, nor to description, nor to informal inspiration. What is 'real' always remains so relative to the specific observer, the focus adopted and the analytical tools used. It is essential that this relativity is constantly borne in mind.

References

Batty M., Longley P. (1994) *Fractal Cities: A Geometry of Form and Function*, Harcourt Brace & Company Academic Press, London.

Berry B.J.L., Goheen P.G., Goldstein M. (1968) *Metropolitan Area Definition: A Re-evalutation of Concepts and Statistical Practice*, U.S. Bureau of the Census, Working Paper, Government Printing Office, Washington D.C.

Cafiero S., Busca A. (1970) *Lo sviluppo metropolitano in Italia*, SVIMEZ, Rome.

Cafiero S., Cecchini D. (1990) Un'analisi economico-funzionale del fenomeno urbano in Italia, in Martellato D., Sforzi F. (eds.) *Studi sui sistemi urbani*, Angeli, Milan, 69-105.

Camagni R. (1989) Cambiamento tecnologico, milieu locale e reti di imprese: verso una teoria dinamica dello spazio economico, *Economia e politica industriale, 64*, 209-236.

Camagni R., Pio S. (1988) Funzioni urbane e gerarchia metropolitana europea: la posizione di Milano nel sistema dell'Europa meridionale, in IRER, *La trasformazione economica della città*, Angeli, Milan, 59-84.

Conti S., Spriano G. (1990) *Effetto città. Sistemi urbani e innovazione: prospettive per l'Europa negli anni '90*, Fondazione Agnelli, Turin.

Coombes M.G., Dixon J.S., Goddard J.B., Openshaw S., Taylor P.J. (1978a) Towards a More Rational Consideration of Census Areal Units: Daily Urban Systems in Britain, *Environment and Planning A, 10*, 1179-1185.

Coombes M.G., Dixon J.S., Goddard J.B., Openshaw S., Taylor P.J. (1978b) Functional Regions for the Population Census of Great Britain, in Herbert O.T., Johnston R.J. (eds.) *Geography and the Urban Environment: Progress in Research and Applications*, John Wiley and Sons, Chichester.

Coombes M.G., Green A.G., Openshaw S. (1985) New Areas for Old: A Comparison of the 1978 and 1984 Travel-To-Work Areas, *Area, 3*, 213-219.

Crevoisier O., Maillat D. (1991) Milieu, Industrial Organisation and Territorial

Production Systems: toward a New Theory of Spatial Development, in Camagni R. (ed.) *Innovation Networks: Spatial Perspectives*, Belhaven Press, London.

DATAR and GIP-Reclus (1989) *Les villes europeénnes*: rapport pour la DATAR, GIP-Reclus, Brunet R. (ed.) La Documentation Française, Paris.

Dendrinos S.D., Mullally H. (1985) *Urban Evolution. Studies in the Mathematical Ecology of Cities*, Oxford University Press, Oxford.

Forrester J.W. (1969) *Urban Dynamics*, MIT Press, Cambridge, Massachusetts.

Hall P., Hay D. (1980) *Growth Centres in the European Urban System*, Heinemann Educational Books, London.

Hall P., Rey T., Gracey H., Drewett R. (1973) *The Containment of Urban England*, Allen and Unwin, London.

IRES (1986) *Rassegna critica dei metodi per l'individuazione dei mercati locali del lavoro,* La Bella A. (ed.) Quaderni, Turin.

IRES (1991) *Le aree metropolitane tra specificità e complementarietà*, Dematteis G., Ferlaino F. (eds.) Dibattiti 2, Turin.

ISTAT (1986) *Classificazione dei comuni secondo le caratteristiche urbane e rurali*, Note e relazioni no. 2, Rome.

ISTAT-IRPET (1989) *I mercati locali del lavoro in Italia*, Angeli, Milan.

ISTAT-IRPET (1994) I sistemi locali del lavoro in Italia, University of Newcastle upon Tyne, University of Leeds, working paper, Rome.

Lowry I.S. (1964) *A Model of a Metropolis*, The Rand Corporation no. R.M. 4125 R.C.

Mees A.I. (1975) The Revival of Cities in Medieval Europe. An Application of Catastrophe Theory, *Regional Science and Urban Economics, 5,* 403-425.

May R.M. (1976) Simple Mathematical Models with Very Complicated Dynamics, *Nature, 261*, 459-467.

Pumain D., Saint-Julien T., Cattan N., Rozenblat C. (1992) *Le concept statistique de la ville en Europe*, Eurostat, Bruxelles, Luxembourg.

Smart M.W. (1974) Labour Market Areas: Uses and Definition, *Progress in Planning, 2*, Part 4, Pergamon, Oxford.

Tagliacarne G. (1973) *Atlante delle aree commerciali d'Italia*, Mondadori, Milan.

Thom R. (1980) *Modèles mathématiques de la morphogénèse*?, Bourgois, Paris.

Thom R. (1988) Rivoluzioni: catastrofi sociali? in Barbieri G., Vidali P. (eds.) *La ragione possibile. Per una geografia della cultura*, Feltrinelli, Milan.

Von Thünen J.H. (1826) *Der isolierte Staat in Beziehung auf Landwirtschaft und Nationalökonomie*, Pythes, Hamburg, Leopold, Rostok.

Wilson A.G. (1970) *Entropy in Urban and Regional Modelling*, Pion, London.

Wilson A.G. (1981) *Catastrophe Theory and Bifurcation: Application to Urban and Regional Systems,* University of California Press, Los Angeles.

16. Towards a Semiotic Theory of the City and the Territory as a Landscape

Carlo Socco

16.1 The City and its Meanings

In her contribution to this book, Denise Pumain examines a wide range of theories and models which have in common the fact that they interpret the city as a *spatially structured socio-economic system.* According to Pumain, it is this particular view of the city which provides the starting point for the interpretation of the city as a complex system, including the paradigms of self-organisation, chaos, evolutionary theory and so on. More than interpretative paradigms, these represent the recognition, or hypothesis, that the system possesses properties and behavioural laws around which it is possible to formulate mathematical models and experiment with different algorithms.

If, however, we look at the vast and varied field of studies concerning the city, we find a number of paradigms at a more fundamental level, i.e. that of the *meaning of the city*. If I speak of 'Venice' or 'Florence' it is unlikely that those listening will envisage these cities as spatial socio-economic systems. In the same way, we may talk about agricultural land, but to say 'the Tuscan countryside' expresses something which is perceived in more immaterial terms, not merely as a place where a certain quantity of alimentary products are grown. The city, likewise, has many meanings other than that underlying the theories examined by Pumain. One, in particular, cannot be ignored by those interested in the physical container we call the 'city' or, in a wider sense, the territory. This is the 'landscape' of life associated with the territory, the expression of the culture and history of the society which lives there. As a cultural expression, it can take forms which give it aesthetic value.

However deeply rooted the habit (at least of the urban planner) of seeing the territory as a complex socio-economic system, as individuals we still react to the quality of the aesthetic image we perceive. The quality of the landscape is always considered an essential and fundamental component of environmental quality. If the importance of the aesthetic qualities of the city is generally accepted, this means that scientific research has to accept the challenging task of understanding how the landscape functions as an aesthetic text.

The urban scientist approaching the city as a structured, socio-economic system filters the information through a spatial interaction theory. Similarly, if we consider the city as a landscape, we have firstly to identify a structural form through which to organise the information used for its interpretation. If our purpose is to interpret the territory as a landscape with aesthetic qualities, we must make use of a general theory which is sensitive to aesthetic values. As, at least for the moment, semiotics would appear to possess the most powerful formal structure for the interpretation of the various manifestations of aesthetic phenomena, we are obliged borrow the theoretical framework from the *semiotics of an aesthetic text*.

Our first task, therefore, concerns the possibility of producing a theoretical interpretation of the landscape *as a text*. Only on the basis of such an interpretation can we get to grips with the complex problem of the *aesthetic enhancement of the landscape*. The usefulness of such a task must inevitably be measured in practical terms. It is necessary therefore that the theory be formalised and translated into an operative tool able to provide us with something more than the romantic approach which still has such a strong influence on landscape aesthetics (Assunto, 1973, Tagliolini and Venturi Ferriolo, 1987).

16.2 A Semiotic Theory of the Landscape as an Aesthetic Text

16.2.1 The Language of Landscape

The landscape, whether urban, agricultural or natural, consists of a system of *signs*, i.e. a system of *significants* (the *syntactical* system) which, by means of a *code*, expresses a system of *meanings* (the *semantic* system). In other words, the landscape provides perceptive information about the territory and can be interpreted as a phenomenon of signification and communication, i.e. a *semiotic phenomenon* (Eco, 1975). The landscape is in fact a complex semiotic phenomenon with a structure which makes it similar in some ways to a text written in an *ostensive language* (Dematteis, 1985)[1]. Seen in this light, the landscape to which we attribute the values recognised in 'Venice', 'Florence' or 'the Tuscan countryside', presents itself as a *text with an aesthetic function*[2].

The first step in this kind of analysis is the interpretation of the landscape as a text. For this purpose, we need to be able to distinguish within the complex of information which makes up a landscape text, those parts which belong to the *expressive level*, i.e. the system of significants, and those which belongs to the *content level*, i.e. the system of meanings. It is necessary therefore to understand the organisation of the formal structure of the information at these two levels in order to arrive at a *landscape grammar*. Lastly, we can analyse the mechanisms by which we give aesthetic value to the landscape. We shall now discuss briefly these various aspects.

---

[1] The phenomenon of semiosis (the basis of the study of semiotics) can be represented by the semiotic triangle (Eco, 1984) whose vertices are: an expression (or significant) *y*, a content (or meaning) *x* and a thing (or reference object) *z*. In common language *y* is a word; in an iconic language *y* can be a drawing or a photograph. In the 'language of things', that is in the language through which we perceive things to have meaning, *y* is hidden within every *z*, in the sense that every *z* is presented directly as the significant of itself, without need for intermediation of a symbol such as a word, drawing, number or cryptogram. Ostensive language is that particular form of language which abolishes the intermediation of symbols, reducing the semiotic triangle to a segment whose end points are *x* and *z*(*y*). Obviously, we need some form of intermediation if we wish to communicate the meaning *x* which we have attributed to a specific referent *z*. The relationship between significant *y* and meaning *x* is the code. In ostensive language, the code links any referent *z* (as substitute for the significant *y*) to the meaning *x*.

[2] Grammatically speaking, by 'text' we mean a sequence of connected statements with cohesion and coherence. More generally, we understand as a text a set of enunciations "made contemporarily on the basis of some semiotic systems" (Eco, 1984, p. 64). We can therefore also consider a text that which we perceive when looking at a square in a historic town, listening to the noise of cars moving through it, smelling their exhaust and following what our companion is reading about the history of the place from the tourist guide book. This is what Sestini (1963) means when he declares that the landscape is made up of a series of environmental phenomena "which are visible manifestations or, more generally, manifestations of the senses. In an even wider conception of landscape, we may include sounds (the roar of running water, the rustling of branches, the sound of traffic along a road etc.), sensations produced by the weather (heat, cold, movement of air etc.) and also smells (the scent of Mediterranean shrubs in flower, the perfume of resin in a pine forest etc.). We cannot ignore the fact that these other sensations make their contribution to the overall impression we receive from a piece of the earth's surface" (p. 10, our translation). This concept of 'landscape of the senses' is further developed by Sestini into the 'rational geographic landscape', which includes the consideration of factors, such as climate or history, not perceived through the senses.

16.2.2 The Level of Expression

Imagine we have three objects, a cypress tree, a field and some hills. These *referents* convey the meanings expressed respectively by the words 'cypress', 'field' and 'hill'. We can now imagine that the cypress stands in front of us in a field with hills in the background. The three objects composed syntactically in this way define a *place*, that is they signify and communicate a *topical content* which we could express in the phrase: 'a lone cypress standing in a green field with a background of hills'. Thus the referents, arranged in space according to a certain topological structure perceived in terms of perspective, convey a content.

The text with an aesthetic function has a particular characteristic that distinguishes it from others: it is ambiguous and self-reflective (this is discussed further in Section 6). If we now move the cypress to one side (changing its position in the perspective structure), the view of the hill opens up before us. If we wanted to translate this into a phrase, we could say something like: 'the hills rise from the green fields alongside a lone cypress'. The topical content is the same (the basic theme of the landscape view and of the phrase which translates it into words has not changed), there has, however, been a change in the syntax: the subject is no longer the cypress, but the hills. This is the result of having moved the tree in relation to the perspective structure of the view. Any such change in the position of the referents in relation to the perspective structure produces a syntactical change, with the consequent modification in the communication perspective.

The syntactical structure of the landscape view can be defined as a perspective structure of 'empty positions' which indicate the syntactical role of the referents which occupy them. The syntax of the landscape is essentially a systematic analysis of the typology of the various possible syntactical structures.

The expressive level is not, however, exhausted with the syntactical analysis of the referents. These objects have a material composition, a texture, colour and geometrical form, i.e. they have physical and geometrical properties which provide important information contents, especially for the purpose of aesthetic evaluation of the landscape text. If we replace the cypress with an oak, we will immediately notice that the slim form of the cypress has been substituted by the bulkier form of the oak. The most visible change is in the physionomic/expressive content, i.e. in the *expressive connotations* which are not specific to the objects of one given place, but general to all things. This applies not only to what we perceive visually, but also to what we hear. Semiotic theory recognises this phenomenon as a process in which information concerning the physical nature of the objects (the material nature of significants), acquires 'pertinence' and hence becomes meaningful (Eco, 1975).

In common language, and also in that normally used by geographers, landscape architects and architects, the word 'form' is used to refer to information concerning the 'material' composition of the significants. Hence, when we speak about the form of a landscape and its connotations, we are referring to what in semiotic terms is a process of 'acquisition of pertinence' by the material composition of the significant. We could therefore use the word *morphology* to describe the systematic study of this kind of information at the expressive level, and reserve the word *syntax* for the analysis at the level examined previously[1].

16.2.3 The Level of Content

The territory is a system of places, each of which has its own topical content. To communicate the content of this system of places, there is no better alternative than a good geographical description, though to provide this we need to know how to use the geographical 'dictionary' correctly. The full significance of a place, however, is far more than that expressed in a *geographical statement*. The meaning generally attributed to a desert landscape, for example, is not limited to that conveyed by the expression: 'a limitless extent of sand'. The desert landscape for many people will carry the *connotation* of hostility. In other words, *the contents conveyed by the topical referents are of a denotative or connotive type*. If we take the three urban landscapes known as 'St. Peter's Place', the 'Rockefeller Centre' and the 'Pentagon', we can provide a denotative description, but it is certain that this does not exhaust all the meanings that current culture has associated with these places. The stratification of connotive meanings of these places is far richer and, above all, more fuzzy: they are 'symbolic'

---

[1] 'Form' is one of the many polysemous words whose content is frequently unclear even to the user. In the present study, it is used to indicate meaningful perceptive information about that which in semiotics is referred to as the 'material of the significant'. When architects speak of the form of a building or of a piece of land, however, they often include the kind of information which in the present article has been called syntax. In other words, there is a tendency to bring together under the heading 'form' a wide range of expressions, without distinguishing between syntax and morphology (generating confusion in the grammatical analysis of the landscape text). It should also perhaps be specified that the meaning adopted here has nothing to do with that normally attributed by semiologists, who speak of form only with reference to the 'formal structure' of the expression and of the content (Eco, 1975).

places. In semiotic terms, they constitute *syntagmatic structures generated by processes of hypercoding.*

The contents of the landscape, therefore, can be subdivided into three main levels: (i) the *denotative contents* which, in normal spoken language, or in the less imprecise language of the geographical sciences, we associate with a visual image of the landscape; (ii) the *connotive contents* of which the denotative contents are the significants and; (iii) the *hypercoded syntagms* constituting the metaphors and symbols of which the landscape is such a rich deposit. A *semantic analysis* of the landscape must take all three levels into account.

From what has been explained so far, we may have the impression that every referent-significant will have a certain denotative meaning, expressed by means of a code, and that on this meaning will be laid a certain 'thickness' of connotive stratifications. In reality, the form of our semantic universe is more complex, since each significant will have several meanings, depending on the context and the circumstances in which it occurs. We can try to make a single meaning correspond to each significant, but only if we use *dictionary-type semantics* as opposed to *encyclopedia-type semantics* which include all possible meanings a given culture associates with the significants[1] .

In examining the contents that a culture attributes to the landscape, it is necessary to examine the dictionary semantics currently used in landscape interpretation. We must also understand how our complete encyclopaedia is structured and how it governs the interpretation of landscapes, including the aesthetic interpretation.

---

[1] What we know about the world is organised in encyclopedia form, that is in the form of instructions for making inferences able to connect the experience of the world with meanings, taking into account the 'frames' (in the sense used in artificial intelligence) of the signification and the communication (Eco, 1984). The dictionary structure, on the other hand, represents one portion of the encyclopedia where the range of possible meanings is limited by reducing the possible contexts (the dictionary may be limited to the conventional and most widespread context, to the context of a specific discipline, a given text etc.). Taking as an example the referent known as 'cypress' in common language; in the field of botanics its significant is *cupressus sempervirens*, and its meaning is that given by its botanical description. In our encyclopedia, however, things are more complicated, as the meaning associated with the referent can change the relative connotative content. Consider, for instance, the differences between the following contexts in which the cypress may appear: next to a farm in Tuscuny, in a Catalan monastery, on a hot August day in Cappadocia, in the garden of a post-modern villa in Hollywood standing by a guitar-shaped swimming pool.

16.2.4 The Aesthetic Function

We know from semiotics that for the aesthetic evaluation of a text we require information deriving from all levels of expression and content (Bense, 1965), as well as from a general level where the various contributions are fused in an individualised narrative text which is based on its own inimitable *aesthetic idiolect* (Eco, 1968)[1]. In other words, we know theoretically that for each individual level there exist independent (and more or less distinct) systems of information units subject to variations of state. It is only by adopting given states that these units bring aesthetic value to the text. The identification of these states and their interdependence is the task of a *semiotic theory of the landscape as an aesthetic text.*

In the space available, it is obviously impossible to develop exhaustively such a vast subject. I shall therefore limit myself to presenting in the following sections a tentative general outline of the theoretical structure. This work is at present undergoing verification so that it can be drawn up in a more definitive version and translated into operative form for practical application (Socco, 1994). In Sections 3 and 4 we shall examine the level of expression, distinguishing between morphology and syntax. In Section 5 we look at the level of content, in Section 6 the ways in which we can interpret the aesthetic value of a landscape, and finally, in Section 7, we draw a number of conclusions.

16.3 The Morphology of the Landscape

We do not have to delve far into what is normally called the *form* of the landscape to discover that it consists of a complex structure of information, which can be divided into at least two different levels. At the first level, it appears as a set of visible features regarding the *material* of which the landscape is composed. The information given at this level concerns the *light, colour and material composition.* If we remove these, the form does not disappear completely; what remains is its diaphanous *geometrical structure* (rather like the reticular shape which CAD has accustomed us to interpreting as solid form). At this second level then, it presents a set of visual characteristics conveying information about the

[1] By idiolect we mean the language of a single speaker. Aesthetic idiolect is synonymous with the poetry of a particular text or author (see Eco, 1975).

shape, texture and dimension of the various parts. These two sets of visual characteristics constitute the *morphological components* of the landscape form[1].

Morphology, as the systematic study of form, has a double function. One is to provide an analysis of the morphological components and their field of variability; the other is to arrive at a synthesis which makes it possible to determine how, and through what combinations of these components, different *expressive connotations* can be derived. The analytical breakdown is the prerequisite for understanding how each component contributes to the final overall effect, which we generally refer to as the *expression* (we employ the term here differently from the way it was used in Section 2). As observed by Arnheim (1954): "expression is the primary content of vision" so that "in front of an open fire ... I do not normally see shades of red, different tones of light, or geometrical forms moving at varying speeds; I see the friendly play of tongues of fire, their sinuous flickering and the vivacity of their colour" (p. 370, Italian transl., 1984). This observation is of strategic importance in identifying the direction which our research should take, if we are to understand how information relating to form can give aesthetic value to the landscape. We shall now attempt to outline, very briefly, the grammar which Arnheim called expression.

The morphological components can take on different *figurations*. The field of variation of these figurations extends between extreme opposites - conditions of light, for example, will range from 'luminosity' to 'darkness', colours from 'bright' to 'pale' (Berlin and Kay, 1969). These extremes, and the whole intermediate range, can be defined in operative terms, making it possible to carry out a rigorous morphological analysis. Colour for instance can be classified by locating its position along the axis of a colour card representing all tones from the brightest to the palest (Itten, 1961). The *expressive connotations* of the landscape will vary according to the figurations of the morphological components. These variations, too, occur in a field limited by extreme opposites, i.e. connotations which give a particular character to the physionomy of the landscape (a kind of *expressive redundance)*, emerging when the morphological components reach the extremes of their range of variation.

---

[1] Form can be subdivided in many ways. The breakdown proposed here is the result of an experiment carried out on a sample of landscape images with the intention of determining which were the salient aspects. This does not mean that others may not prove to be equally effective. As any subdivision is based on pragmatic criteria, it is irrelevant at the theoretical level. What counts is that the overall form cannot be understood, if it is not broken down into simpler phenomena.

We shall see that the form as well, with its effect of expressive redundancy, contributes to the aesthetic value of a landscape text. While we know practically everything about the movement of the sun and the earth, the effect of the setting sun when it tinges the sky red will continue to represent something more than the simple refraction of the sun's rays through the atmosphere. A morphological theory of the landscape, before examining the ways of measuring the aesthetic value, must provide a systematic and operative definition of: (i) the morphological components and their field of variation; (ii) the expressive connotations and their field of variation; (iii) the relationship between the expressive connotations and the figurations of the morphological components. We do not have the space here for a full description of the theoretical model, but we can attempt very briefly to give an outline.

We should imagine that to provide an exhaustive description of the expressive physionomy of the landscape we have devised a list of expressive connotations which can be described in terms of their extreme opposites: 'sunny' vs. 'shady', 'heaviness' vs. 'lightness', 'smoothness' vs. roughness', 'simplicity' vs. 'complexity', 'finite' vs. 'infinite' etc.[1]. We can now assume that through practical experimentation it is possible to establish that the first connotation depends solely on the morphological components 'light' and 'colour', and that we have the connotation 'sunniness' when these two components reach the extremes of 'luminosity' and 'brightness'. On the contrary, we have 'shadiness' when the figurations of light and colour tend towards the extremes of 'darkness' and paleness'. In other words, there exists a matrix which relates the figurations of the morphological components to the expressive connotations, in the same way that there is a relationship between the expressive connotations and their empathic or emotional effects.

As we should expect, especially when dealing with the expressive connotations and empathic effects (rather less with the morphological components), we are moving in an imprecise, 'fuzzy' world which involves the physiological and psychological aspects of perception. There exists a fuzzy code (i.e. based on chance relational structures), which has been analysed by *Gestalt* psychology (Henle, 1961, Kanizsa, 1980, Marr, 1982) and *Einfülung* (Stewart, 1956). This code relates the expressive

[1] The same applies to the expressive connotations as the morphological components already mentioned above. There are many ways of breaking down the complex entity we call expressive physionomy, but some form of breakdown is necessary to understand its working. The connotations proposed here are the result of a study carried out on a sample of landscape images. The practical purpose, which should not be forgotten, is to increase the awareness of the architect or landscape architect.

connotations to the emotive states they generate (Pezzini, 1991), similar to a 'programmed stimulation' (Eco, 1975). The expressive connotation is the net sum of the mixture of the figurations of the morphological components. As in a chromatic mixture, we do not perceive the individual constituent colours but the resulting colour; it is this which determines our emotive states. The landscape painter, however, is interested not only in recording the emotive states stimulated by the landscape, but also in knowing how to compose the colour mix to obtain given states.

The fact that we are moving in an aleatory world does not remove the need to make a systematic analysis, which is, in effect, the only way to arrive at a full understanding of the phenomenon. At present, the problem is not so much whether the morphological components or the expressive connotations are the most effective for this purpose (this could be established with an empirical survey on a statistically significant sample), but the question of the effectiveness of the formal structure proposed here in providing a grammar of landscape form.

We should perhaps add to the above that a grammar of landscape form cannot ignore the fact that the morphology and hence the expressive connotations of the landscape will vary according to the time of day and the season. Every landscape is multifaceted, and this dynamic aspect of its expression must be taken into account. This can be recorded in the morphological analysis by means of a *vector* representing the possible transfigurations of a landscape image (number and frequency of the transfigurations, the 'gap' between one expression and another). The value of the vector (i.e. the degree of expressive variability) constitutes a significant element in the aesthetic evaluation of the landscape: the higher the value of the vector, the greater the expressive redundancy of the landscape and, as we shall see, it is in this direction that one of the features of aesthetic evaluation should be sought.

16.4 The Syntax of the Landscape

Syntactical analysis concerns the structure of a system of significants, i.e. the structure of 'positions' which indicate the syntactical role of the contents which go to occupy them. In a visual message of the spatial type, like that sent by the landscape, the syntactical structure (as we saw in Section 2) is anchored to the *perspective structure* of the view, in the sense that the syntactical roles of the contents depend on their position and visibility within the perspective structure. For example, whatever content occupies the dominant position in the view (e.g. the central position or foreground of the visual field), whether a tree, house, river, or road, it will become the main subject. The syntax of the landscape view has the task of defining the syntactical roles and the typology of the syntactical structure of the various possible landscape views.

But the landscape is something more than that which can be captured in a single view. It is also an overall figuration of the territory (the same type of figuration we previously described as 'Tuscan countryside') and as such has a good probability of stimulating a repertory of landscape images in the visual memory of the individual (Arnheim, 1969). In syntactical terms we could suggest that the overall image offered by a landscape is the sum of the views offered by all possible itineraries through the territory. It is in this sense that the landscape takes on the aspect of a syntactical entity we call a narrative text (Ritter, 1963). In other words, the 'journey' across the territory can be likened to the act of reading a literary text. The fact that semiologists find the metaphor of the journey pertinent to the act of reading (Eco, 1994) would seem to add legitimacy to this concept.

Though we can liken a journey through the landscape to a narrative text, we must nevertheless recognise that we are dealing with a special kind of text. Its narrative episodes, as already observed in morphological analysis, change with time. The passage of time can act on the visibility of the landscape and thus its syntactical structure, transforming the single image into a sequence of images. The syntactical analysis must therefore take into account the fact that the narration of a landscape text develops contemporarily along the trajectories of space and time.

The above is an outline of the basis on which a syntax of the landscape could be developed. This development can take the form of linguistic syntax which, beginning with the smallest unit necessary to communicate a message, i.e. the minimum necessary to construct a phrase (or, by analogy, to produce a landscape view), we proceed to the examination of increasingly complex structures, i.e. the sentence and, finally, the whole text. More precisely, we could divide the syntax of the landscape into the syntax of a *monothematic view* (equivalent to the phrase) and syntax of the *pluri-thematic* view (the sentence). The syntactical elements could be divided into *elements relating to the thematic nucleus* (e.g. the subject, its support and background) (Koffka, 1935) and *complementary elements* (expansion of the subject, support and background).

From a systematic analysis of the possible variations of the syntactical structure of the monothematic and pluri-thematic views, we can obtain the typology of the syntactical structure of the landscape. Operatively, this involves examining all the different positions of the elements of the landscape in the prospective structure and recording the different syntactical landscape types to which these give rise[1].

The syntax of the landscape text concerns the structure of the sequence of views along an itinerary, analysing the typological variations. In brief, the syntactical structure can figure as a sequence of episodes consisting of different *landscape units*[2] encountered along the itinerary and which we can perceive through a certain number of views, as variations on a theme. The different types of structure depend on the type and number of landscape units and landscape views which these produce[3].

Taking into account the fact that each picture will vary with time, it is clear that the syntactical structure of the landscape must define the vector of the spatial sequence of the modifications in the views over time. This can be represented formally as a matrix. The variations in distance between one position and another in the spatial and time sequences have the effect of increasing or decreasing the density (number of images per unit of space) and the frequency (number of images per unit of time) of the views necessary to represent a landscape exhaustively, hence increasing or decreasing the dimension of the syntactical matrix used to 'tell its story'.

16.5 Dictionary and Encyclopaedia

What is the semantic system of the landscape? If we examine the vast theoretical and practical bibliography concerning the landscape, the enormous variety of different meanings (many of them metaphorical and

---

[1] In normal language certain syntactical type distinctions are used to describe different kinds of landscape view. The designation 'panoramic view', for example, indicates a type of structure in which, since there is no foreground subject, the background fulfills the role of the subject. These distinctions are however, approximate and incomplete, in the sense that they do not include the whole range of syntactical types (or subtypes).

[2] A unit of landscape is generally defined as an area of homogeneous landscape, as distinct from simply adjacent areas. More precisely, it can be defined as the portion of territory characterised by a given vector of 'topical elements', organised according to a given 'topological model'. In this context, the concept of landscape unit is basic: no geographical description can do without it. As explained in greater detail in Section 5, a non-arbitrary subdivision of the territory into landscape units requires the use of a geographical type of dictionary-semantics.

[3] A landscape syntax which does not limit the concept of landscape to a set of views of the whole, but also takes into account the important role of detail, will have a more complex structure, interweaving two distinct narrative sequences: one consisting of panoramic-type views, the other containing close-up views.

vague) attributed the subject are likely to leave us with a sense of disorientation. The landscape does not have a single semantic field, but a semantic galaxy! It is in geography that we find a particularly large number of different interpretations (Farinelli, 1981), from the traditional physical types (Sestini, 1963) and historical/political types (Sereni, 1962) to the more recent ecological type (Ingegnoli, 1993). The territory can be considered as a system made up of both natural and artificial subsystems, all providing different types of perceptive information. The kind of information in which we are interested will depend on the importance given to the various subsystems. The system of meanings attributed to this information will be determined by the relevant disciplinary code. In other words, the perceived information can generate pertinence and hence give rise to different segmentalised interpretations of landscape.

Despite the fact that the interpretation of the landscape passes through the filter of a certain number of codes, almost all of them seem to have as a formal model the 'supercode' of scientific language, i.e. a structure which, with statistically predictable margins of error, leads from a significant to a meaning and vice versa. Scientific semiosis narrows the range of possible meanings towards a single meaning. But even the most obstinate geographer cannot refrain from expressing aesthetic judgement on the landscape. When he does this, however, he is abandoning the exercise of scientific semiosis in favour of aesthetic semiosis, which is characterised by the acceptance of multiple equiprobable meanings. The aesthetic text is *ambiguous,* in the sense that the semiotic process triggered opens numerous hermeneutic possibilities, pushing the receiver of the message towards untrodden paths within the semantic universe (Eco, 1962). The meanings tend not to be found at the level of the normal dictionary definition, but favour higher levels of syntagmatic stratification, where there are also the multimeanings which constitute the symbols of our universal encyclopaedia.

The landscape provides the stage for a large part of our human affairs. It absorbs the meanings and the memory of those events; it is this which gives it life. The landscape of our individual world records our joys and sufferings, and is part of ourselves. The same is true at the collective level. Every community, as a cultural entity, is irrevocably tied to its locality as a 'cultural product', i.e. as a significant able to convey the meanings with which the community memory has endowed it. This semantic endowment is not evenly distributed over the landscape or its topical components. Where the greatest accumulations occur, it generates those metaphors and symbols which animate the landscape. The image of such places recalls meanings not contained in the normal 'dictionary definition', and are made up of nebulous areas of content, or aggregations of vague meanings, lost in the childhood of the individual and of humanity. This memory is inexorably connected with the emotional effects caused by the stirring of the deepest layers of our psyche.

The study of landscape semantics, especially the aesthetic aspects, is above all the study of the semantics of symbols[1]; those syntagmatic constructions, both hypercodified and hypocodified, which a culture 'deposits' on the images of the landscape. These therefore become significants which, however much we try to remove their ambiguity through rhetorical operations (Eco, 1981), never lose their capacity to generate new meanings. They are great 'swallowers' of meanings, black holes of the semantic universe of the landscape!

Landscape semantics can be divided into two types. The first are the dictionary-type semantics, largely circumscribed by the geographical sciences (although normal spoken language is making increasing use of this dictionary). These consist of the layer of denotive and connotative contents through which we provide a geographical description of the territory before our eyes. Also included are things the eye does not see: the meanings we associate, through geographical codes, with visual perception, such as the historical meaning we attribute to a landscape feature (although only historical research can reveal its actual content) (Gambi, 1973).

The dictionary semantics of the landscape analyse the topical components and the combined structures through which these components generate landscape units. They provide a way of systematically identifying the landscape units which make up the 'episodes' of a landscape text (see Section 4). Dictionary semantics are in reality geographical semantics and, as the language of the territory is an ostensive language, it takes the form of an atlas - the universal atlas of real world places, including those seen through the eye of artificial satellites.

The second type, encyclopaedia semantics, consist of the syntagmatic

---

[1] A symbol is a particular form of sign which, through an analogical type of code, refers us to an indirect meaning made up of a nebulous area of contents. When, as in the case of a symbol, the code leads from a significant to a vague meaning, the interpretation operates in conditions of 'hypocoding'. The symbol, because of its vagueness, feeds many possible meanings which cause a semantic overloading, i.e. a situation of 'hypercoding'. This does not however cancel out the hypocoding situation. Rhetorical type textual practices, such as metaphors or metanymes, also refer to meanings which are indirect, but precise. In this sense, the metaphorical translation of a symbol involves the task of rendering a vague meaning unambiguous (Eco, 1984). We can state that a symbol remains such, as long as its hidden side has not been revealed.

hypercoded layers of metaphorical or symbolic contents. Semantic loading has been carried out on each unit of the landscape and each topical component (individual, local and international culture, in various epochs), generating a universe of metaphors and landscape symbols (from the Matterhorn as the romantic symbol of the sublime mountain, to Polynesian beaches as a status symbol for consumers of the products of tourist agencies). The systematic survey of this symbolic universe is relevant to the encyclopedia semantics of the landscape. This encyclopedia also has the form of an atlas whose contents, while associated with geographical denotations, includes the connotive deposit laid down by history and that which the processes of semantic fission, operating continually on historical material, have lent to this deposit[1] (Eco, 1968).

This atlas contains not only places in the real world, i.e. those of the geographical dictionary, but also those belonging to the world of fantasy, from Dante's inferno to the Star Wars, where we find a wealth of symbolic landscapes from which we frequently draw to identify or design the symbolic landscapes of the real world. Consider, for example, the influence of landscape painting on garden design throughout the 18th and 19th centuries (Clark, 1959).

The structure of the encyclopaedia, at least according to the definition given by Eco (1975, 1984), permits us to explain how landscape units even of the same basic type can adopt different contents, stimulating different reactions and behaviour, according to the *scenario* in which they occur (Pezzini, 1991)[2].

---

[1] The term 'semantic fission' was coined by Lévi-Strauss. It was taken up by Eco in the semiotic analysis of architecture and other objects to indicate the fusion that history produces between 'prime functions' (e.g. utilitaristic functions) and 'secondary functions' (e.g. symbolic functions). He claims that history produces a separation between the two types of function, and therefore their meanings, which originally were one. But this causes a 'loss of memory' which can affect one type of meaning or the other, or even both. This loss of memory means that not only does historical material, because of its vagueness, tend to take on symbolic meanings, but that these meanings, as they are adopted in new contexts, are subject to a series of redefinitions (Eco, 1968).

[2] In the dictionary interpretation of geography, the range of possible meanings is limited by the conventional limitation of possible scenarios. Aesthetic interpretation on the other hand is freer, in the sense that it may be influenced by a large number of contexts and circumstances not considered significant in the scientific interpretation, so it appears as a subjective interpretation, as opposed to the objective interpretation given by the geographical sciences. In reality, it involves two different forms of the eminently subjective, or cultural, act of semiosis. Aesthetic semiosis moves within, and is regulated by, the complex network of the encyclopaedic code, even when it makes great metaphorical or metonymical leaps. It is this capacity to establish unusual connections from within the global encyclopedia that allows it to acquire a deeper

16.6 The Process of Aesthetic Evaluation

The aesthetic appreciation of the landscape text depends on the morphological, the syntactical and the semantic level. Each level makes its specific contribution, but these merge into each other in the unitary act of perception, giving rise to particular idiolectic forms, similar to those produced by a poetic text. Every landscape is unique, although there are landscape types which seem to have a common poetry: the sublime beauty of Alpine pastures or gentleness of a pastoral scene for the romantics, the urban landscape of *Blade Runner* for those attracted by terror.

The semiotics-inspired investigation of aesthetics has produced several theories. Though in general they are only at early stages of development, they provide some useful indications of possible future directions for research. One proposal, which is particularly helpful at the operative level, is the suggestion that what distinguishes an aesthetic text is its *ambiguity*. A text which is ambiguous, without descending into pure disorder, attracts the attention of the receiver and places him/her in a situation of 'interpretative orgasm' (Eco, 1975). The receiver perceives that the text conveys a content which is not exhausted with the dictionary definition. The beauty lies in that which transpires over and above the immediate meaning of things. In this, the beauty of nature has a great affinity with that of art (Kant, 1790, Adorno, 1970). Faced with such ambiguity, the receiver is pushed to investigate the potential of the text and also of its code. The text attracts attention above all to its semiotic structure and thus becomes *self-reflexive*. As Roman Jakobson (1963) has remarked, language with an aesthetic function is distinguished by these very characteristics of ambiguity and self-reflexiveness.

Attempting to transfer this idea to a landscape text, we can put forward the hypothesis that ambiguity is produced when there is *expressive redundancy* or when there is a *semantic overload*. As suggested earlier, the landscape text takes on expressive redundancy when the morphological components reach the extreme figurations in their field of variability.

---

knowledge of the potential of the encyclopedia itself and therefore extend its limits. Aesthetic judgement is the result of a conoscitive process. The knowledge of the territory involves not only scientific, but also the aesthetic interpretation of the landscape. It is in this sense that we should understand the landscape as a 'cultural resource'. When it is beautiful, the landscape is like an open text with numerous different interpretations. These invite us to take an 'inferential walk' within the global encyclopedia, increasing our knowledge. When the landscape is ugly, it is a text enclosed in immediate denotative meanings and therefore provides no opportunity to improve our culture.

There is general recognition (Arnheim, 1954) of the fact that between the different expressive connotations there are emotive links which, in some way, enhance the connotations. The structure of the relations which link the expressive connotations of the overall expression has a synergetic effect. In particular, the accent is frequently placed on multiplier effect of the presence of opposites (Chen Congzhou, 1990). We can therefore put forward the hypothesis that the *co-existence of extreme opposites produces synergetic relations*.

Although these hypotheses can be supported by a large number of examples concerning both landscapes (Cheng, 1981) and other subjects (Calvino, 1988), full confirmation can derive only from systematic experimentation[1].

We have already briefly mentioned the phenomenon of semantic overloading in connection with the hypothesis of the metaphorical and symbolic nature of the landscape. We now wish to come back to this question. It is an incontrovertible fact that the world around us is populated with metaphors and symbols, and it is equally evident that these are related to our cultural horizon. If we consider the Illuminism and Romanticism movements, which in European culture led to aesthetic values being attributed to landscape, we find a codification of aesthetic categories which at the time were widely shared, at least among the European intellectuals who took the Grand Tour (De Seta, 1982). This gallery of beautiful landscapes is full of symbols alimented by the ambiguity of the classical myths. Today these myths have largely disappeared and their symbols have lost their original evocative power. This is not because the present-day consumer society has ceased to use the landscape to create symbols. It is simply that the classical myths have been pervasively substituted by the kind of contents which advertising attributes to nonessential consumer goods (Grandi, 1992). A systematic survey of

[1] There is a very fine dividing line between 'expressive redundance' which, through its evocativeness, stimulates abductive processes typical of aesthetic interpretation, and 'expressive excess' which makes us grimace. The idea that the aesthetic effect of the former can be achieved through extreme figurations of the morphological components is one which could be worth further research. It is important however in each individual case to know how to fall on the right side of the dividing line. Aesthetic criticism talks at length about counterpositions and the correct balance between 'full' and 'empty', between 'heaviness' and 'lightness'. Just what degree of precision is required in their measurement is demonstrated by Calvino's (1988) classical pages on 'lightness' in "*Lezioni americane*". The theoretical hypothsis on extreme figurations is at the macro level, identifying a general direction. It needs to be segmented and refined in order to construct a good microcode between figurations of the morphological contents of the landscape and its expressive character.

the iconography and literature of romantic landscape would make it possible to reconstruct the gallery of symbolic landscapes and related emotive impact which characterised the earlier period. The same could be done with contemporary films, photographs and written matter through which the tourist industry, ecology movement, advertising and so on help to mould the landscape symbols of modern society.

An encyclopaedic analysis should not limit itself, however, to recording these landscapes and their possible subdivision into categories, but should also record various shades of meaning in relation to concrete communication contexts. As already explained, aesthetic evaluation involves the syntactical as well as the morphological level. The syntactical structure of a landscape view, or of a narrative sequence of views, establishes the relationships between these views. The structure brings together or distances the various contents of the space-time structure of the landscape text, each content taking on a different semantic tone. When this is a complex syntagmatic construction, such as a metaphor or a symbol, the syntactical structure can have a multiplier effect on its evocative potential (take for example the dialectic association so dear to Assunto, 1984, between 'sublime' landscapes and landscapes of 'grace'). But this does not only occur with the most complex syntagmatic constructions. Even simpler denotations can produce a multiplier effect, in particular where the denotative contents are organised in a syntactic structure which gives rise to a scenario not yet recorded in the encyclopedia-atlas. When those contents have a strategic position, even the most common can transform itself into a Joycian epiphany (Eco, 1981); a pale nude can take on unexpected symbolic value, though probably, at least for a Christian, it would not be as evocative as a cross.

It is in this way that aesthetic work can invent symbols in a secularised culture in whose encyclopedia the archetypes of sacredness have been eliminated. In our atlas there will certainly be a chapter dedicated to symbolic places, but aesthetic research would become sterile if we were simply to repropose existing symbols. The researcher knows that symbols are nested in the whole structure of the encyclopaedia. They are not codified images of places, but unusual pathways which can be taken within the structure of the encyclopaedia, inviting further inferential explorations. It is from the *collage* of fragments of landscape in the abstract topology of the syntactic structure that new ambiguities can unexpectedly appear.

We can therefore put forward the hypothesis that *syntactical relations can exercise synergetic effects on the semantic potential of a text*[1] .

16.7 Conclusions

The initial phase of a study is certainly not an ideal one at which to draw conclusions. But even at an early stage such as this, we may be able to reach conclusions of a sort, even if there remain a list of doubts and concepts which are still unclear with missing or weak connections. These are nevertheless all themes which hold a kind of aesthetic fascination for the researcher! In the limited space available, we have been able to touch only briefly on the main features of a subject which in reality extends to wide horizons and deserves to be dealt with in far greater detail. To conclude this tentative survey of the vast field offered by the application of the various semiotic theories to landscape analysis, I should like to add just a few comments.

The filter of semiotic theory allows us to put some order into a complex system of information in which we find ourselves immersed daily and on which we draw in order to 'locate ourselves' in the narration of our existence. It is said that man has an irresistible urge to "construct his life as a story" (Eco, 1994, p. 160). His personal story is interwoven with collective history and held together though the memory. In this narration, life is one with the world scenario; the landscape is an integral part of this story and within it the landscape becomes a narrative text. When, finding ourselves before a natural landscape, we forget for a moment the utility of nature, or that which science tells us about it, the landscape does not lose its meaning. The vacuum is immediately filled by our narrative capacity, in other words, the "possibility of freely exercising that faculty we use both to perceive the world and to reconstruct the past" (Eco, 1994, p. 163). Cancelling the beauty of the landscape does not eliminate this free

[1] The same observations made in the previous footnote in relation to extreme figurations applies to the hypothesis about the symbolic value of the landscape and the synergetic effects of the syntactical relations. It is a 'macro' hypothesis in that it identifies strategic directions for aesthetic research but requires the kind of systematic operative development needed by all experimental translations of theory. By the end of this article, the meaning of the 'code of aesthetic sensitivity' which connects information from an image with the pathemic states it produces should have been made clear. A pragmatic type of semiotic investigation would allow us to record this code and to discover the mechanisms underlying it. In discovering how it works, we can refine it or even overturn it with catastrophic innovative processes (for a discussion of innovation as the product of invention, see Eco, 1975).

cognitive ability, we simply force it to locate our narrative fictions in that landscape of fear which is increasingly the backcloth to metropolitan life. The beauty of the landscape is a public resource, its care cannot be entrusted only to those with artistic flair, but should become a normal professional task of the architect. It is for this reason that the architect should learn the rules for the correct 'reading' and the correct 'production' of the particular form of narrative text represented by the landscape.

In the theoretical literature on narratology, the idea of various 'reader-author' pairs has been put forward (Pugliatti, 1985). The transposition of this idea into landscape terms offers interesting possibilities. We could, for example, apply the distinction introduced by Eco (1979, 1990) between 'empirical reader/author' and 'model reader/author', using it as a basis to distinguish between an 'empirical' landscape architect and a 'model' landscape architect. Whereas the former *uses* the landscape text, reading into it contents not admitted by the narrative structure, the latter *interprets* the landscape as a narrative text, based on a structure revealed by a rigorous semiotic survey.

There are many different ways to walk through the city or the woods, but only one characterises the 'model landscape architect reader'. He traverses the landscape to discover the narrative structure which underlies its text[1]. This interpretative behaviour does not mean that behind a natural walk should be hidden the intention of a Le Nôtre, the essential thing is that the landscape is seen by the one who interprets it as the result of an intentional design, even when this turns out to be totally fictitious. What the landscape reader is looking for is that which is semiotically imprinted on the

[1] It is now generally acknowledged that aesthetics is an important ingredient of architecture and the landscape. When this is not accepted, it is normally due to the belief that aesthetics is still anchored in Romanticism and, therefore, obsolete. A modernization of landscape culture would seem, almost inevitably, to lead us to semiotics. A semiotic survey of aesthetic phenomena offers a reconcilation, bringing us in line with the present-day cultural horizon. It also means, however, that we must establish new rules of aesthetic criticism, teaching us what to search for and how to value it. There is an urgent need for such a rigorous approach to the interpretation of architecture and landscape, where the *empirical author* is frequently confused with the *model author*. This confusion is very evident in what we may consider the high point of landscape architecture: the art of landscape gardening. In the proceedings of the numerous conventions on the subject, the garden is rarely quoted as one of the possible semiotic forms of the aesthetic text. It seems that we want to know everything about gardens, even the private affairs of the gentlewomen who had them built - the only thing which appears not to interest us is the structure of the gardens themselves! The historic garden has become a cult object and, as we know, everything which comes into contact with a cult object becomes sacred, as demonstrated by the *ex voto*. But using gardens to collect *ex voto* has little to do with aesthetic criticism.

landscape as a narrative structure, independently of the empirical landscape authors who have composed it.

The landscape is moulded by many different empirical authors. Firstly there is nature herself, to which everything can be attributed, except the *intention* of writing a text (though this idea pervades religious thought - in this case, however, it is Nature with a capital N). Then there is man who modifies nature for utilitaristic ends (often continuing to think of it with a capital N) and who would look askance at whoever claimed that in this way a narrative text is being created. There is also that select company of aesthetes who have had the natural landscape modified with the express intention of composing an aesthetic text. But can we consider all landscape signs (including natural, intentional or unintentional signs) in all the various historical periods, as belonging to a model author? If we consider that it is appropriate to interpret the landscape as a text, we have no alternative. We must regard the landscape as if it were the product of a strategic design which has led to specific results, sometimes fortunate, sometimes less so. It does not matter if the generator of these strategies is a complex historical process concerning the relationship between man and nature, the essential thing is that as 'model landscape readers' we possess a set of rules for entering into the narrative text and discovering how it is made. To give a theoretical foundation and operative construct to the grammar and the encyclopaedia of the landscape means providing the basic structure of those rules.

References

Adorno T.W. (1970) *Ästhetische Theorie*, Suhrkamp Verlag, Frankfurt.

Arnheim R. (1954) *Art and Visual Perception: a Psychology of the Creative Eye*, Regents of the University of Califorrnia, Berkeley, Los Angeles.

Arnheim R. (1969) *Visual Thinking*, Regents of the University of California, Berkeley, Los Angeles.

Assunto R. (1973) *Il paesaggio e l'estetica*, Giannini, Naples.

Assunto R. (1984) *Il parterre e i ghiacciai*, Novecento, Palermo.

Bense M. (1965) *Aesthetica*, Agis, Baden Baden.

Berlin B., Kay P. (1969) *Basic Color Terms: Their Universality and Evolution*, Berkeley University Press, Berkeley.

Calvino I. (1988) *Lezioni americane*, Garzanti, Milan.

Chen Congzhou (1990) *I giardini cinesi*, Muzzio, Padua.

Cheng F. (1981) *L'éspace du rêve*, Phébus, Paris.

Clark K. (1959) *Landscape into Art*, John Murray, London.

Dematteis G. (1985) *Le metafore della terra*, Feltrinelli, Milan.

De Seta C. (1982) L'Italia nello specchio del *Grand Tour*, in *Storia d'Italia, 5*, 'Il paesaggio', Einaudi, Turin, 124-263.
Eco U. (1962) *Opera aperta*, Bompiani, Milan.
Eco U. (1968) *La struttura assente*, Bompiani, Milan.
Eco U. (1975) *Trattato di semiotica generale*, Bompiani, Milan.
Eco U. (1979) *Lector in fabula*, Bompiani, Milan.
Eco U. (1981) Simbolo, in *Enciclopedia*, Vol. XII, Einaudi, Turin.
Eco U. (1984) *Semiotica e filosofia del linguaggio*, Einaudi, Turin.
Eco U. (1990) *I limiti dell'interpretazione*, Bompiani, Milan.
Eco U. (1994) *Sei passeggiate nei boschi narrativi*, Bompiani, Milan.
Farinelli F. (1981) Storia del concetto geografico di paesaggio, in *Paesaggio: immagine e realtà*, Electa, Milan, 151-158.
Gambi L. (1973) *Una geografia per la storia*, Einaudi, Turin.
Grandi R. (1992) Sistema di consumo e significazione delle merci, in *Semiotica: storia teoria interpretazione*, Bompiani, Milan, 321-336.
Henle M. (1961) *Documents of Gestalt Psychology*, Berkeley University Press, Berkeley, Los Angeles.
Ingegnoli V. (1993) *Fondamenti di ecologia del paesaggio*, CLUP, Milan.
Itten J. (1961) *Kunst der Farbe*, Ravensburg.
Jakobson R. (1963) *Essais de linguistique générale*, Minuit, Paris.
Kanizsa G. (1980) *Grammatica del vedere*, Il Mulino, Bologna.
Kant I. (1790) *Kritik der Urtheilskraft*, Legarde, Berlin.
Koffka K. (1935) *Principles of Gestalt Psychology*, New York.
Marr D. (1982) *Vision*, Freeman, New York.
Pezzini I. (1991) *Semiotica delle passioni*, Esculapio, Bologna.
Pugliatti P. (1985) *Lo sguardo nel racconto*, Zanichelli, Bologna.
Ritter J. (1963) Landschaft, University of Münster, Münster.
Sereni E. (1962) *Storia del paesaggio agrario italiano*, Laterza, Bari.
Sestini A. (1963) *Il paesaggio*, Conosci l'Italia Series, Vol. VII, Touring Club Italiano, Milan.
Socco C. (1994) Appunti di paesaggistica. Per una interpretazione semiotica del paesaggio nella sua funzione estetica (mimeo).
Stewart D.A. (1956) *Preface to Empathy*, The Philosophical Library, New York.
Tagliolini A., Venturi Ferriolo M. (1987) *Il giardino, idea, natura, realtà*, Guerrini e Associati, Milan.

17. Assessing Uncertainty and Complexity in Urban Systems

Silvana Lombardo

17.1 Introduction

The general aim of this work is to discuss some aspects and problems of urban planning. Here, urban planning is viewed as a learning process applied to the continuous control of the evolution of cities and integrated - as a decision aid tool - in the decision-making process behind public policies which manage urban transformation.

The specific aim is to deal with the relationships between some traditional *needs* of urban planning - such as, and above all, the ability to forecast and control - and some problems which seem to frustrate the satisfaction of such needs. These are mainly problems of knowledge: they generate uncertainty as to evolutive processes of city and society and are strictly connected with the new idea of the complexity of all kinds of system (including urban ones), which has now pervaded all sciences.

In the wide discussion concerning the issue of complexity, two directions of thought emerge. These are well represented by a cartoon exhibited at a congress: two scientists are scowling at each other, one of them says: "*Complexity is what you don't understand*", and the other retorts: "*You don't understand complexity*". For the former, then, complexity depends on the available level of knowledge of the system, while the latter claims that complexity is an intrinsic feature of the system and then does not depend on the greater or the lesser knowledge which an observer can have of it.

This work assumes the former point of view. If complexity depends on the level of knowledge and predictability of a system's behaviour, it will present different intensities in different systems and in different points in time in the same system and can be therefore considered as a measurable property of a system.

In this work, then, some steps are proposed in the direction of defining measures of the complexity of an urban system's behaviour and, consequently, of the level of uncertainty connected with actions of forecasting, planning and control.

I shall try to demonstrate that urban planning is not annihilated (as someone claims) by the inapplicability of positive rationality and the consequent impossibility of long term forecasting - which allowed us to rely on the possibility of controlling, in a cybernetic sense, urban systems transformations (Lombardo, 1993a). On the contrary, such aspects widen the meaning and the possibilities of urban planning, provided that we change our way of considering it and use methods adequate to support knowledge.

17.2 Complexity, Knowledge and Planning

What does complexity mean? On this theme, a wide discussion is going on in many scientific fields and there exists a wide historic, philosophical and scientific bibliography. However, I quote here some definitions which are useful for the reasoning.

A physicist (Haken, 1983) defines complex systems as systems whose behaviour cannot be *understood* in a simple way by the behaviour of their elements and whose whole behaviour presents some properties deriving by the *co-operation of their elements*. A planning scientist (Le Moigne, 1984) defines complexity as the property of a system to show behaviours which cannot be wholly *predetermined* even though potentially *envisaged* by an intentional observer of this system. From these definitions some fundamental indications can be drawn as to the nature of complexity and its implications for knowledge and planning.

Complexity is then, above all, a problem of *knowledge*: the behaviour of a system cannot be understood in a simple way and it is caused, among other things, by the working of an internal *organisation* (co-operation of elements). It is a problem of *forecasting*: the behaviour cannot be always foreseen. This generates the *planning problem*: the behaviour, the evolution, the organisation of the urban system, as effects of a plan, cannot be wholly predetermined with certainty.

Let us now focus on the issue of knowledge, as it emerges by the previous statements: the behaviour of a system cannot be, in a simple way, analysed, foreseen and determined *by an observer*. Complexity is not then in the structure of the system, but in the observer or, better, in the coupling of observed system and observing system (Von Foerster, 1981), in which the observer's choices, aims and also ignorance play their role.

If we admit that the complexity of a system is not necessarily a property of the system itself, but is a property of the currently available representation of it, then complexity must be considered as a *way of knowledge of systems* and as a stimulus for the study and the creation of new and different representations. As an example, we can recall that the motions of heavenly bodies appeared endowed with an inextricable complexity until Kepler represented them by means of elliptic trajectories instead of circular ones. Complexity was reduced in the observing system and not in the observed one.

17.3 Organisation and Self-Organisation

As we have said above, the understanding of a system's structure is made arduous by its *organisation*, that is the process which connects different elements, makes them components of a whole and produces the complex unit or system, endowed with qualities unknown at the level of the components if analysed one by one.

Organisation is at the same time transformation and formation (morphogenesis): it is transformation because the elements are changed into parts of a whole, losing some properties and acquiring new ones, it is morphogenesis because it gives form, in space and time, to a new reality - the complex unit or system - whose composition *rules* are not additive, but transforming (Morin, 1977).

The concept of organisation is therefore central as far as complexity is concerned. It allows us to represent, without mutilating them, phenomena perceived as complex. One relevant property of organisation is to guarantee a *relative solidity* of the links among the elements, assuring the system of a certain possibility of *duration*, notwithstanding the occurring of perturbations. But if the perturbations exceed a certain level, a change of organisation and structure can appear.

In this case, the property of *self-organisation* of the observed system reveals itself to the observing system. Self-organisation is the property of a system which reacts to perturbations by maintaining its organisation or reorganising itself in different structures. This last case does not correspond to a simple rearranging of connected elements, but to the production of a new functional organisation, i.e. a different meaning of the relationships between elements.

In general terms, self-organisation is characterised by a state between two extremes: one of rigid, unchangeable order (therefore always foreseeable and controllable), but unable to change without being destroyed; the other of continuous change, without any stability (and therefore always unforeseeable and completely uncontrollable). Such an intermediate state is not fixed, but allows the system to react to the perturbations by means of changes of its organisation. These changes do not correspond to a mere destruction of the previous organisation, but to a reorganisation which makes it possible for new properties, that is new behaviours or new structures, to emerge (Atlan, 1983).

Nicolis and Prigogine (1977) show that self-organisation can be generated by perturbations which take place in the history of the system. They say that the historical path along which a system evolves is characterised by a sequence of stable regions, ruled by deterministic laws, and unstable regions, where the system can choose more than one possible future. *This mixture of chance and necessity makes the history of the system.*

More precisely, he demonstrated that open systems far from equilibrium can undergo spontaneous self-organisation processes, trigged by fluctuations produced in order to fit a changing environment, as a reaction to perturbations of the state of the system.

17.4 Toward Complexity Methods

The study of the evolutive potential of a system does not lead to a reliable and exhaustive forecasting capacity. Scientific laws[1] cannot tell us much about the real space-time trajectory of phenomena; they can simply express a set of possibilities within which actual processes take place (Ceruti, 1985), indicating a range of trajectories, which will be determined partly by chance, partly by planning intervention and partly by the specific characteristics of the systems.

The problem is therefore to transform the discovery of complexity into a *complexity method* (Morin, 1977). How can we design a model which can express a phenomenon without exhausting it, and account for its possible complexities through an intelligible complexity?

---

[1] For which the rules observed in the past will be the same in the future (Von Foerster, 1981).

Let us recall Kepler's work: the building of a complex model contributed to making an otherwise inextricable complexity understandable. The complexity method is above all a method of designing complex models.

In the field of urban system research, the study on complexity methods has taken some stimulating directions, which tend to enrich and innovate, among other things, the methods for simulation-evaluation of processes. In particular, the new models of self-organisation allow us to treat urban system not as a kind of machine ruled by an exogenous deterministic programme, but as a self-organising system[1].

Moving on to experimental issues, Section 5 illustrates briefly the results of some experiments carried out by means of a self-organisation urban model (Lombardo and Rabino, 1989), able to simulate the phenomena and processes described in Section 3 and to lead to an assessment of the complexity of the behaviour of the system. Section 6 presents a methodology, based on some principles of Information Theory, aimed at detecting complexity nodes in a system of planned actions in order to direct, with greater accuracy and effectiveness the use of resources and the actions of control and monitoring. We also describe the project and the expected results of an application of such methodology, presently in progress[2]. In this case, the assessment of the complexity of the planned interventions, provides elements for comparing and evaluating different planned urban transformation processes.

17.5 Self-Organisation, Stability and Uncertainty: Retailing Services in an Urban System

A nonlinear dynamic model, formulated by means of some simple differential equations[3], was applied to the location choice behaviour of

---

[1] The basic work was made by Allen and Sanglier (1979, 1981) and Wilson (1981), followed by Pumain, Saint-Julien and Sanders (1984), Lombardo and Rabino (1983, 1984), Lombardo (1986, 1993b).

[2] The development and the application of such methodology are being carried out in the context of a study, directed by S. Lombardo, for the Ministry of University and Scientific Research.

[3] The equations, derived by the Harris and Wilson model (1978), are the following:

$$W_j = \varepsilon[\Sigma_i e_i P_i\, W_j^{\alpha} \exp(-\beta c_{ij}) / \Sigma_j W_j^{\alpha} \exp(-\beta c_{ij}) - K_j W_j]$$

where:

- i, j are the zones of the urban system;
- W_j is the size of retailing services in j;
- $e_i P_i$ is the retailing services demand expressed by zone i;

retailing services in an hexagonal fictitious system and, later, in the urban system of Rome.

The fictitious urban system (Fig. 1) is made up of 61 hexagonal zones, whose centres are connected by an uniform network with the size of the links assumed equal to 1. A total population of 1,000,000 inhabitants is distributed among the zones, with a density exponentially decreasing from the core zone to the peripheral zones.

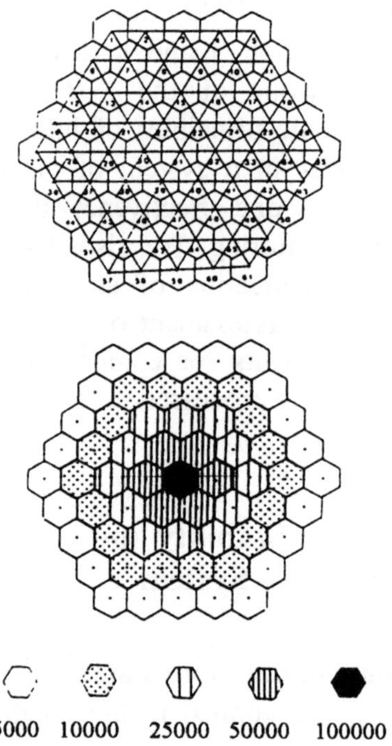

Fig. 1 Hexagonal spatial system: transport network and population by place of residence

---

c_{ij} is the generalised travel cost from i to j;

α is a measure of consumers' sensitivity to economies of scale;

β is a measure of the deterrence of travel costs;

K_j is the cost per unit of supplying the service at j;

ε is a measure of the speed of response of the system to the perturbations.

The first term of the right hand side represents the demand for services in zone j, while the second term represents the supply in the same zone. The trigger for the dynamic behaviour is then the disequilibrium between demand and supply of the service in each zone.

In Fig. 2 service centre size trajectories from the initial state to the equilibrium state are shown for three zones of the hexagonal system. In the same figure, the trajectories of the demand are also represented by a dashed line.

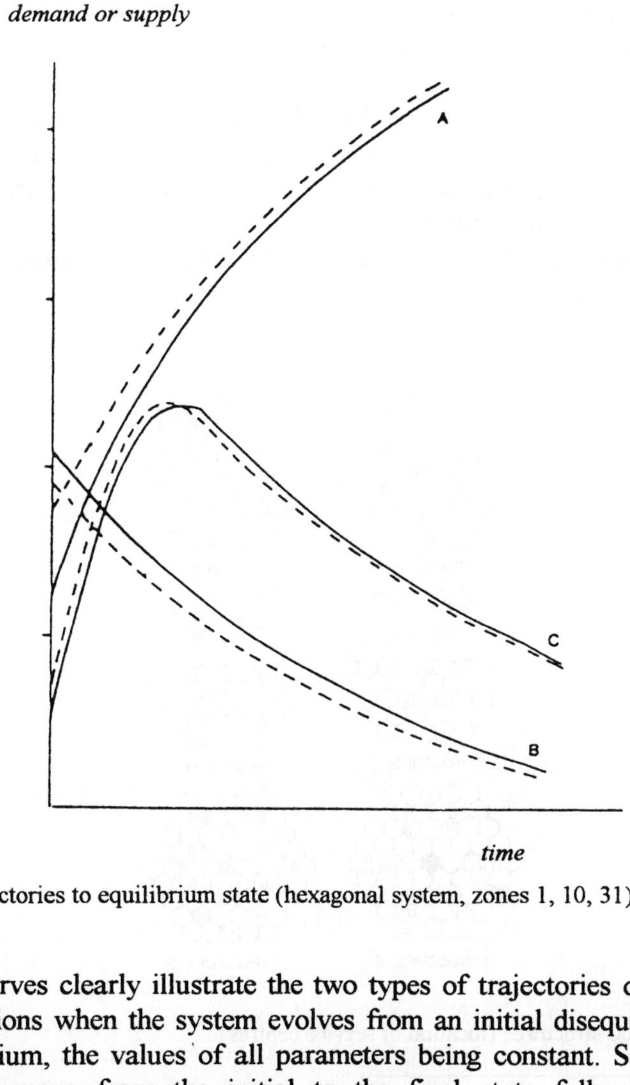

Fig. 2 Trajectories to equilibrium state (hexagonal system, zones 1, 10, 31)

These curves clearly illustrate the two types of trajectories obtained in all simulations when the system evolves from an initial disequilibrium to an equilibrium, the values of all parameters being constant. Some zones (dominant) move from the initial to the final state following quasi-exponential path, the others follow a humped path.

In case A, demand exceeds supply from the outset, so the service centre of the zone grows until the supply meets the demand. On the contrary, in case B, supply always exceeds demand, so the zone declines monotonically to zero. In case C, demand initially exceeds supply, so supply increases towards an equilibrium. In the meanwhile, however, the changes of supply in the other zones reduces demand in the zone by attracting it elsewhere. As a consequence, the supply in the zone exceeds the demand and starts to decline towards zero[1] .

We simulated the reactions of the urban system (and therefore its self-organisation) caused by some perturbations due only to changes of mobility (β), viewed as deterrence of travel cost perceived by consumers. Such reactions are described here by means of the service location patterns, the structures of consumers flows and the structures of market areas, representing the ways in which the system can reorganise itself because of the perturbations.

Figs. 3, 4a and 4b show a set of structures of reorganisation of the analysed system.

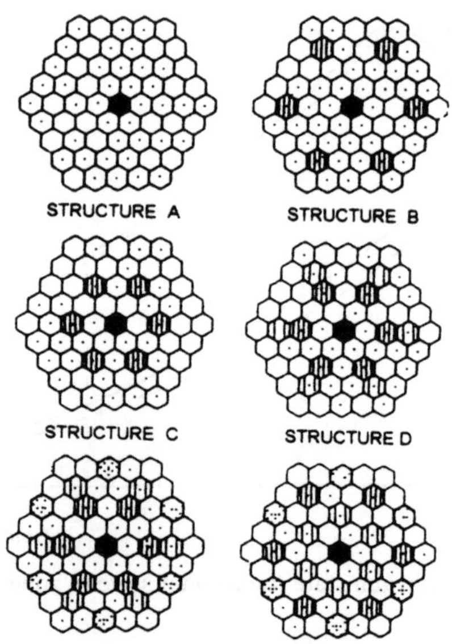

Fig. 3 Spatial structures (location of service centres)

[1] It is worthwhile observing that, in this case, an interaction process is established between demand and supply which recalls the well-known predator-prey interaction. This is not surprising because of the strong similarity between the Volterra model and the present one (see, for instance, Wilson, 1981). But, as in the present model the interactions are nonlinear, the oscillatory trajectories of the Volterra model, which are caused by the linearity of the interactions, do not appear.

Fig. 3 presents the location of the service centres for the six most meaningful equilibrium solutions of the hexagonal system, obtained for β ranging from 0.0 to 5.0 (with α=1.3). As expected, a decrease of consumers' mobility (increase of the value of β) leads to an increasingly diffuse distribution of service centres, and vice versa.

Case A of Fig. 3 corresponds to a single large centre which supplies the whole system. In case B, a decrease in mobility is followed by the emerging of small service centres in peripheral areas, which supply the outer rings. However the core zone still supplies about 75% of the population of the system. A further decrease of mobility leads to a noticeable change of the location structure of services. The large service centre of case A is replaced by a set of seven zones (one is the core zone and the others belong to a peripheral ring). The core zone now supplies approximately 40% of the population, while the remaining 60% is shared uniformly among the other six zones. In Fig. 4a, the main journeys to service are represented for a selected number of the structures analysed and in Fig. 4b the corresponding market areas can be seen, obtained by considering the largest flow coming out of each zone and headed for a service centre.

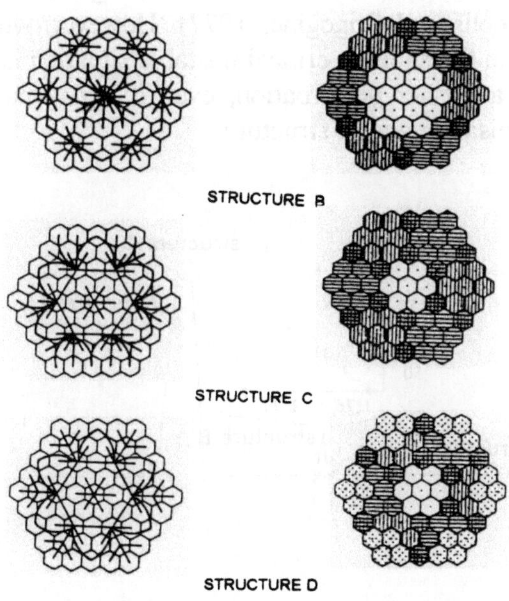

(a) (b)

Fig. 4 (a) Main journeys to service (b) Market areas

Notwithstanding the extreme schematisation and symmetry of the system, it can be noted that the different structures do not correspond to the simple rearrangement of a few elements. On the contrary, they represent *different functional organisations*, that is different structures of market areas, whose elements and interactions change as to their role and meaning. After the transition from structure B to structure C (Fig. 4a), for instance, some zones belonging to the third ring do not depend anymore on the core (first ring). They become quite relevant nodes, to the detriment of core zone and of fourth ring. After the transition from structure C to structure D, beside the emerging of new market areas (Fig. 4b), the flows produce a hierarchical structure of services. Indeed, it can be seen that the centres of the fifth ring gravitate mostly towards the six service centres of the fourth ring and these, in their turn, gravitate towards the service centres of the third ring (Fig. 4a).

The stability of the simulated structures was measured in terms of the time needed by the system to reach a new equilibrium after the perturbations. This analysis (represented in Fig. 5) generated a complex profile with *valleys* and *peaks*. Each *valley* corresponds to one spatial structure and to a state in which the system, although the perturbation become stronger (increase of β, that is a decrease of mobility), holds its organisation. The system then is in a stable region, where deterministic laws work (Nicolis and Prigogine, 1977). However, when perturbations exceeds a certain threshold, a critical unstable situation is produced (in the *peaks*), where a further perturbation, even if very small, can lead to a change of organisation and of structure.

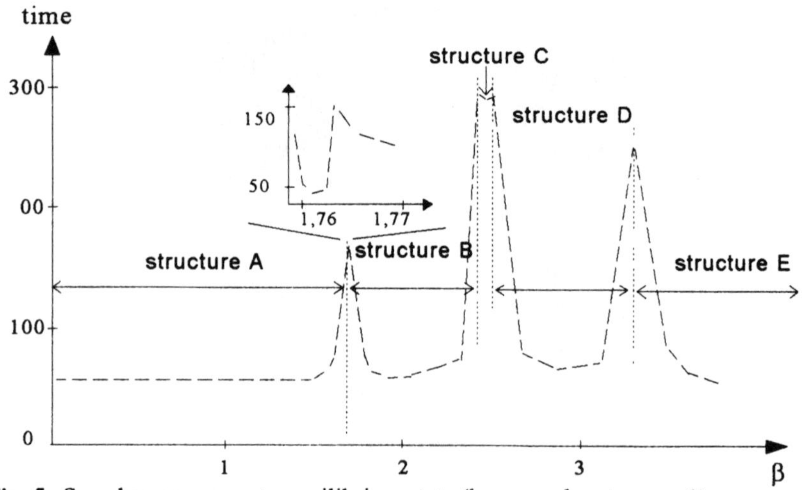

Fig. 5 Speed convergence to equilibrium state (hexagonal system, α>1)

It must be stressed that the results of such analysis can be used as a *measure* of the complexity of the system: narrower the *valleys* and more numerous *peaks* correspond to more unstable and unforeseeable behaviour and, therefore, to a higher degree of complexity (and vice versa).

An experiment similar to that described above was carried out on the metropolitan area of Rome. The different structures of market areas obtained by progressively reducing consumers' mobility are shown in Fig. 6. A decrease of mobility (for β ranging from 0.08 to 0.16) leads to different organisations ranging from 2 to 13 different market areas.

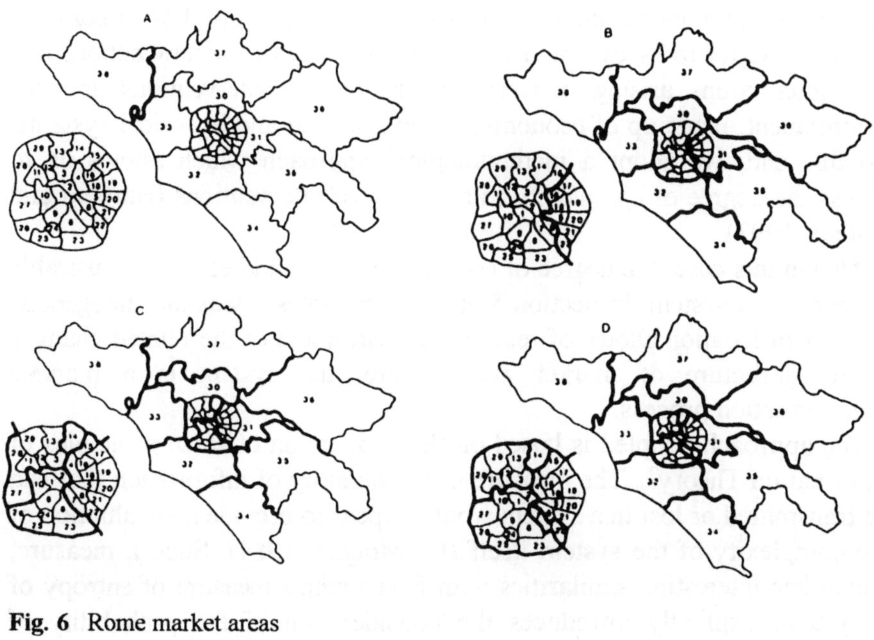

Fig. 6 Rome market areas

17.6 Urban Planning and Uncertainty Assessment: the Location of a Metropolitan Service

This study stems from the debate on the location, within Rome urban area, of a structure devoted to musical activities. It was generally acknowledged that, to meet the needs of a metropolitan area, the response could not simply be a concert hall; it was necessary to consider a project which included a concert hall (Auditorium) in a multipurpose centre endowed with, among other things, structures for teaching, for staging different kinds of performance, for congresses and exhibitions. Such a structure would, therefore, involve a large part of the city in a transformation process.

As far as the choice of location[1] was concerned, the results of the debate did not satisfy most of those involved, which, in this case, included not only public authorities and technicians, but citizens and private entrepreneurs. The aim of the study was therefore to analyse, compare and assess, by means of a *complexity method*, some location hypotheses stemming from the debate, in order to assess the possibility of success of the transformation process underlying each of them.

We consider the transformation process connected with each project as a complex system interacting with an environment made up of other complex systems and try to assess the degree of complexity of the new actions and new interactions arising between the system of interventions and the environment, made up of economic, social, natural and historical systems. To this end, we adopt a methodological approach which allows us to assess the degree of complexity in a system of interventions (Barbera and Butera, 1992).

Also in this case, the degree of complexity is considered as a measurable property of a system. In Section 5 the system analysed was the endogenous process of location choice of retailing activities and of the related changes in the structure of market areas. Here, the system is a planned transformation process.

The approach adopted is based on the application of some principles of Information Theory[2]. The measure of the quantity of information that can be transmitted or lost in a system would appear to provide an evaluation of the complexity of the system itself (Le Moigne, 1984). Such a measure, which has interesting similarities with Boltzmann's measure of entropy of a system, explicitly introduces the consideration of the probability of occurrence of behaviours occurring and their *relative uncertainty* for an observer. This is a fundamental point for planning complexity.

The methodology leads to a quasi-quantitative assessment of the effort that should be put into new actions and new or different interactions with

---

[1] The choice was made only very recently, after a long series of debates, publications and meetings.

[2] In the current literature, different measures have been proposed to assess the complexity of a system: the measure of cybernetic variety (based on the number of possible behaviours), the measure of connection networks (based on the level of performance demanded to the system), the computational measure (based on the complexifying effects of feedbacks), the information measure (based on the quantity of information that can be transmitted or lost in a system).

the economic, social and natural environment in order to make the transformation process (project) work. In other words, it assesses the probability that a specific action will be successful.

Such an approach is usually based on interviews and their evaluation with a simple matrix technique derived through Information Theory, as sometimes used in studies concerning communication processes among animal species (Wilson, 1975).

In brief, the methodology implies the identification of the *actions*, through the detailed analysis, step by step, of the planned transformation process, as well as the identification of the individual, social and institutional *actors* directly or indirectly involved in the process. It is necessary, moreover, to identify the *factors* (transport facilities, the presence of transformable areas, of monuments, etc.) that is, the constraints which potentially influence each action.

The following phase involves the quantitative evaluation of the *intensity of the interactions* between actions, actors and factors. The intensity of such interaction corresponds to the probability of the interaction occurring: it can be interpreted as a probability of communication, which can be measured in terms of *information bits*. We can then compute the values of H (information or complexity) for each action, actor and factor: the computed values of H are a *complexity indicator*. By plotting these values in the form of a histogram (Fig. 7), this can be interpreted as a detailed representation of the system's structure, like the spectrum of a luminous or acoustic source.

This kind of spectrum analysis of the complexity of a project allows us to identify the actions, actors or factors which are most important for the process to work properly. When we compare the spectra of transformation processes derived from different projects, then we can identify the subsystems of actions, actors and factors with the highest level of complexity. These are the subsystems to which we have to pay the greatest attention. If many subsystems show a high level of complexity, it may be wise to modify the project, since the high degree of uncertainty about the behaviour of the system could imply a very low probability of success.

A further evaluation can be made to identify the most critical action, actor, or factor, by analysing the sensitivity to perturbations of each subsystem. This can be made by perturbing the subsystem, i.e. by modifying the values of the interaction matrix within a wide range and analysing the effect on the structure of the complexity of the whole system. The most critical elements will be those with a high level of complexity, which is maintained or increased when the interaction structure is modified. Moreover, this sensitivity analysis allows us to evaluate the overall stability of the planned transformation process.

In the following, we present the indicators derived from the above approach.

Let us define:

j	action,
i	actor or factor,
P_{ji}	probability of communication (interaction) between action j and actor or factor i,
S_{ji}	intensity of the interaction between action j and actor or factor i,
$\Sigma_j S_{ji}$	total interaction of actor (or factor) i with all the actions,
$\Sigma_i S_{ji}$	total interaction of action j with all the actors (or factors),
$\Sigma_{ji} S_{ji}$	total interaction of the system action/actors (or factors),
$PA_{ji} = S_{ji} / \Sigma_i S_{ji}$	probability of interaction between action j and actor or factor i,
$PF_{ji} = S_{ji} / \Sigma_j S_{ji}$	probability of interaction between actor (or factor) i and action j,

we can then build an *indicator of the complexity of action j*:

$$AH_j = \Sigma_i S_{ji} / \Sigma_{ji} S_{ji} (-\Sigma_i PA_{ji} \log_2 PA_{ji})$$

where the first term and the second term (within brackets) of the right hand side represent respectively the intensity and the structure of the interactions between each action and the environment, considered as the whole universe of actors and factors.

Relatively high values of AH_j identify the most critical actions, which must be closely controlled, while low values correspond to action easier to be managed.

Moreover, an *indicator of the complexity of actor or factor i* can be obtained:

$$FH_i = \Sigma_j S_{ji} / \Sigma_{ji} S_{ji} (-\Sigma_j PF_{ji} \log_2 PF_{ji})$$

where the first term and the second term (within brackets) of the right hand side represent respectively the intensity and the structure of the interactions between each actor or factor and the set of actions. FH_i expresses how much an actor or factor i is relevant for the management of the transformation process and allows to rank actors and factors as more or less critical and thus determine whether they need more or less monitoring effort in order the project be successful.

A perturbation modifying the structure and the intensity of the interaction of an actor or factor showing a high relative value of *FH*, will strongly affect the transformation process, in either a positive or negative way.

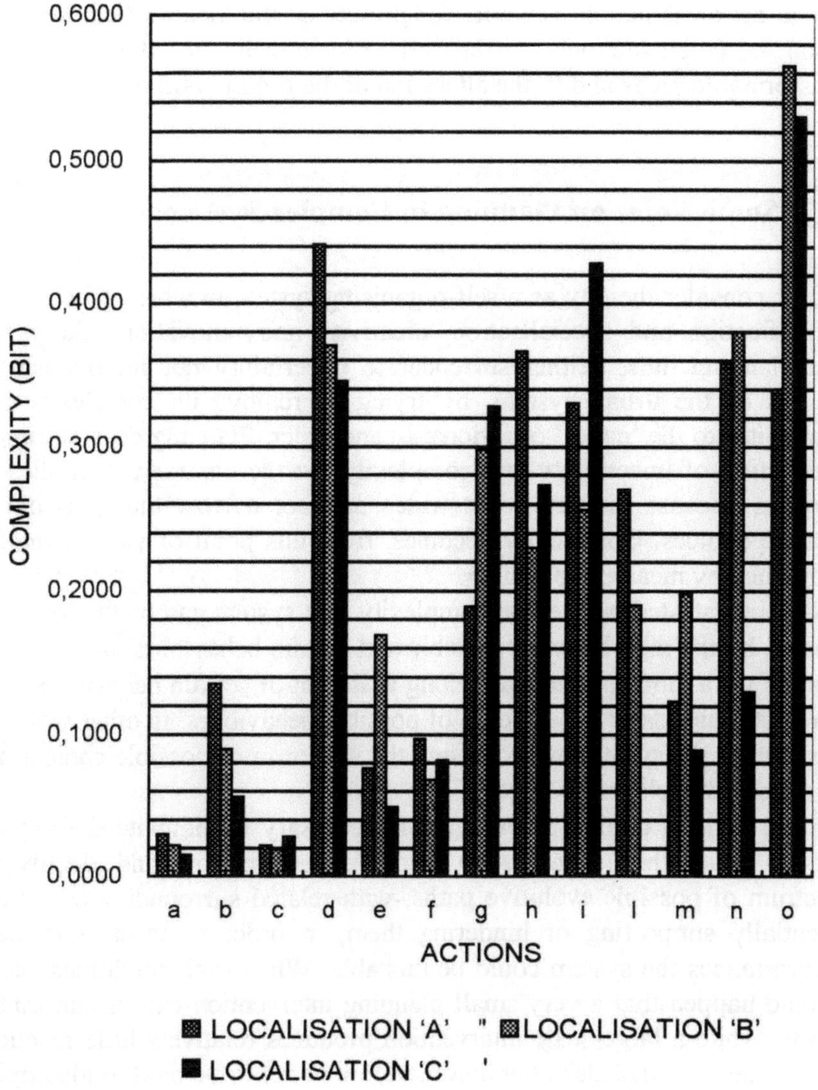

Fig. 7 Complexity (AH_j) of actions (j=a,b,c,...) vs. whole universe of factors

Fig. 7 shows an example of the representation of the complexity level AH_j of the individual actions (a, b, c, ...) belonging to each considered project, versus the universe of factors. Similar histograms can be plotted with respect to the universe of actors, as well as for FH_i indicators. It can be noted that the spectra of the three different location projects (A, B, C) show some substantial differences. For instance, for project A, the most critical action is action d, while, for project B, the most critical action is action o (such actions correspond, respectively, to the finding of transformable areas and to the alteration of the ground outline).

17.7 Some Notes on Planning in Complex Systems

If we consider the city as a self-organising system in which processes of diversification and specialisation, creativity and innovation take place, then planning must neither surrender to uncertainty nor impoverish the concept of the urban system, by trying to remove its complexity and forcing it into the 'cages' of rationality and order. Planning does not imply a reduction of uncertainty and complexity, on the contrary, it implies an increase, because projects must widen and not narrow the spectrum of possible choices. Complexity becomes, from this point of view, a *quality* to produce by means of planning.

As demonstrated before, the complexity of a system can be measured, in general, by the ratio between possible and certain behaviours of the system itself. If we assume projects to belong to the set of certain behaviours, then a project must foster the increase of possible behaviours, in other words, it must enrich the system by increasing the spectrum of possible choices and introducing novelty and variety.

In the context of city planning, it is necessary to simulate the process through which the urban system reacts to perturbations and identify the spectrum of possible evolutive paths, with related surrounding conditions potentially supporting or hindering them, in order to identify in what circumstances the system could be unstable. When such conditions occur, it could happen that a very small planning intervention causes substantial effects, while a large-scale intervention produces relatively little reactions in the system. Particular attention therefore needs to be paid to identifying the system's conditions of criticality.

In general terms, a transition is emerging from controlling and forecasting to the 'interaction game'. The evolutive processes depend on the interaction among three elements: general mechanisms which act as *constraints*; the variety, individuality and contingency of *events*; strategies and choices of *actors* who move between constraints and events in order to build new scenarios and new possibilities (Bocchi, 1985).

References

Allen P.M., Sanglier M. (1979) A Dynamic Model of Growth in a Central Place System, *Geographical Analysis, 11*, 256-272.

Allen P.M., Sanglier M. (1981) Urban Evolution, Self-Organisation and Decision-Making, *Environment and Planning A, 13*, 167-183.

Atlan H. (1983) L'émergence du nouveau et du sens, in Dumouchel P., Dupuy J.P. (eds.) *L'auto-organisation. De la physique au politique*, Editions du Seuil, Paris.

Barbera G., Butera F.M. (1992) Diffusion of Innovative Agricultural Production Systems for Sustainable Development of Small Islands: A Methodological Approach Based on the Science of Complexity, *Environmental Management, 16*, 667-679.

Bocchi G. (1985) *Dal paradigma di Pangloss al pluralismo evolutivo: la costruzione del futuro nei sistemi umani*, Feltrinelli, Milan.

Ceruti M. (1985) *La hybris dell'onniscienza e la sfida della complessità*, Feltrinelli, Milan.

Haken H. (1983) *Synergetics. An Introduction*, Springer-Verlag, Berlin.

Harris B, Wilson A.G. (1978) Equilibrium Values and Dynamics of Attractiveness Terms in Production-constrained Spatial Interaction Models, *Environment and Planning A, 10*, 851-862.

Jantsch E. (1980) *The Self-Organizing Universe*, Pergamon Press, Oxford.

Le Moigne J.L. (1984) *L'Intelligence de la Complexité*, GRASCE, CNRS, 640, Paris.

Lombardo S. (1986) New Developments of a Dynamic Urban Retail Model with Reference to Consumers Mobility and Costs for Developers, in Haining R., Griffith D.A. (eds.) *Transformation through Space and Time*, NATO ASI Series D, Nijhoff, 192-208.

Lombardo S. (1991) Recenti sviluppi nella modellistica urbana, in Bertuglia C.S., La Bella A. (eds.) *I sistemi urbani*, Angeli, Milan, 641-706.

Lombardo S. (1993a) Dalla nuova complessità urbana alla 'città intenzionale' in Proceedings of the International Seminar: *Per la città del XXI secolo. Un'enciclopedia e un progetto*, Giannini, Naples, 359-365.

Lombardo S. (1993b) Innovazione tecnologica e localizzazione, in Lombardo S., Preto G. (eds.) *Innovazione e trasformazioni della città*, Angeli, Milan, 95-129.

Lombardo S., Rabino G. (1983) Some Simulations of a Central Place Theory Model, *Sistemi Urbani, 5*, 315-331.

Lombardo S., Rabino G. (1984) Nonlinear Dynamic Models for Spatial Interaction: The Results of Some Empirical Experiments, *Papers of the Regional Science Association, 55*, 83-101.

Lombardo S., Rabino G. (1989) Urban Structures, Dynamic Modeling and Clustering, in: Hauer J., Timmermann H., Wrigley N. (eds.) *Urban Dynamics and Spatial Choice Behaviour*, Kluwer, Dordrecht, 203-217.

Morin E. (1977) *La méthode. I. La nature de la nature*, Editions du Seuil, Paris.

Nicolis G., Prigogine I. (1977) *Self-organization in non Equilibrium Systems,* John Wiley and Sons, New York.
Nicolis G., Prigogine I. (1987) *Exploring Complexity. An Introduction,* Piper, Munich.
Pumain D., Saint-Julien T., Sanders L. (1984) Dynamics of Spatial Structure in French Urban Agglomerations, *Papers of the Regional Science Association, 55*, 71-82.
Von Foerster H. (1981) *Observing Systems*, Intersystems Publications, Seaside.
Wilson A.G. (1981) *Catastrophe Theory and Bifurcation*, Croom Helm, London.
Wilson E.O. (1975) *Sociobiology, the New Synthesis*, Harvard University Press, Cambridge.

SESSION 3: THE PLANNING OF THE CITY

18. The Nexus Between Analysis and Design: A Case Study of an Uneasy Relationship[1]

Andreas Faludi

18.1 Introduction

A well-known definition of planning is that proposed by Friedmann (1993), who called it "a professional practice that specifically seeks to connect forms of knowledge with forms of action in the public domain" (p. 482). So planning thought must focus on the relationship between research and design, on the knowledge-action nexus. How is this nexus achieved? What are the methodological pitfalls involved? This work is about how the nexus has been conceptualised by Dutch planning.

The present author is not, however, merely an observer of the Dutch scene. Rather, he is a participant with a stake in Dutch planning thought who, at the same time, takes part in international debates. Capitalising on this dual position, this work seeks to interpret Dutch planning against the backcloth of international planning thought. However, it is also a personal account of the author's struggles to come to terms with the knowledge-action nexus as the key problem of planning thought.

Maybe this is just a reflection of personal development, but in the opinion of the author the conceptualisation of the knowledge-action nexus seems have been a three-stage process. The classic period in planning was characterised by the fact that the nexus was ignored. The problem was to obtain knowledge, not how to translate it into action, which was deemed unproblematic. Along with others, the author has made it his business to criticise this last assumption. This has resulted in the so-called 'procedural

[1] This work draws on a study, coauthored with Arnold van der Valk, on 'Rule and Order: Dutch Planning Doctrine in the Twentieth Century'. Unless otherwise indicated, translations from Dutch are by the author. References to Dutch sources have been relegated to the notes.

planning theory' (Faludi 1984, 1st edition 1973) Below, the author's present 'decision-centred view of planning', responding as it does to Dutch debates, will also be discussed.

Although a focus on the knowledge-action nexus is the hallmark of modern planning thought, this is not where the story ends. The concern for how the gap between analysis and design can be bridged has been augmented by attention to how planning issues are being formulated. Much current planning thought is, indeed, about how agendas are formed, how planning discourses become dominant, how the basic paradigms of planning are established. There is much to recommend the reflective quality of the largely academic research involved. Invoking a much-used term from the social science literature on planning and design, this paper invokes the notion of framing (Rein and Schön 1986, Schön and Rein 1994). Attention to how the knowledge-action nexus is being framed is a precondition of planning becoming self-conscious and self-critical, and ultimately also of the long-term sustainability of planning as a professional activity.

The paper not only takes the reader through the three-stage process, considering each period - from ignoring the nexus, to focusing on the nexus, then attention to the way in which it is framed - it also seeks to trace the interrelations between the professionalisation of planning and the development of planning as a discipline. After all, a profession assumes the existence of a discipline, and the discipline of planning has a clear purpose: to serve the interests of planning in its own independent and critical way, as a loyal, but staunchly independent friend (i.e. planning not as perceived by its practitioners, but by academic planning thinkers).

The Dutch have a name for this academic discipline, 'planologie', which Needham (1988) anglicises as planology. More important than the term as such is what it stands for: a social science approach to planning. However, engineers and not social scientists were the first to advocate scientific planning. They were first joined and later replaced by geographers. Geographers were to provide the knowledge base[1]. In fact they regarded planning as applied geography. In the paper quoted above, Needham concedes that, like geography, planology is indeed concerned with the relation between society and space. However, planology differs from geography in being evaluative, optimising, and practice-oriented. As we shall see, at present there is a remarkable consensus amongst Dutch

[1] Van der Valk A.J. (1982) Opleiding in opbouw: Een geschiednis van het Planologisch en Demografisch Instituut, Institute of Planning and Demography, University of Amsterdam, p. 69.

academics about the nature of the discipline. The consensus does not, however, extend to urban designers. Leading designers continue to insist on planning being a matter for experts with a specific gift for the 'creative leap'. Apparently oblivious to the lively debate on design methodology and the underlying epistemological assumptions, these designers are still firmly rooted in the classic period when the knowledge-action nexus was ignored. So planning in the Netherlands is characterised by a bifurcation of the field into planology and urban design (Faludi and De Ruijter, 1985). Albrechts notes that the European profession is similarly divided, with north-western Europe inclining more towards what we describe here as planology, and southern, central and eastern Europe being wedded to a design approach[1].

The term planology (which, with the exception of recent attempts in Italy by Franco Archibugi[2] to introduce it into the international debate, is not used by planners outside the Netherlands) was probably coined by the now largely forgotten pioneer J. de Casseres in lectures at the University of Utrecht under the title "*Foundations of Planology*"[3]. An authoritative academic committee has described planology as involving "scientific and methodological reflection on spatial ordering and planning, forming - on the basis of empirical research - descriptive, explanatory and normative theories"[4]. So, for planology to fulfill its mission, it must critically reflect upon practice.

---

1 Albrechts L. (1994) Plannen in en voor Europa, in Zonneveld W., D'hondt F. (eds.) Europese ruimtelijke ordening: Impressies en visies vanuit Vlaanderen en Nederland, Vlaamse Federatie voor Planologie, Nederlands Instituut voor Ruimtelijke Ordening en Volkshuisvesting, Gent, The Hague, p. 138.

2 See Archibugi F. (1994) Verso uno nuova disciplina della pianificazione, in Archibugi F., Bisogno P. (eds.) Per una teoria della pianificazione, *Prometheus*, 16/17 (special issue) pp. 40-42.

3 De Casseres J. (1929) Grondslagen der Planologie, *De Gids, 93*, 376-394. A 1944 book by P. and F. Bakker Schut is the first one under the title 'Planologie'. See: Bakker Schut P., and Bakker Schut F. (1944) *Planologie, van uitbreidingsplan over streekplan naar nationaal plan*, Noorduijn, Gorinchem. By that time provincial planning agencies had been established under the flag of 'Provinciale Planologische Dienst'. Purists have been complaining ever since about the misuse of the adjective describing practice institutions rather academic pursuits. See: Van den Berg G.J. (1981) *Inleiding tot de planologie: voor iedereen een plaats onder de zon?*, Samsom, Alphen aan den Rijn, Brussels.

4 Sectie Planologie en Stedebouwkunde i.o. (1972) Advies inzake de taakverdeling bij het wetenschappelijk onderwijs in de planologie; translation B. Needham.

18.2 The Classic Period: Ignoring the Nexus

18.2.1 Introduction

The dominant idea during this period, not only in the Netherlands, but also in the United States and Britain, was that plans needed to be based on scientific evidence (Faludi, 1987). Once made, they needed to be implemented without much ado. There was little questioning of the underlying ideas. Obstacles to scientific planning were seen as temporary. First we discuss professionalisation and then disciplinary developments.

18.2.2 Professionalisation

At the root of the formation of a Dutch professional body of sorts, the Netherlands Institute for Housing and Physical Planning (presently known by its Dutch acronym as NIROV) was the feeling that the Housing Act of 1901 (one of the first pieces of legislation in Europe that foresaw statutory planning) was falling short of expectations. Roots in housing and sanitary reform are a common feature of planning, though more so on this side of the Atlantic than in the United States. These issues were discussed by politicians and experts, including economists, lawyers and municipal engineers. Architects, about whose role more below, were still absent. Examples discussed came mainly from Germany, and the term used to describe the field, 'stedenbouw', echoed the German 'Städtebau'. Its present form, 'stedebouw', connotes a design rather than a social science orientation. In the old days, stedebouw was more or less the equivalent of town planning.

It was a German debate, too, which sparked off the first dispute about professional domains[1]. The debate was about whether planning was a science or an art, with implications for the interface between analysis and design. Municipal engineers sought to systematically establish the need for various facilities, expressing them in the form of standards, such as that for beams used in constructing bridges. To engineers, being scientific meant applying well-established standards.

The engineering pioneers came from Delft University of Technology, where J.H. Valckenier Kips had drawn their attention to the writings of German engineers such as Reinhardt Baumeister, Joseph Stübben and A.E. Brinckmann, but also the architects' hero, Camillo Sitte.

---

[1] Fehl G. (1980) Stadtbaukunst contra Stadtplanung, *Stadtbauwelt, 65*, 451-462.

These writers had in common an emphasis on standards, for instance as regards the width of roads, differentiated according to function, the number of rooms per household and so forth. The concern of Valckenier Kips extended to the type, extent and location of facilities, including open space. The calculation of social need expressed in terms of requirements for the provision of facilities was the basis of plan-making. Students at Delft, like Theo Karel van Lohuizen (to be discussed), were amongst the first to perceive this as 'social engineering'. Of course, standards were a common concern in the planning literature at the time. For instance, the C.I.A.M. movement (Congrès International de l'Architecture Moderne) devoted much attention to the establishment of standards. A loose-leaf book with planning standards is still an important source for Dutch planners.

Architects did not dispute the role of engineers nor of standards, but they claimed the final synthesis for themselves. The issue of who is to be in charge remains a constant in debates about professional domains. It reflects a notion of the nature of design which is still common, in the Netherlands as elsewhere. In the wake of Sitte's famous 'Town Planning According to its Artistic Principles' (Collins and Craseman-Collins, 1965), architects under the leadership of Hendrik Petrus Berlage defined planning as an art. They demanded more attention to aesthetics, monumental design and amenity, subordinating social problems (then the domain of the engineers) to three-dimensional form, this being another constant in Dutch debates. The intuition and vision of the master designer were seen as crucial. Whether and how to use research findings was for him to decide. If we are to believe the architect-planner Grandpré Molière, then his information needs were strictly circumscribed. The designer was almost super-human: "For his work he has all necessary knowledge; his technique is flawless, his patience endless, and his spontaneity never lets him down"[1] .

Engineers did not consider themselves competent to deal with aesthetics, nor did they have a method for synthesising the various elements of a plan. Like architects, engineers too, regarded this step as a creative leap. As we shall see, they shared with survey researchers the inability to say much about synthesis. In fact, the internationally renowned proponent of survey research, Patrick Geddes, did not say much about this last step either.

[1] This enthusiastic description of the role of the designer was noted by one of his students; see: Grandpré Molière M.J. (1949) *Woorden en werken van prof.ir. Grandpré Molière (bijeengebracht door zijn vrienden en leerlingen)*, De Toorts, Heemstede, p. 31.

Geddes' museum at Edinburgh, the Outlook Tower, included a bare room, called the Inlook Tower, in which visitors were invited to contemplate what they had seen and draw conclusions accordingly (Boardman, 1978). From this we may infer that Geddes, too, thought synthesis to be an achievement of a 'sound mind'. Indeed, one of his disciples and often his mouthpiece, Victor Brandford, writes about this vital step: "Having made his civic survey, the student retires, let us say, into his meditative cell. He takes with him a carefully built store of mental imagery... of the given city and its inhabitants as evolving towards definite ideals or degenerating towards their negation... (then) the student of sociology re-emerges into the world as civic statesman... The man of action is getting ready, with a programme and policy" (Brandford, 1914, p. 343).

Thus, Dutch planners were not alone in ignoring the issue involved in the nexus between analysis and design. Synthesis as a personal quality complemented the self-image of designers. Groups in the field, other than designers, were the housing and sanitary reformers. Eventually, there was broad agreement about setting up the Netherlands Institute for Housing and Physical Planning, affectionately called 'the Institute'. The Institute aimed to promote housing in the spirit of the Housing Act, and to promote good planning[1]. The ingredients of classic planning thought are present in this programme. They have an empiricist orientation, with the idea of the planner in the role of an umpire impartially weighing various claims on land against each other. Planning is seen as requiring coordination, cooperation and team work, but the final synthesis is considered an art rather than a science and as such the preserve of designers.

The three largest cities each hired survey researchers[2]. Most practitioners were architects though. Where the engineers worried about

---

[1] A study by De Ruijter P. (1987) *Voor volkshuisvesting enstedebouw: Voorgeschiedenis, oprichting en programma van het Nederlands Instituut voor Volkshuisveting en Stedebouw 1850-1940*, Faludi A., Van der Valk A.J. and Van Kesteren G. (eds.), Uitgeverij Matrijs, Utrecht, has reconstructed the programme binding this coalition together. The reconstructed programme comprises two manifests ascribed to the founding director of the Institute, Dirk Hudig. The housing manifesto is about preventing unsanitary conditions and promoting adequate provisions, the opposite of nineteenth-century disorder. It also addresses aesthetics and the role of architects. The planning manifesto stipulates as a key requirement of the classic period that plans must be based on scientific evidence.

[2] After several years at Rotterdam, Th.K. van Lohuizen went to Amsterdam. L.H.J. Angenot, another engineer, succeeded him at Rotterdam, and W.B. Kloos, an architect by training, but a keen researcher, went to The Hague. He would write a Ph.D. on the national survey and join the Government Service for the National Plan in 1941 (see Bosma, 1990).

facts and underlying trends, architects had a knack for synthesising various concerns into one overall plan. Undeniably, all design, including architectural design, requires the formulation of some kind of ordering principle (Cross, 1986) shaping the solution to the problem, and, indeed, designers may be more attuned to this than social analysts. Intuitively, architect planners were also able to anticipate societal trends, so it is little wonder that, apart from the handful of cities which could afford to employ specialist researchers, architects were dominant.

After the Second World War, opportunities for engineers in reconstruction were plentiful. To their disappointment, pioneering engineer-planners found it difficult to lure graduates into following them in their quest for 'scientific' planning by means of surveys. They had to turn to geographers instead. Already before the war, some geographers had gone into survey research. After the war, there were more openings. As we shall see, in due course, geographers would take over from engineers as the standard bearers of scientific planning.

Architect planners formed an Association of Dutch Urban Designers (Bond van Nederlandse Stedebouwkundigen) to look after professional matters like fees. A link was forged with the Netherlands Institute. Professional matters were, however, the exclusive domain of the Association. The Institute has always been a platform for those concerned with housing and planning, and a pressure group watching over policy, similar to the Town and Country Planning Association in Britain, rather than a professional association like the Royal Town Planning Institute. Concurrently with urban designers, researchers set up a Study Circle of Planning Researchers (Studiekring Planologisch Onderzoekers). As the name suggests, this was not a professional body either. Membership was by co-optation and reserved to research directors and the like. Setting up separate associations, designers and survey researchers began to part company.

18.2.3 The Discipline

During the classic period, the view of planning was squarely 'survey-before-plan'. German textbooks had detailed the surveys needed. Surveys came naturally to urban designers. Even Sitte, Berlage's source of inspiration in his artistic view of planning, had advocated surveys. The British influence came later, at the 1924 Amsterdam conference of the International Garden Cities and Town Planning Association, where Patrick Abercrombie talked about regional surveys. Until the fifties, the term 'survey' was common, indicating the extent to which practice reflected foreign influences.

Initially, surveys focused on specific aspects, like housing, but increasingly they were becoming comprehensive. A key person was Van Lohuizen, famous for his role in the survey for the 1934 General Extension Plan of Amsterdam[1]. Before going to Amsterdam, Van Lohuizen had prepared a plan for Rotterdam and its environs. Behind this plan, Van der Valk discerns a fully-fledged programme for the emerging discipline of planology[2]. Concerning, as it does (albeit implicitly), the role of knowledge in design and policy-making, and methods and techniques of applied social science research, this programme is relevant to the topic of this work, analysis and design in planning. The following are the main elements.

1. Planning must be based on evidence. Urban development is never a matter solely of intuition. However, in line with the historic compromise at the foundation of the Institute, designers retain ultimate responsibility for the final product.
2. Planning is not merely town extension. To combat nineteenth-century disorder, and taking a leaf out of the book of Taylor on management, Van Lohuizen demanded a 'scientifically based organisation plan', reflecting the dynamics of urban development.
3. Studying cities gives insight into functional interdependence. The slogan, derived from the 'Regional Plan for New York and Its Environs' by Thomas Adams which Dutch planners keenly read, was: for everything a place, everything in its place! This implied an organic view of communities. Surveys were to cast light on relations between the parts and the whole. These relations were expressed through standards. By means of surveys, the analyst penetrated the 'essence' of the city. Urban form had to give physical expression to this essence.
4. The key to urban growth was industrial development. Industry attracted people and generated traffic, so both research and policy had to focus on the distribution of industry.

Van Lohuizen developed effective research tools. His surveys took the

[1] Van Lohuizen's design counterpart, Cornelis van Eesteren, was then president of C.I.A.M., so there are links with 'functionalist' planning and the Athens Charter; see Giedion, 1967 (1st edition 1941).

[2] Van der Valk A.J (1990) Het levenswerk van Th.K. van Lohuizen 1890-1956: De eenheid van het stedebouwkundig werk, Delftse Universiteire Pers, Delft.

form of maps, diagrams, photographs and reports. He made trend extrapolations to identify future needs. Van Lohuizen and his like were well aware of the fact that the selection of topics was necessarily somewhat arbitrary. Being engineers, they were sympathetic to designers and their need for timely information. In selecting topics for their surveys, they felt no compunction about following design leads (Faludi and De Ruijter, 1985). Indeed, Van Lohuizen himself saw good survey research as a creative task, giving a synthetic view of the real world[1]. So, in fact, the positivism of the pioneers was mediated by a dose of pragmatism, suggesting a more interactive procedure than any purist would condone. However, what is crucial is that the pioneers regarded synthesis as a personal thing. They felt no need to explain the methods used. In reality, they were ignoring the knowledge-action nexus, thus discouraging critical debate on this vital step in planning. Critical debate depends on the use of methods, especially for making the knowledge-action nexus (Faludi, 1987).

Surveys were expensive and could rarely be mounted. By pressing for provincial and national planning, planners believed it was possible to assemble the funds and expertise for proper surveys[2]. As indicated, these surveys increasingly involved a new breed of experts, geographers. Initially, the influx of geographers failed to generate discussion about the discipline. Geographers had a tradition of making regional monographs. Based on detailed surveys, these monographs merely enforced the engineering ethos of exactitude and comprehensiveness. Also, geographers were content with the division of labour represented by 'survey-before-plan'. Over time though, the situation polarised, much as it did elsewhere (Glass, 1959, Broady, 1968). Designers held on to their view that planning was synthetic and that it had to be suffused with 'vision', a property of the individual designer, although a team was needed to implement it. The team shared responsibility for the outcome, but that is as far as designers were willing to go[3].

Geographers started getting edgy: they felt that designers were unaccountable to the public, that the intuitive approach to social problems was inadequate, and the underlying research merely instrumental. Designers paid too much attention to the end-product and not enough to alternatives, in effect imposing their own, albeit implicit views. That such

[1] Van Lohuizen Th.K. (1948) 'De eenheid van het stedebouwkundig werk', inaugural lecture, Delft University of Technology.

[2] Another reason was that planners distrusted small municipalities.

[3] Van der Valk A.J., Het levenswerk, p. 122.

criticisms failed to make their mark was due in part to the fact that geographers themselves were split over the knowledge-action nexus. There were those who (unlike Van Lohuizen and other engineers turned researchers) took survey-before-plan literally. Before contemplating design or policy implications, they wanted to have complete insight into the situation. They were unadulterated positivists in the classic sense and, as such, out of step with state-of-the-art thinking on logical positivism and planning (represented, for instance, by Otto Neurath, founder member of the Vienna Circle, who had worked in the Netherlands before the war, Faludi, 1989b). When they were at all concerned with synthesising survey findings, they relied, not unlike designers, on intuition. This was evident in a leading geographer's discussion of the regional monographs in which most survey research culminated. The author of such a monograph has to make a selection from the assembled facts, emphasising some and neglecting others. Nowadays, this is common fare. Even systems analysts recognise that "like beauty, a system lies in the eye of the beholder, for we can define a system in an infinite number of ways in accordance with our interests and our purpose..." (Chadwick, 1971, p. 42). Descriptions, too, evolved around central themes which do not emerge from the analysis, but are chosen. However, like designers, rather than dwelling on the methods to be used in making such choices, this study stresses that synthesis depends on personal qualities. Again like designers, Chadwick seems to think that this last and most important step of forming insights into the nature of regions is therefore not subject to the canons of science[1].

With such scant attention to the role of scientific method in the business of synthesis, survey researchers had little incentive, and in fact also little justification, to challenge designers and their privileged position in planning, conceived as design involving creative leaps. Being on the whole more junior than the designers, they were not in a position to do so anyhow, so this group acquiesced in the situation as helpmates to master designers.

The second group of geographers was far less concerned with surveys. In their view, geography did not pay enough attention to theory. Some of them were setting themselves up as sociologists[2]. Within the mainstream of academic geography, too, modern concepts and approaches were in the

---

[1] Den Hollander A.N.J. (1968) *Visie en verwoording: Sociologische essays over het eigene en het andere*, Van Gorcum, Assen.

[2] Van Doorn J.A.A. (1964) Beeld en betekenis van de Nederlandse sociologie, Utrecht; Van Paassen C. (1982) Het begin van 75 jaar sociale geografie in Nederland, Institute for Social Geography, University of Amsterdam, Amsterdam.

ascendancy. Eventually, Bours and Lambooy edited a reader on these approaches, with articles dating from 1955-1968[1]. For planning, the most important of the concepts discussed was that of the city-region[2]. Upon his appointment at the Catholic University of Nijmegen, one of the authors, Gerrit Wissink, was to focus on the city-region concept, applying it to Nijmegen.

Applied geography is strong in the Netherlands, so much so that the existence of planology as a separate discipline seems miraculous. Planology focuses on bridging the gap between knowledge and action. In so doing, it is more attuned to the engineering tradition than geography. In the beginning, however, the distinction between geography and planology was by no means obvious. Van Lohuizen had adopted the best of research techniques, and he himself was regarded as quite a pioneer of regional population forecasts[3]. In 1940, he was giving courses on survey research to geographers at the University of Amsterdam. In 1948 he got a chair in survey research at Delft University of Technology.

It is fair to say that, at the end of the classic period, the discipline was suffering from a raging intellectual crisis. It gave no guidance as regards the step from analysis to design. Ignoring the nexus between knowledge and action meant leaving this vital step to the whims of visionary master designers. The only recipe of the classic programme, to do ever more research, was impractical. Nor did geographers have the answer. They were absorbed in the quantitative revolution. They did not give much thought to the step from analysis to design either.

---

[1] Bours A. and Lambooy J.G. (eds.), (1974) *Stad en stadsgewest in de ruimtelijke ordening*, Van Gorcum, Assen, p. 2.

[2] A text book by Willem Steigenga, the first incumbent of the Amsterdam planning chair, was another compendium of state-of-the-art geography applied to planning. Steigenga W. (1968, 1st edition 1964) *Moderne planologie*, Aula, Utrecht. His truly revolutionary focus on the nexus between knowledge and action, defining it as a constructive task, culminating in political decision-making, was not reflected in this book, but in a paper to be discussed below.

[3] Van den Berg G.J. (1991) Over de betekenis van het toegepaste geografisch onderzoek voor het bestuur (afdruk van een voordracht gehouden in 1960), in Tussen ontwerp en bestuur - Prof. drs. G.J. van den Berg: Veertig jaar denken over planologie (Publicatie 12 van de Werkgroep PSVA ingesteld door het NIROV), The Hague, pp. 73-99.

18.3 The Modern Period: Focus on the Nexus

18.3.1 Introduction

Gradually, the view of planning surpassed that of the mere application of knowledge. Some geographers were interested in breaking through this barrier. The classic view of planning cast them in the role of handmaidens, refusing to concede that plan-making proper was the prerogative of designers. Geographers were the ones to bring about the development of planology as a separate discipline. Focussing on the nexus between knowledge and action, they filled the void which Van Lohuizen and his followers had left gaping.

The modern period can be divided into two parts: the heyday of planning during the sixties and seventies, coinciding, as it did, with the rise of the systems approach to planning in the United States and Britain, and the crisis of planning and the response to it during the eighties and nineties. During the heyday, there was optimism as regards state intervention. As before, experts were to provide the scientific base for action. Planners put the systems approach to work, producing increasingly comprehensive national planning reports, achieving more or less total coverage of the country with provincial structure plans and a successful growth-management package providing alternatives to suburban development in so-called 'growth centres' (Faludi, 1992, 1994, Faludi and Van der Valk, 1991, Van der Valk and Faludi, 1992). Unfortunately, since the crisis of the early eighties, none of these achievements has counted for much. A redefinition of planning is in progress, focussing the minds of planners on new issues and increasingly divorcing them from the welfare state and its achievements. Like elsewhere, a project-le approach wedded to boosterism is the order of the day, with waterfront developments emulating Baltimore and Boston stealing the show.

Paradoxically, as regards planning thought, the situation may be said to be the reverse. During the sixties and seventies, planning thought correctly defined the main issue as that of establishing a nexus between knowledge and action. The discipline, however, was still far from appreciating all the implications, in fact it still had technocratic leanings. Since then, academic planning thought has been going from strength to strength. We now examine this aspect, discussing as before firstly professionalisation, then making an analysis of the discipline.

18.3.2 Professionalisation

There were no momentous developments during the heyday of planning. Planning activities expanded, so there were enough jobs to go round, taking the sting out of debates about the role of the profession and its theoretical underpinnings. A new breed of social-science based planners calling themselves planologists went to work in strategic planning, pursuing the systematic approaches considered the hallmark of modern planning[1]. Initially, they were geographers styling themselves as planners. As time went by, new graduates in planology, which since 1972 has been offered as an academic degree, entered the scene.

The membership of professional and quasi-professional bodies increased. The study circle of researchers, now a section of the Netherlands Institute, widened its appeal to include people other than survey researchers. New sections covered planning law and housing. All this went on without much upheaval. It was the crisis of the eighties that brought the rivalry between designers and planologists out into the open. It is a rivalry keenly felt in other countries, too. Designers started the ball rolling[2]. One of them criticised the growing social science influence in planning at the expense of design. Misrepresenting the social science input as surveys of consumer satisfaction, he claimed: "You cannot design a city by describing how it is being appreciated, this is like a cookbook consisting of descriptions of the taste of dishes instead of the recipes. Recipes are important, also in urban design (and architecture)"[3]. So urban design needed visual techniques. The task of the urban designer was to generate an urban morphology capable of absorbing meanings, values and functions. Reacting as they did to the growing role of planologists, not

---

1 The National Physical Planning Agency took the lead, publishing an important paper devoted to methodology. See: Dekker A. (ed.) (1975) Planningmethodiek: Eerste deel van de reeks algemeen ruimtelijk planningkader, Ministry of Housing and Physical Planning, The Hague. In terms of numbers, provincial planning agencies between them were more significant. Taking a leaf out of books like Chadwick (1978, 1st edition 1971), and in particular McLoughlin (1969), planners were beginning to practice their home-grown version of the systems approach. Cities and regions being complex systems was a point well appreciated by planners.

2. De Ranitz J., Wissing W. (1978) Wat is, wat kan, wat doet een stedebouwkundige, ***Stedebouw en Volkshuisvesting, 59***, 221-229; Wissing W. (1981) Heeft de stedebouw nog ruimte?, ***Stedebouw en Volkshuisvesting, 62***, 41-43; Zandvoort F. (1980) Denkraam: Duidelijkheid gewenst, ***Stedebouw en Volkshuisvesting, 61***, 628; Zandvoort F. (1981) Denkraam: Poging tot afbakening, ***Stedebouw en Volkshuisvesting, 62***, 254.

3 Weeber C. (1979) Formele objectiviteit in stedebouw en architectuur als onderdeel van rationele planning, *Plan, 11*, p. 32.

only in research, but also in public participation and in policy-making generally, designers had gone on the offensive[1] .

Nobody spoke out against the arrogation by designers of a central role in planning. Designers also got the ear of politicians. As elsewhere, grand projects became fashionable and international architects were given important commissions. A privately initiated exhibition featuring scenarios for 2050 got much acclaim[2] . Design, in particular on the regional level, continues to be the focus of much attention.

Professional associations took a back seat in these debates. In the Netherlands, they are not the gatekeepers to practice[3] . The professionalisation of planology occurred in a roundabout way. On the international scene urban designers had been monopolising representation of Dutch 'town planners' for quite some time. In so doing, they were helped by the ambiguity of English translations. 'Town planner' can refer to both the planologist as well as the urban designer. So through a system of co-optation, designers were dominating the delegation to the International Society of City and Regional Planners (ISOCARP). Through their international contacts, they became aware of a pending directive from Brussels concerning mutual recognition of qualifications. The Association of Dutch Urban Designers subsequently became a founder member of the European Council of Town Planners, set up in response to this initiative.

Planologists had no answer. Traditionally, the Institute had left the

---

[1] The planning journal 'Stedebouw en Volkshuisvesting' staged a crisis debate in which designers gave planners a whacking. Planners lacked vision, and their disciplinary base was weak. Planners were 'moral theologians'. They were criticized for their pretension of policy being based on scientific analyses. Planning needed imagination combined with the use of 'formal-critical methods'. Scorning intuition, planning had become a paper-shuffling exercise. Constant criticism was deadly for creativity. Taking their cues from politicians, planologists were opportunists, anyhow. Little wonder, therefore, that they were cynical. They knew full well that research did not produce objective knowledge, and that rules were not obeyed. (Presumably, by implication, designers were maintaining high professional standards in the face of political pressure, the common term for this being technocracy.) See Heeling J. (1985) Planologen: de moraaltheologen van Nederland, *Stedebouw en Volkshuisvesting, 66,* 109-111.

[2] Van der Cammen H. (ed.) (1987) Nieuw Nederland: Onderwerpvan ontwerp - Boek I: Achtergronden; Boek II: Beeldverhalen, Government Printing Office, The Hague.

[3] However, the urban designers were hitchhiking quietly on the backs of architects to a sort of recognition. Legislation passed in 1987 concerns the use of the title of architect and urban designer. The legislation is based on the pre-1960 view of the researcher as a helpmate, and the designer as responsible for synthesizing information. The legislation takes note neither of practice, where planologists are holding senior positions in policy-making, nor of the university courses in planology which, since the seventies, have been preparing graduates for policy-oriented roles similar to those of urban designers.

representation of professional interests to the Association of Dutch Urban Designers with which it enjoyed a congenial relationship. The forum for planologists was the Section for Planning Research (Sectie Planologisch Onderzoekers), the successor to the 'study circle'[1]. As the Section was not a professional body, universities concerned about employment opportunities for their graduates felt compelled to make a move. They acted as the midwives of an Association of Dutch Planologists of which current membership (as of 1995) is over two hundred[2]. The background to this was the setting up, in 1982, of fully-fledged four-year planning courses, more or less on the lines of the British example. At least in academia, planning has gained recognition and consequently created opportunities. There is now a core, albeit small, of people with a commitment to planology. In the end, the Netherlands Institute was supportive of the initiative to set up an Association of Dutch Planologists, and there was acceptance, though grudging, by designers as well. Traditional tolerance for diversity had prevailed. Designers made room for planologists on the European Council of Town Planning, and a federation of the two associations was formed in 1995.

18.3.3 The Discipline

The focus on the step from knowledge to action reflects criticisms of the classic approach with its emphasis on surveys and the concomitant neglect of the knowledge-action nexus. The high point of the classic approach was the Second National Physical Planning Report. The research leading up to this 1966 report was conducted under the responsibility of a disciple of Van Lohuizen[3]. Academic critics attacked the report as a blueprint, a

---

[1] The Section had broadened its appeal to all those concerned with planning as such, and not just research. A change of name into Section for Physical Planning ran into opposition even so. The urban designers represented on the Institute's Council were suspicious, and so were some researchers. At that time, planning research was threatened with cuts, and rescinding research from the title of the Section seemed inopportune. The compromise was to change the name into the hybrid 'Section for Physical Planning and Research' (Sectie Ruimtelijke Planning en Onderzoek). Meanwhile, what had become clear was that the Section did not stand up for planology as a profession.

[2] Universities also took an active hand in setting up the Association of European Schools of Planning, or AESOP, the inaugural conference of which was held at Amsterdam in 1987.

[3] His report introduced the notion of 'concentrated deconcentration', shorthand for the strategy of channeling suburban development into designated so-called growth centres. Not surprisingly, the approach to formulating this policy reflected the classic Van

fashionable complaint at that time. Indeed, the report failed to recognise uncertainty regarding, amongst other things, population growth[1]. Another criticism, also a common theme, concerned the lack of opportunity for public participation.

Prominent amongst these critics was Steigenga, who since 1962 had held the Amsterdam chair. He observed that there were no alternatives[2]. Drawing on Foley (1964) and his adaptive (as against the unitary) approach, Steigenga pleaded for flexibility[3]. He took his cues not only from Foley, but from Karl Mannheim, Karl Popper, Melvin Webber, Yehezkel Dror, John Friedmann, Ruth Glass and Paul Davidoff and Thomas Reiner. "*Urban Land Use Planning*" (Chapin, 1965) figured prominently on his reading lists. Steigenga was fascinated by the constructive aspect of planning. A Labour member of the Provincial Legislature, first of South and later of North Holland, he was concerned with rendering political decisions systematic. This was different from the designer emphasis on creativity and intuition. A programmatic paper on 'Social science research and physical planning', written as early as 1956, marks him as a representative of modern planning thought concerned with the organisation and procedures, but above all with the methodology of planning-as-decision-making. Having been a visiting professor at the University of Minnesota in 1954, he was well aware of the latest international developments. In this paper, he defined planning as "the sum total of decisions aiming to create the conditions for a particular type of

---

Lohuizen programme. The analytic concepts invoked were more modern, however. They included central place theory, the rank-size rule, Pareto-distribution and gravity models. These were pieced together into an overall model resting on an analogy between urban development and living beings as 'complex systems'. In fact, this was a continuation of prewar ideas about the city and/or region as an organism. See: Nassuth G.A. (1991) 'De totstandkoming van concepten voor de Tweede Nota', paper given at a workshop of the Working Party for Planning Archives held at the Royal Dutch Academy of Sciences, Amsterdam, 12 June, p. 3.

1 The assumption was that of an exorbitant twenty million inhabitants by the year 2000 for which the Second Report calculated the need for housing and attendant facilities.

2 "Evidently the idea was, and indeed I have nothing but praise for this, that the public at large should be spared the trouble of having to spend too much of their spare time in deliberating, and thus a pre-selection has been made so that not only citizens, but in due course also parliamentarians discussing the budget can economize on their use of time and energy." (Steigenga W., 1973, first published 1966, Inleiding naar aanleiding van het verschijnen van de Tweede Nota Ruimtelijke Ordening, in Steigenga W., *Planologie in beweging: Verkenningen in planologie en demographie, 6*, Institute of Planning and Demography, University of Amsterdam, p. 165.)

3 Steigenga W. (1971) Inzake het planologische model, *Stedebouw en Volkshuisvesting, 52*, 382-386.

social development..."[1] This implied 'social engineering', a fitting task for social scientists.

With this, Steigenga put the cat amongst the pigeons! Defining planning as social engineering, he questioned the pre-eminence of designers. Rather than relying on the creative leap, he recommended the development of models of spatial structure to provide the basis for publicly accountable decision-making. So he went beyond the classic positivism of his predecessors who tried to derive policies straight from the facts.

Steigenga distinguished three phases in planning:

1. the 'descriptive' phase of formulating the problem, including an overall view of the area concerned;
2. the 'analytical' phase of searching for interrelations and underlying trends, involving theory (as in fundamental research);
3. the 'social engineering' phase of synthesising findings into conceptions of societal structure and form as the basis for designing spatial structure.

He called such conceptions planning models. Constructing them was not for technicians because "the technician, the engineer and the architect lack the conceptual apparatus needed to analyse and bring about new forms of society, and to understand the consequences of intervention and to know the effects of their creations"[2]. The role of technicians was that of translating planning models into three-dimensional form. Thus he distinguished between planning as policy-making, in which social science based planners took the lead, and three-dimensional design. About three-dimensional design as such, Steigenga equivocated. Sometimes he seemed to be arguing that social scientists and designers shared responsibility, sometimes that it was the mere translation of policy into concrete form.

Steigenga recognised the creative element in planning, but denied that it was a prerogative of designers. Instead, he traced the lineage of planning models to the Utopian tradition in the social sciences. He was careful, though, not to arrogate political choice to the social engineer. The social engineer was to explore degrees of freedom, "which make it worthwhile to consider various options which then become largely a matter of arbitrary political choice. It is then the role of social research to indicate the ultimate consequences of a certain decision. In this manner we can seek ways to

---

[1] Steigenga W. (1971) Het sociaal-wetenschappelijk onderzoek en de ruimtelijke planning, ***Stedebouw en Volkshuisvesting***, *37*, p. 106.

[2] Steigenga, Het sociaal-wetenschappelijk, p. 109-110.

improve the quality of these decisions, thus counterbalancing arbitrary choice"[1]. Thus, as against the classic thinkers, Steigenga drew a clear line between expert judgement and political choice.

The paper ends by referring to 'Man and Society in an Age of Reconstruction' (Mannheim, 1940) deriving arguments from it for a 'science of social planning' concerned with the application of the sciences in planning. This is of course what was soon to be termed planning theory (Faludi, 1973, Faludi, 1984, first published, 1973). There are other reasons, though, why the present author feels affinity with the work of Staging. He was not only a professed critical rationalist, but also committed to planning as a discipline and to planning education as distinct from education in the more analytical social science disciplines.

The second founding father of planology, Van den Berg, is important for his insistence on participation. In so doing, Van den Berg distanced himself from Van Lohuizen in the fifties[2]. Up until then, North Holland plans had emulated Van Lohuizen's masterpiece, the General Extension Plan for Amsterdam, in preceding design by surveys. In his capacity as planning researcher, Van den Berg proposed discussing fundamental choices before mounting surveys. As adviser to the team, Van Lohuizen was shocked. Research needed to be value-free! "We know now that no research is possible without a necessarily value-laden paradigm, but at that time this was a fashionable gimmick which Van Lohuizen, trained and experienced as he was in practising value-free science, strongly rejected"[3].

Van den Berg went further. Such choices required involvement of the public. Van Lohuizen acquiesced. His paradigm had reached its limits. Initially, consultations were perhaps more a ploy to obtain otherwise unavailable information[4]. However, consultations carried within them the seeds of a more open, democratic approach[5]. As a professor, Van den

---

[1] Steigenga, Het sociaal-wetenschappelijk, p. 111.

[2] In a frank paper on the occasion of Van Lohuizen's one-hundredth anniversary, Van den Berg looks back on this with disciplined emotion. See: Van den Berg G.J. (1991) Over de impuls die Theodor Karel van Lohuizen gaf aan de rol van onderzoek, kennis en reflectie bij ontwerp, planning en beleid, in *105 jaar onderzoek: Voordrachten naar aanleiding van het 100ste geboortejaar van Th.K. van Lohuizen*, EFL-Stichting/NIROV, The Hague, 13-20.

[3] Van den Berg G.J. (1991) Over de impuls, p. 17.

[4] Van der Heiden C.N. (1991) Herwaardering van de grote stad: Strategische ruimtelijke planning in Noord-Holland 1974-1987, Working Papers of the Institute of Planning and Demography, Nr. 139, University of Amsterdam, Amsterdam, p. 41.

[5] Wolff A. (1991) Tussen ontwerp en bestuur - Prof. drs. G.J. van den Berg: Veertig jaar denken over planologie (Publicatie 12 van de Werkgroep PSVA ingesteld door het NIROV), The Hague, p. 48.

Berg expanded upon this view of planning. Drawing on many international authors, he incorporated ideas about social mobilisation into his model[1]. Participation has become standard fare in Dutch planning. It has not delivered what many have hoped for: fundamental democracy. However, the procedures for conducting orderly debates around planning issues are in place, and the right to be listened to is something which action groups cherish.

The above view of planning as publicly accountable decision-making, though conforming to international state-of-the-art thinking, did not go unchallenged. In his inaugural lecture, the Delft Professor of Urban Design, S.J. van Embden, denied that form could be the outcome of team work, let alone democratic decision-making. Much as elsewhere, the standard joke in the Netherlands is that the camel is a horse designed by a committee. Van Embden insisted that designers could only interact meaningfully with other designers who have command of the language of form. That a nonverbal language is needed in planning is borne out by the international design literature (Faludi, forthcoming). There is, however, a twist in Van Embden's argument. He justifies not only the right of designers to their own private language, but also their claim to supremacy. After all, form is said to represent "the totality of the thing which is in need of making"[2].

What followed was non-communication. Designers continued to hold that theirs was a superior role, based on their command of the language of

---

[1] Van den Berg G.J. (1981) Inleiding tot de planologie.

[2] Van Embden S.J. (1964) *Vorm* (inaugural lecture), Delft University of Technology, Uitgeverij Waltman, Delft. 'Totality' is of course a difficult concept. Analyzing the debate, Launspach (a social scientist) sided mostly with Steigenga. However, in one crucial respect he followed Van Embden. He claimed that form indeed represented a 'totality' requiring a logic of its own which only the initiates could appreciate. Thus, Launspach seemed to support the claim of designers for ultimate responsibility in planning, arguing that the social scientist could fulfill his role only by accepting "...that form is one whole, only partially subject to scientific analysis..." See: Launspach J. (1967) 'Stedebouw en samenleving, materiaal voor een theorie van de stedebouw', Ph.D. thesis, VUGA, The Hague, p. 78. Steigenga furiously criticized Launspach for supporting the imperialism of Van Embden's design philosophy. Launspach made it seem as if form had no social function. Research was needed to uncover hidden ideologies behind design concepts. Steigenga unveiled one such ideology behind the assertion that inventiveness was the exclusive property of designers. This reminded him of authoritarian systems, "...where absoluteness of form, whether aesthetic or political, is dominant and where the freedom of the critical-analytic role of thought is limited." See: Steigenga W. (1968) Bespreking van J.A. Launspach, Stedebouw en samenleving, materiaal voor een theorie van de stedebouw, *Tijdschrift voor Sociale en Economische Geografie, 59*, p. 279.

form and their unique ability to grasp its essence. During the remainder of the sixties and seventies planologists put flesh to the bones of the ideas of the pioneers. Steigenga himself passed away in 1974. By that time, judging by his lecture notes, there must have been a text book in the making such as has never before seen the light of day in the Netherlands. The notes show Steigenga to have kept abreast with the international planning literature. They give the ingredients of a synthesis between geography and planning thought. Anyhow, by the time of his untimely death he had imparted his ideas to the first generation of planning graduates, some of whom are now in positions of authority.

The year after Steigenga passed away, one of these graduates, Van der Cammen, published a paper on the process approach. Like Steigenga, Van der Cammen drew on Foley (1964) relating the process approach to the entry of social scientists into the field. Van der Cammen showed the two sides to process planning: planning as cyclical process, and planning as a social process[1]. Another paper by Kreukels gave an overview of new methods, categorising them into formal and behavioural methods. Drawing on Friend and Jessop (1969) and Friend, Power and Yewlett (1974), Kreukels discussed the integration of planning theory with the behavioural sciences[2]. At Nijmegen University, similar themes were discussed[3]. There are frequent references to procedural planning theory, but soon the latter began to evoke unease[4]. Of course, Dutch planning academics also took part in the great Marxist-inspired debates of the seventies, but without making an original contribution.

As the discipline evolved, mechanistic ideas about rational planning came under scrutiny, the notion of planning as starting with goal-setting

---

[1] Planning as a social process is of course central to the idea of planning as an academic social-science discipline. Van der Cammen took note also of the demise of blueprint planning. There is also an inkling in his paper of his later concern with Utopian design. See: Van der Cammen H. (1975) De moderne ruimtelijke planning, een situatieschets, *Stedebouw en Volkshuisvesting, 55*, 462-474.

[2] Kreukels A.M.J. (1975) Stuurmethoden in de planning: Eenoverzicht met bijzondere aandacht voor de netwerkplanning en tegen de achtergrond van procesplanning, *Stedebouw en Volkshuisvesting, 55*, 276-286; Kreukels A.M.J. (1980) 'Planning en planningsproces', Ph.D. thesis, University of Utrecht, VUGA, The Hague.

[3] Ganzevles M.G.J., Van Genugten J.M.O., Linden G.J.J. (1975) Enige beschouwingen over procesplanning, *Stedebouw en Volkshuisvesting, 55*, 251-256; see also: Linden G. and Ganzevles T. (1993) Terugblik op het onderzoek Stadsgewest Nijmegen, in: Dekker A., Ekkers P., Ganzevles T., Muller N. (eds.) Gerrit Wissink: Dertig jaar universitaire planoloog, Nijmeegse Planologische Cahiers, 43, Department of Planology, Catholic University of Nijmegen, Nijmegen, 144-153.

[4] Van der Cammen H. (1979) 'De binnenkant van de planologie', Ph.D. thesis, University of Amsterdam, Coutinho, Muiderberg, pp. 174-184.

being one of them. The idea, emulated in many planning documents, was that planners should deduce alternatives from agreed goals and on that basis recommend the best course of action. Although impracticable, this deductive approach appealed to the taste for 'scientific' policy, whilst simultaneously holding the promise of politicians having a real say. The approach gained official approval[1]. The swing to a more pragmatic approach came in the late seventies. This 'strategic' approach focussed selectively on concrete problems rather than goals[2], another feature of Dutch planning reflecting international debate. For instance, there was a discussion in the Journal of the Royal Town Planning Institute on whether planning is a goal-seeking or a problem-solving activity (Needham, 1971, Faludi 1971, Needham and Faludi, 1973). The break with the deductive approach stemmed from the realisation of the limitations of human knowledge and information-handling capacity and the role of political choice in defining problems. Selectivity also reflected awareness that the scope of planning was more limited than had been hoped for. In fact, the problem-oriented approach was a tactical retreat preceding the more general crisis to be reported below. Anyhow, planning theorists had appreciated such problems for much longer (Friend and Jessop, 1969, Faludi, 1973, 1984, first published 1973, Friedmann, 1973). What people were reacting against was the caricature of systematic planning which over-enthusiastic practitioners had concocted from the literature, not the real thing. Although the vital need to attend to the nexus between knowledge and action had been recognised, planning thought was still far from having gained a secure footing.

What the eighties did witness was a cogent articulation of the procedural and the substantivist positions. Eventually this led to a new consensus between the two camps. The present author, having come to the Netherlands in 1974, was very much part of this. He recast procedural theory in the form of a decision-centred view of planning (see Faludi, 1973, 1984, first edition 1973). This view builds on the IOR-School (Institute for Operational Research), best known for its 'strategic choice approach' (Faludi and Mastop, 1982). According to the decision-centred view, planning stands for rendering ongoing decisions meaningful by

---

[1] This was so much the case that the first part of the Third Report on National Physical Planning, the so-called Orientation Report, published in 1973, culminated in an elaborate statement of the goals of national planning.

[2] Van der Cammen H. (1982) "Methodisch geleide planvorming (1)", *Stedebouw en Volkshuisvesting, 63*, p. 385. An example of this can be found in the so-called Urbanization Report, being Part 2 of the Third National Physical Planning Report, the first version of which was published in 1976.

analysing them in their wider context of choice. In the past pride of place has gone to plans, and decisions have been expected to follow. The decision-centred view puts the onus on planners to make plans relevant to ongoing decision-making. It follows that plan-making must take ongoing decisions as its point of departure, the more so since these decisions form the interface between public policies and the aspirations of private actors (Faludi, 1986a, 1987).

The focus on ongoing decisions relates to the problem of implementation, then as now a source of much concern. Existing literature defined implementation from the point of view of planners. It concentrated on the obstacles which hindered the process of rendering operational the 'beautiful ideas' enshrined in plans. Now, the issue was being defined in a radically different way. Ongoing decision-making was central, and the role of plans that of helping decision-makers, rather than being imposed on them. This has implications for plan evaluation, an aspect which has been thoroughly explored and has led to a distinct Dutch line of research concerned with the performance of plans in assisting with day-by-day decision-making (Faludi, 1989a, Alexander and Faludi 1989, Faludi and Korthals Altes, 1994). Seen in this way, departures from plans do not necessarily indicate failure. Instead of worrying about conformance between plans and the actions which follow, all that can - and must - be insisted upon is that all decisions of a public authority must be well considered. To this extent, the decision-centred view shares the spirit of 'scientific' planning. After all, the methodology of science, too, evolves around the making of well-considered statements about reality.

Accountability, so conceived, relates to a central issue in planning thought: rationality. There has been much discussion of this concept. Suffice it to say that one of the benefits of rational planning is that the reasoning underlying decisions and/or plans is made explicit, thereby rendering them accessible to public scrutiny. This is what accountability means. Rational decision-making presupposes, to invoke a technical term, clear 'definitions of decision situations' (Faludi, 1986a). Of course, defining decision situations is also a learning process, and this extends to social learning. The interactive element comes into it because planning decisions involve many actors. Whether they can form common definitions of the decision situation is crucial. This implies communication, but more is involved. Any honest attempt to do so involves finding out about other actors and their aspirations, which is what social learning implies. This is where social learning takes place, a theme which is strongly represented in the literature, especially in the works of Friedmann (1969, 1973, 1987).

Such social learning bears a crucial relation with the issue, particularly controversial in Dutch planning, of flexibility. It was raised by Thomas *et al.* (1978) in their comparison of Dutch planning with planning in England and Wales. Faludi (1986b, 1987) has extended the scope of comparison to zoning in the U.S. and Australia. Frequent departures from plans make one wonder about the difference between laudable flexibility and despicable opportunism. Now, finally, a cogent distinction can be drawn. Accordingly, flexibility is distinguished by the awareness of the reasons for, and the ramifications of, departures from a previous plan. Such departures need to take place in open fashion and follow recognised procedures. Ideally, not only the immediate action, but the plan too, must be adapted. Thus, and only thus, we may speak of social learning as a consequence of adapting plans during day-by-day decision-making. As against this, opportunism refers to mere violations of previously formulated principles without considering the consequences. Opportunism has the serious shortcoming that it diminishes trust in planning, whereas flexibility, properly conceived, increases it.

Because of the crucial role, indicated above, of agreed-upon definitions of decision situations in rendering planning possible, planners are not the passive recipients of the consensus on which they base their plans. Rather, it is planners who organise consensus on issues regarding their work. This explains why planners have been keen on public participation[1]. Nowadays, other techniques for organising consensus often seem to be taking its place. Be that as it may, the organisation of consensus is a political role. That planning is political is, however, no startling revelation!

As indicated, the original intention behind the decision-centred view had been to clarify the issues in the proceduralist versus substantivist debate of the seventies. An additional concern had been to understand the nature of the rationality in planning as a decision rule, analogous to Popper's rules for accepting or rejecting scientific hypotheses (Faludi, 1986a). Clarifying the issues in the proceduralist-substantivist debate required restating not only what the proceduralist (now decision-centred) view stood for, it also required a clear statement of what its protagonists were against. Contrary to what some critics implied, it was not that planning decisions needed to be based on substantive knowledge. What the protagonists of the decision-centred view were against was the positivistic idea, implied in the approaches of the classic period, that planning required nothing but the thorough study of its object of concern, after which action, design, or policy would spring ready-made from the minds of planners. Rather than ignoring the nexus between knowledge and action, as the classic view did,

---

[1] At present, public participation, although enshrined in legislation, is less popular.

the decision-centred view makes this nexus the focus of its attention (hence the title of this section).

The position opposed by protagonists of the decision-centred view was labelled the 'object-centred' view of planning (Faludi, 1982). To restate this view, the underlying assumption was that analysing the object of concern provided a rock-solid basis for plan-making, in other words the ideal of classic planning thought. In formulating these views, the hope was that substantivists would abjure their positivistic assumptions, leading to a common platform for discussing planning as evolving around choice.

A comprehensive statement of the decision-centred view in Dutch is by Mastop[1]. Defenders of substantive planning theory at Nijmegen University responded with an 'action-oriented approach'. In this, they took a leaf from a landmark report sponsored by the Scientific Council for Government Policy[2]. This report had criticised the planning of the seventies, demanding a more direct involvement of planners with societal action. In describing the ensuing action-oriented approach, we follow Needham (1988). Like the decision-centred view, this approach does not start with the making of plans. However, unlike the decision-centred view, it does not focus on the ongoing decisions of the planning subject either. Rather, its starting point is the bringing about of change, and is therefore 'action-oriented'. It is concerned with how the spatial order can be influenced by spatial measures, and how measures should be chosen and implemented so as to achieve the desired spatial order (thereby taking account of possible side-effects). Needham quotes Wissink[3], arguing that planning must begin with the decisions of private actors shaping the environment, since public measures take effect through their actions. Needham concludes that planology is the study of how the environment can be influenced by influencing the decisions of private actors[4].

The confrontation, such as it was, with the decision-centred view took

---

[1] Mastop J.M. (1987, 1st edition 1984) Besluitvorming, handelen en normeren, *Planologische Studies, 7*, Institute of Planning and Demography, University of Amsterdam, Amsterdam.

[2] Den Hoed P., Salet W.G.M., Van der Sluijs H. (1983) Planning als onderneming, Government Printing Office, The Hague.

[3] Wissink G.A. (1982) ***Ruimtelijke ordening als mensenwerk***, Van Gorcum, Assen.

[4] The Nijmegen School has spawned three books: Needham B., Wissink G.A. (eds.) (1982) ***Ruimtelijke planning en ruimtelijke ontwikkeling***, Van Gorcum, Assen; Muller N., Needham B. (1989) ***Ruimtelijk handelen: Meewerken aan de ruimtelijke ontwikkeling***, Kerckebosch BV, Zeist; Dekker A., Ekkers P., Ganzevles T., Muller N. (eds.) (1993) Gerrit Wissink: Dertig jaar universitaire planoloog, Nijmeegse Planologische Cahiers, 43, Department of Planology, Catholic University of Nijmegen, Nijmegen.

place in 1985[1]. This has laid the foundations for the current consensus[2]. Both approaches now focus on social interaction around public decisions and the actions that follow concerning the environment. In this, both approaches have clear affinities with communicative planning, receiving broad attention in the international literature, most recently from Fischer and Forester (1993) and also Sager (1994). Both appreciate that, as human actors, the addressees of the planners' messages interpret them in the light of their own situations as they perceive them (what Eco, 1979, and Faludi and Korthals Altes, 1994, describe as the 'double construction of texts'). Lastly, both understand that these addressees are in principle free (albeit perhaps to their peril) to negate, subvert and/or contravene plans. This reminds us of the theory of structuration by Giddens (1984).

Designers were unperturbed by such debates. Unlike their international colleagues, Dutch designers have not spawned a body of literature on design methodology. Of course, some publish in foreign journals, but the point is that there is no significant Dutch literature or discourse. Urban design is still a craft. Planning academics sometimes step into this void. A good example is Van der Cammen. A much-quoted paper by this author, published in the early eighties, looked like the final reckoning with seventies-style systematic planning and a re-emergence of design as a dominant force. The paper pointed at the unease with planning as the making of deals behind closed doors. It complained about a dearth of theory about this 'negotiative planning', stating that people were keeping their distance from a practice which they regard as inevitable. Van der Cammen predicted a reaction. Himself a respected opinion-leader, he proceeded to turn his prediction into reality, advancing the notion of 'compass plans' giving direction to negotiations[3]. Compass plans should represent the professional point of view: "Planning agencies would be well

---

[1] Wissink G.A., Needham B., Mastop J.M. (eds.) (1985) Planningmethodologie, SRPO Cahier, 7, NIROV, The Hague.

[2] On various occasions since, the similarities between the two schools have been emphasized. See: Needham B., Dekker A. (1989) De handelingsgerichte benadering van de ruimtelijke planning en ordening: een uiteenzetting, in Muller N., Needham B. (eds.) *Ruimtelijke handelen: Meewerken aan de ruimtelijke ontwikkeling*, Kerckebosch, Zeist, 1-12; Van Marwijk A. (1990) Doorwerken met de Vierde Nota: Een interpretatie van de praktijk aan de hand van Giddens, Working Papers of the Institute of Planning and Demography, no. 144, University of Amsterdam, Amsterdam.

[3] Van der Cammen H. (1982) Methodisch geleide planvorming (2), *Stedebouw en Volkshuisvesting, 63*, p. 455; see also: Van der Cammen H. (1984) Doeltreffende ruimtelijke plannen, Voorstudies 11, Programmeringsoverleg Ruimtelijk Onderzoek, The Hague; De Klerk L.A., Van der Cammen H. (1983) Voorbij de stilstand, *Rooilijn, 16*, 142-146.

advised to reapply themselves to the making of a plan which they can support wholeheartedly, which they themselves have made and which contains a detached interpretation of policy"[1]. Generating discussion, compass plans would also rekindle political interest in planning. This paper was a forerunner to the present interest in Dutch academic debates in framing planning thought and action.

Unfortunately, this paper falls short on the methods of compass planning. In this respect, Van der Cammen is not unlike Van Lohuizen. He merely exhorts planners (and this is where his paper relates to the role of design) to return to the proverbial drawing board and to show the world what the future could look like, if only people had the guts to accept their vision. The author has every right to be pleased with the response he got. In the mid-eighties, when the Fourth Report, one of the series of Dutch national planning documents, was being prepared, planners took a leaf out of his book. His recommendations admirably suited the spirit of boosterism prevailing in the eighties, when glossy imagery was invoked to sell the planners' wares.

Boosterism receives a theoretical underpinning under the flag of 'city marketing'. City marketing is not specific to the city scale. At its most general, Ashworth and Voogd (1990) talk about 'place marketing'. This approach has been used by planners with consummate skill, and not only in the Netherlands. In fact, though, Ashworth and Voogd warn against the four-colour brochures which city marketing often comes down to. Rather than relying on such gimmicks, city-marketing should substitute the traditional focus on ordering of space "by a closer attention to the wishes and needs of actual or potential users" (Ashworth and Voogd, 1990, p. 1). This reminds one of the action-oriented approach. In their emphasis on bottom-up decision-making, the approach of Ashworth and Voogd has affinities with communicative planning, too.

Not only cities and regions, but planning itself was being marketed. In the mid-eighties, an international management consultant turned Minister of Housing, Physical Planning and the Environment challenged planners to relate to up-and-coming, rather than ongoing issues. Drawing on the concept, well known in the management literature, of a product life cycle, he invoked the notion of a policy life cycle. Accordingly, policies went through stages of growing political attention to issues, to be followed by phases of decline, until attention to policies evaporated. The implication was that well-established concerns of national planning at the time, like growth management (Faludi 1992, 1994, Faludi and Van der Valk, 1994),

---

[1] Van der Cammen H., Methodisch geleide planvorming 2, p. 455.

had insufficient appeal to be sustained. Taking heed from this, planners developed a marketing strategy. It led them into exploring hot issues, like the information revolution, for their planning implications. Inevitably, this was at the expense of attention to ongoing growth management (Faludi and Korthals Altes, forthcoming).

Notwithstanding all these efforts at understanding design and how it can contribute to the appeal of planning, the gap between designers and planologists remains. Most attempts to close it have come from planologists. The latter show a distinct concern now for the role of design. An example is a study analysing the entries to a design competition[1]. It put 'design' in the broader context of the societal process by which ideas are formulated and decisions are taken. It paid homage to the role, long recognised in the design literature (for a review see Cross, 1986) of intuition. In addition, it laid bare the categories in which designers think, including their notions of space. The study related these notions to the diagrammatic representation of ideas, pointing out (as Steigenga had done before) the choices involved. At the more abstract regional scale, aggregation, classification, selection and combination of information can be guided by concepts, recipes, basic principles, a leitmotif, or something similar. In fact, the study confirmed what Van der Cammen had based compass planning on (and what the analysis in terms of planning doctrine also shows), that the formulation of concepts for discussing problems and solutions is an important aspect of planning[2].

So, planologists now have much respect for design. Whether designers will respond in kind remains to be seen[3]. A hopeful sign is a study which draws, amongst others, on Habermas[4]. It rejects the designer tactics of 'seduction' and insists on urban design and planology joining forces. The author opts for a radical enlightenment programme, striving for 'communicative rationality'. So far there has been little discussion in the

---

[1] Ekkers P., Mastop H., Dekker A., Raggers J. (1990) Regionaal ontwerp en beleid: Plananalyse stad en land of de helling, Faculty of Policy Sciences, Department of Planology, Catholic University of Nijmegen, pp. 9-23.

[2] Ekkers P., Regionaal ontwerp, p. 21.

[3] As indicated, there is not much design literature in Dutch. The main concern of one recent work is with urban design as an independent field of activity. See: Doevendans K. (1988) Stedebouw en de vormgeving van een speciale wetenschap: Een onderzoek op Loeniaanse leest, *Bouwstenen, 10*, Faculteit Bouwkunde, Eindhoven University of Technology, Eindhoven.

[4] Boelens L. (1990) Stedebouw en planologie - een onvoltooid project: Naar een communicatief handelen in de ruimtelijke planning en ontwerppraktijk, Ph.D. thesis, Delft University of Technology, Delftse Universitaire Pers, Delft.

Dutch planning literature of the work of Habermas, so this is welcome[1]. Works like Sager (1994) and the many studies by Forester (1985, 1989) and Fischer and Forester (1993) are still waiting to be assimilated.

However, this work is not representative of designer thinking in the Netherlands. The distinctness of design, sometimes still wedded to the claim to a superior role in planning, continues to be the main concern. Even if the language of design is distinct from other forms of language, this does not give designers a privileged position. Such a position could only be justified if the designer claim of the sixties of urban form encapsulating the totality of human experience, and of designers having exclusive access to its meaning, held water. However, it should be obvious that urban form cannot encapsulate the totality of human experience, and that neither its analysis, nor the synthetic expression of it, are in any way privileged over other forms of comprehension and action.

In short, therefore, academic planologists have reacted to the design challenge true to form by conceptualising the role of design. In the margin of debates about professionalisation, practice and who should be in the lead, they have also looked upon the whole enterprise of planning. The next section gives an account of their theorising about planning, going beyond reflection on the best methods. Such theorising now extends to the conditions under which planning ideas are being produced and become effective. This is where the concept of framing comes into its own.

18.4 Framing the Nexus Between Knowledge and Action

18.4.1 Introduction

Planning has already been described as a social learning process. One

[1] The exceptions are De Jong M.J. (1986) 'Idee- en consensusvorming in de ruimtelijke ordening', Ph.D. thesis, De Jong, Kollumerzwaag; and Zonneveld W. (1991) Conceptvorming in de ruimtelijke planning: Patronen en processen, *Planologische Studies, 9A*, together with Conceptvorming in de ruimtelijke planning: Encyclopedie van planconcepten, *Planologische Studies, 9B*, Institute of Planning and Demography, University of Amsterdam, Amsterdam. To be sure, works in public administration like: Van der Graaf H., Hoppe R. (1989) *Beleid en politiek: Een inleiding tot de beleidswetenschap en de beleidskunde*, Coutinho, Muiderberg; and Edwards A. (1990) 'Planning betwist: communicatieve strategieën van boeren en natuurbeschermers in de ruilverkaveling Wommels', Ph.D. thesis, University of Amsterdam, Uitgeverij Jan van Arkel, Utrecht, do refer to Habermas.

implication is that patterns of interaction amongst professionals and between professionals and the body politic must be attended to. After all, these patterns condition social learning. In the Netherlands, there has been sustained research into how planning arenas and planning approaches are being constructed and maintained. This last section discusses the relevant works.

This research has been made possible by the professionalisation of academic research and teaching. With it, disciplinary developments cease to merely follow developments in practice. Rather, academic pursuits acquire a momentum of their own. The professionalisation of academic research is the first topic to be discussed.

18.4.2 The Professionalisation of Academic Work

The first generation of planning academics came out of practice. Most of the professors had Ph.D.'s, of course. However, these were in their original discipline, not in planology. Most lecturers had no Ph.D., but various forms of practical experience. The eighties saw the professionalisation of academic work. The Ph.D. was increasingly looked upon as the natural entry-point. Many (but by no means all) university staff gained their Ph.D., and the level of academic work improved. The pioneering generation was replaced by professors with a solid grounding in planology as a discipline.

The professionalisation of academic work has been enforced by sustained efforts to streamline research. The professionalisation of planning in relation to planning doctrine and to the discipline of planology has been one of the themes of such programmatic research. Drawing on the work of a whole team, the book on which this paper is based (Faludi and Van der Valk, 1994) represents the outcome of that programme. There is also much encouragement to publish in English and to participate in international debates. As a consequence, there are a fair number of book-sized publications on the market, along with special issues of international journals, dealing with the Netherlands. There are even two journals published in English, the "*Tijdschrift voor Economische en Sociale Geografie*" and the "*Netherlands Journal of Housing and the Built Environment*". So the Dutch situation is accessible to international scholars. One of the effects of this open-door policy is that there is no dearth of international visitors, from student groups to sabbatical scholars, giving Dutch researchers opportunities galore for participating in international debates.

Indeed, the prospects for academic work in the Netherlands could be described as almost idyllic. Unfortunately, the opportunities for new Ph.D.'s to enter academic work are limited and will remain so until tenured staff appointed in the sixties and seventies retire. Also, further investment in research is conditional upon the continued existence of planning as an academic discipline. Whether this will be the case, depends on forces outside the control of the academic planning community.

18.4.3 Reflections on the Nature of the Discipline

The studies discussed in this section concern the development of planning thought in relation to professionalisation and planning practice. They have culminated in the attempt to give an overall evaluation of the Dutch planning system, including the philosophies behind it, the view of plans and of planning on which it is based, and the performance of plans. The intention has been to step back and reflect, amongst others upon issues discussed in this paper: the conceptualisation of the step from analysis to design, the disparate traditions, but also the role of planning concepts in generating consensus. The intention has been to transcend past disputes. In particular, there is acceptance of the role, emphasised by Van der Cammen in his polemic against seventies-style planning, of visual imagery, such as the Green Heart concept, in shaping policy.

One of the first studies in the series, concerned with the setting up of the Netherlands Institute of Housing and Planning introduced the concept of a programme for housing and planning[1] . 'Programme' stood for that which binds the coalition forming the Netherlands Institute for Housing and Planning together. The study was the first that took its cues from the history and philosophy of science, especially Kuhn and Lakatos. Another study concerned late-nineteenth century planning in Amsterdam[2] . The author sought to interpret planning in Amsterdam in the light of the decision-centred view. However, in order to come to grips with the sequence of events, he could not confine himself to the various plans and (this being a central concern of the decision-centred view) how they are invoked in operational decision-making. Rather, he had to examine a complex of substantive and procedural ideas of professionals. This complex forms the backcloth to their actions. The study referred to this

---

[1] De Ruijter, Voor volkshuisvesting.

[2] Van der Valk A.J. (1989) Amsterdam in aanleg: Planvorming en dagelijks handelen 1850-1900, *Planologische Studies, 8*, Institute of Planning and Demography, University of Amsterdam, Amsterdam.

complex as 'systematic town expansion'. Systematic town expansion encapsulates expert thinking about planning. It signifies intent to do away with the consequences of 'chaotic' private nineteenth-century development. Systematic town expansion relates both to the shape of development as well as to the manner in which order should be achieved. The quest for systematic town expansion was the driving force behind the planners' strategies and actions. So to understand planning requires an appreciation of what systematic town expansion entails. These two studies have laid the foundations for the study of the professionalisation of Dutch planning in a way which seems unparalleled.

The same emphasis on the dynamics of disciplinary developments characterises a study of planning concepts adopted since the twenties[1]. As with other Dutch authors, this study shows the same determination to analyse planning as philosophers of science have done, i.e. as a dynamic process of interaction between the planning community and the outside world in which ideas play a crucial role[2]. Tracing patterns in their formation, it identifies a 'hard core' and 'positive heuristics', together with 'explosive issues', analogous to Kuhn's 'anomalies'. They cause fundamental change in the dominant 'conceptual complex', a notion comparable to that of a paradigm. The study is one of several following on the lines of the sociology of science[3].

All this relates to the theme, already alluded to, of the importance of

---

[1] Zonneveld, Conceptvorming.

[2] Zonneveld, Conceptvorming.

[3] The same author has taken the analysis of 'conceptualization' further, recommending how to deal with planning concepts. He rejects the idea of recipes for the formulation of concepts. Creativity and intuition are indispensable, but so is critical reflection. Zonneveld W. (1992) Naar een beter gebruik van ruimtelijke planconcepten, *Verkenningen, 64*, Institute of Planning and Demography, University of Amsterdam, Amsterdam, pp. 91-92. The focus in the debates on planning concepts has also inspired Faludi and Van der Valk (1994) in their analysis of Dutch planning doctrine. One of the two elements of doctrine is the spatial organization principle for the plan area, and Faludi and Van der Valk make great play of the importance of a central, mobilizing metaphor underlying doctrine, such as the organic metaphor underlying the famous Green Heart concept. (See also Alexander and Faludi, 1990) In the same paper, Zonneveld dissects spatial configurations into areal, nodal and communicative patterns. Areal patterns divide space into more or less homogeneous zones, like urban versus rural areas. Nodal patterns refer to distinct points in space, like Christaller's central points. Communicative patterns are networks, like the system of roads. There is an obvious way of giving diagrammatic expression to each: surfaces for areal patterns, points for nodal patterns, and lines for communicative patterns. These are basic principles for how to represent spatial configurations. Issues of scale, of the symbols to be used, of the type of projection of the spherical surface of the globe onto maps and so forth follow.

overall frames in consensus-building. The formulation of frames is a key to political effectiveness. Hajer (1989) picks up this theme, introducing the notion of the hegemonic project, which consists of a discourse, a system of positions and practices and strategic action. A discourse serves a specific cause, is related to a specific alliance and forms the basis for strategy. The struggle between discourses is the fight between groups "to get their interpretation of the state of affairs dominant. It cannot be discussed usefully in terms only of discourse: it is basically a struggle for hegemony which involves more than ideology alone. It concerns the fight against a dominant hegemonic project, the emergence and formation of alternative projects and the transformation of existing alliances to keep in power" (Hajer, 1989, p. 41). This is similar to Kuhn's scientific revolution, a notion which, together with its counterpart, that of a paradigm, has also inspired the formulation of the notion of doctrine (Alexander and Faludi, 1990).

Discourses use symbols to engender consensus. "Often, they give names to a discourse. It is symbols which give people the sense of inclusion. A symbol functions to give people a feeling that they belong to a discourse" (Hajer, 1989, p. 45). One is immediately reminded of the Green Heart, a planning concept which, indeed, symbolises much of what Dutch national planning stands for.

The next element of a hegemonic project is institutionalisation: the emergence of a system of positions and practices which reflects a pattern of domination. A case in point is the struggle of Dutch planners to establish their position and to institute practices to be able to properly articulate their concerns.

The last of the elements of a hegemonic project is strategic action. It takes place within the context of a discourse and the system of positions and practices. Hajer emphasises that the three elements must be looked at together: "Every political act takes place in a society full of contradictions. Politics is the struggle for hegemony... Actors with different interests form alliances around specific discourses to gain dominance. These discourses form the basis of a hegemonic project. These projects consist first of all of a set of societal actors... with, secondly, their specific societal positions, and, thirdly, the discourse itself, which cements their alliance." (Hajer, 1989, p. 77)

Comparing his notion with that of doctrine (as found in an earlier form in Faludi, 1987), Hajer (1989) acknowledges the similarity, but claims that doctrine can be conservative and that his analysis allows for progressive change. This leads into a discussion of doctrine. Having introduced the notion in the work referred to by Hajer above, Faludi has made it into a key-concept in the analysis of Dutch planning by himself and his associates. The concept originates from a study of the Dutch "*Urbanisation Report*" presenting a growth-management package. The report articulated an image of a desirable future and a way of reaching it. The report was in fact quite effective as a planning document. As the reader will remember, in the early eighties planners (in particular those of an urban design-bent) argued that plans should do precisely this: package attractive images which people would want to follow, thereby ensuring implementation. This had been more or less what the Urbanisation Report had done. The report was also one in a chain of documents. Between them, this succession of documents has generated a definite view of what the country should look like. Decision-makers are imbued with this view. That which provides such self-evident guidance is being described as planning doctrine.

Planning doctrine delineates an arena for discussion and action. Herein lies its importance. By performing its 'framing' role, doctrine enables 'normal' planning. 'Normal' planning involves primarily professional, administrative, and bureaucratic actors. Within the context of agreed-upon values and a generally imaged principle of spatial organisation, professional-bureaucratic debate and political discourse can produce a succession of planning concepts to respond to changing situations. So one benefit of having a doctrine is to reduce the burden of plan-making. Planning becomes cumulative and progressive. This may account for the effectiveness of Dutch planning. By implication, the absence of an agreed-upon doctrine may explain ineffectiveness in planning.

Planning doctrines (such as the Dutch) that have lasted and been successful, have at times displayed significant changes. But, at the same time we are saying that these are the same doctrines throughout. How do we account for this? To answer this question, Alexander and Faludi (1990) invoke once more the analogy between doctrines and paradigms. Lakatos (1974), in relation to paradigms (which he calls: 'scientific research programmes'), distinguishes between negative and positive heuristics. A negative heuristic, the 'hard core' of a research programme, cannot change, whereas a positive heuristic, which encourages development of a 'protective belt' of theories, models and observations around the core, may change.

In planning doctrine, the same distinction exists. Various concepts may be replaced throughout the lifespan of the doctrine. Thus, the emphasis in Dutch planning has gone from concentric development around towns and cities to a policy of controlled dispersal and back to what is called the 'compact-city' policy, all within one and the same doctrine evolving around Randstad and Green Heart. However, the doctrine itself, with its mobilising metaphor as its hard core, is replaced in a different kind of discourse, one even more political and value-oriented. Thus, if ever the Green Heart were abandoned, it is reasoned that this would amount to a doctrinal revolution. The dissolution of doctrine is similar to a situation where there is no agreed-upon doctrine, in that discussion focuses on the doctrine. Alexander and Faludi (1990) call this doctrinal discourse. Rein and Schön (1986) and Schön and Rein (1994) call it frame-reflective discourse.

Doctrinal discourse is exceptional and occurs as 'anomalies' emerge due to the emergence of competing principles (using another term culled from scientific methodology). At least this is what the analogy with Kuhn and Lakatos suggests. Doctrinal discourse is also more political, than 'normal' planning. This is no reason for professionals to stay out of it. Doctrine expresses their deepest concerns about their field and the values they pursue. This is why doctrinal discourse intimately relates to professionalisation and to the formation of what Faludi and Van der Valk in their book describe as the planning community.

The notion of a planning community relates to how planning reflects wider concerns and how that community generates societal consensus. Now, it would be preposterous to assume that the planning community, through the medium of a well-considered doctrine, could generate consensus where it does not exist. Van der Heiden, Kok, Postuma and Wallagh (1992) have pointed out that there are mechanisms other than doctrine responsible for consensus in the Netherlands. Consensus is ingrained in the fabric of society. Whether Dutch planning holds any lessons for abroad remains a moot point, therefore. This is all the more true since land policy in the Netherlands is also unique. That being said, it is possible to argue that planning doctrine has played a role in specifying a pre-existing consensus, focusing it on a particular way of conceptualising the shape of the country, and framing Dutch policies accordingly.

The ultimate challenge is, of course, to turn the new understanding of how doctrines are being shaped, and how they in turn shape action, to good use as planners. Are there such things as discourses on doctrines? The notion of 'frame-reflective discourse' of Rein and Schön (1986) suggests a positive answer. However, can we reasonably expect to shed expert light on such intensely political matters? Or are doctrines, as Kuhn would no doubt argue, incommensurable and thus beyond rational debate? What is the developmental pattern of doctrine anyhow? Are periods of 'normal' planning, as Alexander and Faludi (1990) have put it, in analogy to Kuhn's normal science inevitably followed by doctrinal revolutions?

Certainly in the Netherlands with its well-developed doctrine this is an urgent question. A recent study[1] focuses on the dynamics of doctrinal development, using the Fourth National Physical Planning Report as a case study. It explores the implications of more recent literature by Laudan (1984). Laudan suggests a pattern of doctrinal development which is more evolutionary than revolutionary. The study puts flesh on the bones of an idea which Alexander and Faludi have explored: that of an 'open' doctrine.

18.5 Conclusions

Sometimes calls for 'relevance' of academic research to the immediate concerns of practice must be resisted, so as to make room for research relating to the long-term sustainability of planning. The above line of research is a case in point. Practice is not waiting for an 'open' doctrine. Rather, what practitioners ask from theorists is the re-affirmation of existing doctrine. However, if doctrine does not become more open and flexible, it may be swept away as irrelevant to the changing world. At least this is the upshot of the above discussion of the dynamics of doctrinal development. Be that as it may, the reader will appreciate that the analogy between planning and science, between doctrine and paradigms, continues to be a source of inspiration.

Combining substantive with procedural elements, doctrine also bridges the gap between what was once described as procedural and substantive theory. It will be clear that Dutch planning academics have left that particular conflict behind. In so doing, they have gone a long way towards providing a reflective frame for the ongoing practice of planning. The issues now are how to feed new critical insights back into practice and how, at the same time, to maintain the independence from practice which is the hallmark of any academic enterprise. With pressure now on to supplement meagre university funds through contract research, this poses a dilemma.

---

[1] Korthals Altes W.K. (1995) De Nederlandse planningdoctrine in het fin de siècle: Ervaringen met voorbereiding en doorwerking van de Vierde nota over de ruimtelijke ordening (Extra), Van Gorcum, Assen, Maastricht. For a preliminary publication in English see Korthals Altes (1992).

References

Alexander E.R., Faludi A. (1989) Planning and Planning Implementation: Notes on Evaluation Criteria, *Environment and Planning B: Planning and Design, 16*, 127-140.

Alexander E.R, Faludi A. (1990) Planning Doctrine: its Uses and Applications, Werkstukken van het Planologisch an Demografisch Instituut, 120, University of Amsterdam, Amsterdam.

Ashworth G.J., Voogd H. (1990) *Selling the City: Marketing Approaches in Public Sector Urban Planning*, Belhaven Press, London.

Boardman P. (1978) *The Worlds of Patrick Geddes - Biologist, Town Planner, Re-Educator, Peace-Warrior*, Routledge and Kegan, London, Henley and Boston.

Bours A., Lambooy J.G. (eds.) (1974) *Stad en stadsgewest in de ruimtelijke ordening*, Van Gorcum, Assen.

Bosma K. (1990) Town and Regional Planning in the Netherlands 1920-1945, *Planning Perspectives, 5*, 125-147.

Brandford V.V. (1914) *Interpretations and Forecasts*, Duckworth, London.

Broady M. (1968) *Planning for People*, The Bedford Square Press, London.

Chadwick G.A. (1970) *A System View of Planning*, Pergamon, London.

Chapin F.S. (1957) *Urban Land Use and Planning*, Harper, London.

Collins G.R., Craseman-Collins C. (1965) Camillo Sitte and the Birth of Modern City Planning, London, New York.

Cross N. (1986) Understanding Design: The Lessons of Design Methodology, *Design Methods and Theories, 20*, 409-439.

Eco U. (1979) *The Role of the Reader: Explorations in the Semiotics of Texts*, Indiana University Press, Bloomington, London.

Faludi A. (1973) *A Reader in Planning Theory*, Pergamon, Oxford.

Faludi A. (1982) Three Paradigms of Planning Theory, in Healey P., McDougall G., Thomas M.J. (eds.) *Planning Theory Prospects for the 1980s*, Pergamon Press, Oxford, 81-101.

Faludi A. (1984, 1st edition 1973, reprinted with a new preface by the author) *Planning Theory*, Pergamon, Oxford.

Faludi A. (1986a) *Critical Rationalism and Planning Methodology*, Pion, London.

Faludi A. (1986b) Flexibility in US Zoning: A European Perspective, *Environment and Planning B: Planning and Design, 13*, 255-278.

Faludi A. (1987) *A Decision-Centred View of Environmental Planning*, Pergamon, Oxford.

Faludi A. (1989a) Conformance vs. Performance: Implications for Evaluations, *Impact Assessment Bulletin, 7*, 135-151.

Faludi A. (1989b) Planning According to the Scientific Conception of the World: The Work of Otto Neurath, *Environment and Planning D: Society and Space, 7*, 397-418.

Faludi A. (1992) Dutch Growth Management: The Two Faces of Success, *Landscape and Urban Planning, 22*, 93-106.

Faludi A. (1994) Coalition Building and Planning for Dutch Growth Management: The Role of the Randstad Concept, *Urban Studies, 31*, 485-507.

Faludi A., Framing with Images, *Environment and Planning B: Planning and Design* (forthcoming).

Faludi A., Korthals Altes W. (1994) Evaluating Communicative Planning: A Revised Design for Performance Research, *European Planning Studies, 4*, 403-418.

Faludi A., Korthals Altes W., Marketing Planning and Its Dangers, *Town Planning Review* (forthcoming).

Faludi A., De Ruijter P. (1985) No Match to the Present Crisis? The Theoretical and Institutional Framework for Dutch Planning, in Dutt A.K., Costa F.J. (eds.) *Public Planning in the Netherlands*, Oxford University Press, Oxford, 35-49.

Faludi A., Mastop J.M. (1982) The I.O.R. School: The Development of a Planning Methodology, *Environment and Planning B: Planning and Design, 9*, 241-256.

Faludi A., Van der Valk A.J. (1991) Half a Million Witnesses: The Success (and Failure?) of Dutch Urbanization Strategy, in Faludi A. (ed.) *Fifty Years of Dutch National Physical Planning (Special Issue), 17*, 43-52.

Faludi A., Van der Valk A.J. (1994) *Rule and Order: Dutch Planning Doctrine in the Twentieth Century*, Kluwer Academic Publisher, Dordrecht.

Fischer F., Forester J. (eds.) (1993) The Argumentative Turn in Policy Analysis and Planning, Duke University/UCL Press Ltd., London.

Foley D.L. (1964) An Approach to Metropolitan Spatial Sructure, in Webber M.M. *et al.* (eds.) *Explorations into Urban Structure*, University of Pennsylvania Press, Philadelphia, 21-78.

Forester J. (1985) Practical Rationality in Planmaking, in Breheny M., Hooper A. (eds.) *Rationality in Planning: Critical Essays on the Role of Rationality in Urban and Regional Planning*, Pion, London, 48-58.

Forester J.F. (1989) *Planning in the Face of Power*, University of California Press, Berkeley, California.

Friedmann J. (1969) Planning and Societal Action, *Journal of the American Institute of Planners, 35*, 311-318.

Friedmann J. (1973) *Retracking America: A Theory of Transactive Planning*, Doubleday Anchor, Garden City, New York.

Friedmann J. (1987) *Planning in the Public Domain: From Knowledge to Action*, Princeton University Press, Princeton, New Jersey.

Friedmann J. (1993) Towards a Non-Euclidean Mode of Planning, *Journal of the American Planning Association, 59*, 482-485.

Friend J.K., Jessop W.N. (1969) *Local Government and Strategic Choice*, Pergamon, Oxford.

Friend J.K., Power J.M., Yewlett C.J.L. (1974) *Public Planning - The Intercorporate Dimension*, Tavistock, London.

Giddens A. (1984) *The Constitution of Society: Outline of a Theory of Structuration*, Polity Press, Cambridge.

Glass R. (1959) The Evaluation of Planning: Some Sociological Considerations, *International Social Science Journal, 11*, 393-409.

Hajer M.A. (1989) *City Politics: Hegemonic Projects and Discourses*, Avebury (Gower Publishing Company Ltd.), Aldershot, Brookfield, Hong Kong, Singapore, Sydney.

Kloos W.B. (1939) Het Nationaal Plan: Proeve eener beschrijving der planologische ontwikkelingsmogelijkheden van Nederland, Samsom, Alphen aan den Rijn.

Korthals Altes W.K. (1992) How do Planning Doctrines Function in a Changing Environment?, *Planning Theory Newsletter*, 7-8, 99-114.

Lakatos I. (1974) Falsification and the Methodology of Scientific Research Programmes, in Lakatos I., Musgrave A. (eds.) *Criticism and the Growth of Knowledge*, Cambridge University Press, London, 91-196.

Laudan L. (1984) *Science and Values: The Aims of Science and their Role in Scientific Debate*, University of California Press, Berkeley, Los Angeles, London.

Mannheim K. (1940) *Man and the Society in an Age of Reconstruction*, Kegan Paul, London.

McLoughlin J.B. (1969) *Urban and Regional Planning - A System Approach*, Faber and Faber, London.

Needham B. (1971) Planning as Problem-Solving, *Journal of the Royal Town Planning Institute, 57*, 317-319.

Needham B. (1988) Continuity and Change in Dutch Planning Theory, *The Netherlands Journal of Housing and Environmental Research, 3*, 3, 5-22.

Needham B., Faludi A. (1973) Planning and the Public Interest, *Journal of the Royal Town Planning Institute, 59*, 164-166.

Rein M., Schön D. (1986) Frame-Reflective Policy Discourse, *Beleidsanalyse, 15*, 4, 4-18.

Schön D., Rein M., (1994) *Frame Reflection: Toward the Resolution of Intractable Policy Controversies*, Basic Books, New York.

Sager T. (1994) *Communicative Planning Theory*, Avebury, Aldershot, Brookfield, Hong Kong, Singapore, Sydney.

Steigenga W. (1964) *Moderne planologie*, Aula, Utrecht.

Thomas H.D., Minett J.M., Hopkins S., Hamnett S.L., Faludi A., Barrell D. (1978) *Flexibility and Commitment in Planning*, Martinus Nijhoff, The Hague, Boston, London.

Van der Heiden N., Kok J., Postuma R., Wallagh G.J. (1992) Consensus Building as an Essential Element of the Dutch Planning System, *Planning Theory Newsletter, 7-8*, 115-134.

Van der Valk A.J., Faludi A. (1992) Growth Regions and the Future of Dutch Planning Doctrine, in Breheny M.J. (ed.) *Sustainable Development and Urban Form*, Pion, London, 122-137.

19. Order and Change, Rule and Strategy

Luigi Mazza

19.1 Introduction

This comment on Faludi's paper focuses on three issues: the definition of strategic planning, the concept of planning doctrine, and the uneasy relationship between planning and urban design, that is, between planners and architects. In Faludi's paper the three issues are interrelated in various ways, but to capture these relations adequately it is necessary to recall Faludi's other contributions, in particular his latest book (Faludi and Van der Valk, 1994).

The doctrine concept plays a key role in providing the definition of strategic planning with specific content, making the definition appealing from a disciplinary point of view. It is an analytical and normative tool for describing and evaluating planning activities, and may also be viewed as helping to explain the nature of planning.

The confrontation between architect and planner has always been a feature of planning history. Faludi analyses that confrontation, referring to roles and cultural differences, and examines the relative autonomy of strategic planning with regard to criticisms of planners by architects and urban designers. It is not by chance that Faludi considers strategic planning to be the preferred form of planning, one on which the planners' élite focuses its attention and commitment (Faludi, 1994, Faludi and Van der Valk, 1994). This is an aspect which raises many questions. There has been competition and conflict between architects and planners for too long for it to be considered irrelevant. It is not however easy to resolve because, as Faludi stresses, the positions on each side are far from homogeneous and sometimes not clearly identifiable. Moreover, in relation to planning issues, architects often introduce simplifications that prevent the development of an effective dialogue. From this point of view, it is understandable that many planners choose to delineate for themselves a field in which dialogue is not necessary. Nevertheless, we should ask ourselves why, in so many different countries, architects' proposals and criticisms seem to be so successful, even though poorly argued. A second issue worth considering is the fact that, as environmental transformation is one process, there should be a single view of it, independent of specialisations within the professional arena. It should be possible to demonstrate that the contribution of each expert is consistent with the overall view.

In 1984 in *Critical Rationalism and Planning Methodology*, Faludi first expressed his preference for a decision-centred view of planning, which was also the theme and title of the book, *A Decision-centred View of Environmental Planning* (1987). In a more recent volume, produced in collaboration with Van der Valk (1994) and in the paper discussed here, he again proposes this view. There are however at least two innovative elements, one substantive and one methodological.

In substantive terms, Faludi integrates his decision-centred approach with acknowledgement of the special role played by spatial images (see Faludi, 1996) in framing planning and social learning processes. The key to this acknowledgement is the concept of planning doctrine, which defines strategic planning as the process of formulating decisions aimed at spatial and physical ordering. This permits the traditional distinction between substantive and procedural planning theory to be overcome, as it was historically in the concept of systematic town expansion, which was concerned with the form of the development as well as the manner in which order was achieved (Faludi, 1994).

In methodological terms, the theoretical argument draws on a historical account of Dutch planning which offers a focal point for a number of concepts - planning doctrine, planning concepts, principles, etc. These are strengthened by their reference to specific contexts and concrete planning experience. The concept of doctrine is a key to explaining the successes and failures of Dutch planning, and becomes a real planning tool because it is constructed on the basis of that experience. The definitions of planning doctrine and other concepts are useful for revising the technical language used in planning. Even those who do not share the assumptions behind these definitions should accept them as terms of reference for exploring the meaning of these concepts in other contexts and, on this basis, move towards new definitions. It is in this way that we can acquire knowledge and construct, slowly and gradually, an operational planning language.

19.2 Spatial Ordering and Strategic Planning: The Planning Doctrine

Faludi's approach is in accord with the general post-positivist shift which, in relation to the decision-making process, from Lindblom and Cohen (1979) to Fischer and Forester (1993), has led to the recognition of the key role of discussion in policy development, seeing policy as argument. This approach has also defined the design process, through Simon (1981), as a reflective action, a 'dialogue' with the materials of the situation (Schön, 1983).

According to Faludi, the basic assumption is that "planning is more than the mere application of knowledge. It is decision-making" (Faludi and Van der Valk, 1994, p. xiii-xiv). In other words, it implies abandoning the modernist project approach and technical rationality in favour of reflection on the course of action and an interactive process involving open discussion within all those in some way concerned. According to this approach and on the base of Dutch experience, Faludi offers a fertile definition of strategic planning which is specific enough to possess a strong disciplinary characterisation. It is important to underline that Faludi arrives at this specific definition from a very general one: "Strategic planning is overall planning... We attach no further meaning to the adjective strategic. We say this emphatically because strategy is a troublesome concept" (Faludi and Van der Valk, 1994, p. 2-3).

It is difficult not to agree with such a general definition, which is accompanied by an equally general definition of the strategic plan: "the coordination of many actors, each making decisions of his or her own. Such coordination is continuous, and since all actors want to keep options open, timing is crucial. Rather than a finished product, a strategic plan is a momentary record of fleeting agreements reached. It forms a framework for negotiations and is indicative. The future remains open. Action never flows automatically from a strategic plan. Each decision needs justification in its own right. So the logic behind evaluating strategic plans is more complex than with project plans. A strategic plan needs to be interpreted" (Faludi and Van der Valk, 1994, p. 11).

According to a widely shared opinion, strategic planning is defined as a comprehensive action aimed at formulating a decision. It is therefore essentially a political activity because it aims to build a consensus, "planners organise consensus on issues regarding their work" (Faludi, 1994b, p. 36). It is the ambiguity and openness of strategic planning which give it its strength and flexibility: "what is often considered its vagueness and lack of immediate relevance, are features which are inherent to strategic planning. If it was not abstract and general, it would not be strategic planning" (Faludi and Van der Valk, 1994, p. 3-4); and again: "plan-making must take ongoing decisions as its point of departure, the more so since these decisions form the interface between public policies and the aspirations of private actors" (Faludi 1994b, p. 34-35).

But the general definition, and this is the element of novelty, is immediately clarified in procedural and substantive terms. According to Faludi, strategic planning is not a voluntary or informal action, but a formal one whose specific and well defined objective is spatial ordering: "the *statutory* activity of formulating and applying overall plans to guide the production of the *physical* environment" (Faludi and Van der Valk, 1994, p. 6, emphasis added); moreover, the objective of strategic planning is pursued within the specific context of 'planning doctrine'. In Faludi's terms a doctrine is essentially "a body of thought concerning (a) spatial arrangements within an area, (b) the development of that area, (c) the way both are to be handled" (Faludi and Van der Valk, 1994, p. 18-19), and in order to speak of a doctrine it is necessary to satisfy three conditions: "(a) a planning subject which (b) recognises the relevant planning area, and (c) adheres to the doctrine over time" (Faludi and Van der Valk, 1994, p. 21).

It may be observed that in general the features of the doctrine concept are permanent, not fleeting ones, which characterise, or at least should characterise, effective and comprehensive planning actions. It is therefore not by chance that Faludi traces the roots of research on the doctrine concept developed by Alexander (1990) to authors such as Selznik (1953) and Foley (1963). These features are recognised as characterising a specific form of planning, the originality of the doctrine idea consisting of the definition of a system of links between the procedural and substantive elements of planning actions. Doctrine therefore reconstructs analytically the unity of procedural and substantive components that any reflective practitioner discovers in practice. The idea of doctrine underlines that planning activities involve connecting a vision or a spatial and physical model (of action) with a political will and formal ability to act; it allows us to identify the link between the design and institutional dimensions of planning and claim it as the specific feature of strategic planning. It is this link that architects miss, as they tend to focus on the principles and images of spatial organisation. Doctrine, moreover, stresses the nature of planning as a social construct and collective action, and the need for planning to root its structural choices in the long term in order to be effective.

It is therefore understandable that Faludi takes care to avoid any association of his definition of strategic planning with the very different style of strategic planning connected with business, the 'strategic choice approach', often recommended to planners. He also avoids the contrast which is sometimes made between strategic and operational or tactical planning; his definition of strategic planning is complementary to operational or tactical planning.

I have emphasised the importance of the doctrine concept for two main reasons: firstly because the concept defines strategic planning in disciplinary terms as spatial and physical planning, and secondly because, as a tool for analysing planning activities, it is independent of concepts and categories from other disciplines.

In my own practical experience, spatial and physical projects and policies are the main tools for developing regional and metropolitan strategies. It is far more difficult to get the cooperation of the various bodies concerned, especially firms, needed to develop other types of policy. Spatial and physical policies are more likely to be successful and are those about which public and private firms can (and will) more easily take a stand. This is understandable, since such policies bind firms to the locality, while the internationalisation process is tending to cause a separation (Mazza, 1990, 1994a, 1994c).

19.3 The Nature of Planning and Planning Doctrine

Many of the analytical tools commonly employed in planning studies have in the past come from urban history, architecture and town history, urban sociology and economics. More recently administrative history and policy analysis have added their tools. But until now there has been a lack of analytical tools able to connect the engineering, ethical and ritual roots of planning, which constitute the specific basis physical planning (Mazza, 1993b, 1993c).

The history of planning thought and practice is also the history of periodical polarisations, swinging from technocratic to more liberal approaches, fluctuating between an authoritarian view to a form of planning rooted in popular expectations. We can in fact observe that two of the three main origins of planning, the engineering and ethical currents, tend to alternate; they also integrate the third current, the ritual one, though each time in a different way. In recalling the ritual origin of planning, I am not referring in a literal sense to the city foundation ritual or the traditional interpretations of these rites, for example, the foundation ritual as the symbolic transfiguration of religious and cosmological beliefs into the city plan (Eliade, 1943, Rykwert, 1976, Wheatley, 1971), but to ritual as a tool of the 'purification' of violence (Girard, 1972) and a recomposition of the action-thought, 'ethos-worldview' dichotomy (Bell, 1992).

A foundation ritual may be considered as the purification of the violence perpetrated against the 'natural' order of space. In the course of time a plan repeats this violence in the attempt to transform an existing spatial order into a new one. The violence of the plan has a double and contradictory valence, expressed by the conservation/change dichotomy - the violence of mastering change and denying the possibility of conservation matches the violence of mastering conservation and denying the opportunity of change.

Plans dissolve and conceal violence through their manifest and latent functional motivations; when plans are institutionalised, the violence they express becomes an essential component of their character, and spatial order becomes an institutional order (the Dutch would say a spatial ordering). Descriptions of rite as an action, or as a mechanism which reintegrates the thought-action dichotomy, may appear in a model which describes ritual as "the affirmation of communal unity in contrast to the frictions, constraints, and competitiveness of social life and organisation", and subsequently "as those special, paradigmatic activities that mediate or orchestrate the necessity of opposing demands of both *communitas* and the formalised social order" (Turner, 1966, quoted by Bell, 1992, p. 20-21). The ritual origin of planning is recognisable in the resolution of the dichotomous nature of order conceptions and action dispositions in the symbolic systems of ritual.

Given the assumption that planning has three origins, a planning doctrine can be interpreted as the interaction between them. Values which permeate the engineering and symbolic dimension, urging political action, can be traced back to the ethical origin; the construction of spatial and physical models and technologies can be traced back to the engineering origin; the symbolic form which expresses, legitimises and makes permanent the need for violence implicit in the principles of spatial organisation, can be traced back to the ritual origin. In this perspective, the doctrine concept too, because of its capacity to symbolically recompose the thought-action dichotomy in a constitutive idea of the locality, seems to offer a new analytical and normative tool for describing and evaluating planning experiences.

19.4 Strategic Planning, Project Planning and Urban Design

To construct and clarify theoretically the definition of strategic planning, Faludi defines several analytical tools. Some of them, such as doctrinal discourse, 'normal' planning, the principle of spatial organisation and planning principles, are closely tied to the doctrine concept; others, such as the action-centred/decision-centred view of planning, constitute approaches which are rather different and tend to define strategic planning through convergent aims.

To capture adequately the character of strategic planning, the notion of strategic projects is a key concept. Faludi defines these as "a specific class of projects so grand that each one is considered in its own right. The point is that they are not the outcome of strategic planning. Rather, they set the context within which planning takes place - plans and planning have to adapt to *them*" (Faludi and Van der Valk, 1994, p. 3-4).

A strategic process develops from and sets up around a strategic project. The aim of strategic planning is to formulate decision processes using a strategic project as a catalysing component which defines the context. The recognition of the role of strategic projects is important because it introduces a cooperative relationship between project and plan - a relationship which often seems to be denied by other analytical tools - like the contrasting pairs project plan/strategic plan (Fig. 1), conformance/performance, object-centred/decision-centred planning, technocratic/sociocratic planning (Fig. 2), which are used by Faludi to define strategic planning as the opposition between a product-oriented and a process-oriented approach.

	project plans	*strategic plans*
object	material	decisions
interaction	until adoption	continuous
future	closed	open
time element	limited to phasing	central to problem
form	blueprint	minutes of last meeting
effect	determinate	frames of reference

Fig. 1 Two types of plan (Faludi and Van der Valk, 1994)

If the opposition pairs were reduced simply to the dichotomy of good and bad planning form, for instance sociocratic planning versus technocratic planning, they would lose much of their heuristic value. A project plan is not necessarily a technocratic one, just as a strategic plan is not necessarily a sociocratic one. In a democratic context any strategic plan, like any project which is not the expression of a specific aesthetic aim, is a social construct and involves collective action. As such it may be the product of a selected coalition - progressive or conservative - which develops a particular perspective of change, or of many coalitions which, through confrontation consolidate a cooperative perspective of change.

	technocratic planning	*sociocratic planning*
planning subject	monolithic	coalition
role experts	linchpin	one out of many
centralisation decisions	great	small
plan as product	dominant	relative
form of plan	blueprint	indicative
measure of effectiveness	conformance	performance
scope	comprehensive	selective
notion of rationality	absolute	contextual
planning process	linear	cyclical

Fig. 2 Two forms of planning (Faludi and Van der Valk, 1994)

A planning system is in practice the result of the interaction between many different types of practice and plans: strategic projects and policies, strategic plans, project plans, urban and architectural projects. A good assessment of the functioning of the system depends on the understanding of the role of a plan and its relationship with the planning system. This is also necessary to identify how the roles of plans may be cooperative from the normative viewpoint.

The definitions of strategic projects and strategic plans given by Faludi are a clear contribution both to the understanding of the functioning of the planning system and the construction of a shared technical language; the role distinction introduced by the contrasting definitions of project plan and strategic plan is more problematic (Fig. 1). The distinction is surely, as Faludi suggests, a distinction of plan content, interaction, relationship to time and the effects produced. With regard to these differences, only the strategic plan is consistent with the definition of strategic planning given by Faludi. But project plans and strategic plans do not necessarily compete for the same role, they differ simply in that they fulfill different roles; to justify this assertion, we make a short digression about project plans.

In relation to the contrast between the object-centred/decision-centred view of planning, Faludi argues that the assumption underlying the former "is that analysing the object of concern provides a rock-solid base for plan-making, which has of course been the ideal of classic planning thought" (Faludi, 1994b, p. 37). It is difficult to disagree with this, but the assumption is simply a theory - practice has suggested something quite different. The production of project plans is generally an interaction process using planning analysis as the rhetoric for justifying planning choices; the choices are made through negotiation processes and certainly not following positivistic methods. The project plans of which I have direct or indirect experience have scarcely ever resulted from classical procedures, but tended to take form around one strong proposal from a confused and contradictory flux of conservation and transformation projects continuously recurring in the decision process. The dominant proposal may be also an 'architectural' one, such the 'compass plan' which gives direction to negotiations (Faludi, 1994b).

In Italian practice at least, the production of *piani regolatori* has rarely been the product of classical planning thought, even though planners have been able to vindicate them as a scientifically valid tools and convert their procedures into planning law. The divide between analysis and plan has always in practice been as great as the dividing line between law and praxis (Mazza, 1986, 1987, 1988, 1994b).

If we think of the project plan as the product of classical planning thought, and consider the relationship between project plan and strategic plan in terms of technocratic/sociocratic planning, we must ignore the theory, but so doing we lose the opportunity to understand how different tools and experts interact, consciously or unconsciously, in the process of environmental change. At this point we could ask ourselves why in so many countries it is project plans, despite being very poor technical tools (forms of zoning), which in fact become accepted as the plan with the greatest political, administrative, and technical importance. One answer may lie in the role developed by project plans as tools of social regulation.

In the past, project plans were included in the criticism which considered the state as a mere device in the hands of capitalists, and planning a form of state intervention. Project plans were seen simply a tool of property speculation and social segregation. Since then the role of social regulation associated with the ritual origin of planning, seems to have been kept in the background and obscured by the shared belief that project plans are future oriented.

Project plans in practice have a double nature - the same plan has contradictory and conflictual elements, since it derives from objectives that can be oriented towards conservation and change, tradition and innovation. There is on the one hand an absolute and deontic rationality, on the other a contextual and teleological one; on the one hand a mandatory logic inspired by practices and rules, on the other a logic inspired by principles and strategies. This double nature is also reflected in the plan form, which is a blueprint, but not a homogeneous one. While conservation prescriptions are detailed and determining, the prescriptions for change are indicative, general and partially discretionary. While conservative and exclusion choices are consequently closed to the future, transformation and expansion choices are open; the whole plan approach is at the same time comprehensive and selective. Conservation choices involve an interaction which lasts until they are adopted, but for choices concerning the main transformations, the interaction continues after the adoption, since it is necessary to translate the general plan into detailed action plans and then into operational projects. The implementation process is therefore linear for the conservative choices, and cyclical for choices concerning change; the importance of time is relative and decisive respectively, the effectiveness measured respectively by conformance and performance (Fig. 3; see also Mazza, 1995).

scope	conservation	change
term of reference	tradition	innovation
future	close	open
approach	comprehensive	selective
rationality	absolute	contextual
	deontic	teleological
logic	rule	strategy
	obligation	prevision
	practices	principles
form	blueprint	indicative
prescriptions	determinant	framing
	detailed	general
interaction	until adoption	continuous
time importance	relative	determinant
implementation	linear	cyclical
measure of effectiveness	conformance	performance

Fig. 3 The double nature of project plans

In conclusion, project plans are in practice a blend of order and change, rule and strategy. They are both object-centred and action-centred. They are discretionary and may sometimes degenerate into corruption and long expensive plan variations. It must therefore be stressed that the political and monetary cost of modifying choices inhibits plan variation. It is not so much the incompetence of planners, but the double nature of project plans which explains their frequently poor technical quality and has led to the incomplete institutionalisation of planning. This shows itself at the technical level, but also as a continuous tension between the search for rationality and the need to 'break with form' in order to provide for the future.

Project plans are *above all* rules aimed at reproducing a conservative order - that consisting of property rights and revealing in the physical environmental the social hierarchies and power relations within the community. Project plans are above all rules for ordering, *even though* they are also tools for changing and reconstituting that order. The belief that project plans are future oriented hides their ritual dimension, which imposes spatial order, sometimes through the institutional violence of a police tool. The belief that order is necessary so that a local community can stand as a local state is more or less consciously nourished by planners, who prefer to identify themselves in the heroic role of introducers of change, rather than the severe role of watching over conservation and order. Nevertheless it is also nourished by the support given by plans to urban growth, a function of project plans especially in the past. The blend of conservation and change, rule and strategy, seems to prevent planners themselves from recognising - or allows them not to recognise - that project plans are more important than a desired future transformation of the physical environment. They are a local constitution, or a second level contract inscribed in the first level contract, the national Constitution (Mazza, 1987, 1992a, 1992b, 1993a, 1993d, 1993e; for a different approach see Baer, 1994).

As local constitutions, project plans are a material product which can be (and often is) varied. They tend however not to have time limits and aim to produce determined effects. From this point of view, the attempt to produce flexible project plans is contradictory. In order to be effective, project plans have to be inflexible, like laws which may be interpreted, but not at random. In many countries approved project plans are in fact State law. The search for flexibility is understandable inasmuch as project plans are also a representation of a desired future, and an inflexible representation is counter to the development of processes that the plan should lead and support.

The contradictory blend which characterises a project plan is such that it is almost impossible to foresee how the components may interact and hence predict the consequences of the decisions. Not surprisingly, it is quite difficult to 'read' and evaluate properly a project plan, since planners do not have at their disposal tested protocols of plan analysis. In the attempt to carry out two incompatible functions (a conservative constitutive function and a transformative design function), not surprisingly project plans have at times ended up by paralysing rather than facilitating the process of change. In the last decade their bad performance has underpinned the planning crisis.

Our account of project plans may to some extent explain the recurrent success of architects' proposals in public debate. Planners should be conscious that architects' simplified proposals, even when they present conservation projects, often interpret the role of introducing change far better than planners' plans. The representation of power relations within the city is more easily recognisable in architectural and urban design projects than in plans. But many of these projects though striking, may be inflexible and sometimes unrealistic, hiding their ordering characteristics, consciously or unconsciously, behind an appealing future orientation. Imagination and creativity can be useful tools for concealing spatial ordering and exploitation processes, avoiding the scrutiny of public confrontation! However, while architects' projects may obscure the reality of urban conflicts, they have the merit of making the form of the physical environment more visible and understandable than a detailed report would do.

It would be too simple to trace the responsibility for the ugliness of the built environment produced over the last fifty years entirely back to architects and planners. It is true however that planners have been far too indifferent to the physical features of spatial order. The passionate advice given to planners by Peter Hall (1988) that they should interest themselves again in these essential themes, seems for the time being to be unheeded.

One of the issues that most irritates planners is the reference frequently made by architects to the permanence of the physical form of the city and its autonomy with regard to land-use models (see, for example, Webber quoted by Faludi, 1994b, p. 29-30). It is hard not to find pretentious the idea that diachronic and synchronic interpretations of urban morphology may command urban development plans, in particular when interpretations are not clearly argued. But the long term nature of planning doctrines is also an implicit recognition that, except in periods of rapid urbanisation, the physical environment changes very slowly, or often not at all, forcing land uses to be suitable for their form and location. Admittedly, in recent decades many urban development disasters have been caused by the unwillingness of planners to take into account the knowledge stratified over centuries of environmental transformation. Morphological studies produced by human geographers, land historians, and architects often capture that knowledge better than planning analysts. The recognition of the autonomy and permanence of physical form does not however imply that the design of their change should be put in the hands of experts without any public confrontation; architects should be more alert and careful about this issue.

If strategic planning is a decision-framing process developing from a strategic project, and project plans are considered to be a technically confused, less acceptable face of planning and simply an occasional operational complement to strategic planning, then we could examine the hypothesis of ridding project plans of the planning dimension. In this case dialectics within the planning system would be developed through strategic projects and strategic plans, and operational urban design through architectural projects - the latter possibly referring to zoning maps.

In effect, urban design and architectural projects play many different roles in the planning system. Very general projects extending over a large area may be the visible form of a strategic project, or a project plan unconscious of its constitutive role. They may consciously subordinate their constitutive role to particular speculative and/or aesthetic needs, or, finally, they may be operational projects and part of the strategic plan from which they derive their constraints and limits. But in all of these cases, to refuse or not to consider the constitutive effects of urban design or architectural projects is not consistent with professional integrity and technical responsibility. Such projects may suggest new strategies and help to improve or give precision to both strategic and project plans. Urban design and architectural projects therefore need to be taken into account not only because they are important tools of the urban market, but because they may be helpful to the strategic plan, even when difficult to justify in reasoned terms.

Project plans need to introduce both order *and* change, making them some form of 'non-permanent local constitution', as expressed by Haar and quoted by Baer (1994). This role recognises the ritual origin of planning, i.e. the function of concealing and legitimising the violence implicit in planning choices, which planners seem to be reluctant to recognise. For over forty years - the Schuster Committee report was published in 1950 - the limitations and defects of project plans have been widely stressed. Nevertheless, to exclude them from the planning system seems difficult, since they supply indispensable guarantees and directions to political and economic markets. To modify this approach is not easy - attempts over the last decade at deregulation have usually made things worse, since they have weakened the guarantee system without improving the flexibility of development processes.

With regard to project plans, three hypotheses can be put forward. Firstly, they could be left out of the planning system completely, but their constitutive role would be concealed and the entire development process less transparent. Definition, discussion and evaluation of urban and architectural projects would then be made with reference solely to strategic plans. There would be no detailed or formal description of the existing situation to serve as a safeguard and a second point of reference. Secondly, project plans could be left to those practising outside the academic horizon, since they seem intractable to the scientific approach. The third solution would be attempt to come to a better understanding. Are projects plans really incompatible, for example, with the more advanced methods and techniques and newer forms of plan which correspond to criteria of conformance and performance? Assuming that the planning system should take into account all phenomena involved in environmental transformation processes, the third hypothesis would seem most justified.

19.5 Planning Doctrine and Local Planning

Faludi does not seem to apply the doctrine concept to local planning, though this could be a theme worth exploring. If my understanding is correct, the concept could be helpful as an analytical and normative tool at the local level too. On the basis of my own experience, I would suggest that the doctrine concept is independent of the size of area for which it is developed.

There are cities in which change has been marked by a distinct change in planning principles and the principles of spatial organisation. In two cases where I have practised, the metropolitan city of Turin and the provincial town of Alessandria, I have had a glimpse of the local planning history, and what struck me most was the long term permanence of very simple spatial organisation principles. Those principles were determining factors for the definition of urban change over many decades until the whole framework, apparently quite abruptly, changed completely. In both cases reasons for change and the way it occurred were political rather than technical. To use Faludi's terminology, normal planning came after the development of the doctrinal discourse (Faludi, 1994b, p. 53).

The first great change in spatial organisation in Turin occurred at the beginning of this century when the bourgeois local government of the time decided to abandon the orthogonal grid which had for centuries characterised the urban form and to develop a radial pattern. The justification was that it opened the city to the surrounding countryside and markets (Mazza, 1991). In 1956, the first plan (*piano regolatore*) approved after the second World War, opposed the proposal to develop the city along the north-south axis, which would have changed the spatial organisation once again, and maintained the radial pattern in the extension of almost all suburban areas. Finally, in the early 80s, a new proposal which openly opposed the 1956 plan, attempted to change many planning rules, but did not modify the existing spatial organisation. It would be interesting to establish to what extent the absence of a new principle of spatial organisation, due in part to ideological conflicts within local government (Mazza, 1988), contributed to the failure of the proposal.

Alessandria, in the second half of the last century (after the town wall was demolished) experienced slow growth following a form typical of town expansion projects of the time, but devoid of clear principles of organisation. Urban development continued for about a century, but not until the 60s did the city council chose a project plan which established a new principle of spatial organisation. The new plan radically changed the traditional model of growth. The decision was made as a consequence of the first political change in local government for twenty years; the plan was at the same time their manifesto and the first outcome of the new majority in the council (Mazza, 1976, 1982).

Change in other cities, though it was on the basis of new planning principles, i.e. 'notions about the preparation, form, working of plans' and 'interrelated and durable notions about spatial arrangements, development, how they are to be handled' (Faludi and Van der Valk, 1994, p. 18-19), nevertheless occurred mostly in an incremental way through progressive manipulation, often weakening the original principle of spatial organisation without producing a new one. In other words, despite the development of formal planning, the final outcome was not really satisfying since, in the end, there was no recognisable overall principle of spatial organisation. Historical towns like Lecco, Pinerolo and Trino, or more recent towns like Desio, where I have worked, are to varying degrees examples of towns of this type with very weak spatial organisation principles.

It would be simplistic, however, to attempt to transfer the doctrine concept directly to the local level, especially as it was conceived with regard to national-regional planning, "our emphasis is on national planning" (Faludi and Van der Valk, 1994, p. xiv), Faludi also seems inclined to transfer it to the international scale (Faludi, 1994a). However, if what I recalled briefly above is correct, it would be helpful to understand whether the doctrine concept might be applied in Turin and Alessandria.

A different but not less interesting question is why in other towns the original spatial organisation principle was lost and why it is now so difficult to find a new one. If project plans are a kind of local constitution, they might also be considered a materialisation of the doctrine, i.e. a physical manifestation of the suggestions which the doctrine necessarily expresses in an undetermined way.

19.6 Planning Theory, Planning Practice and Planning History

A relatively recent activity such as contemporary planning needs a professional and disciplinary history to construct a shared identity and a technical language to allow the cumulative development of technical knowledge. Traditional histories of authors and works are useful for compiling a history of the discipline, but they are not sufficient. Similarly, urban and administrative history, or the history of urban design, architecture or landscape are insufficient, because in all of these planning is the context, object or tool and not the subject.

I am not competent to discuss the historical account given by Faludi, but it is not difficult to imagine that Dutch urban historians, economists, geographers and other experts may well give different accounts of the same events. From the planning point of view I do not think however that this really matters. The value of Faludi's paper and of the book co-authored with Van der Valk, is the interlacing of the historical account with a theoretical framework which constitutes an autonomous planning perspective. In other words, planning is recognised as a discipline capable of providing its own interpretation of Dutch events. This is the result of the authors' ability to develop the relationship within urban history, the planning profession and discipline. The concept underlying this relationship is that of doctrine, which is the authors' key to understanding Dutch planning history. It is at the same time constructed and justified on the basis of planning events. Possible corrections of the historical account do not modify the core content of the authors' effort. For these reasons, I think that, even if planning is a contextual activity, the contribution of the paper and book crosses the borders of national experience and can be accepted as a piece of technical planning history offered from within the discipline.

References

Alexander E.R., Faludi A. (1990) Planning Doctrine: Its Uses and Application, *Working Papers of the Institute of Planning and Demography*, no. 120, University of Amsterdam, Amsterdam.

Baer W.C. (1994) Are Plans Really Constitutions?, Proceedings of the 36th ACSP Annual Conference, Tempe.

Bell C. (1992) *Ritual Theory, Ritual Practice*, OUP, New York.

Eliade M. (1940/1943) *Commentarii la legenda Mesterului Manole*, and other essays, Policom, Bucarest.

Faludi A. (1984) *Critical Rationalism and Planning Methodology*, Pion, London.

Faludi A. (1987) *A Decision-Centred View of Enironmental Planning*, Pergamon Press, Oxford.

Faludi A. (1994) European Planning Doctrine: A Bridge Too Far? in Planning for a Broader Europe, Proceedings of the VIII AESOP Congress, August 24-27, Volume 1, Yildiz Technical University, Istanbul, 1-14.

Faludi A. (1996) Framing with Images, *Environment and Planning B: Planning and Design, 23*, 93-108.

Faludi A., Van der Valk A.J. (1994) *Rule and Order: Dutch Planning Doctrine in the Twentieth Century*, Kluwer Academic Publishers, Dordrecht.

Fischer F., Forester J. (eds.) (1993) *The Argumentative Turn in Policy Analysis and Planning*, UCL Press, London.

Foley D.L. (1963) *Controlling London's Growth: Planning in the Great Wen 1940-1960*, University of California Press, Berkeley, California.

Girard R. (1972) *La violence et le sacré*, Grasset, Paris.

Haar C.M. (1955) The Master Plan: An Impermanent Constitution, *Law and Contemporary Problems, 20*, 3, 353-376.

Hall P. (1988) *Cities of Tomorrow, An Intellectual History of Urban Planning and Design in the Twentieth Century*, Blackwell, Oxford.

Lindblom C., Cohen D. (1979) *Usable Knowledge*, Yale University Press, New Haven.

Mazza L. (1976) Problemi della gestione urbanistica in una città di medie dimensioni: il caso di Alessandria, in Clerici A. (eds.) *Contributi per un'indagine sulla gestione del territorio*, Giuffrè, Milan, 65-153.

Mazza L. (1982) Il caso di Alessandria: il progetto di intervento per l'area centrale, *Urbanistica, 74*, 71-84.

Mazza L. (1986) Giustificazione e autonomia degli elementi di piano, *Urbanistica*, 82, 56-63.

Mazza L. (1987) *Dispense del corso di teoria dell'urbanistica*, CELID, Turin.

Mazza L. (1988) Politica amministrativa e pianificazione, *Spazio e Società*, June, 76-79.

Mazza L. (1990) Società locale e strategie economiche: è possibile una convergenza nella politica urbana? in Bagnasco A. (ed.) *La città dopo Ford, il caso di Torino*, Boringhieri, Turin, 108-125.

Mazza L. (1991) Le trasformazioni del piano, in Mazza L., Olmo C. (eds.) *Architettura e Urbanistica a Torino 1945-1990*, Allemandi, Turin, 61-85.

Mazza L. (1992a) An Exercise in Re-constructing a Planning Tool, International Conference on Planning Technologies and Planning Institutions, 8-11 September, Palermo.

Mazza L. (1992b) Descrizione e previsione, in Lombardo S., Preto G. (eds.) *Innovazione*

e trasformazione della città, Angeli, Milan, 181-196.
Mazza L. (1993a) Conservazione e trasformazione: una ridefinizione del piano regolatore, *Controspazio*, 5, September-October, 4-10.
Mazza L. (1993b) Il sapere tecnico comune degli urbanisti, in Palermo P. (ed.) *Urbanistica, politiche e tecnica*, Grafo, Milan, 43-52.
Mazza L. (1993c) Designing a Domain for Planning Theory, introduction to the section one, in Mandelbaum S., Burchell B., Mazza L. (eds.) *Planning Theory in the '90s*, Rutdgers U.P., New Brunswick.
Mazza L. (1993d) Previsione e obbligazione, cambiamento e conservazione: un esercizio di ricostruzione del piano regolatore, *Territorio*, 15, 71-93.
Mazza L. (1993e) Conservazione e trasformazione: una ridefinizione del piano regolatore, in Vila E., Scattoni P. (eds.) *Organizzazione del territorio e gestione urbanistica*, Florence, Annali, 17-32.
Mazza L. (1994a) Piano, progetti, strategie, *Critica della razionalità urbanistica*, 2, 50-55.
Mazza L. (1994b) Pubblico e privato nelle decisioni di urbanizzazione, in Camagni R., Boscacci G. (eds.) *Tra città e campagna: periurbanizzazione e politiche territoriali*, Il Mulino, Bologna, 351-374.
Mazza L. (1995) About the Nature of Traditional Local Plans, Proceedings of the 9th AESOP Congress, Glasgow, August 16-19.
Mazza L. (1996) Difficoltà della pianificazione strategica, *Territorio*, 2, 176-182.
Rykwert J. (1976) *The Idea of a Town, The Anthropology of Urban Form in Rome, Italy and the Ancient World*, MIT Press, Cambridge, Massachusetts.
Schön D. (1983) *The Reflective Practitioner*, Basic Books, New York.
Selznik P. (1953) *TVA and the Grass Roots: A Study in the Sociology of Formal Organisations*, University of California Press, Berkeley, California.
Simon H. (1981) *The Sciences of Artificial*, MIT Press, Cambridge, Massachusetts.
Wheatley P. (1981) *La città come simbolo*, Morcelliana, Brescia.

20. Selected Issues of Urban Planning

Giovanni A. Rabino

20.1 Introduction

As Mazza usefully outlines in this book, a central role in Faludi's paper and, more generally, in the whole of Faludi's thought (see Faludi and Van der Valk, 1994) is played by the concept of planning doctrine. This idea can be described as "a systematic thought concerning: (a) the spatial organisation of an area, (b) the changement of this area, (c) the way both are pursued" (Faludi and Van der Valk, 1994, p. 18-19). These characteristics of systematic setting and pertinence enable doctrine (see Mazza, 1986, De Luca and Las Casas, 1997) to give relevance and consistency to planning, ensuring effectiveness and credibility (which are in fact the main requirements, see Piroddi, 1997).

While there are numerous papers on the relevance of planning doctrine, there are far fewer on the systematic setting, namely the scientific foundation of urban planning. The reason is the antiscientific attitude, which is widespread among urban planners, despite the fact that, for the reason given above, this can undermine the effectiveness of planning. In the present paper I wish to examine the scientific foundation of the discipline with the aim of presenting the most recent developments of this approach.

I intend to present this emerging system of ideas (in my own re-elaboration) as a set of theses. These, which cover the classical themes of the current debate on urban planning, deliberately exclude some important dimensions such as aesthetics and, above all, ethics. This is partly for reasons of space and my own incomplete development of these themes, but mainly because of the lack of resolution of these aspects of the cultural paradigm. Useful elements have been provided for interpretation, but not prescriptive guidelines (Crook, 1994).

Foundations of this paradigm or, to be more precise, postulates of my *weltanschauung*, are:

- in relation to the nature of knowledge, a position definable as critical genetic realism (Piaget, 1970, Popper, 1981);
- in relation to the theoretical and methodological basis of the sciences (including territorial sciences) a synergetic-evolutionary type axiom (Delbruck, 1986, Haken, 1977, Prigogine, 1980);
- in relation to the specific character of the current socio-economic scenario (subject and container of urban planning) an interpretation in terms of the cosmo-creative society (McLuhan, 1970, Laszlo, 1972, Andersson *et al.*, 1993).

Even though necessarily in a synthetic and incomplete way, it could be useful to make some elements of these assumptions more explicit.

As regards the first (the epistemological framework), it seems to me that a good starting point is the mind-brain problem. Almost all neurobiologists nowadays believe (and I share their opinion) that the mind, and probably consciousness, is the expression of the activity of a very large number of neurons (the brain). This biological naturalism (Searle, 1992) means that mental phenomena should not be dealt with as something metaphysical, but at the same time it does not mean that they are *nothing but* mechanisms, in a simple and mechanical sense. Therefore any counterposition between culture and nature is artificial; they are two systems which evolve and coevolve. In this context, although the truth of an argument does not necessarily consist of what human evidence and human means of investigation can show, the conclusions reached in this way are not necessarily the determining factors of this truth. So there is a space between the utopia of classical absolute rationality and the irrational nihilism of hermeneutically closed linguistic systems, for the development of a Popperian style of knowledge which is objective but not dogmatic (true but not certain) and always self-critical in a evolutionary way. And this applies, according to Piaget, to the highest levels of knowledge, where epistemologies and scientific theories co-evolve with retroactive feedback.

Finally, we should mention the problem of scientific progress (that is to say the definition of the best or fittest theories and epistemologies in the evolutive comparison). Personally, I think that the question, which has still not been resolved in the debate among post-Popperian epistemologists (Laudan, 1977), has to be considered within the paradigm of critical realism. This point of view provides a solution to the problem.

As regards the second assumption (the theoretical framework), the natural starting-point is the deep change which is taking place in scientific theories. Most people perceive this as radical enough to suppose a complete break with classical science and the birth of a new paradigm (the complexity paradigm) focussing on self-referencial, hologrammatic principles (Morin, 1990). Even though it is disputable whether it is a true paradigmatic breaking-point, this view is particularly effective in explaining why in the past:

- large interactive systems were analysed in the same way as small ordered systems, mainly because the method formulated had proved itself so effective for simple systems;
- we were convinced that the behaviour of large interactive systems could be anticipated by studying their constituent elements and analysing the microscopic mechanisms individually;
- in default of a better theory, we supposed that the output of the large interactive systems was proportional to the perturbations, hence the dynamics of these systems could be described in an equilibrium state, disturbed from time to time by an exogenous force.

Over the past decade it has become more and more evident that:

- in large interactive systems, global properties can emerge (so-called because they appear only from the whole system), requiring nontraditional methods of analysis;
- the unexpected appearance of these properties derives from the interactions between local subsystem behaviour and the global behaviour (the so-called micro/macro relations);
- this usually involves conditions of irreversibility and disequilibrium. Large interactive systems never reach an equilibrium state, they evolve from one metastable state to another.

In this synergetic (and also intrinsically evolutionary) approach to systems, we should perhaps specify that Darwinist principles still have a place - no longer however as a mere biological metaphor, but in the form of the so-called neo-neo-Darwinist systems theories (see for instance Casti, 1989). Rather at random, and left to the intuitive understanding, we mention some key concepts relating to these theories:

- the concept of window (limited horizon) of observability of the system through the subsystems (with the consequent focus on heuristics, robustness, satisfaction etc.);

- the mechanism of trial and error and differential selection in the evolution of the system (with the consequent importance of coevolution between system and subsystems, the uniqueness and similarity of the subsystems, their phenotype and genotype);
- the blind and imperfect character of evolution (with its serendipity, creative but without any finalism).

We should also specify that today's evolutionism, far from being a monolithic and complete construction, is a living science, with objections, unsolved problems and lively diatribes over basic matters. But everybody now admits that these are internal affairs and that the essential truth of evolutionism is never in doubt (Gould, 1987).

As regards the third and last assumption (the context of urban planning, i.e. the character of our civilisation), we need to consider aspects such as high technology and the media as well as the various social phenomena associated with them. To describe the importance of innovation and the diffusion of its effects, we often hear the use of expressions such as the 'information society', 'technopolis' etc. Many of these linguistic symbols however do not seem to go beyond a mechanistic vision of society. To be consistent with the two previous assumptions, I prefer to speak of the 'cosmo-creative society', since this catches a salient feature of the present civilisation, the explosive spread of creativity in all its manifestations throughout our culture (especially, but not only, in the scientific world). The trigger of this process is coevolution, the positive self-accelerating feedbacks between culture, science and technology (see for instance Cini, 1990).

Firstly, in the relationship between technology and science, there has been an enormous increase in the speed of transfer of scientific advance to technological products and growing scientific added value in these same products. Technology, on the other hand, has been able to provide scientific research with investigation and measurement instruments of greatly enhanced performance. There is a decreasing distinction between science and technology. It is increasingly difficult to differentiate complex technology projects from pure scientific research. Technological progress has also opened immense potential and influenced the style of research due, for instance, to the possibility of computerised data processing. Armed with this accumulation of science and technology, our society, which is already deeply changed but still rapidly evolving, is facing:

- transformation in its demographic structure;
- redefinition of the basis of its organisation;

• and, above all, growth in its culture.

So ours is a more deeply and more widely educated society (and for this reason more complex and varied), which wants to be creative *per se* and express this creativity through science, technology and art. We should note that, being an educated society, it accompanies this with self-critical reflection. It remains to say that the city, as the place where the above system of interactions are manifest (Mela, 1985) is the natural seat of modern civilisation (the common etymology: city ↔ civitas ↔ civilisation is more valid than ever). And as the scale of civilisation is now planetary, the city is everywhere!

It is on the basis of these assumptions, accepted by society in general, even though sometimes in a latent way (and familiar to the Italian urban planning community, but not normally applied), that I approach the following six questions, two related to epistemology, two related to theory and method, and two concerning planning praxis. The paper concludes with a final observation, which derives logically from the reasoning presented.

20.2 Urban Planning: Art, Science or Profession?

Sciences are usually cited in terms of concepts (fractals, neural nets etc.), with which the name of some scientist is connected, while in the arts we tend to cite personalities (Hopper, Warhol etc. and similarly Geddes, Lynch ... Rodwin, Schön ...) with whom a cultural movement or cultural position is associated. It is a small sign that reveals, despite the widely held opposite view, how deeply rooted is the artistic conception of urban planning. So a reflection on the nature (artistic or scientific) of the discipline is not obsolete. Some results, now unanimously accepted, of cognitive psychology concerning types of knowledge (see for instance Bara, 1990) help to clarify the issue. Summarising, we can present a map of the types of knowledge and their interactions (Fig. 1):

• the first type is so-called explicit knowledge (K-explicit). This corresponds to the intuitive concept of knowledge. It is what we know we know, i.e. an aware knowledge, about which we can voluntarily reflect and that can be linguistically expressed. The characteristic representation of this type of knowledge is through logical formalism, in particular classical logic; but since this has revealed limits, other systems of logic have been proposed - default logic, self-deductive logic etc. - or other approaches, such as semantic nets or frames;

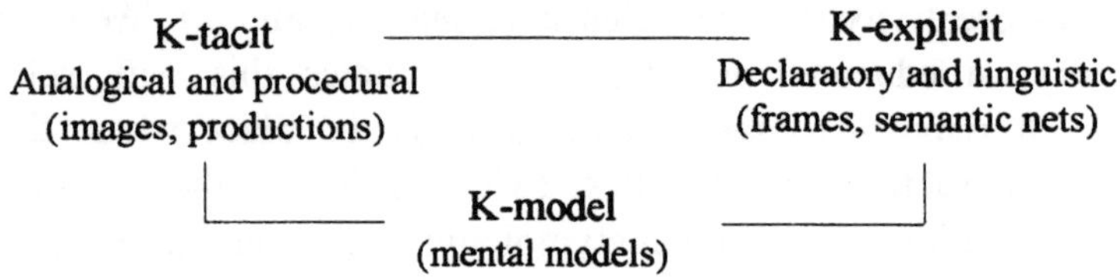

Fig. 1 Types of knowledge and their interactions

- the second type is tacit knowledge. It corresponds to knowing how to interact effectively with the world, even if we are not able to make such knowledge explicit (i.e. describe it directly and reflect on it). Examples are: being able to ride a bicycle, to recognise a wine, to be at one's ease, be able to sing, paint, write poetry etc.; all things that the expert can do well, but can only express verbally approximately, with analogies and metaphors. This type of knowledge is represented through the so-called rules of production, procedural codes which allow us to know how to act;
- the third type of knowledge, model knowledge, is a specific model which integrates the two previous types of knowledge. It is a set of partial configurations of the theoretical knowledge imbedded in K-tacit and K-explicit knowledge; it is the aspect of knowledge which the thinking person is effectively using in a specific moment. The analogical representation of this form of knowledge is the mental model (Johnson-Laird, 1983). This is made up of elements and relations which represent a particular condition of things, already structured in a appropriate way for use.

I should add that the relationship between K-tacit and K-explicit knowledge is complex and elusive, corresponding roughly to the connection between that which is experienced and that which is symbolised.

From tacit knowledge we can try to build a propositional theory and from explicit knowledge to organise procedural codes. However, while K-explicit is by its nature transmissible or teachable (and therefore socialisable), K-tacit knowledge is personal, and learnable only by doing.

On this basis, I think we can formulate the following thesis:

- in urban planning (as in human actions in general) art and science cannot be put in antithetic terms. In every action or professional activity (to include science and art), to proceed by means of mental models inevitably involves merging the two types of knowledge: tacit (which is the essential, but not exclusive, basis of artistic creation) and explicit (which is specific, but not exclusive, to scientific research);
- society, in its growing cosmo-creativity and consequently growing and increasingly aware participation in planning processes (both as informed subject and direct actor), necessarily requires explicit (transmittable) knowledge in order to be able to participate. So the development of scientific skills in urban planning, and achieving a higher K-explicit content is necessary for its own *raison d'être* (and survival as a recognisable academic discipline).

20.3 Technical Languages in Urban Planning

Strong languages (that is to say propositional theories and K-explicit knowledge) do not seem to be particularly appreciated in current urban planning. Not only the systemic and formalised approach (never totally accepted, even at the time of the so-called rational-comprehensive planning), but also the logical systemism of the jurisprudential approach and the technological certainties of physical planning seem to be marginal to prevalent urbanistic thought. The use of certain locutions of technical language or background knowledge subliminally communicates the sense of little importance.

Let us consider the criticisms of the mathematical approach (i.e. the systemic and formalised method) as the case *par excellence* of K-explicit knowledge, not forgetting the general validity of the considerations that follow.

While the radical ideological objections of the seventies have now been exhausted (note: the history of evolutionist thought and of historical materialism is full of consonances and misunderstandings; see Rosser, 1991) and many of the objections formulated on a hermeneutic basis in the eighties having failed (see Vozza, 1990) with the diffusion of critical rationalism, the most virulent attack on the mathematical method has come from post-structuralist deconstructionism (Derrida, 1967). The accusation is that its application in this field does not correspond intrinsically to the principles usually given as justification for its use (or its preference in relation to other languages): its universality (i.e. its validity at any time and in any place); the logical rigour (which make it possible to deduce consistent theories); its objectiveness (which excludes individual and collective cultural bias); the simplification (to read the complexity of the world) and the exactness (which eliminates ambiguities). To be precise, we have to say that the criticism is directed towards a classical rationalistic conception of the method; but it is also a useful exercise for critical rationalism.

As Barnes (1994) argues, from a deconstructionist point of view, the mathematical method does not keep any of its promises because of contradictions from which he claims there is no escape:

- universality is always founded on local assertions;
- logic is unable by itself to justify the use of logic;
- the argument of objectiveness is in itself a judgement of value;
- simplification is obtained only through complexity;
- precision is expressible only through the inaccuracy of ordinary language.

These contradictions are proved in the following way:

- it is shown (by means, for instance, of the Russell's logical paradox and Gödel's theorem of formal incompleteness) that attempts to found mathematics all fail to build a closed self-explanatory system;
- there is therefore always a residual meaning rooted in local institutional practices that arose historically, i.e. in the social dimension of knowledge (Bloor, 1976);
- this determines the inaccuracies, arbitrariness, contingencies etc. that are the basis of the above contradictions.

This reasoning, in an epistemological approach which I accept, seems however to be biased by an inconsistency. We implicitly assume that the analyst, in order to be able to declare failure in the test of self-explanatory nature of mathematics, must have a strong rationality. But the analyst cannot at the same time be free of the declared great cognitive fallacy (of knowledge as a social activity). So, if without the axiom of strong rationality, the analysis confines itself to the proposition of mathematical language as an open programme (see, for instance the developments on Russell's paradox) inspired by principles of logic, precision etc. which are understood in a relativistic way; and if knowledge, without denying its contextualisation in society, appears with naturalistic objectivistic conditionings (reinforced by the same use of strong languages); then instead of an antinomy, we have a virtuous (evolutive) circle that qualifies mathematical language in comparison with other languages.

The conclusion is a solicitation to planners "to cultivate the necessary, cumulative and provisional certainties of technical languages" (Mazza, 1994, p. 158). I personally would put no exceptions on this invitation and, among the strong languages, give an emphasis to the systemic formalised language of mathematics.

20.4 The Myth of Irrationality

Following on Mela and Preto (1990) and Camagni (1988), a useful diagram relating concepts of planning (understood in the broad sense of the capacity to govern the territory through appropriate actions) is given by:

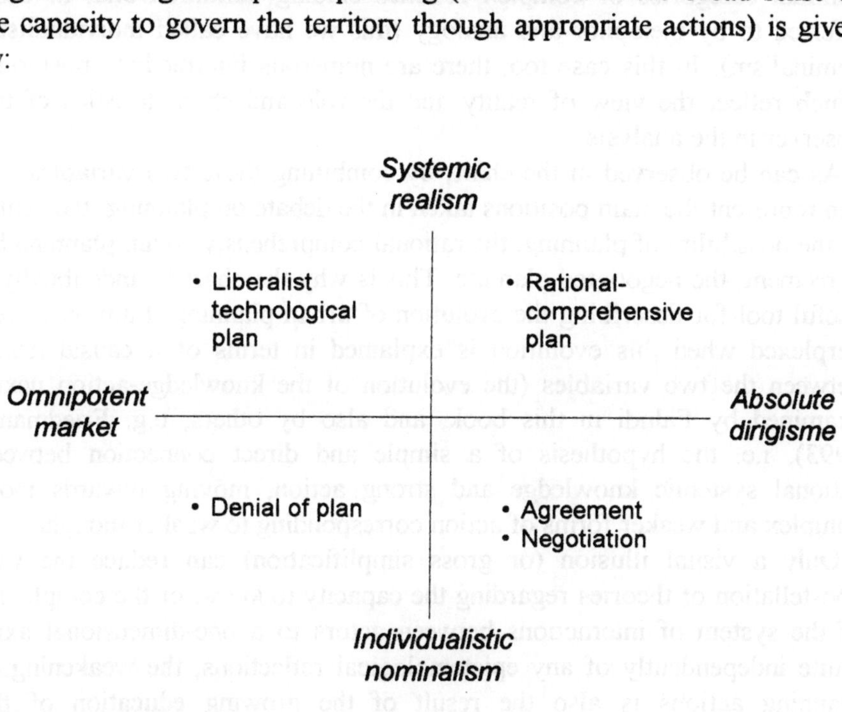

Fig. 2 Concepts of planning

The two variables that define the space of this chart pertain to the forming dimension and the cognitive dimension of the plan. The first considers the degree of coordination among the system of actors. At one extreme there is total reciprocal autonomy, where the government of the territory is left completely to the operational mechanisms of the system (i.e. the market has free play). At the opposite extreme there is complete coordination (by agreement or by hierarchical structure) among the actors, who can together totally guide the system (absolute dirigisme). Naturally, the degree of coordination between decision-makers can vary and be differentiated according to how it is achieved. Therefore, in reality, there is a wide range of intermediate possibilities.

The second variable concerns the degree to which the planned system can be interpreted. At one extreme there is a system made up of physical, economic and social entities with an objective reality for which we have strong explanatory theories describing their essence and interrelations (classic systemic realism). At the opposite extreme we have the weak concept of thought which considers the objects of the plan in terms of nominal categories of complex realities eluding definition and, at best, referred to by metaphor and analogy (this we have called individualistic nominalism). In this case too, there are numerous intermediate positions, which reflect the view of reality and the role and characteristics of the observer in the analysis.

As can be observed in the chart, by combining these two variables, we can represent the main positions taken in the debate on planning: the denial of the possibility of planning, the rational-comprehensive plan, planning by agreement, the negotiated plan etc. This is why the chart is undoubtedly a useful tool for describing the evolution of urban planning. I am, however, perplexed when this evolution is explained in terms of a causal nexus between the two variables (the evolution of the knowledge-action nexus examined by Faludi in this book, and also by others, e.g. Friedmann, 1993), i.e. the hypothesis of a simple and direct connection between rational systemic knowledge and strong action, moving towards more complex and weaker forms of action corresponding to weaker thought.

Only a visual illusion (or gross simplification) can reduce the vast constellation of theories regarding the capacity to know, or the complexity of the system of interactions between actors to a one-dimensional axis. Quite independently of any epistemological reflections, the weakening of planning actions is also the result of the growing education of the population in planning matters. People are taking on more direct involvement, and making it necessary to redefine (reinvent?) forms of coordination, also to innovate the role of the urban planner, who previously has acted as the benevolent prince's right arm! I should like to point out that this is a general phenomenon gaining vast importance. It is currently affecting the international political order as well as the organisation of Italian politics. The mechanism outlined - cultural growth leading to role crisis - is providing a necessary and suitable terrain for the autonomous development of urban planning practice and academic reflection with the latter, as observed by Faludi, in this book, being gradually professionalised.

It should be added that the illusion referred to above has operated historically in the opposite way. Faced with the transformation of society and the crisis in planning, there was a tendency to reduce the causal factor to a single axis (rationality/irrationality) overlooking the variety of epistemological paradigms and including all questions, ontological, semantic etc. in this simplified dichotomy. In this regard the debate of the eighties on rationality is significant. As observed by Reade (1985), planners did not refer to rationality in the limited sense adopted for instance by economists, but they reasoned on the basis of functional rationality and substantive rationality (among the exceptions, see Faludi), touching on, but never delving deeply into, the epistemological issues that this raises. For this reason the debate turned out to be inconclusive and left as an aftermath a vague (yet painful) sense of irrationality, that still pervades all concepts of planning except the traditional one.

An alternative to this state of things is to deal more thoroughly with epistemological matters, as well as theoretical and methodological questions. In fact advances have been made in knowledge about the nature of rationality by scholars such as Rapoport, Tversky and Harsanyi in disciplines like psychological economics and mathematical game theory (see Barry and Hardin, 1982).

20.5 Revealed Creativity

The thesis of this section is that creativity (the act, whether individual or social, displaying varying degrees of ingenious inventiveness and leading for instance to the urban project or plan), which has long been the centre of scientific attention, is a phenomena whose mechanisms are now being clarified and explained scientifically. This does is not imply that we are nearing a mechanical reproduction of the creative act (even though that may happen in the long run), but that even the most hidden component of 'knowing how to do things' - tacit knowledge - which permits the planner or designer to considers himself such, is becoming explicit knowledge.

In the past, with respect to creativity, there were two opposite attitudes (not connected on levels of erudition or by an opposition of the two cultures, humanities and science) (see Melucci, 1994):

- one, mysterious, which emphasised the uniqueness of the creative genius (Einstein, Leonardo etc.), the fortuitousness of the event (Newton's apple, Archimedes *eureka* etc.) and the connection with the irrational (the association of genius with disorderliness, the stereotype of the absent-minded professor etc.);
- the other, rationalising, which underlined the normal character of creative activity (the sweat of research, the mediocrity of much work of famous geniuses), the regularity of the event (the creative act as inevitable in given historical, cultural and economic settings, the possibility of 'learning' creativity (the lateral thought of Gardner, 1983) and, even though highly complex and not yet fully understood, the mechanistic nature of the phenomenon (creation appears at the end of a process, almost naturally, as an inevitable consequence of certain passages).

This is the context of the diatribe about the procedure of systemic planning (by McLoughlin and Chadwick). While this approach is criticised for its supposed claim to absolute control over the system (an unfounded criticism, given the space left for creativity in defining alternative policies), it is defended too weakly as being a merely informative - analytical and evaluation - phase of the planning process. In reality, it had begun to address the problem of clarifying the creative mechanisms in planning.

The latest evolutionary theories of creativity cast some light on the problem. In the differential selection process (mutation of genotypes and environmental selection) these theories reconcile the two aspects of the creative act. It is the serendipity of invention, which derives not only from problem-solving (generally speaking, the phase of selecting the most suitable from among various alternatives), but also from the factors operating in problem formulation (the phase of generation of alternatives through suitable mechanisms). It is in the problem-solving process, made into a social activity through the clarification of the mechanism involved, that systemic planning has found a precise reference, as shown for instance by the particular attention given to the nexus between simulation models, performance indicators and evaluation methods (Bertuglia, Clarke and Wilson, 1994).

Problem formulation is the subject of current studies in the field of creativity. If we reject the idea that the creative act is only the new combination of old ideas (that it does not follow simple mechanisms like random mutation or ideas such as gene-crossings), but involves the expansion of a field of endeavour through ideas that do not emerge simply by following the usual rules, it nonetheless does not suffice to rashly break the rules. A careful examination of creative work reveals the presence of constraints (various kinds of metarule). It is the interaction between the representations of the problem (i.e. the field of endeavour) and these metarules that generates a series of possibilities which may eventually produce a radical change, or creative invention. Artistic and scientific creativity thus seem to belong to the properties of the large interacting systems cited in the introduction.

These problem-solving processes are now, in simple versions, captured in software programmes - computers are at the dawn of creativity (Matthews, 1994).

20.6 Technological Innovation and Ecological Planning

Among the current topics being debated among Italian urban planners, two of the most important are the relationship between territorial systems and technological innovation, and the ecological orientation of planning (both underlie many other matters being discussed, from the amendments of normative land-use laws to the evaluation of large infrastructural projects, renewal and rehabilitation policies, and so on).

Regarding the former, among a large number of studies and projects, the Megaride Charter 94 (Beguinot, 1994) stands out. This contains, very briefly, the following proposal:

- to invert the logic of the relationship between technological innovation and territorial transformation (from innovation as an exogenous and uncontrolled factor of transformation to innovation that is functional to transformation);
- to build a liveable city (a city of beauty, of peace and of science) specifically by means of innovation.

Regarding the theme of ecologically-oriented planning, almost the entire urban planning community has made an effort, though some more specifically (see Magnaghi, 1990), coming up with a proposal that can be outlined as follows:

- to pursue sustainable development as an alternative to destruction, sacking and impoverishment of the territory caused by the search for

unlimited economic growth (IUCN, 1980);
- to this purpose, to aim for a form of local development which takes into account all specific and unrepeatable features of a place and therefore values all differences.

The two proposals seem similar in many ways; they are certainly not conflicting. Both emerge from the complexity paradigm (in contrast with the classical paradigm) cited in the introduction. Nonetheless, between the two schools of thought there is an evident contrast. The thesis put forth here is that the conflict arises from the lack of examination of complex thought in both proposals. They attribute, more or less explicitly, a positive or negative moral value to technological innovation. This moral value derives from criteria of judgement of a society that still thinks to a large degree according to the canons of classical rationality. To be precise, it is acknowledged that the scholars involved in the Megaride Charter 94 have sought motivations for their proposal in the science of complexity (see for instance Rabino, 1993) but:

- the arguments on complexity often lend themselves to ambiguous interpretations; that is, they tend to be interpreted according to classical rationality (see the proclaimed principle of simplifying complexity or the controllist nature of planning that the proposal seems to reveal);
- the science of complexity is used more as an analytical method for criticising the present urban situation than as a planning tool (the objectives of the proposal emerge in general as an antithesis to the status quo determined by the usual way of interpreting the situation, rather than a new way of perceiving problems).

At this point, it is fair to say that, in the most recent studies (for instance, Beguinot, 1997), care has been taken to avoid the kind of misunderstandings mentioned above. As far as the ecologist planning school is concerned, it should be said immediately that it is difficult to recognise in the whole ecology movement any single clear theoretical basis. Completely different cultural positions, such as bourgeois naturalism, eco-industrial technocratism, total romantic ecologism, neo-Malthusianism, one-worldism etc. appear to be grouped together. This lack of foundation has been denounced within the ecology movement itself (O'Connor, 1988) as the germ of its own destruction.

As a result of this confusion at the theoretical level, we can ascertain in authors belonging to this school:

- the persistence of a classical mentality (for instance, the enunciation of the principle of sustainable development in terms of functional extremals - minimising, constraining etc. - which is typical of absolute rationality);
- the logical shortcuts, free inductions, metaphorical jumps etc. in deriving indications for planning from the principles of complexity, often containing errors, such as the counterposition between local and global (which in complex thought are in fact always connected through the hologrammatic principle).

It is therefore no wonder that misunderstandings occur between the two schools.

20.7 Strategic Planning and Evolutionism

Another theme of the current urban planning debate is strategic planning. Derived from economics and management, like many other topics (from urban marketing to networking, the problem of sustainability itself and technological innovation), strategic planning is characterised by the following elements (Gibelli, 1993):

- it gives priority to perspective and scenario analysis;
- it is dynamic and flexible with respect to the selection of objectives and implemental choices;
- it identifies opportunities and challenges in the outside environment, and the strong and weak points inside;
- it operates in an openly pragmatic dimension;
- it proposes systemic analysis, uses learning processes and repetitive revisions and prefers negotiating interaction rather than conflicting opposition;
- it promotes consultation and extended participation;
- it attributes strategic importance to the implementation phases of the plan;
- its predominant role is one of persuasion and marketing.

To emphasise the complex and organic view that underlies these features of strategic planning would be superfluous: from the irreversible dynamics of the environment, to the limited rationality of the actor, and his flexible and satisficing behaviour. What I wish to emphasise is that these characteristics, going beyond pure contextualisation and mere biological metaphor, introduce into planning an environment, actors and mechanisms typical of evolutionary complexity; for instance:

- the co-evolution between the environment and the planning process (the role of persuasion, consultation and participation, the importance of the implementation phase etc.);
- the internal structure (genotype) of the actors that changes in relation to the interaction between its behaviour (phenotype) and the environment (identifying the strong and weak points, preferring negotiating interaction etc.);
- the centrality of trial and error in the mechanisms (learning and revision processes, scenario analysis, openly pragmatic operative dimension etc.).

What has been said should not be surprising since evolutionism seems to be the theoretical reference which dominates modern economics (Nelson and Winter, 1982) and management (Hannan and Freeman, 1989). The message of this chapter for strategic planning, and urban planning in general, is to root its investigations directly in such theoretical foundations instead of adopting them from other disciplines. In this way, I feel the analysis could be more correct and productive; certainly more original.

An addendum is appropriate here. The above might seem to imply that evolutionism is in opposition to mechanicism (and that I have abandoned the latter for the former). In fact, evolutionary theory is a more general theory that includes mechanicism as a special case. In such relationships, the principles which define the conditions under which the general rule degenerates to the particular case are particularly important (e.g. Bohr's theorem for quantistic and classic systems and the K.A.M. theorem for chaotic and regular systems). Between evolutionism and mechanicism the correspondence principle is the theorem of evolutionarily stable strategies which, generally speaking, states that: in a constant environment certain behaviour strategies are better than all the others and cannot be improved (Maynard Smith, 1982). These behaviours, seen from the outside, may seem like mechanical relationships between the factors by which they are determined. On the other hand, real systems, though evolutive, must be fairly constant (otherwise we would not be able to talk about them). Mechanicism is therefore certainly an approximation, but it often remains the best available cognitive hypothesis.

20.8 Conclusions

Throughout this entire paper I have argued in favour of scientific - but not scientistic - urban planning: I propose a scientific approach, renewed as a consequence of the (justified) criticism raised by the so-called weak school of thought, but not succumbing to the anti-scientific vein of much post-modern thought.

Space does not allow, and it was not in any case the intention of this paper, to make a detailed presentation of the operational principles of the proposed style of planning (even though some general principles have been proposed in Sections 6 and 7), but the interested reader can find an initial attempt in Rabino, 1996, concerning the planning of the transportation subsystem and a post-modern transport model. Although only a tentative approach, the soundness, validity and promise of this direction of work seem to be widely shared.

References

Andersson Å.E., Batten D.F., Kobayashi K., Yoshikawa K. (1993) *The Cosmo-Creative Society*, Springer-Verlag, Berlin.

Bara B.G. (1990) *Scienza cognitiva. Un approccio evolutivo alla simulazione della mente*, Bollati Boringhieri, Turin.

Barnes T.J. (1994) Probable Writing: Derrida, Deconstruction and the Quantitative Revolution in Human Geography, *Environment and Planning A, 26*, 1021-1040.

Barry B., Hardin R. (1982) *Rational Man and Irrational Society*, Sage, Beverly Hills.

Beguinot C. (eds.) (1994) *La carta di Megaride 94. Città della pace - città della scienza*, Giannini, Naples.

Beguinot (1997) Città, innovazione, programmazione, in Bertuglia C.S., Vaio F. (eds.) *La città e le sue scienze: la programmazione della città*, vol. 3, Angeli, Milan, 135-146.

Bertuglia C.S., Clarke G.P., Wilson A.G. (eds.) (1994) *Modelling the City: Performance, Policy and Planning*, Routledge, London.

Bloor D. (1976) *Knowledge and Social Imagery*, University of Chicago Press, Chicago.

Camagni R. (1988) Lo spazio della pianificazione, in Gibelli M.C., Magnani I. (eds.) *La pianificazione urbanistica come strumento di politica economica*, Angeli, Milan, 61-71.

Casti J.L. (1989) *Paradigms Lost: Images of Man in the Mirror of Science*, Morrow, New York.

Cini M. (1990) *Trentatrè variazioni su un tema. Soggetti dentro e fuori la scienza*, Editori Riuniti, Rome.

Crook P. (1994) *Darwinism, War and History*, Cambridge University Press, Cambridge.

Delbruck M. (1986) *Mind from Matter? An Essay on Evolutionary Epistemology*,

Blackwell Scientific Publications, London.
De Luca G., Las Casas G.B. (1997) Dalla razionalità alla ragionevolezza, il contributo delle scienze cognitive all'argomentazione in urbanistica, in Bertuglia C.S., Vaio F. (eds.) *La città e le sue scienze: la programmazione della città*, vol. 3, Angeli, Milan, 193-208.
Derrida J. (1967) *L'écriture et la difference*, Editions du Seuil, Paris.
Faludi A., Van der Valk A.J. (1994) *Rule and Order: Dutch Planning Doctrine in the Twentieth Century*, Kluwer, Dordrecht.
Friedmann J. (1993) Toward a Non-Euclidean Mode of Planning, *Journal of the American Planning Association, 59*, 482-485.
Gardner H. (1983) *Frames of Mind: The Theory of Multiple Intelligences*, Basic Books, New York.
Gibelli M.C. (1993) La crisi del piano fra logica sinottica e logica incrementalista: il contributo dello strategic planning, in Lombardo S., Preto G. (eds.) *Innovazione e trasformazioni della città*, Angeli, Milan, 207-239.
Gould S.J. (1987) *An Urchin in the Storm. Essays about Books and Ideas*, Norton and Co., New York.
Hannan M., Freeman J. (1989) *Organizational Ecology*, Harvard University Press, Harvard, Cambridge, Massachusetts.
Haken H. (1977) *Synergetics*, Springer-Verlag, Berlin.
IUCN (1980) *World Conservation Strategy*, IUCN (International Union for the Conservation of Nature).
Johnson-Laird P.N. (1983) *Mental Models. Towards a Cognitive Science of Language, Inference and Consciousness*, Cambridge University Press, Cambridge.
Laszlo E. (1972) *The Systems View of the World*, Braziller, New York.
Laudan L. (1977) *Progress and its Problems*, University of California Press, Berkeley, California.
Magnaghi A. (1990) *Il territorio dell'abitare*, Angeli, Milan.
Maynard Smith J. (1982) *Evolution and the Theory of Games*, Cambridge University Press, Cambridge.
Matthews R. (1994) Computers at the Dawn of Creativity, *New Scientist*, 1955, 30-34.
Mazza L. (1986) Giustificazione ed autonomia degli elementi di piano, *Urbanistica*, 82, 56-63.
Mazza L. (1994) I linguaggi della pianificazione, *Urbanistica*, 102, 153-158.
McLuhan M. (1970) *Culture is our Business*, McGraw-Hill, New York.
Mela A. (1985) *La città come sistema di comunicazioni sociali*, Angeli, Milan.
Mela A., Preto G. (1990) Alla ricerca della strategia perduta, in Curti F., Diappi L. (eds.) *Gerarchie e reti di città: tendenze e politiche*, Angeli, Milan, 127-154.
Melucci A. (1994) *Creatività: miti, discorsi, processi*, Feltrinelli, Milan.
Morin E. (1990) *Introduction à la pensée complexe*, E.S.F. Éditeur, Paris.
Nelson R.R., Winter S.G. (1982) *An Evolutionary Theory of Economic Change*, The Belknap Press, Cambridge, Massachusetts.
O'Connor J. (1988) *Capitalism, Nature, Socialism*, Santa Cruz.
Piaget J. (1970) *L'épistemologie génétique*, Presses Universitaires de France, Paris.
Piroddi E. (1997) Credibilità ed efficacia del piano, in Bertuglia C.S., Vaio F. (eds.) *La città e le sue scienze: la programmazione della città*, vol. 3, Angeli, Milan, 209-222.
Popper K.R. (1981) *Objective Knowledge. An Evolutionary Approach*, Clarendon Press, Oxford.

Prigogine I. (1980) *From Being to Becoming*, Freeman, San Francisco.

Rabino G.A. (1993) Per una scienza della complessità urbanistica, in *Atti del convegno internazionale: Per il XXI secolo - una enciclopedia e un progetto*, Giannini, Naples, 385-395.

Rabino G.A. (1996) Complessità, scienza della complessità e modello dei trasporti, *Le strade*, volume 98, no. 4, 299-306.

Reade E. (1985) An Analysis of the Use of the Concept of Rationality in the Literature of Planning, in Breheny M., Hooper A. (eds.) *Rationality in Planning. Critical Essays on the Role of Rationality in Urban and Regional Planning*, Pion, London, 77-97.

Rosser J.B.Jr. (1991) *From Catastrophe to Chaos: A General Theory of Economic Discontinuities*, Kluwer Academic Publishers, Norwell.

Searle J.R. (1992) *The Rediscovery of the Mind*, MIT Press, Cambridge, Massachusetts.

Vozza M. (1990) *Rilevanze: epistemologia ed ermeneutica*, Laterza, Bari.

21. New Conditions and Requirements for Urban Government

Francesco Indovina

21.1 Social Actions and Policies: A Question of Interaction

This contribution aims to stress the interaction between policies and social actions in the management of urban change. The intention is to interrelate the positive aspect of the 'science of the city' with the rules of urban planning, and can be summed up as the *government of urban change*. The background of the following analysis is the capitalistic system and its consequent social relations. Even though capitalism is the main production system in the world, it still contains contradictory elements which have engendered a great many criticisms. To avoid possible misunderstanding, it is important to clarify the content of the terms used (because of their multiple meanings and the overlap of different ideologies).

Policies refer to those decisions undertaken by public officials and turned into operative actions. They may involve particular sectors (housing, transport, etc.), but they are not necessarily specific; on the contrary, implicitly or explicitly, correctly or not, they tend to underline *general interests* representing the legitimate choices made by institutional bodies. The basic elements in policy-making are therefore the public official, namely the source or the authority, and the general interest, namely the content.

Policies concerning the management of urban change (a mobility policy, for example) have a clear territorial content, but others too can have direct effects on the city. A policy which aims, for example, to restrict illegal activities, such as prostitution, smuggling or drug trafficking, to defined areas (creating a sort of 'free trade area') can affect the physical nature of a district, causing degradation, for instance. Occasionally policies are illegitimately used to pursue specific interests. However, such policies do not change the above situation just because they are illegitimate. It should be pointed out that policies tend to be 'dressed up'; even when they favour particular interests, this is hidden under the cover of being for the general good. This implicitly confirms the 'generalistic' feature of policies.

Social actions are those activities which the members of a community carry out, individually or as a group, to achieve their specific and lawful objectives. Consequently, actions, whether carried out by single individuals, bodies, organisations or other concerns are characterised by *partiality*. In this sense, even those actions undertaken by public officials (i.e. in the public interest) can be considered partial. For social actions the source is more complex than for policies (it can be the owner of building land, or a factory, a public body, a trade union, a committee of citizens, etc.), but the content is easier to single out.

Social actions can also have some general content when they are the result of the expression of collective actions with goals valid for the whole community (the inhabitants of a city, of a district etc.). Social subjects use this kind of policy to influence the political decisions of institutions. They are mostly processes of self-organisation which often provoke conflict (social or urban) and are specifically intended to achieve objectives neglected by the institutions. Social actions can arise from various elements: the absence of institutions; a favourable conjuncture which unites social subjects to achieve mutually held objectives; the democratic character of a given situation and the social subjects' direct involvement in such a democracy. Mainly they represent the acknowledgment of the rights of citizenship (Ferrajoli, 1993), which are at present being increasingly ignored.

By *government* (of the town) we refer to the complex and coordinated activity of the body ruling the town, i.e. the local authority. As we have explained above, this takes place through the application and coordination of policies with a specific territorial content or a direct effect on the territory. It affirms the *intentionality* (Indovina, 1994) related to the dynamics of the city. In fact, the public role can only express itself through a clear intentionality, i.e. some clear concepts about community living, development and change which as a whole take the form of a 'city design' and of precise objectives. The definition of the contents is a matter of politics. Here we can only give a few indications: a city which would encourage general and collective interest, development, social justice, balance and community living. The only way to attain a mutually held design is through democracy, i.e. responding to social subjects, their needs and collective requirements.

Public authority decisions have a very important function, arising from the need to reconcile contrasting interests, manage complex situations in which the right balance would not otherwise be found, intervene in relevant matters and, also very significant, modify spontaneous tendencies directing them towards determined objectives. The local government's duty consists of the application of policies (in this context an urban plan can be considered as a policy) directly produced by the local authorities and the coordination of initiatives taken by other institutions.

Referring to *urban change* we intend to underline the nonstatic nature of the city, which changes continuously, though differences may be noted in the intensity, rhythm and speed of change, in different periods and places. Such processes can bring about progress or regression, but nothing remains stable. This means that even though an authority's activity can determine the trend of this change, a lack of activity would not stop the evolution - this would occur spontaneously, possibly engendering nonpositive results on urban life. In other words, the government's non-intervention can be considered an *abdication* rather than a lack; the definition and direction of change in this case passing from collective, public hands to private ones.

21.2 The Strength of Social Actions

Social actions undoubtedly produce transformation. They can make a community dynamic, allowing it to exploit opportunities and technology, develop creativity and give birth to new forms of organisation of everyday life. We have to identify those social actions with the ability to create innovation, modify the way the community lives and, in general, promote dynamism in the city. Actions are brought about by and for private interests, but are legitimated by institutions, i.e. they are *regulated* by the system of rules accepted by the collectivity. The power of an initiative is private, but its assertion is ruled by the collectivity. It would seem, therefore, that there can be no contradiction between the application of these actions and the general interest. However, this is not so. The rules are general in character, defining the content but rarely the 'where' and 'when', nor do they consider all the possible but often unpredictable effects. In this context, the question of the effects should be stressed. In fact, although general conditions favour social initiatives, their application can cause negative phenomena, despite regulation.

The consequences of partiality, i.e. social actions which represent the interests of a specific sector, social group or individual can be regarded as the first negative effect. Even a highly simplified view of society affirms that the general good of all is the result of the free expression of many interests. It has never been considered beneficial, in any political, cultural or historical context, that the city should be the result of the domination of a single interest. On the contrary, the urban community has always been seen as the product of intentions which go beyond individual interest. In recent history however, the city has been nearly exclusively the expression of the requirements and interests of industrial production, provoking such a negative effect from a social, hygienic and moral point of view, that a long phase of *urban reformism* (Indovina, 1995) has been necessary. Social actions, because they are partial, need to be limited and corrected, that is diverted towards the wider interest.

Within the dynamism created by social actions, the strong tend to dominate the weak, not only in terms of social class (which is already serious), but in many other areas. This represents a second negative effect. Since social actions are characterised by competition (for resources, space, values, etc.) or by conflict with general social 'rules' and the collective good, a rebalancing intervention by public authorities is needed. The prevalence of a kind of social Darwinism, the survival of the fittest, is completely unacceptable as it risks leading to weak cultural values.

Finally, the third negative effect results from the application of social actions which compromise the 'raison d'être' of the city, i.e. the urban experience of a collectivity, which includes history, morphological conditions, habits in the use of space, etc. It has been found that legitimate social objectives can cause variations in the organisation of space which end up by conditioning the nature of the city. This assertion should not be considered as a contradiction of what was said above about the basic role of social actions in determining urban evolution. In fact, in facing change there should nevertheless be a conscious preservation of positive elements. Consider, for example, the recent tendency to 'import' forms of retail organisation from other countries with different urban cultures. In the European, and particularly, Italian experience, the street serves many functions including a commercial one; the shops located along urban streets determine a specific way of using the city and vitality which adds character and prestige. The introduction of shopping malls (brought about by the domination of real estate interests) changes this characteristic, reducing urban quality. Although supermarkets are justified from some viewpoints (low prices), they trigger a reduction in the diffusion of commercial activities in the city.

The negative elements resulting from social actions need to be controlled and coordinated. A conscious mediation of the future of the city is necessary, and hence the intervention of local government. One of the reasons for the recent worsening in urban conditions is the undervaluing of this function. This does not mean that we should repropose simplified, old fashioned schemes and plans, nor that the public authority's role should be preponderant (changes in technical, economic, social and cultural conditions have to be taken into account), but that there is an urgent need for a *direction* in processes of transformation. The city is the product of a design for civilisation, bringing together many values and perspectives, but undergoes tensions due to transformations brought about by the changes in the social mechanism which, in turn, are influenced by the city.

Because of its complex and contradictory character, its potential for influencing the 'pattern of life' and the intense economic, social and cultural relations of the city, urban planning needs strong motivation and goals to be able to produce an integrated series of actions capable of facing the emerging problems.

21.3 The Need for Government

From a normative point of view, the fact that social actions have at the same time a highly dynamic content and a high rate of partiality is highly relevant. While the positive aspect should be preserved, the complete and unconditional exercise of any partial interest cannot be accepted. The risk is the general deterioration of urban environment. Social processes and the city influence each other, but this mechanism stresses the leading role of public authorities. In fact, if technical, economical, organisational, social and cultural changes were not managed (a technical term is used intentionally), their spontaneous evolution would tend to undermine the role of the city, not because it would be overwhelmed, but because it would be made unusable by implosion, with negative social effects. Many people now assert that the quality and functionality of the city are fundamental for attracting new economic initiatives.

A balance is needed between the use of urban government and the free expression of economical, cultural, social and technological development. Initiatives should not be repressed, but they may need to be constrained when the general interest is threatened. The management of urban change should favour social actions, correcting and avoiding their undesirable effects; at the same time, it has to defend the balance of the urban organisation, offering new opportunities and directions. The government's action should aim to *allow* and *forbid* but, above all to create new and suitable conditions for social actions which are in harmony with principles of social justice, and capable of ensuring the preservation of the urban heritage for future generations.

Among economic scholars who have investigated globalisation (for example, the Lisbon Group, 1995) many have stressed the emergence of extremely negative consequences from the free expression of competition and the lack of management of such processes. Many imbalances are strengthened, new ones may emerge, but brute power prevails, with the result that benefits tend to concentrate rather than spread in society and instability becomes the rule. Curiously enough, even though many scholars point out the limits of competition, politicians and legislators defend it from monopolistic initiatives, cartels, etc. and public interference. They exhalt the potential of competition for creating more favourable conditions for everyone, ignoring the structural mechanisms which cause imbalance, in reality they only determine the dimension of the ring where the strong can overwhelm the weaker elements.

The control of this political phenomenon is not only an economic concern, but also involves the urban environment and its management, although to a slightly lesser extent. Social initiatives take place in a competitive regime with effects similar to those in the economy. Naturally, some differences exist. We shall emphasise those emerging in government. Although economic intervention is becoming weaker and weaker (it has to be so, it seems), and the hope of a world government is sustained by the attainment of globalisation (improbable, though a lot of goodwill exists), as far as the city is concerned, and luckily for urban civilisation, a 'government of change' has always existed. Urban development has been subject to different forms of public control, even though not always adequate. The theory and methodology is rarely discussed, but its application has very often provoked contrasts between social actions and public government. There is evidence that, throughout the history of civilisation, it has been the existence of public management of territorial evolution which has allowed the city to survive as a place of 'sociality' and progress. This government has always been characterised by the dialectic between single and collective interests. In fact, the interrelation between these interests has favoured the preservation and development of the city, though contradictory and sometimes unbearable.

Policies, therefore, are the instruments used by the collectivity, through its institutions, to manifest its intentions for the future of the city. These intentions are expressed by the function defined as 'government of urban change', but the policies it produces for specific sectors are introduced today for the future, to generate new conditions suitable for activating social actions. Planning does not only have to 'solve problems', it should embody a 'corrective' measure for the long term, making it possible to avoid contradictions between current and future intervention. In other words, the solution of emerging problems has to be coherent with the future perspective (Crosta, in this volume).

21.4 The Quality of Government

Before analysing the latest situation, it is useful to make a few observations about the quality of urban government. In particular, we should like to give a brief explanation for the preference in this paper for the expression 'government of urban change' rather than 'urban planning'.

Because of its static nature, its limited contents and inability to establish the operative conditions for social actions, the statutory 'Town Plan' is not really a suitable tool to determine the whole evolution of the city (or surrounding area). As such plans have, however, frequently been the main element of urban government, they have often been assigned functions which have made them unnecessarily complicated, baroque and largely ineffective (both from the point of view of specific planning tasks and urban government in general). When attempts have been made to incorporate aspects like income, difficulties have arisen and mechanisms (such as equalisation) have imposed a strong 'conditioning'. The plan has also run into difficulty when attempting to give specific indications about the 'destiny' of the city, because of the impossibility of controlling the necessary elements. A plan is essential, but should be just one of a series of instruments.

An urban plan has to take into account the particular characteristics of a community - its economic features, and social and cultural processes - but these simply represent a conceptual basis (though precise and spatialised) for producing an urban design. The problem is that, in an attempt to make a concrete contribution to the urgent need for urban management in the face of wide-ranging urban change, the plan has been invested with a whole range of tasks which it is simply not possible to tackle successfully with such a limited tool. It has been pointed out, correctly, that it is impossible to plan a city without considering, for example, the local situation in relation to the national and international economy.

This increased range of tasks has even had a semantic result, transforming the expression 'urban design' into 'urban planning' (the latter includes urban design, but is enriched with economic and social features). Today urban design as an activity is reclaiming its independence, as more attention is being paid to urban form and aesthetics. The risk is that dealt with separately, urban planning could become an empty shell, while urban design is an arbitrary act, unconnected with the overall plan.

To avoid these negative effects, it is necessary to reduce the number of tasks tackled by planning, but to enlarge its responsibility in specific areas. In other words, it should deal with the organisation of space, taking into account the direction of urban evolution, namely *public intentionality*. In this way, the plan would be connected to reality and to the community's political decisions about its destiny, but would enjoy greater independence. In other words, it would become just one of the policies of the government of urban change, possibly even a driving force, while being integrated within the global design.

At this point, it should be evident that the plan needs a strong political content and a clear division of roles. It is the local authorities' responsibility to establish the overall goals, but this is a political choice *from* and *for* the city and is determined through normal political procedures. It is not a technical process, even though substantial technical contributions can be provided. A plan is the territorial embodiment of the goals and should be capable of interpreting them without transforming them. This is not an automatic process and it is necessary to examine the proposals of the plan to assess their correspondence with the objectives.

The whole set of policies which interpret public intentionality (each in its own field) represents the governing of urban change. In this context the urban plan no longer conditions other policies: the choices are, in fact, made operative by the various policies. They become a programme to be implemented and are no longer an obstacle to social actions. Moreover, the government function includes the important task of coordinating the policies of the various public institutions to ensure they are coherent and relevant. Well designed policies can favour the attainment of objectives, giving a direction to the action and the use of public resources. The aim is not the creation of new 'opportunities' (a modern term) for private citizens, but the determination of precise limits within which they can reach their goals and any appropriate modification of private interests.

The characteristics of a high quality government of urban change are (very briefly):

- *clarity of direction* (intentionality), the statement of a clear project for the future, a whole which is made precise by policies and a steady control of processes;

- *precise policies*, which have to match the objectives and available means;
- *coordination* of the operations which contribute to the achievement of the objectives;
- *coherence* of the whole;
- *control* of the results and, if necessary, correction of the actions.

21.5 Some Observations about the 'New Conditions'

21.5.1 Introduction

In the present phase, some new conditions are rendering the government of urban change more complicated. The following analysis considers the most relevant ones.

21.5.2 An Increase in Complexity

Hierarchical relations, which previously characterised the organisation of space, reflecting the structure of society, have been overlain by network relations. These have given birth to a new structure with multiple relations. The passage from hierarchical to network relations is the result of both territorial and social interactions. As many scholars assert, in this new context the 'bearers of interests' cannot be easily distinguished, since each is involved in a great number of interests (sometimes even contradicting one another).

But it seems quite possible that this phenomenon has been overestimated. Even in the past, the bearer of a single interest was rare. After all, the expression of individual interests has always been filtered by a cultural process; this is the means through which the individual's and the community's identity was built. The weakening of identity cannot be the consequence of the multiplication of interests. We should perhaps look instead at the mechanisms which produce identity. It seems undesirable to have individuals of weak and fragmented identity, since there are numerous negative consequences, mainly because the process of disaggregation cannot be regulated. As with the 'sorcerer's apprentice', the forces released cannot be controlled.

The substitution of spatial hierarchies with network relations is, in part, only due to a change of viewpoint. Network mechanisms already existed in the organisation of space. Individuals are active in many social activities which may be contradictory or even in competition with each other. These considerations bring about different positions as far as 'government' is concerned (not only the governing of the city, but at any level and in any context, disciplinary and political). We can identify three positions which result from the awareness of the above mechanisms.

The first supports the limitation of government because of the increasing difficulty of understanding social processes and the interests manifested in social actions. The idea is that fewer rules means more competition. According to this position, the market mechanism is the best means of choosing between different options. In relation to urban planning, this would signify the conclusion of a long path, the end of urban intervention and of the planner as controller of urban dynamics. Attention would focus exclusively on form and aesthetics, in other words planning would give way to design, it would be a victory of architecture over town planning. After a period of success, this position has been seen in the right perspective, from a disciplinary and political point of view.

The second position, supporting some limitation of control, would eliminate any connection between government actions and social actions. According to this theory, the inscrutable nature of social dynamics means that they need independent regulation. It does not seem that this position has produced anything new as far as the town planning is concerned. It stresses the preceding position, considering the city as the result of architectural design (the sum of a multitude of designs). This position, in general and as far as the territory is concerned, seems to exalt arbitrariness, attenuated only by the 'farsightedness' of the designer.

The third position proposes an *increase* in the action of the government, considered necessary because of growing complexity. It is considered that a government suitable for the situation should be provided with more effective instruments of analysis and control, since the increased complexity can provoke greater entropy or disorder. Many speculations have exalted this condition of growing disorder: it has been defined as ineliminable, full of opportunities and, even, of freedom ('extremely individualistic' expression, Ilardi, 1995). However, to avoid the establishment of a world of terror rather than one of 'reason set free', regulation is necessary together with the ability to 'listen' to the need (expressed sometimes through forms of conflict) for new identity.

The awareness of the need for more effective government of urban change brings us to some observations about the instruments available, the necessary evolution towards greater flexibility and also better specification of the areas of intervention.

21.5.3 Reduction in Resources

The action of government can be considered as an activity of redistribution. A possible reduction in resources, therefore, has a considerable influence on this function. The reduction in resources is the result of several mechanisms: a greater concentration of wealth; restrictive budget policies aimed at lessening deficits; growing costs (for structural reasons) of the many social services; and growing investment in technology, which will produce savings in running expenses only in the distant future. As a consequence, local authorities have fewer resources for investment and the government of urban change.

The reduction in resources stresses the need for a change of direction, thus *intervention* should be replaced by a process of reorganisation. In fact, in the past intervention was seen to represent the solution to urban problems, but the effects on public expenses were ruinous (the recent legal trials have confirmed that this policy was often connected with political corruption). In some sectors, fewer resources and budget cuts could to some extent lead to innovation in government. This new situation requires operative creativity; urban organisation becomes a variable rather than a given, and intervention becomes a process of reorganisation (not embodied only in action).

However, it should also be remembered that cuts in resources triggers a reduction in public services, with serious consequences for the prestige of the city. As a result, social policies become necessary, but there are fewer funds available. In this way, 'natural' social injustice (in the sense of the specific connotation of a given social order) is fed and the function of social compensation (Indovina, 1995) loses its importance - a function the city has fulfilled in the past.

21.5.4 New Settlement Characteristics

The third and last condition to be considered concerns the features of modern urban settlement, territorial differentiation and the new requirements for the city. The prevalence of decentralisation and diffused settlement patterns (Indovina, 1990) has brought about a modification in the nature of towns and the use of space. Sprawling urban areas, created principally by cultural and economical factors, little by little have turned into 'diffused cities', characterised by discontinuity (despite the strong tendency to the merging of existing settlements), by low-density building and a widespread urbanisation of the territory. This kind of settlement causes very high collective costs, the cost of private mobility compensating for the lower housing expenses.

The new phenomenon of the 'urban user' (Martinotti, 1993), people using the city but not resident, puts pressure on services and poses problems of urban management. This evolution has been the effect of changes in production processes and in the location of economic activities, together with the reduced quality of city living. The high density city is still the main location for many activities, but it tends at the same time to drive people away because of the environmental problems (traffic and pollution) and rent costs (higher than those of the diffused city).

New technological discoveries can definitely help improve urban conditions. The problem is that public authorities have great difficulty in exploiting this opportunity, partly due to their limited resources, and partly to cultural and political resistance to innovation. This tends to widen the gap between private and public use of these technologies, causing negative effects on the city.

A further factor is the rapidly growing mixture of ethnic communities, cultures and religions. This variety could be the basis for the creation of a 'golden melting pot', but unfortunately the resulting spatial separation, due to distrust and defence of identity, causes tensions. Connected with this is the emergence of the problem of security. The perception of urban 'danger', often amplified by the mass media, ends up by triggering a spiral reaction. Certainly, criminality and violence have spread, but their diffusion has also been encouraged by the way the city is used, changes of habit and the fear of violence. Solutions such as protected residences, alarm systems, etc. and a reduced use of the city by women, elderly people and children do not seem to solve the problem. Instead, they provoke a subdivision of urban life, which has a negative effect on the quality of city life, above all, its safety.

These phenomena, and others not discussed here, have created new problems concerning the way the city responds to its citizens' requirements and even the status of citizenship itself. The need for a new system of government has emerged, requiring a coordinated system of policies.

21.6 Some Observations about Models

The review of the various models proposed (Pumain, in this volume) is an interesting exercise, because it underlines the efforts made recently to give a more attentive interpretation of urban phenomena. It seems, however, that very often the basis for the construction of these models is not represented by urban mechanisms, but analogies 'imported' from other disciplines and different contexts. In many cases the model does not even seem to be constructed from reality, it consists of an application. This in itself is not a criticism (many useful results in all disciplines have derived from similar adaptive procedures), nevertheless these very intense research efforts have yielded modest results in the understanding and interpretation of phenomena, and in the operative field. Probably, this accounts for the growth and decline of enthusiasm for certain models. The development of the 'science of modelling' as a self-referential branch is found in many disciplines, especially the social sciences, not only those dealing with the territory. Formalisation and modelling represent a 'requirement' of scientific research.

The features of the city today and of territorial organisation, described above, have had consequences on the research and the instruments used. We have underlined the conflict existing between the decisions which make a city or a territory dynamic and their capacity to engender processes which diffuse the urban effect. In other words, the result is an interaction whose 'rules' are difficult to identify. In theory, it should be possible to determine them from the specific kind of interaction implemented, even though its extension and the subject involved cannot be known. The study of decisional processes and strategies can give useful indications which can also be used to create some reasonable model. For interpretative purposes, strategic games (Cecchini and Indovina, 1992) provide a basis - relations between social actions, and relations between social actions and policies can be assumed to be acts of 'competition' or 'collaboration' relating to nonrandom strategies. If urban policies or planning are considered to be administrative *actions* rather than mere *acts*, they need a strategy to attain their objectives, to use their forces better than their adversaries, to exploit their counterpart's advantages and disadvantages. The relevance of mechanisms of interaction and mutuality has led to the development of game simulation techniques (computerised or otherwise), which have given interesting results (Cecchini and Indovina, 1989, Cecchini and Taylor, 1987).

For representing some urban phenomena, e.g. those for which the 'proximity' to other parts of the territory is important (urban degradation and rent values, for example), models based on cellular robots can be useful. In this case too, while the didactic results have been considerable, the interpretative results for the moment are interesting, but not easily applicable.

Models are useful only if they widen knowledge and enrich interpretation (not simply because they turn what was already known into elegant formalisations). The city is in continuous evolution. The changes are so evident that their definition does not require particular effort. On the other hand, their interpretation is more complicated (the identification of specific causes, trend, results, etc.). However, a starting point is to give the 'science of the city' a statute adequate for the purpose; no instrument should be snubbed, every effort should aim at the goal. Nevertheless, often the 'spectacles' (prejudices and ideologies) through which urban phenomena are observed have more influence than the instruments used (this is the reason for the dialectic between researchers).

References

Cecchini A., Taylor J.L. (eds.) (1987) *La simulazione giocata*, Angeli, Milan.
Cecchini A., Indovina F. (eds.) (1989) *La simulazione*, Angeli, Milan.
Cecchini A., Indovina F. (eds.) (1992) *Strategia per un futuro possibile*, Angeli, Milan.
Gruppo di Lisbona (1995) *I limiti della competitività*, Manifestolibri, Rome.
Ferrajoli L. (1993) Cittadinanza e diritti fondamentali, *Teoria politica, 3*, Bologna.
Ilardi M. (1995) *L'individuo in rivolta*, Carta and Nolan, Rome.
Indovina F. (eds.) (1990) *La città diffusa*, DAEST, Venice.
Indovina F. (1994) Intenzionalità e innovazione nella pianificazione, *Critica della Razionalità Urbanistica*, 2, Naples.
Indovina F. (1995) 'La città che verrà', Report at the International Seminar *'La ciutat a la fi del mil.lenni'*, Barcelona, May 1995.
Martinotti G. (1993) *Metropoli*, Il Mulino, Bologna.

22. Town Planning and Fluctuations of Rules Covering Time and Space

Giuseppe Longhi

22.1 Introduction

Françoise Choay in her paper '*Le regne de l'urbain ou la mort de la ville*' ('The reign of the urban or the death of the town') observes that: "The town has a spontaneous overflow that seems to be the consequence of an out of control cataclysm. We have to admit the disappearance of the traditional town and ask ourselves what will take its place, that is, about the kind of conurbation and 'non-town' which seems to be the destiny of Western affluent societies". In short, she asks this essential question: "is urban place a synonym for town?" (Choay, 1993, p. 26 and p. 38).

This leads to other questions concerning how to approach analysis and design within the present urban context and what role town planning can play. Françoise Choay claims that technology has played an important role, pointing out that the exponential rise in speed (of technological innovation as well as physical and nonphysical movement), has produced an enslavement of urban space to the logic of the network, creating a growing confusion between the constituent elements of the town: *urbs* and *civitas*.

Carlo Maria Cipolla (1989), the economic historian, in his book: 'The Three Revolutions' confirms Françoise Choay's concern, adding some interesting arguments. He begins with the consideration that, before the merchant revolution, Europe belonged to the barbarians, but the subsequent revolutions (in trade, knowledge, and industry) have all had their epicentre in Europe and have been urban revolutions. At this point, Cipolla poses these questions:

- as the latest revolution originated not in Europe but in Asia (the centre of

gravity of innovation and progress has now shifted toward the Pacific area, after moving from Europe to the Atlantic coast of the U.S.A. during the industrial era), will Europe revert into a land of barbarians?
- as the new technological processes challenge the classical rules of agglomeration, can the latest revolution be considered an urban revolution?

These questions warn us that European town planning has to think about its future existence in a space which, unfortunately, is becoming peripheral. The first advice of Choay (1991) is to give up the terrorism by which the architect-planner-demiurge exerts his choice on the people and to face the present problems with restraint. Many planners accept this advice as coming at the right time. In fact, those responsible for the planning of Lyon's metropolitan area, in the introduction to the plan, comment "if we are extraordinarily able, with our plan, faced with great conceptual effort and investment, we shall only be able to maintain the present situation" (SEPAL, 1988).

The rapid change in the logic of space has raised fundamental questions about town planning. In industrial society, town planning responded to the problems of a society whose rules were gigantism, hierarchical organisation and concentration, with the myths of utopia and hygiene, but now this patrimony is being doubted as other more intangible problems emerge.

The nature of town planning has therefore become problematical: is it possible to explain the definition of urban science as a postulate of objective and neutral space? Is the project explainable as a regularising event? To define town planning as a science setting the rules of the town is contradictory, while to define it as urban modelling is limiting.

This paper does not attempt to give a complete answer to these complex questions, but in the first part we propose some critical considerations about the limits of the basic paradigms of town planning, and in the second part we explore more recent planning experience showing the limits and the diversity of urban thought in the present period of change. In the third part we consider some initial elements of the lexicon of the plan, as the fruit of conceptual renewal.

Because of the complexity of the argument, the paper is constructed like an index, with a bibliography showing the different positions concerning the central problem of the renewal of town planning.

22.2 The Limit of the Consolidated Paradigms

22.2.1 The Question

The paradigm of the modern movement: speed = chaos, makes it seem as if the crisis of the classical idea of planning is due to the rapidity of technological innovation, but the effects of technology are more complex (Braudel, 1986). We also need to consider the change in rules affecting the use of space caused by incessant innovation.

The main questions concern:

- the change in the classical rules of agglomeration;
- the effects of technology in the building of towns. By imposing miniaturised and standardised processes, it has destroyed the classical rules of urban design;
- the mechanical system of intervention, whose 'externalities' have been undervalued, producing negative consequences for the environment.

This means we need to critically rethink both consolidated 'traditional' and recent paradigms: agglomeration and zoning, as well as the network idea.

22.2.2 The Agglomeration Paradigm

This paradigm is explained by Jean Paul Lacaze (1992). Density makes people happy: there probably doesn't exist any serious scientific research which supports this statement, openly declared as real dogma. The myth helps to justify the personal aesthetic preferences of the planner and the demands of investors: it is based on a mix between neighbourhood liveability, which requires a concentration of public space and land use density, although correlation cannot be established between these two elements.

Beyond the controversy, we have to admit that space considered as distance commonly enters into current analysis, but that space considered from an organisational point of view poses many irresolvable problems. (It is possible to find very good treatises regarding agglomeration economies, inspired by economic theories, particularly those regarding costs, for example Camagni (1992), but it is more difficult to find exhaustive studies about organisation and the urban effects of those economies. The work of

Davezies (1992) is one of the better research projects in this field (and the source of the main arguments used in this article).

According to neoclassic economic theory, the town minimises location costs and permits specific advantages linked to the density of factors available (such as the circulation of information, the variety of markets, labour, capital, and other production factors). But the town also produces externalities and indivisibilities that economic market theory doesn't take into consideration. The indivisibility of urban organisation has become a dark area for the economy, in which exchange value is no longer represented only by explicit commercial or monetary exchange, but depends largely on management and urban policy.

This fact is very important, because it means economics and politics should not be two separate worlds. They should live together within an economic theory where the externality idea is taken into consideration (Klaassen, Molle and Paelinck, 1981). Even more difficult is the question of the character and operation of economies of agglomeration. What are the costs and benefits of urban concentration?

In this field there are still many unresolved conceptual difficulties. Alonso (1971) stated that we don't know if the urban employee's tie is an intermediate consumption, i.e. a cost, or a final consumption, i.e. a benefit. One more difficulty is that externalities are linked to the model of urban management, so a town's economic efficiency is not easily assessed on the basis of 'general laws' concerning dimension, density or shape. To explain the relative efficacy of different policies, it is more useful to use econometric models which can evaluate the positive and negative externalities generated.

This uncertain background is increasingly affected by the globalisation of the economy and the need for structural change in order to remain competitive. Planners and public managers often ask economists: what kind of urban form or, more generally, urban organisation is best to support economic growth? We must admit that economists seem unable to give a clear answer, but it is significant that the emphasis is shifting from economies of scale and questions of urban form to management.

For the modern urban economy, the quantities and costs of production factors are less important than organisation factors, such as logistics, the limitation of risk of error or production break-down, co-ordination between actors and managers, and the rapid circulation of goods and information. Organisation is replacing the minimisation of factor costs as the crucial element in efficiency.

A delay in interpretation can be deduced from the kind of descriptive literature prevailing over the last fifteen years: at the end of the 70s urban decline was the main recurring theme, in the middle of 80s it was re-urbanisation or urban renewal, while at the end of 80s the central theme was urban congestion. This last phenomenon may be interpreted in two different ways:

- it is intrinsically linked to urban concentration, which tends to produce negative externalities;
- it is the price of inadequate political action concerning the structure of urban areas. The increasing demand for agglomeration economies from industry has produced a late response with urban policies inclined to produce positive externalities.

The first explanation is linked to the theory of the 'city life cycle', which claims that towns grow and decline in a cyclical way, with a series of congestion/decongestion effects producing a net balance of urban externalities close to zero. There is also the associated theory of the natural cycle of urban decline/renewal, proposed at the beginning of the 80s by many authors (influenced by the work of Hoover and Vernon, 1959). This has led to a 'weak' urban policy, and stimulated policies aimed at regional equilibrium, supported by public opinion strongly conditioned by the idea of the 'unliveable town'.

The second explanation (Mera, 1989) is connected to the kind of urban policies adopted. We could ask, in fact, whether the 'urban-concentration' phenomenon of the late 80s was linked to previous disinvestment processes. Inevitably, the new phase of concentration led to congestion because 'metropolitan renewal' was not foreseen or accompanied by appropriate infrastructure policies. If this explanation is correct, it should be hoped that the authorities at least had a clear idea and more solid doctrine about macro-economic urban management.

In fact, the strong deregulation which has inspired government policy since the mid 80s has strengthened the natural trend of social and economic factors towards concentration and polarisation. However, as already stated, this process of polarisation, resulting from weak public intervention in the macro economic field, produces a set of externalities which demand strong public intervention in urban areas.

This is one of the paradoxes caused by current public policy. The abandonment of intervention in town planning and in the macro economy is bringing to light the need for planning and the formulation of new forms of regulation. It is not yet clear whether the resulting balance of these adjustments will mean less politics or more!

22.2.3 Zoning

Zoning is an integral part of the scientific paradigm of town planning, but today this principle is being strongly contested (Dupuy, 1991). We should not forget that the practice of zoning, linked to the myth of hygiene, originated in the industrial cities of Europe and was intended to mitigate the effects of the 'infernal factories'. But we should not undervalue the validity of zoning as an instrument of mediation. The town planner, facing multiple conflicting interests, is forced to mediate to attain his aims. By limiting, under strict political control, building and renewal on urban sites, he reassures landowners, or at least makes clear their property situation regarding the plan (Dupuy, 1992).

Built development now spreads over a vast area and is very different from the traditional pattern of settlement of the industrial town. Factories are located outside the city, shopping centres attract people away from urban centres, and the use of spare time is influenced more by tastes and trends, than by proximity. These elements raise doubts about the idea of 'mass standards'. What does standardisation mean, expressed as land lots, in a world based on quality and differences?

The idea of restricting people and activities within a defined space network through zoning (the subdivision of urban territory established by law) has proved ineffective compared to the new relationship models - logistics seem a better answer than zoning to the needs of the modern conurbation. Was zoning only appropriate for the industrial era? Probably not, we may see town planning linked for some time yet to the zoning principle because it is strongly rooted in professional culture and practice. Moreover, zoning plays a decisive role in public housing policies, in transport and telecommunication network boundaries, and has gathered new conceptual strength in the Californian 'new urbanism' concept (the proposals for the core of this project are described in Calthoper, 1995, Katz, 1993, Duany and Plater-Zyberk, 1993).

22.2.4 The Network Paradigm

The enormous extension of networks (water, gas, sewerage, electricity, telephone, railways, and so on) constructed, above all, since the middle of the last century, has been used by town planners to avoid the constraints imposed by built-up land and estate interests and made possible the creation of new building areas with good services and accessibility, away from the more expensive areas (Dupuy, 1991).

Settlement based on a network logic may become a real alternative to zoning only when the technology has reached full development. Services will as a result be universally distributed, and from this point of view the spread of car ownership undoubtedly plays a decisive role. We shall then have the perfect synergy of transport and communication networks, permitting the rapid and efficient transfer of goods, people, information and energy. The network could take on a social meaning that goes beyond the technical horizon and individual interests, aiming at the general opening up of urban space.

At this point zoning would undergo a 'gentle revolution' (Dupuy, 1992). Boundaries, limits and perimeters would not disappear, but their importance would become relative compared to that of the networks, which cross or ignore them. Networks offer more in terms of social meaning, public service and individual choice. But it is not certain that networks will permit an approach to the town that differs greatly from the functionalistic approach. Networks emphasise the social and economic links rather than a political idea of space, unlike the classical idea of the town in which the prevailing desire is to create a place with political meaning. These considerations lead from critical reflection about the role of the 'architectural objects' in the town to consideration of overall civic design and the 'psychological' quality of life (Unal, 1992).

The era of networks emerged from the crisis of functionalism. Many architects prefer the network approach, which is apparently more human and flexible than the functionalistic approach. It is based on the transposition in space (or rather, the virtual transposition) of functions previously strictly assigned to specific zones. It is a mechanism aiming to press a group of actors to promote a given project. According to Marion Unal, this explains the symbolist fashion in town planning, bringing about metaphors destined to mark the 'memory places' in space, the 'mythical crossings' which are able to establish other kinds of links among the citizens. It follows the claim to have found presuppositions for a new urban project, i.e. inclusion in a network and pragmatism, with the purpose of maximum efficacy.

Speaking about networks means speaking about efficacy and movement, noting, at the same time, the new kinds of natural selection among towns. So, today, there exist some towns able to build up networks of actions, and others which have remained merely onlookers, hoping to be involved. But we must remember that citizens with the same rights, duties and opportunities are living in these towns. In other words, networks, while aiming to stimulate the local economy, extend across space and help to mould it, with great indifference to the social and political structure.

The biggest problem is represented by the difference of interests between economic needs and social demands. Consequently, local politicians, faced with network strategies not formulated in legal terms, are in the difficult position of having to 'manage' the dichotomy between the indifference to economic fluxes and the stability of social places, underlined by Manuel Castells (1989). Hence the increasing uncertainty in local development policies and the legitimate worry of forces of democracy responsible for social cohesion, since the strategies of the economic world act with increasing indifference to local politics.

In brief, the network philosophy seems to represent a factor of probability, rather than a guarantee of positive outcome at an urban scale. The physical result of the uncertainties and contradictions is the overvaluation of city centre space, traditionally occupied by the middle class and overloaded with political symbols, while suburban space, planned for the needy, grows in a mono-functional way without any political function. Recent urban developments have been promoted in town centres, increasing the lack of balance between centre and suburbs. The interventions in suburban areas have generally been heterogeneous, built in the wake of infrastructural networks, unlinked with urban concepts of space or image of the man/society relationship. We should reformulate both the needs/ideals of citizens concerning their town, and the professional's role in project design. This needs to be based on the principle that town development should enhance the 'psychological quality' of its inhabitants' life, in other words, a concept representing the culture, behaviour and images of its inhabitants (Fabris and Mortara, 1986, Martinotti, 1988, 1994).

We should realise that the psychological dimension of cities, historically expressed by the Utopians, is today largely inadequate. How long does it take to put into a concrete form a new idea of city living and not simply an organisation principle?

22.3 Renewal of Codes and Planning Experience

22.3.1 The Transition

The above questions remind us that we are facing the classic problems of the transition between one cycle and the next. We are living in the 'post-modern' era, the turning point between the declining industrial civilisation and the rising electronic civilisation. This transition can be interpreted through recent planning experience and the process of revision of town planning thought (Longhi, 1994). Although this process does not give absolute or certain answers, it would seem that there are two streams of thought concerning the nature of the transition. These claim that it consists of:

a. a modification of the classical paradigm;
b. the emergence of a totally new paradigm.

The first stream, supported by Lyddon (1991) (see Table 1), and Lacaze (1990) (Table 2) starts from the presupposition that the historical urban system, even though subjected to considerable changes and innovation, is not being radically overturned. It is claimed that the key problem is the shortage of resources available to municipalities/ communities (GLC, 1985) due to the adoption of classical Keynesian policies. The plan can no longer be imposed as previously and has become a lobbying tool of policies, resources and agreement.

According to this logic, the municipality does not passively regulate estate rights, but is called upon to operate actively with a higher level of organisation and management ability. The key features of the plan are strategy and flexibility, with the municipality working as a civic manager. The priority is the improvement of the competitive position of towns, in a network relationship system, which is increasingly replacing the hierarchical organisation of space and social relationships.

This leads to the application of strategic planning (Camagni and Gibelli, 1992), whose aims are evolving towards:

- overcoming the strictly technocratic approach to decision-making, attempting to make the process of specification of objectives more democratic;
- a widening of the temporal horizons by means of a continuous planning process.

The second stream of thought can be further subdivided into two subgroups consisting of those who emphasise:

- the radical new opportunities for urban transformation offered by technological change: a trend synthesised by Mitchell (1995) (Table 3);

FROM	TO
Garden city	Inner city
Expansion	Conservation and renewal
Population and employment	Stability, changes in social structure and understanding of the structure of work
Simplistic notion of planning as enlarged architecture	Understanding the city as a social and economic system
Creating and controlling whole environments	Acceptance of diversity and 'happy accidents'
The 'end state' master plan	Flexible plan
'Top down' planning	Encouraging self-help initiatives
Planning product according to design rules	Planning process as a result of participation
Control by plot ratio	Urban impact analysis
Separation of land use for health reasons	A mixture of uses for social diversity
Confidence in the computer and quantitative methods	Mistrust of model based planning
Quantitative methods	Qualitative concerns
The planner as the only professional involved in planning	Corporate view and product from a wide range of disciplines
The pursuit of existing but simplistic new images	The discovery of order in existing diversity
Industrial technology	Electronic technology
Cheap energy	Expensive energy
Central system	Quest for decentralisation
Consensus and agreed definitions	Roles of experts questioned
Municipality provides services	Municipality acts as civic entrepreneur
Urban governance: worst first	Municipal marketing: invest in success

Table 1 25 years of towns and cities; a summary of the changes

- the radical replacement of classical 'mechanical' paradigms with new 'biological' paradigms of planning and design (Morita, 1991).

The former group believes that the technological process leading to greater dispersion of urban development and increasing use of telecommunications will 'destructure' the traditional habitat in favour of a system of 'planetary villages', whose survival depends on high accessibility to advanced telematic systems. This society of the future will be characterised by new means of production based on automation and the physical diffusion of workplaces. A new 'fourth' sector, consisting of communications, education, free time and health, will be added to traditional sectors of the economy.

The latter, most fully expressed by Edgard Morin (Salat, 1992, Morin, Kern, 1994, Longhi, 1992), points out the changing scientific paradigms, marking the passage from the rules of 'Divine Proportion' of man as represented by the classical geometry of Vitruvio or Leonardo da Vinci, to the nonequilibrium produced by the complicated interlacing of 10 thousand million neurons, represented by Jean Pierre Changeaux's neural man (Morin, 1990).

Method	*Object*	*Idea of town*	*Main factor*	*Reference value*	*Professional knowledge*	*Decision criteria*
Strategic planning	To modify structures of urban space	Economic centre	Time	Efficacy returns	Engineers Economists	Technocratic
Urban design	To build new neighbourhoods	Built space	Space	Aesthetics Culture	Architects/ Town planners	Autocratic
Advocacy planning	To improve way of life	Social relationship space	Man	Space appropriation usage values	Sociologists	Democratic
Management planning	To improve service quality	Services network system	Services	Ratio cost/efficacy	Management experts	Managerial
Communication planning	To attract new firms	Global image	Symbolic	Recognizability	Architects Communication experts	Ad-hoc

Table 2 Town Planning Methods

Electronic agoras	
from	**to**
Vitruvian man	Neural man
Spatial	Antispatial
Focused	Fragmented
Synchronous	Asynchronous
Voyeurism	Engagement
Contiguous	Connected

Soft City	
from	**to**
Real estate	Cyberspace
Face-to-face	Interface
On the spot	On the net
Enclosure	Encryption
Physical transaction	Electronic exchanges
Territory	Topology
Street maps	Hyperplans
Community standards	Network norms
Plan	Programmed places
Surveillance	Electronic panopticon
Electoral politics	Electronic polis

Recombinant Architecture	
from	**to**
Façade	Interface
Model	Decomposition-recombination
Stacks	Servers
Office towers	Working with the net
Homes	Telecottages
Circulation	Telecommunication
Trading floors	Electronic trading system
Bookstores	Bit stores
Galleries	Virtual museums
Theatres	Entertainment networks
Schoolhouses	Virtual campuses
Hospitals	Telemedicine
Prisons	Electronic supervision
Going to market	Sorting lists

Table 3 City of bits.

The neural man metaphor imposes on the planner a new project paradigm marked by uncertainty, by an infinite range of combinations and by the possible integration between physical and biological rules.

22.3.2 Codes and Plan Solutions

The variety of codes (Longhi, 1994) is empirically expressed by means of a very wide range of plan solutions, which may be divided into: 'back to the past', 'foundaries of excellence', 'continental programme', 'a few equilibrium solutions'.

I. Back to the past

The post-modern age is characterised by a strong reaction against modern age excess, rather than a real understanding of towns. The lack of certainty has led to a way of thought which aims to return to an imaginary past; hence the eclecticism of urban neoclassical, neo-technical and neo-modern shapes.

The break between the industrial and the electronic world is producing strange towns - ultra-modern at a technical level, but ultra-'retro' at a formal level. According to this school of thought, the first towns of the 21st century's electronic revolution should be planned for the future while employing shapes of the past, just as the cities of the industrial revolution were camouflaged as baroque cities, the first cars camouflaged as carriages, or the first trams as stage-coaches. This is the approach of the planners of the Californian 'New Urbanism', who are combining the telecottage and high speed transport with shapes borrowed from the Victorian age.

II. 'Foundaries' of excellence

The difficulty of reinterpreting the whole town has led to a series of polarising projects aiming to concentrate urban innovation in 'places of excellence', often disused harbours and derelict ex-industrial sites situated close to the town centre. These projects show clearly the gap between the world of business and the social reality of the city. They illustrate the brutal contrast between the few people and places who benefit from evolution and renewal and the rest of society, and the city excluded from these processes (Longhi, 1988, 1990a).

III. Continental programmes

The age in which we are living poses the question of how to go beyond metropolitan and regional space, that is, how to reach a new equilibrium between local and global (meaning at the planetary level). This opening of the spatial dimension makes the city part of infinitely wider system; a need felt in Japan, the U.S.A. and the European Community, but expressed through different philosophies (Longhi, 1990b).

Japan: through the MITI (Ministry for Industry, Commerce and Planning), Japan began in the late 70s to undertake a radical revision of its idea of urban space. Beginning with the supposition that the new technologies produce polarising effects, in the urban area of Greater Tokyo MITI has developed a project for renewing the smaller cities. These previously depended on the metropolitan area because of the externalities produced, but were neglected by the innovation process. The idea is that they should develop an independent dialogue with the system of world cities. So in the smaller Japanese cities new local specialisations are being encouraged and preferential links and complementarity established with the international city network.
MITI's program is at the same time a local, national and international programme. An example of this policy is represented by Antwerp, a European harbour city in decline saved by the Japanese, as it has became their 'network-headquarters' in the Old Continent.

U.S.A.: the United States are aiming at worldwide computer literacy, making practicable the metaphor of a language which destructures traditional rules of space. But this raises the real doubt as to whether the new revolution will be urban at all. The fate of the town is left to the contradiction of liberal policies, avoiding the question of what the 'town' of those excluded from the Internet will be like.

Europe: in 1994 the E.C.C. launched the project: 'Towards a more liveable city' (E.C.C., 1994), which reproposes the principles of the Japanese policy, in the light of the possible acceleration of telematic connections promoted by the American policies. This project is based on three questions:

- what will the new 'agora' be like? i.e. how can we manage the transformation from the hierarchical and functionalist urban plan to an urban 'milieu' based on understanding, coexistence and the development of common values?
- how can we combine the global and the local level? i.e. how can we convert our present cities with 'areas of excellence' surrounded by large areas of new poverty into integrated spaces able to develop network relations, based on the re-evaluation of differences, and counterbalance the power of the global economy?

- how can we develop biologically sustainable cities? i.e. how can we transform the 'disconnected' town (with the ability to compete with other towns and with multinational industries whose future is conditioned by external forces) into a local holistic self-supporting system, which is part of an interactive network of local systems?

IV. *A few equilibrium solutions*

This multifaceted scenario is contradictory, because the speed of the technological and social evolution has not yet been metabolised by a mature planning idea. Nevertheless there are some projects which probably reach satisfying levels of equilibrium between the physical, technical and social problems (Research Committee for Designing Integrated Regional Information Systems, 1985). The Kawasaki plan, for example, which has joined the physical level with the telematic level, thanks to the fusion of hard and soft projects, and the Emscher Park plan (Muller, 1991), which has translated the complex long term evolutionary logic of historical territory into a mature management system.

22.4 Elements for a New Project Lexicon

22.4.1 Introduction

The range of possible scenarios, though technically rich, seems to be prisoner of a series of questions: how can the existing context survive under the push of innovation, which in the mature industrial world is linked to a shortage of resources? how can we adapt the existing context to new technological scenarios?

The technocratic anxiety is justified. The lack of work, for example, justifies the push to try experimental projects, hoping they will produce some form of urban renewal and maintain the town as the centre of gravity for employment. Another question is to what extent we are facing a new kind of town, as stated by Françoise Choay: '*the equilibrium between urbs and civitas is definitively compromised*'. Despite this uncertainty, the experiences considered prefigure a new project lexicon whose most important elements are: the revision of the idea of *civitas* and of space, and the urgency of renewing knowledge.

22.4.2 A Revision of the Idea of Civitas: The Street Address and the Internet Code

Can the idea of *civitas* be left to passing opportunities and the 'schizophrenia' represented by the fact that the modern citizen possesses both a street address, representing a fixed physical place, and an Internet code, completely indifferent to the traditional geographical terms of position? We are living in an incomplete revolution. All previous revolutions have produced forms of organisation and social revolution clearly identifiable with a physical place. The merchant revolution produced the organisation and physical revolution of the town council. In that era society was self-regulated, the rules were established in its centre, the city centre. The industrial revolution produced multiple centres (i.e. the business centre, the industrial centre, the station, and so on), but since then, society has begun to be artificially constrained in hierarchical structures linked to economic production rather than to civic values, marking the passage from the civilisation of the city to the degradation of the conurbation. The latest technological revolution is not concerned about new forms of social organisation: we know that the citizen is immersed in a chaotic system of opportunities, but no 'political' translation of the scientific word 'chaos' has yet been offered.

Do we therefore have to passively put up with the idea of a town theatre of 'balkanised' social relationships proposed to us daily by the media? Probably not, since we are living in a transition period and are just beginning, with difficulty, to have an idea of some possible 'political' translations of the term *civitas*. There is a clear contrast between the stability of periods of historic civic organisation and the interactivity of new technological systems, which permit the citizen to be protagonist of decisions in real time and according to nonhierarchic models (Virilio, 1985). This dichotomy is now translated in extreme proposals: the 'panoptic' and the 'forum'.

Over the last few years we have seen the insistent proposition of the 'Bentham's panoptic' (the famous building from the centre of which the Prince secretly controlled the life of his subjects), a disquieting physical anticipation of Orwell's images. If the complexity produced by new technologies causes confusion among people, what better solution than a hyper-concentration of knowledge and power, able to manage in real time the needs of subjects whose only function is to react passively to a refined system of orders (Zaboff, 1988)!

The experience of the 'forum' is based on very different presuppositions.

It has been tried in the recent plans of Emscher Park and Kawasaki. These projects reject the stability of the traditional town council as a place of command and planning management, in favour of a refined system of relationships, able to manage the projects in the various stages of evolution, from inception to realisation, with the active participation of citizens. The central assumption of this approach is that the new town should not simply be the result of orders imposed *ex-ante*, but the fruit of a complicated interweaving of responsibility, in which the citizens are put in the position of making decisions, thanks to an active educational policy.

These two philosophies tell us that the new strategic element in town construction will be knowledge, but that this can provide either an opportunity to repropose the old oligarchic system, or a new frontier of democracy. The choice is up to those who propose projects and plans.

22.4.3 The Revision of Space Rules

The decline of traditional interpretations of space makes it unfeasible to repropose former administration levels or technical instruments, but there has to be an intensification in the drawing up of new strategies and collective projects (I.Re.R, 1991). Consequently, if the complicated Italian urban system wants to have some role in the Europe of Maastricht, it has to take greater advantage of incentives offered by the E.C. programmes, widening the possibilities of intervention and redefining its role within the great urban cluster of which the open European region will consist. The last Italian document to define the complex connection between place and the national system, 'Project for the 80s', dates back to the end of 60s, and was the final outcome of the 'regionalist' trend which began after the second World War!

The Italian urban system has to pursue the ambitious aim of developing a complex project, the result of agreement on heterogeneous forces: the urban specificity, the metropolitan central position, and the regional network (De Roo, 1992). The idea of the enlarged region does not mean the recomposition of administrative spaces, but represents a space of functional balance with moving boundaries. It is *integrative,* because it links economic and social networks, which both tend to close themselves in local ghettoes, or their own local territory; it is an *open structure*, because it diversifies the scale of social identity and of economic organisation, and because it secures territorial fluidity from the local to the universal.

Consequently, the regional network cannot be likened to the metropolis/region, where all flows are centred on one node. The urban 'regional' cluster should be thought of as a fine-mesh net, characterised by multiple connection points and by multiple scales of reasoning. In this system the historical town becomes a *go-between space*, whose essential quality is its organisation, which ensures links among local networks, maintains territorial coherence, and allows the passage from local to international. The historical town therefore should exalt the social matrix of urban complexity, ensuring cosmopolitanism and diversity by means of social mobility, and opposing a closed urban civilisation, reduced to fragmentation and the formation of 'ghettoes'. The historical town must be understood as the link between an open and integrative regional structure and the space of civic identification.

These arguments raise the issue of the re-interpretation of local self-government, as a stimulus to informal co-operation between different entities, without necessarily producing new structures. This would seem to be the most useful way of giving full meaning to urban space as the place of political negotiation and social contract.

22.4.4 The Revision of Plan Principles

As we have said, the network is a recurring metaphor in urban policies both within and outside Europe. The aim of most planning interventions is either to integrate towns in network relationships or create a specialisation, creating a town or city which becomes known as an important node. If this is the condition for economic survival, these policies have to face the contradictions of the network concept, and especially the new territorial and social disequilibrium produced by the growing lack of distinction between economic and political decisions.

The central problem of plans for our post-industrial towns is the difficult integration between economic and social time. The town is not changing in real time, according to the floating model given by the short life cycle of new technologies or by liberal economic schemes. Consequently, we should be careful of plans for collective organisation, which aim to transform residents into 'political actors'.

Once we have abandoned the illusion of the plan as a pre-established set of interventions promoted by landed interests, we find it becomes a complex series of actions, in which organisation is central and a condition of the plan's effectiveness. The aim is to create a synergy, increasing the probability of meeting citizens' hopes as well as economic interests, using the managing ability of Public Administration. In other words, the planning process is no longer linear, like Taylor's organisational model, consisting of a series of orders given by the Public Administration, but takes on a more complex form, inspired by the Toyota organisation model. The Public Administration defines the strategy and develops a multiplicity of co-ordination activities among the actors involved in this strategy. Great attention must be paid to the production and distribution of knowledge, which becomes the main function of intervention, and to the form of representation of civic power, moving from town council to 'forum', a permanent place of meeting, comparison, testing and, hopefully, the realisation of projects.

22.4.5 The Urgency of Renewing Knowledge

The acceleration of change, its ubiquity and the complexity of choices make the rapid diffusion of knowledge essential, above all, so that people can make conscious choices about proposals by technicians and administrators. This need has been assimilated by the new plans, which include an educational aspect. Françoise Choay (1994) hoped for a deep revision of teaching systems to achieve the structural transformation of *crafts and professions connected to the environment*. This revision includes:

- a continuous updating of information for local politicians;
- new training for architects to give them management skills: this would include a far more abstract theoretical package, new media techniques, project paradigms inspired by biological rules, a new technological tradition (i.e. imposed by the decline in the use of reinforced concrete);
- the revaluation of artisans, as is happening in Japan, the most advanced country in the virtual techniques field, where craft skills are protected and defined 'national treasure' by a special law - a useful lesson for Italy, which has destroyed its technically-based educational system;
- the promotion of associationism and its cultural modernisation, to enable the jump from the defence of special interests to a wider vision, making citizens more aware and giving them a new culture so that they can actively interpret the concepts of public space and town.

22.5 An Assessment of the Italian Experience

The Italian urban reality is complex: on the one hand, the new has undoubtedly met with strong resistance at a scientific and administrative level, on the other hand, we can see the blossoming of a series of experiences promoted by a group of scientists and professionals, 'civic agitators' who have, with originality, been rethinking our towns according to new scenarios.

Among these experiences, the first, and perhaps the most important for its operational results, has been the 'Cabling Lombardy' project, the outcome of cooperation between local authorities, experts and firms (Regione Lombardia, 1987). The merit of this project is to have faced a central issue of the metropolitan region: the delivery of high added-value services and its impact on the urban format.

This project has been followed by a series of others involving medium-sized towns, mostly with high living standards. Among the more active promoters of this second stage is the RESEAUX group, with important contributions by Franco Morganti and Paola Manacorda (1989). One of the projects is 'Telematic Trentino' (Ciborra and Longhi, 1989), promoted by the Chamber of Commerce, the University and the Union of Industrialists. This experience posed the problem of achieving a balance between promoting telematic communication and re-stimulation of civic unity.

At the academic level, we should mention the work of Corrado Beguinot's group (1988, 1994), which has codified all aspects regarding innovation processes and the town in the "Encyclopaedia of the Cabled City" in order to arrive at a new town planning 'Charter'. The eco-planning concept too has been considered (Longhi, 1992), although results are still far from the research of the international scenario (Morita, 1991, Choay, 1994).

At present, effort is being concentrated on creating 'virtual towns', involving an interactive relationship between civic authorities and citizens. Such projects have been launched in Bologna, promoted by the Decentralisation Office, in Venice, promoted by the University Institute of Architecture, and Milan, with the most strongly marked open civic network, promoted by a small group of experts from the Polytechnic.

But, as Priscilla de Roo (responsible for the planning sector of DATAR) would say, the impression is, on the whole, that ours are 'neglected towns' within a 'neglected Europe': 'neglected towns' because most entrepreneurial and intellectual forces have not really understood the urgency of entering the network of new relationships which has characterised the international texture of towns since the mid 80s, and because political forces have not understood the new 'rules for belonging' characterising post-industrial society; 'neglected Europe' because the towns of the old continent are now facing a historical crisis, caused by the transfer of the central relationships and interchange from the Atlantic basin (Europe/US eastern seaboard) to the Pacific basin (the US western seaboard/South East Asia).

This transfer, as we are reminded by the plan for metropolitan Lyon, has made European administrators and experts aware that 'the optimism in planning of the roaring 30s' is now just an unproductive memory for planners of future urban realities. As regards the Italian situation in particular, we may put the question: has the effort of chasing industrial paradigms foreign to our culture, made us so tired that we have renounced the challenge of a new complex situation? Or, returning to Carlo Maria Cipolla's doubt, will we be able to accept the cultural challenge from the Pacific area without having to assimilate the previous cultural revolution coming from the Atlantic area?

The task is, obviously, fascinating but difficult. From this point of view, we understand how strong the temptation is for Italians to isolate themselves in an illusory 'Mediterranean dimension' and give up the challenge. To avoid this, it would be useful to reconsider the positive results of our planning culture, to reflect upon the new kind of plan emerging, the rules of space, and the urgent problem of renewing knowledge. It is certain that in the planning process some basic precepts must be established: the supremacy of the Public Administration, the active participation of citizens in the process of plan formulation, and the consideration of historical towns as the fulcrum for all spatial designs. But, these elements must be re-interpreted according to the new reality.

22.6 Conclusions

In this article, as stated in the introduction, we have reflected on the spatial effects of the latest waves of technological innovation. Considering the speed of these innovations and resulting confusion, we have made an effort to look at the situation in a systematic way in order to identify the new rules of planning. This should not be interpreted as the presumption to produce a new theoretical synthesis, but simply a contribution to the understanding of that ever-changing nebula of fragments which is now colliding with our territorial reality.

From this point of view, the current process of technological renewal is pushing us to revise the general spatial rules and assess our first empirical experiences. We have pointed out the danger of the present tendency to consider the project as an autonomous intervention, unlinked to the more general processes of transformation, and have emphasised the usefulness of considering the project as an 'intersection' of differing knowledge and the importance of rapid re-thinking. These considerations have led us to underline the present 'neglect' of towns in the urban context and to hope for their re-insertion in the complex network which is creating a 'world city' thanks to the rapid and indispensable renewal of knowledge.

References

Alonso W. (1971) The Economics of Urban Size, *Papers of the Regional Science Association, 36.*

Beguinot C. (eds.) (1988) *L'Enciclopedia della città cablata*, Giannini, Naples.

Beguinot C. (eds.) (1994) *La Carta di Megaride*, Giannini, Naples.

Braudel F. (1986) *Una lezione di storia*, Einaudi, Turin.

Calthoper P. (1995) *The Next American Metropolis: Ecology, Community*, Princeton Architectural Press, Princeton.

Camagni R. (1992) *Economia urbana*, La Nuova Italia Scientifica, Rome.

Camagni R., Gibelli M.C. (1992) Per una pianificazione strategica in Lombardia, in I.Re.R., Aggiornamento del piano territoriale della Lombardia, I.Re.R., Milan.

Castells M. (1989) *The Informational City*, Basil Blackwell, Cambridge, Massachusetts.

Choay F. (1991) L'urbanistica disorientata, in Gottmann J., Muscarà C. (eds.) *La città prossima ventura*, Laterza, Bari.

Choay F. (1993) Le règne de l'urbain et la mort de la ville, in the catalogue of the exhibition 'La ville, art et architecture en Europe 1870-1993', Centre Georges Pompidou, Paris.

Choay F. (1994) Penser la non ville et la non campagne de demain, *Intellos, 4*, 23-30.

Ciborra C., Longhi G. (1989) *Telematica e territorio nella terza Italia, il caso del Trentino*, Angeli, Milan.

Cipolla C.M. (1989) *Le tre rivoluzioni*, Il Mulino, Bologna.

Davezies L (1992) A la recherche de la ville globale, *Action et Recherches Sociales, 4/92-1/93, monograph, Où va la ville?*, 43-58.

De Roo P. (1992) La métropolité (ou l'invention de l'urbain), *Action et Recherches Sociales, 4/92-1/93, monograph, Où va la ville?*

Duany A., Plater-Zyberk E. (1993) *Towns and Town Making Principles*, Rizzoli International, New York.

Dupuy G. (1991) *L'urbanisme des réseaux*, Colin, Paris.

Dupuy G. (1992) Relire Cerda pour aménager la ville d'aujourd'hui, *Action et Recherches Sociales, 4/92-1/93, monograph, Où va la ville?*, 67-74.

E.C.C. (1994) City Action RDT Programme: Toward a Better Liveable City, E.C.C., Bruxelles.

Fabris G., Mortara V. (1986) *Le otto italie*, Mondadori, Milan.

G.L.C. (1985) The London Industrial Strategy, Greater London Council, London.

Hoover E.M., Vernon R. (1959) *Anatomy of a Metropolis*, Harvard University Press, Cambridge, Massachusetts.

I.Re.R (1991) *La città metropolitana*, I.Re.R., Milan.

Katz P. (1993) *The New Urbanism toward an Architecture of Community*, McGraw-Hill, New York.

Klaassen L.H., Molle W.T.M., Paelinck J.H.P. (1981) *Dynamics of Urban Development*, St.Martin Press, London.

Lacaze J.P. (1990) *Les métodes de l'urbanisme*, Que sais-je?, no. 2524, Presses Universitaires de France, Paris.

Lacaze J.P. (1992) L'urbanisme entre mythe et réalité, *Action et Recherches Sociales, 4/92-1/93, monograph, Où va la ville?*, 21-30.

Longhi G. (1988) La città del 2000 mette radici: la riqualificazione urbana hi-tech, *Il nuovo cantiere*, 10, 8-12.

Longhi G. (1990a) Tecnologie e pianificazione, *Il nuovo cantiere*, 2, 20-23.

Longhi G. (1990b) Struttura della metropoli, mutamenti tecnologici e prospettive della convivenza, in Totaro M. (ed.) *Città e diritti di cittadinanza*, Angeli, Milan, 49-63.

Longhi G. (1992) Verso l'ecopiano, in Catalogo della XVIII Triennale, La vita tra cose e natura, Electa, Milan, 226-234.

Longhi G. (1994) Il nuovo ordinamento delle autonomie locali a fronte dei cambiamenti negli indirizzi di progettazione territoriale, in I.Re.R. *Il nuovo ordinamento delle autonomie locali in Lombardia*, Angeli, Milan, 191-226.

Lyddon D. (1991) Planning in the Age of the Information City, *Planning Administration, 28*, 111-115.

Martinotti G. (1988) *Milano ore sette: come vivono i milanesi*, Maggioli, Rimini.

Martinotti G. (1994) *Metropoli*, Il Mulino, Bologna.

Mera K. (1989) An Economic Policy Hypothesis of Metropolitan Growth Cycles, A Reflection on the Recent Rejuvenation of Tokyo, *Review of Urban and Regional Development Studies, 1*, 85-90.

Mitchell W.J. (1995) *City of Bits, Space, Place and Infobahn*, MIT Press, Cambridge, Massachusetts.

Morganti F., Manacorda P. (1989) Nuovi servizi telematici per Reggio Emilia, Report of RESEAUX, Milan.

Morin E. (1990) *Introduction à la pensée complexe*, ESF, Paris.

Morin E., Kern A.B. (1994) *Terra-Patria*, Raffaello Cortina Editore, Milan.

Morita T. (1991) Japan's New Concept for Sustainable Development in Urban and Regional Planning: Ecopolis, Ecobusiness and Ecological Life-Style, in *Planning Administration, 28*, 44-47.

Muller R. (1991) IBA Emscher Park, Cologne.

Research Commitee for Designing Integrated Regional Information System (1985) *Creation of an International Scientific and Cultural City: Kawasaki*, Japan Association for Planning Administration, Tokyo.

Regione Lombardia (1987) Lombardia cablata, Regione Lombardia, Milan.

Salat S. (1992) Complessità. La spirale e il cubo, in XVIII Triennale, *La vita tra cose e natura*, Electa, Milan, 290-294.

SEPAL (1988) Lyon 2010: un projet d'agglomération pour une metropole européenne, Communauté urbaine de Lyon.

Unal M. (1992) Réseaux urbaines et utopies humaines: deux approches inconciliables?, *Action et Recherches Sociales, 4/92-1/93, monograph: Où va la ville?*, 75-84.

Virilio P. (1984) *L'espace critique*, Bourgois, Paris.

Zaboff S. (1988) *In the Age of the Smart Machine*, Heinemann, New York.

[illegible] R. (1988) *La città senza [illegible]*, [illegible], Milan.
[illegible] (1983) *The New Heritage [illegible]*, McGraw-Hill, New York.
Ha[illegible] J.H., Mohn W., [illegible] (1988) *[illegible] Development*, [illegible] Press, London.
La[illegible] (1989) *Les [illegible] de l'informatique*, Que sais-je? no. 2224, Presses Universitaires de France, Paris.
[illegible] (1982) [illegible] et nature, [illegible] Monograph, [illegible], 31-50.
[illegible] (1988) [illegible] la [illegible] urbana [illegible], 8-12.
[illegible] (1996) [illegible], *Il nuovo cantiere*, 3, [illegible].
[illegible] (1990) [illegible] della [illegible] tecnologie [illegible], in [illegible], [illegible], [illegible] XVII Triennale di Milano, [illegible], Milan, [illegible]-225.
[illegible] (1989) [illegible] nella [illegible] di [illegible], in [illegible] *Il nuovo [illegible]*, [illegible], Milan, 194-225.
[illegible] (1991) [illegible] the Information City, *[illegible] Administration*, [illegible].
Martinotti G. (1988) *Milano ore sette: [illegible] di [illegible]*, Maggioli, Rimini.
Martinotti G. (1994) *[illegible]*, Il Mulino, Bologna.
[illegible] (1988) An [illegible] of Metropolitan [illegible]: A [illegible] on the [illegible] of [illegible], *[illegible] Development Studies*, [illegible].
[illegible]
[illegible] (1990) [illegible], [illegible].
[illegible], Edizioni [illegible], Milan.
Morita T. (1996) Japan's New Concept for Sustainable Development in Urban and Regional Planning, Ecopolis, Ecobusiness and Ecological Life-Style, in *Planning Administration*, [illegible].
[illegible] (1994) *Les [illegible] à Paris*, [illegible], Cologne.
Research Committee for Designing Integrated Regional Information System (1995) *Creation of an International Networked Regional Information* [illegible], Japan Association for Planning Administration, Tokyo.
Regione Lombardia (1987) *Lombardia [illegible]*, Regione Lombardia, Milan.
Sal[illegible] S. (1992) [illegible], in XVIII Triennale, *[illegible]*, [illegible], Milan, 200-204.
[illegible] (1986) [illegible] projet d'une [illegible], [illegible], Communauté urbaine de Lyon.
[illegible]
[illegible]
[illegible], New York.

23. China in Search of a New Planning Paradigm

Tunney F. Lee

23.1 Introduction

Viewed through Western eyes, urbanisation in China is chaotic, unbridled and destructive of the environment. The magnitude of new construction is unlike anything seen by Western eyes since the postwar reconstruction of Europe. New central business districts, office complexes, industrial parks, shopping centres, hotels, highways and tracts of suburban villas have been built with no end in sight. For the Chinese, this explosive development is viewed both as a sign of progress and the way out of poverty. Whereas in the West, growth is seen as something to be controlled, ordered and contained, China is buoyantly encouraging it.

"In China today we are witnessing the emergence of the world's largest market economy, as well as a building boom of unprecedented scale. Yet while supply-side economics are creating new opportunities in the business environment, there is a short supply of anything resembling high quality architecture, or even decent construction, in the built environment" (Chen, 1993, p. 35).

Whereas in the West automobiles and urban sprawl are seen as environmentally destructive, China plans to build three million automobiles per year by the year 2000. Americans see the city as a dangerous place, while the Chinese flock to cities to seek out economic opportunities.

How can we understand these divergent perceptions? Is this a case of cognitive dissonance between knowledge and action? Are the normative planning theories of the West relevant to understanding the forces at work in China today? That is, do Chinese planners know how to measure and analyse the consequences of rapid uncontrolled growth, but simply not have the power to implement their policies? Will the Chinese repeat all the mistakes of the West, passing through the whole urban cycle before value systems change, and before urban sprawl and car ownership are controlled or channeled?

These questions suggest some of the difficulties in understanding China's urbanisation in terms familiar to Western planners. Perhaps it would be more productive to re-state these questions in the following form: Given that China is undergoing modernisation at an unprecedented rate, what kind of city planning can hope to give coherence to this process? Are some of the methods being discussed at this conference applicable? If so, in what ways?

The meanings and metaphors which breathe life into the particular forms of development of Chinese cities are significantly divergent from the European models. In order to answer the questions above, we firstly examine the historical factors which have contributed to the changing form of the city over past centuries. The examination is divided into four parts.

Firstly, in Section 2 we examine the city as a manifestation of the harmonious and hierarchical society of traditional China, pointing out some significant contrasts with European medieval cities. Section 3 describes the painful transition undergone in the 19th century when the city became an instrument of European exploitation. This period saw the growth of the coastal entrepôts and migration of population from the country to the city. China's population grew dramatically, but stark contrasts developed between the urban life and the desperately poor standard of living in rural areas.

The impact of Socialist policies adopted from the 1950s onwards are discussed in Section 4. An attempt was made to achieve a better balance between urban and rural areas, and strict limits were imposed on city population. The initial approach to planning and the form of the newly planned cities bore a strong Soviet influence. In the more recent period, however, there has been a reversal of urban policies and emphasis is now placed above all on economic development. In the last section, we ask whether it is possible to curb the present free-for-all attitude to development and, if so, what lessons should be learned from the experience of the West and which techniques are the most appropriate to apply. We reflect on the importance of finding new ways of understanding city growth and more flexible approaches to planning. The potential of advanced technologies is stressed, and it is suggested that GIS, as well as modelling techniques, could play a significant role in determining urban policies for the future.

23.2 The City as a Manifestation of the Harmonious and Hierarchical Society

The contrasts between the Chinese and the European city in medieval times were many and profound. European cities became worlds in which relatively democratic forms of government were established. These worlds were, in fact, both very free and highly democratic compared with rural society, which was dominated by serfdom and feudalism (Bairoch, 1988).

Unlike the European city, the Chinese city served mainly as an administrative centre for the empire and never achieved independent status. There never developed in China the independent, democratic and mercantile character which became the basis for the organisation of the modern capitalist Western city and the institutions and practice of city planning. Contrary to the notion that 'city air makes you free' cities in China over the last few centuries have been more oppressive than the countryside. Peasants were petrified of the city and wary of being ensnared in its strange and arbitrary laws. Merchants were restricted and scorned.

Despite great differences in the spoken language, the common written language along with a shared culture and shared beliefs and rituals meant that Chinese cities did not develop separate political systems nor did they adopt different religions. Commenting on the reasons for the lagging of China in science, Clayre noted that "one partial explanation is China's lack of political variety. Whereas the diversity of Europe meant, at a crucial time, that ideas not acceptable in one place could be cultivated in another - for instance the Galilean notions of astronomy could be discussed in Holland when forbidden in Italy - the unity of China made for a single orthodoxy" (Clayre, 1985, p. 216). The result was that although individual cities developed very distinctive characteristics and specialisations, they did not experiment with different political ideas or different ways of organising society.

The classical urban tradition lasted from 1500 B.C. almost to the present, and was based on a concept of the ideal Chinese City. "It should be square, regular, and oriented, with an emphasis on enclosure, gates, approaches, the meaning of the directions, and the duality of left and right. Creating and maintaining religious and political order was the explicit aim. Ritual and place were fitted together. They expressed, and indeed were believed actually to sustain, the harmony of heaven and men, which it was disastrous to disturb" (Lynch, 1981, p. 13).

A further reason suggested for differences in the density of city development was the ecology associated with the dominant grain crops in

China and Europe. "The difference between rice and wheat is crucial. Like the potatoes and maize of the pre-Columbian cities, rice is more favourable to urbanisation than wheat. Rice supplies around 3,600 calories per kilogram, as against 3,400 for wheat, which eases transport problems. But the great advantage of rice is the higher yield: one hectare of land yielded 1,600 kilograms of rice, as opposed to only 600 of wheat. Therefore in terms of usable output, the ratio in favour of rice is 3 to 1. Furthermore in many parts of China farmers could produce two harvests per year. This means that the area needed to provision a city in a rice-producing region would have been three to six times smaller than that for a city in a wheat-producing region" (Bairoch, 1988, p. 355).

The consequence was that the level of urbanisation in traditional China became the highest in the world. As noted by Marco Polo and other European travelers, Chinese cities were large, rich and well organised and governed.

"It becomes more and more clear that the rise of agriculture all but inevitably carried in its wake a process of urbanisation. For, while it is true that urbanisation could not get under way without the concentration of population and the surplus of food resulting from agriculture, it is equally true that the emergence of agriculture set in motion forces that sooner or later led to the growth of cities" (Bairoch, 1988, p. 94).

The kind of society that emerged in China was directly related to the form of agriculture that developed in the great river plains of the Yellow and Yangtze Rivers. The amount of rainfall was volatile and unpredictable, but of a seasonal character; that is, the rains came in the spring and summer, while there was no rain at all in the fall and winter. The need to store, move and control water over a large territory became the essential duty of the state.

Success depended on the creation and maintenance of a national system of dikes, dams, canals, artificial lakes and reservoirs. For thousands of years and through innumerable dynastic changes, the top of the Chinese social order had been the scholar-official who manned the imperial bureaucracy and whose responsibility was to organise, coordinate and direct public works. At the apex, the Emperor as the Son of Heaven had to mediate between a capricious heaven and the people.

The contrast with Europe is striking. As noted above, the productivity of rice per acre is much higher than the yield of subsistence farming in medieval Europe, where the state had a minor role in improving and sustaining productivity. Success depended more on rainfall, rotation of crops and balance among different agricultural activities such as dairying, stock raising, etc. As a result, a centralised bureaucracy and despotism never emerged as it did in China.

Supporting this physical form was an elaborate and complex set of rituals and rules of siting. The rules of siting known as *feng shui* controlled and influenced all aspects of siting cities, villages and buildings. Derived from Chinese religion and philosophy, it is part of a system that comprehensively describes and includes the individual human being in the cycles and movements of nature. The universal application of *feng shui* has contributed to the harmonious fitting and siting of the built environment into the natural environment. "The cosmic model upholds the ideal of a crystalline city: stable and hierarchical – a magical microcosm in which each part is fused into a perfectly ordered whole. If it changes at all, the microcosm should do so only in some rhythmical, ordered, completely unchanging cycle" (Lynch, 1981, p. 81).

This model has tremendous power and has appeared in Korea, Japan and Southeast Asia. Its last influence was in the nineteenth century when the Nguyen Dynasty adopted it for on the plan of Hue, their capital of Vietnam.

23.3 The Painful and Unfinished Transition

In 1793, China was at the greatest extent of her territorial limit and, for the last time, complete and consistent within her own society and set of values. Her cities were among the largest and best organised in the world. When George III of Great Britain sent Lord Macartney with three ships loaded with the latest inventions: clockwork toys, a planetarium, telescopes, scientific instruments and Wedgewood porcelain and candelabra, Macartney was dismissed by the Emperor with the famous edict: "You, O King, live beyond the confines of many seas, nevertheless, impelled by your humble desire to partake of the benefits of our civilisation, You have despatched a mission respectfully bearing your memorial ... Swaying the wide world, I have but one aim in view, namely to maintain a perfect governance and to fulfill the duties of state: strange and costly objects do not interest me ... We possess all things. I set no value on objects strange or ingenious, and have no use for your country's manufactures" (Singer, 1992, p. 99).

Although Macartney's embassy was seen as a failure, he had already observed in his journey through China that there were serious problems of poverty and stagnation. The Emperor Qianlong presided over a deteriorating and declining country which had been the victim of its own success. The system had so completely relied on maintaining the appearance of harmony through ritual and beliefs, that it could not withstand the dynamic thrust of an expanding Europe. In the past, the Empire had absorbed foreign invasion by asserting its superior governmental system and better developed culture, but now the Europeans posed a different problem with their newly developed systems and a culture supported by vastly superior military technology.

To make things worse, China was caught in a period of stagnation. The Qing Dynasty had brought one of the longest periods of peace in many centuries. But the consequences of changes in the New World were daunting. High nutrition foods from the Americas came via Europe; new crops of potatoes, yams, and corn could be grown on marginal lands not previously exploitable. New strains of rice that were more productive and new methods of planting increased the yield further. The net result was the rise of the population from 110 million in 1500 to 420 million by 1850. This had been achieved by using up almost all available cultivable land and expansion into marginal lands.

China was facing a Malthusian brake: insufficient food resources to feed its growing population. The same Imperial system that created the agricultural successes of the past now stood in the way of its solution. Peasant rebellions began to break out all over China as the pressure of taxes on an increasingly impoverished population exacerbated the inequities of the system. Despite the orderly appearance of the overall system, local patterns of land ownership were plagued by irrationalities and fragmentation of holdings. There was a need for an agricultural and industrial revolution which was not forthcoming. What happened instead was that the newly industrialised European countries came seeking markets.

"Already by the 1790s, a worker in an English textile mill could, due to mechanisation, produce 120 to 160 times as much cotton thread as an Indian or European traditional craftsman. During the same period, because of the use of coke, with a population of eight million, England turned out some 100,000 metric tons of cast iron, or probably as much as, if not more than, all of India with a population of nearly 200 million. Within about twenty years (around 1810-20) this industrial change brought about a complete reversal in the flow of trade between Europe and those regions in other parts of the world with advanced technological cultures, particularly Asia" (Bairoch, 1988, p. 397).

By 1842, the British had gained by force what they could not gain by diplomacy in 1793. China tried to delay the encroachment but decisively lost every battle, each time giving concessions of cities and land to various

European powers as well as the newly industrialised Japanese. These treaty ports were 'concessions' to the foreign powers by which land was leased in perpetuity. China did not lose sovereignty (except in Hong Kong) but in effect, the treaty ports became foreign territory.

The treaty ports were built to facilitate trade. Port facilities, warehouses, offices, government buildings, barracks and public works were built for the fast growing trade which rose tenfold between 1840 and 1913. Residential areas, with their churches, clubs, cricket fields, botanical gardens, race courses, and so on, seemed to be transplanted from the home country to the treaty ports. Due to cheap labour, the expatriates were able to enjoy a life-style modelled on that at home, but with a considerably higher standard of living.

The Chinese too flocked to the new cities, attracted by the economic opportunities and also freedom from the restrictions of traditional Chinese society. Shanghai grew from 200,000 to 800,000 from 1850 to 1900. By 1938, it had reached 3.5 million and had become the leading city in Asia. It provided the base for China's film industry, book-publishing and financial markets. It had also become corrupt and was dominated by crooked businessmen and gangsters.

The Qing Dynasty ended with a whimper in 1911. The boy Emperor abdicated, leaving a country so weak and devastated that it would be several decades before China unified under a single government. The Nationalist government struggled to take control of a vast country and population torn by regional conflicts and poverty. Nevertheless, the ensuing years saw continuing growth of the population and major cities. But the attempt to establish a system of administrative control over the rural areas proved elusive. This administrative system left fundamental problems in the countryside unsolved, and in many rural regions life was little different from what it had been under the Qing. "In China's cities and towns, the contrast was stark. Medical care became more sophisticated, new hospitals were built, schools and college campuses featured sports grounds and laboratories. New power stations brought electricity to urban China, cinemas established themselves as part of urban life; radios and phonographs appeared in the richer homes. For the wealthier Chinese, life could be very comfortable indeed" (Spence, 1990, p. 368).

However, life for the peasants was disastrous. The fragmented and irrational land-ownership patterns, the landlords in collusion with corrupt officials, the lack of education, the low standard of health and hygiene, the uncertainties of the weather and market all exacerbated the conditions for peasant discontent. The Japanese invasion further disrupted and savaged the country, leading to the triumph of the Communist revolution in 1949.

23.4 The Socialist City

The Revolution was fought first in the countryside; the cities were not taken until the last stage of the civil war. The tactics and strategies had been worked out in Mao's long exile in rural Yenan. The process of class struggle and land reform had been tried and honed in the areas held by the Communists. What had not yet been achieved was the administration of the cities.

In traditional Chinese society, merchants were the lowliest of the four occupational classes: scholar-officials, peasants, artisans, and merchants. Trade was strictly regulated and heavily taxed by the bureaucracy. Merchants were considered parasites who did not labour with their minds or hands, but only made profits.

Reinforced by their years of working in the countryside, the peasant cadres who entered the cities were appalled by the corruption and decadence they found there. The Chinese Communist Party (CCP) had serious problems in administering the liberated cities and soon established street committees to deal with problems of public security and moral order. Campaigns were launched against prostitution and opium addiction which had evaded attempts at control by the Qing and Nationalist governments. A Shanghai newspaper wrote: "Shanghai is a nonproductive city. It is a parasitic city. It is a criminal city. It is a refugee city. It is the paradise of adventurers" (Spence, 1990, p. 518).

These concerns prevailed among the leadership of the CCP and were influential in the development of urban policies. Faced with a still growing population and unimaginable poverty, the CCP adopted the Stalinist model of development. This consisted of the establishment of heavy 'producer industries' that would serve as the engine to pull along the rest of the economy. To implement this, the USSR sent technical assistance to build infrastructure, such as hydroelectric dams, steelworks and factories for producing machine tools and tractors, in fifty cities along and at the junctions of the up-graded or newly built railway lines.

A corollary of this policy was the move away from the coastal areas and, in particular, the down-grading of Shanghai. This was partly due to the fear of invasion but, more importantly, also prompted by the desire to achieve a better balance between rural areas and urban areas. A further reason was the distrust of the treaty port cities, thought to be corrupted and contaminated by Western bourgeois ideas.

The first period of reconstruction ended in a rift with the Soviet Union over the speed of agricultural collectivisation. Mao's concerns harked back to the simple existence of Yenan. He disdained city life. His ideal was the omnicompetent man of the soil, a combined farmer, craftsman and militia soldier in a self-sufficient countryside (Fairbank, 1974).

The resulting commune movement was pervasive and covered all of rural China. Even urban communes were formed in cities with major industries at the core. Unfortunately, the experiment was to end in disaster. The attempt to build industries in the countryside (e.g. backyard steel furnaces) produced very little that was usable and also siphoned off labour and raw materials. Agricultural production fell and a severe famine killed millions of people.

In the aftermath of the disastrous Great Leap Forward and with the strain of increasing urban population, strict limits were set on the size of cities. Drastic measures, including the issue of ration cards, identity cards, internal visas and the allocation of jobs and housing, were used to control movement into the city. Groups of people, including workers and professionals, were also forcibly moved to remote rural areas and inland cities. For several decades, China was the only developing country that was successful in controlling city population.

The fifty cities built with Soviet aid were all located around rail lines and old centres. The usual plan was based on Miliutin's Nizhni Novgorod or Stalingrad with industries lined up along the railroad tracks and separated from the residential areas by a green belt with community facilities. The old centre was incorporated into this overall fabric.

In the older cities, new boulevards and squares were created through and around the older neighbourhoods. Most significantly, the seat of government and the capital was re-established in Beijing and the Forbidden Palace of the Qing dynasty was chosen as the place to announce the new People's Republic. The space in front of Tiananmen was enlarged to become Tiananmen Square, still the political centre of China.

In the decades since 1949, when the present government came to power in China, urban and regional development had not followed an economically rational or reasonable course of evolution. China had trodden a circuitous path, without fully exploiting its development potential or harnessing its economic resources. For a long time, China stagnated at low rates of economic growth, with an increasing gulf between its level of economic development and that in advanced economies or even in certain developing economies (Yeung, 1993).

After the trauma and complete chaos of the Cultural Revolution and the succession struggle after the death of Mao, China set out on a deliberate course to catch up with its more successful neighbours - Japan, Korea, Taiwan, Hong Kong and Singapore.

The first stage was the abolition of the collectivisation of agriculture and the introduction of the 'Family Responsibility System'. The reforms gave the initiative back to the family unit, setting quotas to be sold to the government and allowing any surplus to be sold on the private market. Agricultural production doubled in a few years. Again agriculture provided the impetus for the economy to grow and reform.

A consequence of the increased productivity was the increase in savings and the release of surplus labour. This had two effects. One was the growth of township and village enterprises. Built on the earlier structure of the communes, rural industry grew to employ 113 million workers in 19 million enterprises. The second effect was the migration of the surplus labour into the booming coastal areas.

The urban policies of the first three decades of the People's Republic were ultimately all reversed. The strategy favouring the inland provinces to balance development was now abandoned to allow the coastal regions to grow as rapidly as they could. The strict control of city population was relaxed to allow rural workers to migrate to areas of high economic growth. Urbanisation was now seen as both desirable and necessary for economic development. All manner of enterprise was encouraged.

Foreign investment was sought rather than prohibited. Fourteen coastal cities were declared open cities and Special Economic Zones (SEZ) were established to attract foreign investment and to conduct foreign trade. These zones evolved into new cities planned to implement the economic reforms and to experiment with new ideas before being introduced to the rest of China. The SEZ's were an immediate success, conducting 90% of China's foreign trade and accounting for half of the total foreign investment.

As cities like Shenzhen and the other SEZ's prospered, market reforms spread to existing cities. Shanghai, which had been neglected for decades, was chosen to lead the development of the entire Yangtze Valley, an area with a population larger than the United States. In Shanghai's Pudong New Area alone there are fifty buildings over twenty stories being planned or in construction.

In the early days after Liberation, under the Russian influence, city planning institutes were set up at each level of the bureaucracy. These produced Master Plans, but they were often merely formal exercises with very little input from decision-makers and other officials. As a result, the plans were often ignored by powerful politicians and organisations in quest of their own goals.

With the advent of market reforms there was still further deterioration in the role of the planners. Master plans contained no implementation schemes or costs, so they were routinely violated by the over-riding drive towards economic development. Since mayors are judged by their economic performance, the incentive is to encourage and foster development wherever there is a willing investor.

It is no surprise that the new buildings and complexes being built are a reflection of the free-for-all nature of development. Developers from overseas, especially Hong Kong and Taiwan, bring their own sense of marketability, building replicas of what rents and sells in their home markets. Local developers aim for a progressive image, which is often based on glossy magazine pictures of skyscrapers in the West. They demand buildings that are 'new, strange and distinctive'. The result is visual chaos as they pay little regard to consequences for the environment or their own Chinese heritage.

23.5 What Now?

At this point in time, matters are chaotic and out of control. The degradation of the environment, the lack of coordination among sectors and levels of bureaucracy and the rapidly deteriorating quality of life demands a whole new look at the role of city planning. How will China develop its cities and devise the kind of city planning that fits its particular historical and contemporary circumstances?

Undoubtedly, China's current problems, whose origin can be traced back hundreds or even thousand of years, are not easily characterised by labels developed in the West (Huang, 1990). China could certainly learn many lessons and techniques from Western experience. However, many of the institutions and values that have emerged from the Western city and its traditions of capitalist welfare democracy have no analog in Chinese society (e.g. mayors in China have always been appointed. There are no municipal elections of any kind). If some of these institutions and values were to develop in China, then new variations of planning techniques and theories which could emerge, utilising modern technological and ideological thinking within the Chinese context.

Since Liberation, much justified criticism has been leveled at the policies of the CCP. However, what has been forgotten is that many of these policies laid the foundation for what was to follow. Land reform was necessary for the Family Responsibility System to work. The land which had been sub-divided again and again resulted in holdings of so many small pieces that families could not work them efficiently. When collectivisation was ended and the land re-assigned to families, they were allocated as efficiently sized plots, which made modern agriculture possible.

Mao and the CCP created the infrastructure for the new China. "He had cleared the cancerous complexities of inter-household exploitation within the villages, which in the past had been one of the most fundamental obstacles to China's modernisation" (Huang, 1990, p. 250).

What propelled the revolution was the reaction to the failures of both the Qing Dynasty and the Nationalist government. People responded to the ideals of Socialism for its promise of a basic concern for the welfare of all people. Socialism was defined as State control of the means of production and the running of a centralised command economy.

Certain ideas from the socialist experience persist in a positive way. Housing is still considered essential for the wellbeing of society and set-asides for low-cost housing are required of all large private developments. Schools and public facilities are also considered essential elements of a healthy residential area and are mandatory in new developments.

It is possible that the traditional reverence for the environment may re-appear in a modern form. China is an enthusiastic supporter of environmental legislation and was an early adopter of the Rio Declaration and Agenda 21. China was the first of the major countries to develop its own Agenda 21 and a plan for its implementation. Clearly, this may be only rhetoric, but the rhetoric creates a standard by which action can be measured.

Chinese economic planners are beginning to adopt some of the new views on science as the philosophical basis for market reforms. "For the last 10 years, the so-called new scientific methodology - such as control theory, catastrophe theory, and dissipative structure theory - was introduced progressively and extensively to the study of social science and greatly enlightened people's outlook and thinking. This change is the result of China's reforms on the one hand and the driving force for the reform on the other hand. The significance of the new view of science, or rather the new way of thinking, is profound. It is a powerful tool that helps us in analysing the issues concerning social-economic development and reform" (He, 1989, p. 637).

The problem is that the reforms in conceptual thinking in the economy have not been incorporated into traditional city planning practice. While the forms of city planning remain, the rapid growth of the cities and the imperative towards economic development have left the traditional plans in the dust.

But there are certainly indications pointing towards a new way of thinking which could liberate planners from the unimplementable Master Plans that are still part of the mandatory output of the bureaucracy. There are signs of a shift towards a more flexible and less dogmatic view of plannig. Instead of painting utopian pictures of the ideal city, a more selective approach considers focusing on aspects of the city which are pivotal to change. Understanding how the city grows and changes would help enormously in the complex process of city development.

Instead of trying to plan the whole economy down to its every detail, a more modest planning policy might be to just plan major sectors or aggregate categories or particular aspects of the economy (Lavoie, 1989). One example is the recently announced *Anju* (meaning comfortable and spacious) housing programme which aims to build modern, affordable housing for several million families in the major cities. Abandoning the governmental role of financing, designing, building and managing social housing, the government is instead opening up all these functions to many different groups, including local groups, private developers, etc. The government provides incentives in the form of land lease concessions, tax breaks, guaranteed returns, etc. The response so far has been encouraging and may lead to more such strategic interventions.

More productive uses of the new information technologies are also being sought. The central government is highly tuned into the latest developments in computer and multimedia ware. Because China has had a very low level of telecommunications infrastructure, it is possible to leap-frog present technologies altogether. For example the telephone service will directly adopt digital, cellular systems rather than wiring the cities with copper wire. Major trunk lines connecting cities to each other and the outside world will use fiber optics with vast capacity for data transfer. Several Internet providers have een set up and their potential for facilitating faster information exchange within the country and internationally is enormous, despite worries about censorship.

Similarly, China is under-developed with respect to land information of all kinds - plot sizes and location, ownership, land use, growth and change, etc. Because relatively little information is yet recorded, the application of geographical information systems, cadastral systems and the like will be much easier and faster than more traditional formats used elsewhere. The computerisation of information will also enable the development of formalised models to measure the effects of policies such as increased car ownership, energy consumption, loss of farm land, pollution of water supplies etc.

This new technology will certainly not cause any direct change, but may influence the course of development by opening communication in ways that will change the internal decision making structures. Optimistically, the new tools will create a new dialectic, making the public aware of the consequences of short term decisions. Greater public awareness will then influence officials to make better decisions in the new complex and dynamic context.

"Decisions about urban policy, or the allocation of resources, or where to move, or how to build something, must use norms about good or bad. Short-range or long-range, broad or selfish, implicit or explicit, values are an inevitable ingredient of decision. Without some sense of better, any action is perverse. When values lie unexamined, they are dangerous" (Lynch, 1981, p. 1).

China today is in a crisis of values. With the demise of socialism and the cynicism which lies in its wake, she is left rudderless without a moral compass. The traditional is seen as irrelevant to the modern world. The fascination with the dizzying rewards unleashed by the market economy has been overwhelming. The new adage: 'To get rich is glorious' has replaced the Maoist idea of serving the people.

When the dust settles, it will be clear that all the techniques, information, theories and understanding will not produce good and livable cities until the Chinese people has transformed and reinvented its values. That process is underway, but is dependent on the further reform of the economy and the beginnings of political participation. A strong economy will produce the self-confidence and political maturity needed to adapt and use the new urban sciences and technologies to realise the universal value so well stated by Louis Sullivan: "All men in their native powers are craftsmen, whose destiny it is to create ... a fit abiding place, a sane and beautiful world."

References

Bairoch P. (1988) *Cities and Economic Development*, University of Chicago Press, Chicago.

Chen N. (1993) Interview reported in Beginnings, 35, Dept. of Architecture, Chinese University of Hong Kong.

Clayre A. (1985) *The Heart of the Dragon*, Houghton Mifflin, Boston.

Fairbank J. (1974) *China Perceived*, Knopf, New York.

Fairbank J., Reischauer E. (1989) *China - Tradition and Transformation*, Allen and Unwin, London.

He W. (1989) The Impact and Influence of the New View of Science on China's Reform, *The Cato Journal, 3*, 637-640.

Huang R. (1990) *China - A Micro History,* M.E. Sharpe, Armonk, New York.
Lavoie D. (1989) Economic Chaos or Spontaneous Order, *The Cato Journal, 3,* 613-633.
Lynch K. (1981) *Good City Form,* MIT Press, Cambridge, Massachusetts.
Singer A. (1992) *The Lion and the Dragon,* Barrie & Jenkins, London.
Spence J. (1990) *In Search of Modern China,* W.W. Norton, New York.
Yeung Y.M. (1993) Urban and Regional Development in China: Recent Transformations and Future Prospects, Occasional Paper no. 22, Hong Kong Institute of Asia-Pacific Studies, The Chinese University of Hong Kong.

24. The Paradoxical Nature of Territorial Change: Science Parks and the Case of Trieste

Sandro Fabbro

24.1 Introduction

Despite the enormous difficulties arising during the establishment and development of a Science Park, and despite the numerous failed attempts (Castells and Hall, 1994), the appeals for 'technopolitan' policies continue unabated. The object of this report is not to question the reasons underlying this phenomenon, nor is it to judge its timeliness. The aim here is to show that the difficulties and failures of this policy may be due in part to over-simplistic conceptualisations of the territorial context in which they are located and an underestimation of the significance of self-organisation factors and autonomous local decision-making processes.

First of all, in Section 2 we examine the nature of 'technopolitan' policies based on the establishment of Science Parks. Section 3 then highlights the very complex, but also paradoxical nature of projects whose aim is to stimulate innovation and change, in a socio-economic and territorial sense. It is claimed that if this contradiction is ignored, it can become the source of many difficulties. A critical reconstruction is made in Section 4 of the Trieste Science Park, focusing not so much the successes obtained at the structural level and in terms of international renown, as the strategic problems encountered in the attempt to establish a pole of attraction and, in particular, the planning and implementation of the larger-scale project for territorial change involving reindustrialisation, innovation of the regional industrial economy and so on. Section 5 concludes the discussion with some observations on the case study as well as some more general issues (of method and merit) which are raised by the analysis.

24.2 Science Parks and Territorial Development Policies

What exactly is a policy aimed at the establishment of a Science Park (SP) in an economically advanced context? Various definitions can be given according to whether an SP and its territory are considered mainly as 'infrastructural systems' or 'networks of actors'. For example: (i) the territory and SP can both be considered complex infrastructural systems; (ii) the territory may be conceptualised as a local network of actors who make strategic decisions and the SP considered an infrastructure which is the operative result of these local decisions; (iii) the territory can be conceptualised as an infrastructural system, while the SP is conceptualised as a particular network of actors; (iv) both territory and SP can be conceptualised as complex networks of actors.

In the first case the territory and SP are the outcome of a policy, generally laid down by a national government, aimed at obtaining results in terms of national performance (in economics, politics, international prestige etc.). This is essentially a 'top-down' policy, exogenous and indifferent to the characteristics of the territory itself. Notable examples are the science cities founded by the ex-U.S.S.R., the major Japanese research centres, Big Science and Tsukuba, and Taedok in South Korea (Castells and Hall, 1994). In the second case the conceptualisation of the territory is more problematic, but the Science Park is still seen as an infrastructural solution to local development problems (an area equipped for a faster and more effective diffusion of technological innovation or to attract high-tech firms from outside or even to promote the establishment of new local firms). This is essentially a traditional endogenous development model. The British Science Park at Cambridge, where an urban planning decision favoured the establishment of a large number of high-tech firms (Beveridge, 1994), could be considered an example. By contrast, the third case infers an instrumental view of the territory, the SP being considered the real engine of local development in terms of decisions and promotion, and seen as exclusively 'corporate' and 'business-oriented', while the territory, as a whole, is an external economy subordinate to the SP strategy. In the last case, both territory and SP define themselves as actors of development and innovation through mutual interaction. Examples are the 'innovative milieux', such as 'technology districts' (Maillat *et al.*, 1995), where the self-organising abilities of the local actors, together with an autonomous capacity for learning and the existence of networks and synergies, produce a continuous innovation process. This fourth type also represents a framework for those initiatives which, regardless of the context, aim to establish a new territorial development model. It must be recognised, however, as pointed out by Castells and Hall (1994), that this model is 'culturally specific'.

This model of the territory and of the SP derive from:

a. observations from studies of local systems and industrial districts and also studies of networks and synergies (territorial, industrial, inter- and intra-organisational etc.);
b. the epistemological shift from the classical systemic paradigm to the 'paradigm of complexity'[1] in the 'self-referential' and 'constructivist' versions (Watzlawick, 1988) applied to the social sciences and, more recently, geography and the territorial sciences (with reference to the Italian debate, see among others Mela and Preto, 1990, Dematteis, 1994, Cavallaro, in this volume).

These findings appear to coincide with the studies on SPs and some scholars seem inclined to develop this new approach (among others, Bruhat, 1990, 1992). The traditional view of the SP as an infrastructural system is being replaced by its interpretation as an actor, or network of actors, which help to promote an autonomous strategy of territorial innovation. In other words, the Science Park is seen not so much an instrument, with which to pursue some external objective, as a set of actors of change. The territory itself is not considered an *a priori* element, but an outcome of the interaction between these actors, and therefore as a social construct of these actors who require enacting images of their environment to pursue their own strategic actions (Weick, 1977). The territory therefore loses its traditional connotation as a system of relationships between 'things'- whose control is carried out by some other external decision-maker - to become a 'system of actors', who generate constructive images of their environment and consequently take the strategic decisions[2].

---

[1] With the most recent developments of the systemic view (we should like to mention, among others, J.P. Dupuy, J.L. Le Moigne, N. Luhmann, H. Maturana, E. Morin, F. Varela, Von Foerster), attention is clearly diverted from the interaction between the structures of the system to the endogenous change involving these structures. Indeed, while the classic systemic view considers the constraints and possibilities to be due to the structural properties of the system, the latest developments see them as the cognitive properties through which the system processes the change.

[2] The context here is a 'self-referential' and 'constructivist' view of the organization of complex systems which normally refers mainly to biological systems. This view, however, also prompts meaningful analogies with the evolution of social, economic and political systems. In this respect, we must acknowledge the contributions of economists

24.3 The Paradoxical Nature of Territorial Change

If the implementation of an SP involves a project for the whole territorial context and also involves changes which profoundly affect a broad system of relations between economic, social and institutional actors, then one can consider an SP policy, or at least a certain type of SP policy, as an intentional policy for urban and territorial change (see Drewett, Knight and Schubert, 1992). From this point of view, territorial change can be considered to be morphogenetic, because it is based on new forms of interaction and learning, on new synergies among local development actors. Although it is held that SPs can be planned, and consequently 'governed', so that territorial changes (socio-economic, technological, cultural, etc.) can be achieved, it seems that little consideration is given to the profound nature of these postulated changes. There is a tendency to ascribe thaumaturgical powers to certain expected effects, deliberately and emphatically using complex concepts such as 'synergy', 'network', 'integration', 'interaction' etc. to hide those difficulties and uncertainties which seem ever-present in the planning and implementation of SPs.

If the success of a Science Park is measured in terms of the 'system changes' which it generates, the problem lies in discovering which strategies need to be implemented to achieve these system changes. By definition, they are due to complex effects which are often intrinsically nonprogrammable (Elster, 1983a, 1983b). System changes, and therefore everything which in some way lies hidden beneath the thaumaturgical concepts mentioned above, can be linked back to the unexpected, indirect, side-effects of a particular project, as described by Hirschman (1967). However, a further problem arises here. As effects of this type are fundamentally unintentional, a direct and intentional attempt to achieve them implies an intrinsic paradox. If the change is not planned, the problem is not likely to be solved, on the other hand, if the change is planned, the problem runs the risk of being perpetuated in other forms.

A possible solution to this paradox can be found if we adopt a self-reflective view of change as something which stems from a circular creative process. This involves the 'subtle art of re-structuring' world

---

such as A.O. Hirschman, anthropologists such as G. Bateson, sociologists such as N. Luhmann and many scholars in the fields of social and organizational learning. Worth noting too is that, in this context, the concept of self-organization has a very different role from that in other approaches to the complexity of systems (Allen, in this volume). Indeed in this case, self-organization is not so much a natural and intrinsic property of the system as a 'project' to be realized through a process by a 'self-reflective' system.

views (Watzlawick, Weakland and Fisch, 1974) or the 'learning to learn' in which Bateson (1972) identifies a level of learning which sets off system changes and not just changes of state. It could also be related to ways of implementing projects which, according to Hirschman (1967), entail the unconscious underestimation of the difficulties and the activation of forces internal to the project itself ('internal' and essentially unconscious behaviour is considered crucial to the success of the project)[1]. In these terms, the paradox represents an expression of the conflict between intentional plans and casual interactions and events, whose practical solution and management requires great cognitive ability and the capacity for abstraction on the part of the actors involved.

24.4 Trieste: from Industrial Decline to a Scientific and Technological Pole, a Controversial and Bumpy Path

24.4.1 Case Study Methodology

The Trieste Science Park constitutes an emblematic case both for contextual and for methodological reasons. It was the first and largest scientific research area to be established in Italy and was seen as a solution to the crisis of a city afflicted by a serious decline in port and industrial activities and a marked decline in the main socio-demographic parameters. With regard to the methodological reasons for choosing this case study, the Trieste Science Park has a history of over 20 years and lends itself to a long term evaluation.

The assessment was carried out using: (a) a 'longitudinal' analysis (historical and developmental) applied not so much to the social and economic structure, as to the learning process involved in the SP strategy (a sort of 'backward mapping', Browne and Wildavsky, 1983); (b) a SWOT analysis applied to the hypothetical scenarios which can currently be put forward.

The greatest problem is that the Trieste Science Park represented an infrastructural investment by the national government in a city which,

---

[1] Friedmann (1987) appears to maintain something similar in treating the double-loop learning theory of Argyris and Schön (1978). Friedmann considers this true learning, since it involves a re-structuring of the individual's relationship with the world. This re-structuring, however, is inflicted upon [the self] with reluctance, [since] unless not doing [so] it [would] entails a penalty or, even worse, failure of the project itself.

given its geographical position and its turmoil-stricken history, continues to claim support from the State. This means that the SP, ultimately, is likely to pose an obstacle to its own survival. A strategic reorientation is becoming necessary in order to avoid the failure of the whole project. It requires a radical redesigning of the Science Park strategy and of its specific 'milieu'. This will be a complex process, needing a restructuring of the image and the role of the Science Park and of the territorial context.

The questions posed, therefore, are the following:

a. how and why does the shift from an 'infrastructural' and exogenous design to a self-organising conceptualisation of the Science Park and territorial change occur?
b. what internal changes (regulatory, cognitive, managemental) and new external relations and synergies does the new perspective entail?
c. what difficulties will be faced by the process of 'internal' change and the new territorial strategy?

24.4.2 Main Characteristics of the Trieste SP

The Trieste Science Park, called the 'Area', was established under the auspices of the Ministry for University and Scientific & Technological Research (M.U.R.S.T.) and was formally instituted in 1977 by the national law which financed the reconstruction of the Friuli area of the Friuli-Venezia Giulia region after the 1976 earthquake. The explicit intent was to found an SP which would represent a national reference point for the promotion of research in advanced technology.

The laboratories and services which make up the Area are situated on the Carso plateau within the municipality of Trieste (230,000 inhabitants), the capital of the Friuli-Venezia Giulia region, located in the extreme north-east of Italy. The Area covers a total of 160 hectares of land and currently around 800 persons employed. They are housed in over 40,000 m^3 of laboratories, offices and facilities made available by the Consortium which manages the Area, whose members include the autonomous Friuli-Venezia Giulia Region, the Province and Municipality of Trieste and Udine, the National Council for Scientific Research (CNR), the Universities of Trieste and Udine and other public and private institutions. The Area is also home to the 'Elettra' synchrotron and light laboratory, whose promotion and management are the responsibility of the Trieste Synchrotron Society, presided over by Nobel prize-winner Carlo Rubbia. A promotion campaign is currently under way to develop the use of the machine through contracts with firms and laboratories from around the world.

The socio-economic and territorial context of the SP is made more difficult by a number of factors, including the delicate relations with the closely neighbouring countries and different ethnic groups, the natural features of the site, the lack of large metropolitan cities nearby, the serious economic and demographic decline currently afflicting the city of Trieste, and the absence of an entrepreneurial-industrial tradition in the city. There is now a much more lively and dynamic environment, from an entrepreneurial-industrial standpoint, in the Friuli part of the region, which has never entertained good economic relations with Trieste. The last two centuries of history have moulded not only the city, but the typical Triestine character which has had a strong influence on the evolution of the Area.

The history of the Trieste Science Park goes back over 30 years. This can be divided into four chronological phases which, for the sake of simplicity, correspond to the four successive decades: (i) the 60s, characterised by the embryonic stages of the scientific pole and the establishment of a favourable national and international context; (ii) the 70s, characterised by intra-regional dynamics which allowed the concrete institution of the SP; (iii) the 80s, characterised by the establishment of the facilities and by the beginning and then rapid growth of activity; (iv) the 90s, during which numerous and complex strategic and operational problems have been brought to light. A radical re-structuring of the entire scientific pole is now being carried out within an entirely different planning framework.

24.4.3 The Origin of the Idea

During the 1960s, thanks to: (i) the position of Trieste in the East-West relations established after the end of World War II; (ii) the postwar compensation which Trieste claimed from the State; and (iii) the new peace-based science policy, Trieste succeeded in being chosen as the location for the International Centre for Theoretical Physics under the patronage of the U.N.O. and the I.A.E.A. in Vienna. Headed by Abdus Salam, a Pakistani and subsequent Nobel prize-winner, the Centre specialised in the training of scientists for developing countries and revealed its strategic position in North-South world relations. Trieste has become a destination for Nobel prize-winners and noted scientists. One of the challenges this poses is how best to exploit their knowledge.

The opportunity for developing a scientific pole in Trieste seems to have stemmed from a perception of the city as strategically 'centred' in a specific historical and geo-political context deriving from the Second World War, and therefore has nothing to do with either geographical centrality (i.e. based on the infrastructure or distances from other urban centres) nor with the linear development of endogenous opportunities. Indeed the idea is neither a manifestation of current 'metropolitan' potential, nor is it rooted in local tradition. It appears to be the projection of opportunities deriving from the new world order, and on the perceptive and constructive capacities of certain individuals who were particularly sensitive and skilled, and who perceived the 'virtual' centrality of the city as an opportunity.

24.4.4 The Seventies and the Start of Project Implementation

In the 1970s the strengthening of the Friuli area of the region was accompanied by the decline of the Trieste area. There was post-earthquake reconstruction in Friuli and the founding of the University of Udine, which added to the list of traditional sources of conflict between the Friuli and Venezia Giulia parts of the region. In compensation for its diminished political, economic and functional importance, the city of Trieste requested the government to establish a 'Research Area' and to strengthen its science-oriented institutions. At more or less the same time, however, the city declined the proposal (part of the Osimo agreements between Italy and the then Yugoslavia) to establish a large industrial area on the Carso plateau. Underlying the refusal was not so much the question of environmental conservation, but anti-Slav resentment and a culture disinclined towards industrial investment.

The Science and Technology Research Area was formally established in 1978. The project was vague with regard to the strategic aspects of the technology research and local and regional development; there were however clear indications of significant commitments in this direction. For years the Area remained an empty 'container', an institutional formula which was, however, earnestly invoked as a readily-available solution for the crisis of Trieste and for the city's reduced role in the regional context. The globalist and universalist standpoint of the 60s thus became integrated with or, rather, made way for an outlook which was more confined to the local and regional context. The result was the emergence of a new image for the area prior to another important strategic shift in orientation: the 'universalist' vision involving international actors in North-South cooperation projects, gradually gave way to a far more local vision, in which the motivations behind specific courses of action also became more pragmatic.

The Government contributed generously (in financial terms) to the reconstruction of Friuli without interfering in the destination of the resources. For its part, the Region Authority attempted to mediate between the areas within its confines and ensured an allocation of national resources which would guarantee regional unity. Triestine public institutions took advantage of the situation, successfully obtaining resources with which finally to launch the SP project, which had originated in a different context and with different aims.

24.4.5 The Eighties: Location Choice and Plant Establishment

The initial location envisioned for the park was a central position with respect to the region, but eventually a site in the suburbs of the city of Trieste was chosen. This choice of site meant that the Area represented more a solution to the particular crisis of Trieste than a regional development project. In addition, the decision to locate the Area on the Carso limestone plateau appears to have been based only to a minor degree on strictly technical or financial advantages (external economies, greater access to support services, a previously established State-owned settlement, etc.). These, in a cost-benefit evaluation, may not have proved of greater value than the disadvantages of the location. The real reasons for the choice were:

a. to bring new verve and political prestige to the city of Trieste, inducing significant structural transformations with the consolidation and expansion of science-based activities already established in the city;
b. the noteworthy landscape aspect of the Carso which, according to the view of scientists, represents a primary requirement.

The plots of land set out for the future expansion of the Area were very large because the Area's first and foremost aim was to attract large-scale laboratories which could take advantage of the on-site technical and research services on a tenancy basis. The expected effects in terms of economic development depended on a chain of activities deriving from the establishment of the 'growth pole' through a hierarchical 'filtering down' process.

A debate arose regarding the nature of the research activity to be carried out in the Area: was it to specialise in scientific research or become a technological-production park? Environmental opposition helped to break the stalemate by coming out against setting up production plants on the Carso. However, 150 hectares still seemed a very large allotment of land for the exclusive purpose of scientific research. It was more appropriate for a large industrial area, so how was the Area to be filled?

The first laboratories that set up in the Area were from the public sector and, through intense diplomatic negotiations and a firm Government commitment, two outstanding projects were attracted to the area: the Italian Synchrotron and the UNIDO Biotechnology Centre. But what was the point of a synchrotron if it could be equalled or made obsolete by foreign competitors in few years, with its market still to be established? Biotechnology research for the southern hemisphere in partnership with a similar centre in New Delhi did not imply any direct productive application of the results in Italy. Indeed, the strategic aspects were scarcely considered. With regard to the expected territorial effects, the dominant approach seemed to be: ‘first of all, we must grow by attracting research initiatives of any type, then we shall tackle the technological fall-out and the creation or attraction of new high-tech ventures’. All the management efforts were directed toward achieving an acceptable dimension, in the least time possible, in order to avoid the failure of the project.

The dominant image of the park was a quantitative-infrastructural one, strategic and identity aspects seeming to be neglected. The site was viewed both as an advantage, due to its ‘physical nearness’ to the city (e.g. the locational aspects), and a possible obstacle, as it could raise problems of ‘cultural distancing’ (a feature of the industrial dynamics in the rest of the region and the north-east of Italy in general).

24.4.6 The Nineties: What Are the Strategic Prospects?

By the end of the 1980s, State finance for the Research Area was depleted and the national financial and institutional crisis meant that research expenditure was reduced. The factors on which the SP had founded its stability and success were faltering and influencing the expansion of the scientific and advanced educational institutions of the so-called Science City.

The Elettra synchrotron was finally completed. It proved an extraordinary achievement, in technological as well as scientific terms, making the Trieste Science Park unique because the potential market area spanned the whole of Europe (the only other synchrotron at the time was in

Berkeley). At present, however, only a few of the many potential light beams are functioning and the completion of other synchrotrons elsewhere is only a few years away.

Today much work is being carried out. There are numerous installations and several hundred people are employed on the site. The Science Park area is host to many research laboratories, but very few high-tech firms. Questions are emerging about the justification of the investments amounting to several hundred billion lire, the size of the Area, the construction of the synchrotron, etc.

Evaluated in terms of structures, the results achieved are extremely significant at national and also international level. Nevertheless, two important aspects remain as yet far from satisfactory: (i) the achievement of the aim of influencing regional innovation still appears quite remote; (ii) the structures available on site do physically exist, but they lack the necessary strategy and 'autonomous intelligence'.

The causes of these problems can be traced back, first of all, to the 'genetic' and evolutive characteristics of the organisation and in particular the system of actors involved. With regard to the 'genetic' characteristics, it is a system subject to the bureaucratic and centralistic regulations typical of state institutions. This affects the regulation of the 'invisible college' of scientists and is the source of the lack of interest in local territorial innovation problems. In addition, it should be said that the main behavioural pattern is characterised by opportunistic tactics. Notwithstanding its important results in quantitative and structural terms, this behaviour constrains any serious attempt to promote a strategic orientation for the existing structures.

To summarise, the combined effects of an organisation highly dependent on State finance and regulation, an evolution process based exclusively on an opportunistic view and the prevailing 'infrastructural' image of the scientific pole, appear to have limited the capacity to effect significant changes in the local territorial and socio-economic system.

The organisation itself seems to acknowledge that with regard to: (i) the turbulence caused by the financial and institutional crisis of State and various other uncertainties; (ii) the need to give new meaning and perspective to the structures available and, therefore, to some degree the organisation as a whole, means that there is an urgent need for a serious reorientation of strategies.

This impulse has been translated into an attempt to relaunch applied research in addition to basic research, and to switch from a State-dependent regulation system to a more local autonomous regulation system through agreements with the regional actors. This process of restructuring is not, however, revealing itself to be an easy task. First of all, there is a lack of the necessary vocational and structural resources in the city (e.g. systems of small and medium-size firms with an innovative orientation). This means that the proposed policy could generate conflicts of interest with existing local and sectorial arrangements, transforming what should be a game with a positive solution into a game with zero or even negative solution. Secondly, the solution requires some sort of 'social construction of the market' and on systems of small and innovative firms whose origins are generally endogenous and spontaneous.

It is therefore legitimate to question whether such a reorientation is realistic, i.e. whether it is in fact possible to 'relaunch' the structure, providing it with greater autonomy and self-sufficiency, simply by assigning new objectives and releasing the structure from bureaucratic ties. There is nevertheless a definite change in the image of the SP - reference is now being made explicitly to 'territorial systems', 'industrial districts', endogenous potential, synergies and so on. New images deriving from the debate of the 80s on the new local development paradigms have been adopted, apparently without much criticism. But there would appear to be an intrinsic paradox, since an attempt is being made to consciously generate and regulate processes which are essentially spontaneous.

A turning point has therefore been reached. As the original aims of the SP and its relations with the local area have proved strategically inadequate, the SP needs to reformulate the underlying motivations and radically redefine its nature and identity. This inevitably means modifying the previous 'world views'. It is no longer sufficient for the local context to be seen in the simplified form which dominated previous phases. A more realistic and complex view is needed and at the same time co-operation has to be encouraged between local actors.

24.4.7 Some Final Comments

The analysis of the Trieste Science Park has shown that the Park has developed in quantitative terms, but that there is almost no autonomous governance nor consideration of the local and regional territory. Put simply, there is a complete lack of territorial strategy. Above all, the Science Park has been seen as a large, exogenous and 'neutral' infrastructural investment by the State. This has entailed serious problems for the effectiveness of the SP policy as a whole since: (i) the structures which do exist continue to lack 'autonomous intelligence'; (ii) the aim of profoundly influencing the future of the city and regional development remains largely unachieved; and consequently, (iii) the survival of the project itself is being threatened.

The underlying causes can be traced back to the nature of the SP itself (the fact that it is subject to State regulation, that it consists of large laboratories concentrating on basic research, and characterised by the 'academic' culture of the scientists and researchers). The instability and 'threats' to its existence, at least in the current Italian context, have contributed to increasing uncertainty about the future of the project. It has been recognised that there is a need to make a significant change in strategy, in the nature of the organisation and the 'enactment' of relations with the local 'milieu'.

What has emerged is a strategy in which the science and technology pole would form the main focus of a network of local and regional centres for the diffusion of technologically-advanced reindustrialisation and innovation. Despite being generally accepted (although more by default than conscious decision), this strategy has revealed itself intrinsically paradoxical. Apart from calling for a radical restructuring of the existing socio-economic structure of Trieste, this strategy also entails the 'invention' of an innovative industrial milieu which, by definition, is something which depends largely on spontaneous and random interactions.

It should also be noted that the reorientation entails a drastic revision by the SP of its methods, behaviour, culture and the image of the local context, in other words, the entire development policy. All this implies the need for profound and painful change from a hierarchical, exogenous and heterodirected scientific pole into a nonhierarchical, endogenous and more independent entity, capable of self-organisation and cognition in terms of strategic learning. The outcome of this process is by no means certain. It seems that the reorientation could be pursued using a 'dualistic' and 'paradoxical' action strategy: 'dualistic' because it would be directed towards the restructuring of the cognitive capacity and organisation of the organisation itself as well as the external environment (e.g. through the construction of networks and synergies to implement new strategies), 'paradoxical' because the change should be pursued by acting on different and contradictory logical and operative levels.

24.5 General Conclusions

The Trieste case study can help us to understand certain problems relating to the strategic planning and management of ongoing urban and territorial change. We shall therefore attempt to draw some conclusions of general interest and relevance.

Firstly, the analysis has confirmed that if an SP is seen as a multidirectional infrastructural element, constructed and expanded without a profound reflection regarding the more 'subjective' components and territorial context, then not only will it not produce the expected change, but it also runs the risk of failing to achieve its fundamental aims. An SP can also be conceived as the structured result of complex interactions and synergies between territorial actors and is, therefore, a territorial 'actor' in its own right, capable of promoting new, more complex and widespread interactions.

The problem faced by the SP as an actor is therefore that of conceptualising and managing its more specific and sectorial activities for a market which often requires, by definition, to be 'invented'. A further problem entails the activating of system change at both local and regional level, which is vital to the success of the SP.

From a more general standpoint, the Trieste Science Park analysis reveals that:

a. territorial projects and policies which aim to achieve 'strategic change' tend to rely on heavy infrastructural investment, and often fail to take into account the more complex aspects and cultural context which represent the only guarantee for activating authentic development and processes of change. The success of these processes also depends on being able to identify (even in embryonic form) the organisational nuclei capable of autonomous learning, i.e. those which can take on a leading role and establish the 'structures' needed for implementing the project (Hjern and Porter, 1981). There must therefore be coherence between policies for change, the promotion of organisational nuclei and the construction of the wider context (the 'innovative milieu'); in other words, the three policies are mutually interdependent and should be developed in parallel;
b. in addition, a policy for development and change can spring from different factors and conditions, both exogenous and endogenous. It does not, however, appear to be capable of developing either spontaneously or through simple incremental growth, since the latter is associated with opportunistic and tactical behaviour, concentrating on structural growth and neglecting strategic problems. There appears to be a point beyond which incremental growth can no longer be considered acceptable, because it places the success of the project in jeopardy;

c. a threat to the survival of the project as a whole and fear of failure can force the organisation to search for a different strategy (even though creating new difficulties and paradoxes), both at the level of the single organisation and at the wider level of actor networks. The change, therefore, must also involve learning processes and entail the formulation of possible scenarios at higher levels of abstraction, as well as the construction of networks of actors. This is what has been defined as 'learning to learn' or, to use an expression which is perhaps more explicit in this context, 'strategic learning'[1].

Furthermore, this enables us to claim that:

a. such a policy probably has a dualistic and paradoxical nature: dualistic because it entails addressing change both within the organisation (regarding its strategic behaviour) and outside (the interaction with other actors of the 'milieu'), paradoxical because, although it cannot be directly investigated, because it is linked to collateral, indirect and unexpected effects, it cannot be left to chance, so we must intervene in some way. But since such a policy cannot be designed on the basis of a comprehensive *a priori* rationalisation, it implies creating an openness within the organisation to unexpected combinations of the various logical, cognitive and operational levels. There are particular ways of learning which the organisation must learn to develop. These derive from a greater disposition to thinking and communicating in 'paradoxical' terms. This can be the outcome of the cumulative experience of history, but can also be the painfully gained fruit of threat and defeat;
b. territorial policies for change imposed from outside risk confining themselves to the production of structures without independent strategic intelligence. If they are seen in self-organising terms, the policies must centre on organisations (agencies, firms, formal and informal networks)

[1] It should be remembered here that there is a certain correspondence between Bateson's concept of 'learning to learn', used in the context of his theory of communication, and the concept of 'double-loop learning' elaborated by organization researchers like Argyris and Schön (1974, 1978), a concept which entails the overcoming of organizational norms which are incompatible with the needs for change "through a defining of new priorities and a weighing of the norms, or better through a restructuring of the norms themselves with their associated strategies and theses." (Argyris and Schön, 1978, p. 24). On the basis of this correspondence we have, in the course of this study, taken the liberty of extending Bateson's concepts to territorial organizations and of using the intuitively more effective concept of 'strategic learning' as a synonym of 'learning to learn'.

within the local territorial system. The task of these organisations is to assist the process of change with continuous reference to themselves and their own context, in order to redefine, widen and possibly modify the interaction and learning processes (Butera, in this volume). The strategic plan, the policies and organisations which implement it and the policy for the construction of new inter-organisational networks and synergies, therefore, are exactly the same thing. A policy for change is, simply, the self-implementation of a complex organisation which, in producing itself, also produces networks of territorial change and vice versa;

c. the 'territorial system' can therefore be reconsidered not so much as a system of structures naturally endowed with self-organising capacities, or as a subjective, abstract and virtual entity, but as the contextual specification of a certain type of actor and a certain 'milieu' of networks both in a structural and in a virtual and organisational sense; that is, as a system whose 'subjective' and 'objective' components merge and define themselves simultaneously;
d. at this stage, neither the 'infrastructural' image of the territory (appropriate for the most traditional forms of Government intervention), nor an all-network image of the territory or the random and spontaneous development of interactions can aid the extension to the urban field of concepts of self-organisation which entail forms of strategic learning. This approach seems relevant not so much in the hierarchical context, nor in those dynamic systems which evolve spontaneously and have a strong endogenous creativity, but in a certain type of intermediate situation (which is in fact quite frequent) falling between the hierarchy and the market. Seen from this point of view, territories and cities could also be considered as aggregates of complex and self-referential organisations which plan, in the sense that they 'think' and learn, how to generate their own identity, their own strategic action and the 'structures' needed for the implementation of those plans.

This contribution does, to some degree, support the ongoing debate on strategic planning since it highlights how, in certain territorial and urban contexts, the pragmatic thrust to 'change', 'innovate' and 'realise concrete things' can lead to a vicious circle, in particular when those local 'actors' able to learn, interact with and face the paradoxes of change are not identified. This seems to point to a concept of the strategic plan as a 'learning' tool for territorial organisations, and implies the need for structures which permit reflection about the plan and also its implementation. It also implies the need for the strategic plan to critically re-elaborate the territorial images (Faludi, in this volume), according to new principles and structures of space organisation (Mazza, in this volume). In this way the strategic plan could encourage the restructuring of norms and hence give rise to more effective planning policy and intervention.

References

Argyris C., Schön D. (1974) *Theory in Practice: Increasing Professional Effectiveness*, Jossey-Bass, San Francisco.

Argyris C., Schön D. (1978) *Organizational Learning: A Theory of Action Perspective*, Addison-Wesley, Reading, Massachusetts.

Bateson G. (1972) *Steps to an Ecology of Mind*, Chandler Publishing Company, San Francisco.

Beveridge L. (1994) Cambridge Science Park, *Nature*, 368, 6467, 170-171.

Browne A., Wildavsky A. (1983) Implementation as Mutual Adaptation, in Pressman J.L., Wildavsky A. (eds) *Implementation*, University of California Press, Berkeley, California.

Bruhat T. (1990) *Vingt technopoles, un premier bilan*, La Documentation Française, Paris.

Bruhat T. (1992) Evaluating the French Science & Technology Parks Experience, Paper for the Workshop on Science and Technology Parks Impact Evaluation, CEC DG XII, Bari, March 26-27.

Castells M., Hall P. (1994) *Technopoles of the World. The Making of 21st Century Industrial Complexes*, Routledge, London.

Dematteis G. (1994) Sistemi locali e reti globali: il problema del radicamento territoriale, Paper for Incontri pratesi sullo sviluppo locale, IRIS, Artimino, September (mimeo).

Drewett R., Knight R., Schubert U. (1992) The Future of European Cities, the Role of Science and Technology, report for the Monitor-Fast Program, CEC.

Elster J. (1983a) *Explaining Technical Change*, Cambridge University Press, Cambridge.

Elster J. (1983b) *Sour Grapes. Studies in the Subversion of Rationality*, Maison des Sciences de l'Homme and Cambridge University Press, Cambridge.

Friedmann J. (1987) *Planning in the Public Domain: from knowledge to action*, Princeton University Press.

Hirschman A. O. (1967) *Development Projects Observed*, The Brookings Institution, Washington D.C.

Hjern B., Porter D.O. (1981) Implementation Structures: a New Unit of Administrative Analysis, *Organisation Studies*, 2/3.

Maillat D., Lecoq B., Nemeti F., Pfister M. (1995) Technology district and innovation: the case of the Swiss Jura Arc, *Regional Studies, 29,* 3, 251-263.

Mela A., Preto G. (1990) Alla ricerca della strategia perduta, in Curti F., Diappi L. (eds) *Gerarchie e reti di città: tendenze e politiche*, Angeli, Milan, 127-154.

Watzlawick P., Weakland J.H., Fisch R. (1974) *Change,* Norton, New York.
Watzlawick P. (1988) *Invented Reality*, Norton, New York.
Weick K. (1977) Enactment Processes in Organizations, in Staw B.M., Salancik G.R. (eds.) *New Directions in Organizational Behaviour, St Clair Press, Chicago.*

25. Strategic Planning in Italy and the New Local Authority Act: The Master Plan for the City of Venice

Mariolina Toniolo

25.1 Introduction

Established planning doctrine in Italy tends to picture urban planning simply as a form of urban design plus legal control. Just recently strategic planning has come to be mentioned with growing frequency in planning literature, not though as something actually being practiced in our country, but as a brand new approach which could hopefully be introduced. In the author's opinion (together with Gibelli, 1993), the strategic approach is probably the most promising way of making planning more effective, something other than the mere prescription of desirable goals. It is therefore important to understand why this promising approach has not so far been adopted in Italy.

One explanation may come from cultural factors, namely the very Italian emphasis on the legal aspect, along with the distrust of Public Administration, due to its lack of ability and lack of responsibility for goal achievement. Another explanation is possibly the type of education Italian planners are given, trained as they are only in Architecture or Engineering schools. This is very specific to Italy and, along with the above mentioned cultural attitudes, tends to result in plans that are just a combination of design and legal prescriptions. In fact, many planners have been brought up to believe that a plan is finished when zoning and norms are defined.

It is true that in recent times university programmes in planning, where new approaches are considered and taught, have been established both for graduate and undergraduate students. Nevertheless, the physical environment remains the only field for which a specific planning education is provided. This is quite different from the United States, and it is not by chance that strategic planning was born there. Rachelle Alterman (1992) provides an interesting comparison between the kind of education a planner is given in Europe and in North America. In the former, the teaching concerns mainly technical issues, while in the latter management techniques and conflict resolution abilities are included. Italy is probably an extreme example of the European bias in this respect!

Coming back to the question of why strategic planning tends to find little acceptance in Italy, academic education cannot provide a full explanation. Strategic planning uses skills that can be learned from experiences which can also be gained outside a strictly academic environment or schools specifically dedicated to the teaching of planning. A further explanation may therefore relate to the institutional framework. The chronic instability affecting all levels of government in Italy until quite recently was a major factor in reducing politics to day by day survival, making any strategic perspective impossible. Local governments frequently consisted of heterogeneous coalitions, containing representatives of the various elected parties on a strictly proportional basis. These coalitions were only brought together by the need to ensure a majority within the City Council and rarely shared long term goals on specific issues. It was impossible, therefore, to define common objectives or establish a cooperative implementation strategy. If this institutional explanation is correct, then the 1993 Local Authorities Act which removes such instability may create more favourable conditions for applying strategic planning in Italy.

This paper stresses the importance of the institutional factor in influencing urban planning. In the author's opinion, the widespread idea that strategic planning has never been implemented in Italy ignores certain innovations currently being adopted in the Public Administration. In fact, in this respect an unexpected discovery is made and reported with the same astonishment probably felt in the past by the discoverers of new worlds! Like the tales they wrote upon returning from their voyages, this too may include some exaggeration but the core is absolutely true. It is, after all, the report from a journey the author is making into what for her is a new world.

We can begin, however, by admitting that the poor reputation of Italian bureaucracy is not undeserved. The Municipality of Venice, reminiscent of its Byzantine origins, is no exception. Nevertheless, some sections of its administration are adopting some innovations that would seem to provide a favourable basis for strategic planning, even though those involved are probably unaware of it. Something similar was detected by Kaufman and Jacobs (1987) a few years ago with respect to Public Administration in the United States. There, too, it was commonly understood that the logic of strategic planning - which was then more closely identified with its corporate origins - was unfamiliar to the public sector, to which it was recommended as something new (Bryson, Freeman and Roering, 1986). The authors carried out some interviews with local authority planners and in fact found them quite familiar with the strategic approach, although it was used in ways appropriate to the Public Administration context. Many of those interviewed spoke of it as a widespread *de facto* approach within the public sector long before professional planners had incorporated it formally within a theory. The situation in Italy seems similar.

In the case of the Venice Municipality, it emerged that strategic planning was not really an established practice, but that there was a receptiveness to this way of thinking. In the author's opinion, favourable conditions have been established by the new Local Authority Act (no. 81 of 1993). We describe in the following sections the innovative features of this law and how it could make an important difference as far as public authority planning is concerned. It should be stressed here, though, that the legislation alone will not automatically bring change. The inherent potential for innovation has to be understood and exploited by the new local authorities.

The case of Venice is taken as an example, not because it is a particularly good one, but because it is the one the author knows best. We are interested in investigating the opportunities provided by the new conditions, rather than how far they were exploited in this specific case. Venice was among the first Italian cities to elect a Mayor and City Council under the new legislation. The new Mayor was elected just a few months after the law came into force and has now been governing the city for over two years. The study concerns the author's own area of work, the drawing up of a new Master Plan for the city. Other examples could undoubtedly have been provided in other fields, but this case is interesting because the contents and procedure of the Master Plan is strictly codified by Italian law. Italian planning laws were conceived within a strictly rational-comprehensive framework, so it is a paradox that a strategic approach may be applied to the preparation of the Master Plan. Our thesis is that the new local authorities elected under 1993 law can afford to use even very traditional planning tools in a new perspective.

In Section 2 of this paper we provide some definitions which help to clarify what planning in general, and strategic planning in particular, involve in this context. The new Local Authorities Act is then briefly explained in order to show how it affects the ability of local authorities to act strategically. Section 3 gives some examples, all taken from the preparation of the new Master Plan for the city of Venice. Attention is drawn to the method rather than the actual results. Section 4 draws some conclusions.

25.2 Strategic Planning and a New Style in Local Government

25.2.1 The Meaning of Planning under Different Institutional Conditions

It is always important for definitions to be made clear. The meaning of strategic planning has varied over time and space following a kind of path. A number of successive stages can be identified (Gibelli, 1995), hence the need to state the specific meaning referred to.

It is outside the scope of this work to provide any contribution to the theoretical debate. The definitions given here are simply for the sake of clear understanding, without any implications concerning their correctness. There is hopefully no orthodoxy to be observed in this field, and we accept that other definitions may be equally valid in other instances.

First of all, the meaning of planning. In Italy, this word is generally used with specific reference to the physical environment. In economics and in social issues, the word planning took on negative associations during the Cold War, so in those fields the word 'programming' tended to be used instead. In the United States this confusion did not arise, so the word 'planning' is applied to many fields other than land use regulation. In fact, it is used in such a broad sense as to become meaningless (Alexander, 1981). In this paper the word 'planning' will be used in the English sense, referring to any field, not just the physical environment. This is the only meaning appropriate for strategic planning as its very nature denies an *a priori* restriction within the boundaries of any specific field (Bryson, 1995).

References in Section 3 to city planning therefore regard not just its physical layout, but how the city works. Recent planning theory in Italy stresses that urban design is the only promising approach to urban planning. This viewpoint must be taken seriously as it is espoused by those planners who made the first attempts at interdisciplinary co-operation. It was the failure of this experience that led to the suggestion that planners would be better off using the tools they can fully master. By confining planning to urban design, any strategic approach is of course ruled out. We attempt to show how, under the new institutional conditions, Italian planners could possibly afford to make another attempt at a broader approach to planning.

My own recent experience suggests that, within the Public Administration, disciplinary segregation may be overcome more easily than within the academic environment, thanks to the existence of civil servants who have developed planning skills in several fields. It is interesting to note that these people often do not have an academic background in planning, and often don't even think of themselves as planners. This situation has some similarities with the findings of Kaufman and Jacobs (1987) in their survey concerning the U.S. The civil servants referred to have the kind of skills that Alterman (1992) indicates as typical of American rather than European planning education, namely the ability to negotiate and manage conflict.

25.2.2 Strategic Planning: Meaning and New Perspectives

As far as strategic planning in the public sector is concerned, the aspect we stress here is the implementation. With respect to what we call prescriptive planning, the difference lies in the drawing up of an explicit implementation strategy and the monitoring of performance, allowing for the plan to be adjusted over time. Implementation does not follow the making of the plan, but is part of the plan itself. In a strategic perspective, the procedure is not less important than the plan design.

The consequences are, first of all, the need for an open attitude to cooperation with those involved, other than the planning authority, the flexible nature of the plan itself, and the relevance of time in implementation.

Existing Italian laws still assume the local authority to be the only body with the right and obligation to plan the urban environment. The contribution of private bodies to the implementation of the plan is confined to complying with the plan after it has been approved. The plan determines *what* can be done on each parcel of land, but not *when*. Attempts to introduce a time perspective into urban planning were seldom successful and have recently been abandoned.

The existing legal framework for town planning in Italy dates back to 1942. Since then many changes have been introduced, but none have radically altered the prescriptive nature of planning and lack of time reference. It is therefore no wonder that the procedure has become increasingly discredited. Growing dissatisfaction has been due to the lack of visible output.

With a delay of twenty years, Italian planners in the 1980s discovered Lindblom's (1959) point of view. Urban change was expected to come from the implementation of a limited number of important projects and the long term perspective was often abandoned. As such projects usually depend on private investments, during the 1980s planning practice became increasingly involved in negotiation with private agencies. In so doing, some public officials were found to have acted illegally and cases of bribery detected. Sometimes it became apparent that the existing legal framework had no clear provision for public-private negotiation on development issues. In any case, 'negotiated planning' has now taken on a bad name in Italy, implying something less than correct. As a reaction, some planners have preferred to go back to 'safe' prescriptive planning, while others (together with Forester, 1989, Kaufman, 1978, Marcuse, 1976) think that flexible, negotiated planning should be practiced openly, under the control of public opinion, and be part of a comprehensive long term strategy. The author shares this second point of view.

The cooperative attitude to other bodies by the public administration, suggested by strategic planners, may also mean greater participation in the plan-making process by citizens. Until quite recently, urban plans in Italy were drafted in a strictly secluded environment. They were disclosed to the city only after approval by the City Council. Existing urban planning laws still dictate a complicated bureaucratic procedure for citizens wanting to suggest any change in the plan after its approval. The aim of this provision was to avoid the leaking of information to people who might try to influence the plan's design for their own interest. In fact, what happened was that certain powerful interests managed to get the information they wanted in spite of all the secrecy, while the majority of citizens remained uninformed. Strategic planners have to take the risk of debating the plan before it is finished. This helps the implementation process by finding an agreement with possibly conflicting interests, whether land owners or grassroots movements. It also gives the plan what Benveniste (1989) calls 'the trigger', through a loud announcement.

Another point is whether the strategic approach allows urban plans to have wider scope and a long term perspective. There may be good reasons for believing it does not. One is the fear of repeating the disappointments of the old rational-comprehensive approach. However, the use of more efficient tools could provide a sufficient safeguard against this risk. Recently, more sophisticated reasons were put forth in support of the incremental approach to planning. It has been argued (Sager, 1990) that the only form of planning which allows fully informed citizen participation is incrementalism, since wide-ranging plans can lead to public opinion manipulation by experts. In the Italian experience, though, and probably in northern Europe too, there would seem to be less danger from technocratic manipulation of public opinion than from the narrow consideration of the issues and lack of a clear vision of the context.

The NIMBY attitude is spreading fast. Many things have changed since Davidoff (1965) called for planners to openly fight in the defence of the neighbourhoods that were threatened by authoritative planning. Now grassroots movements are strong and vocal. The problem is that they often forget the general interest. This is certainly one of the hardest challenges for present day planners, as Popper (1992) points out. The public administration is increasingly put in the uneasy situation of having to beg for approval of projects that are vital for the whole community (such as the location of public facilities) or in the interest of the weakest social groups (the mentally ill, the homeless, or people recovering from drug addiction).

If planning requires agreement by the community over common goals, the risk now is of having no planning at all, because of the lack of accord among different local interests. The only possible escape is negotiation in local communities to find an acceptable solution to common problems in a sufficiently wide, if not global, perspective.

25.2.3 The New Local Authority Act and Consequences for Urban Planning

In the past, local authorities in Italy (both municipalities and provinces) were ruled by governments that were elected by the General Council. The Council, in turn, was elected by the citizens according to a strictly proportional rule on a party basis. Coalitions were formed to chose the Mayor (or President of the Province) and his or her Government. Quite often, in order to get the Council's approval for his coalition, a Mayor had to accept in his government councillors who did not fully share his views on important issues. So related aspects of local government could be determined by councillors with very different views. This led to conflicting strategies. Local elections were usually held every five years, but the local government did not always last that long. Often the majority within the Council changed before the five-year term expired and a new government, often with a new Mayor, came into office. Only few Mayors in Italy enjoyed enough consensus within the Council to make long term plans. Many local governments spent most of their time merely struggling for survival.

This institutional framework obviously affected urban planning. Only projects with limited scope could reasonably be carried out. This perhaps provides an explanation for city-wide planning becoming less and less popular in recent years. Also, the attitude among department heads (the *assessori*) was more competitive than cooperative, so planning could seldom involve the joint efforts of more than one department. Each carried on with its own plans, physical planning being the exclusive responsibility of the town planning department.

The 1993 Local Authorities Act fostered deep innovations in this respect. Citizens now vote not only for the city councillors, but for the Mayor him/herself. The Mayor is free to chose the people who are to assist in the government. They are submitted to Council approval, but the Council itself can be dissolved and new elections called if no government is formed within a given time. As a result, a Mayor who uses his powers properly can have a very homogeneous group of people making up the local government. Under the new law, elections are held every four years, but there now seems a far greater likelihood of the government lasting for the whole term. It is true that the Council can force the Mayor to resign, but most councillors have little interest in doing so because the Council itself would be dissolved and new elections held. A Mayor and his government can therefore plan their work over a four year term having a fairly good chance of being able to accomplish it.

In addition, the new electoral rules give the Mayor much more prestige. The Mayor is elected directly by the citizens in a two-round poll. After the first poll, the two highest scoring candidates confront each other in a two-week campaign that makes them and their programmes highly visible. Surveys have shown that many more citizens know the name of their Mayor and have an opinion about his/her policies than previously. One consequence is that this person can afford negotiation with other public or private subjects from a more favourable position (Fisher and Ury, 1991).

To sum up, the new law gives Local Authorities more cohesion, duration and prestige. What this paper intends to show is that, under such conditions, urban planning can change from a self sufficient activity to a cooperative one, with wider scope and a more flexible set of instruments. These can include not only the traditional physical planning tools such as zoning, but tools previously outside the control of the Planning Department, as other departments may cooperate in the pursuit of common goals. Although it is true that four years is a short term for planning major changes at the city level, we shall discuss later how the problem was tackled in Venice with the new Master Plan.

Along with the major innovations in regulations affecting Local

Authorities, some changes made recently in the legislation concerning civil servants and their relationship with elected officials could also affect the approach to planning. Such innovations were so numerous that their final outcome is uncertain. On the whole, they will probably tend to link the upper levels within the civil service hierarchy more closely to the government in power, though granting them more responsibility in the face of the law and some personal guarantees. Again, this change could be used by the Mayor and the government in different ways. If they seek personal power in a narrow-minded way, they will probably use the control they have over civil servants to reward those who serve them most devotedly. But they can also use incentives to reward those officials who help achieve objectives of the government.

25.3 The Case of the New Master Plan for Venice

25.3.1 Master Plan and Strategic Plan: Different but Integrated

The new Mayor of Venice was elected at the end of 1993, immediately after the new legislation had become effective. During his campaign, the Mayor had been explicit about his idea for the city's future (Cacciari, 1989). On the basis of this idea he formed the coalition now supporting his government within the City Council. The philosophy supporting his programme foresaw the possibility of reconciling the two apparently conflicting issues regarding the future of Venice: the safeguarding of the city's unique physical characteristics and cultural heritage, and the development of an economic base not totally dependent on tourism. The assumption was that modern technology, and the shift of the urban economy from the production of material goods to information exchange, would allow such a reconciliation. As a consequence, much stress has been put on the re-building of a diversified economic base for the city, along with control of the impact of tourism.

Another aspect of the Mayor's programme was to draft a new Master Plan, since the one in force was more than thirty years old. The Master plan was to be instrumental in achieving the Councils' objectives outlined above. Now the question is: how effective can a Master Plan be in promoting goals whose scope goes beyond the mere design of the physical environment?

In fact existing laws in Italy define in detail what a Master Plan should be, but assumes that it only concerns transformations of the physical environment. Within this framework, it is difficult for a Master Plan, operating as a self-sufficient tool, to provide any positive support for a social and economic development programme. It can only be expected not to hamper it. This is why even one of the strongest supporters of strategic planning in Italy (Mazza, 1995) contends that a Master Plan is not an appropriate tool for the purpose, since it cannot be expected to produce anything but land use regulation. In the case of Venice, an attempt was made to incorporate the Master Plan, along with other instruments, in a comprehensive strategy. This was made possible by the cooperative climate that the new law created within the local government.

The Master Plan affords many questions. In order to give an overview of the method that was followed, we describe three of the issues faced:

i. the relationship between general goals, the overall structure of the Plan and its individual parts: how this relationship was dealt with, especially with regard to the time sequence through which the planning process was to be carried out;
ii. the Plan for the Marghera Harbour industrial zone, as an example of the flexible use of different tools and of cooperation among different participants, both public and private;
iii. the way access to the historic core of the city is to be regulated, with recourse to an integrated bundle of regulations, agreements and city design projects.

25.3.2 Time Sequence in the Drawing up of the New Master Plan

The first issue mentioned above, concerning the time sequence for the Plan, was necessarily conditioned by the very short time interval within which the local government could act before facing new elections. Drawing up a Master Plan for a city like Venice in four years is an ambitious goal, and even more so if it has to be done by the City Planning Department itself, rather than entrusting it to a private practice, as many other cities do. This decision was crucial however for linking the new Plan with its implementation. It required a reorganisation of the Planning Department, which was to be supported by the scientific consultant, Leonardo Benevolo, who agreed to work in close collaboration with the Department.

In addition to the time constraint, there were other reasons for ruling out the traditional top-down approach involving a sequence of steps: the analysis of the problem, the definition of general goals, then more specific ones, followed by the sketching of the main outlines and, eventually, the detailed definition of the Plan. It was soon clear that, for several reasons, some parts of the city needed to be planned before others. For the industrial zone of Marghera Harbour, as we shall see in the next section, the reason was the urgency of launching a policy to fight industrial decline. Being relatively isolated, with few interactions with the rest of the city, there was little harm in planning it before the full definition of the new Master Plan. The Mayor's programme provided sufficient guidelines to ensure consistency with the rest of the Plan. Evidence of the deep employment crisis in the area was such that any effort to reverse it could count on widespread political support.

For other parts of the city, the production of a detailed design at an early stage proved essential for the opposite reason. They were so crucial in the new functional scheme that it was necessary to draw up detailed plans in order to check whether the overall project would actually work. In some cases private investments that were instrumental to the Plan as strategic goals could not wait. A general Plan was also drafted of course, (Benevolo, 1996), but this did not come as the first step. It was carried out after the plans for some parts of the city had already been drawn up and approved by the City Council. The general Plan was therefore submitted to the public for debate as a sort of mid-term adjustment for the city's planning policy. Far from making it useless, this probably gave confirmation that it was really to be implemented. Previous experience has shown that even when a general Plan was adopted, if it could not be implemented within the planned time interval, the whole project could be deprived of any effectiveness, because the next administration could easily reverse it.

25.3.3 The Plan for Marghera Harbour Industrial Area

As previously mentioned, Marghera Harbour industrial zone is currently facing a deep crisis. Over the last fifteen years, the number of jobs has decreased from forty to fourteen thousand. Founded in 1910, the industrial zone was the most important manufacturing centre in the region, based on heavy manufacturing and port-related sectors: chemicals, oil refineries, metals and shipyards. Its present day decline largely depends on global trends. In addition, most of the industries are state-owned and were driven close to bankruptcy by financial crisis in their controlling groups, regardless of actual economic performance of individual plants.

Many old industrial zones in Italy are now facing a similar crisis. A solution in some cases has been the conversion of factory premises to offices, shopping centres or other nonindustrial uses. In the case of Marghera, such a solution did not seem feasible for several reasons. First of all, it seemed important to safeguard manufacturing employment in the area for social reasons and also to diversify the city's economic base, which already depends too much on consumption and tourism-related activities. In addition, the area is too large (approx. 2000 hectares) and too polluted. Only manufacturing could raise enough investment to be able to decontaminate it. The objective is therefore to encourage conversion from heavy industry to more promising innovative production activities.

This kind of goal would have been outside the scope of traditional town planning, even though the design of a new physical layout can be instrumental to industrial policy. Several strategies were put forward, including the development of a new physical plan for the area. In the past the Municipality had bought a considerable amount of land in Marghera. One parcel was offered to the Port Authority to accommodate a part of the existing harbour now in the historical city core. Another part is being offered on a 60-year lease to new investors, selected according to explicit priorities including employment levels, on condition that they help to decontaminate the site. Some promising applications are presently under scrutiny. The Municipality expects to have sufficient economic return to be able to buy some more sites from other industries likely to close in the future, and so continue with the same policy. The hope is that the coming of new enterprises may attract further activities, resulting in less need for the Municipality's intervention in the future.

The European Union recently included Marghera among the areas to be funded under Objective 2. This allows a number of actions including the retraining of workers and the promotion of new co-operative forms of self-employment. More funds from the EU derive from the project URBAN, which is targeted at deprived residential areas. This will allow some initiatives in the nearby housing areas, where unemployment exposes young people to the risk of drug addiction. The new Plan was designed to be part of this comprehensive strategy.

The area needs an improved layout to match prospective change. It will require better roads with convenient access from the residential neighbourhood, along with more parking places and green areas. The edges of what is now the industrial area may be converted to office buildings and other related uses in order to break down the split between the industrial area and the rest of the city. New industries replacing the declining ones will have a different physical form. The plan gives flexible guidelines for the conversion in order not to interfere with the timing of the projects, always difficult to anticipate.

Flexibility is in fact the key word underlying the whole plan and applies also to the institutional relationships. In the past, it seemed questionable whether the Municipality even had the right to be involved in the planning of the area. There was a long legal controversy between the Municipality and the Port Authority over this point. The new Administration chose to solve the problem through voluntary agreement. The whole plan was drawn up with the Port Authority's approval, proving in the end easier, and above all faster, than fighting a legal procedure. This choice may appear obvious (see Susskind and Ozawa, 1984), but in the Italian context it is quite innovative. It is probably not by chance that it was taken by a City Council whose prestige was reinforced by having a Mayor elected by the citizens themselves.

25.3.4 Regulating Access from the Mainland to the Historical City Centre

The number of tourists visiting Venice is so huge that tourism itself is now considered by many Venetians as a mixed blessing, if not a calamity. Estimates for the year 1996 show that about ten million people visited the city, the majority of whom did not stay overnight. Almost all visitors arrive in the historical city through two access points, the railway station and the car terminal. They then move to St. Mark's Square, which is the focal point of the visit, generating such a huge flow that it seriously interferes with the movement of inhabitants and commuters.

In many city centres traffic congestion is being reduced by forbidding private cars to move through the central area. In Venice, the historical core has already been a pedestrian area for centuries, so the solution proposed involves providing different access points to the city, each targeted at different types of user. This would segregate tourists and people coming to the city for other purposes. The City Council wanted to give priority to inhabitants and commuters by dedicating the fastest routes to them, leaving the tourists to reach the city centre by slower but more interesting itineraries across the lagoon. Three departure points were therefore identified on the mainland, where tourists could leave their cars or bus and take a boat.

This was not a new idea. A similar provision had been tried in the past, but not adopted by a sufficient proportion of visitors. Reconsidering the reasons for its failure, it became clear that it was not due to a shortage of demand, but to the lack of an integrated set of measures for its implementation. The public boat service had not been as frequent as required, and the interchange points offered little comfort. Arrival points in the historical city were set too far away from the destination of the majority of tourists, St. Mark's Square. Little, if any, publicity had been made and last, but not least, tourists' cars and buses had not been forbidden to park in closer locations.

Each of the above aspects depended on a different authority, including various departments within the Municipal Government which had different views on the subject and supported different interest groups. This is why no consistent policy ever took off. This time, the policy to set up interchange points on the mainland is fully shared within the City Government. This means they can make a comprehensive plan including appropriate transport and provision of facilities, as well as integrated regulatory measures.

All is still at the planning stage. Implementation has not yet begun, so no conclusion can be drawn for the moment about its success. Still, some hope comes from the fact that potential investors have not hesitated to offer finance for the project, though it involves large sums.

25.4 Conclusions

The issues outlined above are just examples to show how the Master Plan may become part of integrated strategy to change the city, not just its physical shape but its way of functioning. The Master Plan alone does not have such power. It can however become a very powerful tool if it supports, and is supported by actions that previously lay beyond the scope of town planning.

The limitation of planning in Italy to mere city design has arisen from awareness by planners of the limits inherent in their tools. This is a step forward with respect to the attitude widespread in the past, when planners tended to envisage profound transformation of city life resulting from their plans. But, although more realistic, the new attitude leaves many questions, and probably the most urgent ones, unanswered.

Local authorities have control over a wide range of instruments which can be used to stimulate the willingness of other bodies to work towards common goals. The authorities often have people on their staff able to use them appropriately, but lack of cooperation between different departments within the same administration has tended to weaken public policy in Italy.

Town planning could count only on tools strictly belonging to the discipline itself, i.e. zoning regulations and little more.

In the preceding sections, an attempt has been made to show that this lack of inter-departmental cooperation depended largely on the fact that Local Governments were made up of weak coalitions, consisting of parties and individuals with different goals and perspectives. We have also explained how the new Local Authorities Act has triggered a deep change by laying down conditions for more homogeneous and stable local governments.

After change at the Government level, staff have often shown an unexpected ability to cooperate. This is probably more widespread among civil servants, who are often shifted from one Department to another, than among University staff, who are pushed into specialisation by academic regulations. It has to be stressed, though, that the new law can produce favourable conditions for change, but for these to be exploited, they have to be used cleverly. Their potential needs to be fully grasped above all by the Mayor himself, who has to use his freedom to choose a valid and cooperative team.

The reform still has to be completed. A strong need is now felt to make civil servants, especially at the executive level, more responsible for the outcome of their actions. In addition, the decentralisation of many powers from the central State to local authorities is a high priority in the Italian political agenda. The examples here, taken from direct experience, nevertheless bear witness to the new climate that is emerging in some of the recently reformed Italian local authorities. The plans referred are still in the process of being implemented, which means that there are not yet any results. No conclusions can therefore be drawn, except than there are reasons for hope.

References

Alexander E.R. (1981) If Planning isn't Everything Maybe it's Something, *The Town Planning Review, 52*, 131-142.

Alterman R. (1992) A Transatlantic View of Planning Education and Professional Practice, *Journal of Planning Education and Research, 12*, 39-54.

Benevolo L. (ed.) (1996) *Venezia, il nuovo piano urbanistico*, Laterza, Bari.

Benveniste G. (1989) *Mastering the Politics of Planning*, Jossey-Bass Publishers, San Francisco.

Bryson J.M. (1995) Approaches to Strategic Planning, Report presented at the seminar 'Verso politiche urbane condivise. Approcci strategici alla pianificazione e alla

gestione urbana', Milan Polytechnic, 16-17 March (mimeo).
Bryson J.M., Freeman R.E., Roering W.D. (1986) Strategic Planning in the Public Sector: Approaches and Future Directions, in Checkoway B. (ed.) *Strategic Approaches to Planning Practice*, Lexington, Massachusetts, 65-85.
Cacciari M. (1989) Idea di Venezia, *Casabella*, 557, 42-58.
Davidoff P. (1965) Advocacy and Pluralism in Planning, *Journal of the American Institute of Planners, 31*, 331-338.
Forester J.F. (1989) *Planning in the Face of Power*, University of California Press, Berkeley, California.
Fischer R., Ury W. (1991) *Getting to Yes: Negotiating Agreement Without Giving In*, Penguin Books, Harmondsworth.
Gibelli M.C. (1993) La crisi del piano fra logica sinottica e logica incrementalista, in Lombardo S., Preto G. (eds.) *Innovazione e trasformazioni della città*, Angeli, Milan, 207-239.
Gibelli M.C. (1995) Tre famiglie di piani strategici: verso un modello reticolare e visionario, Report presented at the seminar 'Verso politiche urbane condivise. Approcci strategici alla pianificazione e alla gestione urbana', Milan Polytechnic, 16-17 March (mimeo).
Kaufman J.L (1978) The Planner as Interventionist in Public Policy Issues, in Burchell R.W., Sternlieb G. (eds.) Planning Theory in the 1980's, Centre for Policy Research, New Brunswick, New Jersey, 179-200.
Kaufman J.L., Jacobs H.M. (1987) A Public Planning Perspective on Strategic Planning, *Journal of the American Planning Association, 53*, 23-33.
Lindblom C. (1959) The Science of Muddling Through, *Public Administration Review, 19*, 78-88.
Marcuse P. (1976) Professional Ethics and Beyond: Values in Planning, *Journal of the American Institute of Planners, 42*, 264-274.
Mazza L. (1995) Piani ordinativi e piani strategici, in Sartorio G., Spaziante A. (eds.) *Il Piano Regolatore Generale nelle legge urbanistica del Piemonte*, Turin, Regione Piemonte, 45-52.
Popper F. (1992) The Great LULU Planning Game, *Planning, 14*, 15-17.
Sager T. (1990) *Communicate or Calculate: Planning Theory and Social Science Concepts in a Contingency Perspective*, Nordplan, Stockholm.
Susskind L., Ozawa C. (1984) Mediated Negotiation in the Public Sector: The Planner as Mediator, *Journal of Planning Education and Research, 4*, 5-15.

26. The Art of the Science of the City

Angela M. Spence

26.1 Introduction

The nature of a society influences not only the character of its towns and cities, but also the way that urban problems are perceived and how they are dealt with. A series of factors came together after the second World War to permit a significant change in the approach to urban planning, especially in north-western Europe. The need to replace slum housing and reconstruct war damaged cities provided the opportunity to find a more rational way of organising the built environment. The adoption of theories from locational economics and social science, among others, transformed town planning from a design-based craft to a 'scientific' activity (for a detailed analysis, see Batty, 1994).

Armed with the newly developed techniques, planners adopted a systematic and supposedly objective approach to the task of analysis, evaluation and plan formulation. Confidence in their role as technical experts reflected the general faith at that time in logical positivism. The systems view of the city provided a framework, while a range of tools, including mathematical models, was formulated for the study of urban phenomena and investigation of policy alternatives. The underlying assumption was that, through the application of a rational process of analysis and evaluation, it was possible to arrive at an optimum solution.

The application of this process to the real world proved to be rather less straightforward. To derive specific objectives from abstract goals was not as easy as it seemed and the urban systems rarely behaved as predicted. Many well-intentioned schemes (such as the slum replacement and city centre redevelopment of the 60s and 70s) produced disappointing results and some, in the long term, created worse problems than those they were designed to resolve. One of the fundamental shortcomings of the approach was that the city was seen in simplified functional terms, not recognising that urban systems possess numerous complexities not present in the kind of systems from which the techniques had been derived.

The inability of the early attempts at 'scientific' planning to live up to its promise undermined faith in the planner as an expert. But among urban analysts themselves there is still a noticeable lack of consensus about how to deal with these complexities. While it is clear that a co-ordinated approach to the management of cities is required, there are few indications of the form it should take.

Increasing deregulation and cuts in public spending are limiting the possibility for direct investment by municipalities and putting greater reliance on the private sector. It is no longer realistic, or even desirable, to envisage a return to comprehensive *planning* in the sense understood earlier this century. We do however require comprehensive *understanding* of the city and also a long term perspective, if we are to be able to allocate planning decisions and reconcile the conflicting objectives (Harris, 1994).

The difficulties of urban planning are compounded by the changing world context. In Europe, the pressures of global competition are making economic survival a major preoccupation. Cities are having to restructure their economies and attempt to create new jobs. The severity of environmental problems has also given urgency to finding an ecologically sustainable pattern of urban life. A further major problem, the phenomenon of social exclusion and of 'degraded' districts within cities, seems likely to get worse, unless social inequalities and unemployment are reduced, and better ways found of coping with immigration.

These three aspects: social, environmental and economic need to be dealt with jointly, but the relationships are not simple (Wulf-Mathies, 1996). Economic growth does not necessarily resolve the problem of social inequality (in fact it frequently seems to increase the gap), and industrial competitiveness is often in conflict with environmental objectives. The current challenge of urban planning is to achieve an adequate level of economic well-being without compromising social or environmental goals and, above all, to understand how all three are influenced by the physical arrangements of cities. But what is the appropriate approach and what kind of tools are required?

Other papers in this volume have identified some of the features of the present situation: the emphasis on planning as mediation and programme management, the growing use of computerised data-bases and the search for new ways of dealing with urban complexity through the adaptation of theories from other disciplines. Optimism has been expressed about the capacity of the new analytical tools to help us understand the complex interrelations and nonlinearities inherent in urban systems, but opinion seems to be divided as to whether the current difficulties in their practical applicability represent a strictly a technical limitation, which can be overcome in time, or a more fundamental limitation of their appropriateness for the task. Views on the subject reflect the continuing 'clash of cultures' described by Faludi.

It is argued in this paper that the use of sophisticated analytical techniques is not in itself sufficient to ensure the definition of adequate policies, and that the approach to urban planning and analysis "should be more in style than in the kit bag of tools" (Lee, 1994, p. 38). We conclude that the full implications of the 'science of complexity' are far more radical than generally suggested, offering not simply a series of new tools, but a completely new approach to the comprehension of the world around us. An essential first step is to go beyond the misleading counter-position between the formalised 'scientific' approach and the intuitive style of thought. A fruitful way of coping with complexity is to integrate the two strategies and, in doing so, to 'tap into' far wider sources of knowledge. In other words, it is important to learn the art of using the sciences of the city.

26.2 Aspects of Contemporary Urban Form and Structure

The evolution of urban form provides a reflection of the way we live and how we conceive our place in the order of things. As stated by Hall, "The planning of the city merges imperceptibly into the problems of cities, and those into the economics and sociology and politics of the cities, and those in turn into the entire socio-economic-political-cultural life of the time. There is no end, no boundary to the relationships" (Hall, 1988, p. 5). In other words, the complexity of the city reflects the complexity of human life.

According to the classical conception, the city is both the physical manifestation and the tool of man's aspiration to achieve higher levels of civilisation. To be considered 'civilised', a city should provide not only for the basic material needs - decent living conditions and opportunities for work - but also possibilities for social and cultural exchange, and a setting in which the community can express and reinforce its human and spiritual values. In different periods of history and in different societies, the emphasis given to these different aspects has varied, but any marked imbalance leads in the long term to negative consequences. However, it would seem that when 'corrective action' is taken, there is often a swing to the other extreme.

It was the dreadful physical conditions of life in the nineteenth century, particularly severe in the industrial towns, which has led to this century's overriding concern with hygiene, space and orderliness. Although new housing construction and the application of planning standards have provided better living conditions for the majority, modern urban life has been impoverished in many other respects. Recent studies (Kunstler, 1994, Rogers, 1996) have identified a series of problems deriving from the form and structure typical of contemporary urban development (found in its most pronounced form in the United States, but is also common in Britain and other countries of northern Europe, whereas in southern Europe a more traditional form is still predominant). We examine briefly some of these features.

The physical segregation of different types of activity. The tendency to specialisation, already a feature of economic life, is increasingly reflected in the use of urban space. One of the characteristics of recent development is for parts of the city to 'specialise' in one predominant type of function (business districts, shopping centres, residential estates, industrial areas, etc.) and even further specialisation often occurs within a single function (e.g. residential areas which cater for a specific income bracket). This can occur both intentionally, in new development, and 'spontaneously' where existing areas with previously mixed land use are affected by a gradual separation process (an example is described in Duncan and Tamás Sikos, 1995).

One of the more extreme results of residential segregation is the phenomenon of lower income groups becoming trapped in 'ghettos', where the downward spiral of unemployment, crime and environmental decay reinforce the problem of social marginalisation. It has been pointed out that, as well as reducing the variety and richness of local experience, the geographical separation of different ethnic and income groups can have a serious effect human relationships within the urban community, since excessive segregation feeds fear and misunderstanding, contributing to the build up of social tension (Lozano, 1990).

Low density development, due to the relatively generous space provision, results in a highly diffused pattern of urbanisation. The combination of low density and highly segregated land-uses increases the difficulty of achieving ecological sustainability. Large areas of land are 'consumed' and there is a relatively inefficient use of the infrastructure (transport, utilities etc.). The physical separation of residential, shopping and workplaces, especially in a society dominated by the private car, though possibly reducing local congestion, generates a higher overall volume of traffic and pollution. The difficulties of adapting sprawling cities to more sustainable patterns of mobility are considerable, as the residential density is rarely sufficient to permit viable systems of mass transit.

The rationalisation of service provision, which has led to the reorganisation of many services (from shops to health centres and schools), results generally in larger, but fewer outlets and a new location pattern. This process has radically affected the urban hierarchy and also accessibility. Many small and medium towns have become dependent on larger centres for all but the lowest level services, some have been virtually 'put out of business' by large out-of-town shopping centres.

This trend has also tended to produce inequalities within the population in terms of choice and opportunity of access to services. The location of many services outside residential areas and (in the case of shopping centres) often completely outside the city, combined with the lack of public transport, causes serious difficulties for some sectors of the population, especially the elderly and non car-owners. Rural areas, small towns and even certain city districts are particularly disadvantaged by the lack of services (Smith, 1991).

Poverty of the townscape. Whereas sustainability has become a major issue, and episodes of violence have brought the most extreme forms of social marginalisation to the public notice, rather less attention is generally paid to the 'urban habitat'. As pointed out by Kunstler (1994), the modern urban environment is incomparably poorer than the traditional city in terms of those factors which contribute to the quality of 'character'. While there is little doubt that the modern city represents a highly complex entity especially in terms of flows of traffic, information and goods, from the point of view of the organisation of space, a major shortcoming is its lack of complexity (Berna, 1996). Building has become anonymous, taking little regard of history and ignoring the importance of public space. Many city centres are no longer a focus for the urban community and are visited, if at all, only for utilitarian purposes. The vitality, as well as the cultural and civic life of cities, as well as being crucial from a social point of view, can play an important economic role. In a world where increasing weight is given to the quality of life, cities which can attract the highly mobile skilled professionals have a competitive advantage. The paradox is that "as we are ever more enmeshed in the electronic web, we increasingly insist on a sense of place" (Morris, 1994).

We have outlined above just some of the shortcomings which have emerged in the long term from the so-called 'diffused' form of urban development. The environmental and social costs have in fact lent increasing support for a return to a more traditional pattern with greater mix of activities and higher densities (Rogers, 1996). Considerable efforts and resources are being dedicated in many cities in Britain, for example, to providing more city centre housing, better public transport, improving urban design quality and encouraging business activities to move back to the centre. It is proving very hard, however, to recreate many of the qualities which have been lost.

As pointed out by Faludi in this volume, northern Europe has in general shown a greater propensity than the south to an institutionalised and 'rationalistic' approach to town planning. Already fifty years ago Britain instituted a comprehensive planning framework (though the last two decades have seen increasing deregulation, with market forces being allowed greater play). This has nevertheless not mitigated the kind of problem described above, in fact, we argue below that the approach to intervention has in some cases exacerbated the problems.

In southern Europe, including Italy, where planning intervention has in general been far less systematic, the majority of cities still retain a more traditional form and role: city centres, for example, have a mixture of functions, a wide variety of users and represent the focus of urban life. However, the changes which have occurred in recent decades are putting Italian cities under great strain and forcing municipalities to take action. While it is essential to find solutions, especially to the severe problems of traffic congestion and pollution, the kind of experience described above suggests that an over-reaction to functional defects can lead to the loss of valuable qualities.

The irony is that while efforts are being made in Italy to encourage a more 'scientific' approach to urban analysis and to make planning intervention more systematic, Italian cities serve as a model for many qualities that North European cities are now trying to recreate. In view of the difficulties now being faced elsewhere, there is an obvious interest in understanding how the form of urban development described above has become so prevalent and what lessons there are to learn.

There are certainly powerful economic forces and business interests which have pushed in this direction. One of the root causes of current problems is undoubtedly the preoccupation of contemporary society with material needs, and the systematic disregard for community responsibility in favour of individual interest and short-term economic expediency (Lozano, 1990). But the form and nature of the city would also seem to have been affected by the way in which urban problems have been analysed and tackled.

In the second part of this section, we indicate how the adoption of a 'rationalistic' approach can affect urban form and structure as a result of:

a. the kind of tools used for analysis, and hence the way in which the city is perceived,
b. the way the urban problems are defined, and therefore the kind of 'solution' adopted.

Firstly, the emphasis on *statistical and mathematical representations* results in a focus on quantitative factors, such as population and employment data, transport flows, neglecting many of the less easily expressed factors. This can lead to the neglect of many qualitative aspects and, in particular, of the way the urban ingredients come together, i.e. the city as a 'three dimensional' place. As stated by Smith (1994), one of the weaknesses of urban policy is that too often it has been based on "the manipulation of data that explains far less than it pretends to, judging the urban experience on economic, physical or statistical factors that may move in directions quite separate from urban dwellers' cultural and emotional wishes" (p. 63).

A second weakness is the tendency to *codify standards and practices*, leading to uniformity and the adoption of routine solutions without regard for local factors. This applies to the approaches both to analysis and policy implementation: the tendency to use standard classifications and visions of the city or, for example, conventional practices, such as land-use zoning (although initially justified by the need to separate heavy industry and housing, it has continued to be systematically applied, even where there is no real incompatibility between activities).

One of the most serious defects, however, is the lack of understanding of the complex dynamic behaviour of many aspects of the urban system. This can lead to: (i) the adoption of mistaken assumptions relating to cause and effect, e.g. in transport planning, new roads were designed on the assumption that the extra capacity would resolve the problem of congestion (in reality, it frequently generated new demand and therefore the benefit was only temporary), (ii) the tendency to assume linear trends, and hence insufficient flexibility of design to permit adaptation. Many city centre malls and housing projects of the 60s and 70s have already had to be completely rebuilt due to changing requirements and tastes, (iii) the failure to realise the possible long term consequences of a policy, e.g. the substitution effects of new out-of-town shopping centres (a characteristic of such processes of change is that they are not gradual. They seem to reach a breakpoint, then accelerate, both in the positive sense, e.g. a new business development which is suddenly 'takes off', and the negative sense, the 'domino effect' when a process of decline spreads rapidly and overtakes a whole district. Once the threshold has been reached, external intervention has little influence).

We now examine this last series of defects in greater detail. It involves not only the tools employed for urban analysis, but also the approach to planning intervention, and in particular the process of *problem definition and problem solving*.

The complexity of an urban system makes it hard to analyse rigorously, but also hard to know how to intervene when some dysfunction is diagnosed. An interesting example of the kind of self-defeating chain, typical of modern land management, is given in this volume by Allen. He describes how changes in agricultural practice on the Argolid Plain in Greece (policies supported by the EC and intended to protect rural agriculture) eventually managed to destroy land fertility, making it almost impossible for farmers to make a living. How was it possible for an apparently logically-derived policy, based on systematic analysis, to have an outcome precisely the opposite of that intended? (This, incidentally, is a question relevant not only to land management or urban planning, but many other areas of human endeavour!)

Two factors which seem to provoke this kind of chain are *economic pressure*, pushing producers to seek maximum exploitation of resources, and *the (inappropriate) use of technology*, which makes it possible. But a third, highly significant factor, is the process of problem definition. The tendency is for attention to focus on the immediate problem or symptom, and for efforts to concentrate on modifying the factor(s) perceived as the direct cause. The 'solution' thus derived may provide temporary relief, but in the long term there is the risk that either the original problem will reappear, or that a whole new series of problems will emerge.

This process could be likened to the allopathic approach to medicine. After diagnosing the symptoms of the malady, a strong dose of the counteracting remedy is administered. While possibly alleviating the immediate symptoms, it risks bringing about a number of side-effects which then need to be tackled with further 'cures'! An example from the field of urban planning is the approach to city centre congestion. Typical solutions involve limiting car access, introducing payment for parking and transferring business activities to the periphery, i.e. reducing the demand for trips. This may successfully eliminate traffic problems, but risks reducing the vitality and attraction of the centre, eventually causing its decline (and often resulting in its replacement by a new suburban centre which is highly functional and able to cater for large volumes of cars, but without the variety and 'complexity' of a traditional town).

This suggests that the conventional 'problem-solving' approach needs to be rethought. Returning to the medical analogy, we should perhaps adopt something more similar to a naturopathic approach and act 'from inside', identifying the minimal intervention and using the qualities of the system itself to effect a re-equilibrium (or assist it to evolve to a new pattern). This kind of approach means:

a. adopting a sufficiently wide perspective to see the whole picture (and hence identify the 'real' problem);
b. acquiring a better understanding of the relationships within the system;
c. using greater creativity in the search for solutions (perhaps totally redefining the problem or attempting an unconventional approach);
d. encouraging the co-operation of elements within the system (individuals, groups, firms, etc.) in the search for solutions and means of successful implementation.

The aim should be not just to eliminate the immediate problem, but recreate a new balance. As pointed out by Toniolo in this book, this is likely to involve actions which go beyond purely physical planning. In the case of the traffic congestion problem mentioned above, the objective should be to 'restore the quality and attraction of the centre', not simply to 'eliminate traffic problems'. This could perhaps be achieved by finding alternative forms of access, encouraging staggered working hours, ensuring the possibility of local housing for city centre employees, etc. rather than reducing the range of activities.

Most European cities are in fact now paying more attention to the quality of urban design, promoting a broader 'mix' of activities and, above all, making efforts to achieve sustainability. But, as explained above, there is a danger that if the approach continues to be one of 'problem elimination', further long term problems are likely be created. The pro-environment policy adopted in some cities, for instance, of keeping air pollution below a given maximum, may eliminate the peaks, but risks causing higher overall pollution levels, by encouraging the diffusion of activities. Similarly, if we make the creation of an attractive city our primary aim, treating it predominantly as a question of aesthetics, we risk a purely cosmetic solution. This is the fate of many urban improvement schemes, which provide the right physical 'ingredients', but become little more than tourist attractions or up-market shopping districts, virtually unused by the local population.

Whatever the most appropriate form for the city (a more traditional structure, a diffused form, one with strong network connections, etc.), it is significant that the focus is now shifting to questions such as urban identity and image, accessibility, the reciprocal influence of social change and the physical environment, as well as the prickly issue of sustainability. It is therefore important that the tools and approaches available are able to express these aspects. In the next section, we discuss some of the problems of dealing with analysis and decision-making in an urban system, with reference to the tools and concepts deriving from the new 'science of complexity'.

26.3 An Approach to Complex Systems

An eloquent description of the method of enquiry is given by Bronowsky, likening it to that of the portrait artist: "We are aware that these pictures do not so much fix the face as explore it, each line that is added strengthens the picture, but never makes it final. We accept that as the method of the artist, but what physics has done is to show that it is the only method to knowledge. There is no absolute knowledge, all information is imperfect. We have to treat it with humility." (Bronowsky, 1977, p.353).

The study of complex dynamic systems is revealing a picture of the world which departs radically from that described by classical science; it is a world of instability, nonlinear change, paradox and above all unpredictability. The meaning of the word itself ('complex' = braided together), provides an effective image of the many interwoven threads which make complex systems so difficult to unravel. They create 'surprise', in the sense that they do not react in the way we would expect. In fact, the closer we get, the more they seem to elude analysis.

The attempt to fit the social sciences into the straitjacket of formal analysis and linear causal processes has always been controversial. Considerable interest has therefore been raised by demonstrations that at least some kinds of 'non regular' and apparently chaotic behaviour seem to fit recognisable patterns and can even be expressed in formal terms. In this section, we look at the problem of providing a description of complex phenomena, the nature of the processes of change (patterns of growth and transformation) and how to deal with unpredictability, considering to the possible role the tools of the so-called 'science of complexity'. As detailed descriptions of these tools are provided in other papers in this volume, we concentrate on the discussion of their applicability to urban analysis and planning.

Classical science encouraged belief in the possibility of being able to provide a complete description of phenomena. This led to the idea that even highly complex systems could be explained by increasing the level of complication of the analysis. When the fit of a model was found to be imperfect, this was assumed to be a technical deficiency, which could be resolved by adopting more sophisticated approaches or techniques. Lack of success, instead of being understood as the impossibility of encompassing reality in any single analytical system, was frequently explained away by the 'only if' syndrome (if only we had more money/time/resources, etc.). The approach to urban modelling seems at times to have fallen prey to this attitude.

One impact of the new ideas emerging from the study of complex systems is that it is 'loosening the grip' of old myths: the idea that everything can be understood, predicted and hence managed. From this follows the recognition that not all aspects of human thought and behaviour are reducible to mathematical language or can be represented through logical and symbolic procedures. In describing social systems there are a number of fundamental problems. Before even facing the problem of measurement, there is the difficulty of deciding which phenomena to include (this already implies a choice and hence a value judgement), then how to match them to the established classification. A further constraint is the fact that this framework is inevitably static, whereas reality is fluid.

These limitations do not of course mean that such analytical procedures are irrelevant to a highly complex system such as the city, but that any form of model must be accepted as partial, simply providing an 'indicator'. The aim of the model is to capture in a vivid way some aspect of a structure or behaviour, not to reproduce reality. An approach to the building of a more complete picture is discussed in the next section.

The application of any tool inevitably condtions the way the world is perceived, just as the language used influence the way it is represented and the questions which can be asked. The 'paradox effect' confirms that contrasting viewpoints can provide equally valid insights (the city is both node and network, it contains both predictable and unpredictable phenomena, etc.), it is therefore necessary to approach from all sides, making use of a variety of tools, some formal, some not. Empirical investigation is fundamental, helping to clear away 'myths' and subjective bias, but we must be prepared to accept other sources of knowledge.

The need for plurality of approach includes the sources of information. While GIS, for example, certainly provide rapid access to a very large quantity of data, they nevertheless offer, as specified by Van Geenhuizen and Nijkamp in this volume, a "stylised picture of reality using the fruits of modern computer technology" (p. 723). The danger is in using them as an exclusive source of information. As stressed by Occelli, the real problem is not simply to provide information, but contribute to meaningful knowledge. The main difficulty in the future seems likely to be in finding the balance between too little and too much information. A phenomenon already recognised in the field of business, but increasingly relevant elsewhere, is that of 'information overload', where insufficient selectivity results in confusion (a survey in a local planning authority in Britain revealed that 75% of the staff involved considered that the use of the GIS did not improve the *quality* of decisions!).

Recent studies of decision-making and the thinking process suggest we need to distinguish between different contexts. Whereas analytical thinking (and the use of quantitative information) is appropriate is certain situations (e.g. the solution of some traffic problems), it may not always capture the full complexity of an issue. In cases where noncognitive considerations are important, intuitive thinking deserves attention, because it offers an overall view of a situation. This is especially so where the future behaviour of complex systems is concerned.

A fundamental feature of classical science was the belief in the predictability of the behaviour of natural phenomena. Although it was accepted that in the social sciences human choice made the situation more complex, it was generally felt that some underlying pattern exists. The strategy normally adopted consists of attempting to narrow uncertainty to a minimum, or at least of expressing it in terms of probability. Whereas this may provide an acceptable approximation for relative stable systems, in times of rapid change the approach becomes seriously inadequate.

The constant innovation and instability typical of so many areas of modern life are changing values and attitudes, producing reactions which are no longer rooted in the past. This makes it very difficult to model or predict behaviour. Even 'relaxed' assumptions, such a bounded rationality, are based on the unrealistic idea that people are homogeneous and adopt a systematic strategy. One of the characteristics of the contemporary world would seem to be that choices are based on widely divergent criteria, not easily reduced to a single strategy.

One of the most valuable features of the study of complex systems is that it has provided descriptions of nonlinear modes of change. We briefly mention four such modes, all describes in greater detail in Casti (1994).

One of the simplest, but perhaps most common forms of change is that represented by logistic growth. The examination of a great many processes, including economic, social and political activities has shown that they frequently follow an S-shaped curve. A further example of nonlinear change is the feedback effect, i.e. the accumulative advantage (or disadvantage) which causes one system or factor to gradually draw apart from the others. Despite very similar starting conditions, it will begin to pull ahead, success feeding on success (or the opposite, decline leading to further decline). In certain kinds of complex system, catastrophic behaviour may appear. This is marked by stable periods in which behaviour is relatively predictable, and disturbed periods in which dramatic changes can occur in response to very small external stimuli. The system reaches a 'critical point' at which there may be a jump to a new level or type of organisation which, especially in more complex systems, can be completely unpredictable. Lastly is the butterfly effect, where a single apparently insignificant action sends reverberations throughout the whole system, with dramatic repercussions even at a considerable distance.

Convincing examples have been presented in other papers in this book of urban phenomena which fit these patterns. Most authors have admitted, however, that although it is possible to recognise them in historical data, it is far harder to anticipate situations where they are likely to occur, and almost impossible to make precise predictions. This dilemma was stated succinctly by Handy (1994) "life is best understood backwards, but has to be lived forwards" (p.10). So how should such analyses be used? As suggested above in relation to models, they can nevertheless serve as 'indicators', helping for example to avoid the assumption of linear growth, by showing that even well-established current trends are unlikely to be permanent. Experience in other cities could be used to provide a kind of 'early warning system' by identifying: (i) factors which may act as triggers to radical changes in the system, (ii) the vicinity of a critical threshold where behaviour could change drastically. They could also help to identify those actions which, though small, can send reverberations throughout the whole system. This information can be used in two ways: (i) to avoid an action likely to have widespread negative consequences, and (ii) to recognise the minimum action needed to create the maximum effect, by identifying possible catalysts of change and choosing the right moment to implement a policy.

These processes of change underline the importance of timing. By implementing a decision at the right moment, its chances of success can be considerably increased and its costs (both social and economic) reduced. The 'right' policy carried out at the wrong time may encounter resistance and waste resources.

A great many of the decisions affecting urban systems include elements which involve subjective value judgements. It is also very often necessary (both for individuals and institutional decision-makers) to make choices without access to full information. This can occur either because the necessary information is not available, or because it involves knowledge of the future. The former dilemma can (in theory) be rectified, but the latter is more problematic, especially where highly dynamic systems are concerned. Even if it is possible to arrive at a reasoned evaluation at a given moment in time, it is likely that within a relatively brief period the situation will have changed (through an unexpected modification in an existing element, or the introduction of a completely new one), completely invalidating the previous evaluation. It is therefore quite possible that an apparently illogical choice will turn out to be more appropriate than a 'reasonable' one. How can we cope with such a dilemma?

Rather than eliminating uncertainty, an alternative strategy for coping with the unexpected is that of accepting it, "an appreciation of process makes uncertainty bearable... we must be free to change and assimilate new information, making the future as we go along" (Ferguson, 1986, p. 108). This implies promoting flexibility, i.e. adopting systems which are amenable to change, avoiding irrevocable decisions and over-specialisation, but also making greater use of intuition, which is the theme we explore in the next section.

26.4 Sources of Knowledge

"What is sought can only be found where there is illumination. ....Two things matter: the size of the circle of light that is the universe of one's inquiry, and the spirit of one's inquiry. The latter must include an acute awareness that there is an outer darkness, and that there are sources of illumination of which one as yet knows very little." (Weisenbaum, 1984, p. 127).

In many ways town planning as a professional activity has suffered from overspecialization. There is often little integration between the many disciplines concerned, with a wide separation not only in terms of areas of interest, but also in the kind of approach employed. One of the most unfortunate aspects of the debate over recent decades has been the continuing antagonism between the proponents of the intuitive (or design) approach and those supporting the use of more formal analytical tools and methodologies, including mathematical models. While the designers are accused of presumption in their unwillingness (or inability) to explain the mechanism of their 'creative' leap, urban scientists are criticised for their over-reductive view of the world and neglect of the qualitative aspects.

The consequent division into opposing camps, described so vividly by Faludi in this volume, has limited the opportunity for a better understanding the fundamental complementarity of the two approaches.

The antagonism is heightened by the influence of a society which channels individuals towards developing one skill to the exclusion of the other, creating the 'two cultures'. But if we examine the process by which scientific knowledge itself has advanced, we find that logical analysis and rigorous verification procedures are frequently accompanied by a creative leap akin to that of artistic inspiration.

The well-known study of creativity by Arthur Koestler (1965) sheds light on the way in which some of the major scientific breakthroughs have come about. He suggests that the process which foments a new discovery involves the bringing together of normally unconnected areas of knowledge or 'matrices'. It is the synthesis of different dimensions of experience which enables man to achieve deeper understanding. The creative leap can occur when conventional ways of thinking are abandoned - a new level of understanding is made possible through perception from a different angle. If we look at this process in further detail (see Fig. 1), we find that two very different modes of thought are involved, one rational and 'conscious', the other intuitive.

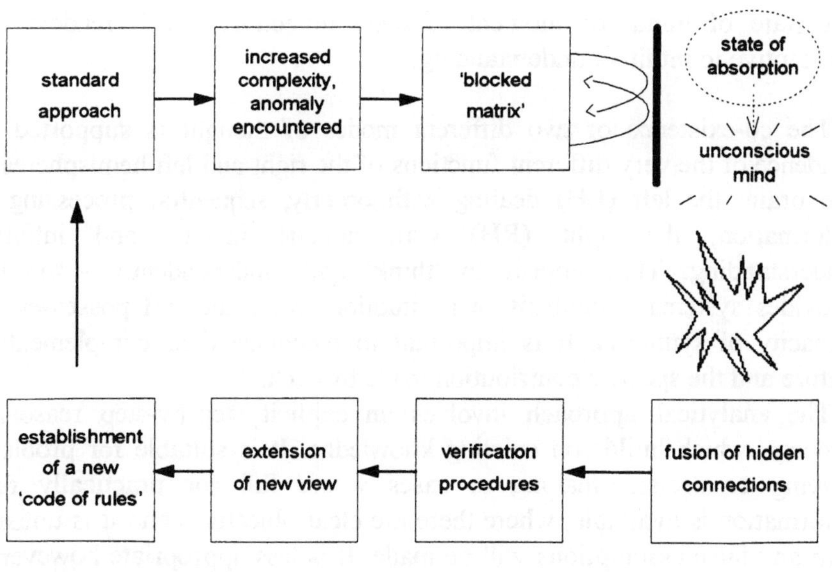

Fig. 1 Representation of the creative process (based on Koestler, 1965)

In the initial stage, proceeding by means of systematic investigation, the researcher is faced with some form of anomaly which is inexplicable in terms of conventional schemes of analysis or established precepts. This creates a 'blocked matrix' causing thought to 'run in circles' until released by a flash of intuition.

Documentary evidence reveals a fascinating picture of how this may come about - from Kekulé the chemist who 'saw' chains of molecules in a dream, to Poincaré whose ideas started coming after drinking black coffee late at night, and Gutenberg's association of the wine-making process with printing. The understanding seems to arrive unexpectedly and from a deeper part of the mind. It would appear that awareness of the problem permeates through to the unconscious and that the connection which produces the solution is triggered by a flash of intuition. According to Koestler (1965), this is able to occur when the rational mind is temporarily 'switched off' and the inhibiting effect of the usual logical framework is no longer operative. The necessary conditions for arriving at the new intuition would therefore appear to be:

- a state of 'ripeness', i.e. preparation of the mind through systematic application to a problem or absorption in a subject;
- the occurrence of circumstances, including a chance event, or even an accident, which permit the coming together of ideas or unexpected associations; and/or
- a state of mind, or moment of deep reflection, which renders one receptive to intuitive understanding.

The co-existence of two different modes of thought is supported by evidence of the very different functions of the right and left hemispheres of the brain: the left (LH) dealing with orderly, sequential processing of information, the right (RH) with holistic images and intuitive understanding. They appear to 'think' quite independently - the LH provides systematic analysis of a situation, while the RH possesses the capacity of synthesis. It is important to recognise their complementary nature and the specific contribution made by each.

The analytical approach involves an explicit step-by-step reasoning process, which builds on existing knowledge. It is suitable for problem-solving or decision-making in cases where full (or practically full) information is available, where there are clear objectives and it is unlikely that any false assumptions will be made. It is less appropriate however in situations which are 'untidy', such as those involving difficult predictions, or sensory and emotional aspects which are difficult to formalise or express verbally. Here, an intuitive response can be more effective, as it has the capacity to integrate many different considerations (some of which the conscious mind may not even be aware of).

Intuition is not infallible (its accuracy depends upon the refinement of perception), but it is increasingly accepted that it should be taken seriously. Knowledge of this deeper level of awareness has long existed, but in recent times it has been devalued and frequently misrepresented. Ever since the Cartesian revolution, the mind has been identified above all with rational thinking. Another tendency is to confuse intuition with random guesswork, a lucky hunch or mere fancy. True intuitive perception represents transcendent reasoning, "the brains capacity for simultaneous analysis we cannot consciously track" (Ferguson, 1982, p.114). It is the insight, recognised by ancient philosophies, which permits direct access to the 'implicit reality'.

Whereas analytical procedures permit preparatory exploration of the field, providing the 'groundwork' (and are also necessary for the subsequent verification of new ideas), intuition is a necessary ingredient if we wish to encompass the full complexity of living systems.

26.5 Some Brief Conclusions

The new discoveries of science are revealing aspects of the world around us too rich for analysis. The inevitable conclusion is that the kind of complexity typical of urban systems is not amenable either to complete description or reliable forecasting. Although the new tools deriving from the science of complexity provide useful 'indicators' of typical structures and processes of change, they have limited predictive power when applied to social systems. Nevertheless, integrated with a range of other techniques, they can help to build up a more complete picture. As for all tools, it is essential to have a clear idea of their specific strengths and weaknesses, and to pay sufficient attention to the stages *before* and *after* the application. Creativity should permeate the whole process. The real 'art' lies in knowing which features to investigate, what information to use, which tool to apply and, above all, how to interpret the results. Giving meaning to the results "requires that we bring every scrap of knowledge, imagination and insight to the task, and there is nothing to help us here - no books, machines, only ourselves" (Gould, 1986, p. 9).

It is for this reason that the analyst needs to be open minded and use his full experience of the city, constantly refining his perceptions. In the present context, where so many fundamental features of urban life are changing, we have to be prepared to rethink many accepted ideas, classifications and standard policy 'solutions'. A more open approach would also be beneficial in relation to disciplinary and institutional integration. One of the major obstacles to effective urban planning is in reality not the lack of precise information or highly refined models, but the absence of appropriate co-operation. Transport plans are still made with little reference to land-use, administrative divisions often make it impossible to co-ordinate policies for a 'functional region'. Equally serious is the distance, pointed out by Hall (1994) between the academic world and planning practice. Although many interesting approaches are being developed in research institutes, most studies are of a highly theoretical nature, and the practitioner remains without useful tools to apply in the everyday context.

Within Europe itself there is considerable scope for the exchange of experience in many key areas of urban policy. We must be able to assess the qualities of different solutions, but find the most appropriate for the specific requirements of a given application. A worrying feature of both analytical approaches and policy proposals is the influence of 'fashionable' ideas.

Exchange of information is essential, but too much data can also cloud the view. The idea that with enough information man can predict the future is one of the deeply rooted illusions of modern man. While possibly true of mechanical systems, the basic quality of human systems is their uncertainty. Man's strategy for living has always taken the lack of full knowledge of the future into account. In other areas of life this is instinctively understood. The solution lies partly in the acquiring confidence in insight, and partly in creating flexible systems which are able to adapt to changed circumstances.

One of the features of the positivist view is that: "almost every human dilemma is seen as a mere paradox ... that can be untangled by judicious application of cold logic" (Weisenbaum, 1984, p. 252). This reinforces the idea that the analyst can remain morally aloof and ignores the fact that there can be genuinely incompatible value systems. Decision-making in the urban context is not a purely technical process as it is bound to involve fundamental human dilemmas.

The city is the 'object' which the urban analyst is trying to understand, but it should also be a source of inspiration. The capacity to understand and to perceive possible policy options does not depend only on intellectual ability, but also how the city is 'lived', i.e. what is learned from the urban experience - it is in other words a reciprocal process. So, at this point we come full circle: the richer, denser, and more varied the city, the greater the possibility it has of being the source of creativity necessary for its own survival.

References

Batty M (1994) The Anglo-American Modeling Experience, *Journal of the American Planning Association, 60*, 1, 7-16.

Berna L. (1995) Una città troppo semplice per una vita troppo complessa, *Paesaggio urbano, 4*, novembre-dicembre, 5-8.

Bronowsky J. (1977) *The Ascent of Man*, B. C. Associates, London.

Casti J. (1994) *Complexification*, Harper Collins, London.

Duncan J.E., Tamás Sikos T. (1995) The Application of the Shift-Share Method to Retail Change in Boston City 1945-1993, in Fischer M.M., Tamás Sikos T., Bassa L. (eds.) *Recent Developments in Spatial Information, Modelling and Processing*, Geomarket Co., Budapest, 248-286.

Ferguson M. (1982) *The Aquarian Conspiracy: Personal and Social Transformation in the 1980s*, Paladin, London.

Gould P. (1986) Allowing, Forbidding, but not Requiring: A Mathematic for a Human World, *Biomathmatics, 16*, Springer-Verlag, Berlin, 1-21.

Hall P. (1988) *Cities of Tomorrow*, Blackwell, Oxford.

Handy C. (1994) *The Empty Raincoat*, Hutchinson, London.

Harris B. (1994) The Real Issues Concerning Lee's Requiem, *Journal of the American Planning Association, 60*, 1, 31-34.

Koestler A. (1986) *The Act of Creation*, Hutchinson, London.

Kunstler J. (1994) *The Geography of Nowhere: the Rise and Fall of America's Man-made Environment*, Touchstone, New York.

Lee D.B. (1994) Retrospective on Large-Scale Urban Models, *Journal of the American Planning Association, 60*, 1, 35-40.

Lozano E. (1990) *Community Design and the Culture of Cities: the Crossroads and the Wall*, Cambridge University Press, Cambridge.

Morris D. (1994) The Return of the City State, *Utne Reader, 65*, 78-81.

Rogers R. (1995) The Imperfect Form of the New, (Reith Lectures) in *The Independent*, February 13.

Smith G.C. (1991) Grocery Shopping Patterns of the Ambulatory Urban Poor, *Environment and Behavior, 23*, 86-114.

Smith S. (1994) Saving our Cities from the Experts, *Utne Reader, 65*, 59-75.

Weisenbaum J. (1984) *Computer Power and Human Reason, from Judgment to Calculation*, Penguin, New York.

Wulf-Mathies M. (1996) Le mille e una Europe, *Dossier Europa*, 19, EEC Commission.

SESSION 4: THE METHODOLOGIES OF THE URBAN SCIENCES

27. Design and Use of Information Systems for a Sustainable Complex City

Marina van Geenhuizen, Peter Nijkamp

27.1 Introduction

The city is the functional appearance of modern civilisation. It is the heartland of the economy, the cradle of creativeness and innovation, the source of science and culture. It is the meeting place of a modern society, and the nodal centre of a gradually globalising network society. The city is essentially the place where individuals shape a contemporary sense of community and where a post-modern urban space determines individual lifestyles (see Leontidou, 1992).

It is noteworthy that despite uniformities in urban architecture and urban shape all over the world, the intrinsic feature of a modern city is the provision of opportunities for individualistic behavioural patterns. This is in agreement with the view of Fukuyama (1992) on the free self-determining man in a liberal society which has emerged after the collapse of centrally governed states in recent years.

The existence of free individuals in the post-modern city also provokes many questions on the sustainability of such cities, as there is no unifying force or behavioural code which would ensure continuity under changing conditions and in conflicting situations. This is essentially the 'social dilemma' of contemporary cities: rational behaviour of individuals (or households or segmented groups in an urban society) does not necessarily mean that overall urban viability or survival is guaranteed, as there is an intrinsic discrepancy between individual and collective interest. The existence of social externalities in particular (both costs and benefits) creates a case for public intervention in the form of urban management and policy (see Fokkema and Nijkamp, 1994). A prerequisite for balanced and effective urban policy-making is the availability of up-to-date and reliable urban information systems which respect individually-oriented interests in the city and which are tuned to the needs of urban policy-makers.

In recent years, the awareness has grown that up-to-date and customised information should constitute the basis for urban analyses and policies (see also Batty, 1994, Harris, 1994, Wegener, 1994), whether computerised information systems or modelling results. Especially in a dynamic urban context, it is clear that planners and policy-makers require reliable information in order to justify proposals and actions in an uncertain world. There exists however the danger that the current information technology will take the lead in urban decision-making, making policy more engineering driven than quality-based.

The present work will offer a description of the architecture and use of user-oriented urban information systems. First, a sketch will be given of some key evolutionary trends in contemporary cities which lead to the notion of the 'complex city'. Then the challenges of governing the complex city will be outlined. Attention is next focused on the need for adequate information systems. It will be shown that modern geographical information systems (GIS) in combination with decision support systems (DSS) including multiple criteria analysis (MCA) provide a powerful tool for decision-making in a complex city. The architecture and substance of such information systems will be described, with particular emphasis on the management of sustainable urban development. A key question, of an empirical and testable nature, concerns the kind of information needed to support a certain policy decision. The intrinsic quality of information for urban policy-making can essentially be subject to scientific testing.

27.2 The City and the Home of Man

In our century urbanisation has become a world-wide phenomenon. Urbanisation presupposes an intricate and efficient organisation of space and time for all human activities. Mainly due to the division of labour, mankind has managed to benefit significantly from economies of scale in geographical concentrations. Especially after the introduction of mechanised production (leading to industrialisation), large scale concentration of human activities - and consequently urbanisation - became a widespread phenomenon. Admittedly, numerous old annals also witness the existence of relatively large towns in ancient civilisations, but such cities were mainly organised according to strict hierarchical political (if not military) principles in both spatial and social terms. As these cities derived their size predominantly from administrative centralisation, they often did not survive when the central power of political regimes collapsed.

Urbanisation in Europe after 1300 was initially purely based on gains from trade. Although some mechanisation took place (e.g. windmills), most towns remained rather small (fewer than 20,000 inhabitants). Only when economic power was very strong or combined with administrative power did towns exceed the above limit.

The industrial revolution in the last century radically changed this situation. The enormous economies of scale that could be achieved compared to the previous rural society as well as the introduction of motive power and energy in urban and inter-urban transport allowed the emergence of large cities. Therefore, it is plausible to claim that modern urbanisation is almost literally fuelled by the large scale introduction of fossil fuels. This also suggests that the transition of developing countries to a higher stage of development may cause a massive rise in energy consumption.

Modern urban systems are increasingly faced with environmental problems, ranging from air, oil and water pollution to more intangible externalities such as noise, lack of safety, destruction of 'cityscape' or visual pollution. There is a wide variety of sources generating these urban environmental problems, such as demographic factors, socio-economic development, inefficient energy consumption, inappropriate technologies, spatial behaviour (travel) patterns, and most important of all, inappropriate and/or badly enforced environmental policy measures. An improvement of the current unfavourable situation clearly requires a mobilisation of all forces.

Nowadays, the notion of sustainable cities is receiving growing attention. The notion of sustainable development has gained much popularity in recent years. The political formulation of this notion is most clearly described in the publication 'Our Common Future' (the Brundtland Report 1987) as follows: "a process of change in which the exploitation of resources, the direction of investments, the orientation of technological development and institutional changes are made consistent with future as well as present needs" (p. 46). It is clear that the idea of sustainable development is much broader than environmental protection. Furthermore, sustainable development is not a predetermined end state, but a balanced and adaptive evolutionary process (Nijkamp and Perrels, 1994). Sustainability refers in this context to a balanced use and management of the natural environmental basis of economic development. A basic underlying principle may be that the stock of natural resources should not be depleted beyond its regenerative capacity.

Sustainable development has of course a global dimension, but there is clearly also a close interaction between local and global processes. Cities are open systems impacting on all other areas and on the earth as a whole. Therefore, an urban scale for analysing sustainability is certainly warranted. At the same time, it should be realised that in an open spatial system cross-boundary flows of resources and even of pollution and waste play an important role. Thus, sustainability requires a conception and spatial scale of analysis that accounts for the openness of regional or urban areas (see Breheny, 1992).

Especially in the European context, the reinforced focus on the city seems warranted, as European cities are facing a stage of dramatic transformation as a consequence of the move towards the completion of the internal market. There is both increased competition between cities within the European community and between European cities and cities in other economic blocks in the world. However, the aim to make European cities more competitive in economic terms may be at odds with the aim of environmental sustainability.

Although sustainable urban development has become an important issue in social science research, the theoretical underpinnings and the critical success parameters of actual urban sustainability policies are still feeble. Moreover, there lacks a multidisciplinary approach to urban sustainability, combining the various actors, their activities and institutions, their different contributions to economic growth, as well as their resource consumption and the negative externalities caused.

Now the question is: how can urban planning policies be used in order to contribute to sustainable urban development. How can solutions be found for urban congestion problems, to raise the quality of life, or reduce the enormous (hidden) unemployment? How can effective instruments be applied? Do these instruments only serve environmental-technical aims or are they also related to other policy fields, such as transport, housing and physical planning? And last but not least: how can we design information systems which are consistent under changing urban circumstances and flexible with respect to complex urban dynamics?

A necessary condition for implementing an effective planning system for urban economic and environmental management is the development of a system of suitable, i.e. policy relevant, indicators. For example, the OECD (1991) has drawn up a list which combines a number of economic and environmental indicators with an emphasis on the latter. However, it appears to be extremely difficult to operationalise such an indicator system for urban sustainability. This means that precise empirical evidence on urban environmental quality and on the implications for both household and company behaviour is not always available.

In the meantime, a new discipline has appeared, 'urban ecology', which aims to design principles for sound urban environmental policy. Examples of such principles, which also require tailor-made information, are:

- minimise space consumption in urban areas (e.g. underground parking);
- minimise spatial mobility in cities by reducing the geographical separation between working, living and facility spaces;
- minimise urban private transport;
- favour the use of new information and telecommunication technology;
- minimise urban waste and favour recycling;
- minimise urban energy use (e.g. combined heat and power systems, district heating etc.).

Such principles are comparable to those formulated in the so-called 'Gaia' concept (see Lovelock, 1979). The fulfilment of such principles will of course require an effective multifaceted urban policy, which covers a broad range of aspects of current city life.

An interesting illustration of concrete attempts at achieving sustainable cities can be found in the Danish 'Green Municipality Project' in which various cities in Denmark work together, with the aim of generating awareness and policy actions at the local level, in order to pave the road for economically and ecologically responsible development of cities. Various pilot projects have been initiated in the meantime, focusing attention on life styles in the city, health care, education/information, landscape, clean technology, water management, energy policy, transport and built environment.

Despite the above and other initiatives, it has become clear that the road towards sustainable urban development is paved with stumbling blocks! A few caveats for achieving an operational integrated policy for sustainable cities will be mentioned here.

- the profile of a sustainable city is not unambiguous and appears to generate a lot of debate;
- the many very different policy sectors need to be considered in a sustainable city policy, for example, industry, transport, energy and recreation;
- the measurement of indicators and identification of bottlenecks have seldom been translated into sustainability policy measures and their implementation;

- changes in urban land use involving substitution between different activities (e.g. parking place into office space) provoke much discussion on the trade-offs between the socio-economic and environmental implications and evaluation of such changes;
- financial budgets of cities impose severe constraints on the flexibility and feasibility of new urban environmental plans;
- small-scale improvements in the direct living environment of urban inhabitants are often much higher valued than strategic urban development plans;
- transport in the city appears to lead to many externalities which are, however, extremely hard to cope with;
- an integrated urban environmental policy (the so-called 'chain' concept, going from the source to the final link) has not yet become a widely accepted idea.

The previous discussion indicates that information on urban sustainability is of critical importance for a balanced urban policy.

27.3 The City: A Complex System

Urban (or metropolitan) regions are essentially large production and information processing systems, encompassing the core economic activities of a country and acting as the focal point of an interurban, regional or national network. Their evolutionary paths do not only reflect stages of fast growth and stagnant maturity, but also obsolescence and decline.

Currently, we face a major change from an economic system based on mass manufacturing of uniform products, division of labour and hierarchical control, to a system based on flexible modes of production, leading to a network economy. This transition includes competition between 'old' locations of production based on local availability of natural resources, access to complementary resources and output markets, and 'new' locations. Regarding the new locations, the emergence of multilayer networks points to a large number of possible sites for local network interaction. What seems to be different in new locations is that they focus far more on accommodating and attracting creativity and knowledge, providing education in cognitive skills, through creative organisations and cultural facilities. Such cities or smaller towns also, almost invariably, provide modern communications, including high speed rail and large capacity telecommunication services.

The shift towards a network economy is only one change in a long sequence of major changes that have affected the location and evolutionary path of cities. A dynamic urban economic system striving for sustainability faces specific characteristics influencing its behaviour (Nijkamp, 1990a). Among those which appear to particularly affect the potential of the city-system, ensuring the maintenance or improvement of the technological and market position of economic actors in an increasingly competitive economy are:

- the carrying capacity which is limited. This is concerned with land and resources (physical and human);
- multifunctionality. This generates benefits to various activities within the urban territory (urban symbiosis). It strongly influences the incubation potential and underlying local learning processes;
- interaction or communication networks. Through these networks a city-system is linked to other cities or regions. This characteristic affects the potential for adoption of new technology from elsewhere, and the potential for development of new output markets.

Spatial (urban) dynamics are thus the result of internal and external responses by various actors (institutions) in an urban system. In this respect the self-organising capacity is of crucial importance. It provides, for example, consensus among the actors in the system, as well as a certain level of coordination between them. The self-organising capacity of an urban system is also closely related to permanent and dynamic learning processes based upon various local amenities (see Camagni, 1991).

It is commonly acknowledged that conventional city centres, but also other metropolitan areas tend to lose part of their innovation and incubation potential to peripheral areas (see Davelaar and Nijkamp, 1992, Van Geenhuizen and Nijkamp, 1993). Such an outward shift seems to be partially the result of the rise of (new) bottleneck factors in metropolitan areas. They also result from severe congestion in the urban traffic systems.

The previously mentioned characteristics of dynamic urban systems clearly offer various possibilities for a policy aimed at improving the competitive position. For example, one may focus on the technological production factors, by means of an emphasis on the inputs necessary to generate technology and economic change in the city (such as knowledge and capital). Similarly, urban policy can focus on the improvement of 'seedbed conditions' (e.g. the presence of university research facilities, the availability of cheap and flexible incubator premises), and on the improvement of conditions for networking (e.g. meeting points for key actors, the quality of the traffic and telecommunications infrastructure).

As previously indicated, there is a growing realisation that the sustainability debate needs to focus on urban areas. Large cities are the major consumers of natural resources and the major producers of pollution and waste. But it is important to recognise that cities are also the major sources of new technology and economic growth.

A common element of urban change processes is inertia (or lack of resilience) in adjustment mechanisms. This may lead to uncoordinated behaviour in various types of urban policy, such as infrastructure policy and industrialisation policy. For instance, the interaction between the production system and a given infrastructure system requires adjustments which are close to instantaneous. Changes in the capacity constraints and reallocations are however filtered through a time-consuming decision process. Hence, investment and relocation decisions are often delayed in relation to warning signals representing under and over-utilisation of existing amenities in the urban system.

Cities in our modern world are thus intrinsically dynamic systems which may exhibit various types of regular (i.e. linear) and irregular (i.e. nonlinear) evolutionary patterns. Apart from external shocks which may change the dynamics of the urban economy, the city also has endogenous growth caused by innovative behaviour, competitive strategies, inter-urban migration, residential relocation and other key forces (see Nijkamp and Reggiani, 1994). Consequently, urban dynamics are not easy to predict, as the possibility of nonlinear evolutionary patterns precludes simple straightforward forecasts of the future. This implies that nonlinear dynamic models are appropriate tools to map the evolution of city systems, as the development pattern of cities may exhibit bifurcation phenomena, catastrophic events and even chaotic behaviour. Complexity does not mean that the development of the city is hard to understand on subjective grounds. On the contrary, complexity - in the sense of a nonlinear evolutionary development of a phenomenon - is an objective and researchable feature of a city, even though the empirical validation of complex urban models is far from easy.

The previous observations mean that strategic urban policy is a field fraught with uncertainties, as a result of (i) the semantic insufficiency of urban model specifications and (ii) the lack of evolutionary orientation of existing urban data bases. This will be further discussed in the next section.

27.4 Policy Analysis in the Complex City

The use of dynamic nonlinear models for depicting urban evolution offers many opportunities to grasp the driving forces - and impacts - of urban dynamics, in particular the less foreseeable consequences of various determinant parameters of urban dynamics. To a large extent, such complex urban models can also be used to make conditional forecasts of the future of cities, although such experiments are more in the nature of dynamic sensitivity analyses regarding key parameters than actual forecasts.

This means that the relevance of complex urban models for long range urban policies is fairly limited. As a result, there has been a shift away from the use of urban forecasting models to the design of urban scenario experiments.

Scenarios can be conceived of as hypothetical sequences of events within a particular time-perspective, based on explicit assumptions (see Masser, Sviden and Wegener, 1992). As a policy tool, they serve to give insight into alternative choices and potential impacts. Scenarios are different from forecasting methods in that they may include both quantitative statements (e.g. based on statistical extrapolation) and qualitative insights about the future. In addition, scenarios usually have a higher complexity by incorporating a large number of influences, leading to various (contrasting) potential developments. The marked differences between forecasting models and scenarios analysis (see Zwier *et al.*, 1995) are shown in Table 1.

Scenario writing involves the complex task of creating, registering, discussing, synthesising, storing and presenting information on future development processes. One of the ingredients which may be necessary within a multidisciplinary approach is expert opinion. The use of expert opinion is increasingly popular in the design and testing of scenarios. For example, it may be used to bring about the convergence of a large number of diverging scenarios. An illustration of the use of expert panels for generating and synthesising information and sustainable urban transport strategies can be found in Nijkamp, Rientstra and Vleugel (1996), where a so-called Spider approach has been developed to present expected and desirable urban development options.

It should be noted that scenario analyses can in principle be combined with nonlinear dynamic models in order to test the dynamic robustness of qualitative system's changes, since the interest is more in qualitative regime switches than in precise numerical predictions.

The use of scenario analysis to depict and to analyse dynamic uncertainty is also compatible with modern tools for urban policy analysis, notably MCA. This offers the possibility of including qualitative information of a structural nature, besides quantitative measurements. Especially in the area of programme evaluation, such methods may be helpful when the data base is more qualitative in nature, as discrete jumps in a spatial system can also be taken into consideration. As shown in Nijkamp Rietveld and Voogd (1991) and Nijkamp and Blaas (1994), scenarios offer a useful way to include qualitative shifts in the structure of a dynamic (urban) system in policy evaluation methods. Also fuzzy information and rough set data, as well as risk statements based on value function approaches can be incorporated. This means that the relationship between information (un)certainty and policy evaluation deserves more attention. This will be further discussed in Section 5.

Forecasting models	Scenarios
- Focus on quantified variables	- Focus on qualitative pictures
- More emphasis on detail	- More emphasis on trends
- Results determined by status quo	- Results determined by future images
- From present to future	- From future to present
- Deterministic analysis	- Creative thinking
- Closed future	- Open future
- Statistical-econometric tests	- Plausible reasoning
- From simple to complex	- From complex to simple
- From quantitative to qualitative	- From qualitative to quantitative

Table 1 Differences between forecasting models and scenario analysis

27.5 Spatial Information and Decision Support in the Complex City

It goes without saying that in view of the uncertainty of urban evolution, the well-known statement that 'information is power' applies; in other words, urban information is urban policy power. Therefore, it is no surprise that in recent years information systems and policy analysis have become twins (see Nijkamp and Scholten, 1993). The main question is whether the available information is of a sufficient quality on which to base policy decisions.

Strategic urban and regional planning (including land-use and transportation systems planning) requires the design and use of tailor-made information systems. Such spatial information systems (incorporating also data management systems and simulation models) may become effective planning tools if the following conditions are satisfied:

- the presence of a flexible and user-friendly urban information systems model for strategic urban process planning and policy making, which is fairly robust under various development conditions of the spatial system concerned;
- an integration of recent software advances, notably in the field of geographical information system (GIS), with modern tools for evaluating urban and regional development projects or programmes;
- the provision of a flexible decision support framework (including monitoring mechanisms, early warning systems, cost-benefit analysis, multicriteria analyses and simulation experiments) which is able to assist urban planners and decision-makers in making strategic choices in an uncertain environment.

In recent years, elements of this approach have been developed separately for various purposes (for instance, decision support and expert systems for urban and environmental planning, GIS for spatial reference), but a uniform, integrated and coherent information systems model framework suitable for pro-active urban planning and policy making is lacking. The development of GIS can stimulate advanced information systems technologies by incorporating sophisticated planning tools (software, models) in strategic phases of planning (see Fedra and Reitsma, 1990, Nijkamp, 1990b, Scholten and Stillwell, 1990). The need for a co-evolutionary pathway between technical methods (including GIS) and planning is becoming increasingly evident.

Information has undoubtedly become a key variable in urban planning. In the postwar period, many countries have experienced an information explosion. The introduction of computers, micro-electronic equipment and telecommunication services have paved the way for such an explosion, not only for scientific research, but also for the transfer of information to a broader public and for urban planning or policy purposes. Several reasons may explain this information explosion in urban planning and policy making (Nijkamp and Rietveld, 1983):

- our complex society needs insights into the mechanisms and structures

determining intertwined socio-economic, spatial and environmental processes in the city;
- the high risks and costs of wrong decisions in public policy require a careful judgement of all alternative courses of action for the city;
- scientific progress in statistical and econometric modelling has led to a clear need for more adequate urban data and information monitoring;
- modern computer software and hardware facilities (e.g. decision support systems) have provided the conditions for a quick and flexible treatment of data regarding all aspects of policy analysis in the city;
- many statistical offices have produced a great deal of data in electronic form which can be usefully included in appropriate information systems.

It is therefore essential to integrate the evaluation software with the monitoring and modelling instruments into tailor-made decision support models or expert systems. An expert system is defined here as a computerised system which utilises specialised knowledge about a particular application area to help decision-makers solve ill-structured problems. As such, expert systems can be considered as a subset of conventional decision support systems. This subset is characterised by the fact that it concerns knowledge-based systems (see Leary, 1987, Sagalowicz, 1984).

Effective and accessible information systems are nowadays vital to economic performance and strategic decision making. According to a recent estimate, more than half of all jobs are already directly or indirectly related to the information and service sector. This figure is likely to grow in the near future (Naisbitt, 1984). It is increasingly believed that advanced infrastructures for information exchange and services (so-called 'electronic highways') will be as dominant in the last decade of the twentieth century as the waterways, rail and road transport infrastructures were in previous centuries.

The rapid development of digital and electronic technologies, for instance, in the form of digital recording and transmission of sound and pictures, optical fibres for the high speed transmission of information, super-fast computers, satellite broadcasting and video transmission, offers new potential for sophisticated voice, data and image transmission. Clearly, in this field the development of hardware and software has to run parallel in order to ensure maximum benefits for the city.

In conclusion, modern information systems have become an indispensable part of policy preparation in cities.

27.6 Geographic Information Systems

As mentioned above, in the past two decades a new tool in information handling, Geographic Information Systems (GIS), has emerged. These provide a special type of information, namely spatially referenced data. Spatial data sets are frequently heterogeneous and often comprise data sources with different scales, coordinate systems, accuracies, and areal coverage. Management and analysis of spatial data cannot be accomplished by simply applying typical data base management systems designed to handle only numerical and textual information. In the geographical field, the trend toward advanced information systems has led to the design and use of GIS. A GIS offers a coherent representation of a set of geographical units or objects which - besides their location - can be characterised by one or more attributes. Such information requires a consistent treatment of basic data, from their collection and storage to their manipulation and presentation.

Such information systems may be extremely important for the planning of scarce space, not only on a global scale (e.g. monitoring the development of rain-forest), but also on a local scale (e.g. physical planning). Within this framework, spatial information systems are increasingly based on topology and combined with pattern recognition, systems theory, statistics and finite element analyses.

Since the mid 1980s, major technological advances have affected costs, speed and data storage capacity of computer hardware in general, the advent of stand-alone workstations and PCs capable of running GIS applications, and also progress in the software, have made GIS technology easily affordable and accessible to local, regional and federal agencies as well as to small and medium sized companies in a variety of fields, such as environmental management, economic development, marketing and public service delivery.

Although GIS have reached a certain stage of maturity, current systems have realised only a modest portion of their potential. Despite widespread recognition that spatial analysis is central to the purpose of GIS, the lack of integration of spatial statistical procedures and spatial models is a major shortcoming (Goodchild, 1991, Clarke, 1990, Openshaw, 1990a, Birkin *et al.*, 1987).

The ability of GIS to store, handle and analyse spatial data is usually seen as the characteristic which distinguishes them from information systems developed to serve the needs of business data processing and computer aided design (CAD) systems or other systems whose primary objective is map production.

Important fields of current GIS applications are natural resource management (agriculture, forestry), land use planning, environmental assessment, environmental protection and economic development, urban and regional planning, public utilities planning and management (telephone, water, gas and electrical utilities), transportation planning, the siting of sports and health facilities, distribution planning and market analysis at different spatial scales. Each field is supported by specific disciplines. Almost any discipline using information with a spatial component is a potential candidate for GIS applications. Applications in the private sector (geo-marketing and locational analysis) are increasing in importance.

Spatial data sets provide usually two types of information:

- data describing the locational positions of objects in spatial or space-time systems (so-called positional data or spatial attributes eventually incorporated in topological systems or relationships);
- data describing nonspatial attributes of the objects recorded (so-called attribute or thematic data).

Spatial referencing can take several forms. The reference may locate a single data point to an exact location in a spatial system (e.g. the position of a house). It can be a set of references locating a more complex entity in space, e.g. the route of a road, or it can be a reference to an area (e.g. an administrative area). The primitive elements of spatial data are referred to as objects. Point, line and area are primitives. The elements of the object classes are characterised by attributes (nonspatial information). Object pairs (dyads of objects) combine any type and are especially important in spatial analysis. In this work the term GIS is used as a generic term for the whole field of spatially referenced data systems, including land use information systems (LIS).

To cope with the above mentioned features of spatial data, the database management systems (DBMS) used in a GIS may thus include various types of data, i.e. numerical data and nonnumerical data, as well as regional and nonregional data. A DBMS should namely fulfil certain practical requirements. The following are especially relevant for the user:

- to store spatial attributes (locational data), specify their types (point, line, area or combination) and coordinate systems;
- to link layers to spatial attributes and specify the main layer defining attributes along with its range of values;

- to store nonspatial attributes and relate them to spatial attributes;
- to represent complex objects (i.e. objects characterised by spatial and nonspatial attributes from several layers);
- to enable the exchange of complete geographic data sets;
- to support or to be suitable to support both raster and vector data;
- to specify the topological relationships between objects;
- to support spatial analysis operators;
- to provide spatial index structures for handling point data in any dimension, polyline and polygonal data structure;
- to allow the implementation of an advanced graphic user-interface.

There are different ways in which spatial data is stored for data processing. These may vary considerably from one GIS to another and may be quite significant in terms of specific technical considerations. A general way is to organise geographic data as a hierarchy of different components: data layers (also termed overlays) representing different themes, resolutions, orientations, zones, attribute values, locations and coordinates. A GIS may be therefore be seen as a computer-based information system which attempts to capture, store, manipulate, analyse and display spatially referenced and associated tabular attribute data for solving complex research, planning and management problems. The system may embody:

- a **database of spatially referenced data** consisting of locational and associated tabular attribute data;
- appropriate **software components** encompassing procedures for the interrelated transactions from input via storage and retrieval, and linking manipulation and spatial analysis facilities to output (including specialised algorithms for spatial analysis and specialised computer languages for making spatial queries);
- associated **hardware components** including high-resolution graphic displays, large-capacity electronic storage devices which are organised and interfaced in an efficient and effective manner to allow rapid data storage, retrieval and management capabilities and to facilitate the analysis.

Efficiency in this context means that the system uses the minimum number of information resources to achieve the output level the system's users require. Effectiveness means that the system performs the intended function and that the users get the information needed in the right form and in a timely fashion.

The function of an information system is to assist a user in solving complex research, planning and management problems and thus to improve the user's ability to evaluate policy issues, to compare policy alternatives and hence to facilitate decision-making. Data processing in an information system may be viewed as a number of interrelated transactions from input, storage and retrieval, to the data manipulation and analysis. A GIS may be considered as a subsystem of an information system which itself has five major component subsystems including:

- data input processing;
- data storage, retrieval and database management;
- data manipulation and analysis;
- display and product generation;
- a user interface.

Data input covers all aspects of the transformation of spatial and nonspatial information from printing or digital files into a GIS database. To capture spatially referenced data effectively, GIS should be able to provide alternative methods of data entry. These usually include digitising (both manual and automatic), satellite images, scanning and keyboard entry. The data may come from many sources such as existing analogue maps, air photography, remote sensing (data from satellites and from sensors from other platforms), existing digital data sets (for example, from other GIS), surveys and other information systems. Often prior to encoding, the data requires manual or automated processing, including format conversion, data reduction and generalisation of data, error detection and editing, merging of points into lines, edge matching, rectification and registration, and interpolation.

Database management functions control the creation of, and access to, the database itself. For the storage, integration and manipulation of large volumes of different data types at a variety of spatial scales and levels of resolution, a GIS has to provide the facilities available within a DBMS. Most commercial GIS (such as, for example ARC/INFO) have a dual architecture. The nonspatial attribute information is stored in a relational database management system and the spatial information in a separate subsystem which enables the user to deal with spatial data and spatial queries. Such an architecture however, reduces the performance, because objects have to be retrieved and compiled from components stored in the two subsystems. This problem is not easy to solve. Spatial data processing is performed with vector or raster data, or a combination of these two formats. Vector representation allows more object-oriented manipulation, but the raster data also has advantages, especially in terms of efficiency and data preparation. Rapid, precise and accurate data integration is a key issue in successfully merging these geometric data structures within GIS.

The most important distinguishing feature of GIS over a mere computer mapping system is the ability to **manipulate and analyse spatial data**. The manipulation and analysis procedures usually integrated in a GIS are often limited, however, to simple spatial operations, such as:

- geometric calculation operators such as distance, length, perimeter, area, closest intersection and union;
- topological operators such as neighbourhood, next link in a polyline network, left and right polygons of a polyline, start and end nodes of polylines;
- spatial comparison operators such as 'intersects, 'inside', 'larger than', 'outside', 'neighbour of', etc.;
- multilayer spatial overlay involving the integration of nodal, linear and polygone layers;
- restricted forms of network analysis.

Product generation is the phase where final products from the geographical information system are created. The displays and products may take various forms such as statistical reports, maps and graphics of various kinds, depending upon the characteristics and on the media chosen. These include video screens for an animated time-sequence of displays similar to a movie, laser printers, ink jet and electrostatic plotters, colour film recorders, micro film devices and photographic media.

The final module of a geographic information system consists of software capabilities which simplify and organise the interaction between the user and the GIS software via, for example, menu-driven command systems.

In the analytical functions of a GIS, the combination and selection of information on the basis of the geographical component is central. From the point of view of model construction and, more generally, from that of quantitative-empirical methods, the specific methods of GIS analysis supplement existing analytical techniques. They are however, by no means a replacement of such methods. Models are stylised representations of reality, based on the logic of mathematics. GIS offers also a stylised picture of reality, using the fruits of modern computer technology and cartography and, above all, the integration of spatial and attribute data.

The central role of GIS in modern urban and spatial planning can be depicted by means of visual representation taken from Grothe, Scholten and Van der Beek (1994) (see Fig. 1).

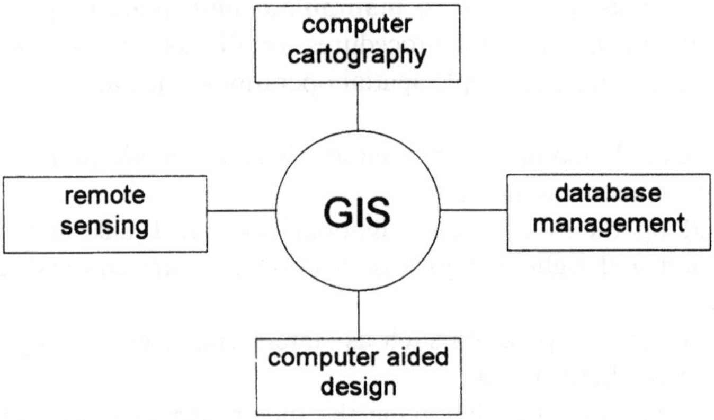

Fig. 1 The central role of GIS in spatial planning

Birkin *et al.* (1987) wrote of the need for a marriage between the model-based methods and GIS techniques to provide adequate tools to assist decision-makers, e.g. techniques for data transformation, the synthesis and integration of data, updating of information, forecasting, impact analysis and optimisation. They conclude that the two approaches have barely come together because of different historical traditions and research foci.

Most current geographic information systems are strong in the domains of data storage and retrieval, and graphic display, but their capacity for more sophisticated forms of spatial analysis and decision-making is still limited. This lack of analytical and modelling ability is widely recognised as a major deficiency (see Goodchild, 1991, Clarke, 1990, Openshaw, 1990a, 1990b). As mentioned above, the analytical possibilities usually refer to polygon overlay with logical operations, buffering in vector maps, interpolations, zoning, and simplified network analysis. The capacity to deal with location-allocation problems, optimal land use allocation and management, e.g. routing vehicles for delivery of goods and services, for example, is currently limited to simplistic types of analysis. Therefore, in the next section we pay some attention to spatial analysis.

27.7 Geographical Information Systems and Spatial Analysis

Spatial analysis - which in its widest sense can be considered to be the analysis of spatial and nonspatial information concerning phenomena in spatial or space-time systems as an aid to their description, explanation and prediction, offers a wide range of methodologies and procedures relevant to GIS research. Spatial analysis is more general than statistical analysis of nonspatial information, because it requires access not only to attributes, but also to locational and topological information.

From a technical point of view, we can consider four ways to link GIS and spatial analysis (see Openshaw, 1990b, Goodchild, 1987, 1991):

- to develop generic spatial analysis tools to be integrated into appropriate GIS as standard operators (full integration of spatial analysis tools into GIS technology);
- to write user-friendly interfaces to special statistical software packages for spatial data (loose coupling of GIS and spatial analysis) made necessary by the lack of adequate spatial analysis technology in statistical software packages;
- to develop a basic spatial tool box for inclusion in standard statistical packages (SPSS-X, SAS, GLIM, GAUSS or RATS) to provide a set of independent portable spatial analysis macros which can co-exist with a targeted GIS (close coupling of GIS and spatial analysis);
- to embed GIS procedures within spatial analysis or modelling frameworks which attempt to exploit the unique capabilities of GIS technology to devise new and more relevant analytical procedures (full integration of GIS procedures into spatial analysis and modelling frameworks).

The first strategy involves deciding which spatial analysis operations should be included and assumes there are functions which are on a sufficient scale, general and generic, to justify the efforts (Openshaw, 1990a). Given the tremendous efforts needed and the nature of the GIS software industry, this strategy seems unlikely to be realised in the near future. The second strategy is relatively simple because the availability of export facilities in GIS makes data transfer easy, but there is a lack of GIS-relevant standard spatial statistical packages. Concerning the third - most likely - strategy, the unresolved problem is the nature of the user-friendly interface for the linkage. Possibilities range from source code through subroutine libraries, high-level languages, commands and menus to virtual reality. The final strategy represents the opposite of the first. Here GIS technology is embedded within spatial analysis and modelling frameworks by utilising GIS databases and access at the subroutine level.

Advanced spatial analysis modules may be critical elements for the next generation of more intelligent geographic information systems based on principles of artificial intelligence.

From the above, it can be concluded that not all users wish or need to make similar use of the main GIS functions (Table 2). It is clear that the further development of GIS should take advantage of this differentiated demand. It is also clear that potential users should specify more precisely which type of GIS best serves their interest.

On the basis of what has been established above, a multitrack development in GIS seems to offer the best solution for optimum use. Consequently, the most important directions of development are an 'integrated environment', a user interface to overcome the different command languages of the various packages in use, advanced interfacing with appropriate spatial analytical procedures, the development of a 'small and beautiful' tool for decision-makers, and a highly accessible information and query system for the public.

Type of user	*Information demand*	*User demand*	*Type of GIS*	*Development*
Information specialist	Raw data	Analysis Flexibility	Large Flexible	Links to other packages
Preparer	Raw data and Macrolanguages	Analysis	Compact	
of policy	pre-treated data (=information)	Good accessibility	Manageable	Interfaces to other packages
Policy decision-maker	Strategic information	Good accessibility to users Weighting and optimisation models	'Small and beautiful'	User-friendly interface Key information
Interested citizen	Information	Good accessibility to users	'Small and beautiful'	User-friendly interface

Table 2 Types of user demand for a geographical information system (GIS)

27.8 The Use of Urban Information Systems

Urban information systems - and hence also GIS - are tools for improving the quality of urban planning and decision-making. The requirements of such systems depend on the specific nature of each planning problem, but various generic criteria can be distinguished (see also Fig. 2). The axes of the octagon measure the intensity of such criteria. The outer extent of the envelope reflects the highest possible value, in reality an urban information system will take an intermediate position between the envelope and the origin, as is suggested by the dashed line.

Systematic and coherent insight into the complex pattern and evolution of an urban system requires the design of an up-to-date, accessible and comprehensive urban information system. Information systems for urban (and regional) planning should contain structured data on the development of real-world patterns, their properties (stability, for example) and their links. Frequently, however, the use of information systems as decision aid tools in urban and regional development planning has been neglected. Far more attention needs to be paid to the design and development of information systems reflecting socio-economic processes, so as to achieve a better representation of urban systems and a better adaptation to the needs of planners.

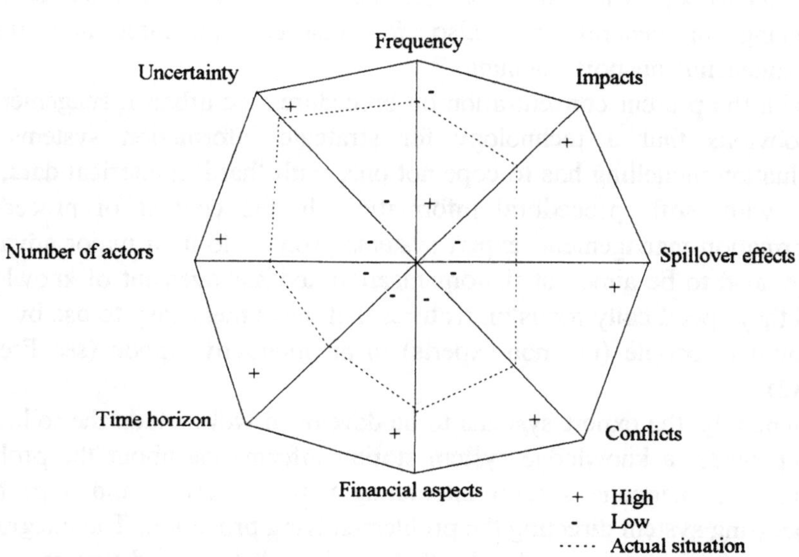

Fig. 2 The demands on information systems caused by the nature of choice problems

Information systems have generally been designed for specific purposes, e.g. transport data for transport policy, housing data for housing market policy. Clearly, there are various reasons (institutional and technical) why their integration is hard to achieve in current planning practice. On the other hand, a lack of integration means an enormous waste of effort. It would already be a significant step forward if national or regional statistical bureaux were authorised to provide uniform rules for data collection and standard classifications of urban socio-economic activities, even if these would concern information systems beyond their own responsibility. Experience in technical and medical sciences has demonstrated the power of uniform rules and classifications, and there is no logical reason to prevent the creation of a standard frame of reference for the design of spatially-oriented information systems. This would also increase the polyvalence of such systems and provide a more appropriate basis for coherent urban sustainability policies.

In the current era of political, social and economic restructuring and technological progress, cities and regions are increasingly facing structural changes which have many major impacts, e.g. on the urban housing market and the urban transport system. In most cases, an integrated view of such developments is missing because of the lack of data, inappropriate information systems or partial modelling efforts. The lack of such tools also precludes pro-active and strategic urban planning, which is necessary in order to cope with these changes. This holds for urban (and regional) planning in general, but also for land-use planning and traffic management/transport planning.

With the present concentration on procedures for urban management, it is obvious that a technology for strategic information systems for evaluation modelling has to cope not only with 'hard' numerical data, but also with 'soft' procedural information. In the context of procedural information management, expert systems promise to be a major advance. They tend to be aimed at the organisation and management of knowledge and they specifically focus on features that make them easy to use by 'non computer' people (i.e. non-experts) in an interactive mode (see Freska, 1982).

Generally, the expert systems to be developed will include the following components: a knowledge system storing information about the problem domain, a language system interacting with the users, and a problem processing system directing the problem-solving processes. The integration of these three systems is also denoted as a knowledge-based system.

Frequently, information systems for urban and regional planning have been developed in close connection with multiregional models. Such models are an extension of traditional econometric modelling and aim at providing consistent and coherent information on a complex spatial world, so as to identify the main driving forces and the mechanisms of a complicated multiregional system (see Issaev *et al.*, 1982). The aim of coherence and consistency will, in general, lead to a rejection of economic models that do not take into account the openness of a region or city. Thus, without consideration of interregional and national-regional links, there is no guarantee of consistency for the spatial system as a whole. Usually, there are various kinds of direct and indirect cross-regional linkages caused by spatio-temporal feedback and contiguity effects, so that regional developments may have a nation-wide effect. National or even international developments may also exert significant impacts on a spatial system. This is especially important because such developments may affect the competitive power of regions and cities in a spatial system; a national policy on innovation, for instance, may favour areas with large agglomerations. The diversity in an open spatial economic system requires coordination of information activities at all (national, regional and urban) levels.

The previous discussion has shown that the explosion of available information in combination with the modern computer hardware and software facilities provides many possibilities for coupling different information systems. Nevertheless, the actual performance of informatics in urban planning is not always very impressive. In this respect, urban (and regional) planning agencies might learn useful lessons from multiplant and multiregional corporate organisations, which have generally been better able to solve the organisational problems of dealing with very large diversified databases. In this regard, the organisational aspects of internal communication within urban planning agencies deserve much more attention, especially during the implementation stage of urban plans. In the next section, we will focus on the evolutionary aspects of information in a complex city.

27.9 Management Models for Complex Cities

The management of complex cities includes many different policy (planning) activities and information needs, marked by uncertainty and a long-range horizon. By considering conventional policy cycles, we can distinguish several stages:

- problem diagnosis (including causal analysis);
- policy-making (including impact analysis and scenario design);
- resource programming and allocation;
- communication, co-ordination and participation;
- control of implementation;
- monitoring and evaluation (*ex-post*).

In Fig. 3 the successive management activities are linked with various planning outputs into a process including a sequence of different loops. These planning outputs are concerned with different spatial levels within cities (for example, ranging from the street level to the urban level) and different time horizons, i.e. long range policy options, middle range strategic programmes and short range action projects.

In a similar vein, various 'domains' of planning output, such as normative policy options, strategic programmes, public intervention and action projects are linked up with various clusters of problem areas and policy issues within an urban sustainability framework. This leads to the creation of a sequence of learning loops, including processes of transformation, new trajectories and new scopes (Fig. 4).

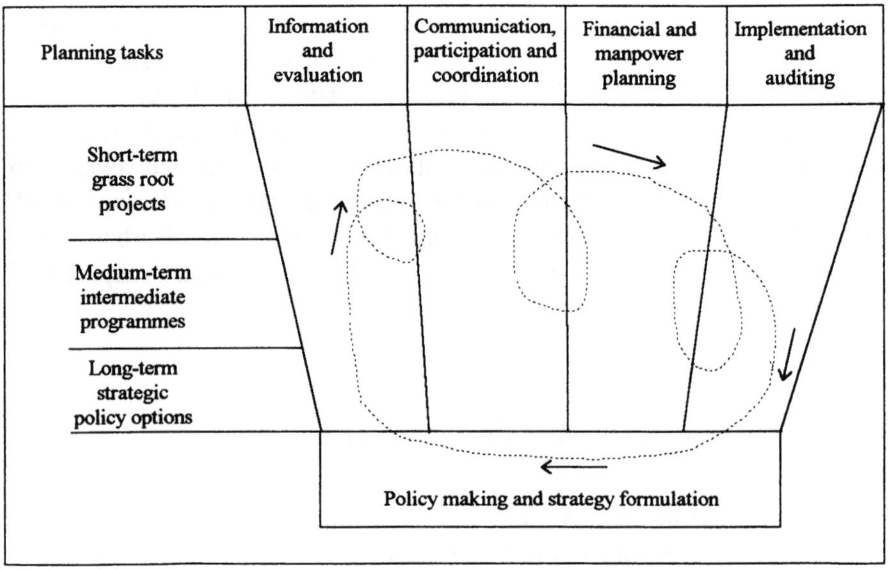

Fig. 3 Urban management activities and multilevel planning outputs

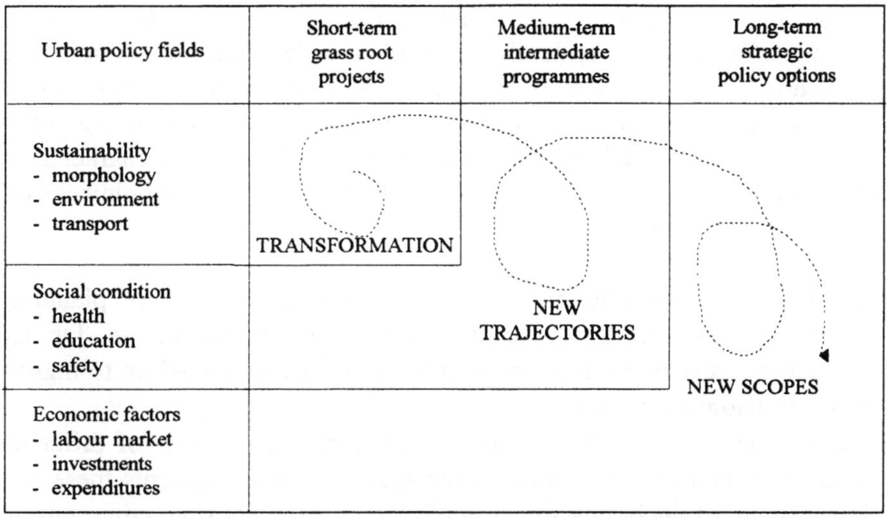

Fig. 4 Sequential learning and decisions loops in urban management

It stands to reason that learning and decision loops in city management involving multilevel, multifacet and multiperiod features, have a comprehensive need for information.

27.10 Design of an Urban Information System Architecture

From the discussion in Section 9, it has become clear that the type, amount and degree of precision of information depends on the specific nature of a given urban policy question. In the establishment of an urban information system two aspects can be distinguished, i.e. the **architecture** of the database and the **organisation** of the system. The architecture of the database is concerned with the type and structure of the data, whereas the organisation of the system includes the actual data input, data processing and control of the database. In the latter, various different actors are involved, such as the controlling institution, data suppliers and data users.

Spatial information systems have certain characteristics which deserve close attention before a design can be developed. They have to support decisions which are closely connected with the various fields incorporated in the urban system and connected with planning methodology. Regarding the former, urban planning entails **multiple components**, e.g. fields such as housing, employment and industry. As regards the latter, present planning conditions are typically based upon a decentralised (bottom-up) model and a large amount of flexibility. In addition, urban planning is **multilevel** in nature because it includes various spatial scales, e.g. sub-local, local and regional. This multicomponent and multilevel character of urban information systems can clearly be mirrored in the eight dimensions discussed below (see also Fig. 2):

- the degree of **centralisation**: to take account of the shift of planning towards decentralisation, the information system should as far as possible be developed from the lowest major planning level up to higher levels **(bottom-up** approach);
- the level of **comprehensiveness**: this concerns the number of different core components (or modules) in the system. In most systems these are fairly standard, including for example, population, housing, environment and economic sectors. To deal with specific planning problems (such as traffic congestion) or opportunities (such as innovative potential) specific additional information can be inserted;
- the degree of **integration** between the various components; this concerns the production of synchronised and standardised information on the various planning fields at each planning level. A major condition is the use of **uniform definitions** of planning subjects (entities and their attributes) over time and space;
- the degree of **local detail**: this varies with the specific planning field (population, economic sectors), as well as with the spatial level of data collection. For example, a high degree of detail is necessary for local economic planning. The lower the level of detail, however, the higher the chance that confidentiality issues on information arise;
- the definition of the **city** or **city-region**: the city is preferably defined in a functional sense, including the daily urban system or urban labour market area. However, spatial aggregation and disaggregation is necessary in a flexible way, for example, for reasons of comparative research (national, international) and for reasons of focusing on specific local issues. Accordingly, a uniform definition of the basic (lowest level) spatial unit is important;
- the time-dimension: this dimension concerns the length of the retrospective observation period and the frequency of updating. The length of the observation period may be different for the various planning fields. Certain demographic projections can only be produced by means of data on a sufficient number of past years (e.g. from 1980). Data

including a time-dimension may be longitudinal in nature, which means that the same objects or panel (persons, enterprises) are measured at various points in time or continuously. The alternative, which has a number of disadvantages, is the transversal method, in which different objects (e.g. in various samples) are measured at a small number of points in time. In general, annual updating is preferable. This does not imply that the full detail should always be used in projections or modelling, but rather that lack of detail is not a restriction;

- the statistical coverage: this concerns the methods of data collecting, for example, censuses, panel surveys, sample surveys and case studies. In this respect, there is a trade-off between the accuracy and precision needed in projections, modelling, scenario analysis, etc. and the availability and costs of data collection;
- the extent of computerisation: a high level of computerisation is particularly relevant concerning the input and output of data. This ideally involves on-line connections to the administrative databases for the major users. Strong computerisation enables a high degree of flexibility in planning, as the monitoring of developments and evaluation of urban policy is possible.

The supply of information does not usually completely satisfy the need for information within a planning framework of complex cities. Various problems may emerge concerning the **matching** between the demand and supply of information, namely:

- under-information: a level of information which is too low in relation to complex urban policy questions;
- over-information: a level of information which is too high in relation to structured decisions;
- misinformation: a qualitative discrepancy between demand and supply of information;
- incoherent information: too little co-ordination between information relating to multiple sectors, levels or time periods (e.g. mismatch of definitions of planning subjects; mismatch of levels of data aggregation);
- uncertainty concerning information of a stochastic or fuzzy nature;
- non-updated information: the necessary information is only available for the past;
- scale discrepancy: the necessary information is not available on the appropriate spatial scale;
- non-accessible information: the information does exist, but is not accessible due to secrecy, confidentiality or high cost;

Fine-tuning between supply and demand of information is therefore necessary. We discuss this further with a specific view on the information needs for planning and managing complex cities.

First, we focus on the architecture of an urban information system. The major types of datasets are discussed and the rationale behind their selection is explained. Particular attention is given to the relation of the datasets with the planning fields, the planning activity (e.g. analysis of past trends, project appraisal) and the data input into the system. Then, a systematic description of the headings and substance of various important examples of datasets in a mature urban information system will be given. This section will be concluded with an inventory of critical factors that influence the success of an urban information system.

The **architecture** of urban information systems is based on the following aspects of management and planning:

- the type of field, reflecting the priority given in policy (major, medium, minor);
- the approach of analysis of planning fields, i.e. quantitative (forecasts) and qualitative (scenarios);
- the planning activity, e.g. forecasting trends, project appraisal.

As far as the first is concerned, the major priorities are usually standard, e.g. population, housing, employment, and the environment. However, various other fields may receive high priority if identified by a specific diagnosis on socio-economic problems in the planning area. In general, data on planning fields with a relatively low to medium priority are optional in an urban information system.

By taking the above aspects of urban planning into account, three datasets of an urban information system can be distinguished in a so-called CBD structure (Fig. 5):

- Core data (C)
- Basic data (B)
- Distinct project data (D).

The aim of the core data is to provide a reliable statistical representation for the analysis and forecasting of the major planning fields in the region or city, namely socio-economic development and environmental development (including land use). The data includes the demography, housing and industrial sectors and constitutes the permanent core of the information model. Consequently, in this part of the information model, data input is on a continuous or very regular basis (yearly). As reliability of core data is very important, complete coverage of the relevant population is generally achieved through public records, such as the Municipal Population Register or Chamber of Commerce Registers.

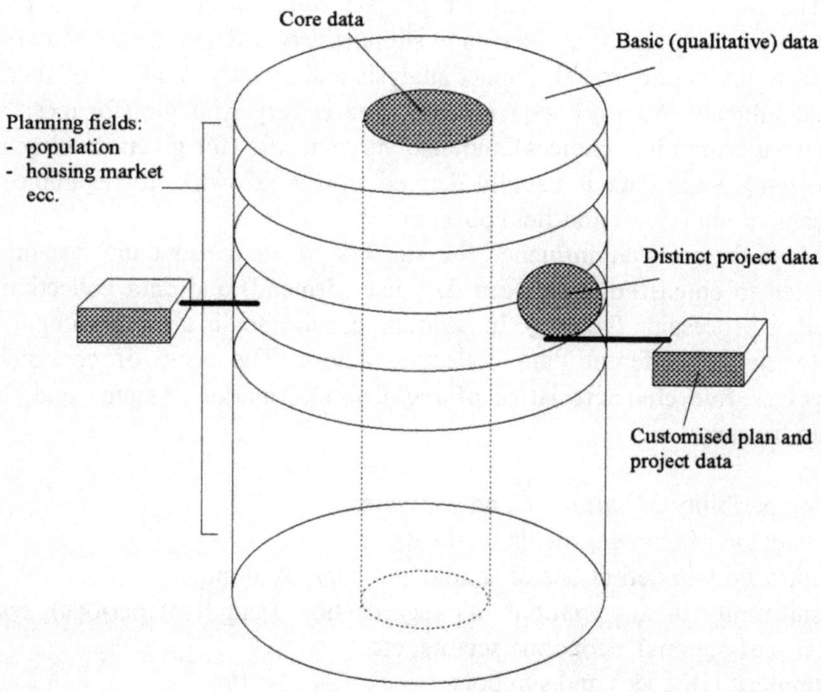

Fig. 5 Architecture of a general urban information system

Basic data is of two types. The first covers the same major planning fields as the core data, but the approach is different, i.e. it is **qualitative** in nature. A good example is information on the economic strength and weakness of cities in a competitive urban setting. The second type includes fields which fall outside the main focus of urban management and planning, but may contribute to an understanding of the phenomena described by core data. This data does not have immediate priority in analysis and forecasting, but may become of strategic importance under certain conditions. A good example is data on the finance of public sector bodies, such as income from local taxes. Such data is **optional** in an urban information system, because not directly necessary or indispensable in management and planning activities.

The two types of basic data have different input characteristics. Qualitative data is usually collected for a short period for a given policy issue. Non-priority data is collected on a permanent or semi-permanent basis, but usually with a low frequency and with less detail.

The third set of data, **distinct project data**, usually concerns major urban planning fields, but serves to support very different types of activity, namely project appraisal, impact analysis and scenario analysis of specific field subjects. As a consequence, the data is very strategic (for meso and micro urban policy choices) and also very specific (for given development projects). Such data is usually derived from field work, for example, by means of interviews and field observation.

Most factors that influence the **success** of an urban data system are related to **consistency** between data use (demand) and data collection as well as processing (supply). In addition, consistency is also very important between the different data collecting actors. The issue of consistency involves five **characteristics** of regional information systems and their data, namely:

- compatibility of hardware and software;
- coverage of planning fields by the data;
- uniformity in definitions of spatial and nonspatial units;
- uniformity in aggregations, as regards time (length of periods), space (size of regions), economic sectors, etc.;
- similarity of codes and symbols.

The need for communication (transfer) between (information) systems of data producers, processors and users requires **compatibility of hardware and software**. A second issue is the **coverage of the planning field**. It is very important that the data covers the subjects in sufficient width and depth (detail). In addition, when the planning field is concerned with subjects which are 'latent', such as quality of life, the variables in the data base should be valid indicators. A further important aspect is the statistical validity of the data. This matters, for example, when sample surveys are used as data sources, and there is a large non-response or a high level of 'hidden' events (non-registration).

A third important characteristic of data concerns the **definition** of the spatial and nonspatial **units** and their **attributes**. Each relevant actor should ideally use the same definition of the labour force, unemployment and hazardous waste, for example. The definition issue applies also to cities or city-regions and time periods. The definition of cities (and spatial subdivisions) in the data system should correspond with the spatial units in planning. The data system should also contain a sufficient number of past years to enable forecasting or extrapolation of trends with regard to the period selected in the planning. In general, it is also advisable to collect data at the lowest level of **aggregation** possible. When aggregation is necessary, there should be consistency in the method used. This concerns particularly spatial levels and time periods, but also economic sectors. A further issue of consistency concerns the use of **codes** and **symbols** by the different data producers and users. A sufficient level of similarity in this use is necessary in order to transfer and integrate data from different systems in a smooth way.

To sum up, the success of data systems depends on:

- consistency in systems and data between different data collecting actors;
- consistency in systems and data between demand (data users) and supply (data collecting and processing actors).

The systematic use and design of general urban information systems may become a strategic tool in the management and planning of complex cities. In the next section we focus on specific urban data needed within the framework of measuring urban sustainability and implementing an urban policy for sustainable growth.

27.11 Information Systems for Urban Sustainability

Data on urban sustainability is needed to study the quality of the urban environment and to develop policies aimed at achieving sustainable cities. A suitable data system will provide indicators on the current state of the urban environment and changes occurring, the behaviour of the urban population and changes in these habits, as well as basic indicators that measure the ways in which cities cause environmental problems. In addition, it will provide data directly helpful in the design and implementation of an urban sustainability policy.

An urban sustainability data system can be viewed as a **specification** of the general urban information system outlined in the previous section. This section will focus on the structure of such an urban sustainability data system, as well as the criteria which are to be fulfilled in the selection of the indicators in question. The structure of an urban sustainability data system will follow that of the CBD structure of a general urban information system described above, containing:

- **core** indicators: these provide the minimal essential information to measure urban sustainability quantitatively;
- **basic** indicators: these support core indicators in a qualitative sense (e.g. in providing an explanation) or in a relatively low priority field;
- **distinct** project (area) indicators: these are concerned with major urban sustainability fields in terms of environmental impact assessment (EIA) and scenario analysis; they may also apply to particular areas where specific (non generic) sustainability problems or conditions are found.

Indicators of urban sustainability are still being developed, and are influenced by the current policy views and the definition of sustainability. For example, a major point of political divergence concerns the relationship between economic growth and sustainable development (see Haughton and Hunter, 1994). 'Deep ecology' writers stand at one extreme, adhering to the total incompatibility between continued high levels of economic growth and the advance of sustainable environment. At the other extreme are those who believe in the abundance of nature and the capacity of mankind to find new solutions to emerging environmental problems. In the latter view, economic growth is necessary for sustainability, as it alone can provide the financial and technological capacity required to solve environmental problems.

Urban sustainability can be conceived in various ways. In a narrow sense, the concept focuses merely on the state and change in the natural urban environment, whereas in a broad sense it also includes the well-being of the resident population, largely in connection with the social and economic system.

It should be emphasised that there is still a **lack of knowledge** regarding the systems and mechanisms in the natural environment and the human (behavioural) causes behind environmental problems. This situation inevitably hampers the design of suitable indicators. For example, little is known about threshold values (and tolerances) in natural systems, such as the (concentration) values above which the regenerative capacity of natural bodies will decrease. Similarly, it is still largely unknown which indicators in urban natural and social-economic systems can serve as key parameters, describing the qualitative condition of these systems.

Despite different viewpoints and shortage of knowledge, a set of **criteria** can be established for the design of urban sustainability indicators. The criteria concerned with the scope of the indicators and their policy function are as follows:

- covering the key areas of environmental sustainability, namely natural resources, natural capacity to absorb waste and pollution, human capacity to contribute to environmental sustainability, and human well-being in a social and economic sense (core indicators);
- directly measurable in an unambiguous way and leading to a consistent data system (core indicators);
- similar to (or at least compatible with) existing practice in European cities (core and basic indicators);
- providing detailed explanation of core indicator values (basic indicators);
- closely linked to continuous urban evaluation and monitoring systems (core and basic indicators);
- can form a basis for advanced urban policies and urban plans (long term strategic as well as short term detailed local plans) (core and basic indicators, distinct data);
- can be presented ('translated') in such a way that they can perform a public information function (core, basic and distinct data).

We list here some suggested core sustainability indicators.

Core sustainability indicators in an urban framework preferably encompass the following four classes of data:

1. Natural resources for production and consumption:
   - production from agricultural land and fishery waters in the urban area;
   - consumption of water and energy (including gasoline) by the resident population; consumption of raw material, energy and water by the urban industry.

2. Natural capacity to absorb and recycle pollution and waste within the urban area:
   - production of solid waste, air and water pollutants, hazardous waste and radiation by the resident population and local industry;
   - quality of urban air, climate, water bodies (rivers, lakes, coastal zones), soil and ground water;
   - vitality of urban green (parks, forests);

- number of waste disposal areas, amount of spots of (strongly) contaminated soil.

3. Human capacity to produce renewable and clean energy, save energy use, recycle and remove waste:
   - urban production of wind and solar energy, urban production of combined power-heat, integrated urban heating system;
   - urban capacity for recycling organic and nonorganic waste, for sewage treatment and waste incinerators;
   - number of companies credited to eco-management and auditing schemes (European Standard).

4. Human well-being in a social and economic sense:
   - 'quality of life' in the urban area as measured by statistics such as the number of people living below poverty line, number of crimes, number of car traffic accidents, size of unemployment, number of public facilities/services at certain distances, and as measured by surveys on satisfaction among the resident population;
   - urban social and economic 'vitality', measured by the number of urban social-cultural activities, number of voluntary organisations active in the urban area, number of newly established companies and their survival rate.

We shall now illustrate **basic** urban sustainability data by means of data on urban energy use for private transport.

From a policy point of view it is highly desirable that differences in fuel (petrol) consumption can be explained in terms of the pattern of the built environment, the functional structure of the city, and behavioural (lifestyle) patterns of the resident population. There is still need for major research in this field, particularly on the functional structure of cities and on the behavioural side - the relationship with urban density of population is quite clear (see Kenworthy and Newman, 1990). This type of supportive research needs to take into account the internal variation in density within cities and the functional interdependence between residential, working, service and recreational districts.

The following basic data may be required in order to find an explanation to different urban travel behaviour in terms of purpose, distance, frequency and mode:

- urban form and function: population density (overall, per quarter), housing density and housing type per quarter, indicator for functional homogeneity (or mix) per quarter and per town, in relation to neighbouring towns;

- traffic infrastructure: length of roads, cycling and walking lanes, public transport services (routes and service schedules), car parking facilities, road congestion;
- life-style of resident population: age and income, valuation of time, car ownership.

The above indicates that basic data can be derived from general urban core statistics (e.g. on urban form and traffic infrastructure) and from specific surveys on individual characteristics of the residential population, including activity registration in diaries.

Distinct data is closely related to the implementation of urban policy. One major use of distinct data is in **environmental impact assessment (EIA)** for a particular development project. EIA will be discussed in more detail in the remaining part of this section.

As a proactive policy tool, EIA provides a framework for the prior assessment of the potential impacts of urban (or other) policy development in such a way that adverse environmental effects can be eliminated or minimised before the development commences. Most recently, EIA in the developed world has been extended beyond the project level to the level of policies, plans and programmes. The use of EIA in these earlier stages of the planning process has been named strategic environmental assessment (Haughton and Hunter, 1994).

EIA has become an accepted part of planning procedures of many developed countries, its primary function being to include those impacts on the environment which cannot be measured in monetary terms. In addition, the formal adoption of EIA into environmental protection legislation has proceeded rapidly in developed countries in the past few years. In many developing countries, however, there remain serious obstacles to the operation of formal EIA, such as a lack of political will or awareness of the benefits of EIA, inadequate legislative frameworks, insufficient manpower and lack of environmental data.

In all policy and planning systems using EIA, **prior screening** of projects is required to determine whether these should be subject to a full EIA procedure. The formulation of criteria for such screening is rather difficult as it is necessary to strike a balance between ensuring environmental protection and the time and expense of imposing a full EIA procedure. Screening criteria may include the physical size (scale) of the project, the environmental characteristics of the area in question, and the physical and process characteristics of the project. Once projects have been selected, an EIA system and its written output (an environmental impact statement) ideally cover the following types of item:

- the main features of the project, including the residues and wastes that it is likely to create;
- aspects of the environment likely to be significantly affected by the project;
- likely significant effects of the proposed project on the environment;
- measures envisaged to reduce harmful effects (including alternative plans and the reasons for their rejection);
- assessment of the compatibility of the project with environmental regulation and land use plans;
- a nontechnical summary of the overall assessment.

It should be emphasised that the prior assessment of '**scoping**' of potential impacts enables the identification of major issues to be included in the EIA. This is of utmost importance as it directs the (often expensive) collection of data and determines the focus of analysis on those impacts considered to be of most concern. At the same time, EIA allows for a prioritisation in valuation of project effects in the final assessment. For example, it is possible to assign different **weights** in the final assessment to economic (cost-benefit) analysis, effects on the environment and on the safety of the resident population. In a correct EIA procedure, 'subjective' selection criteria and assignment of weights are often based on reports or expert consultation. The results are then made explicit in a written statement.
Aside from the above 'subjective' or 'selective' elements in the EIA procedure, the adequacy of EIA depends to a large degree on the **quality** of the information used in the processing of results on likely effects. The information includes ideally the following (see EC Directive on EIA, 1985):

- data on the project in terms of site, design and size, particularly its external effects on human beings, fauna and flora, soil, water, air, climate and the landscape, material assets and the built cultural heritage;
- data on the site (area) and the environment of implementation of the project, particularly the sensitivity of the environment for specific effects (baseline conditions);
- impact models, forecasting and evaluation techniques, regarding short and long term effects, primary and secondary effects;

- information on measures and alternatives that can prevent (or reduce) harmful effects;
- information on established environmental regulations and land use plans which may affect the project.

The EIA is subject to all the problems of **matching** between demand and supply of information already indicated in Section 9. Of particular importance is the lack of accuracy of impact predictions caused by shortcomings in the prediction models and techniques, and in data on environmental quality (Haughton and Hunter, 1994). The availability and quality of baseline information on the status of the environment is crucial for making accurate predictions of environmental change. It is therefore essential to have reliable and continuously updated environmental data on the urban area for the core database (as indicated earlier in this section). From this permanent dataset, information can then be made available to EIA practitioners and decision-makers in relation to specific projects and plans. In addition, **post-development monitoring** of actual environmental impacts is needed, in order to increase the understanding of environmental processes and tolerances.

Despite the above problems and inadequacies, EIA can play a substantial role in the decision-making and planning process in urban areas. Firstly, it contributes to the principle of 'prevention is better than cure' and secondly, it allows greater involvement of the local community in decision-making, as well as an increasing awareness of environmental sustainability issues among developers, local communities and urban policy-makers. From a scientific perspective, it allows us to answer the intriguing test question - based on empirical facts - whether sustainable urban policy decisions are sufficiently substantiated by the available information on urban developments.

27.12 Retrospect

It goes without saying that the phenomenon of a complex city needs a new spectrum of analytical approaches based on nonlinear dynamic modelling in order to trace the complex trajectories of urban development. Examples of such methodologies are niche models, ecologically-based models and chaos models. These models can help to map out the complex dynamics of urban systems. The main problem, however, concerns their empirical validity. There is therefore a strong need for well-designed information systems, a *sine qua non* for empirical application. This work has demonstrated the potential offered by modern information systems in a spatial setting.

Information plays a central role in managing and studying the contemporary complex city. One may even argue that, without a sufficient data base with a long time series, it is not even possible to speak of a complex city, as the test of nonlinear evolution would be absent. Consequently, an information base has two functions: it permits tests on the existence of hard-to-predict nonlinear dynamics (including chaos and bifurcations), and also on the justification of a given policy decision (irrespective of the political choices involved).

Modern cities represent a multifaceted and dynamic phenomenon. Planning for sustainability requires analytical insight into the driving forces of urban dynamics and into viable - often nonlinear - trajectories for the city's future. In view of possible unstable dynamics, scenario experiments based on backcasting may be designed. There is also a need to depict the current reality of the city by means of a modular urban information system which combines traditional quantitative statistics with flexible qualitative descriptions of the key components of the urban system. It is suggested that such information systems should be structured according to the CBD model (Core - Basic - Distinct data) to allow integrated planning of the city. The approach sketched in this study attempts to pave the way towards a sustainable urban future, by outlining the design of information systems which are sufficiently flexible with regard to nonlinear evolutionary phenomena that are an intrinsic feature of the complex city.

References

Batty M. (1994) A Chronicle of Scientific Planning, *Journal of the American Planning Association, 60,* 1, 7-16.

Birkin M., Clarke G.P., Clarke M., Wilson A.G. (1987) Geographical Information Systems and Model-Based Location Analysis; Ships in the Night or the Beginnings of a Relationship?, WP-498, School of Geography, University of Leeds, Leeds.

Breheny M. (ed.) *Sustainable Development and Urban Form*, Pion, London.

Camagni R. (1991) *Economia urbana*, NIS, Rome.

Clarke M. (1990) Geographical Information Systems and Model Based Analysis: Towards Effective Decision Support Systems, in Scholten H.J., Stillwell J.C. (eds.) *Geographical Information Systems for Urban and Regional Planning,* Kluwer, Dordrecht, 165-175.

Davelaar E.J., Nijkamp P. (1992) Operational Models on Industrial Innovation and

Spatial Development, *Journal of Scientific and Industrial Research, 51*, 273-284.
Fedra K., Reitsma R.F. (1990) Decision Support and Geographical Information System, in Scholten H.J., Stillwell J.C. (eds.) *Geographical Information Systems for Urban and Regional Planning*, Kluwer, Dordrecht, 177-188.
Fokkema T., Nijkamp P. (1994) The Changing Role of Governments, *International Journal of Transport Economics, 21*, 127-145.
Freska C. (1982) Linguistic Description of Human Judgements in Expert Systems are in the "Soft" Sciences, in Gupta M.M., Sanchez E. (eds.) *Approximate Reasoning in Decision Analysis*, North Holland, Amsterdam, 297-305.
Fukuyama F. (1992) *The End of History and the Last Man*, Hamilton, London.
Goodchild M F. (1987) A Spatial Analytical Perspective on Geographical Information Systems, *International Journal of Geographical Information Systems, 1*, 327-334.
Goodchild M.F. (1991) Progress on the GIS Research Agenda, in Harts J., Ottens H.F.L., Scholten H.J. (eds.) *EGIS 91, Proceedings Second European Conference on Geographical Information Systems, 1*, EGIS Foundation, Utrecht, 324-350.
Grothe M., Scholten H., Van der Beek M. (1994) GIS: Noodzaak of Luxe?, *Nederlandse Geografische Studies*, Amsterdam.
Harris B., (1994) The Real Issues Concerning Lee's "Requiem", *Journal of the American Planning Association, 60*, 1, 31-34.
Haughton G., Hunter C. (1994) *Sustainable Cities*, Jessica Kingsley, London.
Issaev B., Nijkamp P., Rietveld P., Snickars F. (eds.) (1982) *Multiregional Economic Modelling Practice and Prospects*, North Holland, Amsterdam.
Kenworthy J.R., Newman P.W.G. (1990) Cities and Transport Energy: Lessons From a Global Survey, *Ekistics*, 344/345, 258-268.
Leary M. (1987) Development Control: the Role of Expert Systems, *Town Planning Review, 58*, 331-342.
Leontidou L. (1993) Postmodernism and the City, *Urban Studies, 30*, 949-965.
Lovelock J. (1979) *Gaia: A New Look at Life on Earth*, Oxford University Press, Oxford.
Masser I., Sviden O., Wegener M. (1992) From Growth to Equity and Sustainability, *Futures, 24*, 539-558.
Naisbitt J. (1982) *Megatrends*, Warner Books, New York.
Nijkamp P. (ed.) (1990a) *Sustainability of Urban Systems*, Avebury, Aldershot.
Nijkamp P. (1990b) Geographical Information Systems in Perspective, in Scholten H.J., Stillwell J.C.H. (eds.) *Geographical Information Systems for Urban and Regional Planning*, Kluwer, Dordrecht, 240-252.
Nijkamp P., Blaas E. (1994) *Impact Assessment and Evaluation in Transportation Planning*, Kluwer, Dordrecht.
Nijkamp P., Perrels A. (1994) *Sustainable Cities in Europe*, Earthscan, London.
Nijkamp P., Reggiani A. (eds.) (1994) *Nonlinear Evolution of Spatial Economic Systems*, Springer-Verlag, Berlin.
Nijkamp P., Rientstra S., Vleugel J. (1996) *Expert-Based Scenarios for Sustainable Transport*, Kluwer, Boston.
Nijkamp P., Rietveld P. (eds.) (1983) *Information Systems for Integrated Regional Planning*, North Holland, Amsterdam.
Nijkamp P., Rietveld P., Voogd H. (1991) *The Use of Multicriteria Analysis in Physical Planning*, Elsevier, Amsterdam.
Nijkamp P., Scholten H. (1993) Spatial Information Systems: Design, Modelling and Use in Planning, *International Journal of Geographical Information Systems, 7*,

85-96.
OECD (1991) *Environmental Policies for Cities in the 1990s*, Paris.
Openshaw S. (1990a) A Spatial Analysis Research Strategy for the Regional Research Laboratory Initiative, Regional Research Laboratory Initiative Research Discussion Paper no. 3.
Openshaw S. (1990b) Spatial Analysis and Geographical Information Systems: A Review of Progress and Possibilities, in Scholten H.J., Stillwell J.C.M. (eds.) *Geographical Information Systems for Urban and Regional Planning*, Kluwer, Dordrecht, 153-163.
Sagalowicz D. (1984) Development of an Expert System, *Expert Systems, 1*, 2, 137-141.
Scholten H.J., Stillwell J.C.M. (eds.) (1990) *Geographical Information Systems for Urban and Regional Planning*, Kluwer, Dordrecht.
Van Geenhuizen M., Nijkamp P. (1993) Urbanization, Industrial Dynamics and Spatial Development, Tinbergen Institute, Amsterdam.
WCED (1987) *Our Common Future* (The Brundtland Report), Oxford University Press, Oxford.
Wegener M. (1994) Operational Urban Models: State of the Art, *Journal of the American Planning Association, 60*, 1, 17-30.
Zwier R., Hiemstra F., Nijkamp P., Montfort K. (1995) Connectivity and Isolation in Transport Networks, Tinbergen Institute, Amsterdam.

28. An Information System for Planning Mobility and Transport in the Metropolis: Proposed Methodology for Setting Up an Integrated Tool for Decision Support

Roberto Tadei, Marco Dellasette

28.1 Introduction

In this chapter we investigate some aspects of the theoretical design of an Information System for use at the metropolitan level, giving particular emphasis to data concerning mobility and transport. Firstly, we shall outline the basic features and requirements of such a system.

An Information System of this kind is characterised by its *transverse* and highly *intersectorial* nature. It also needs to be *logically integrated*, although it may be made up of elements which are physically independent. In addition, it is important that the information required should be easily obtained at source without creating additional work for the administrative authority or causing conflict with established work routines. It should therefore, as far as possible, be designed in such as way that existing data gathering and processing procedures can be maintained. The system must also guarantee a high degree of *transparency* and *ease of use*.

The use of *advanced mathematical models* will be necessary for the analysis of transport demand and supply, the demand/supply interactions and for the evaluation of the various scenarios relating to demand and supply alternatives. In order to take account of the intersectorial nature of mobility and transport information, the Metropolitan City should be considered as a system made up of a number of interrelated parts. The principal parts will be the various subsystems of the city: the housing market, the job market, the service sector, land-use market and transport.

Although the *scale of reference* used here is the metropolitan level, both the regional and local scale must be taken into consideration for two reasons. The first reason is 'systemic', since intervention in any of the subsystems of the city is bound to impose objectives and constraints at the regional and local level and, vice versa, other objectives and constraints from these levels will impose themselves at the metropolitan level. The second reason is operative, since the information necessary for the analysis and planning of mobility and transport does not derive solely from the metropolitan level, but also from the regional and local levels.

This means that *different institutional levels* are involved in setting up an Information System of this kind. There will be not only Regional and Local Authorities, but also the various companies and bodies whose operations affect the metropolis and interact with the mobility and transport system. The methodology for representing the pattern of mobility and transport, and hence the relative information system, will therefore necessarily have an inter-institutional character. Fig. 1 shows the kind of open Information System conceived here, responding to the modern requirements of transversality, physical independence, modularity and efficiency (Dellasette, 1993).

This chapter is structured in the following way. In Section 2, the various subsystems of the metropolitan system are identified and the main interactions between systems and the general architecture of the metropolitan system are described. In Section 3, we propose an integrated system for the planning of mobility and transport, able to manage jointly the analysis of the demand, the planning of the supply and the management of transport services. Section 4 introduces the main mathematical models necessary for the Information System, explains their use and discusses some procedures for the evaluation of the demand scenarios and supply alternatives. In Section 5, the computing aspects of the Information System are outlined. Lastly, in Section 6, some conclusions are drawn and a brief summary given of the main characteristics that an advanced Information System for mobility and transport should possess.

28.2 The Metropolitan City as a System

28.2.1 Introduction

The metropolitan city can be conceived as a system. By this we mean that we can consider it to be an entity made up of a set of interrelated parts: basic industries, housing, services and so on, distributed over a wide area and related to each other through various forms of communications and transport (Bertuglia *et al.*, 1987b). Due to the very large number of parts and inter-relations, the city must be considered a complex system (Bertuglia and La Bella, 1991).

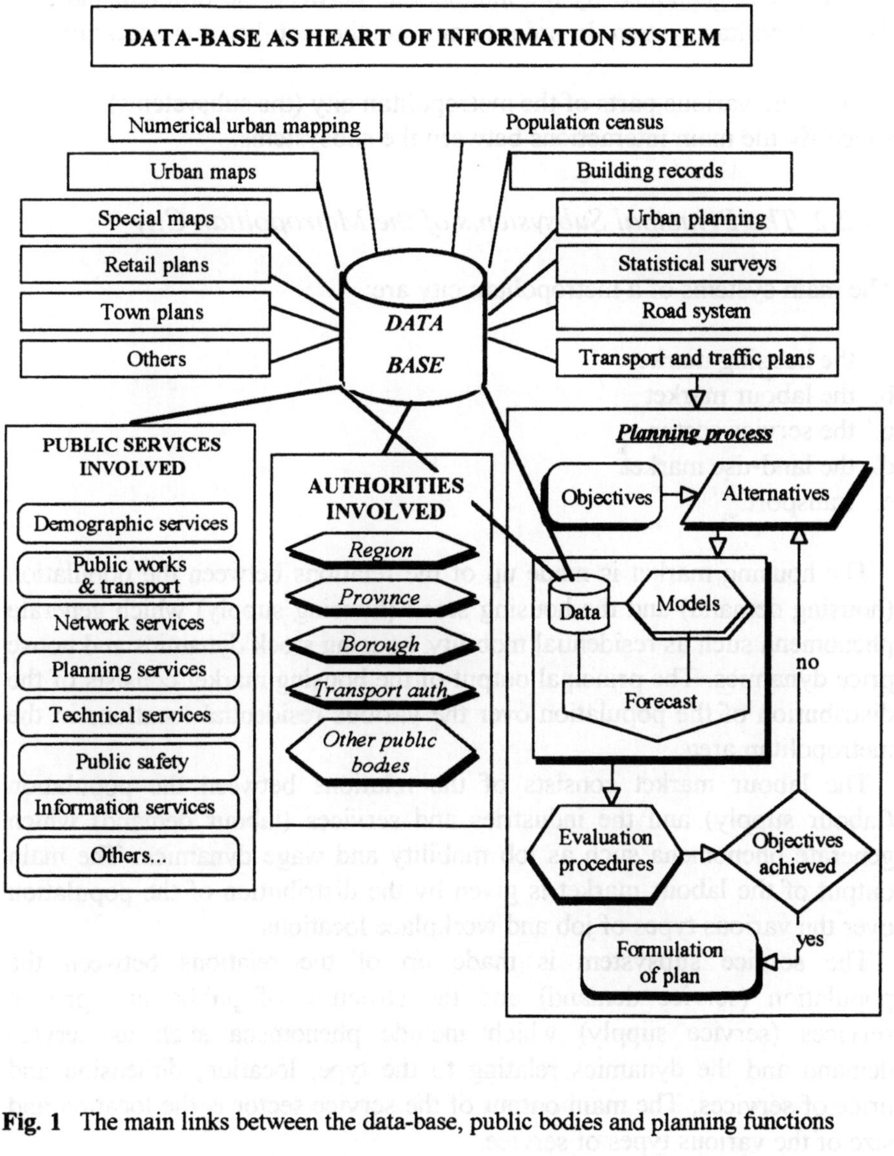

Fig. 1 The main links between the data-base, public bodies and planning functions

It is important to recognise that the city, as a complex system, possesses the characteristic that any action relating to one part or relation will generate effects on other parts or relations in a way which is extremely difficult to predict, analyse or control without adequate tools. In addition, many of the effects are likely to appear with a certain delay. The system therefore has some inertia and nonlinear internal dynamics which need to be taken into consideration.

Mobility is generated by the interactions between the different parts of the metropolitan system. In order to analyse the mobility it is necessary to:

- define the various parts of the metropolitan city (the subsystems);
- identify the main interactions between the subsystems.

28.2.2 The Principal Subsystems of the Metropolitan City

The main systems of a metropolitan city are:

a. the housing market
b. the labour market
c. the service sector
d. the land-use market
e. transport.

The housing market is made up of the relations between the population (housing demand) and the housing stock (housing supply) which generate phenomena such as residential mobility, housing stock dynamics and house price dynamics. The principal output of the housing market consists of the distribution of the population over the various residential locations in the metropolitan area.

The labour market consists of the relations between the population (labour supply) and the industries and services (labour demand) which generate phenomena such as job mobility and wage dynamics. The main output of the labour market is given by the distribution of the population over the various types of job and workplace locations.

The service subsystem is made up of the relations between the population (service demand) and the structure of public and private services (service supply) which include phenomena such as service demand and the dynamics relating to the type, location, dimension and price of services. The main output of the service sector is the location and size of the various types of service.

The land-use subsystem is made up of the relations between the land required for housing, industry and services (land-use demand) and the land available (land supply) which generate phenomena such as land rent dynamics.

The transport subsystem consists of the relations between the population and the industrial and service activities which create phenomena such as transport demand and supply, demand/supply interactions relating to the transport of goods, journeys made by individuals (for business, shopping, leisure, school etc.).

28.2.3 *The Main Interactions between Subsystems*

The interactions between the above subsystems and generation of the relative variables are shown in Fig. 2.

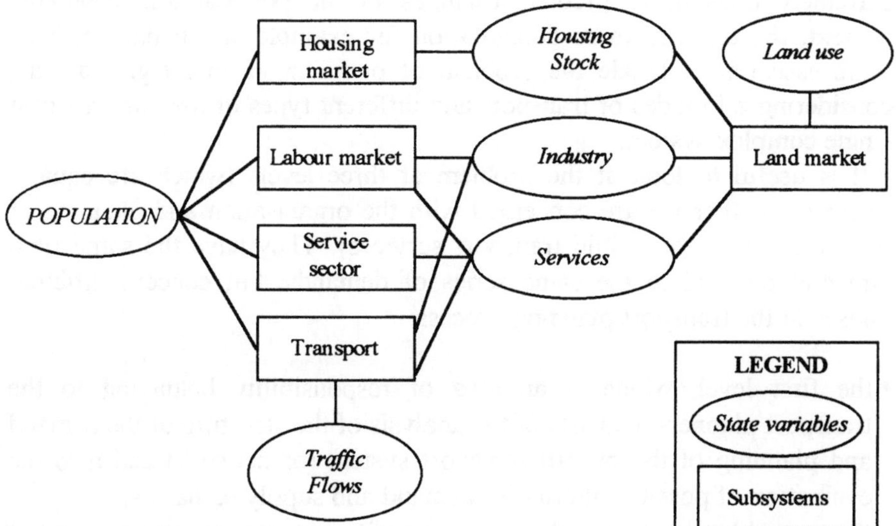

Fig. 2 The general structure of the system

The system represented is a *dynamic system* (Bertuglia *et al.*, 1990a, 1990b), which differs profoundly with respect to the classic equilibrium models, in that the balancing between demand and supply does not occur immediately. The supply can therefore be greater than the demand, or vice versa. This disequilibrium reflects the inertia present in real situations, such as the housing or the labour market.

28.3 An Integrated System for the Planning of Mobility and Transport

The planning of transport and the infrastructure of large urban areas is a highly complex task to which considerable attention has been dedicated by urban scientists and transport specialists. The problems associated with the environment (pollution levels and congestion) make it essential to examine the overall efficiency of the transport system, considering not only its effectiveness in transporting people and goods, but also the energy consumed (and hence emissions produced), the space needed for parking, roads and the exchange between modes of transport.

The demands of present day mobility require a transport system which can provide a high degree of flexibility and autonomy of demand. The application of modern technology is making available new kinds of vehicle and operating systems. It is a scenario in which future developments are extremely difficult to predict. Changes in the political and economic context, the trend towards deregulation for example, are making it even more essential to tackle the problem of planning in an integrated way, considering all modes of transport and different types of user as part of a single complex system.

It is useful to look at the problem at three levels (which are equally relevant whether we are concerned with the organisation and planning of the road system or public transport services). They have the same final aim and respond to the same series of demands, but concern different phases of the transport planning process:

- the first level, which is an area of responsibility belonging to the transport planners, consists of the analysis of the structure of the demand and planning of the overall transport system (or network) and also the evaluation of possible alternative demand and supply scenarios;
- the second level concerns the behaviour of transport users and consists of all those local factors which condition the choice of transport mode and route;
- the third level, which concerns the management of the system, involves the monitoring of the impact of the policies adopted (on both the system and the community). This ranges from the evaluation of the services provided to the provision of information systems for users.

It is obviously important for the data-bases within the information system to correspond to the various needs outlined above. If such data-bases have a common source and are totally integrated, they can provide operative reference scenarios and evaluation methodologies which are of direct use to the public transport authorities, the bodies responsible for providing funds and those which manage the infrastructure and provide other related services.

Only if there is total integration between the different levels, is it possible to achieve an economical system for the location, processing and maintenance of the data relating to the demand and supply of transport, i.e. through the setting up of a common information system for the various bodies involved in the planning and management of mobility. With an integrated system, it is possible to provide a complete set of modules for the simulation of the transport networks (public and private), the provision and control of public transport by the relevant authorities, and the optimisation and management of services by the transport companies.

Fig. 3 represents a possible level of integration of the common data-bases with the principal bodies involved and their responsibilities (Crotti, Dellasette and Villa, 1993).

It is highly desirable for there to be an interface between the data-bases described above and any Geographical Information Systems or GIS (Clarke, 1990, Nijkamp, 1990), since this will give access to more detailed information about the elements making up the various subsystems and provide additional support for the analysis of land-use, the relationship between transport and other urban activities. Another important advantage is that it is possible, with the aid of a GIS, to carry out analyses at different scales, by integrating data and output from the models.

28.4 Modelling Aspects of the Information System

28.4.1 Mathematical Models

The mathematical models integrated in the Information System consist of a series of independent modules which, using the 'raw' information from the data-base, make it possible to carry out various analyses, simulations and evaluations, hence enriching the basic data with additional information. In Fig. 4 we show the principal modules which could be included in an Information System for transport and mobility.

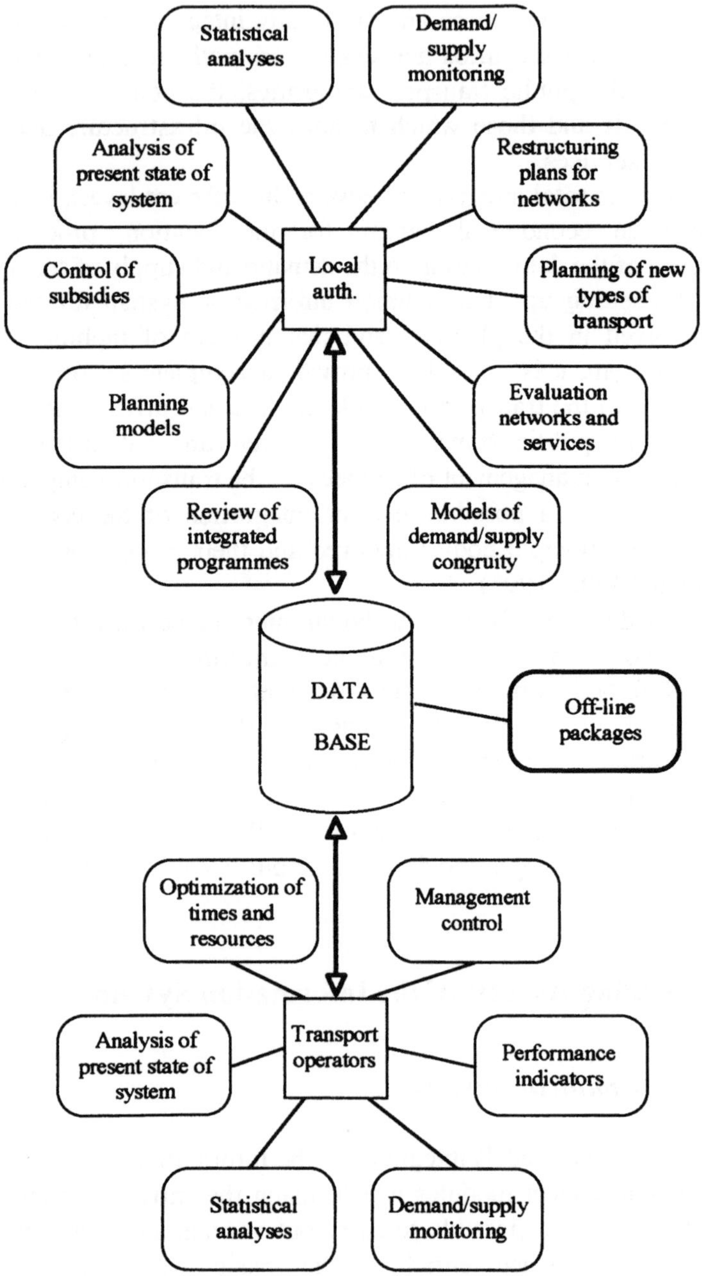

Fig. 3 Integration of data-bases with the main bodies involved and their responsibilities

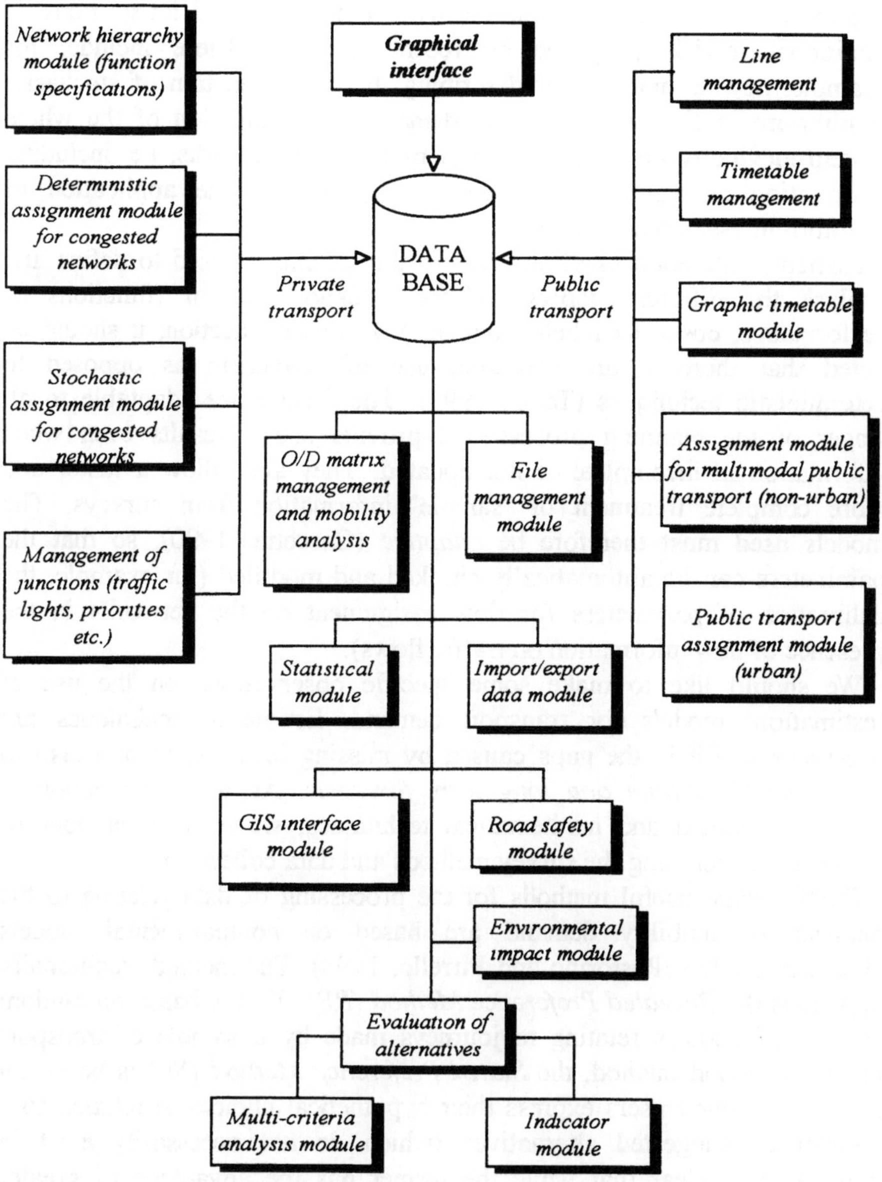

Fig. 4 Principal modules for the Information System

In order to define the necessary models, it is useful to refer to the four-stage process generally applied to transport planning (Bianco and La Bella, 1991): trip generation, distribution, modal split and assignment. It is important to take into account the main extensions and integrations of traditional models proposed by recent research. These include, for example, new techniques for describing the delay function of stochastic equilibrium in the case of elastic demand. An evaluation of the whole system inevitably involves the use of multimodal networks, i.e. including information on public and private transport, and the application of equilibrium assignment models.

Particular attention must be given to the techniques used to define and analyse the different aspects of the transport system (functions of performance, costs, user behaviour etc.). In this connection, it should be noted that there is an increasing use of stochastic as opposed to deterministic techniques (Tadei, 1993). The former are adaptable to all stages of the planning process and provide useful results even when information in incomplete or not updated. They also allow a faster and more complete treatment of 'sample' information from surveys. The models used must therefore be *adaptive* (Cascetta, 1990), so that the parameters can be automatically checked and modified (for example, the calibration of parameters for flow assignment on the networks in the presence of new information on traffic flows).

We should like to make some specific observations on the use of 'estimation' models for transport demand. Estimation techniques are necessary to fill in the gaps caused by missing information and also to make *short, medium and long term forecasts*. As well as mentioning certain statistical and mathematical techniques, we make a number of comments concerning the survey methods and data collection.

Firstly, some useful methods for the processing of data relating to the analysis of mobility demand are based on compartmental models (Postorino, 1993, Postorino and Pirrello, 1993). The method traditionally applied is the *Revealed Preference Method (RP)*. This is based on random surveys of choices relating to journeys made by a sample of transport users. A second method, the *Stated Preference Method (SP)* is based on surveys in which users express their hypothetical choices in relation to a number of suggested alternatives (which do not necessarily exist in reality). It is clear that while the former has the advantage of greater reliability, the latter method makes it possible to evaluate possible future scenarios. Recently, other methods have been put forward in which RP and SP are combined.

The *evaluation* phase, which is frequently neglected in the definition of the mathematical aspects of transport models, is in fact extremely important. It could, in a sense, be considered the fifth stage of the planning process. By applying reliable and complete methods of evaluation to the traditional planning process, the Information System goes beyond a simple simulation model to become a *decision support tool* (Bertuglia and La Bella, 1991). As such it is able to provide detailed qualitative and quantitative information on the performance of the transport system. We can, for instance, define indicators to assess the environmental impact, social implications, costs, efficiency and efficacy of a transport system (Tadei, 1989, Tadei and Williams, 1994).

Models for the evaluation of noise and pollution are particularly relevant in the present-day context (Painho, Gy Fabos and Gross, 1987). Recent experiments in Italy and other countries have revealed that these factors are strongly dependent on parameters which can easily be obtained from the classical simulation models described above (e.g. traffic composition by vehicle type, categories of driver behaviour, vehicle age, type of road surface etc.). On the basis of this information, it is possible to apply estimation techniques to derive estimates of total exhaust emissions.

With reference to road safety, the analysis and monitoring of traffic information combined with data on road accidents can constitute a basis for the identification of 'black spots' on the road network and help formulate policies to improve safety conditions (Villa, 1992).

In relation to public transport, it is necessary for the information relating to services in urban areas (running at fixed intervals) to be treated separately from suburban or nonurban services (running at fixed times). The models used have to take into account these different operating conditions and use algorithms for the selection of alternatives which, in the former case, consider the frequency of the service (hypertrip algorithms) (Nguyen and Pallottino, 1986) and, in the latter case, the actual times ('attractor' type algorithms) (Dellasette, 1995b). This permits a correct evaluation of the journey times, possibilities of connecting services, and so on.

In designing the Information System, it is important that maximum use is made of all *existing data* that is relevant and can be easily integrated into the system. As far as possible, it is preferable that existing software and systems of work should be retained. In deciding which tools and data to include, the following criteria should be respected:

- ease of integration into the proposed Information System and other calculation modules
- use-friendliness
- homogeneity of input/output information with the Information System
- compatibility with the chosen hardware system
- possibility of integration and/or personalisation.

28.4.2 Main Applications of Mathematical Models

In this section we describe the main applications of mathematical models in constructing an Information System for mobility and transport.

From a practical point of view, the principal purpose of such models is to produce 'value-added' to data and information. This can be achieved in different ways:

- the transformation, synthesis and integration of data
- the updating of information
- forecasting
- impact analysis
- optimisation.

We now look at each of these in turn.

a. The transformation, synthesis and integration of information.

It is fairly common for available data to be incomplete, or for there to be a large amount of data at different spatial levels which is difficult to integrate. In situations such as these, mathematical models can be of great help. Firstly, they can provide a systematic structure which makes it easier to analyse or synthesise the data. This kind of analysis produces added value through the generation, for example, of efficiency or efficacy indicators. Models can also be used to fill gaps in data. A problem typical of spatial data is that it tends to be available at very different levels of detail. The Census, for instance, provides some highly detailed information on households, but other information (for example, relating to income) exists only for large-scale geographical aggregations. In this case, too, mathematical models may help to complete some of the information at lower levels of aggregation.

b. Updating of information

It is often said that, from the moment it has been recorded, data is already out of date! Continuous data collection, however, would incur impossibly high expenditure in terms of cost and resources, hence the usefulness of models able to update existing information. The microsimulation models (Clarke and Holm, 1987) fall into this category

c. Forecasting

While updating brings our information from past to present, forecasting involves the examination of possible future changes. Information on future trends are a fundamental requirement for the organisation of all kinds of public and private service (from education and health to retailing and leisure activities). In relation to transport, the enormous expense involved in providing the infrastructure means that accurate forecasting is particularly important for establishing the economic feasibility of investment. This requires the forecasting of transport demand, which depends on demographic dynamics and the distribution of jobs. The methods used for this purpose range from straightforward extrapolation techniques to sophisticated econometric models.

d. Impact analysis

One of the questions to which models can help provide a reply is: "What would happen if ...?". In other words, they can investigate the effects on the system of a given action, which may be a planning intervention or a variation in one of the variables in the system. Taking into account that many such effects have a long-term impact, it is useful, and frequently essential, to be able to carry out a dynamic impact analysis (Tadei, 1990). By this we mean a continuous monitoring process which will provide feedback to guide the readjustment of policies.

Continuous monitoring is certainly the best way of analysing and controlling the impact of policy measures, but unfortunately the necessary data is not always available (and the costs of constant updating of existing data prohibitive). For this reason, impact analysis often becomes an exercise in comparative statics, i.e. a comparison of the situation in the system at two different moments (e.g. in two successive years).

e. Optimisation

Optimisation methods provide a way of arriving at a solution which maximises (or minimises) the value of an objective function, satisfying at the same time a number of constraints or conditions. Obviously, both the objective function and the set of constraints need to be clearly defined beforehand. The problem could consist, for example, of the choice of location of hospitals in a region in such a way as to maximise the equity of distribution of services to the population, respecting the capacity limits of the hospitals and their total budgets.

There are in fact two main ways in which optimisation methods can be used. Firstly, they can be adopted as a technique for dealing with a specific problem, providing a solution which complies precisely with the stated objective of the decision-maker. Secondly, they can be used in a more general way, providing a point of reference or serving as a standard for evaluating to what degree the current situation differs from the optimum.

Modelling approaches of the kind described above can be integrated within an Information System (Nijkamp and Scholten, 1993). Alongside the modelling functions, however, it is necessary to have other functions such as information retrieval, data analysis and systems for numerical or, where appropriate, graphical representation. We close this section with a general observation concerning the acquisition of data for an Information System. Data gathered without clear idea of its final use risks not only being a useless exercise, but may also hinder and delay the decision process. It is therefore necessary to have clear objectives from the outset and to organise the data collection operation accordingly.

28.4.3 Evaluation Procedures for Demand and Supply Scenarios

A great deal could be said about the question of evaluation. We refer the interested reader to Bertuglia, Rabino and Tadei (1988, 1989, 1991, 1992), Bertuglia and Rabino (1990), Bertuglia, Clarke and Wilson, (1994). In the present work, we intend to concentrate on two aspects of the problem:

- the benefits for the decision-maker of the use of scientific methods of evaluation
- methods which are suitable for evaluation in the field of mobility and transport.

We begin by observing that the evaluation of policy interventions is always the responsibility of the decision-maker. The role of the expert consists simply of the presentation of a series of possible alternatives and an assessment of the advantages and disadvantages of each (with an explanation of the methodology used). He or she does not, in other words, in any way replace the decision-maker, who continues to carry the weight and honours of the decision itself. The expert is able however to provide decision-support, in particular by rationalising the decision process.

There are numerous benefits to be derived from the use of scientific methods of evaluation, among these are the following:

- the reduction of uncertainties both in the analysis of the problem and the formulation of the decision;
- the rationalisation of the decision process and improved understanding on

the part of the decision-makers;
- the possibility of using evaluation methods not only as a decision-making instrument, but also as a way of stimulating discussion between the various bodies involved and hence obtaining wider consensus;
- the possibility of taking into account diverging points of view in a rational and well-documented fashion.

Despite these advantages, it is important to recognise that scientific methods of evaluation also have their limitations and it is unlikely that they will completely satisfy all the requirements of a given decision process. We retain however that, notwithstanding their imperfections, they remain useful tools which it would be a mistake to reject out of hand. It should be borne in mind that they are *decision-support* and not *decision-making* instruments.

We now briefly describe two evaluation methods suitable for adoption in an Information System for mobility and transport, then give some indications as to how they can be integrated.

The first, *multicriteria analysis*, is a technique involving the evaluation and ranking of alternatives according to a series of weighted criteria (Nijkamp, Rietveld and Voogd, 1991, Voogd, 1993, Nijkamp and Blaas, 1994). The most important and most delicate phase is the establishment of the criteria and their relative weighting. For the evaluation to be valid, the set of criteria needs to be complete and meaningful. An additional problem, especially when they consist of abstract concepts, is the 'translation' of the criteria and their impact into measurable quantities, i.e. the assignment of quantitative or qualitative values which make it possible to compare the alternatives.

The second method involves the use of *evaluation indicators*. These can be seen as indicators of the 'performance' of a given action or policy in a particular set of circumstances. In other words, they provide a way of measuring its efficiency or effectiveness. The indicators can be calculated either directly from the data already available or arrived at through the application of mathematical models (Tadei, 1989, Tadei and Williams, 1994, Bertuglia, Clarke and Wilson, 1994).

To what extent can performance indicators be interfaced with multicriteria analysis? If we can devise a series of indicators which express the efficiency and effectiveness of an action, we possess a measure of their impact on the various alternatives and hence a way of making a comparative analysis. This constitutes a first, intuitive step towards the integration of the two methods. The advantages and disadvantages of this integration can be analysed in theoretical terms, but above all requires adequate empirical checks.

We maintain that one of the objectives of the design of an Information System should be the creation of a methodology and an operative system for the calculation, evaluation and visualisation of appropriate performance indicators. The construction of a set of indicators for transport/land-use requires:

- a detailed knowledge of the different forms of mobility;
- an evaluation of transport provision and the quality of the service;
- the identification of a wide range of indicators of the economic, social and environmental impact;
- the creation of a data-base which can be integrated into the general information system or GIS.

The most important innovative features of this kind of evaluation indicators are:

- the fact that they represent an operative integration of the transport and land-use/economic models (as they are based on the spatial and functional inter-relations between the different urban activities and between these activities and the transport system);
- the approach adopted for their construction, which is: (i) flexible in terms of required data, the classification of variables and zoning, (ii) user-orientated, i.e. easy to use and visualise and (iii) policy-oriented, i.e. able to analyse and evaluate the impact of scenarios and policy alternatives;
- their theoretical and methodological content and systemic nature.

The Information System should be designed in such a way that it can easily be used as an analytical instrument, even by non-specialists. It therefore requires an *intelligent system* for the calculation, evaluation and visualisation of indicators or evaluation criteria. An Information System structured in this way becomes a powerful decision-support tool and also guarantees total transparency in the decision process, since it is possible to follow the procedure followed, step by step. Interpretation can be facilitated by the creation of modules which allow one to keep a trace of the operations carried out with the various models.

Finally, it is essential that the contents of the evaluation indicators are 'communicable' to decision-makers, other bodies involved and the public in general. For this purpose, *multimedia facilities* can be a useful way for informing the public of the planning and management strategies adopted by the transport authorities.

28.5 Computing Aspects of the Information System

The heart of the Information System is the data-base which serves the various modules for the processing of the data. This should include all information relative to transport, the road and rail networks and other services (Dellasette, 1995a). The *data base structure* should be of a standard type in order to permit access via the most commonly used languages and operating systems. The management system, consisting of the *interactive graphical interface* for the visual editing of data, must allow the simple and rapid input and maintenance of data. It should not be necessary for the personnel responsible for the updating of the data-base to have detailed knowledge of traffic engineering or the calculation algorithms used for the simulation models. They should only have to deal with easily recognisable transport network diagrams and familiar geometrical values.

The data-base is the *integrating element* of the whole system, through which all the models, algorithms and software packages are linked. Fig. 5 shows an example of the structure of a data-base for transport.

The large quantity and diversity of data included in such an Information System, including the procedures for updating data and the calculation of the algorithms for obtaining the indicators, makes it extremely complex, especially as both the data and the activities concern a number of different authorities. The design therefore needs to be carried out in a rigorous fashion with the use of formal conceptual models. For this reason, it is suggested that use is made of a technique (Bruno, 1995) which formalises:

- the *structure of the data*: using a system to describe the classes of data and the relations between them (relations of association, heredity or composition);
- the data *elaboration processes:* including the activities carried out (e.g. updating processes, calculation of algorithms) and the rules activating these activities (sequence, synchronisation, choice, confluence, repetition etc.);
- the *agents and resources* involved: describing their characteristics and role in the organisation.

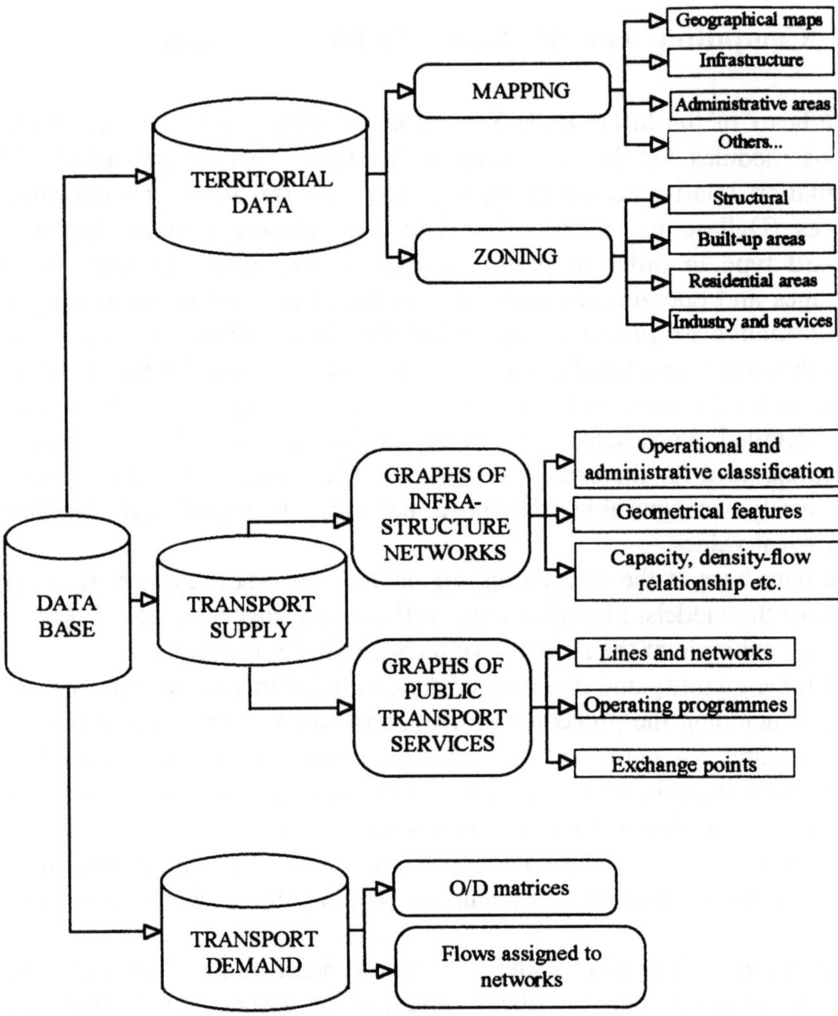

Fig. 5 Structure of a transport data-base

The above process has been successfully applied to the analysis of company activities in various fields. It has the advantage of possessing a graphic and simulatable language which, in Italy, is already being standardised by UNINFO (UNI Informatica). An important benefit of applying such a system is that it offers the possibility of validating the project, allowing a rapid and effective process of *quality control.*

As far as the data is concerned, the Information System must be based on an *open and distributed data-base* which can incorporate existing databases and is able to provide different users with *integrated and aggregate information*. The architecture of the system needs to be structured in function of the processing capacity required of the various algorithms, and takes into account the geographical location of the different authorities likely to use the system. For this reason it is suggested that it is organised as a *network*. The Information System must allow access by a number of other bodies (local and national authorities, statistical institutes etc.) as well as the regional authorities and other information systems (through the use of protocols). It should be noted that certain bodies providing information may also be consumers of processed information (e.g. Chambers of Commerce, transport network managers etc.).

28.6 Conclusions

In this chapter, we have described the main characteristics which we consider an advanced Information System for mobility and transport in a metropolitan city should possess. We now briefly summarise these features, dividing them under four headings: (a) modelling aspects, (b) questions concerning evaluation, (c) information/computing and (d) other aspects.

(a) Modelling aspects

- as well as the classical elements of transport modelling, the mathematical models should be able to consider social aspects of transport demand. They therefore need to be able to express not only quantitative, but also qualitative data;
- the generation, distribution, modal split, assignment and flow models should be stochastic, so that they are able to provide a more realistic representation of user behaviour;
- adaptive type models should be used for the assignment of flows for both public and private transport to allow the calibration and rebalancing of demand and supply, even in relation to future traffic;
- assignment models for public transport outside the main urban area should be based on 'fixed time' graphs, whereas classic 'frequency' graphs can be used for city transport.

(b) Questions concerning evaluation

- the aim should be to develop an integrated system for decision support;

- systemic performance indicators should be constructed for the demand and supply of transport and the demand/supply interaction. These should include indicators of environmental impact (air and noise pollution) and road safety. The Information System should also be able to deal with information relating to the 'sustainability' of the city, i.e. data concerning the quality of the urban environment and policies aimed at the self-sustaining development (Fusco Girard, in this volume, Van Geenhuizen and Nijkamp, in this volume);
- multicriteria analysis should be applied to the evaluation of alternative planning and management policies;
- an intelligent (automatic) system should be used for the calculation, analysis and visualisation of indicators.

(c) Information and computing aspects

- an 'object oriented' structure should be used for the analysis of the requirements of the Information System;
- the data-base should be open, standard and accessible to all modules and commonly-used external programmes;
- the Information System should make maximum use of existing information systems, in other words, it can be spatially dispersed and organised on a network basis with the use of protocols for access;
- unless it is decided to give the possibility of access by all operators to all the information, it is necessary to have a synthesis of the data which interest each user;
- it is necessary to be able to aggregate the information, passing from micro to meso and to macro level (e.g. aggregating Census data from local sections to provide information by district and by planning zones);
- the Information System should have a data dictionary containing a description of all information available (constituting a sort of 'identity card'). This operation, which can be very complex, is essential if the various data-bases are to be effectively integrated and the system to be functionally homogeneous;
- the system must be designed in such a way as to incorporate the output of models and indicators in the data-base;
- the system should be subjected to a detailed verification procedure to satisfy quality standards.

(d) other aspects

- the Information System should be conceived as an instrument for multimedia consultation by the public;
- maximum use should be made of existing resources, both in terms of

software and working systems adopted by users. New systems which are radically different from existing ones can prejudice the success of the Information System.

A great many of the features listed above are already in use in current Information Systems. Others represent interesting areas for further research and experimentation. From the methodological point of view, we feel that the more widespread adoption of Information Systems of this kind could have a significant influence on the approach to urban problems and, in particular, the planning of mobility and transport.

The setting up of Advanced Information Systems should also help to encourage the diffusion and standardisation of mathematical models. The systematic approach to the collection of information means that all the data necessary for such models is available and hence constitutes an incentive for their more continuous use. Greater experience in the use of mathematical models would encourage the development of new models for the updating of data, monitoring and other functions.

A well-designed interactive information base with analytical and evaluation functions can make a positive contribution to the acquisition of a deeper understanding of urban phenomena, and make it possible to meet the planning objectives more rapidly. They may also serve as a stimulus to the establishment of new decision-making procedures.

References

Bertuglia C.S., Clarke G.P., Wilson A.G. (eds.) (1994) *Modelling the City. Performance, Policy and Planning*, Routledge, London.

Bertuglia C.S., La Bella A. (eds.) (1991) *I sistemi urbani*, Angeli, Milan.

Bertuglia C.S., Leonardi G., Occelli S., Rabino G.A., Tadei R. (1987a) An Integrated Model for the Dynamic Analysis of Location-transport Interrelations, *European Journal of Operational Research, 31*, 198-208.

Bertuglia C.S., Leonardi G., Occelli S., Rabino G.A., Tadei R., Wilson A.G. (eds.) (1987b) *Urban Systems: Contemporary Approaches to Modelling*, Croom Helm, London.

Bertuglia C.S., Leonardi G., Rabino G.A., Tadei R. (1990a) A First Example of an Integrated Operational Model, in Bertuglia C.S., Leonardi G., Wilson A.G. (eds.) *Urban Dynamics: Designing an Integrated Model*, Routledge, London, 367-394.

Bertuglia C.S., Leonardi G., Wilson A.G. (eds.) (1990b) *Urban Dynamics: Designing an Integrated Model*, Routledge, London.

Bertuglia C.S., Rabino G.A. (1990) The Use of Mathematical Models in the Evaluation of Actions in Urban Planning: Conceptual Premises and Operational Problems,

Sistemi Urbani, 12, 121-132.

Bertuglia C.S., Rabino G.A., Tadei R. (1988) Mathematical Model and Plan Evaluation, *Sistemi Urbani, 10,* 237-261.

Bertuglia C.S., Rabino G.A., Tadei R. (1989) I modelli matematici e la valutazione dei piani, in Barbanente A. (eds.) Metodi di valutazione nella pianificazione urbana e territoriale: teoria e casi di studio, Proceedings of the International Seminar Capri-Naples 1988, Quaderno 6, IRIS-CNR, Ragusa Grafica Moderna, Bari, 47-71.

Bertuglia C.S., Rabino G.A., Tadei R. (1991) La valutazione delle azioni in campo urbano in un contesto caratterizzato dall'impiego di modelli matematici, in Bielli M., Reggiani A. (eds.) *Sistemi spaziali ed approcci metodologici,* Angeli, Milan, 97-143.

Bertuglia C.S., Rabino G.A., Tadei R. (1992) Review of the Main Conceptual Issues Facing Contemporary Urban Planning, *Sistemi Urbani, 14,* 151-171.

Bianco L., La Bella A. (eds.) (1992) *Strumenti quantitativi per l'analisi dei sistemi di trasporto,* Angeli, Milan.

Bruno G. (1995) *Model-based Software Engineering,* Chapman and Hall, London.

Cascetta E. (1990) *Metodi quantitativi per la pianificazione dei sistemi di trasporto,* CEDAM, Padova.

Clarke M., Holm E. (1987) Microsimulation Methods in Spatial Analysis and Planning, *Geografiska Annaler, 69B,* 145-164.

Clarke M. (1990) Geographical Information System and Model Based Analysis: Towards Effective Decision Support System, in Scholten H.J., Stillwell J.C.H. (eds.) *Geographical Information System for Urban and Regional Planning,* Kluwer, Dordrecht, 165-175.

Crotti A., Dellasette M., Villa M. (1993) Proposta di sistema integrato per la pianificazione, progettazione, gestione e valutazione di reti e servizi di trasporto, Proceedings of the ANIPLA Conference 'Automazione e sistemi di trasporto', Trieste (mimeo).

Dellasette M. (1993) Strumenti e tecnologie informatiche per la redazione dei P.U.T., Proceedings of the Seminar Regione Piemonte / A.I.I.T. 'Il D.L. 285/1992 Nuovo Codice della Strada: pianificazione del traffico e della mobilità: problemi e adempimenti degli Enti Locali', Turin (mimeo).

Dellasette M. (1995a) Pub-Line - Road-Line: sistema integrato per la pianificazione e la gestione dei trasporti, in Guariso G., Rizzoli A. (eds.) *Software per l'ambiente,* Pàtron Editore, Bologna, 243-248.

Dellasette M. (1995b) Nuovo modello di assegnazione ai cammini 'attrattivi' per il trasporto pubblico ad orario, Proceedings of the IV National Convention S.I.D.T.: 'Il trasporto pubblico nei sistemi urbani e metropolitani', Turin.

Nguyen S., Pallottino S. (1986) Assegnamento dei passeggeri ad un sistema di linee urbane: determinazione degli ipercammini minimi, *Ricerca Operativa, 38,* 28-47.

Nijkamp P. (1990) Geographical Information System in Perspective, in Scholten H.J., Stillwell J.C.H. (eds.) *Geographical Information System for Urban and Regional Planning,* Kluwer, Dordrecht, 240-252.

Nijkamp P., Blaas E. (1994) *Impact Assessment and Evaluation in Transportation Planning,* Kluwer, Dordrecht, Boston.

Nijkamp P., Rietveld P., Voogd H. (1991) *The Use of Multicriteria Analysis in Physical Planning,* Elsevier, Amsterdam.

Nijkamp P., Scholten H. (1993) Spatial Information Systems: Design, Modelling, and Use in Planning, *International Journal of Geography Information Systems, 7,*

85-96.
Painho M.O., Gy Fabos J., Gross M. (1987) Multiobjective Programming Models in Landscape/Land Use Planning, *Research Bulletin, 715*, Massachusetts Agricultural Experiment Station, Amherst, Massachusetts.
Postorino M.N. (1993) Il metodo delle Stated Preferences nell'analisi dei sistemi di trasporto, Quaderno scientifico, Facoltà di Ingegneria di Reggio Calabria.
Postorino M.N., Pirrello P. (1993) La stima dei parametri di scelta modale attraverso l'uso congiunto di dati RP ed SP, Quaderno scientifico, Facoltà di Ingegneria, Reggio Calabria.
Tadei R. (1989) Indicatori di performance e valutazione multicriteri, Proceedings of the CNR Education Programme 'Tecniche e modelli per la programmazione regionale', X Corso: Processi di sviluppo dei sistemi urbani: modelli e strumenti di governo dell'economia e del territorio, Capri (mimeo).
Tadei R. (1990) Modelli dinamici di assetto del territorio, Proceedings of the CNR Education Programme 'Tecniche e modelli per la programmazione regionale', XI Corso, Modelli di analisi e politiche regionali per gli anni '90: nuove tecnologie ed innovazione nel governo del sistema sociale, economico e territoriale, Capri (mimeo).
Tadei R. (1991) Indicatori di performance e valutazione multicriteri, in Bertuglia C.S., La Bella A. (eds.) *I sistemi urbani*, vol. 2, Angeli, Milan, 735-758.
Tadei R. (1993) Modelli di interazione localizzazione-trasporti con particolare attenzione agli aspetti economico-finanziari, rapporto finale Progetto Finalizzato Trasporti Due-CNR, 1st year, Rome (mimeo).
Tadei R., Williams H.C. (1994) Performance Indicator-based Planning with a Dynamic Model, in Bertuglia C.S., Clarke G.P., Wilson A.G. (eds.) *Modelling the City. Performance, Policy and Planning*, Routledge, London, 82-104.
Villa M. (1989) Indagine sui dati della sicurezza stradale, 45ª Conferenza del Traffico e della Circolazione di Stresa, Stresa, (mimeo).
Villa M. (1992) Metodologia per la classificazione e la previsione di standards di sicurezza della viabilità ordinaria, *Vie e Trasporti*, 587, 53-56.
Voogd H. (1993) *Multicriteria Evaluation for Urban and Regional Planning*, Pion, London.

29. Urban Information Systems and Sustainable Urban Development

Luigi Fusco Girard

29.1 Introduction

Regional Science has recently formulated a number of new methodological approaches and some quite sophisticated models, many of which relate to the understanding of a systemic approach to urban realities and the search for tools for its development, beginning with the improvement of decision-making processes. These are, however, often theoretical constructions which are difficult to apply, firstly because they are unable to fully reflect the real complexity of the urban system and secondly due to the lack or scarcity of urban and regional indicators.

The urban information systems discussed by Van Geenhuizen and Nijkamp in this book currently represent one of the most important tools with which to monitor and manage the level of sustainable development of the city and of the territory. The recent regulations in Italy which are creating a decentralised pattern by attributing new responsibilities to local authorities, i.e. to municipal, provincial and metropolitan governments, in the formulation of development strategies, have met limitations to their practical implementation due to the lack of an adequate system of indicators for providing the three following elements:

1. improvement of the knowledge of the existing urban set-up and living conditions vis-à-vis the following dimensions: employment, environmental integrity, urban health, energy etc. Indicators are needed not only to permit constant monitoring, but also to control the timing and spatial trends of the different urban budgets (the energy budget, environmental/natural resources budget, the economic budget and the social budget) both in quantitative and qualitative terms at an

intraurban, urban and metropolitan scale;

2. improvement of the decision-making processes of various actors (public and private) through the possibility of evaluating alternative actions. Indicators are needed to express and control the achievement of social, economic and environmental targets so as to allow the improvement of those agents' actions;
3. improvement of communication processes in cities, i.e. improvement of interactions between the various actors in order to integrate the different viewpoints and stimulate cooperative processes. The communications dimension is often overlooked, although it is as relevant as the construction of an adequate decision support system. It requires the ability to represent the complex systemic urban reality in terms which are intelligible, also for lay persons, so as to create consensus in a very pluralistic reality through dialogue and cooperation. The indicators therefore must neither be too general nor too specific, but highlight the link between quantitative and qualitative aspects as well as the comparison with threshold values.

The approach by Van Geenhuizen and Nijkamp perfectly reflects the above points. Therefore, I intend in this work to make some considerations based on his approach. I wish in particular to stress the relationships between communication processes, decision-making processes and urban indicators and show how they should pursue two types of goal: to improve decisions and, at the same time, communications, actions and interactions. Improving decisions and communication processes is of basic importance in order to achieve not only sustainable but also self-sustainable urban development.

As we shall see later, self-sustainability depends on the level of urban organisation, i.e. on the degree of coordination of the system of public, social and private institutions and, therefore, on that critical element that can be defined as 'institutional capital'. The general objective of this work is, therefore, to identify some guidelines and elements which can contribute to the formation of the institutional capital (rules, organisation and knowledge) essential for the development of self-sustainability, since it relates the natural urban system to the economic and social systems. Information systems, indicators and evaluation models are capable of enhancing the degree of organisation of such institutional capital, which is badly needed by the urban/ metropolitan reality.

29.2 Complex Cities, Decision-Making Processes and Communication Processes

Our cities are the places where we find the contradictions of modernity, the place where reality appears more and more fragmented, where different cultures, styles of life, interests and ethnic groups meet and clash. The city is the theatre of many diversities and differences, the place where multiplicity is most evident. But it is also the place where this multiplicity of interests, objectives and values gives rise to the deepest conflicts. It is therefore in the city that the need is most deeply felt for reducing the many contrasts and exclusions in order to bring conflicts to an acceptable level and foster more solidarity and cooperative behaviour.

Because of its high population density, in the city all elements tend to be interlinked. Quoting Commoner (1977), it is the place where everything is connected with everything else. This means that in the city the levels of interdependence relating each component of the urban system to the others are higher than elsewhere. The cultural and natural capital is connected with the human capital in a bundle of relationships that go through the social capital produced by the institutions. However, cities are not only the place where pluralities/multiplicities are most evident and interdependences are maximised, but also where the change characterising this age of transition from modernity to post-modernity is most rapid and where new ideas and courses of action are being searched for and tested.

The biggest change, the closest interdependences and the greatest plurality mean, in brief, that the city expresses the maximum systemic complexity. Therefore, the tools of analysis applied in the city in order to manage and govern it must be able to face such complexity. Indeed, this complexity must be organised. The risk must be avoided that this hyper-complex system 'spins out of control' jeopardising its staying power, thus exacerbating the internal disequilibria among its components due to the exponential increase of interdependences. This means, in particular, that scientific tools of urban and government analysis must be able to face the following issues:

1. uncertainty due to change, i.e. to the above-mentioned dynamic dimension (which, among other things, implies the use of fuzzy approaches);
2. multiplicity, i.e. the issue of multidimensional indicators and their ability to highlight the distribution aspects on the territory;
3. conflict deriving from the above interdependences and repercussions.

The conflict originates from and, in turn, is increased by what is generally defined as 'rational' behaviour whose logic relies on the matrix of economic calculation, as noted by Schumpeter, 1954, ruling out the possibility of including in an individual's choice the social, economic and environmental externalities affecting other individuals. Individual rationality clashes with what could be considered 'social rationality', because it gives rise to growing conflict with all the other members of society and the environment. This, in turn, causes further social divisions, new marginalisations and, by producing new environmental decay, causes waste and an inefficient utilisation of resources. The negative results of individual rationality, which becomes social and environmental irrationality, make urban and territorial development absolutely 'unsustainable'. On the contrary, in order to sustain the development of the urban system, we must activate individual and collective decision-making processes capable of incorporating the general interest. A richer, multidimensional approach is needed to overcome the present communication crisis.

Van Geenhuizen and Nijkamp, however, very conveniently consider self-sustainability, i.e. the self-organisation capacity, as a critical element of city development. But self-sustainability, multiplicity, general interest, the multicriteria approach, coordination and consensus are closely interrelated and strongly dependent on the communication levels that can be activated and, therefore, on the information system available.

29.3 'Horizontal Communication' and Information Systems

The nature of current urban problems leads us to ask ourselves not only about the decisions, but also the goals underlying urban development (Zamagni, 1995). The choice of goals means choosing the values that give shape and meaning to the development of cities in order to make them 'sustainable'. Therefore, the choice of values is a necessary and unavoidable step.

It is indeed easy to observe that whereas tools for the rational choice of means are already available, the instruments needed for choosing the goals, i.e. the values, are far less developed. To choose values means to build a sufficient basis of consensus. Therefore, one has to choose values that can be shared and are indeed shared by the greatest number of subjects. Since in our multicultural and pluralistic society there are no goals whose value is 'evident', that is, shared by everybody from the very beginning, then such goals must be derived from the comparison of many different points of view. A suitable process is needed for communication between institutions and society, within social groups and between individual subjects and the community. In other words, goals are not given, but must be chosen through a construction process that involves the different social components. In this respect one might speak of a social construction.

From the recognition of the multiplicity of viewpoints concerning urban/territorial development issues, one can shift, when there is support from an information system to their integration, i.e. to an overall perspective that represents a shared reference framework from which the general interest can be derived: the 'vision of the city' expressing what is good for the city and its territory. In order to achieve this, we require an information system that permits us to achieve 'horizontal communication' through participation (as envisaged in some recent Italian laws) creating a counterbalance to 'vertical communication'.

The contribution of Van Geenhuizen and Nijkamp to this book deals with this topic by focusing on the relationship between urban complexity and information systems. The 'social dilemma' they describe in the introduction, i.e. "the intrinsic discrepancy between individual and collective interest" (p. 617) is the right premise to their proposal for the management and government of the 'complex city' in a sustainability perspective. For the complex city, we need new development policies supported by an appropriate system of indicators and by decision-support techniques based on multicriteria analyses and an adequate information system.

The general framework of the urban information system is well described by Van Geenhuizen and Nijkamp in Sections 10 and 11 of their work. It could be useful however to make some further observations referring to typically Italian urban situations. In the following, we wish to point to some elements that should be reflected in the construction of the urban information system with reference to:

- the relationship between the complex city, the identification of its general interest and sustainability;
- the implication of sustainable development prospects for urban policy tools;
- the integration between information systems and decision-support methods in urban science and, more in general, in the activities of the different local institutions. Such an integration should give rise to an effective way of managing and governing cities.

29.4 Conflict, General Interest and Urban Development

As already noted in Section 2, cities are places of plurality and differences. Because of such pluralities, urban/metropolitan areas appear nowadays more and more suspended between evolution and involution, development and decline, wealth and poverty, private wealth and public poverty, restructuring and destructuring, reintegration and disintegration. Contrasting features coexist and cause cities to be a 'mosaic' of different realities of which some are declining, some evolving and others are in equilibrium.

Innovation processes coexist with processes of involution in a very intricate web of order and disorder. The city is the place where the possibilities of choice offered to each person are multiplying, but also where the capacity for orientation is reduced due to the anonymity of urban space which often lacks visual/perceptive differences and points of reference able to help define a mental map. Maximising the freedom of choice implies greater difficulty in choosing but, at the same time, the need to make choices. The city dweller is freer, but increasingly obliged to choose in order not to be marginalised from social, economic, cultural processes and in order to govern/direct such processes (Zamagni, 1995).

The city is the place with the greatest potential for the production of information, innovation and, thus, wealth, but it is also the place where social marginalisation, exclusion, unemployment and poverty are most severe. In cities, the scale and agglomeration economies have turned into diseconomies, environmental crisis and waste of resources. The high population density and the very high level of interdependence have led to a communication crisis, and often to the dissolution of the sense of community, citizenship and of belonging. Exchanges of information are the maximum possible, but the real interaction among individuals is very limited. Those who are part of an activity circuit are more and more protected, while those who are excluded are increasingly isolated and contribute the growing area of poverty and marginalisation, i.e. those groups of persons who are unwilling to invest in the future since they are obliged - by need - to worry only about the present. The level of conflict is therefore increasing and its external impacts affect all the other subjects and groups because of the interdependences. This increases the already considerable fragmentation and hinders the formation of a sufficient consensus for carrying out the necessary changes.

In cities everything (land, nature, cultural heritage) tends to become a commodity. Attention to exchange values rather than use or other values, reflects the triumph of the development model based on the priorities of an economy unable to reconcile the individual citizen's interest with the community's general interest, i.e. an economy which has always neglected the social dimension of the individual (Daly and Cobb, 1990). This has produced environmental decay and social marginalisation, due to the over-utilisation of environmental resources and underemployment of human capital. An abundant labour force and low environmental quality mix and concentrate in urban/metropolitan areas. This is especially pronounced in the South of Italy, giving rise to a growing conflict with nature and future generations.

However, it is the richness of the heritage of history, culture and nature specific to the city which could become the source of collective identity, the tool of communication between past, present and future generations. At the same time, it could represent the starting point for local, decentralised and participatory development (Benevolo, 1993). It is in the city that citizens can rediscover the elements that make them into a community. This heterogeneous reality is the greatest wealth of the city. In other words, plurality is a basic value of the urban reality and must be defended and enhanced to prevent it from becoming an element of involution, disorder and crisis. It is necessary that this pluralistic principle should be combined with an integrating principle, which is, as we shall see, an autopoietic principle capable of embodying the different components into a systemic and cooperative logic. Therefore, the city must be our starting point in facing the main problems of our society, i.e. marginalisation/poverty and the environmental crisis.

New urban development needs to be more consistent with the environment and the people, and oriented towards a conception of the general interest extended in space and over time, i.e. able to fulfill the needs of the marginalised and future generations, as well as the needs of nature.

29.5 From the Two Issues of the Urban Crisis to Sustainable Development

In this and the following two sections, we present the results of a research project financed by Italian Ministry of University and Research.

The first issue in the urban crisis is represented by the wealth/poverty dualism, i.e. by marginality and unemployment. The average rate of unemployment in Italy is 12%, but its geographical distribution (7.5% in the North, 11% in the Centre, 20% in the South), shows that the phenomenon of marginalisation is most widespread in the Mezzogiorno (where half of the unemployed are less than 25 years old). These levels of marginality, i.e. this dualistic model not only characterises the relationship between the North and the South of Italy, it is an urban phenomenon which is worsening in many cities.

In addition to being a social cost, it represents an economic cost, because it implies losing tax revenue, paying subsidies to the unemployed and having to meet a heavier burden of social services. Moreover, it gives rise to negative externalities, an illegal economy, higher death rates and environmental decay.

In such a context, it is clear we must foster the urban economy, but economic growth alone does not necessarily absorb unemployment and, hence, meet society's most urgent needs. European experience (CEEA, 1994) has shown that economic growth can in some circumstances destroy jobs. Too often, the benefits of economic growth have accrued to those already participating in the activity circuits. Therefore, since the market by itself may be able to give rise to new growth, but not create new employment, we must identify a type of public intervention that can orient the market. This means answering critical questions about the goals: growth for whom? for doing what?

The job issue highlights firstly the need for a set of statistical data of the urban information system which would make it possible to closely monitor labour policies and, thus, sustainable urban development. A second critical issue that, together with unemployment, represents the negative side of urban development, is the environmental crisis. This has been caused by the over-utilisation of natural, environmental and energy resources, and the production of pollution and urban waste at a much greater rate than the capacity for disposal/regeneration. We could, in addition, claim that the cultural and architectural heritage of historic city centres has suffered from its utilisation predominantly by the lowest income groups, resulting for example in insufficient maintenance. The overall result is a worsening quality of life.

Between poverty and environmental degeneration there is a close interdependence, one leading to the other in a reciprocal and cumulative process. A certain type of development leads to marginalisation, exclusion and poverty; this in turn brings about worse conditions of environmental decay that eventually inhibit or discourage economic activities. The overconsumption of environmental resources, including the cultural heritage, causes conflict between present and future generations, which adds to the conflict between the employed and the unemployed. If, as in fact is the case, economic processes over-utilise irreplaceable natural resources, irreversibly depleting them, this means that present growth takes place at the detriment of future generations, who will have fewer resources for the fulfillment of their own needs. Future scarcity will reduce the possibility of overcoming future poverty. It is a matter of lack of justice if the level of consumption of this generation inhibits the capacity of future generations to meet their requirements.

All urban sciences should deal with these two critical aspects, and identify new approaches so as to reduce the conflicts, and encourage new urban development policies able to achieve the integration of economic growth, social equity and protection of the environment, i.e. policies for 'sustainable urban development'. The latter can be defined as development able to manage economic activities so as to reduce poverty and cause no permanent depletion in the stock of resources required for present and future economic activities. The definition of sustainable development from a Paretian viewpoint, expressed as a type of development capable of meeting the needs of the present generation without jeopardising the fulfillment of the needs of future generations (Commission mondiale sur l'environnement et le développement, 1987) reveals that such development is characterised by a long run logic. Moreover, this definition is also concerned with the social dimension of development, which is extended to include an intergenerational perspective. Therefore, the notion of sustainable development focuses on the most marginal social groups, which implies the commitment to reduce poverty.

The basic condition of sustainability is the conservation of the stock of resources that form the natural and man-made capital at the disposal of the present generation and on which the welfare level is based. In other words, any present change can impair the future generations' welfare. If this happens, then compensation is required. Such compensation is represented by the transfer of a given aggregate of natural and man-made capital from the present to future generations (Pearce, Markandya and Barbier, 1989, Turner, Pearce and Bateman, 1994). This implies the introduction of a third key criterion in development in addition to equity and efficiency: environmental sustainability.

Sustainable development implies the production of goods and services in such a way as to preserve the environment, without using it destructively or wastefully. This means that the energy required by the residents, the production system and the transportation system must be used in the best way possible (Nijkamp and Juul, 1989, Nijkamp, 1990). Thus, sustainable development involves three dimensions, i.e. the values stated above (equity, utility and nature). These need to be balanced, and therefore conservation, social equity, or economic growth should be dealt with jointly (Giaoutzi and Nijkamp, 1993, Nijkamp and Perrels, 1994).

The notion of sustainable development is characterised, however, by a wide range of interpretations about 'what, who, how and when to sustain'. There is no doubt that the objectives of economic efficiency, social equity and the safeguard of environmental/ecological integrity are frequently in contrast. In this light, development is sustainable if it is capable of maintaining a dynamic long-run equilibrium between all these heterogeneous components. Such an equilibrium is continuously threatened by pressures from the different social players, both public and private, which form the social system.

Since it is a social construct, sustainable development explicitly recognises the multiplicity of actors who are often in competition with one another. We must therefore understand how to reduce such conflicts and to promote cooperation between public and private players. As it is impossible to maximise the three dimensions simultaneously, some choices have to be made and these inevitably involve value judgements. This requires something more than mere technical tools and experts; in order to face and solve the problems of choice, a participation process is necessary. Sustainable development implies the development of participation. But choices of this kind cannot be made from above or imposed, they can only be taken at local level. This leads us to recognise the local nature of sustainable development and its intrinsic consistency with the subsidiarity principle.

This principle, which is also recognised by the treaty of Maastricht, states that institutions should be as close as possible to the subjects whose needs it is their duty to meet. Only when this is not possible at the lowest level, should higher level institutions take over, i.e. at the provincial, regional level, etc. In conclusion, urban sustainability policies must, right from the very beginning, aim at the human factor and 'fight' urban poverty. The welfare of the urban poor must not be left to private charity, nor must it be the object of redistribution measures by public institutions which do not affect growth mechanisms (Musu, 1995). As noted in Section 5, urban marginalisation is not only an extremely important social issue, it is also a factor causing a waste of resources and inefficiency at a time when the economy is increasingly vulnerable. It is a factor that also hinders economic development, since those who are unable to satisfy even their most basic needs are giving rise to a conflict which is deeper than the traditional capital/labour dichotomy, having a negative impact also upon the rest of society.

Therefore, within the framework of sustainable development, there should be policies whose aim is to reduce the widening gap between opulence and want, wealth and poverty, by including the deprived in the labour circuit. At a time when information, knowledge, know-how and imagination are becoming basic development resources, we can aim at the human factor without resorting to the slow two-stage policy of economic growth and redistribution. Moreover, urban sustainability policies must preserve the natural and cultural heritage. No prospect for sustainability exists, if the present levels of unemployment and decay remain.

All of the above are elements of sustainable development, which the urban information system will have to reflect.

29.6 The Sustainability of Cities as a Private, Public, Social and Natural Economy

In order to tackle the two main issues confronting cities, i.e. employment and the environmental crisis, we need to have a 'good' urban economy, one which consists of a system of innovative enterprises integrated with the social and ecological environment. Moreover, we also need a 'good community economy' and a 'good economy of nature'.

If the environment and labour are recognised as priorities, it could be easily assumed that a free market would exacerbate rather than improve the status quo. We require public institutions which intervene to 'steer' the market, but which operate efficiently and effectively.

Unfortunately, in reality, we too often find public institutions which intervene in many sectors, but without being able to guide the market and fight against urban marginality. What is needed are integrated development strategies that guarantee new jobs, environmental protection and social promotion. Such integration could be achieved by producing goods and services with strong externalities and co-fruition, i.e. public goods such as education, training, transportation, communications, rehabilitation of the environmental/cultural heritage, housing, etc.

Therefore, for the sustainable management of urban areas, good institutions are needed, namely, a public economy which gives absolute priority to the effective and efficient functioning of educational and training institutions fostering investment in human capital, since creativity, invention, skills, culture, etc. are becoming important development factors. Development and high education levels are strongly correlated; where the education level is low, poverty prevails. Knowledge, know-how and training are nowadays the real riches since they enlarge the field of personal choice, the capacity to interpret and evaluate reality and thus 'react', i.e. to take decisions as to the allocation of resources. Investment in human capital is therefore the first issue to be solved if we are to decrease social marginalisation. This is achieved most of all through good educational institutions.

Secondly, a good urban economy, i.e. a sustainable urban economy, is an economy that enhances the role of the 'third sector', between the state and the market, and embodies elements of cooperation, solidarity, benevolence and unselfishness; this includes nonprofit making institutions, nongovernmental organisations, associations, etc. The community economy is that sector of the economy concerned with the social and community dimension of human experience, i.e. a rationality which reflects not only individual interest, but also the good of others. This means referring to a wider idea of what a human being is (not just an isolated person, but a member of a community of individuals and other subjects) and what utility means (not restricted to the benefits of commercial exchange, but seen in a wider perspective that includes reciprocities, intrinsic values, symbolic values, etc.).

The household economy, i.e. the economy that concerns the production of goods and services outside the market, is an important component of the economy of the third sector from the above viewpoint. The first investment in human capital formation is that which takes place within the family (providing care, security, welfare, etc. in a nonmonetary form and outside the market) where the welfare functions of many different subjects are integrated.

Finally, a good urban economy is a 'natural economy' that causes no stress to the self-reproductive/regenerative capacities of natural resources on which all production, exchange and consumption processes rely.

Therefore, urban sustainability is an issue that concerns the system of enterprises, the system of public institutions, the system of social institutions and the natural/environmental system. It is necessary to have an information system capable of appropriately representing the complexity of the urban system with multidimensional profiles that not only highlight the production of public goods, but also social and natural wealth in order to be able to predict the impact of alternative actions.

29.7 Sustainability, Self-Sustainability and Communication

The notion of self-sustainability, already briefly mentioned in Section 1, is not the same as the concept of sustainability described in Section 5. A self-sustainable system is one that draws self-regulating capacity from itself, i.e. from its organisational structure. This means that it coordinates its own activities over time without needing external support from other agents.

Without external support in the form of resources, finance, incentives, knowledge, coordination capacity, etc., if the system nevertheless manages to achieve equilibrium between the three goals, it can be defined not only as sustainable, but also self-sustainable. Such a system is able to reproduce conditions of sustainability over a long period, in a 'self-perpetuating' circuit, because it continuously produces and transfers knowledge, a resource which is essential for identifying the best organisational patterns.

An example of a self-sustainable system is the ecological system characterised by autopoietic processes. All autopoietic systems are self-sustainable. The self-sustainability of the urban system depends upon the relationships between the different components, their coordination and communication. In order to achieve urban self-sustainability, it is first of all indispensable for the actions of all those involved to be coordinated. This therefore concerns the existing organisational set-up, institutions, and the rules that govern the interaction or communication among the different subjects. It is generally assumed that when the components, functions and responsibilities are highly fragmented, those with responsibility for coordination are the public institutions. Their organisational activities should help to bring conflictual situations back to higher levels of competitiveness and cooperation. But the sustainability of urban development is not only an issue concerning public institutions, it also regards the private sector and enterprises.

Of course, it is necessary that public institutions should be in a position not only to sustain but also to be self-sustained by operating under conditions of effectiveness and efficiency. But it is even more important that the system as a whole becomes self-sustainable, i.e. capable of maintaining and reproducing itself (Zeleny, 1994). This is possible through interconnection between the natural/ecological system, the economic system and the social system constructed by the institutions. In general, these are the rules of the game that a society sets in order to shape interaction processes through social, economic and political exchange (North, 1990). This reflects the values and ideals of a community, that is to say, its culture. These, in turn, shape society and therefore modify ways of thinking, values and ideas. Indeed, these are more important than the values which people think of, or recognise as their own (Zeleny, 1994).

The set of rules and organisational capacities that govern and coordinate the interaction and the behaviour of human, natural and man-made capital can be defined as institutional capital (Zeleny, 1992, 1994, Berkes and Folke, 1992). Institutional capital can deeply change the interactions among the above forms of capital, which are all essential for promoting sustainability. Therefore, it can give rise to forms of sustainable and also self-sustainable development if it first of all guarantees suitable functioning of the natural and social systems. In fact, the natural system is able to sustain the three other 'overlapping' systems (economic, public and social) to the extent to which it is itself self-sustainable. If it loses this characteristic, the system will collapse and no longer be in a position to sustain or support other sectors, because it also loses its capacity for self-regeneration. The resources that form the natural system are those that represent the heritage of common goods which are recognised as valuable.

The third sector (of cooperation, free production of goods and services, households, associations, etc.) is another example of a self-sustainable system that, in turn, sustains the private and public economies. The more this sector is developed, the more it is viable. It is therefore necessary that the private economy becomes in its turn self-sustainable and does not erode other social or natural systems by passing on to them its unbalancing external effects. The sustainable enterprise is an autopoietic one in which the firm's interest and the general interest coincide. The decisions of the enterprise give rise to a rationality that consists of the integration of economic/financial accounting, ecological accounting and social accounting (Zoppi, 1994).

The self-sustainability notion refers in both cases to the capacity for fostering cooperation (in cases of conflict), maintaining/preserving the system of common resources or goods while the various subjects making up the system pursue different goals/interests (Zeleny, 1981). This implies the capacity for organising oneself, but 'together with the others', and therefore requires interaction with other subjects (Luhmann, 1986). Such an interaction derives from the existing level of horizontal communication and from the recognition of the mutual interest in preserving a heritage of common resources, whose value is recognised by everybody. Moreover, the ability of a system to sustain itself is related to its capacity for renewal and transformation over time. This depends on the degree of autonomy of each of its components and the ability to respond to external stimuli which, in turn, depends on the capacity for continuously reproducing its knowledge (Maturana and Varela, 1980).

A further characteristic of self-sustainable systems is that of managing their internal conflicts within the boundaries of the system itself, that is, without passing them to other systems (Galtung, 1986). Good communication among all the components is a basic requirement (Prigogine and Sanglier, 1987), since it enables each component to 'in-form' and 'de-form' the others, i.e. change them and adjust to them in a creative process of adjustment/readjustment and co-evolution (Zeleny, 1992). The equilibrium reached is not stable and 'given', but is a dynamic equilibrium which is lost, reconstructed and needs to be continuously reproduced over time despite fluctuations, instabilities, disequilibria, disorder and crises (Nijkamp and Munda, 1994).

Urban sustainability policies are required for the four above sub-sections and also need to be interrelated to each other. Therefore, it is necessary to have a system of indicators that adequately reflect and express the various principles of sustainability. These should make reference to and promote self-sustainability, firstly by highlighting the common patrimony of resources which 'connects' all subjects and from which sustainability originates. Secondly, this system should explicitly represent the third sector, which is intrinsically self-sustainable because it is autopoietic. The latter is often neglected by information systems, though it is a good indicator of viability and sustainability conditions (Nijkamp and Fusco Girard, 1997).

29.8 Communication at the Strategic Level: Indicators for the Choice of Development Goals

The notion of sustainable or, rather, self-sustainable urban development represents the general interest of the city and the territory. It is not something univocal and 'given', but an idea that must be 'constructed' through communication in order to achieve consensus (Nijkamp and Perrels, 1994). The various public and private players possess different ideas of the general interest, therefore they have to reach a consensus on the general contents of the notion of sustainable development, seeing it as a common good that they wish to achieve, taking into account a composite set of interests, objectives and often competing or conflicting values. This means that it is necessary to have a communication process which can compare the various points of view expressed by each social player.

The availability of a system of indicators is indispensable for such a process. Through these the targets of each actor are made explicit, and structured, distinguishing the basic objectives from the instrumental ones that make the former operational. The result of this process, based on dialogue and participation, is the formulation of a new set of objectives/preferences, where some will have been replaced and others added, and where the original priority ranking may also have changed. From a multiplicity of viewpoints expressed by the various actors, a more general standpoint is derived.

In order to simplify this process, instead of starting from the interests, one should start from a more general level, that is to say, the values. To reason on this basis fosters interpersonal communication and, therefore, greater participation, facilitating a strategic approach. The construction of this more general viewpoint represents a rational element of reference against which the different outlooks on the city's future can be compared and evaluated: the ecological city, the city of culture, the city of the arts, the technological city, the city of services, the industrial city, the city of the sea, etc. (Fusco Girard, 1995). It is, therefore, the result of a participation process in which everyone confronts the others, makes his/her own values clear, learns about other values and sees himself/herself as part of a whole.

In Italy, recent regulations make provision for participation, but they have not yet been completely structured and still need the definition of rules and procedures at local level. The activity of comparing and choosing goals that express the general interest of the city and give shape to the notion of sustainability, is also an evaluation process. Goals must be formulated, then reformulated again and again, in a process which attempts to construct consensus on the notion of general interest and a common set of values/objectives.

At this stage, indicators and evaluation are the tools of a communicatory rationality (Habermas, 1979), rather than the tools related to instrumental rationality through which the best means are selected for achieving 'given' goals. In other words, the urban information system at such a level should permit a description and critical evaluation which makes it possible to arrive at a set of values/objectives capable of taking into account the various standpoints and expressing a shared notion of the general interest.

Information systems and evaluation are interrelated. They are the tools for a meaningful social construct able to guide change and construct a pact with the different actors/subjects. Once the most relevant goals have been identified, i.e. the values able to give a meaning to development, they should be ranked. Technical procedures such as the one formulated by Saaty (1990), based on a 9-score scale, can simplify the problem of priority ranking by paired comparisons of different values/targets.

29.9 Communication at the Tactical Level: Urban Indicators for the Choice of Policy

The strategic stage is the one - as already explained - at which attention is focused on what the city wants to become. At this stage we focus on the set of general values, such as the basic values of history, nature, and labour. The information system is useful for understanding the implications when, for example, the value of nature has to be stressed or when development has to start from the enhancement of cultural capital or human capital (i.e. labour).

At the tactical level, we tackle the problem of 'how' and 'when' this is to be achieved, that is to say, through what type of implementation or time priorities. The local information system should enable us to apply evaluation methods such as multicriteria, multigroup, quantitative or qualitative approaches as support to decision-making processes, not only to produce priority ranking of possible alternatives, but also to work out the most suitable strategy of cooperation between the private and public sector.

Such multicriteria evaluation methods (formulated by Keeney and Raiffa, 1976, Nijkamp and Spronk, 1981, Zeleny, 1982, Voogd, 1983, Steuer, 1986) recognise the limitations of decision theory derived from microeconomics - from the inability to take externalities into consideration to the need to explicitly recognise multidimensional aspects and the intangibles. There are three basic approaches to ranking alternatives:

a. the maximisation of overall utility;
b. the minimisation of the 'distance' from the ideal point;
c. the maximisation of cost/benefit ratio.

The first group includes methods based on multi-attribute utility theory, in which the preferability criterion relies on the estimate of a utility function for each criterion, transforming the levels of a given indicator into corresponding utility or disutility values. Even when referring to the same criterion/objective or value, these functions can vary from one social group to another. In the second group we find methods comparing the 'distance' of different alternatives from an ideal solution in an n dimensional space, weighing each dimension by means of a vector w that expresses its importance. The third group includes evaluation analyses based on welfare theory: social cost-benefit analysis, cost-result analysis, and financial analysis.

A list of some multicriteria evaluation methods which can be used as a support to decision-making processes is given in Table 1. These methods help integrate the fight against urban marginalisation and unemployment with actions aiming at the efficient use of energy, the upgrading of human, natural and cultural capital, and economic revitalisation and thus contribute to formulating cooperation and integration processes between public and private actors. Their application should also involve fuzzy approaches (Munda, 1995).

29.10 Indicators of State, Pressure and Response for Urban Sustainability

If urban sustainability implies a priority commitment to reducing urban marginality, i.e. unemployment/underemployment/poverty, and allocating energy and available resources in a reasonable way, what is the impact on the information system? If urban sustainability depends on the four subsystems (enterprises, public institutions, communities and the natural system), what sort of indicators do we need in order to be able to highlight the effectiveness and efficiency of institutions operating in these sectors? We should remember that special attention also needs be paid to the indicators capable of taking into account the relevance of the third sector economy and particularly of households.

If indicators are necessary for achieving a communicatory rationality through dialogue, i.e. 'horizontal communication' with and among citizens, in addition to an instrumental rationality by adopting techniques of decision support and evaluation of alternatives at a tactical stage, how should the urban information system be designed?

It should be possible to use the indicators for the construction of 'value budgets' to supplement the annual financial budget. If the intrinsic values on which urban sustainable development strategies are based are human capital, man-made capital and natural capital, the system of urban indicators should be structured so as to permit budgets to be drawn up for each (see Fig. 1). They might also take into account damage to environmental/natural resources (water, air, etc.), to man-made resources (housing, urbanisation) and to human resources (health, etc.), and thus show the real costs of proposed development (i.e. identify the net benefits).

Since it is possible to recognise many different interpretations of sustainable development, it is understandable that there is not yet an established consensus as to what indicators would make it possible to evaluate and control urban sustainability (CREDOC, 1981, OCDE, 1991, Adriaanse, 1993, Button, 1993). Traditional indicators such as per capita consumption or production, per capita income, the employment rate, etc. are insufficient because they only reflect the produced wealth, i.e. man-made capital and not natural, human and social capitals.

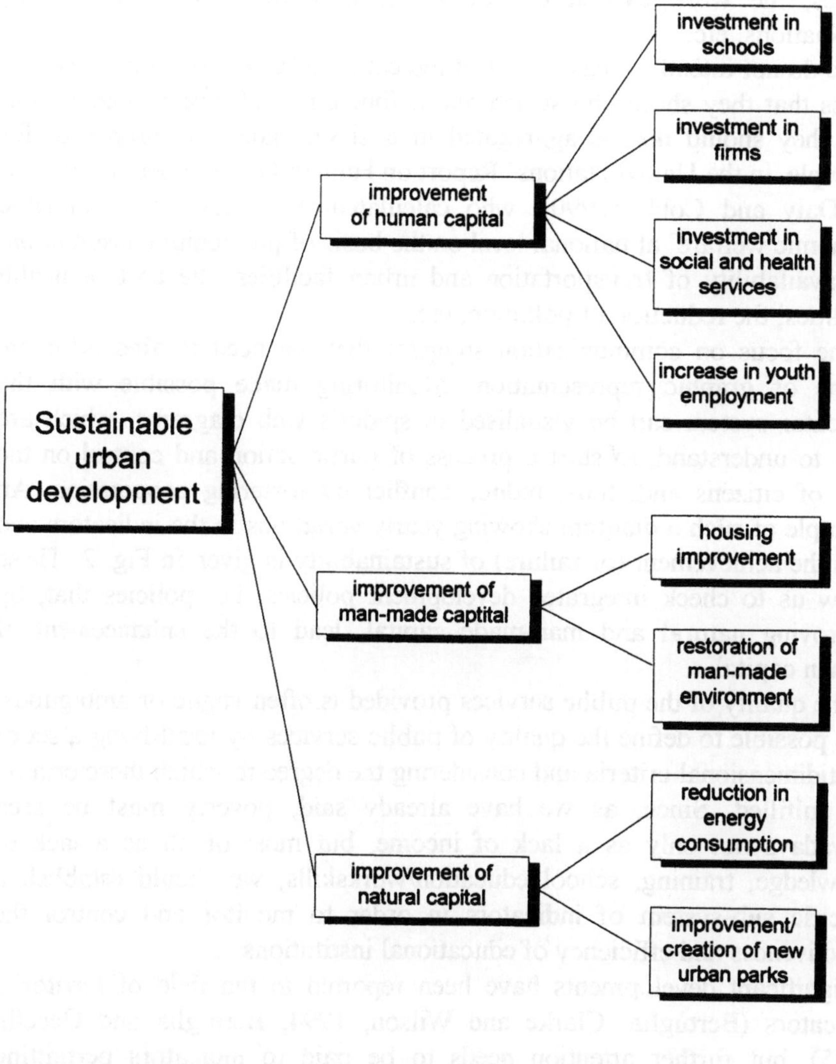

Fig. 1 The various forms of capital coordinated by institutional capital for drawing up a value budget

Sustainability, and in particular self-sustainability, depends on these four types of capital and the way in which they are combined, that is, on their mutual relationship. Therefore, we need not only monetary indicators, but also indicators that can be expressed in physical units concerning the natural system's resources. Most of all, it is necessary to obtain indicators of social capital, concerning the type and functioning of institutions, the organisation of local communities, the type of participation in public bodies, the work carried out in the third sector, within households, associations, etc.

We do not intend to make a list of indicators of urban sustainability, but stress that they should be structured at four different levels, pointing out that they should not be aggregated in a single index, as proposed, for example, in the United Nations' Report on Human Development (1992), or by Daly and Cobb (1990), who calculated the index of 'sustainable economic welfare' at national level on the basis of per capita consumption, the availability of transportation and urban facilities, the cost of health facilities, the reduction of pollution, etc.

The focus on communication suggests that we need to find adequate forms of graphic representation. Monitoring made possible with the indicator system can be visualised in spider's web diagrams, which are easy to understand, to start a process of participation and control on the part of citizens and, thus, reduce conflict by fostering cooperation. An example of such a diagram showing yearly variations in the indicators and thus the achievement (or failure) of sustainability is given in Fig. 2. These allow us to check integrated development policies, i.e. policies that, by improving natural and man-made capital, lead to the enhancement of human capital.

The quality of the public services provided is often vague or ambiguous. It is possible to define the quality of public services by identifying a set of multidimensional criteria and considering the degree to which these criteria are fulfilled. Since, as we have already said, poverty must be seen nowadays not only as a lack of income, but most of all as a lack of knowledge, training, school education/workskills, we should establish a specific sub-system of indicators in order to monitor and control the effectiveness and efficiency of educational institutions.

Significant developments have been reported in the field of territorial indicators (Bertuglia, Clarke and Wilson, 1994, Bertuglia and Occelli, 1995), but further attention needs to be paid to indicators permitting evaluation of urban marginalisation and environmental decay, i.e. indicators of state and pressure (Adriaanse, 1993). These should be integrated with the so-called response indicators that denote not only the state of the urban environment (level of poverty, pollution, etc.) and the pressure imposed upon it, but the costs that have to be met in order to reduce marginalisation, pollution, excessive densities, energy consumption, etc.

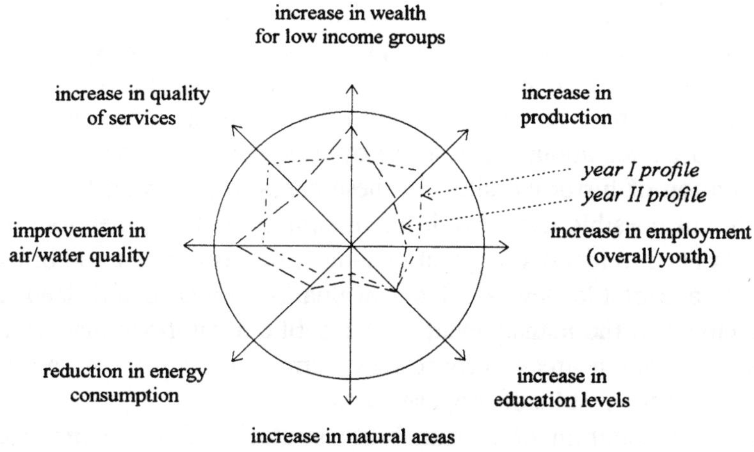

Fig. 2 Example of a spider's web diagram

It is also necessary to possess data concerning the cost of reducing air, water and soil pollution as well as the cost of facilities and infrastructure. The lack of such indicators makes it impossible to construct, implement and monitor urban policies which are sustainable from a social and environmental standpoint, i.e. capable of ensuring social and ecological viability which is a basic element of economic sustainability. The definition of indicators to express the distribution and characteristics of poverty, the types of marginalisation, or the regeneration capacity of essential factors such as air, water, soil, etc. represents an interesting and still open field of research.

29.11 Complex Values, Multicriteria Evaluations and Fuzzy Approach to Urban Choices

A good urban information system should permit the application of different methods of evaluation, including quantitative and qualitative multicriteria methods (also using fuzzy approaches) as decision support tools (Munda, 1995). In order to identify actions, programmes or policies suitable for achieving sustainable development targets in a coordinated way, it is necessary to predict behaviour as a response to given actions and to monitor the results. In other words, to pass from the general principles of sustainability to their concrete implementation, it is indispensable that all development proposals undergo specific evaluation processes.

A good field for experimentally testing the integration between the information system and the evaluation methods is represented by the problem of choice about the best use of large derelict urban areas and the rehabilitation of historical sites. In these cases, the conflict level between the different possible uses and different subjects is high, therefore there is an urgent need to identify ways of dealing with this conflict and bringing it back to acceptable levels. Such evaluation tools could lead to an improvement in the management of the public institutions that operate in urban/metropolitan areas because they permit an *ex-ante* and an *ex-post* evaluation of the results of proposed actions.

The implementation of sustainability strategies should start from the enhancement of those resources whose intrinsic values are the highest. In addition to human capital, there are the natural and cultural resources. Natural resources are valuable because they 'sustain' the various activities of the private, public and social economy. They include the raw materials needed by man (which have an exchange value), but also water, air and soil, as they permit life in its various forms, and supply energy to the various components of the ecosystem. Natural resources are valuable also because they are a source of the energy that, as noted already by Leopold (1966), connects soil, flora and fauna in a single ecosystem and sustains biodiversity. The natural resources therefore have an instrumental/market value and, in addition, a value in themselves: this is an intrinsic value, independent of their direct use, deriving from the autopoietic processes (air and water regeneration, conservation of living species in their respective ecologic niches, etc.). It has been referred to as the 'primary value of nature' or 'natural surplus value' (Turner, 1992).

Cultural/monumental/artistic resources also have a twofold value: an instrumental and an intrinsic value. Artistic production is characterised by the value it has in its own right, independently of its use, i.e. an 'intrinsic' or a 'capacity' value, which is a 'complex value' (Fusco Girard, 1987). In the city one finds various combinations of human capital, man-made capital (in particular architectural/cultural resources) and natural capital (consisting of natural and man-made green areas, open spaces, etc.). They should be linked by closer interdependence. Especially in central areas where pollution, congestion and decay due to over-utilisation are greater than elsewhere, it is necessary to rehabilitate the architectural heritage, expand the green areas and relocate some activities to satellite/peripheral areas.

Culture/history, nature and transportation are, therefore, the three areas in which one has to operate in order to guarantee sustainability conditions through the creation of networks. This means relying on the intrinsic values of culture and nature, linking them with the transportation/communications network in order to reduce the energy/environmental crisis and increase job opportunities. The activities of rehabilitation, conservation and the development of abandoned/unbuilt sites have a high employment capacity and, at the same time, the ability to improve the quality of life and generate conditions for the location of new activities, therefore they are consistent with the promotion of sustainable development.

Town planning is the activity through which it is possible to govern the territory. It permits intrinsic values to be changed into use values by directing demand, redistributing it over time and space to further sustainable development strategies (Nijkamp, 1991). When the aim is to achieve sustainability, first of all the above values must be recognised. Indeed to plan for sustainability means maximising the 'complex value' of urban resources, i.e. both the use value and the use-independent value (Forte and Fusco Girard, 1996). It can achieve its goals by means of a continuous process of evaluation that takes places on several levels.

The *ex-ante* evaluation of the intrinsic value of the territory leads to the choice concerning the selection of the most suitable use, consistent with the intrinsic values. This is possible by predicting the impacts of each function and, thus, by making an evaluation. The third type of evaluation involves the identification of the use intensity of the selected function. This should be the consistent with the maximum carrying capacity of the resource, which depends on its specific characteristics. The intensity of use must also be acceptable and convenient to the private individuals in the case of public/private sharing of transformation or rehabilitation activities (Forte and Fusco Girard, 1996).

A fourth kind of evaluation concerns the way in which qualitative and quantitative changes can be actually implemented in space. There are four types of 'complex evaluation' because a multiplicity of elements, which are often multidimensional, conflictual, quantitative, qualitative and fuzzy, need to be compared. This requires a specific type of urban information system.

29.12 Conclusions

Urban sciences should contribute concretely to the solution of the critical issues raised above by integrating theoretical approaches with operational applications. Some form of urban pact is required to reconcile employment creation and environmental protection. It should be an agreement making it possible to promote new urban development strategies which achieve a compromise between wealth and poverty, i.e. self-sustainable development strategies.

An urban system will never be self-sustainable if it is unable to foster and guarantee communication among its different components (public, semi-public and private actors). When communication within the system is limited, the system needs to be 'sustained' from the outside. The process of urban self-organisation must be furthered by improving institutional 'infrastructures', i.e. that particular form of capital which coordinates and organises human, natural and man-made capital. Institutional capital leads to the construction of a 'good society' characterised by a strong civil conscience, which is a basic requirement for self-sustainable development.

Urban indicators are a basic tool for tackling the evaluation of resources and monitoring, and can be used as decision support systems with special reference to the issues of employment and the environment. But they also have to encourage communication processes in order to improve coordination among the different actors. Evaluation, communication, coordination and sustainability are thus closely interrelated to the availability of an integrated system of urban indicators. Such a system supplies the energy that makes the urban system really autopoietic.

References

Adriaanse A. (1993) Environmental Policy Performance Indicators, Dutch Ministry of Housing, The Hague.

Benevolo L. (1993) *La città nella storia d'Europa*, Laterza, Bari.

Berkes F., Folke C. (1992) A Systems Perspective on the Interrelation between Natural, Human-made and Cultural Capital, *Ecological Economics, 5*, 1-8.

Bertuglia C.S., Clarke G.P., Wilson A.G. (1994) *Modelling the City, Performance, Policy and Planning*, Routledge, London.

Bertuglia C.S., Occelli S. (1995) Gli indicatori territoriali, con particolare riferimento a quelli di performance spaziale: inquadramento storico, presupposti concettuali, problematiche operative, qualche esempio, in Campisi D., La Bella A. (eds.) *Il governo della spesa pubblica e l'efficienza dei servizi*, Angeli, Milan, 313-448.

Button K. (1993) *Urban Indicators and City Management*, OECD, Paris.

CEEA (1994) Crescita, competitività, occupazione, Brussels.

Commission mondiale sur l'environnement et le developpement (1987) *Notre avenir à tous*, Edition de Fleuve, Montreal.

Commoner B. (1977) *Il cerchio da chiudere*, Garzanti, Milan.

CREDOC (1991) Indicateurs sur la qualité de la vie urbaine et sur l'environnement, Paris.

Daly H.E., Cobb J.B. (1990) *For the Common Good*, Green Print, London.

Fusco Girard L. (1987) *Risorse architettoniche e culturali: valutazioni e strategie di conservazione*, Angeli, Milan.

Fusco Girard L. (1995) Lo sviluppo delle aree metropolitane, in *Lo sviluppo sostenibile delle aree metropolitane: quali strategie? quali valutazioni?*, CESET, Florence, 11-33.

Fusco Girard L. (1996) Valutazioni e sostenibilità dello sviluppo, in Forte F., Fusco Girard L. (eds.) *Principi teorici e prassi operativa nel progetto del piano d'uso del suolo*, Maggioli, Bologna.

Galtung J. (1986) Toward a New Economics: On the Theory and Practice of Self-Reliance, in Ekin P. (ed.) *The Living Economy*, Routledge, London, 125-140.

Giaoutzi M., Nijkamp P. (1993) *Decision Support Models for Regional Sustainable Development*, Aldershot, London.

Habermas J. (1979) *Communication and the Evolution of Society*, Beacon Press, Boston.

Keeney R.L., Raiffa H. (1976) *Decisions with Multiple Objectives: Preferences and Value Trade-offs*, John Wiley and Sons, New York.

Leopold A. (1966) *Sand County Almanac*, Ballantine, New York.

Luhmann N. (1986) The Autopoiesis of Social Systems, in Geyer A., Van der Zouwen J. (eds.) *Sociocybernetic Paradoxes*, Sage, London, New York, 172-192.

Maturana H.R., Varela F. (1980) *Autopoiesis and Cognition: The Realization of the Living*, Reidel, Dordrecht.

Munda G. (1995) *Multicriteria Evaluation in a Fuzzy Environment*, Physica Verlag, Heidelberg.

Musu I. (1995) Tra ricchezza e povertà: le nuove sfide dello sviluppo, in *Una buona società in cui vivere*, Edizioni Studium, Rome, 116-134.

Nijkamp P. (ed.) (1990) *Urban Sustainability*, Gower, London.

Nijkamp P. (1991) Evaluation of Environmental Quality in the City, *International Journal of Development Planning Literature, 6*.

Nijkamp P., Perrels A. (1994) *Sustainable Cities in Europe*, Earthscan, London.

Nijkamp P., Munda G. (1995) Urban Sustainable Development: Indicators for a Co-evolution of the Economy, the Community and the Environment, in *Lo sviluppo sostenibile delle aree metropolitane: quali strategie? quali valutazioni?*, CESET, Florence, 223-257.

Nijkamp P., Spronk J. (1981) *Multiple Criteria Analysis*, Gower, London.

Nijkamp P., Juul K. (1989) Urban Energy Planning, Brussels (mimeo).

Nijkamp P., Fusco Girard L. (1997) *Le valutazioni per lo sviluppo sostenibile della città e del territorio*, Angeli, Milan.

North D. (1990) *Institutions, Institutional Change and Economic Performance*, Cambridge University Press, Cambridge.

OCDE (1991) *Environmental Indicators*, Paris.

Pearce D.W., Markandya A., Barbier E.B. (1989) *Blueprint for a Green Economy*, Earthscan, London.

Pearce D.W., Turner K. (1990) *Economics of Sustainable Resources and Environment*, Harvester-Weathsheaf, New York.
Prigogine I., Sanglier M. (1987) *Laws of Nature and Human Conduct*, Brussels.
Saaty T.L. (1990) *The Analytic Hierarchy Process*, RWS Publisher, Pittsburg.
Schumpeter J.A. (1954) *Capitalism, Socialism and Democracy*, Allen and Unwin, London.
Steuer R.E. (1986) *Multicriteria Optimization: Theory, Computation and Application*, John Wiley and Sons, New York.
Turner K. (1992) *Perspective of Mantaining and Investing in Natural Capital*, Stockholm.
Turner K., Pearce D.W., Bateman C. (1994) *Environmental Economics*, Harvester-Wheatsheaf, New York.
UNDP (1992) *Human Development Report*, Oxford University Press, Oxford.
Voogd H. (1983) *Multi-Criteria Evaluation for Urban and Regional Planning*, Pion, London.
Zamagni S. (1995) Soggetti e processi per una nuova progettualità in Italia, in *Una buona società in cui vivere*, Edizione Studium, Rome, 201-222.
Zeleny M. (1981) *Autopoiesis, a Theory of Living Organisation*, North Holland, Amsterdam.
Zeleny M. (1982) *Multiple Criteria Decision Making*, McGraw-Hill, New York.
Zeleny M. (1992) The Application of Autopoiesis in System Analysis, *International Journal of General Systems, 21*, 2, 145-160.
Zeleny M. (1995) L'impresa autosostenibile, in *Lo sviluppo sostenibile delle aree metropolitane: quali strategie? quali valutazioni?*, CESET, Florence, 36-57.
Zoppi S. (1995) Le strategie di sviluppo delle aree metropolitane in una prospettiva di sostenibilità: lavoro e impresa sostenibile, in *Lo sviluppo sostenibile delle aree metropolitane: quali strategie? quali valutazioni?*, CESET, Florence, 58-75.

30. Towards the Complex City: Approaches and Experiments in Spatial Economics

Aura Reggiani

30.1 Introduction

The complexity issue has recently come to the fore, driving scientific research, especially in the analysis of spatial-economic systems, towards 'universal' principles and methodological instruments/tools. For example, the recent conceptual attention to the network system, identified as a 'complex spatio-temporal system' (see e.g. Reggiani and Nijkamp, 1995a), emphasises the interpretation of spatial-economic systems as *evolutionary processes with hierarchical/multilayer organisation* (with inter-dependencies among the various sub-systems). In this context, urban systems, which result from the complex interaction between *slow* and *fast* dynamics of the various components, can be considered the expression of the maximum systemic complexity (Fusco Girard, in this volume) as well as being self-organised systems with various types of evolutionary *(non)regular* behaviour (Van Geenhuizen and Nijkamp, in this volume, Reggiani and Nijkamp, 1995b). This implies the need to examine further 'principles' linked to the complex evolution of spatial-economic systems, such as the stability concept, associated with the more specific concept of *sustainability*[1], and hence the notion (more significant from the planning viewpoint) of dynamic urban *fragility/resilience*[2]. Once we have established, at least for some areas, the typology of the 'complexity/fragility/resilience' relationships, we also need to investigate reactions to (un)stable (endogenous and exogenous) change, in order to

[1] See e.g. Beckenbach and Pasche (1994) who define sustainability as a stable domain/ corridor in a dynamic spatial-economic system.

[2] We define resilience as the capacity of the system to resist to internal/external perturbations.

improve understanding of the *structural tendencies* of the urban system. A clear example is offered by the transport network, whose nonlinear evolution provokes a perturbation on the whole urban network, as well as on the *intercity network*.

It seems that only two resolutions ensure the sustainable evolution of networks: a) the increase of system-resilience; b) decrease in the *variety* of internal/external behaviour and hence its complexity. Consequently, in order to evaluate the sustainable evolution of the urban system, it is necessary to obtain more 'insight' not only into the *high complexity* resulting from the interaction of a variety of components, but also the *high uncertainty* associated with the dynamic behaviour of such subsystems (see, among others, Fusco Girard, in this volume).

Which approaches or methodological tools are able to measure/quantify or at least to formalise/typify these concepts? It is firstly worth considering the possible contribution of the methodological approaches recently adopted in regional science and spatial economics. A brief review will be given in the next section, assessing their potential for analysing and modelling urban complexity. Section 3 then offers some attempts to identify, by means of classical models, the structural tendencies of a systemic component, such as transport evolution in a 'complex-city'. For this purpose, the Italian metropolitan areas to be covered by the future high speed train network will be considered, interpreting this network as a 'complex-city'. Section 4 ends the analysis with some concluding remarks.

30.2 Towards Complexity Theory: Approaches and Methodologies in Spatial Economics

30.2.1 Introduction

Spatial economics can offer a substantial contribution to the research into approaches oriented towards modelling complexity. From a philosophical perspective, complexity concerns the mapping of a system's nonintuitive behaviour, in particular the *evolutionary patterns of connections* between interacting components of a system whose behaviour is hard to predict in the long run (see Casti, 1979). Systems theory in general, and network theory in particular, may help to define the *analytical* (and hence measurable*)* *complexity* of the system, without however capturing the inherent *behavioural complexity*. As stated by Mannermaa:

"The concept of complexity is, however, very 'complex' (Ploman, 1984; Williams, 1985). It has sometimes been attached to 'us' (researchers, people) and sometimes to 'reality' (objects, things) (Flood, 1987). One way of clarifying it might be to divide it into ontological and semiotic complexity (Csàny, 1989). Ontological complexity means the inherent complexity of 'reality' in natural processes (randomness in thermodynamic processes), in human beings and in societies, as well as in the relationship between the researcher and the object of research. Semiotic complexity, on the other hand, refers to the complexity of the models which we have in our minds or which we have constructed (the length of a computer program)" (Mannermaa, 1955, p. 31).

Semiotic complexity can be therefore be identified by means of the following typologies[1] (see e.g. Casti, 1979, Nijkamp and Reggiani, 1993, Reggiani and Nijkamp, 1995a), which clearly characterise a network and consequently a city, conceived as a network-system.

a. Static Complexity

This phenomenon refers mainly to the *structure* of the system, and in particular a configuration where the components come together in an intricate and interrelated way. In this case, the following can serve as indicators of static complexity:

- hierarchical structure (number of hierarchical levels)
- connectivity (manner in which the components or subsystems of a process are connected)
- variety of components (number of components and output)
- strength of interactions (quantitative measure of the interaction)

b. Dynamic Complexity

This phenomenon refers to the *dynamic behaviour* of the system. Dynamic complexity can be measured in terms of:

- computational complexity (number of steps necessary for computing the dynamic pattern of a particular function)
- evolutionary complexity (ability of a system to produce irregular and/or chaotic behaviour)

This second measure can, for example, be based on the identification of

[1] Part of the present chapter is based on Reggiani and Nijkamp (1995c).

strange attractors, fractal phenomena, etc. In this case, it is evident that these are nonlinear dynamic forms, and hence methodologies able to map the related evolutionary patterns are needed (for a review see Nijkamp and Reggiani, 1992a, 1992b, 1993). An analysis of the findings reported in scientific literature suggests that the high complexity and richness of nonlinear motions (like, for example, chaotic motions, see La Bella, in this volume) is a characteristic of simple analytical systems (with two or three equations and a small number of parameters). Consequently, it seems that by increasing the number of equations and parameters governing the system, the possibility of identifying irregular and chaotic motion is confined to a limited domain of parameter space. In other words, complex chaotic dynamics are less frequent when the system becomes analytically more complex. We may therefore reasonably assume the hypothesis of *incompatibility between analytical and evolutionary complexity*. If the robustness and frequency of this phenomenon could be demonstrated, more insight would certainly gained into the driving principle governing the evolution of real systems.

As with the distinction between static and dynamic complexity, if we wish to understand the equilibria of the city, it is essential to distinguish *static and dynamic urban sustainability* (see, among others, Camagni, Capello and Nijkamp, 1996). Here the time horizon plays a fundamental role, giving rise to possible *intermittency*[1] phenomena. The methodologies or approaches useful for the study of network complexity and sustainability can be neatly divided into four fundamental avenues of research which have recently received a great deal of attention in spatial economics, i.e. nonlinear models, chaos models, bio-ecological models and neural network models. These methodological approaches will now be briefly described.

30.2.2 Nonlinear Models

Nonlinear models, developed in order to map out nonperiodic and irregular behaviour, raised a great deal of interest in the 1970s and 1980s, along with the concept of (un)stable equilibrium, mainly as a result of the interest in catastrophe and bifurcation theory. Multiple equilibria, jumps and bifurcations in network evolution can be analysed by means of nonlinear dynamic systems (for a historical/methodological review, see Campisi, 1991, Camagni, in this volume, Diappi and Reggiani, 1997,

---

[1] We define *intermittency* as the phenomenon of irregular/chaotic movements in contrast to regular/stable behaviour.

Nijkamp and Reggiani, 1995a, Pumain, in this volume). However, the formal limit to the ability to determine stability solutions for a complex analytical system (one with a large number of dimensions and connected parameters) has still not been overcome.

One solution to this problem is to focus on the sensitivity of the system to perturbations, i.e. the *structural stability* of the system, rather than on stability solutions for the differential/difference equations governing nonlinear models. In the investigation of structurally unstable systems (like some components of the city-system), this means identifying those qualitative changes in their behaviour and structure which cause a significant impact on the whole system. Various studies have examined the *morphogenetic aspect* of complex dynamics in real world systems: the evolution of cities (see Mees, 1975, Rosser, 1992), regional development (Andersson, 1995, Andersson and Batten, 1988, Wilson, 1981) and dynamic choices in spatial systems (Fischer, Nijkamp and Papageorgiou, 1990). In this context, it should be noted that when time-series data is lacking, empirical applications can be substituted by simulation experiments where the parameters play a fundamental role (since their speed of change leads to very different qualitative configurations).

Among the nonlinear models, we should firstly mention the numerous studies and experiments linked with *spatial interaction models*, based on the dynamic models developed by Harris and Wilson (1978). The main relevance of spatial interaction models is their analytical structure, unifying - both at a static and dynamic level - most of the models adopted in urban and regional science (see Reggiani, 1990). The stochastic versions of dynamic spatial interaction models in particular (see e.g. Vorst, 1995, Nijkamp and Reggiani, 1992a) constitute the first attempt to model dynamic-stochastic factors in urban evolution.

Secondly, there is the branch of research which analyses randomness in the dynamic choice process. This has led to the investigation of *dynamic logit* models as well as their connections with spatial interaction models (see Leonardi, 1983, Nijkamp and Reggiani, 1992). Logit models are based on random utility theory (see Domencich and McFadden, 1975), thus offering a strong economic-theoretical feature to spatial choice problems. In particular the 'time-nested random utility theory' approach, applied to residential mobility, introduces the dynamic evaluation of the decision maker. It can also be demonstrated that fast dynamics or delay effects in a logit-type decision process can lead to chaotic irregularities (Nijkamp and Reggiani, 1992a). In conclusion, these results are significant in the sense that they integrate economic theory (such as random utility theory) with dynamic-spatial processes of urban evolution (see also

Camagni, in this volume).

30.2.3 Models of Chaos

The 'adventure' of chaos theory in the 1980s has led to great interest among scientists in the various spatio-temporal typologies of unstable dynamics produced by *nonlinear deterministic* systems and hence endogenous mechanisms. Chaos models applied to urban evolution can identify those components that seem to 'escape' prediction and are, therefore, able to destabilise other components or even the whole city-system. A review of the numerous experiments and applications of chaos theory in economics and urban/regional science can be found in Nijkamp and Reggiani (1992a, 1993, 1995).

It is interesting to note that most of the applications refer to two prototype chaos models: May's model (1976) and Lorenz's model (1963). The former represents a logistic difference function, where the state variable is expressed in relative terms. An important result is the demonstration of the analytical link between the May model and the dynamic logit model, showing the possibility of generating (macro) irregularities without having introduced external perturbations (see, for example, Reggiani 1990, Sonis, 1992). This result has also opened an interesting debate in spatial economics on the formal structure of dynamic models: whether they should be discrete or continuous. It is known that dynamic difference models adopted in applications where time data are available in discrete form (as in urban dynamics) can introduce unpredictable and chaotic behaviour for particular parameter values and initial conditions[1]. Dynamic continuous models on the other hand frequently have stability characteristics (for a discussion of this issue, with reference to industrial market experiments, see Thill and Wheeler, 1995).

This brings us to the Lorenz model, which does not offer methodological links with economic science, due to its physical foundations (see Nijkamp and Reggiani, 1992a). This model is nevertheless interesting, since it shows the possibility of producing chaos in a dynamic continuous system. Experiments with the Lorenz model in the urban environment have been carried out by Nijkamp and Reggiani (1992a) as well by Zhang (1991), who demonstrated that chaotic phenomena decreased with an increase in the number of parameters, and hence the analytical complexity of the system.

[1] Thanks to the Poincaré-Bendixson's theorem (see e.g. Haken, 1993), it has been shown that chaotic motion is not possible in one/two dimensional continuous systems.

The above findings confirm the incompatibility between behavioural complexity and analytical complexity. They also underline the *relevance of the constraints*, given their fundamental role in governing particular values of the parameters and/or different connections/organisations in the system from which irregular and complex behaviour may arise. However, although chaos models may provide substantial support in identifying irregularities and uncertainties in certain components of the city-system, the problem of modelling the *global evolution*[1] of the system, i.e. the dynamic and complex interaction of its multilayer interacting components, still remains unresolved. This area has been explored recently in spatial economics by investigating the potential of evolutionary theories and, in particular, the network concept.

30.2.4 Evolutionary Models Based on Bio-Ecological Theories

We have already stressed that spatial systems, and hence the city, are dynamic 'man-made multicomponent' systems with a high degree of variability and uncertainty. Research in spatial economics has focused on *evolutionary networks*, since these constitute a clear example of spatial multicomponent organised structures. One approach is the bio-ecological approach, first adopted in economics by Malthus (1798) in his investigation of population dynamics.

The debate in ecology and biology between proponents of continuous evolution ('gradualism', Darwin, 1859) and discontinuous evolution ('saltationalism', Wright, 1931) has been joined by various economists, for instance, by Marshall (1920) and Schumpeter (1934), as well as by Boulding (1978), who takes a position between these two main streams of research, arguing for the possibility of a mix of continuous and discontinuous processes (for a review, see Nijkamp and Reggiani, 1992b, Rosser, 1991).

The formalisation of evolutionary processes in a spatial/economic context is mainly based on bio-ecological models which permit the analysis of competition/complementarity/substitution in a network. Economists have generally adopted one of two main bio-ecological models: (a) for one-dimensional systems, the logistic equation, particularly its discrete version, i.e. the May equation, which is able to produce instability (as explained in Section 2.2), and (b) for multidimensional systems, the Lotka-Volterra equations (see Pumain, in this volume). These, being made up of

[1] It is well known that chaotic phenomena are confined to limited spatio-temporal domains.

interrelated logistics, are also able to map bifurcations and irregular movements.

An approach which integrates these two types of model is *niche theory* (for a theoretical/methodological review, see Nijkamp and Reggiani, 1992b). The formulation for modelling the evolution of a hierarchical self-organising system is the following:

$$\dot{y}_i = \alpha_i y_i (C_i - \sum_{j=1}^{N} \beta_{ij} y_j) \tag{1}$$

where the variable y_i represents the population of a species i (i=1, 2, ... j, ... N), α_i is the growth rate of population y_i, the constant C_i is a function of the carrying capacity for the ith species and the β_{ij} coefficients represent the competition coefficients between species (measurable from the niche overlapping).

It is evident from expression (1) that the case of only one species leads to the logistic equation, while the multiple interactive species system (1) produces a series of Lotka-Volterra type equations. The discrete version of system (1) therefore offers the possibility of modelling discontinuous processes (see the previous subsection). Niche theory has the important methodological function of *unifying nonlinear models* (such as logistic functions) and *chaos theory*, by offering 'universal' evolutionary principles to economic and social science.

Applications of niche theory to urban and regional evolution can be found in the models by Allen and Sanglier (1981) and Camagni, Diappi and Leonardi (1986) which investigate urban dynamics, the studies by Gambarotto and Maggioni (1996) and Johansson and Nijkamp (1987) with applications to industrial dynamics, and finally in the experiments by Grübler and Nakicenovic (1995) and Nijkamp and Reggiani (1995) concerning transport systems.

Niche theory seems to be extremely useful in the study of network evolution. In this context, the city is interpreted as a niche, whose complex evolution is determined by the effects of cooperation/competition between the various sub-niches. We may also identify the evolution of wider 'city-networks' by means of its connections with other interactive networks. The result is the so-called *'multilayer-niche' system* whose main parameters (carrying capacity C_i and growth rate α_i) can also be considered dynamic functions, determining a complete evolutionary interaction as well as an increase in complexity (see Nijkamp and Reggiani, 1995b and Reggiani and Nijkamp, 1995c, for methodological experiments related to transport systems and innovation diffusion networks respectively).

It is evident however that this approach, although satisfactory from the methodological viewpoint, may provoke empirical problems due to the lack of space-time data and the inherent computational complexity. While the adoption of geographical information systems can surely solve certain data problems (see Fusco Girard, Van Geenhuizen and Nijkamp, Tadei and Dellasette, in this volume), it is necessary to explore other methodological approaches able to map network complexity, including the uncertainty produced by the system itself. A typology of these approaches, conceptually different from the previous ones, is based on those derived from Artificial Intelligence, in particular neural network models.

30.2.5 Neural Network Models

Neural network (NN) approaches are distinguished from the previous models by their capacity to *generalise* from experience without fixing *a priori* any behavioural rule/model among variables. Applications of neural networks are nowadays abundant in many disciplines, especially transport and spatial economics (for a review, see Reggiani, Romanelli and Tritapepe, 1995), where it is necessary, as in the case of the complex city, to adopt tools able to map the connectivity, communication, adaptivity, control and prediction patterns.

NNs are characterised by the property of *self-organisation,* therefore such models are able to independently establish the parameters of the phenomenon being studied and emulate the *stochastic* decision events suggested by behavioural theory. Consequently, it is not necessary to add a random variable when using a utility function related to the decision process (as in the case of discrete choice models). In this respect the decision process can be imagined as an NN, where the decision-maker's choice is a result of the network adaptation. NNs thus represent 'a useful tool for exploring the microfoundations of economic processes based on adaptive and emerging rationality' (Dosi, 1993).

NNs were developed to reproduce the structure and functioning of nervous system which, given the large number of components linked by interactions and feedbacks, is undoubtedly highly complex dynamic system. NNs are therefore well suited to representing spatial/economic systems, which can also be interpreted and studied as complex systems. As stated by Holland (1988) "The global economy is an excellent example of a nonlinear adaptive network... Nonlinear adaptive networks provide a substantial extension of traditional economic theory. The basic components of the latter are rational fixed agents, who operate in a linear, static, statistically foreseeable environment. On the contrary, nonlinear networks take into account the continuous nonlinear interactions occurring between a great number of agents, whose characteristics change in time" (quoted in Fabbri and Orsini, 1993, p. 239).

In the following subsections we shall briefly describe the main characteristics of NNs, by illustrating in particular the 'back-propagation' model, which is the one most used in spatial/economic applications.

i. General characteristics of the neural network

NNs were conceived with the aim of studying the organisation of the human brain, by simulating the biological processes occurring during the learning phase. In this process, information transmission among the nervous cells (*neurons*) is spread by means of electrochemical signals produced by spatial connections (synapses) between *axons* and *dendrites*. An axon is a single long fibre of the nervous cell (ending with ramifications), while dendrites are ramifications around the nucleus. The electric stimuli, which reach the neuron by means of synapses, polarise the cellular membrane giving rise to the so-called *action potential*. This phenomenon firstly appears near the nucleus, then spreads over the adjacent parts of membrane, giving rise to electric impulses propagating along the axon. The impact of these stimuli on other neurons depends, however, on the (dynamic) characteristic of the synapses, which may be *inhibiting* or *activating*. This underlines the fundamental role of synapses in the dynamics of the nervous system (see Fabbri and Orsini, 1993).

Even though the brain's anatomy is not fully known, NNs have been conceived as a simplified reproduction of a highly interconnected network, inspired by the neural structure of the human brain, whose nodes represent neurons and connections synapses. Each node represents a logical unit processing and transmitting the information received from other neurons, while each connection exerts an excitatory or inhibitory effect on its own terminal neurons.

This structure enables NNs to capture some typical characteristics of *intelligent behaviour*: firstly, their ability to 'learn' a functional relationship, by training themselves on the basis of experience (see Openshaw, 1992), secondly, their capacity to self-organise, and finally, their ability to generalise, i.e. to create new 'output' in accordance with previously studied examples. These properties provide NNs with flexibility and capability when facing complicated and noisy information, as in the case of a complex network/city.

The NN literature (see e.g. Kosko, 1992, Maren, Harston and Pap, 1990, Reggiani, Romanelli and Tritapepe, 1995, Terna, 1992) suggests the following methodological implications for spatial/economic science. Firstly, there is evident potential in the field of *nonlinear dynamic systems and/or in the presence of stochastic factors*. In these cases the NN approach, by means of inner representations (see Openshaw, 1992), creates a functional approximation, similar to a regression model, overcoming the limits inherent in the linear models usually adopted. Secondly, it is interesting to point out that NNs *are suitable in a forecasting context*, given their ability to generalise, i.e. to elaborate new situations for different spatio-temporal scenarios. It should be noted that this characteristic depends on the chosen training set and the architectural configuration of the network (the number of hidden levels, the number of neurons on these levels, etc.) (see e.g. Fischer and Gopal, 1994, Mussone, 1994). The first point presupposes the choice of a representative sample in order to get an 'unbiased' distribution of observations, while the optimal configuration is reached only by means of experimental methodologies (see also Malliaris and Salchenberger, 1992).

In the next subsection, we briefly describe the basic theoretical components of a network typology well known in scientific literature, the back-propagation network. This type of network has been applied in many different sectors, particularly the transport field, where it is used for the construction of O/D matrices, the forecasting of vehicle-flows, the simulation of user behaviour, etc. The NN literature on transport applications indicates that the back-propagation algorithm has been unanimously adopted.

ii. The back-propagation multilayer-feedforward approach

The NN structure can be linked to a weighted and oriented graph, consisting of a set of nodes and weighted links (Fig. 1). This structure implies the need to identify the functional relationships (connections) between neurons in order to define the most suitable architecture for a given problem. In general the network architecture is subdivided into layers, each layer containing a different number of neurons. Neurons are linked in three ways ('forward', 'backward' and 'lateral') leading to so-called multilayer, single-layer and bi-layer architectures (Maren, 1990). A multilayer-feedforward network is characterised by a layer of input neurons (1st layer), connected to a second layer of neurons (2nd hidden layer), which is then connected to the layer of output neurons (3rd layer).

The properties of this type of network are total connection between neurons and unidirectional connections from input to output layers (without feedback), which ensures convergence to a solution. For these reasons, this NN typology is regarded as being very simple from a computational viewpoint.

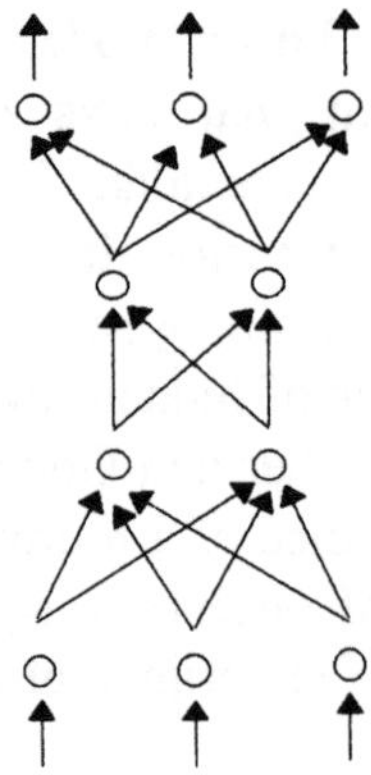

Fig. 1 A back-propagation network

The behaviour of an NN depends on the *activity* of the single neuron, i.e. on the generation and transmission of the action potential occurring when the sum of the perceived stimuli exceeds a given threshold. Each neuron therefore has its own activation state determined by the activity of the neurons linked to it (Fig. 2).

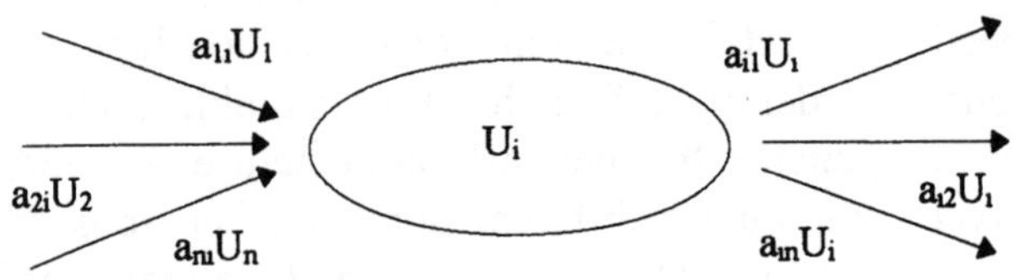

Fig. 2 A neuron activity model

Generally the following activation rule is assigned:

$$U_i = f\left(\sum_{j=1}^{n} a_{ij}\, U_j + s_i\right) \tag{2}$$

where U_i is the output of the unit under analysis, i.e. the activation of the ith neuron, s_i is its threshold[1] ('bias'), a_{ij} the weight of synaptic connections between the jth neuron and the neurons linked to it; U_j the output of each network unit and f the activation (transfer) function. The transfer function can assume different forms (see e.g. Kosko, 1992, Maren *et al.*, 1990). The most well-known function, also adopted in the back-propagation network, is the logistic/sigmoid function:

$$U_i = \frac{1}{1+e^{-B_i}} \tag{3}$$

where
$$A_i = \sum_j a_{ij}\, U_j + s_i \tag{4}$$

The main characteristics of this function are: continuity, increasing monotonicity and differentiability at all points with positive derivatives throughout (Fischer and Gopal, 1996).

A further step in NN behaviour concerns the learning phase. This is a fundamental phase in which the network adapts or organises itself each time it examines a new sample. In other words, during the learning phase the network changes its synaptic weights in order to produce the desired output. During this process the weights of the connections are subject to further adjustments governed by a specific rule, such as Hebb's rule:

$$\Delta a_{ij} = \gamma\, U_i\, U_j \tag{5}$$

where Δa_{ij} is the difference in the weights of the connection between neurons i and j, while U_i and U_j are the activations of neurons i and j, and γ is a learning coefficient which regulates the sensitivity of the algorithm to the difference between the output provided by the network and the expected output. A back-propagation network analyses the input-output

---

[1] The symbol s_i is introduced in order to differentiate the threshold value of the jth neuron. This value, as well as the values of the weights a_{ij}, is automatically defined within the back propagation learning procedure.

set by means of the 'generalised delta rule'[1], which is based on the minimisation of the above difference or error. The error term can then be formulated as follows:

$$\varepsilon = \frac{1}{2} \sum_{j=1}^{N} (O'_j - O_j)^2 \tag{6}$$

where ε is the total error, O'_j the expected output, O_j the resulting output and N the number of neurons at the output layer. Several algorithms - like the error back-propagation - can be used to modify the weights of the connections. The total error is back propagated along the connections among the neurons; it is subject, for example, to Hebb's rule which assigns a new weight $a_{ij}(t+1)$ at time $t+1$ as follows:

$$a_{ij}(t+1) = a_{ij}(t) + \psi \Delta a_{ij}(t) \tag{7}$$

where $a_{ij}(t)$ is the weight related to the connection between the ith and jth neuron at time t, ψ is the learning coefficient (fixed a priori) and Δa_{ij} is formulated as follows:

$$\Delta a_{ij}(t) \propto - \frac{\delta \varepsilon}{\delta a_{ij}(t)} \tag{8}$$

where the function ε, defined in equation (6), is represented in an n-dimensional space (where n is the number of connections). It is then possible to find the minimum value of such a function by means of the 'descent of gradient' technique in order to reach the convergence of the algorithm. Other formulations of the error ε are, however, available (see e.g. Fischer and Gopal, 1996, Reggiani and Tritapepe, 1995).

It should also be noted that in this learning phase particular attention should be paid to the choice of all the 'strategic' elements which can influence an 'optimal' network - for example the training (random or sequential) choice, the objective error choice and the initial conditions of parameters (weights, coefficients, etc.). In particular the parameter's choice/evaluation is usually undertaken empirically. Once the NNs have 'learned', they can be used to simulate choices under uncertain conditions as well as for simulating adaptivity, selection and evolution in interactive

---

[1] For an illustration of the generalized delta rule see Khanna, 1991.

contexts (Fabbri and Orsini, 1993).

A limit concerning the use of NN is given by the data-set. If the data-set is small, the variability of NN results increases, on the contrary, if the data-set is large the variability of NN results decreases so much that it becomes independent of the particular type of model adopted. This constraint shows the impossibility of using NNs in extremely simple systems, even though they may be complex in their functioning. An empirical example is the *city-network system,* such as the system constituting the major Italian metropolitan areas. Given the complementarity relationships between these cities, *this system can be conceived of as a high-order complex city*, where the possibility of choosing between three competitive transport modes (road, train, air) leads to a well-integrated spatial system (the so-called 'integrated city-system', Lambooy, 1993).

30.3 Experiments Relating to a Complex City

30.3.1 Introduction

In order to verify the approaches illustrated above, some experiments relating to a modal split analysis on routes between seven major Italian cities have been carried out. The aim of the application was to provide an estimate of the future demand for transport in the presence of the new high-speed (H-S) train (see Reggiani, Tritapepe and Cisbani, 1995a). The seven areas considered are those served by the new Turin-Milan-Rome-Naples line. This system of urban areas has been interpreted as a *complex city* where the three modes 'road', 'train' and 'air' are in competition. Fig. 3 shows the prevalent mode for each link. It should firstly be noted that an NN approach and the related software has been developed to explore this methodology (Reggiani and Tritapepe, 1995). However, given the small number of examples under analysis it was not possible, as previously pointed out, to generalise the results. In view of the limited amount of data, it was decided to use the classical choice models, more precisely the *nested logit models*, to analyse the structural trends of this complex city, in the light of new transport scenarios.

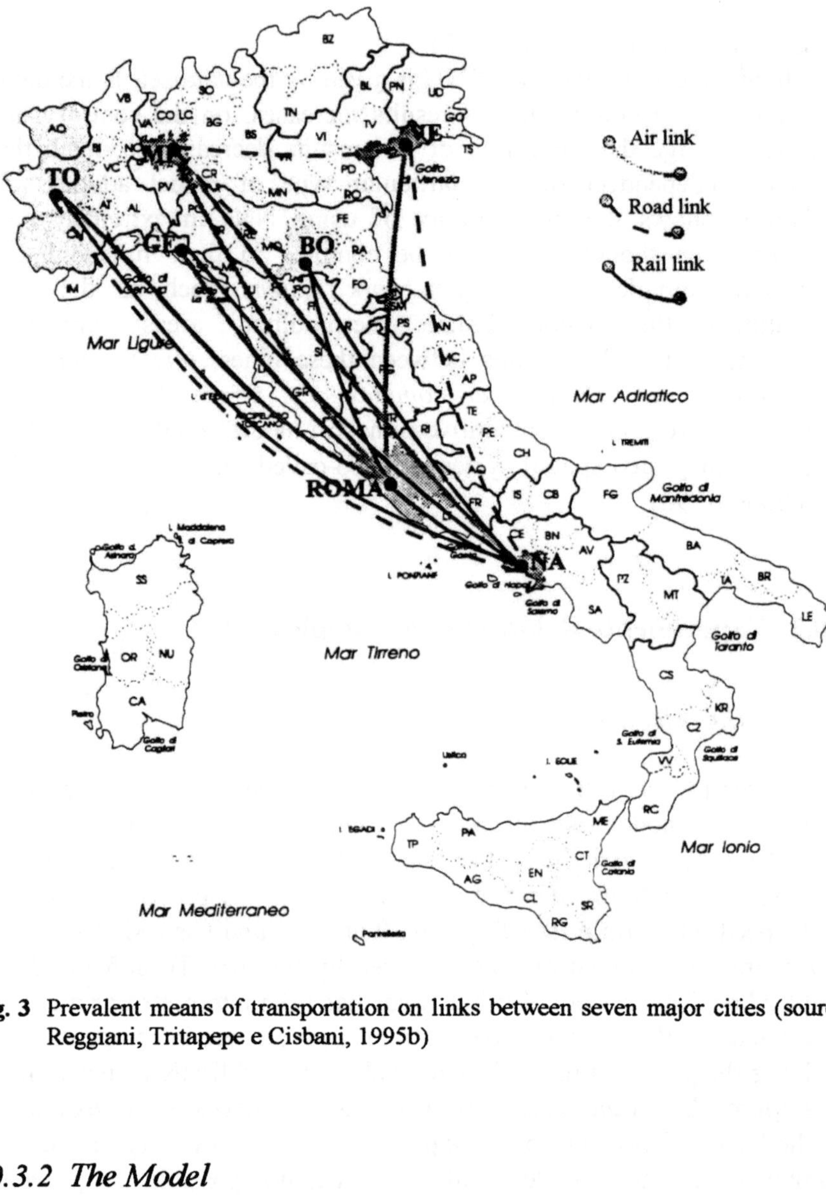

Fig. 3 Prevalent means of transportation on links between seven major cities (source: Reggiani, Tritapepe e Cisbani, 1995b)

30.3.2 The Model

The model adopted for the experiments, i.e. the nested logit model, belongs to the wide category of discrete choice models (static and nonlinear) based on random utility theory (see Section 2). The nested logit model, widely used in transport economics, offers the possibility of modelling sequential choices, such as the destination and mode choice. From a theoretical viewpoint the analysis involves the calibration of a nested logit model in order to obtain an estimation of the commuters travelling from the origin i (i=1, ... ,7) to the destination d (d=1, ... ,7) by using the mode m (m=1,2,3) (Fig. 4).

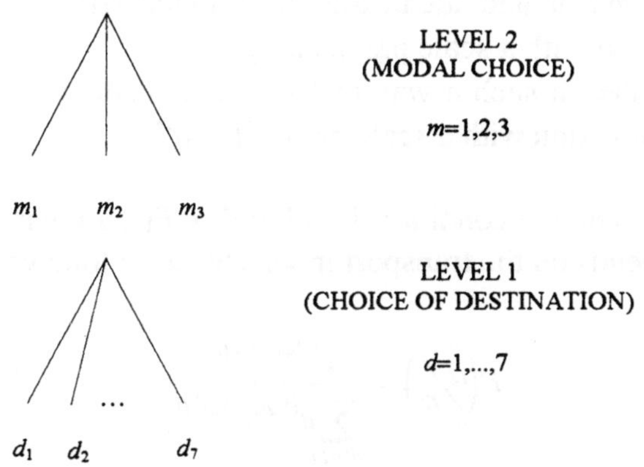

Fig. 4 The nested logit model adopted

The total utility u_{dm} of the joint choice (destination and mode) can be formulated, in the framework of random utility models, as follows (Ben-Akiva and Lerman, 1985):

$$u_{dm} = v_m + v_d + v_{dm} + \varepsilon_m + \varepsilon_d + \varepsilon_{dm} \tag{9}$$

where v_m represents the systemic component of the utility related to the transport mode m (m=1, ... ,3); v_d represents the systemic components of the total utility related to destination d (d=1, ... ,7); v_{dm} represents the 'residual' of the systemic component of the utility related to the combination (d,m); ε_d and ε_m represent the random utility components associated with the destination and the mode respectively, while ε_{dm} represents the 'residual' of the random component of the utility.

If we assume that ε_d is distributed with variance = 0, it is possible to rewrite formulation (9) as follows (Ben-Akiva and Lerman, 1985, Domencich and McFadden, 1975):

$$u_{dm} = v_m + v_d + \varepsilon_m + \varepsilon_{dm} \tag{10}$$

Furthermore, it is possible to derive the nested logit model, if we assume that:

a. ε_m and ε_{dm} are independent for $d \in D$ and $m \in M$;
b. ε_{dm} are Independent and Identically Distributed (IID) according to a Gumbel function with a scale parameter μ^d;
c. ε_m are distributed in such a way that $\max(u_{dm})$ is distributed according to a Gumbel function with a scale parameter μ^m

We can then calculate the conditional probability *P(d/m)* that the choice of destination *d* depends on the transport mode choice as follows:

$$P\left(\frac{d}{m}\right) = \frac{e^{(v_{dm}+v_d)\mu^d}}{\sum_{d' \in D} e^{(v_{d'm}+v_{d'})\mu^d}} \tag{11}$$

The subsequent step involves determining the natural logarithm of the denominator of expression (11), by considering the constant $1/\mu^d$, in other words the inclusive value v'_m which reads as follows:

$$v'_m = \frac{1}{\mu^d} \ln \sum_{d \in D} e^{(v_d + v_{dm})\mu^d} \tag{12}$$

Finally we can calculate the marginal probability *P*(*m*) of choosing the transport mode:

$$P(m) = \frac{e^{(v_m + v'_m)\mu^m}}{\sum_{m' \in M} e^{(v_{m'} + v'_{m'})\mu^m}} \tag{13}$$

The nested logit model adopted is a two-layer hierarchical model. It is also possible to derive multilayer nested logit models, for example, by considering the route choice (besides the destination and mode choice) (Leonardi, 1983, Reggiani, 1985, Reggiani and Zaccarin, 1989).

30.3.3 Data-Set

As previously explained, the competition analysis between the transport modes was performed for the seven cities, considering all three modes, road, rail and air (Fig. 3). However, the available information, based on the subdivision of Italy into zones, had some drawbacks: firstly the intra-zonal flows were missing, secondly, the centroid of each zone had to be considered as the origin/destination point, with the possibility of bias for flows relating to large areas.

The socio-economic attributes adopted in the utility function of the nested logit model (11)-(13) were the following: *residential population*, *total employment* and *added value* (calculated as the difference between the added value in the destination zone and in the origin zone). *Distance*, *cost* and *travel time* were considered as transport attributes. As far as costs and time were concerned, the exit/entrance cost/time (for each mode) were considered, as well as the actual travel cost/time. Finally a dummy variable was constructed, assigning the value 1 to the zone including the seat of local government, 0 otherwise.

The choice of the above attributes was determined by the available information. It should also be noted that the flows were not disaggregated by social class or travel purpose (e.g. study, work, shopping, entertainment, etc.). Consequently, the model included an average utility function, identical for each individual facing the destination/transport mode choice in each of the seven areas.

30.3.4 Calibration of the Nested Logit Model

The calibration of the nested logit model was carried out for the main flow-lines illustrated in Fig. 3 using the data-set previously described. The results are shown in Tables 1, 2 and 3.

Table 1 shows the parameter values for the attributes associated with the utility function of the nested logit model at the 1st layer as well as the asymptotic *t*-values. We can observe that the parameter associated with the travel cost/time has a negative sign, thus the travel cost/time can be interpreted as a 'generic cost' in the utility function of the destination choice.

As far as modal choice is concerned (2nd layer; Table 2), we should point out that there is no 'mean cost', since in this phase the related parameter is not significantly different from zero. This result would seem reasonable since we are likely to be dealing with high income commuters for whom travel time is more relevant than cost. This is supported by modal choice figures for the Milan-Rome line: 61% of users choose 'air', 21% 'rail' and 18% 'road'.

Variable	Coefficient	Standard Error	Ratio t	Prob$\|t\|>x$
cost	$-0{,}35277\cdot10^{-3}$	$0{,}3643\cdot10^{-4}$	-9,684	0,00000
time	$-0{,}73040\cdot10^{-2}$	$0{,}7796\cdot10^{-4}$	-93,695	0,00000
dummy	-0,96109	$0{,}6817\cdot10^{-1}$	-14,099	0,00000
egression time	-0,61316	$0{,}2042\cdot10^{-1}$	-30,028	0,00000
v. a.	$-0{,}26272\cdot10^{-8}$	$0{,}2946\cdot10^{-9}$	-8,917	0,00000
pop.	$-0{,}41699\cdot10^{-6}$	$0{,}1080\cdot10^{-7}$	-38,593	0,00000
egression cost	0,76511	$0{,}2406\cdot10^{-1}$	31,802	0,00000
DUMMYC	-1,0805	$0{,}4617\cdot10^{-1}$	-23,403	0,00000
DUMMYCE	1,5781	$0{,}8881\cdot10^{-1}$	17,770	0,00000
DUMMYTE	0,11411	$0{,}2990\cdot10^{-1}$	3,816	0,00014
DUMMYP	-0,28053	$0{,}2562\cdot10^{-1}$	-10,951	0,00000

Table 1 Logit model parameters: level 1, destination choice (source: Reggiani, Tritapepe and Cisbani, 1995b)

Variable	Coefficient	Standard Error	Ratio t	Prob$\|t\|>x$
average time	$0{,}19214\cdot10^{-8}$	$0{,}3393\cdot10^{-3}$	-5,663	0,00000
access time	0,12298	$0{,}1114\cdot10^{-1}$	11,042	0,00000
access cost	-0,14271	$0{,}1585\cdot10^{-1}$	-9,003	0,00000
DUMMYCA	0,71376	0,1036	6,893	0,00000
DUMMYTA	0,94407	$0{,}4274\cdot10^{-1}$	22,088	0,00000
DUMMYTM	2,4192	0,1016	23,814	0,00000
DUMMYCM	0,20810	$0{,}5550\cdot10^{-1}$	3,749	0,00018
IV	-0,18031	$0{,}1266\cdot10^{-1}$	-14,242	0,00000

Table 2 Logit model parameters: level 2, modal choice (source: Reggiani, Tritapepe e Cisbani, 1995b)

However, given the ‘economic’ significance of cost, an additional nested logit model was calibrated adding ‘mean cost’ to the other attributes. The new parameter values are shown in Table 3.

Due to the lack of statistical significance of the ‘mean cost’, we can conjecture that by varying this attribute, the choice probabilities of the three modes will remain more or less constant. On the contrary, when the ‘travel time’ and ‘mean time’ are associated with significant and negative parameters, we may predict a decrease of flows as travel or mean time increase.

Variable	Coefficient	Standard Error	Ratio t	Prob $\|t\|>x$
average cost	$0,84504 \cdot 10^{-4}$	$0,1077 \cdot 10^{-3}$	0,785	0,43262
average time	$-0,18181 \cdot 10^{-2}$	$0,3653 \cdot 10^{-3}$	-4,978	0,00000
access time	0,11980	$0,1177 \cdot 10^{-1}$	10,181	0,00000
access cost	-0,13497	$0,1841 \cdot 10^{-1}$	-7,331	0,00000
DUMMYCA	0,63690	0,1396	4,563	0,00001
DUMMYTA	0,94875	$0,4315 \cdot 10^{-1}$	21,989	0,00000
DUMMYTM	2,4121	0,1019	23,674	0,00000
DUMMYCM	0,16691	$0,7569 \cdot 10^{-1}$	2,205	0,02744
IV	-0,18384	$0,1346 \cdot 10^{-1}$	-13,660	0,00000

Table 3 Logit model parameters considering mean cost: level 2, modal choice (source: Reggiani, Tritapepe e Cisbani, 1995b)

30.3.5 The Forecasts

As explained above, the aim of the modelling experiment was to forecast the impact of the new H-S train on modal split, bearing in mind that the introduction of the new train will result in a decrease in travel time and an increase in travel cost. Even though the impact will be felt on the whole national network[1], the present application concerns predictions related only to the city-network consisting of the metropolitan areas directly affected by the high speed route (see Fig. 3): Turin, Genoa, Milan, Venice, Bologna, Rome and Naples. The nested logit experiments do not take into account time/cost variations for the Milan-Venice and Genoa-Rome lines since the introduction of the H-S train on these links is still uncertain.

Table 4 shows the forecasts based on the calibration phase. It is evident that the new line will give rise to an increase in rail flows especially in the areas of Milan, Bologna, Rome and Naples. An interesting comparison can be made by graphical representation of the market shares for the three modes on each link before and after the H-S train's introduction (Fig. 5 and Fig. 6).

Figures 5-6 show that after introduction of the H-S train, rail travel is estimated to capture 43% of the trips between the seven metropolitan areas. This result highlights the behaviour of high-middle income

[1] Concerning the predictions - by means of nested logit models - related to the matrix of 67×67 areas (corresponding to a subdivision of the whole of Italy into zones), see Reggiani *et al.*, 1995b.

commuters who seem sensitive to the time factor. By decreasing journey time, it is possible for one mode to dominate the others. Obviously these results could be modified by introducing other relevant attributes in the utility function, e.g. frequency, safety, comfort, congestion, etc. This underlines once again the need to design modular information systems (Van Geenhuizen and Nijkamp, Tadei and Dellasette, in this volume) able to provide data suitable for analyses seeking to identify structural trends in the city-system.

Origin	Destination	Train Flux		Road Flux		Air Flux	
		actual	predicted(AV)	actual	predicted(AV)	actual	predicted(AV)
Turin	Rome	305	417	410	402	589	434
Turin	Naples	103	149	103	106	52	53
Milan	Venice	1079	417	1334	998	62	86
Milan	Bologna	1012	1069	1585	1298	1	83
Milan	Rome	743	1680	659	901	2180	2336
Milan	Naples	424	566	325	238	584	316
Venice	Milan	1240	1028	1334	1062	66	99
Venice	Rome	191	454	158	259	605	515
Venice	Naples	39	173	99	69	49	73
Genoa	Rome	275	232	302	289	369	410
Genoa	Naples	84	81	49	77	65	54
Bologna	Milan	1169	1260	1585	1106	2	0
Bologna	Rome	498	692	258	440	336	209
Bologna	Naples	150	262	125	116	54	91
Rome	Turin	338	388	410	329	592	583
Rome	Milan	833	1544	659	683	2181	1974
Rome	Venice	145	263	158	248	607	751
Rome	Genoa	263	160	302	121	361	297
Rome	Bologna	549	785	258	525	328	75
Rome	Naples	3662	3153	3141	2775	259	382
Naples	Turin	122	189	103	121	45	156
Naples	Milan	457	857	325	256	578	272
Naples	Venice	31	107	99	91	55	142
Naples	Genoa	80	77	49	45	62	38
Naples	Bologna	135	424	125	194	56	56
Naples	Ancona	3867	3405	3141	2923	289	267

Table 4 Results of forecasts with nested logit model (source: Reggiani, Tritapepe e Cisbani, 1995b)

30.4 Concluding Remarks

The present work has analysed, from both a methodological and empirical viewpoint, approaches to mapping the complex dynamics of a city-system. In order to test the capacity of a nested logit model to provide transport forecasts, the impact of the introduction of the new high-speed rail system on travel choices in seven major cities in Italy linked by the route was investigated. Clearly the nested logit experiments do not solve the problem of modelling uncertain and complex realities (which are better captured by the theoretical approaches/methodologies illustrated in Section 2). They are however able to underline structural trends of the transport component resulting from external interventions/perturbations (such as the introduction of the H-S train). Experiments on the sustainability or viability of these new structural forms could then follow.

The general problem of the empirical limits inherent in the theoretical approaches investigated in Section 2 remains an open research question. The small data set for our case study does not guarantee the suitability of an NN approach (even for a comparison with the nested logit predictions). The lack of time data would also hamper empirical applications of nonlinear models, especially the ecological models which are - theoretically at least - able to solve dynamic problems of competition, complementarity and substitution in components of the complex-city under analysis.

As far as the methodological aspect is concerned, currently the only 'operational' instrument appears to be the *decodification* of system complexity by means of analytical tools such as spatial interaction models and, in particular, logit models and their derivations. These have the significant advantage of providing connections with economic theory, as well as providing good performance, which has been confirmed in a number of applications. It is evident however that in order to analyse the complex and multidimensional city, with uncertain and complicated relationships between variables, it is necessary to discover new methodological instruments which are more adaptable to real economic behaviour and also more powerful from the mathematical and computational viewpoint.

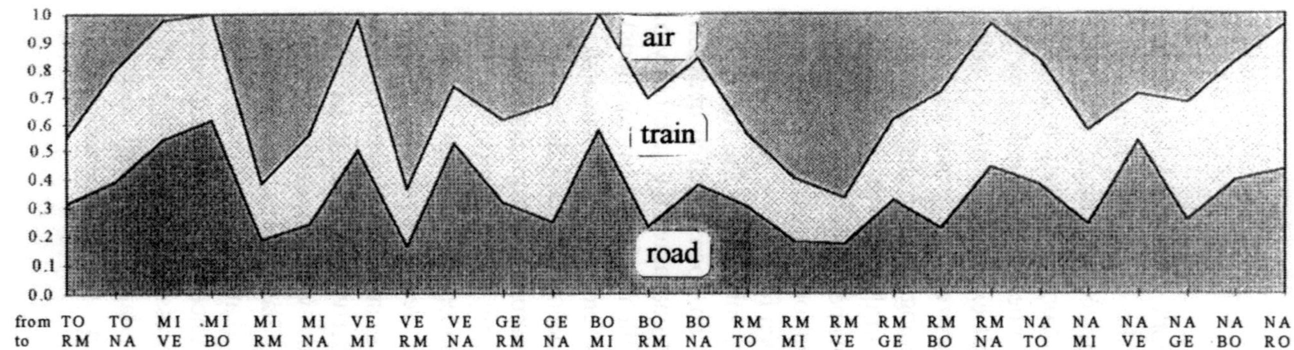

Fig. 5 Market share before HS train introduction (source: Reggiani, Tritapepe and Cisbani, 1995b)
(TO=Turin, MI=Milan, GE=Genoa, VE=Venice, BO=Bologna, RM=Rome,NA=Naples)

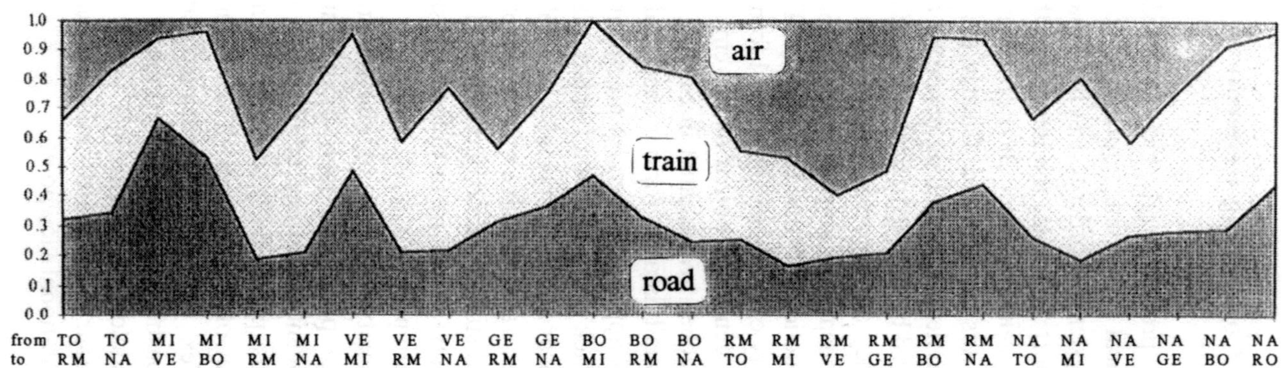

Fig. 6 Market share after HS train introduction (source: Reggiani, Tritapepe and Cisbani, 1995b)

Acknowledgements

This project was carried out by CESPRI/Bocconi University as part of the Progetto Finalizzato Trasporti 2, Contract no. 96.00098.PF74 related to the High-Speed Train Project (commissioned by the Italian State Railways). The author would also like to thank Tommaso Tritapepe and Luca Cisbani for their contributions to last part of Section 2 and Section 3 respectively.

References

Allen P.M., Sanglier M. (1981) Urban Evolution, Self-Organisation and Decision-Making, *Environment and Planning A, 13*, 167-183.

Andersson Å.E. (1995) Creation, Innovation and Diffusion of Knowledge: General and Specific Economic Impacts in Bertuglia C.S., Fischer M.M., Preto G. (eds.) *Technological Change Economic Development and Space*, Springer-Verlag, Berlin, 13-33.

Andersson Å.E., Batten D.F. (1988) Creative Nodes, Logistical Networks and the Future of the Metropolis, *Transportation, 14*, 281-293.

Beckenbach F., Pasche M. (1994) Nonlinear Ecological Models and Economic Perturbation - Sustainability as a Concept of Stability Corridors, Models of Sustainable Development, International Symposium, Paris.

Ben-Akiva M., Lerman S.R. (1985) *Discrete Choice Analysis: Theory and Application to Travel Demand*, MIT Press, Cambridge, Massachusetts.

Boulding K.E. (1978) *Ecodynamics: A New Theory of Social Evolution*, Sage, Beverly Hills.

Camagni R., Capello R., Nijkamp P. (1994) Sustainable City Policy: Economic, Environmental, Technological, in Van den Meulen G., Erkelens P. (eds.) Urban Habitat: Technische Universiteit Eindhoven, 35-57.

Camagni R., Diappi L., Leonardi G. (1986) Urban Growth and Decline in a Hierarchical System, *Regional Science and Urban Economics, 15*, 145-160.

Campisi D. (1991) I fondamenti della modellistica urbana, in Bertuglia C.S., La Bella A. (eds.) *I sistemi urbani*, vol. 2, Angeli, Milan, 508-551.

Casti J. (1979) *Connectivity, Complexity and Catastrophe in Large-Scale Systems*, Chichester, John Wiley and Sons, New York.

Csàny V. (1989) *Evolutionary Systems and Society: A General Theory*, Duke University Press, Durham and London.

Darwin C. (1859) *On the Origin of Species*, Penguin, Harmondsworth.

Diappi L., Reggiani A. (1997) Evolutionary Models in Spatial Systems: Qualitative vs. Quantitative Approaches, in Camagni R., Senn L. (eds.) *Ten Years of Regional Science*, Angeli, Milan.

Domencich T.A., McFadden D. (1975) *Urban Travel Demand: A Behavioural Analysis*, North Holland, Amsterdam.

Dosi G. (1993) Preface, in Fabbri G., Orsini R., *Reti neurali per le scienze economiche*, Muzzio, Padua, IX-XI.

Fabbri G., Orsini R. (1993) *Reti neurali per le scienze economiche*, Muzzio, Padua.

Fischer M.M., Gopal S. (1996) Learning in Single Hidden Layer Feedforward Network Model, *Geographical Analysis, 28*, 1, 38-55.

Fischer M.M., Nijkamp P., Papageorgiou Y.Y. (1990) *Spatial Choices and Processes*, North-Holland, Amsterdam.

Flood R.L. (1987) Complexity: A Definition by Construction of a Conceptual Framework *Systems Research, 4*, 3, 177-185.

Gambarotto F., Maggioni M. (1996) An Ecological Approach to Regional Development Policies, *Regional Studies*.

Grübler A., Nakicenovic (1991) *Evolution of Transport Systems: Past and Future*, RR-91-8, International Institute for Applied Systems Analysis, Laxenburg.

Haken H. (1983) *Synergetics*, Springer-Verlag, Berlin.

Harris B., Wilson A.G. (1978) Equilibrium Values and Dynamics of AttractivenessTerms in Production-Constrained Spatial-Interaction Models, *Environment and Planning A, 10*, 371-388.

Holland J.H. (1988) The Global Economy as an Adaptive Process, in Anderson P.W., Arrow K.J., Pines D. (eds.) *The Economy as an Evolving Complex System*, Addison Wesley, New York.

Johansson B., Nijkamp P. (1987) Analysis of Episodes in Urban Event Histories, in Van den Berg L., Burn L.S., Klaassen W.H. (eds.) *Spatial Cycles*, Gower, Aldershot, 43-66.

Khanna T. (1991) *Fondamenti di reti neurali*, Addison-Wesley Italia Editoriale, Milan.

Kosko B. (1992) *Neural Networks and Fuzzy Systems*, Prentice-Hall, Englewood Cliffs, New Jersey.

Lambooy J.G. (1993) The European City: From Carrefour to Organisational Nexus, *Tijdschrift voor Economische en Sociale Geografie, 84*, 4, 258-268.

Leonardi G. (1983) An Optimal Control Representation of Stochastic Multistage Multiactor Choice Process, in Griffith D.A., Lea A.C. (eds.) *Evolving Geographical Structure*, Martinus Nijhoff, Dordrecht, 62-72.

Leonardi G. (1987) The Choice-Theoretic Approach: Population Mobility as an Example, in Bertuglia C.S., Leonardi G., Occelli S., Rabino G.A., Tadei R., Wilson A.G. (eds.) *Urban Systems. Contemporary Approaches to Modelling*, Croom Helm, London, 136-188.

Lorenz E.N. (1963) Deterministic Non-Periodic Flow, *Journal of Atmospheric Sciences, 20*, 2, 130-141.

Malliaris M., Salchenberger L. (1992) A Neural Network Model for Estimating Option Prices (mimeo).

Mannermaa M. (1995) Alternative Futures Perspectives on Sustainability, Coherence and Chaos, *Journal of Contingencies and Crisis Management, 3*, 1, 27-34.

Maren A., Harston C., Pap R. (1990) *Handbook of Neural Computing Applications*, Academic Press, San Diego.

Marshall A. (1920) *Principles of Economics*, Macmillan, London.

May R.M. (1976) Simple Mathematical Models with Very Complicated Dynamics, *Nature*, 261, 459-467.

Mees, A. (1975) The Revival of Cities in Medieval Europe: An Application of Catastrophe Theory, *Regional Science and Urban Economics, 5*, 403-425.

Mussone L. (1994) Le reti neurali artificiali nei trasporti, *Trasporti e Trazione, 2*, 56-72.

Nijkamp P., Reggiani A. (1992a) *Interaction, Evolution and Chaos in Space*, Springer-Verlag, Berlin.

Nijkamp P., Reggiani A. (1992b) Spatial Competition and Ecologically Based Models,

Socio-Spatial Dynamics, 3, 2, 89-109.

Nijkamp P., Reggiani A. (eds.) (1993) *Nonlinear Evolution of Spatial Economic Systems*, Springer-Verlag, Berlin.

Nijkamp P., Reggiani A. (1995) Nonlinear Evolution of Dynamic Spatial Systems: The Relevance of Chaos and Ecologically-based Models, *Regional Science and Urban Economics, 25*, 183-210.

Nijkamp P., Reggiani A. (1996) Space Time Synergetics in Innovation Diffusion: A Nested Network Simulation Approach, *Geographical Analysis, 28*, 1, 18-37.

Openshaw S. (1992) Modelling Spatial Interaction Using a Neural Net, University of Leeds, Leeds.

Ploman E. (1984) Reflections of the State of the Art and the Interest of the United Nations University in the Field of Complexity, Report presented at the "Club of Rome Conference", Helsinki, July.

Reggiani A. (1990) Spatial Interaction Models: New Directions, Ph.D. Thesis, Free University, Amsterdam.

Reggiani A., Nijkamp P. (1995a) Competition and Complexity in Spatially Connected Networks, *System Dynamics Review, 11*, 1, 51-66.

Reggiani A., Nijkamp P. (1995b) Multilayer Networks and Dynamic Transport Systems, *Geographical Systems, 2*, 39-57.

Reggiani A., Nijkamp P. (1995c) Modelling Complex Networks, *Discussion Paper TI-5-95-226*, Tinbergen Institute, Amsterdam.

Reggiani A., Nijkamp P. (1996) Towards a Science of Complexity in Spatial-Economic Systems, in Van den Bergh J., Nijkamp P., Rietveld P. (eds.) *Recent Advances in Spatial Equilibrium Modelling: Methodology and Applications*, Springer-Verlag, Berlin, 359-378.

Reggiani A., Romanelli R., Tritapepe T. (1995) Approccio reti neurali e modello logit: un'analisi comparata con riferimento alla mobilità pendolare all'area metropolitana milanese, Proceedings of the Annual PFT2 Meeting, Genoa.

Reggiani A., Tritapepe T., Cisbani L. (1995a) La modellizzazione di flussi di trasporto in Italia: sperimentazioni relative ad analisi di previsione, CESPRI (Bocconi University) Report, Milan (mimeo).

Reggiani A., Tritapepe T., Cisbani L. (1995b) Studio della ripartizione modale attraverso il modello nested logit ed il modello di rete neurale: un'applicazione a sette aree metropolitane italiane, Proceedings of the 15th AISRe Conference, Siena, vol. 2, 680-703.

Reggiani A., Romanelli R., Tritapepe T., Nijkamp P. (1995) Neural Networks: An Overview and Applications in the Space-Economy, Discussion Paper TI 95-40, Tinbergen Institute, Amsterdam.

Reggiani A., Zaccarin S. (1989) Urban Mobility in the City of Venice: An Application of the Nested Logit Model, *Ricerca Operativa, 50*, 49-73.

Rosser J.B. (1991) *From Catastrophe to Chaos: A General Theory of Economic Discontinuities*, Kluwer Academic Publishers, Dordrecht.

Rosser J.B. (1992) Approaches to the Analysis of the Morphogenesis of Regional Systems, *Socio-Spatial Dynamics, 1*, 2, 75-102.

Schumpeter J.A. (1934) *The Theory of Economic Development*, Harvard University Press, Cambridge, Mass.

Sonis M. (1992) Dynamics of Continuous and Discrete Logit Models, *Socio-Spatial Dynamics, 3*, 1, 35-60.

Terna P. (1992) Connessionismo, reti neurali e nuova intelligenza artificiale: l'interesse

per gli economisti, *Sistemi Intelligenti, 3*, 379-419.
Thill J.C., Wheeler A.K. (1995) On Chaos, Continuous-Time and Discrete-Time Models of Spatial System Dynamics, *Geographical Systems, 2*, 2, 121-130.
Vorst A.C.F. (1995) A Stochastic Version of the Urban Retail Model, *Environment and Planning A, 17*, 1569-1580.
Williams T. (1985) A Science of Change and Complexity, *Futures, 17*, 3, 263-268.
Wilson A.G. (1981) *Catastrophe Theory and Bifurcation*, Croom Helm, London.
Wright S. (1931) Evolution of Mendelian Population, *Genetics, 16*, 97-189.
Zhang W.B. (1991) *Synergetic Economics*, Springer-Verlag, Berlin.

31. Mathematical Models for Simulating Urban Mobility Systems: State of the Art and Lines of Development

Ennio Cascetta

31.1 Introduction

Mobility (or transport) systems in an urban area may be defined as the set of elements comprising the mobility demand between various points in the area, the infrastructure supply and corresponding transport services. In this sense, mobility systems should be seen as sub-systems of the city which interact with the location of activities and urban functions in both directions: on the one hand, the configuration of urban activities determines mobility demand, and on the other, the accessibility provided by the transport system supply conditions location choices and thus the system of activities.

The key role played by the mobility system with respect to the city's layout, development and its very existence has long been recognised. By the same token, there has for some time been a consolidated technical tradition in the design and construction of some of the city's important components (such as roads and metropolitan railways).

However, it is only in recent years, with the emergence of the systemic approach to socio-technical problems, that the mobility system has been quantitatively analysed and designed as a whole, explicitly taking into consideration the large number of constituent elements, as well as their internal and external interdependencies. Indeed, in the last few decades there has been a considerable development in mathematical models designed to simulate the key elements of urban mobility systems. Such developments concern both the theoretical analysis and the formulation of a system of mathematical models based on a consistent behavioural-topological paradigm, in addition to the applications of such models to very different contexts, and the relative computational tools.

In this work, we firstly define the mobility system and describe its relations with the system of urban activities (Section 2). Section 3 briefly illustrates the 'standard' paradigm of mathematical models used to simulate the elements of urban transport systems in a large number of applications. Section 4 covers the chief focus of recent research and some of the most promising lines of development regarding the modelling of complex trip chains in urban areas (trip-chaining), the dynamic analysis of transport systems and dynamic assignment to networks. Finally, Section 5 analyses various fields of application of the previously described models.

31.2 The Mobility System and its Relationships with the Activity System

This section describes the chief elements comprising the system of mobility and urban activities, and the main interactions between them. Such elements and relations are diagrammatically represented in Fig. 1.

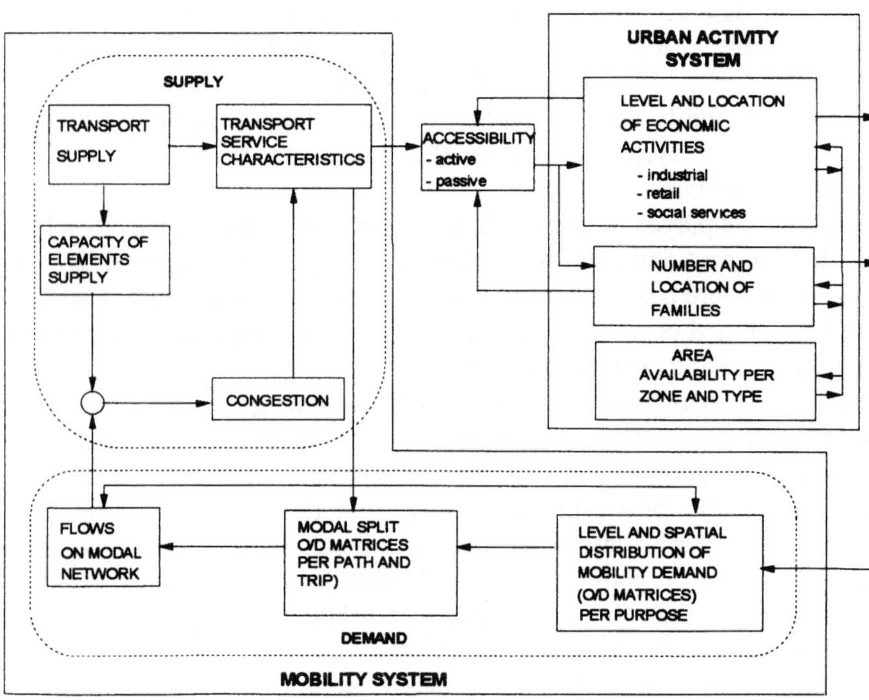

Fig. 1 Relations between transport and activity systems

The system of activities in the urban area may be broken down into three subsystems or 'markets' described by:

- households resident in each zone of the area in question, according to category (defined by income bands, life cycle, composition etc.);
- economic activities located in each zone, according to sector (various divisions for industry, business services, services for households etc.) and type, with relative economic indicators (such as added value) and physical indicators (for example, number of employees);
- surface areas (or volumes) with facilities available in each zone by type (space occupied by industry, apartment blocks, shops etc.) and relative market prices.

Obviously, the components of the activity system interact in varying ways. For example, the number and type of resident families in various zones depends on work opportunities and their distribution, and thus on the subsystem of economic activities. The location of certain types of economic activities (trade, services to families, education, health etc.) in turn depends on family distribution. Lastly, the presence of resident families and economic activities in each zone depends on the availability of compatible premises (houses, shops etc.) and on relative prices/conditions of use.

It is beyond the scope of this work to describe in detail the decisional mechanisms which govern the functioning of single urban subsystems, namely the choice of zone and type/size of family dwellings, the choice of entrepreneurs to locate and start up economic activities and of public decision-makers to locate social services, to earmark areas for development or to set land-use constraints. It is worth noting that, in most of these mechanisms, a substantial role is played by the relative accessibility between the various zones as allowed by the mobility system.

The spatial distribution of families and activities is the basis for mobility demand which arises from the need to use various urban functions in various places. Households, or rather, their components, are the users of the mobility system and make 'mobility choices' (holding of driving licence, car ownership etc.) and 'travel choices' (frequency, time, destination, mode, route and 'chaining') to satisfy their own needs. The overall result of such choices is transport demand, or the number of people who move between the various origin and destination zones in the city, for various purposes, at various times, with the various modes available.

All the trip dimensions are affected in different ways by the transport service characteristics (time, cost, reliability, comfort etc.) of each transport mode available between origin-destination pairs. Hence, destination choice is affected by the generalised transport cost incurred as well as by the activities present in the zone; choice of trip start-time is affected by the time foreseen to reach one's destination; the choice of transport mode is influenced not only by the individual's or the household's socio-economic characteristics (income, licence holding, car ownership, etc.), but also by times, costs, reliability of alternative modes and so forth.

Service characteristics depend, in turn, on transport supply, that is, the set of infrastructures (roads, car parks, railway network etc.) and organisational elements (road traffic regulations, public transport lines, schedules and fares) which allow trips to be undertaken in the urban area. The transport supply system has a limited capacity, that is the various elements which comprise it and thus the system as a whole, can permit a limited number of users to travel in a pre-established time period.

Transport demand distributed among the various modes and different paths produces traffic flows on the network. This results in vehicle flows on trunk roads, parking demand at car parks and loading on public transport lines etc. When flows approach the capacities of the various elements (or exceed them for limited periods) congestion is caused - link performance depends on the user flow on that link. This phenomenon, besides other nonsecondary externalities (such as noise, air pollution and visual intrusion) may significantly modify travel times, reliability and fuel consumption, i.e. the transport service characteristics supplied by the various modes. Congestion often has a transversal effect on the various transport modes: road congestion, for example, severely affects ground public transport service characteristics.

Furthermore, transport service characteristics determine the relative accessibility of city zones, making it easier or more difficult to reach one zone than others ('passive' accessibility), or to reach all the other zones from one zone ('active' accessibility). As has been stated, both aspects of accessibility condition location decisions for households and economic activities, and hence the real estate market. Besides, the accessibility of the residence zone also affects certain user characteristics, such as car ownership. For example, the choice of residential zone takes into account the active accessibility of the zone to the workplace and household services (shops, education etc.), the location of economic activities also considers the zone's passive accessibility with regard to potential clients, while the location of public services takes into account (or should do so) passive accessibility for users.

In the system described, there are a number of feedback loops inducing dynamics with different length periods. The more 'internal' loops, i.e.

those involving fewer elements and with shorter term dynamics, concern the distribution of origin-destination demand for a certain mode (for example, the car) on certain paths. Such a distribution produces flows on the network which, because of congestion, determine travel time and other characteristics of the various possible paths, and thus affect user path choices.

There are more 'external' loops in which trip distribution among the various time bands, possible destinations and alternative modes produce the demand for each mode and time band. This induces traffic flows which, through congestion, modify service characteristics, which in turn affect choice of departure time, destination and transport mode. Finally, there is an external cycle over a longer period in which the location of activities leads to transport demand. The latter, once again through flows and congestion, affects accessibility relative to the various areas in the city and hence locational choices for families and businesses.

The working conditions of an urban transport/land-use system therefore change over time, evolving through temporary equilibria which are continually perturbed by the overlapping of various factors with dynamics of different periods: economic and demographic variables, new building or development schemes and changes in transport supply.

From the above, it may be inferred that urban mobility and activity systems, especially in the presence of high congestion, are 'complex systems' with many elements of a different nature (social, technical, economic) which interact closely in both a direct and nonlinear way. In the last few decades, increasingly sophisticated mathematical models have been proposed to study and reproduce the main variables and reciprocal relationships. Research and applications have developed in particular along two main lines. The first concerns the modelling of mobility subsystems, assuming the activity system as exogenously given. The second focuses on the explicit study of the relations within the activity system, and between the latter and the transport system.

It may be stated that mathematical models of transport systems, the focus of this work, have until now received greater attention from the scientific community, and there exist a large number of case-studies. Transport/land-use interaction models, due to the greater breadth and complexity of the phenomena and the difficulty in finding data for calibration and testing, are still in a phase of theoretical definition and there have been few professional applications.

31.3 The Standard Model of Urban Transport Systems

31.3.1 Introduction

This section deals in greater detail with the mathematical models which have been developed and applied to simulate the mobility system, referred to below as the transport system. The conceptual basis of such models derives broadly from systems theory which, as is well known, first requires definition of the system in question with a view to simulating its behaviour in response to hypothesised measures. At the same time, it requires the identification of the external environment, that is, of the elements which interact with the transport system but are not specifically being studied.

The external environment of a transport system consists of the *activity system*, an expression of land use and the socio-economic activities performed therein. The activity system, which includes the location of residential buildings and productive activities, determines the mobility needs that the transport system must serve.

The transport system may be broken down further into two closely interacting subsystems:

- the *demand (sub)system*, consisting of users with characteristics which affect their mobility behaviour;
- the *supply (sub)system*, consisting of those physical components (vehicles, technologies and infrastructures) as well as organisational and regulatory components (traffic management, line and service structure, schedules, fares etc.) which determine the transport services provided and their characteristics.

A transport system model thus generally consists of three (sub)models:

- a *demand model* which simulates the chief characteristics of mobility as a result of the activity system and supply;
- a *supply model* which simulates levels of service provided and various external impacts in relation to services produced and to loads determined by user behaviour;
- a *demand/supply interaction model* which simulates reciprocal conditioning between demand and supply.

In current applications, the dominant hypothesis concerning the system's time structure is that transport demand stays roughly constant for a sufficiently long period (time bands, typical days etc.) to permit study of the system in 'regime' conditions, that is, within-period staticity. Removal of this hypothesis has a considerable effect on demand and supply models, leading to within-day dynamic models (see Section 4.2).

Reference will be made below to the general structure of currently adopted models which assume the system's behaviour to be static, both from one day to the next and within each day. For more exhaustive and in-depth treatment of such models see Cascetta (1990) and the commented bibliography contained therein. On the other hand, dynamic models, which are still being researched, will be briefly described in Section 4.

31.3.2 Supply Models

The supply of each transport mode is generally simulated with a network model consisting of:

- a graph representing the topology of transport services provided;
- functions which express, for each link, the dependence between the level of service provided (times and costs) and user flows on network links;
- relationships expressing, for each link, the dependence between impact indicators (noisc and environmental pollution, energy consumption etc.) and flow characteristics, such as speed and density.

Construction of the supply model (see Fig. 2) first requires *delimitation of the study area* and its subdivision into traffic zones, or *zoning*. The above steps allow the area in question to be defined and discretised, assuming that most mobility occurs between the zones (and not within them) and that the trip origin and destination are concentrated at a characteristic point of each zone, termed the centroid. External centroids are also considered for representing interchange and crossing mobility.

The subsequent *graph extraction* (for each of the transport modes considered) involves identifying the chief inter-zone connections and representing their topological structure in graph form, with nodes and links. Nodes represent the significant events which define the trip in space and time. Links represent connections between nodes, as allowed by the transport supply in question. For example, a road network for private traffic generally consists of a graph in which the nodes represent intersections or diversion points and the links represent the roads between two nodes. In the case of public transport with scheduled services (buses, trams, trains), the nodes represent access points to services (stops, stations etc.) and the links represent connections between access points allowed by the lines, access/egress connections to services, or operations at terminals.

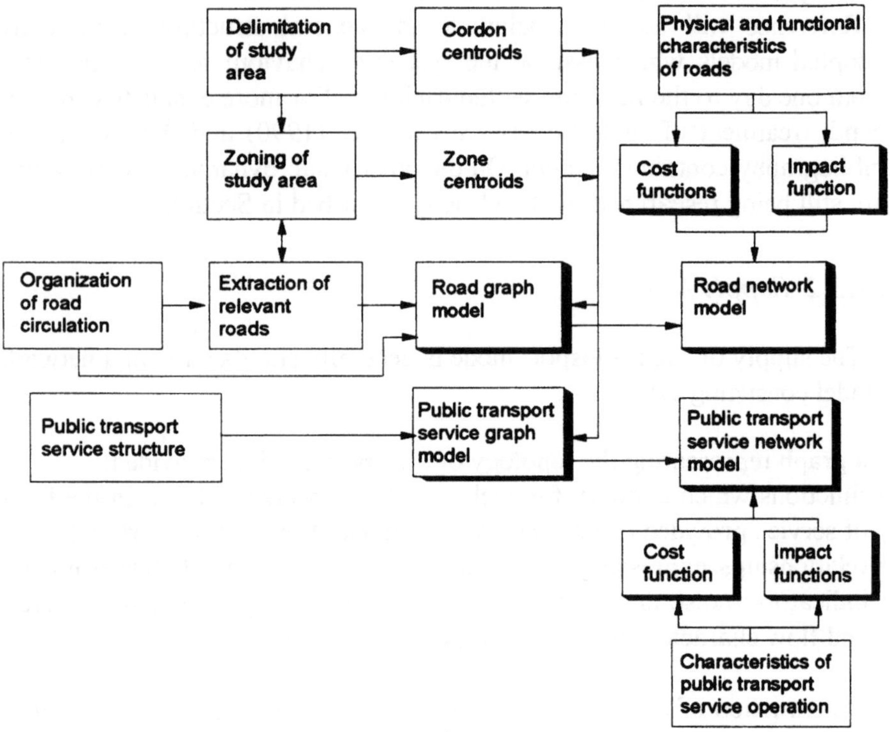

Fig. 2 Functional representation of the components of the transport supply model

Complete network model specification requires several parameters to be associated with each link, making it possible to define the generalised transport cost of link transit, as perceived by users making travel choices. The link cost is generally expressed as a linear combination of several factors, such as transit time (travel or waiting), monetary cost, comfort etc. Moreover, for each link, consideration is taken of the parameters necessary for estimating impacts which arise from the transport system structure, such as air and noise pollution, energy consumption, accidents and so on.

The choices of a user who has decided to move between two different traffic zones with a given mode, are represented graphically as elemental paths, i.e. sequences of consecutive links without circuits which connect centroid nodes in each zone. If the graph represents scheduled transport services, the concept of path may be generalised to include path sets (hyperpaths) with partial overlaps. The user chooses between these sets during the trip, according to various behavioural strategies (see Cascetta and Nuzzolo, 1986, for a discussion of various hypotheses concerning user behaviour).

Let:

- n be the number of links of the graph,
- m the number of relevant paths for mobility between all the origin-destination pairs in the area,
- **A** the link-path incidence matrix with a_{lk}=1 if arc l belongs to path k, a_{lk}=0 otherwise,
- **F** the path flow vector ($m\times 1$), that is the number of those using each path in the reference time period,
- **f** the link flow vector ($n\times 1$),
- **c** the link cost vector ($n\times 1$),
- **C** the path cost vector ($m\times 1$), that is, the cost incurred by users on each path.

Network consistency can supply the following relations:

$$\mathbf{f} = \mathbf{A}\mathbf{F}$$
$$\mathbf{C} = \mathbf{A}^{\mathrm{T}}\mathbf{c}$$

Due to reciprocal conditioning among users on the same physical link (congestion), average link cost depends on user flow (that is, the number of users who move on the link in the time interval considered). *Flow-cost functions* express the dependence between the various cost parameters of each arc and the flow along it, and possibly the flows relative to other links, according to the link's physical and functional characteristics:

$$\mathbf{c} = c(\mathbf{f})$$

Specification of the cost functions depends on the system in question. For example, in the case of road transport it is necessary to distinguish the costs for car travel (by road category) from the corresponding costs incurred by pedestrians or bus passengers. Similarly, cost functions may be specified for links representing public transport systems or transfer nodes (stations etc.). The simultaneous presence on the network of a number of user categories characterised by different costs (such as cars and buses) may be represented by introducing different links with different characteristics.

The supply model is thus generally defined for a congested network, that is by the graph and cost functions. If the link costs are constant and (roughly) independent of arc flows, the network is said to be non-congested. The impacts relating to an element of a transport system also depend on its working conditions and hence on flow vectors. The chief impact functions generally considered relate to air and noise pollution, energy consumption and accidents.

31.3.3 Demand Models

Transport demand is the result of the users' choice behaviour. This is usually described by origin-destination matrices, whose elements represent the number of users (defined according to socio-economic characteristics, such as income class and trip purpose) travelling between each origin and destination pair in an assigned reference period (time band, average day etc.) by a given transport mode/service. It is also necessary to simulate path choice behaviour, possibly expressed by mode and/or user category.

Demand models usually include, as an input, the configuration of activity locations and users' socio-economic characteristics. They simulate mobility demand as a function of the vector of transportation costs on the various paths connecting the *o-d* pairs in the study area. Denoting:

d the demand vector, obtained by re-arranging the *o-d* matrix
SE the matrix of socio-economic characteristics, which expresses the relevant attributes of the activity system (e.g. number of workplaces by production sector). Its dimension equals the number of zones by the number of key characteristics.

The demand model, or function, can be expressed as:

$$\mathbf{d} = d(\mathbf{SE}, \mathbf{C})$$

A transport demand model may be expressed as a system of sub-models relating to the different levels, including general aspects of *mobility* (such as driving licence possession and car ownership) and subsequent *travel* choices, relating to frequency, timing (within-day, etc.) and space (destination) as well as the characteristics (mode, services and path) of each *trip* within the overall travel activity.

The traditional approach to modelling travellers' demand involves simulating trip choices sequentially, though their reciprocal interdependencies are also taken into account. In particular, choices concerning whether, when and why to travel are simulated by an *emission* (or frequency) model relative to each origin and reference period, differentiated by travel purpose and, possibly, by different socio-economic user categories. Subsequently, a *distribution* model simulates trip destination choice and a *modal split* model simulates choice of transport mode. Lastly, a *service* (or *path*) *choice* model simulates how the trip is undertaken.

We therefore obtain a *partial share model* (called a four-stage model system), which is, in general, different for each user category considered and specific to a time band (if this choice dimension is not simulated). If we denote by:

$d_{od}(s,m,k)$	the number of users who travel between origin o and destination d, for purpose s using mode m and service (or path) k,
$d_{o.}(s)$	the number of users who travel from for purpose s, resulting from the emission model,
$p(d/so)$	the probability of choosing destination d, resulting from the distribution model,
$p(m/sod)$	the probability of choosing mode m, the resulting from the modal split model,
$p(k/sodm)$	the probability of choosing service (or path) k,

The four stage demand model can be expressed as:

$$d_{od}(s,m,k) = d_{o.}(s)\, p(d/so)\, p(m/sod)\, p(k/sodm)$$

where, for ease of notation, dependence on socio-economic characteristics and/or service level provided has not been explicitly indicated. It is worth noting that the results of each model are affected by the underlying models. For example, the demand distribution among traffic zones depends on the characteristics of the various modes available for a given destination.

Each of the submodels may be defined either according to a *descriptive* approach, which aims to reproduce the relationships observed between mobility variables and explanatory variables **C** and **SE** or with a *behavioural* approach in which the model specification is obtained from explicit hypotheses concerning user behaviour.

In behavioural models it is increasingly common to apply *random utility theory* to user choice from a discrete set of alternatives (Ben-Akiva and Lerman, 1985) because of the explicit interpretation of phenomena and efficiency in using information for the specification and calibration phases. In models based on random utility theory, each user chooses the alternative with maximum *perceived utility*, represented by a random variable, the sum of a deterministic term (with zero variance) called *systematic utility*, and a stochastic term (called a *random residual*), which is a random variable with zero mean. Whereas the systematic utility of each alternative expresses its average utility among all users, the random residual expresses interpersonal and intertemporal variations, as well as the inevitable modelling errors. Systematic utility is a function - usually linear - of the attributes of the alternative (such as times, costs, regularity etc.) and the decision-maker (such as socio-economic category, driving licence ownership etc.). The choice probability expression of the generic alternative depends on random residual distribution.

The traditional structure of passenger mobility demand models may be developed and enhanced in various respects by using the potential of disaggregate behavioural models. Such models allow us to define integrated models of mobility and travel choice, for example by also considering choices relating to driving licence or car ownership which naturally condition travel choices and, in particular, modal choice.

An example of a complex system of mobility and trip models is shown in Fig. 3 and described in detail in Cascetta, Biggiero and Nuzzolo (1995). A further example of an urban demand model system calibrated in various Italian cities is described in Cascetta and Nuzzolo (1988).

Whatever the type of model and the applicative context, there remains the problem regarding the choice of functional form, the variables used (specification) and the estimation of unknown parameters (calibration). Calibration techniques vary according to model type (aggregate/ disaggregate) and the information available, but their treatment is beyond the scope of this work. Finally, it should be mentioned that there are some very efficient methods for combining different sources of information (surveys, traffic counts, previous estimates etc.) to improve direct demand estimation obtainable through surveys or the estimation of demand model parameters.

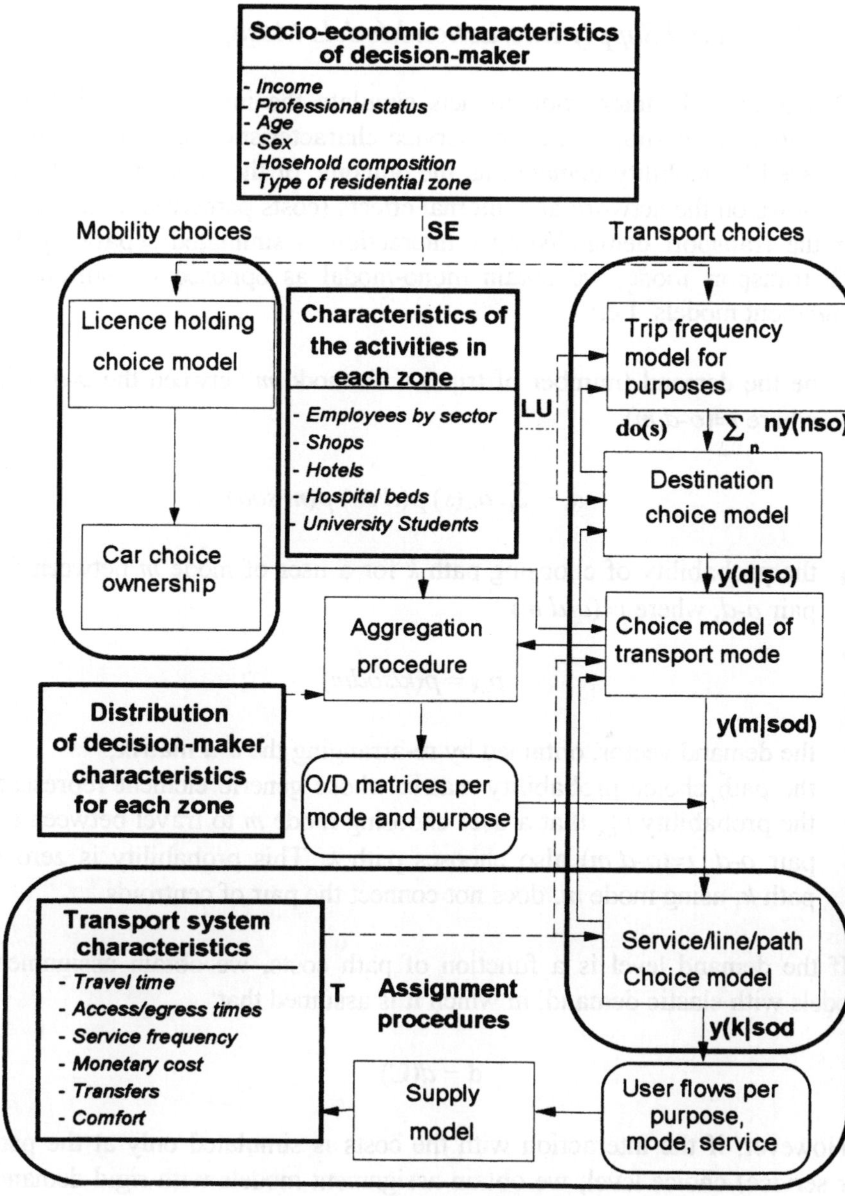

Fig. 3 A disaggregate model system of transport demand

31.3.4 Demand/Supply Interaction Models (Assignment)

Demand/supply interaction models simulate the interrelations between transport system supply, i.e. the service characteristics, and user choice, expressed by mobility demand, as the outcome of the interaction between user flows on the network and internal effects (costs perceived by users).

If the transport demand/supply interaction is simulated separately for each transport mode, we obtain mono-modal as opposed to multimodal assignment models. Let:

d_i be the demand (number of trips) with mode m between the o-d pair, where i=(o-d,m):

$$d_i = \Sigma_s\, d_{o.}(s)\; p(d/so)\; p(m/sod)$$

$p_{i,k}$ the probability of choosing path k for a user of mode m between the pair o-d, where i=(o-d,m):

$$p_{i,k} = p(k/sodm)$$

d the demand vector, obtained by re-arranging the o-d matrix;

P the path choice probability matrix whose generic element represents the probability $p_{i,k}$ that a user choosing mode m to travel between the pair o-d, i=(o-d,m) also chooses path k. This probability is zero if path k, using mode m, does not connect the pair of centroids.

If the demand level is a function of path costs, we obtain assignment models with elastic demand, in which it is assumed that:

$$\mathbf{d} = d(\mathbf{C})$$

However, if the interaction with the costs is simulated only at the path (or service) choice level, we obtain assignment models with rigid demand, in which it is assumed that:

$$\mathbf{d} = \mathbf{d}_o$$

$\mathbf{d}_o$ being the constant demand (either current demand, estimated directly by means of sample surveys or through the use of emission, distribution and modal-split models, or a forecast of future demand for a given scenario).

The flow on each path may be expressed as the product of demand and the path choice probability or, in vectorial notation:

$$\mathbf{F} = \mathbf{P}\mathbf{d}$$

In general, the systematic disutility of a generic path is measured by the perceived generalised transport cost, which is a function of travel time, monetary cost and other attributes. Therefore, path choice probabilities may be expressed as a function of path costs by a model based on random utility theory, in which systematic disutility is assumed equal to path costs:

$$\mathbf{P} = P(\mathbf{C})$$

The path choice model is defined by the hypotheses mode on random residual distribution. For example, a Gumbel distribution hypothesis leads to the Logit model, while the multivariate normal hypothesis leads to the Probit model. A particular case is that of deterministic utility, i.e. of zero random residuals, which leads to the choice of only minimum generalised cost paths. The assignment models described below are defined as either stochastic or deterministic according to the path choice model adopted.

Simulation of the demand/supply interaction involves defining the reciprocal relationship between path costs $\mathbf{C}=\mathbf{A}^{\mathrm{T}}\mathbf{c}$, path choices $P(\mathbf{C})$, link flows $\mathbf{f}=\mathbf{A}P(\mathbf{C})$ and link costs $\mathbf{c}(\mathbf{f})$. In the case of a non-congested network *constant cost assignment models* are obtained, formally expressed by the map of *transport network loading*:

$$\mathbf{f} = \varphi(\mathbf{c}) = \mathbf{A}P(\mathbf{A}^{\mathrm{T}}\mathbf{c})\mathbf{d}$$

In the general case of a congested network, different *variable cost assignment models* are usually possible. Traditionally, an equilibrium configuration between demand and supply is considered representative. In particular, the path flow vector which determines the supply costs is assumed to coincide with the path flow vector arising from the demand model, and likewise for the path cost vector (for the description of a more general dynamic process approach, see Section 4.3).

We thus define a link flow configuration which determines a link cost configuration, in which no user can reduce the perceived disutility of the choice adopted. The equilibrium assignment models thus obtained for a transport network are formally expressed by the fixed point problem:

$$\mathbf{f}^* = \varphi(c(\mathbf{f}^*)) = \mathbf{A}P(\mathbf{A}^{\mathrm{T}}c(\mathbf{f}^*))\mathbf{d}$$

where demand is defined by the relations introduced above for cases of elastic rigid demand. Under some fairly general assumptions concerning the path choice model and the demand function, the existence and uniqueness of equilibrium arc flows are ensured under hypotheses of continuous and/or monotonic cost functions.

The formulation of equilibrium assignment as an optimisation problem, both in the stochastic and deterministic case (or of variational inequality for deterministic equilibrium), allows the definition of solution algorithms. In this case a series of constant cost assignments (network loading) is generally conducted, updating the link cost vector at each iteration. For path choice models, which are more commonly used in practical applications, algorithms allowing network loading without explicit path enumeration have been developed. This makes it possible to obtain overall equilibrium assignment algorithms which are also efficient on large networks (Sheffi, 1985, Cascetta, 1990).

It is fairly straightforward to extend models of this type to public transport networks and, in particular, to origin-destination pairs served by partially overlapping and/or competing lines. In this case, the path actually followed represents a series of successive choices. Prior to the beginning of the trip, the user chooses a set of 'attractive' lines. During the trip at each intermediate stop, the user chooses between the lines initially defined attractive (Cascetta and Nuzzolo, 1986).

31.4 Some Directions for Research and Development

We now present some of the areas of current research and development, which concern in particular the in-depth study of behavioural mechanisms underlying travel choices and, more generally, the interconnections between trip chains and activities, as well as the dynamic phenomena which may occur in a transport system.

31.4.1 Demand Models with Explicit Simulation of Activity/Trip Chaining

The classical approach to modelling urban mobility demand (described above) simulates the sequence of choices relating to the individual origin-destination trip, with regard to frequency, destination, mode and path, but is independent of the trip or daily activity chains within which it occurs (*trip-based models*). One of the aims of research in the field of urban mobility systems is to represent jointly the choices relative to a whole trip chain or, even more broadly, the whole daily mobility programme of a decision-maker. A number of models have been proposed for this purpose (for a systematic review of the literature, see Ben-Akiva, Bowman and Gopinath 1994).

Some of the simpler models begin by modifying classical trip-based models to represent a tour or trip chain which begins and ends at home (*trip-chaining models*). The usual choice criteria are supplemented, implicitly or explicitly, by criteria relating to intermediate or secondary trips, while respecting constraints of internal consistency. The functional structure of such a trip-chaining model (Cascetta, Nuzzolo and Velardi, 1994) is shown in Fig. 4. More sophisticated models aim to simulate trip chains within the overall programme of the individual's daily activities (*activity-based models*).

It is generally possible to define a system of trip-chaining demand models connected to activities if one or more of the following requirements are satisfied:

- trips are always associated with purposes or activities performed at the origin or destination;
- space and time constraints are respected with regard to trip chains;
- models incorporate interdependencies and interactions with the household;
- models incorporate inter-period dynamics, i.e. dependence on activity/mobility decisions taken in previous periods.

As may be observed, models of this type allow us to quantitatively evaluate the influence of many factors which directly condition urban mobility demand. It is, for example, possible to represent the effect of the time budget upon the possibility of undertaking activities/trips, the greater probability of undertaking trip chains according to the relative accessibility of the residence and primary destination zones (there is, for instance, a greater likelihood of undertaking secondary trips for someone living in a peripheral zone with low accessibility to services and shops and working in a central zone with high accessibility to services and shops), as well as the effect of chains upon modal choice and, in particular, upon car use. The results so far obtained in this area of research are extremely promising in terms of their significance and capacity to represent the complex phenomena underlying the structure of urban mobility. It is likely that such model systems, thanks also to the computational power now available, may be applied in the near future.

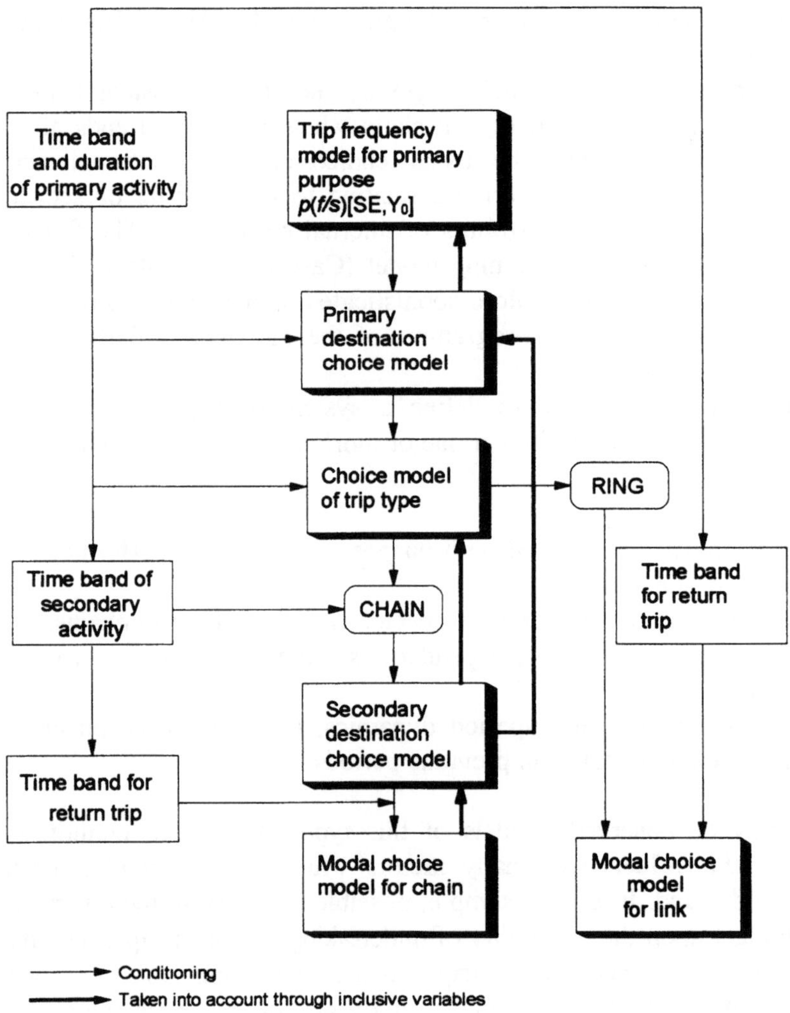

Fig. 4 Structure of a trip-chaining model system

31.4.2 Within-Period Dynamic Supply and Demand Models

Models of within-period dynamics allow us to simulate within-day variations of the system state expressed in terms of flow characteristics, such as arc flows and queues, as well as possible oversaturation phenomena. In such models, explicit consideration may be taken of variations in demand (due to departure time choice) and/or supply (due to the presence of real-time control systems, accidents etc.) within the reference period (whole day, rush-hour etc.). The use of these models also allows for possible en-route modifications in response to traffic conditions encountered and/or route information or path direction systems.

The extension of static supply and demand models to the within-day dynamic case is usually not straightforward. Generally discrete time is adopted, with the reference period subdivided into intervals, although representation in continuous time is theoretically possible (but a numerical solution requires in any case some form of discretisation).

As well as the aggregate approach (macrosimulation), it is possible to adopt a disaggregate approach (microsimulation) in which each user's behaviour is explicitly simulated. Disaggregate models can be more sophisticated and accurate, but are not generally suited to large scale applications because of the computational power required. They are chiefly used for detailed analysis of small areas, following the application of aggregate models.

The specification of aggregate demand models with within-period dynamics also requires simulation of choice behaviour on departure time. Moreover, path choice simulation requires both pre-trip and en-route behaviour to be represented. This is particularly complex since, unlike for the static case, link flow characteristics cannot be uniquely defined.

The dynamic loading map of a transport network expresses the relation existing between link flows **f** and path flows **F** in the various intervals:

$$\mathbf{f} = \Phi[\mathbf{F}]$$

Such a map is not linear, nor can it be defined analytically, as in the static case. Furthermore, the time for crossing a link depends both on link characteristics, such as speed, and on the moment of arrival at the origin node of the link, and cannot be defined as a function only of link flow. On the other hand, link characteristics **r** are generally a function of flow on the link itself (in the corresponding interval):

$$\mathbf{r} = r(\mathbf{f})$$

Once the link flows are known and the flow characteristics determined, the average cost **C** of each path may be defined for each departure interval:

$$\mathbf{C} = \Gamma[\mathbf{r}]$$

The three relations introduced above define, at least formally, the supply model for within-period dynamic models. To solve it, the dynamic loading map has to be made explicit. Various approaches to this problem have been proposed, but none are entirely satisfactory (for a bibliography, see Cascetta and Cantarella, 1993).

31.4.3 Dynamic Process Models for Traffic Assignment

Dynamic process models of demand/supply interaction explicitly simulate the evolution of the system from one period to the next (day-to-day dynamics). They may be seen as a generalisation of the equilibrium paradigm, inasmuch as they explicitly simulate user learning and updating mechanisms, including any routine behaviour. They therefore permit a large class of problems to be analysed, including the effectiveness of dynamic systems of traffic control and management of information for users (for an extensive presentation of such models, see Cantarella and Cascetta, 1995).

Deterministic process models, based on nonlinear dynamic systems theory, simulate the system's evolution with a set of recursive equations which explicitly defines the predicted link cost vector for a generic reference period, starting from the costs incurred in previous periods. A simple deterministic dynamic process model, whose fixed points coincide with configurations of system equilibrium, is illustrated below. Let:

- y^t be predicted link costs on day t
- x^t link flows occurring on day t,
- α the probability of a user reconsidering (but not necessarily changing) the previous day's choice,
- β the weight assigned by users to costs incurred the previous day in forming predicted costs.

The model is specified by the following recursive equations:

$$y^t = \beta c(x^{t-1}) + (1-\beta)y^{t-1}$$
$$x^t = \alpha\varphi(y^t) + (1-\alpha)x^{t-1}$$

It can be demonstrated (Cantarella and Cascetta, 1995), under the assumption that one and only one equilibrium configuration exists, that there is only one fixed-point attractor. Its stability is ensured by low values of parameter α, which somehow measures user inertia in changing path choice in a generic period, and β, which measures the influence of recent experience on memory formation. Moreover, the system is more stable for low demand levels and for high perception errors. However, realistic parameter values can bring about the non-stability of the equilibrium configuration, determining an evolution towards another type of attractor (periodic, fractal or quasi-periodic).

Rigorous statistical analysis is made possible by *stochastic process* models in which it is generally assumed that arc flows on day t are represented by a random variable X^t with distribution dependent on x^{t-1} and y^t, and the expected value:

$$E\,[\,X^t \mid (x^{t-1}, y^t)] = \alpha\varphi(y^t) + (1-\alpha)x^{t-1}$$

A Markov process may thus be obtained. Under fairly general hypotheses, this may be shown to be regular (Cantarella and Cascetta, 1995), i.e. it admits one and only one final probability distribution of states and converges towards it regardless of the initial state. Such a model is straightforward to solve using Montecarlo simulation techniques which allow a pseudo-realisation of the process to be obtained.

31.5 Fields of Application

31.5.1 Design and Evaluation of Transport Systems

Mathematical models for simulating urban mobility systems have various fields of application with different time horizons (short, medium and long term). The technical characteristics of the systems, the objectives and decisional constraints influence choice of methodology, as well as the assumptions and approximations required for the specification of the model to be adopted. Three major applications may be identified, though not necessarily all present in each context:

- the functional design of transport systems, i.e. the definition of the technical and functional supply characteristics and demand management measures;
- impact analyses (externalities);
- financial/economic analyses, i.e. the calculation of user benefits, traffic revenues and operating costs.

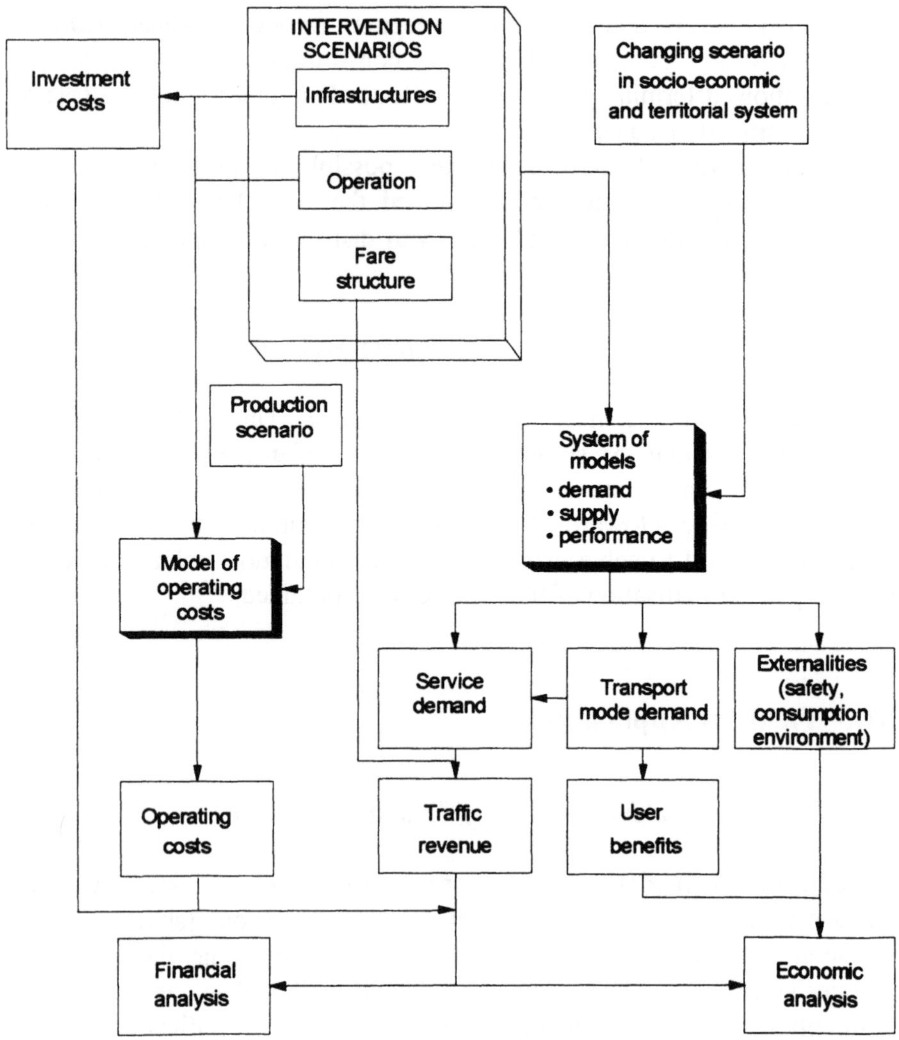

Fig. 5 Design and evaluation of an urban transport system

These applications also involve other disciplines, but the procedure generally adopted for the design and evaluation of an urban transport system consists of the following phases (see Fig. 5):

- analysis of the current situation in terms of internal and external effects;
- construction of project scenarios according to policy objectives and constraints, taking account of current problems or possible future problems;
- impact assessment;
- analysis of functionality and the financial benefits of each scenario.

Transport planning objectives and procedures and, more generally, network infrastructure and public services, have recently been the focus of much theoretical discussion. The modern vision of planning as a evolutionary process requires the construction of monitoring and simulation systems able to forecast the effects of decisions (Decision Support Systems), making use of the kind of methods and techniques described above. It is important in this context to distinguish between the activities of design and evaluation, which are mainly technical, and the decision-making and implementation processes which are a political responsibility. Below is a description of the chief types of application and their implications.

31.5.2 Strategic Planning: The Urban Transport Plan

The strategic or long-term planning of an urban transport system involves determining future interventions which usually include investment in infrastructure and technologies, modification of the service structure, and possibly by the redefinition of fare policies and demand management tools. If strategic planning is carried out within a social framework, the result is the production of an urban transport plan. For those operating in a business framework, the outcome is a business development plan. In strategic planning, the system studied consists of the whole transport system, and all the phases shown above are necessary. Particularly important are the assessment of effects both inside and outside the sector and the economic evaluation of the plan considered (traditionally by means of cost-benefit and/or multicriteria analysis), as well as the evaluation of financial resources required for its implementation.

Current measures and the time horizons adopted are encouraging a growing use of transport/land-use interaction models (or, at least, the evaluation of the plan's spatial impacts). This is consistent with the precepts of modern planning where the two systems, land use and transport, are studied simultaneously and not sequentially (Bertuglia and La Bella, 1991).

31.5.3 Techno-Economic Feasibility Studies of Transport Projects

This type of analysis is belongs to the 'planning by projects' approach. It is based on defining a reference framework for identifying the links to be created in the system (plan) and on the subsequent assessment of alternative projects to determine their profitability, the priorities and methods of construction. A techno-economic feasibility study of transport projects usually entails the formulation of alternative hypotheses for a given link (road, railway etc.). Such hypotheses involve the definition of the link's functional and performance characteristics (the route, connections, operating capacity and performance, type and characteristics of technologies), as well as those of the operating structure and fare structure.

The various alternatives, including that of non-intervention, are then assessed from the functional, economic and financial point of view, within the context of scenarios hypothesised for the area and transport system in question. The time horizon is usually long term. The system is analysed and modelled with various degrees of spatial and functional detail, according to the significance of the link considered. In any case, the system is simulated with reference to mobility demand and the supply of all transport modes.

31.5.4 Tactical Planning: Urban Traffic Plans and Public Service Reorganisation Plans

The short/medium-term tactical planning of an urban transport system usually concerns limited resource interventions, not modifying the infrastructure, but reorganising transport supply and services, fare policies and demand control. From the business angle, tactical plans involve the restructuring of services and/or the definition of commercial or fare policies. One of the most interesting aspects is the impact assessment and the financial analysis of operating costs and traffic revenues for the operators, plus the (simplified) economic analysis. In practice, to make such an assessment, a given socio-economic scenario is adopted for the whole mobility system and a given level and spatial distribution of mobility demand (global *o-d* matrices) assumed. Variations in modal split and flow assignment ensuing from the alternatives are then explicitly simulated.

31.5.5 Operations Management Programmes

Short-term operations management usually primarily concerns a business context and involves defining how particular aspects of each mode are to be managed, using existing resources. It includes traffic light regulation and detailed traffic plans, the design of public transport lines and schedules, and the organisation of factors required for the provision of transport services (for example, the assignment of vehicles to lines and travelling staff to shifts).

The system usually adopted considers only the mode in question, assuming modal demand to be fixed. If necessary, the supply model may be supplemented by detailed simulation models. The design phase may also be carried out with the support of operations research techniques and, given the large number of alternatives possible (often defined by continuous variables), does not only include pre-established scenarios.

References

Ben-Akiva M., Lerman S. (1985) *Discrete Choice Analysis: Theory and Application to Travel Demand,* MIT Press, Cambridge, Massachusetts.

Ben-Akiva M., Bowman J.L., Gopinath D. (1994) Travel Demand Model System Architecture with Daily Activity Schedules, M.I.T. Internal Report (mimeo).

Bertuglia C.S., La Bella A. (eds.) (1991) *I sistemi urbani*, Angeli, Milan.

Cantarella G.E., Cascetta E. (1995) Dynamic Processes and Equilibrium in Transportation Networks: Towards a Unifying Theory, *Transportation Science, 29*, 4, 1305-1329.

Cascetta E. (1990) *Metodi quantitativi per la pianificazione dei sistemi di trasporto*, CEDAM, Padua.

Cascetta E., Biggiero L., Nuzzolo A. (1995) *Passenger and Freight Demand Models for the Italian Transportation System,* Presented at WCTR 1995, Sydney.

Cascetta E., Cantarella G.E. (1993) Modelling Dynamics in Transportation Networks: State of the Art and Future Developments, *Journal of Simulation Practice and Theory, 1*, 65-91.

Cascetta E., Nuzzolo A. (1986) Uno schema comportamentale per la modellizzazione delle scelte di percorso nelle reti di trasporto pubblico urbano, Proceedings of the IV Conference of PFT-CNR, Turin.

Cascetta E., Nuzzolo A. (1988) Un modello di equilibrio domanda/offerta per la simulazione dei sistemi di trasporto nelle aree urbane di medie dimensioni, in

Bianco L., La Bella A. (eds.) *Strumenti quantitativi per l'analisi dei sistemi di trasporto*, Angeli, Milan, 15-53.

Cascetta E., Nuzzolo A., Velardi V. (1994) A Time of the Day Tour Based Trip Chaining Model System for Urban Transportation Planning, in Proceedings of the 21st PTRC European Forum, Warwick.

Sheffi Y. (1985) *Urban Transportation Networks*, Prentice-Hall, Englewood Cliff, New Jersey.

32. A New Perspective for Methodologies in Spatial and Urban Analysis

Sylvie Occelli

Foreword

Public Consultation on the Master Plan for a district in the Turin metropolitan area, February, 1994:
A Citizen: "I should like to know why the Plan contains so many new housing programmes when the Local Authority complains about the increasing difficulty in coping with current expenditure."
Local Government Officer: "Well, there are some State funds which are still unused..."
Another Citizen: "Could you explain why that sports facility has been located so far from the city centre, near such a busy, unsafe road?"
Planner: "Well, you know, town planning is a free-ride activity."

32.1 Introduction

This work addresses some questions which, in the view of the author, are particularly challenging. These questions stem from the recognition of the great contrast between the evident complexity of urban problems and the limitations of the tools available for dealing with them. Bearing in mind the inescapable need for a set of well-suited tools, we offer a tentative comment about the role of methods for the analysis, planning and control of urban systems.

The wide and diversified panorama represented by the relationship between information (i.e. knowledge apparatus), practical tools and the 'planning activity' provide a convenient background to our discussion. Although to be exhaustive it would be necessary to carry out a far deeper examination, we feel that it is nevertheless worthwhile to make an attempt. This is for two reasons:

a. the emerging demand for new ways to conceive and carry out *actions* in urban systems (we refer here to the exchange quoted in the foreword. Although it might seem provocative, it in fact bears witness to the reality of such a demand, particularly in the Italian context)
b. the great potential offered by the development and diffusion of both hardware and software systems for processing and retrieving 'information'.

Within this rather limited brief, we try nevertheless to present a wide conception of methodologies in urban and spatial analysis. These are not considered simply as analytical devices characterised by various formalisations, but as knowledge tools whose application should provide above all an 'information gain'. Our discussion is divided into three parts. In Section 2 we make some observations about the links between analytical tools and planning problems, outlining a framework which gives some indications of the requirements of a new perspective on the role of methodologies. Building upon this discussion, Section 3 concentrates on the features of this new perspective, hinting briefly at three main topics which could constitute a future research agenda: (i) the rationalisation, operationalisation and diffusion of tools for the production of information; (ii) learning and creativity; (iii) the relationships between analytical tools and planning visions. In the final section we emphasise some aspects of the accessibility to analytical tools and the question of communication, which will be an important aspect for the application of these tools.

32.2 Analytical Tools and Planning Problems

32.2.1 Introduction

The works in this book by Van Geenhuizen and Nijkamp, Rabino, and Fusco Girard in particular have already introduced a significant feature of the connection between analytical tools and planning problems: that of the role of information, communication and, more generally, knowledge. In this respect, three different although complementary aspects have been identified:

a. the technical and analytical requirements of an information system (see Van Geenhuizen and Nijkamp, in this volume);

b. the aims of an information system and related system values (see Fusco Girard, in this volume);
c. the epistemological implications of the role of information in relation to the broader issue of the foundation of a 'science' of the city (see Rabino, in this volume).

We now wish to come back to the role of knowledge and focus on some themes which seem crucial in the relationship between analytical tools and planning. In order to provide an adequate framework for the discussion, it is useful to consider the major categories involved in a planning process. It is convenient for this purpose to adopt the categorisation proposed by Khakee (1994), who identified three main dimensions:

a. the *normative* dimension which posits both the objectives and the legitimacy of planning. This ultimately depends on the socio-economic principles and division of power underlying the working rules of a given system;
b. the *institutional* dimension which dictates the organisational procedures of the planning activity and how these vary as new organisational needs emerge;
c. the *methodological* dimension which regulates the ways in which both theoretical and practical knowledge intervene in plan-making.

This categorisation could of course serve a number of different purposes: the need to revise or modify planning practices currently in use, or an examination of the role of the components belonging to each category, as well as the relationships between the categories (see Fig. 1 based on Khakee, 1994). A major point to be underlined is the increasing awareness of the many interactions existing between the three categories, and the fact that these relationships are not unidirectional. Acknowledging this means recognising that some of the relationships are able to activate (make easier) and/or prevent (inhibit) the implementation of planning actions pertaining to the domain of one or more categories.

The existence, nature and intensity of these relationships make up what can be called the domain of existence of the demand for knowledge in planning activities. This same domain also constitutes the field of application for methodologies for urban and spatial analysis. Stated in this way, the concept of knowledge demand in planning may appear trivial, whereas the field of application of urban and spatial methodologies boundless. We can go into this in more depth if we refer to the ideas developed in Rosen (1985) and assume that, relative to a given planning problem (or, alternatively, to a given planning action):

a. each category has its own 'information pool' (e.g. information level). This is not given once and for all, but due to the dynamic nature of planning process needs to be constantly adjusted, creating a continuous demand for knowledge;
b. the 'information levels', which are interrelated and super-imposed in various ways, also have their own autonomous capacity for adjustment, depending on changes in the other information levels as well as on exogenous influences or random events;
c. the interactions established between the information levels give rise to a plurality of 'visions'. These are neither comprehensive, complete nor definitive, but sectoral, partial and changeable.

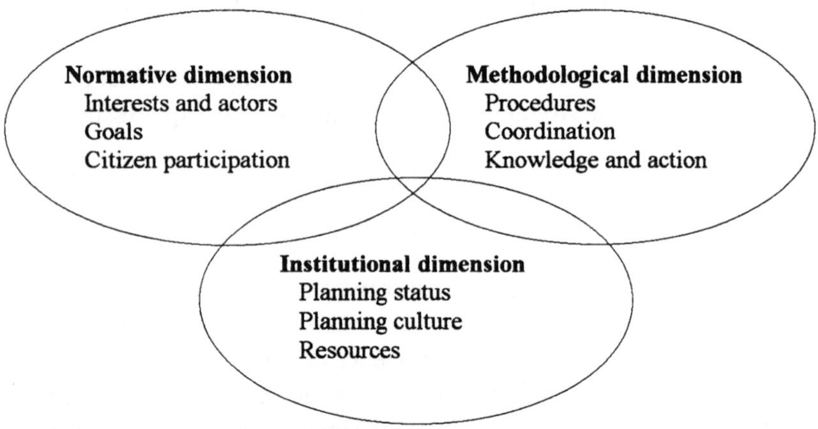

Fig. 1 The main dimensions in the planning process (Khakee, 1994)

The above features are all facets of the intrinsic complexity of the planning process. With reference to this process, we believe that the formation of such visions is becoming a major point of discussion also from the slightly different point of view of plan-design. Although criticisms and revisions have accompanied the theoretical development of urban and regional science over the past two decades (see, among others, Massey *et al.*, 1976, Guelke, 1978, Batty, 1979, 1983, Crosby, 1983, Bahrenberg, 1984, Bertuglia *et al.*, 1987 and the contributions by Pumain and Rabino, in this volume), a current assessment of both the role and relevance of methodologies raises a number of issues which are quite different from those of the past. The major differences can be ascribed to two general trends:

a. the changing context within which methodologies are applied and, consequently, the new requirements which need to be met by information providers;
b. the emergence of new factors in the provision of information.

32.2.2 The Context of Application of Urban Methodologies

It is generally agreed that methodologies must comply with one fundamental imperative, that of providing information (knowledge) and not only an increasing quantity of information, but more and better information (some justifications for this have been provided in Rabino's work, in this book). An implication of the above discussion is that the provision of information matters not only for the methodological dimension, but is becoming increasingly important for the normative and institutional dimensions too. The reasons for this are manifold and can be explained by the social, economic, institutional, technological and cultural transformations which are permeating all advanced economic societies (Alonso, 1991, Hall, 1991, Vernon, 1991, Amin, 1994). Needless to say, these transformations have also been fostered by the establishment of new societal 'values'. Some of these, identified by Maestre (1994), are concerned with the concepts of identity, autonomy, social and cultural innovation, and the environment.

In general terms it could be argued that for each category these changes raise three main questions:

(i) how knowledge about a system's present and (expected) future state can be obtained;
(ii) how individual and system behaviours are triggered;
(iii) how the organisation of activities is modified.

A simple but effective way of describing the terms involved is to apply the concept of multilevel structure. According to this concept, three general structures can be identified (Mesarovic, Macko and Takahara, 1970). These are shown graphically in Fig. 2.

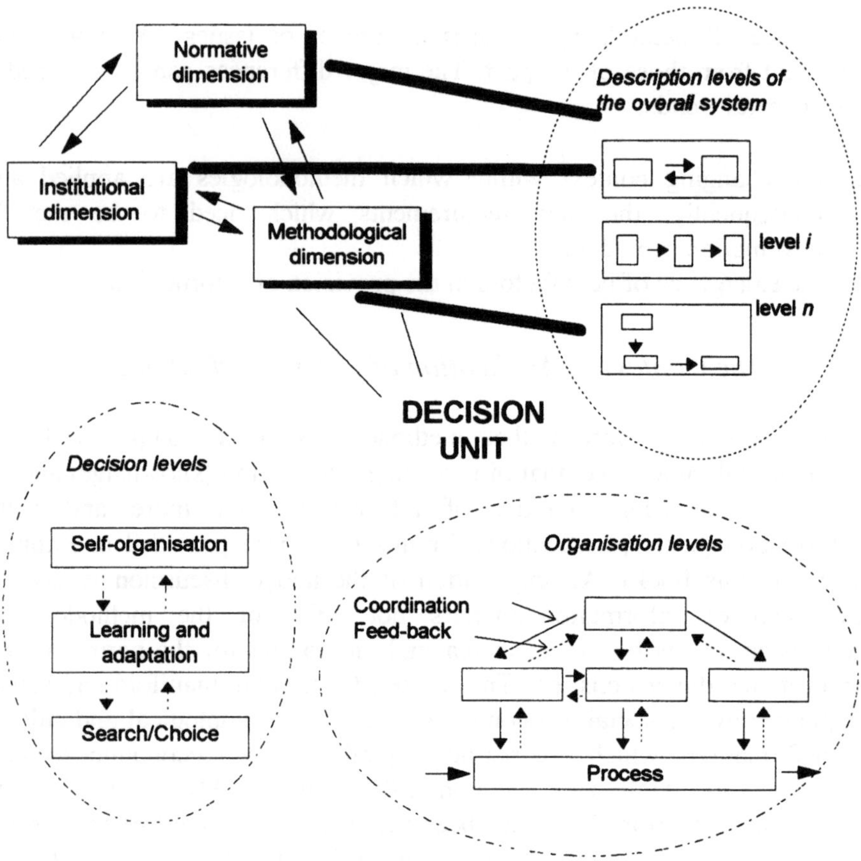

Fig. 2 A multilevel description of the different dimensions of the planning process

a. The first structure refers to the 'description of the system' and concerns the fact that whenever an individual has to face a given problem he/she has to, or at least tries to, acquire knowledge of the whole system, thus formulating his/her own description of the different categories - normative, institutional and methodological - as well as the relationships between them. In this process several descriptive levels, those considered relevant in the observation of the system, can more or less consciously co-exist. The stratification depends not only on the observer and his knowledge, but also on the aims of the observation. Each descriptive level corresponds to a given concept (abstraction) about how the system works. What is considered as an element (a system) in one stratum becomes a relationship (a subsystem) in another. In other words, a stratification defines an ordering (or hierarchy) in the 'descriptive form' of the system. The understanding of the system increases as we cross the various levels - going down to a lower stratum provides a more detailed explanation of the functioning of the system, going up to a higher one allows us to improve our understanding about the meanings of that functioning.

b. The second structure relates to the decision process in which everyone (whether decision-maker or private individual) faces situations characterised by uncertainty and lack of relevant information. The problem can be conceptualised as a functional structure consisting of different levels corresponding to distinct sub-decision problems. A brief outline of these levels is as follows:

- a first level at which the selection of the option is made, relative to the available information and uncertainty conditions;
- a second level pertaining to both learning and adaptation processes. At this level, the uncertainties about the system behaviour, i.e. the effects of adopting a given option, are assessed. The assumptions made about the sources of uncertainty, empirical observations (data) and communication established between the agents involved in the decision process play a fundamental role in such an assessment;
- a final level is that of self-organisation. Here the structure, functions and strategies which will be considered by the other levels are established so that the goal-achievement process can be carried out as well as possible.

Of course, in this functional articulation there is also the possibility that the 'signals' sent by the higher level will be changed or adjusted according to the reactions determined (or the constraints existing) at the lower levels.

c. A third structure relates to the ways in which the subjects (subsystems) belonging to a broader category (a more general system) organise themselves. This structure, which can be considered a complexification of the previous one, is characterised by the existence of a family of interacting sub-systems, some of which have a leading role in the decision-making process, insofar as they have controlling power or influence over other subsystems. As the decision-making units, which can be located at every level of the whole structure, have freedom of action in the choice of their decision variables, conflicting situations can occur. These are determined both by the nature of the goals of the different decision-makers and the working of the system.

It is worth mentioning in passing that this general structure is the basis of a typology of decision-making systems which can be observed in animal as well as human organisations (one-level/one-goal systems, one level/multigoal systems, and multilevel/multigoal systems).

Building upon the above ideas, we could argue that the kind of changes making up the domain of application for methodologies for urban and spatial analysis lead to:

i. a multiplicity of descriptions of the system, which are often difficult to compare;
ii. increasing difficulty in identifying precisely the goals which should steer the changes. This implies a different attitude towards the future as a result of the awareness that uncertainty permeates most current situations and their evolution is unpredictable;
iii. an increasing number of relevant actors, including the appearance of completely new ones.

32.2.3 The Basic Elements in the Provision of Information

Information provision consists of three main elements which are closely interrelated and should be taken into account jointly:

a. what information to provide;
b. for whom - or to what ends- information should be provided;
c. how to provide the desired or expected information.

The question of 'what information?' implies in turn two aspects. The first is substantial and relates to what constitutes information in relation to the problem in hand. The second deals with the nature of information, that is with its relevance relative to expectations and the purposes of acquiring knowledge. As far as the former is concerned, a general though to some extent elusive conception claims that 'information' is anything answering a question (Rosen, 1985). From this point of view, an increasingly important class of questions is that involving the driving forces behind urban (spatial) change and their effects. In this respect, an essential requirement of information is the capacity to provide the indications necessary for the description and interpretation of the processes of urban (spatial) change.

Furthermore, as already stated, in considering what constitutes information, we cannot ignore the question of its relevance. As far as planning is concerned, it is widely agreed that what ultimately matters is the availability of information which in a given context allows us, or makes it easier, to make a decision in a transparent, timely and efficient way, and arrive at an effective solution in terms of policy actions, programme or designs. However, it often happens in reality that even though certain information is known to be valuable for a given decision, the situation itself, often caused by an emergency or spurred by political urgency, prevents the information from being acquired. More often than not, existing information proves to be incomplete or simply not accessible. Even if it exists, it is often necessary to take into account so many implications (economic, social, environmental, etc.) and foresee so many effects (Wellar, 1990), that it is impossible to make a clear assessment of the situation. Such a lack of information then becomes an excuse for the unsuccessful outcome of the decision.

A gap therefore exists between the legitimate need to acquire (or update) information of an intrinsically comprehensive and pervasive nature - spurred by knowledge expectations about the future - and the awareness that in many situations existing information is inadequate for coping with the contingent preoccupations. To bridge this gap, we have to turn to the second determinant in the provision of information, relating to *for whom, or to what ends,* we need to provide information.

Some major aspects of this issue are discussed by Fusco Girard in this volume. Adopting different premises, other authors (see Preto and Occelli,1994) underline the significance of both goal-formulation and planning policy content, recognising the importance of scientific knowledge as a basic ingredient capable of conditioning the definition of goals and ways of achieving them. Wellar (1990), discussing the failure of environmental policies, is even more peremptory about the possible alternatives, arguing that the only way of bridging the gap is to:

a. revisit the whole relationship between information and planning policies. He claims that to improve policies and their outcome, it is essential to strengthen data resources and management ability in order to guarantee a constant stream of information to decision-making;
b. re-activate consulting and participatory processes between the main actors involved, policy-makers, technicians (planners and practitioners) and the public, to set policy goals and contents as well as their priorities, identify the kind and mix of variables (the control variables) which should be represented in these policies and to establish ways and means of carrying out, adjusting and monitoring policy, control variables and planning processes. This latter point leads to the third element of our discussion which concerns *how to provide information.*

We require two main sets of tools, which appear even more inseparable than in the past: technical tools and formal-analytical tools.

As far as the former is concerned, we are seeing a steady growth in computing power which is underlining the importance of the ability to manage, process and provide graphical visualisation of both socio-economic and morphological data. The wide diffusion of computing facilities in many different contexts (academic, governmental, and professional) will surely open up possibilities which would have been unimaginable few years ago. Expert systems (Batty and Yeh, 1994), neural networks (Fischer, 1994), GIS (Goodchild, 1991, Batty, 1993, Van Geenhuizen and Nijkamp, in this volume) represent some of the more innovative 'domains' within which the marriage between technical and analytical tools makes it possible to target the production of information to the realisation of so-called 'decision support systems'.

In addition, the more recent marriage between computers and telecommunications is making available new possibilities for the analysis, control and design of urban systems, The scope of future applications is still largely unexplored (Batty, 1995). It would seem, however, that it is with respect to analytical tools that new potential will be brought to light. In other words, conditions would appear to exist to innovate the traditional role of analytical tools in the planning process.

32.3 A New Perspective of the Role of Analytical Tools in Urban Analysis

32.3.1 Introduction

In chapter 2 an effort was made to shed some light on the principal aspects which characterise the position and general purposes of analytical tools in planning today. We have tried in particular to show that these aspects are determined to a large extent by information production as well as by the 'visions' which are currently produced through the interactions between the information levels embedded in the different planning categories.

The underlying conception is one which departs from the idea of methodologies as simple technical tools and warrants a more focussed perspective, including both the meaning and functions of these tools. This conception has many points in common with the general debate on the evolution of systems analysis. A major theme of this debate has been developed by Maestre (1994), for whom the new approach to systems analysis should be, at least at the individual level, an attitude resulting from learning, from a dialectic between a sanctioning practice and a modelling activity, which is not however limited only to models.

Needless to say, the features of this new perspective are not easily identifiable. In fact, they derive more from the widespread dissatisfaction with experience in the various application fields, than an effort to undertake a deeper revision of its fundamental premises (see, among others, Batty and Hutchinson, 1983, Batty, 1994, Harris, 1994). Some features however are beginning to emerge and relate to the following general themes, which might also be cornerstones of a future research agenda:

a. the rationalisation, operativity and diffusion of the means of information production;
b. the learning and 'creative' abilities;
c. the relationship between analytical tools and planning 'visions'.

32.3.2 The Rationalisation, Operativity and Diffusion of the Means of Information Production

As already mentioned in 2.2, the growth in desktop computing and the development of information technologies will undoubtedly have an increasing impact on the methodologies for urban analysis. Although so far the development has concentrated mainly on the 'hard' content of methodologies (i.e. the computing power), in the near future it seems likely to have a greater influence on the 'soft' component (i.e. the procedures and applications). It is sufficient to recall the plethora of new commercial software (electronic spreadsheets, statistical and mapping packages) developed in the early eighties, which made possible many data processing and information retrieval operations which were unthinkable only a few years earlier.

In this respect, two main consequences can be mentioned. The first is the feasibility of integration of two major areas which were previously considered distinct methodological fields (Couclelis, 1983):

a. methods and techniques for the description and investigation of the 'substantive' components of planning (this includes for example the techniques of spatial analysis, urban and transportation modelling and input-output methodologies);

b. methods and techniques focussing on the treatment of the 'procedural' components of the planning process (i.e. evaluation methods and, more generally, models of the decision process).

It is worth observing that the reasons for this integration are probably to be found in the maturity achieved during the long-standing debate on the planning process itself. This is no longer seen as a one-directional path consisting of successive steps, but as a set of phases that have to be revisited many times, allowing changes and adjustments to be made at each step of the process. The 'decision phase', although very important, is not simply the final step which marks the end of the process, but is one of many (this latter argument also relates to the discussion on knowledge pitfalls hinted at in 2.3).

A second set of consequences is associated with the more technical question of compatibilty and results from the increased potential for interfacing procedures (or packages) originally developed to perform specific and/or specialised tasks, most of which were not specifically developed for spatial or urban analysis. This provides new ways of encompassing in an overall framework methods which are technically different but functionally complementary. They also make it possible to link more efficiently the fundamental contents of the methods themselves, that is:

a. the data, i.e. codified information;
b. indicators, i.e. information with a higher added value;
c. graphical representations (visual images).

Decision support systems, for example, among which we should include performance indicators (Clarke and Wilson, 1987a, 1987b, Bertuglia, Clarke and Wilson, 1994), offer many possibilities for building fully integrated information architectures.

The above arguments would even support the view that technological developments in the production of information will be so pervasive that the 'hard' component will take over the 'soft' one, making the contributions of the latter irrelevant. The exact effects on the future of methodological approaches in the urban and spatial fields are not easy to predict. On the one hand, it seems likely that methods, as traditionally conceived, will tend to lose some meaning. They will also probably be filtered by 'procedures' or 'interfaces' of various kinds so as to be practically invisible to users, who are generally more interested in the results than in the specific characteristics or technicalities of the methods themselves. On the other hand, it is also true that innovation in hardware could allow methods to be applied more efficiently at scales which could not be tackled earlier, or only with prohibitive costs. In addition, it would permit an evolution of existing methods, by making possible alternative means of implementation (for example of the application of computer vision to the analysis of multidimensional data structures.

Finally, we should not disregard the possibility that the diffusion of computing and telecommunications (personal computers, workstations, multimedia systems) among a wider and more diversified public could contribute to a less diffident attitude towards these techniques, which still are so little understood by non-specialist users. It is also evident that the diffusion of hardware will not be enough. The cultural and organisational context, and in particular the so-called 'information environment', within which exchanges of users' experience can take place, is an essential condition in fostering this attitude (see Craglia, 1992).

32.3.3 Learning and 'Creative' Abilities

Three kinds of analytical tool are most frequently used: (a) indicators, (b) models (by models we mean all those methods which are relevant for urban analysis) and (c) graphical representations. Comparing the logical and operational assumptions behind these tools in terms of their potential for information provision, we find that a basic common goal can be identified (Occelli, 1995): the creation of a knowledge gain from given starting information conditions (from this point of view it could be possible to establish complementarity between the tools themselves).

There is also a substantial similarity between the tools as far as 'the analytical path' leading to the provision of this information gain is concerned. The most important of these analytical steps can be briefly summarised as follows:

- firstly, we need to define the area of application of the tool, make a hypothesis for describing the phenomena being studied and formulate the 'observables', i.e. the concepts which are considered to be relevant;
- we then need to identify the conditions under which the observables (and their descriptions) are linked with the entities used for their measurement and the rules adopted for describing the changes in these entities;
- finally, we must assess to what extent the measured entities are related to the original domain of application.

Modelling is the most effective analytical apparatus, and in fact the only one which allows us 'to test our interpretative hypothesis about reality' (Haines-Young, 1989). Besides, it permits us to obtain information about the system state, which in turn is capable of yielding 'an image' (i.e. a manifold configuration of images) about future system situations. Through modelling we can test existing theories or hypothetical constructs and also create new ideas and innovative concepts. More particularly, it is worth recalling that modelling is not only an activity characterised by a progressive abstraction but, as shown by many authors (see, for example, Casti, 1984, Rosen, 1985), it is also an activity which requires a number of analytical phases, all equally relevant in the knowledge acquisition process (see Fig. 3, where we have tried to indicate the content of these phases, showing their relationship to indicators and graphical representations).

It is the knowledge path involved in the use (application) of the analytical tools that gives sense to the methods. We could even argue that the potential of the analytical path underlying the modelling activity is due to the fact it encompasses two major components (Vickery, 1983):

- a 'rationalising' component, which also inherently selects and orders the information contents as well as the means by which it is provided. In a planning context this can be considered as belonging to activities mainly oriented towards 'problem-solving';
- a 'creative' component which, besides yielding information with a higher knowledge content, can bring about additional information to be used in the formation of new images (or interpretative hypotheses). This concerns the so-called 'problem-making' activities, those aimed at foreseeing and/or anticipating future situations likely to lead to problems which need to be dealt with by planning.

In this respect, the arguments already put forward by Cole (1983) more than fifteen years ago should not sound surprising. Commenting on the state-of-the-art of modelling and systemic thinking in planning, he posited "to stop modelling would be like asking people to stop thinking" (p. 417) and "putting models into society is fundamentally more important than putting society into models" (p. 418).

From the point of view of methods, the impact of new technologies and particularly those concerning communication cannot be underestimated. Both the rationalising and creative components of methods will be deeply affected. Moreover, it is likely that the final outcomes will not only be a consequence of greater computing capability or the establishment of new communication networks. The new information technology will also have a major and beneficial influence on the 'analytical path' which underpins the knowledge acquisition process.

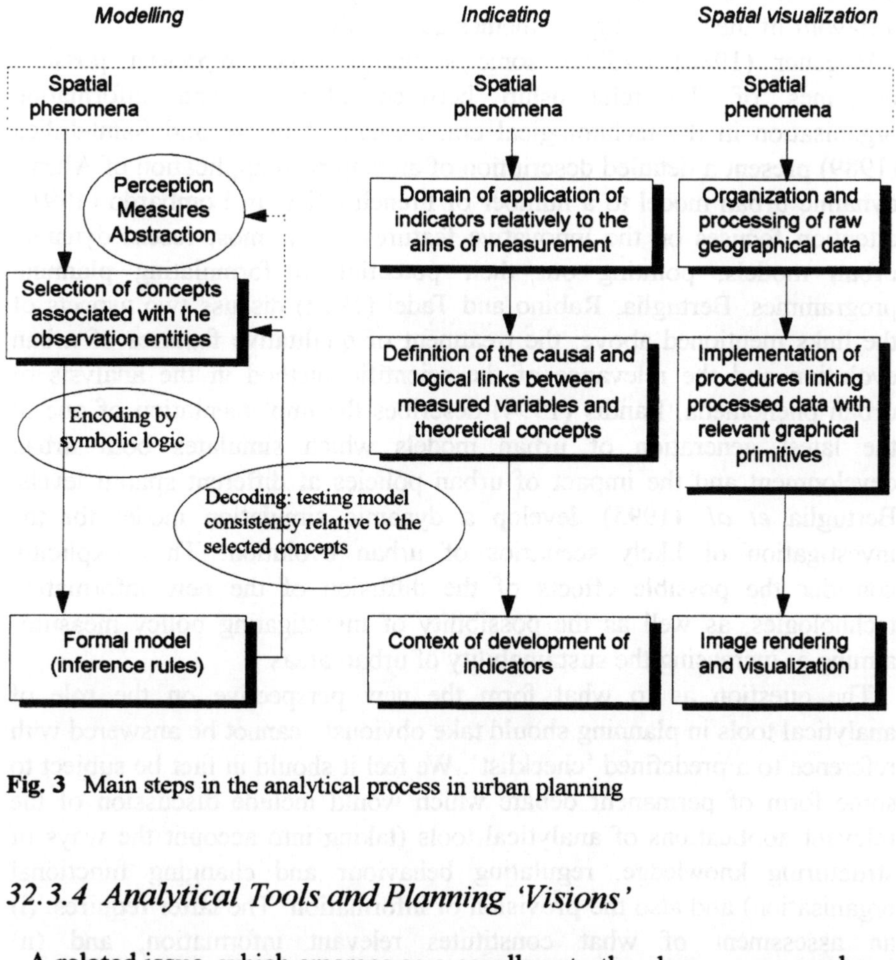

Fig. 3 Main steps in the analytical process in urban planning

32.3.4 Analytical Tools and Planning 'Visions'

A related issue, which emerges as a corollary to the above, concerns how a 'new perspective' on the meaning and functions of analytical tools applies to the overall planning process. This is not an easy question and it would be presumptuous to claim to be able to give an exhaustive answer. Nonetheless, it is one which should not be neglected if, recalling the foreword to this chapter, we wish the 'opinion' which ultimately leads to planning choices at least to be communicable and socially agreed upon.

Although often overshadowed, this issue has been dealt with in recent papers, some of which have provided a number of interesting suggestions. It is worth emphasising the efforts being made to clarify the various links existing between the use of analytical apparatus, the analysis of urban structures and phenomena and the set of operations, procedures etc. which make up planning activities in practice. It would be impossible to review all the relevant contributions, so we mention just a few studies which give a flavour of the wide range of themes dealt with.

Wegener (1988) outlines some scenarios which represent possible outcomes of the relationship between planning and information organisation in the technological era. Pumain, Sanders and Saint-Julien (1989) present a detailed description of an empirical application of Allen's dynamic urban model to a number of French cities. In Lombardo (1991), attention focuses on the innovative features of the most recent dynamic urban models, pointing out their potential in formulating planning programmes. Bertuglia, Rabino and Tadei (1992) discuss two aspects of the links mentioned above: the treatment of qualitative features of urban evolution and the relevance of the scientific method in the analysis of urban phenomena. Landis (1994) describes the implementation of one of the latest generation of urban models which simulates both urban development and the impact of urban policies at different spatial levels. Bertuglia *et al.* (1995) develop a dynamic simulation model for the investigation of likely scenarios of urban evolution. They explicitly consider the possible effects of the diffusion of the new information technologies, as well as the possibility of investigating policy measures aiming at increasing the sustainability of urban areas.

The question as to what form the new perspective on the role of analytical tools in planning should take obviously cannot be answered with reference to a predefined 'checklist'. We feel it should in fact be subject to some form of permanent debate which would include discussion of the relevant applications of analytical tools (taking into account the ways of structuring knowledge, regulating behaviour and changing functional organisation) and also the provision of information. The latter requires: (i) an assessment of what constitutes relevant information, and (ii) consideration of the main agents involved in both the provision and use of the information and (iii) identifying the means of obtaining it.

A salient feature of this new perspective relates to what Harris (1994) identifies as the creation of an environment which favours the development of planners' innovative thinking and actions. We could add that this should apply to all those involved in the planning activity, whether they belong to the normative, institutional or methodological dimension. In such an environment, the 'images' (or planning visions) which, as time goes by, are formed about a given urban issue would be exchanged, refreshed, debated, and maybe selected or refined, but never dismissed a priori. This argument bears some similarity to the ideas discussed in Rabino's contribution to this book, to the extent that in such an environment it would be easier to accumulate and regenerate what he calls 'explicit knowledge'.

A further point not to be overlooked, concerns the fact that the adoption of a new perspective about the meaning and functions of analytical tools implies a revision of the ways in which planning visions are currently created and updated. In particular, the following aspects seem particularly important:

a. the comparison (assessment) of the different images;
b. the communication of the images, considered as a process by which the semantic meaning underlying a vision is transmitted, received and then used to foster actions (in this regard, decision support systems allowing an interactive interpretation of means and ends could be very useful, Aitken and Rushton, 1990).

32.4 Concluding Remarks

On the basis of the above discussion, there appear to be two compelling points affecting the use of knowledge tools in today planning activities:

a. the increasing awareness of the need to improve the interpretation of the evolution of urban processes, and take account of the large number of different actors involved (relative to the normative, institutional and methodological dimension) and the context within which they occur (see, for example, Jayet,1993);
b. the awareness that the understanding of these processes, and hence the possibility of their management and control depends on the information available. This is conditioned by the 'knowledge path' which is undertaken (this point resembles the argument of Casti, 1986, who alleged that the complexity of a plan should be the same - or, at least, as similar as possible - as the complexity of actions necessary to guarantee its control and management).

In this context, we have tried to put forward a conception of analytical tools which goes beyond a narrow definition in terms of technical devices. Besides making it possible to improve the efficiency of their instrumental and functional role, this conception is likely to introduce an educational role, insofar as analytical tools could favour the formation of a more careful attitude towards the ways of coping with planning problems. This latter brings up questions concerning the cornerstones of town-planning culture, as pointed out for example in Genestier (1990) and by Mazza, in this volume, in his discussion of the contents of its 'doctrine'.

Although basically optimistic, this conception is not able to completely solve some nontrivial problems related to the accessibility to 'knowledge tools' and communication within the planning context. Both these aspects will have to be specifically dealt with in the near future. As far as the question of the *accessibility to the 'knowledge tools'* is concerned, it is worth observing that its solution seems likely only partially to result from the development of hardware (i.e. the availability of more powerful and flexible desktop computing) or from the availability of sophisticated (and expensive) system architectures that only large companies will be able to acquire. Intensifying the exchange of ideas, experience and know-how between the academic, research and government environment, as well as with the public, will be of utmost importance.

In this sense, it is evident that those methods which have already achieved a high level of consolidation both theoretically and empirically, should become standard tools and adopted in everyday practice. But to achieve a consolidation in the use of operational tools we require a greater accumulation of experience and know-how in the various areas of urban analysis and planning practice than has so far been possible. (As far as models are concerned, we could ask why so many of the existing commercial applications concern transport, rather than the other urban sub-systems).

From this point-of-view, the *communication* of information, as an activity fostering exchange, appears to be an essential pre-condition. Nonetheless, the effectiveness of communication requires, as stressed by Butera in this book, the existence of reciprocal understanding, otherwise the knowledge gain is lost. A major implication of the thesis stated in this work is that the experience obtained in the course of the application of an analytical tool provides a kind of knowledge which should favour such reciprocity of understanding. But although it may be possible to achieve this between specialists, it would seem more difficult and controversial in a context where the main actors are not technical experts, but political decision-makers and citizens. From this point of view, the question of 'communication' in all its many and different facets constitutes, in its own right, a new challenge for developing more effective knowledge tools for urban and spatial analysis.

References

Aitken S.C., Rushton G. (1990) Perceptual and Behavioural Theory in Practice, *Progress in Human Geography, 17,* 378-388.

Alonso W. (1991) Europe's Urban System and its Peripheries, *Journal of the American Planning Association, 57,* 6-13.

Amin A. (1994) Post-Fordism: Models, Fantasies and Phantoms of Transition, in Amin A. (ed.) *Post-Fordism,* Blackwell, Oxford, 1-40.

Bahrenberg G. (1984) Spatial Analysis: A Retropective View, in Bahrenberg G., Fischer M.M., Nijkamp P. (eds.) *Recent Developments in Spatial Data Analysis,* Gower, Aldershot, 35-50.

Batty M. (1979) Paradoxes of Science in Public Policy: The Baffling Case of Land Use Models, *Sistemi Urbani, 1,* 1, 89-122.

Batty M. (1983) On Systems Theory in Urban Planning: An Assessment, in Batty M., Hutchinson B. (eds.) *System Analysis in Urban Policy-Making,* Plenum Press, New York, 423-448.

Batty M. (1993) Using Geographic Information Systems in Urban Planning and Policy-Making, in Fischer M.M., Nijkamp P. (eds.) *Geographic Information Systems, Spatial Modelling, and Policy Evaluation*, Springer-Verlag, Berlin, 51-72.

Batty M. (1994) A Chronicle of Scientific Planning: The Anglo-American Modeling Experience, *Journal of the American Planning Association, 60,* 7-16.

Batty M. (1995) The Computable City, Keynote Address at the Fourth International Conference on Computers in Urban Planning and Urban Management, Melbourne, Australia, July 11th-14th (mimeo).

Batty M., Hutchinson B. (eds.) (1983) *System Analysis in Urban Policy-Making,* Plenum Press, New York.

Batty M., Yeh T. (1991) The Promise of Expert Systems for Urban Planning, *Computers, Environment and Urban Systems, 15,* 101-108.

Bertuglia C.S., Clarke G.P., Wilson A.G. (eds.) (1994) *Modelling the City: Performance, Policy and Planning*, Routledge, London.

Bertuglia C.S., Leonardi G., Occelli S., Rabino G.A., Tadei R., Wilson A.G. (1987) *Urban Systems: Contemporary Approaches to Modelling,* Croom Helm, London.

Bertuglia C.S., Lombardo S., Occelli S., Rabino G.A. (1995) The Interacting Choice Processes of Innovation, Location and Mobility: A Compartmental Approach, in Bertuglia C.S., Fischer M.M., Preto G. (eds.) *Tecnological Change, Economic Development and Space,* Springer-Verlag, Berlin, 118-140.

Bertuglia C.S., Rabino G.A., Tadei R. (1992) Review of the Main Conceptual Issues Facing Contemporary Urban Planning, *Sistemi Urbani, 14,* 151-171.

Castells M. (1989) *The Informational City*, Blackwell, Cambridge.

Clarke G.P., Wilson A.G. (1987a) Performance Indicators and Model-based Planning: 1. The Indicator Movement and the Possibilities for Urban Planning, *Sistemi Urbani,* 9, 79-127.

Clarke G.P., Wilson A.G. (1987b) Performance Indicators and Model-Based Planning: 2. Model-Based Approaches, *Sistemi Urbani*, 9, 137-169.

Casti J.L. (1984) On the Theory of Models and the Modelling of Natural Phenomena, in Bahrenberg G., Fischer M.M., Nijkamp P. (eds.) *Recent Developments in Spatial Data Analysis,* Gower, Aldershot, 73-92.

Casti J.L. (1986) On System Complexity: Identification, Measurement, and

Management, in Casti J.L., Karlqvist A. (eds.) *Complexity, Language and Life: Mathematical Approaches*, Springer-Verlag, Berlin, 146-173.

Cole S. (1983) Models, Metaphors and the State of Knowledge, in Batty M., Hutchinson B. (eds.) (1983) *System Analysis in Urban Policy-Making*, Plenum Press, New York, 407-448.

Couclelis H. (1983) What Reasons for Rationality? In Search of a Future for Rational Methods in Urban Planning, in Batty M., Hutchinson B. (eds.) (1983) *System Analysis in Urban Policy-Making*, Plenum Press, New York, 449-474.

Craglia M. (1992) GIS in Italian Urban Planning, *Computers, Environment and Urban Systems, 16*, 543-556.

Crosby R.W. (ed.) (1983) *Cities and Regions as Nonlinear Decision Systems*, Westview Press, Boulder.

Fischer M.M. (1994) From Conventional to Knowledge-based Geographical Information Systems, *Computers, Environment and Urban Systems, 18*, 4, 233-242.

Genestier P. (1990) Recherche urbaine, aménagement urbanistique: entre doctrines et aphories, *Sociologie du Travail, 32*, 339-352.

Goodchild M.F. (1991) Geographic Information System, *Progress in Human Geography, 15*, 194-200.

Guelke L. (1978) Geography and Logical Positivism, in Herbert D.T., Johnston R.J. (eds.) *Geography and the Urban Environment*, John Wiley and Sons, New York, 35-62.

Haines-Young R. (1989) Modelling Geographical Knowledge, in Macmillan B. (ed.) *Remodelling Geography*, Basil Black, Cambridge, 22-39.

Hall P. (1991) Three Systems, Three Separate Paths, *Journal of the American Planning Association, 57*, 16-20.

Harris B. (1994) Some Thoughts on New Styles of Planning, *Environment and Planning B, 21*, 393-398.

Jayet H. (1993) Territoires et concurrence territoriale, *Revue d'économie régionale et urbaine, 1*, 55-75.

Khakee A. (1994) A Methodology for Assessing Structure Planning Processes, *Environment and Planning B, 21*, 441-451.

Landis J.D. (1994) The California Urban Futures Model: A New Generation of Metropolitan Simulation Models, *Environment and Planning B, 21*, 399-420.

Lombardo S. (1991) Recenti sviluppi della modellistica urbana, in Bertuglia C.S., La Bella A. (eds.) *I sistemi urbani*, vol. 2, Angeli, Milan, 641-706.

Maestre C.J. (1994) Emergence de l'approche systémique, *Analyse de système, 20*, 1-113.

Massey D., Minnis W., Morrison W.I., Whitbread M. (1976) A Strategy for Urban and Regional Research, *Regional Studies, 10*, 381-388.

Mesarovic M.D., Macko D., Takahara Y. (1970) *Theory of Hierarchical Multilevel Systems*, Academic Press, London.

Occelli S. (1995) Modellistica urbana ed analisi quantitativa: considerazioni metodologiche ed aspetti applicativi, paper presented at the seminar 'Modelli matematici e progetto urbano', Milan, 6-7 March (mimeo).

Preto G., Occelli S. (1994) Zonizzazione territoriale ed ambiti spaziali delle politiche: 1. Considerazioni teorico-metodologiche, IRES, WP 103, Turin.

Pumain D., Sanders L., Saint-Julien T. (1989) *Villes et auto-organisation*, Economica, Paris.

Rosen R. (1985) *Anticipatory Systems*, Pergamon Press, Oxford.

Vernon R. (1991) The Coming Global Metropolis, *Journal of the American Planning Association, 57*, 3-6.
Vickery G. (1983) The Poverty of Problem-Solving, in Batty M., Hutchinson B. (eds.) *System Analysis in Urban Policy-Making,* Plenum Press, New York, 17-30.
Wegener M. (1988) Information Technology, Society and the Future of Planning, in Giaoutzi M., Nijkamp P. (eds.) *Informatics and Regional Development,* Avebury, Aldershot, 42-55.
Wellar B. (1990) Politicians and Information Technology, *Computers, Environment and Urban Systems, 14,* 1-4.

33. Towards an Environmentally Compatible Mobility in the Region of Stuttgart

Frank C. Englmann, Walter Scheuerer, Rainer Carius, Bettina Oppermann, Sabine Köberle, Ortwin Renn

33.1 Introduction

Environmentally compatible mobility now seems to be generally accepted as an aim of sustainable urban and regional development. It is likely to become even more important with the growing awareness that it is not possible to continue with the present pattern of mobility. Today's environmental problems can be shown to be the direct result of the predominance of personal motorised transport systems. This means that we need to adopt a new orientation towards more environmentally compatible ways of achieving mobility. The problem is that while the translation into practical planning terms is still in its infancy, the pressure created by the problems, due to the cuts in financial and personnel resources in the public sector, is increasing rapidly.

Partly for these reasons, the state of Baden-Württemberg, located in the south-west of Germany, decided to create a new political and administrative body for its main urban agglomeration, the Region of Stuttgart[1] . In June 1994 a parliament was duly elected. Along with other bodies, the parliament is responsible for the regional development plan and the regional transportation plan. In order to draw up the latter, a engineering consortium was hired. In the same period a research project entitled 'Towards an Environmentally Compatible Mobility in the Region of Stuttgart' was proposed by a group of institutes belonging to the University of Stuttgart. In July 1995 the transportation committee of the Stuttgart regional parliament decided to contribute to the funding of this project. Other main contributors are the Ministries of Science and

[1] The Region of Stuttgart has 2,341,000 inhabitants and covers an area of 3,654 km^2.

Research, of the Environment, of Transportation and of Economic Affairs for the state of Baden-Württemberg. The main condition for the funding was that the research project and drawing up of the regional transportation plan should be linked.

In Germany there is often a long delay between the decision concerning an infrastructure investment and its implementation. This is partly due to long legal hearings. As German law provides for government decisions to be contested in court, single citizens or groups of citizens are able to delay the implementation of decisions on infrastructure investments.

A member of the Green Party convinced the Minister of Transport of Baden-Württemberg (a member of the Christian Democrats) to set up a workgroup called the "Stuttgart Transportation Area". It consists of the representatives of various associations, groups and firms such as the German Automobile Club, the regional Chamber of Industry and Commerce, Daimler Benz, trade unions, various state ministries and groups of environmentalists. The workgroup is involved in the drawing up of both the regional transportation plan and the research project referred to above.

In Section 2 we present the organisation of the studies for the development of the regional transportation plan for the Region of Stuttgart and in Section 3 the overall structure of the research project "Towards an Environmentally Compatible Mobility in the Region of Stuttgart". Sections 4 and 5 describe in detail two of the three sub-projects. Section 6 contains some concluding remarks.

33.2 The Organisation of the Studies for the Development of the Regional Transportation Plan for the Region of Stuttgart

The project for the development of the Regional Transportation Plan (RTP) for the Region of Stuttgart has a conventional structure. The first step will be the creation of a database for the year 1995 containing structural and traffic data. After that traffic analyses are to be made of passenger and freight traffic, as well as the traffic situation in 1995. Later on, forecasting models will be developed for passenger and freight traffic and scenarios produced for the year 2010. These will lead to traffic and impact forecasts for the year 2010 and finally to recommendations on strategies and policy measures.

The innovative feature of the development of the RTP lies in the project organisation. Fig. 1 shows how various groups and stakeholders are involved in the planning process. Hopefully, this will lead to better results and faster planning decisions because it introduces additional information into the planning process and makes an attempt to establish consensus among the various groups and stakeholders. Apart from the consignee for the studies - a consortium of engineering bureaux (RTP experts) - three groups were formed. The enlarged project group includes important decision-makers in transportation infrastructure, such as the administration of the counties in the Region of Stuttgart, the city of Stuttgart itself, the German Railway Company and others. The Stuttgart Transportation Area workgroup involves social groups concerned with transportation problems. Finally, in the co-ordination group, an exchange of the results of the studies for the RTP and the research project is being organised.

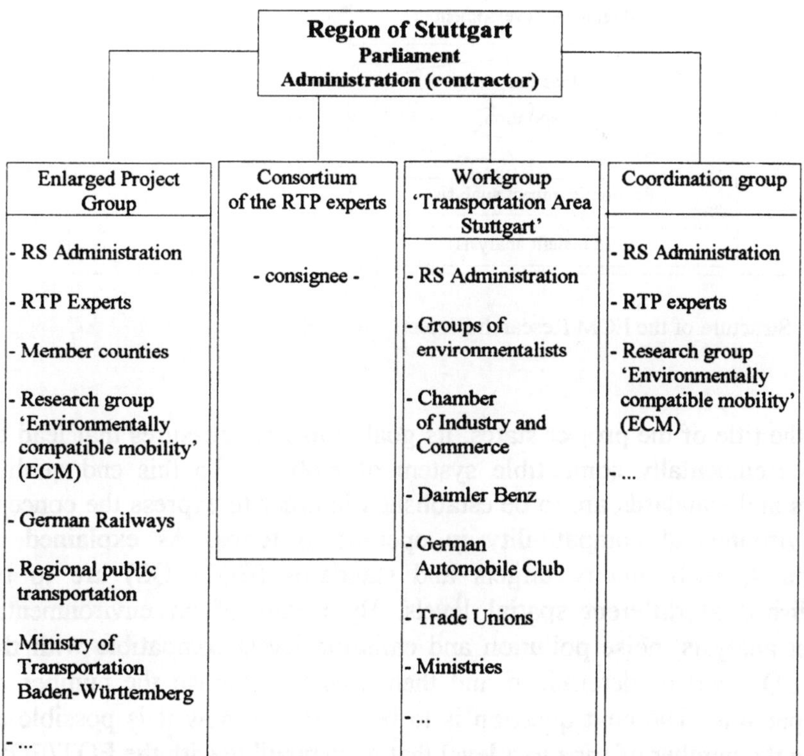

Fig. 1 Organisation of the Studies for the Regional Traffic Planning (RTP) for the Region of Stuttgart (RS)

33.3 The Project 'Towards an Environmentally Compatible Mobility in the Region of Stuttgart'

The overall structure of the research project is shown in Fig. 2.

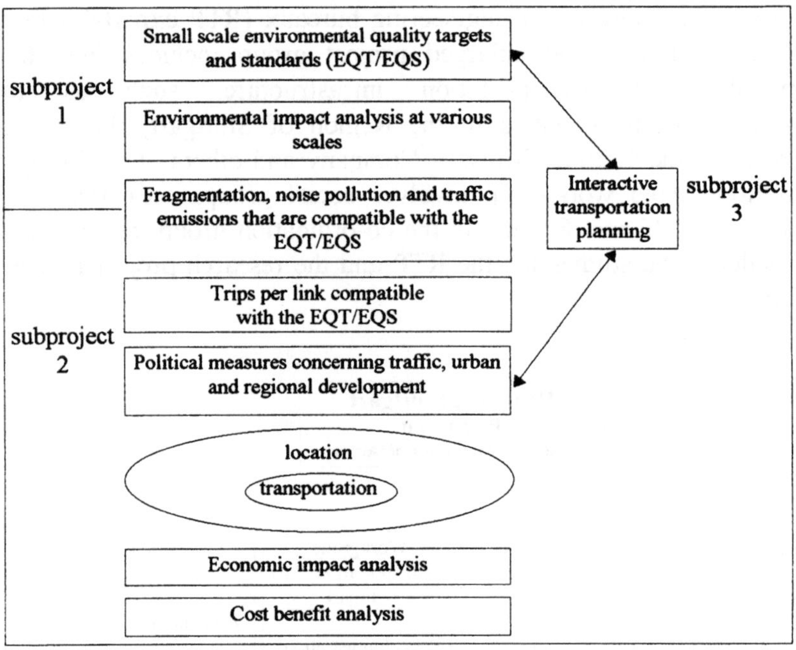

Fig. 2 Structure of the ECM Research Project

As the title of the project states, its goal is to find measures that lead to an environmentally compatible system of mobility. To this end quality targets and standards are to be established in order to express the concept of environmental compatibility in operational terms. As explained in Section 4, such quality targets and standards (EQT/EQS) are to be established at different spatial levels. By means of an environmental impact analysis, noise pollution and emission levels compatible with the EQT/EQS will be determined and then used to establish the number of trips per link. The next question is to be tackled is how it is possible to reduce the number of trips to a level that is compatible with the EQT/EQS. For this purpose it is planned to construct a transportation model for the Region of Stuttgart. This new model will represent an improvement on existing modelling approaches with respect to:

- the analysis of price and administrative/political regulations;
- a stronger subdivision in time and space;
- simulation of pedestrian and bicycle traffic in a spatially limited area.

As transportation and land use are strongly interrelated, current land use will be endogenised. Only urban and regional planning measures concerning land use will be treated as exogenously given. Because of high transaction costs, land use will usually be out of equilibrium. For this reason the master equation approach will be used for the location level of the model, whereas it will be assumed that the traffic level will be in user equilibrium (see Haag, 1986, 1990).

The location level will contain land, labour and housing markets. Hence the effects of various policy measures on employment and income as well as on land prices and housing rents in the Region of Stuttgart will be determined endogenously. The economic impact analysis will also be used in the cost-benefit analysis of the policy measures. (The structure of sub-project 2 is shown in Fig. 3).

As those involved in the project are well aware that environmental quality targets and standards cannot be determined in a scientific 'ivory tower', both the EQT/EQS and the policy measures will be discussed with the Stuttgart Transportation Area work group in the light of the economic impact analysis. This process of interactive transportation planning differs from that organised by the administration of the Region of Stuttgart for the regional transportation plan (see Fig. 1). These organisational differences result from political decisions taken by the administration of the Region of Stuttgart. The Academy of Technology Assessment of Baden-Württemberg, responsible for the organisation of the interactive planning process for the research project, is also prepared to organise similar participation processes for the regional transportation plan (RTP). It remains to be seen how these two participation processes will interact.

With respect to the RTP, the main inputs of the research project will be the submission of research results related to:

- EQS/EQT (after discussion with the 'Stuttgart Transportation Area' work group);
- environmental impact analysis of the transportation system in 1995;
- development of an improved model for forecasting future passenger traffic;
- development of a bicycle traffic concept for the Region of Stuttgart;

- effects of road pricing;
- environmental, urban and regional sustainability of the Regional Transportation Plan for the year 2010.

Due to the interdisciplinary character of the research, various research institutions from the University of Stuttgart and also other institutions are involved in the project. They are listed in Fig. 4.

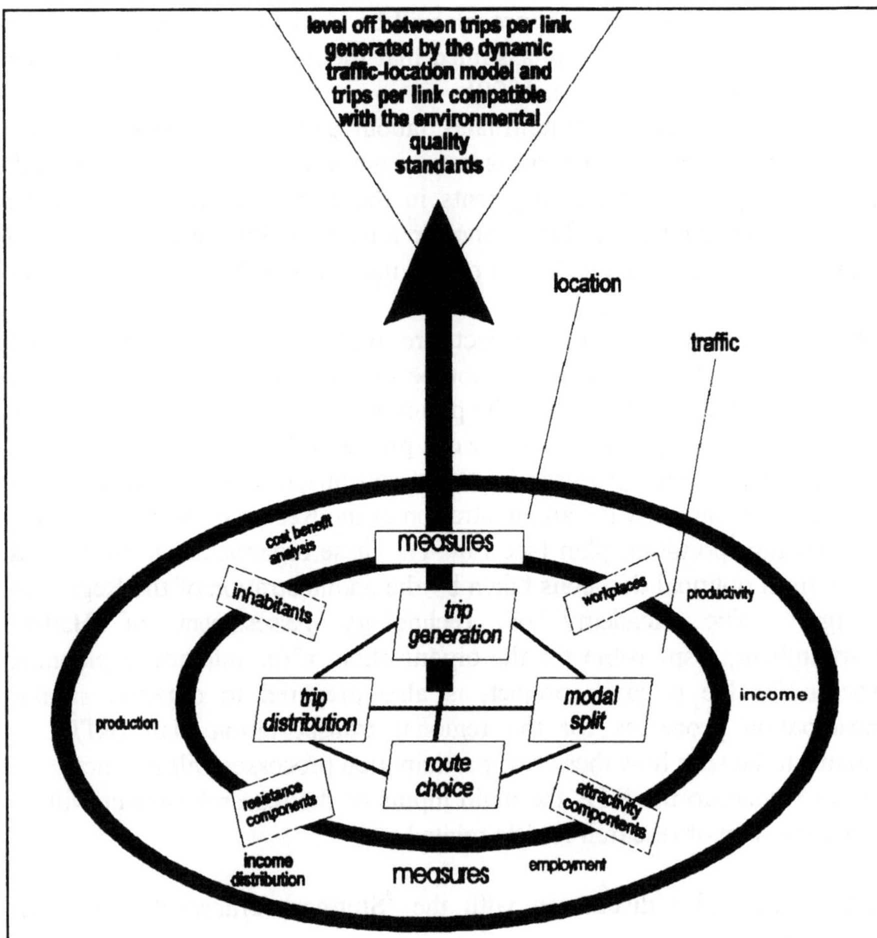

Fig. 3 The Dynamic Integrated Transportation-Location Model

Subprojects	Research Institutes
1. Environmental Quality Targets and Environmental Impact Analysis	• Institute of Landscape Planning and Ecology • Urban Planning and Design Institute
2. Dynamic Integrated Transportation and Location Model	• Institute of Road and Transportation Planning and Engineering • Institute of Railway and Transportation Engineering • Steierwald, Schönharting and Partner, Consulting Engineers • Institute of Social Research - Department of Economics • Institute of Applied Economics, Tübingen • Steinbeis Transfer Centre for Applied Systems Analysis • Institute of Regional Development Planning Urban Planning and Design Institute
3. Interactive Transportation Planning	• Centre of Technology Assessment in Baden-Württemberg

Fig. 4 Research Institutes Involved

The planning area is that of the Region of Stuttgart although some of the representative areas of interest selected in fact affect a much larger area. These areas of interest are subject to political debate and concern important spatial interfaces, such as the city/hinterland, core/periphery, centre/inner city (see Fig. 5). The integrated dynamic transportation and location model will be calibrated and validated with the results of the representative areas, and these areas then tested for their suitability for the participation procedure. Furthermore, the environmental impact analysis will be carried out for these selected areas in a more detailed manner. To sum up, the studies on representative areas pursue the following goals:

- data acquisition;
- calibration and validation of the model;
- detailed examination of some questions of special interest;
- gaining experience in the implementation of concepts through planning procedures;
- testing of new procedures of participation;
- mediation between scientific study and practice;
- ensuring practicability.

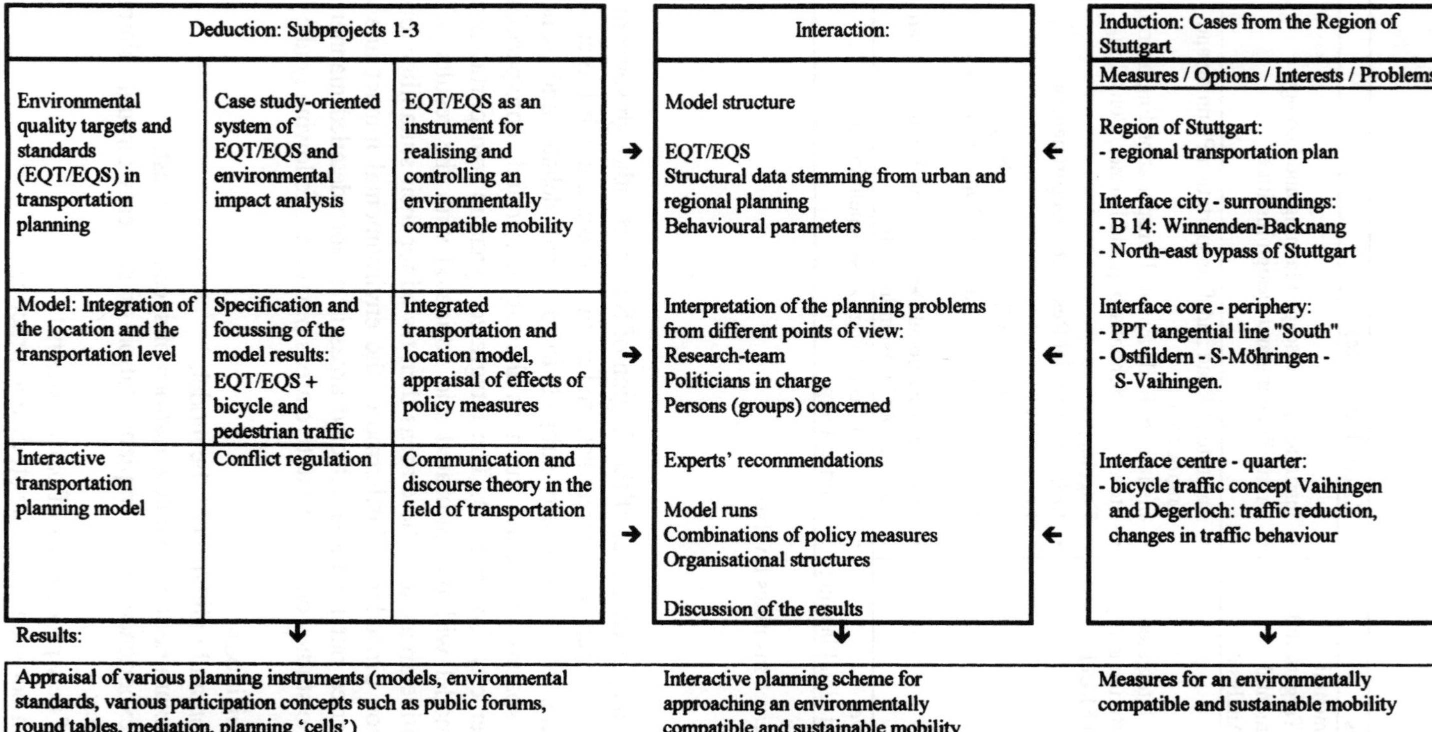

Fig. 5 Deduction, induction, interaction

Fig. 5 shows how deductive and inductive processes are interconnected in the research project. As the main topic of this work is the planning aspects of the research project, the dynamic integrated transportation-location model will not be described in greater detail. The two following sections are devoted to a description of sub-projects 1 and 3.

33.4 Environmental Quality Targets/Environmental Quality Standards and Environmental Impact Analysis (subproject 1)

The compatibility of mobility is closely connected with the environment. The effects of traffic on the environment (e.g. noise and exhaust emissions from motor vehicles, sealing of road surfaces, fragmentation of areas, etc.) will be quantified. In the case of sub-project 1, emphasis is placed on three issues. (i) definition and spatial assignment of environmental quality targets and environmental quality standards (EQT/EQS), their evaluation and utilisation as planning instruments (Kliemstedt and Fürst, 1990); (ii) databases and GIS for integrated spatial analysis of the effects of traffic on the environment and the visualisation of results; (iii) Impact analysis at different spatial scales - regional, landscape and local.

33.4.1 Definition and Spatial Assignment of EQT/EQS

Indicator concepts will be used to represent EQT/EQS, and will be applied to the state and sensitivity of environmental factors in the planning area. These factors are:

- climate
- soil
- groundwater
- surface water
- species and bio-types
- landscape view.

A number of traffic dependent substances and impacts are used as indicators (see Fig. 6). Possible traffic-induced impacts will be analysed and environmental quality standards assigned to certain areas dependent on the specific land-use.

		CO_2	Nitrogen oxides (NO_x), near surface ozone	Heavy metals (Pb, Cd, Zn)	Organic substances (e.g. benzole)	Noise pollution and bad smell	Fragmentation of areas	Waste of material and land
Frame conditions	**Global**	Reduction of 50% by 2020						Longer life cycle/car, higher percentage of recycling
	State	Reduction of 10% by 2005 (1987)	NO_x reduction of 30% by 1998, 60 % by 2005 (1987); BlmSchG §40, EG-Guideline 92/72, VDI 2310	Limits for soils and plants, reduction	Limits for soils and plants, reduction	Reduction of noise to 70 - 71 dB(A) for cars, to 76 - 78 dB(A) for trucks	Portion of unfragmented areas (BFLR)	
Scale of modelling and test	**Regional**	Reduction of traffic-induced CO_2 emissions. Aims of community strategies (e.g. traffic 50%, heating systems 50%)	Critical loads or limits for soils, dependent on soil characteristics and rate of percolation; emission after VDI 2310, BlmSchG §40, EU-Guide line 92/72	Inclusion of surrounding areas (topography; state of road surfaces, age of roads, consideration of background pollution, movement in the air e.g. dust particles)		Agreement to more severe limits than DIN in some areas e.g. recreation areas	Appearance of sensitive species	Waste of land, sealing of surfaces of new roads and buildings
	City/ Landscape						Recreation areas	Reduction of parking areas (parking places/ number of inhabitants)
	Local (type of road and surrounding land use)		Observation of limits at sufficient tolerances, BlmSchG §40, VDI 2310, EG-Guideline 92/72	Accumulation and movement in soils and plants, raised in settlement areas; recommended limit (BUND) $3 \mu g/m^3$ to $10 - 15 \mu g/m^3$		DIN 45641 and 45645/1 'TA Lärm', VDI 2058; mapping of bad smell areas	Procedures for the calculation of ecological standards	Area of parking places (m^2/place)

CO_2 carbon dioxide, Pb lead, Cd cadmium, Zn zinc.

Fig. 6 Example for the derivation of environmental quality standards

The global objectives, e.g. the reduction of CO_2 emissions are considered to be framework conditions. The intention is to test EQT/EQS as planning instruments for evaluation and planning procedures (e.g. suitability analysis, impact analysis) and establish whether they are sensitive enough to express the estimated changes (e.g. traffic density, volume, traffic network).

33.4.2 Database and GIS

One of the difficulties faced by planners is that the analytical tools usually available are not adequate for the scale of the issues they have to deal with. Large volumes of data make processing complex and integration between types of data is difficult. Many of these problems can be overcome however by using computer-based systems for handling geographical or spatialised data, i.e. Geographic Information Systems (GIS). A GIS is an information system designed to work with data referenced by spatial or geographic coordinates (Bill and Fritsch, 1991). In general terms a GIS is any computer-based system for input, storage, analysis and display of spatial information.

Summarised, the role of database and GIS in the project is to:

1. provide a data structure for storing and the efficient management of ecological or socio-economic data for large areas;
2. enable aggregation and disaggregation of data between different scales;
3. locate study plots and/or environmentally sensitive areas;
4. support statistical analysis of ecological or socio-economic distributions;
5. improve the extraction of information from remotely sensed imagery;
6. provide input data/parameters for modelling.

The main tasks are the registration of all relevant data within the project, the digital processing of spatial data and the mapping of related environmental factors together with a suitability analysis.

33.4.3 Impact Analysis at Different Spatial Scales

The analysis of the effects of the transportation system on the environment will be carried out for scenarios at three different spatial scales: regional, landscape and local level. The regional scale is defined as the whole planning area, the landscape level as parts of the region, and the local level is a large scale view of about 1:500. The choice of variables and the appropriate spatial and temporal resolution form the structural basis for the analysis and have a fundamental influence on the nature and reliability of results. Existing models for various spatial and temporal resolutions will be tested for their ability to simulate the effects of traffic on environmental factors. The analysis of environmental impacts associated with traffic is related to:

- air quality and dispersion of emissions (e.g. nitrogen oxides, particles);
- noise and acoustic dispersion;
- soils and groundwater accumulation of residuals from motor vehicles, infiltration and percolation;
- biological effects, habitat destruction and creation of barriers to animal migration, fragmentation of areas;
- visual effects, views of highways, roads and related facilities.

Spatial data must be processed with a suitable resolution or derived from other available data. The ecological effects of traffic emissions at the regional level generally have to be modelled by static models along roads because of the lack of specific data for the whole planning area. Models are linked with GIS and the key operation at this scale is the digital overlaying of maps and their analysis in a GIS (Hübler and Otto-Zimmermann, 1989).

At the smaller landscape scale process-oriented dynamic models are used to simulate the movement and accumulation of traffic-induced pollutants and also to calibrate the coarser models at a regional level. Until now the simulation of the dispersion of traffic emissions has normally been extremely unrefined (highly sophisticated models also have been used, but only to cover very small areas). To simulate the dispersion of traffic emissions at the landscape level, appropriate dynamic models have to be used, considering weather conditions, topography and land use, in addition to the 'traditional' parameters (Vogel, Fiedler and Vogel, 1992). Since the receptors of the pollutants react in different ways, it is necessary to model each one independently (e.g. air, plants, soil, water). For example, noise pollution plays a different role in different areas because the maximum (permissible) noise pollution varies for different types of land use (e.g. housing areas, industrial areas). A classification of noise sensitivity over the whole area has to take this into consideration.

Generally, the EQT\EQS are defined and assigned for large spatial units. This is effective for a general view, but for a detailed assessment it is also necessary to consider the density of development, structure, height of buildings etc. The effect of traffic emissions differs according to the structure of an urban area, so a small-scale view with averaged emission values is not appropriate. The application of large-scale models, which can take into account the detailed building structure, is necessary to be able to make relevant distinctions in urban planning. However, this close-up view can only be achieved in certain small areas because of the lack of specific input data. Of particular interest is the determination and comparison of the effects of different planning scenarios for certain types of area, e.g. new building areas or change of land use in town centres. Various micro-scale simulation models for the dispersion of noise and air pollutants have already been developed. Our objective was to develop a detailed spatial and temporal description of pollution caused by traffic for given planning situations. The first requirement was therefore to link the GIS with 3D CAD-systems for fluid-dynamics and acoustics (the results were to be visualised with three dimensional graphical tools).

Finally, the whole model, including the modelling of environmental effects as well as the transportation and location model, will be tested and applied. This procedure will make it possible to determine the consequences of political measures for the planning area with regard to the environment. The application of EQT/EQS will be tested with respect to their transferability to other regions.

33.5 Interactive Transportation Planning (subproject 3)

The aim of our research project is to integrate social aspects into planning procedures. This integration is based on methods of extended participation, which can help to resolve conflicts if used appropriately.

Personal experience of concrete environmental problems helps in the recognition and understanding of the more abstract concepts associated with the environment. Therefore, environmental protection at regional and municipal levels is a wide field of learning for citizens participating in local politics. Problems with developing and implementing a consistent environmental policy partly result from the lack of standards for environmental quality. The legal framework supports the implementation of ecological aspects through existing procedures in regional and urban development. Often they contain not only objective descriptions of the actual situation but political aims as well, although these have to be discussed in society. Also there are objectively formulated systems deduced from existing legislation or generally accepted principles.

The problem with public planning is the need to show profound knowledge in assessing options as a prerequisite to political assessment of the planning projects. But knowledge is not enough to find a democratic and ethically legitimate solution. This asks too much of politics and administration. To leave the necessary decisions solely to experts contradicts to the normative basis of democracy, to leave these decisions to the market of political powers may lead to scientific facts being disregarded and costs being exceeded. What is needed is a strategy that combines fair decision-making with competent problem-solving. Such a strategy does not only call for a legitimate procedure, but an open discussion with the public as well. The former should be a transparent and comprehensible process of decision-making, the latter an adequate involvement of all interested parties (Haller, 1990). The dialogue, however, needs to be open to both sides, i.e. it should never be restricted to one-way information from the official side or even to marketing strategies, but has to provide the right to participation by all interested parties. Adequate openness of the decision is vital for an acceptable participation procedure. Without such compromise every discussion will end up in frustration.

Participatory decision-making procedures can take various forms. For needs such as ours, it would seem that well structured formal procedures are best, as spontaneous changes of political positions cannot be anticipated in the planning process. This does not prevent action groups from being included or the application of unconventional methods of participation.

The best approach for transportation planning would appear to be the instrument of *mediation*. This includes all participation procedures characterised by the use of a neutral mediator. The aim is to find a solution agreed upon by all participants. By communicating, all parties search for a consensus on the basis of mutual understanding. The mediator should neither pursue his own interests, nor have the power to decide on the matter. Mediation procedures are decision-oriented negotiation processes and should be distinguished from mere hearings or discussions. However, they are not arbitrations or court trials designed to produce an authoritative final decision.

Mediation procedures have been used since the beginning of the 1970s, especially in the US, but also in Japan and Canada. Bingham (1986) documented 132 procedures in the US which had been invoked to reach concrete agreement in matters of conflict. Of these 99 were conflicts concerning sites and 33 concerning agreement on political measures. If success can be measured by the reaching of agreement, 78% of the procedures were successful (79% of those concerning sites, 76% political dialogues). Mediation procedures have now become an established instrument in the environmental and planning policy of the US (Bacow and Wheeler, 1987).

For Germany, mediation procedures in environmental planning are still new political territory. The Wissenschaftszentrum Institute of Berlin has carried out a two-year study on the problem of waste disposal. During the search for a site for a waste dump site in the canton of Aargau (Switzerland), a mediation procedure helped municipalities and citizens to agree on a site (Renn *et al.*, 1993). The ongoing mediation project on waste disposal in the northern Black Forest region has completed its first two stages, determining the volume of final waste and the technical procedures to apply, and is now in the process of determining the site.

33.5.1 Relationship to the Project

Environmentally compatible mobility combines the transport demand with an ecologically-oriented utilisation of the infrastructure. Harsh measures and instruments of 'renouncement' can only be applied if underlying values have been accepted and there is a real chance of success. Models can be very useful tools for assessing the consequences of proposed actions, but an integrated transportation and location model is not regarded as the 'finished product', simply a starting point for a process of mutual learning where scientists, planners from different institutions, road-users, and citizens gather new experience to develop an environmentally and socially acceptable concept of mobility.

The exchange of ideas on the measures necessary to support environmentally compatible mobility and the analysis of interests and fears are the two pillars of the process of mutual learning which may lead to changes in the behavioural assumptions in the model. Most important in this part of the project is:

- clarifying the social desirability of the model's objectives;
- demonstrating to interest groups the environmental consequences of a proposed plan;
- empirically testing the assumptions of the model on behavioural changes in the population;
- discussing measures and assessing their chance of success under given conditions.

Mediation offers a solid platform to all participants of the planning procedure.

In the survey, information will be collected from all professionals involved in the planning and modelling process, and several interviews will be conducted with experts to analyse the consent and dissent among the participants in the research project. This should allow a constant improvement of the quality of planning and modelling. For example, practitioners constantly check the environmental consequences of the plan, the practical applicability of the model, its critical factors and assumptions as well as derived measures. This communication between practitioners and modelling experts enables a closely interconnected perspective of development and application.

By combining the analysis of individual interests, perspectives, expertise and competence, the development of transportation planning remains open. Willingness for open discussion and learning is vital to the project's success.

33.5.2 Implementing Mediation in the Transportation-Location Model

Models can help predict the state of the environment under given conditions and test the effectiveness of measures prior to implementation. In urban and land-use planning, models can also be used to develop scenarios. The aim in this case is not to test a specific measure or policy, but to evaluate possible combinations of measures corresponding to alternative conceptual strategies.

Using the transportation-location model in mediation can be helpful, but is highly demanding in terms of communication capacity for the public as well as for the scientists, who are responsible for a clear and comprehensible description of the scenarios. Mediation also provides input to the model, by validating the assumptions relating to different groups of population. It is a mutual exchange and iterative learning process. The advantage for modellers is that they can gain deeper insight into the practical obstacles to implementation. Measures can be simulated and analysed for their effectiveness as well as their theoretical validity. By combining different interests, perspectives and expertise, the transportation planning process remains open. To ensure constructive and continuous participation, the results of mediation procedures have to be implemented step by step, enhancing the participant's motivation and readiness to be involved.

33.5.3 *Research Procedure*

The project should be checked for feasibility and the conditions to be observed. Answers to the questions: what room the project offers for action, who will be participating in the planning procedure and who receives the results of mediation, are arrived at in three steps.

Firstly a feasibility study clarifies whether the concept of mediation is applicable, or whether other methods of participation might be more suitable. Examples of other participative urban planning projects can be analysed for their structural similarity, and the mediation concept adapted to the special characteristics of urban planning and transportation planning.

A concrete project structure, including time-schedule, is then outlined. The main aim is to find out the motivations of the participants and room for action, including the mission and mandate of such a 'Round Table'. Finally, to achieve fruitful discussions, it is necessary to formulate the issues as clearly as possible for the participants. Working out options in form of scenarios is a part of this step.

33.5.4 *Carrying out Mediation*

At the Centre of Technology Assessment a similar planning project is currently being carried out on waste disposal in the northern Black Forest region. The similarity of the concepts allows transfer of communication or cooperation models to solve problems in other subjects or areas.

One question concerns the combination of participants at Round Table sessions. When there is a large number of participants, an 'inner circle' may be set up, including for example the city and regional authorities, the railway company, scientists and one or two interest groups or environmental associations. In a larger circle, experts on different subjects from the above institutions, e.g. representatives of the urban planning, environmental protection and transportation departments could be included. A third circle involves the broader public. In this case a 'planning cell', i.e. a random selection of persons, could be called upon.

33.5.5 *Application and Examples*

To achieve mediation between planners, researchers and interest groups, a concrete mandate needs to be given. Since the beginning of 1994, the Stuttgart Transportation Area workgroup has been meeting interested parties - political, economic and environmental protection groups - in order to discuss objectives for traffic planning in Stuttgart. The following institutions were involved: ADAC Württemberg, Auto Club Europa, the Federal Department of Transportation, Daimler Benz AG/Mercedes Benz AG, the Federal union association Deutscher Gewerkschaftsbund in the Region of Stuttgart, the municipal authorities Gemeindetag, Landkreistag and Städtetag, the union Gewerkschaft Öffentliche Dienste, Transport und Verkehr, the Craft Association Handwerkskammer Stuttgart, the union IG Metall in Stuttgart, the Chamber of Industry and Commerce for the Region of Stuttgart, UMKEHR Stuttgart, the Ministry of the Environment in Baden-Württemberg, the Region of Stuttgart, the Department of Transportation in Baden-Württemberg, the association Verkehrs und Tarifverbund, and the Ministry of Trade and Commerce in Baden-Württemberg.

In setting up the workgroup, it is hoped that greater consensus on the aims, projects and measures of regional transportation policy in the future will be achieved. One of the main aims is the development of a utility analysis that includes not only the advantages of the current traffic system, but also meets the demands of the future. This can be split up into different steps:

1. the development of utility profiles for all interested parties;
2. the systematic presentation of different utility profiles;
3. comparison of the utility profiles of planners and researchers;
4. the coordination of the utility profiles.

To arrive at a traffic utility profile, it is necessary to establish the intentions of each group. This can be carried out in four stages:

1. theoretical opening of the social function of transportation;
2. sample survey of road users;
3. questioning the survey group;
4. analysis of content and formation of group.

These are followed by the presentation, coordination and discussion of the different utility profiles in a mediative discourse.

33.5.6 Public Relations Work

As institutions, e.g. municipalities, have initiated changes in behaviour in the field of the environment, over the last few years the provision of advice has gained importance. In waste disposal, for example, user-oriented advice models were developed. The first mobility advice centre opened 1993 in Bremen. There are many different forms, measures and instruments of environmental advice, teaching and education. In contrast to the mediation research, the focus here is not the understanding of small-scale communication patterns, but the analysis of effective mass strategies.

The first task of public relations work is to acquaint the public at large with information on the background and range of possible actions. In order to encourage changes in attitude and thus in behaviour, information needs to be fed through several channels and to be user-friendly. A better image for public transportation can be an important component of such a concept. Social psychology and environmental psychology can contribute a great deal to effective communication of these aims.

33.6 Conclusions

The research project covers a wide range of rather diverse topics. As pointed out above, the project was largely applied, but an important part consists of basic research, since it was essential that the different scales involved should be interconnected. For example, environment quality standards are postulated at a meso level, whereas the different emission types come from a micro level. Because of the interactions at the micro level, the resulting emissions at the meso and macro level are not simply the sum of emissions on a micro level. The same holds for the integrated transportation-location model. The measures to reach environmental quality standards are taken at the meso level, but the decisions about single journeys are made at micro level by individuals, households or enterprises. These decisions lead to certain traffic densities with possible congestion (traffic jams), which again feed back to the decisions made by individual economic agents in the planning area. At the same time there are effects on the macro level which again depend upon accessibility and environmental qualities of the region as a whole. Thus, the research project has tried to contribute to basic research on the interrelation between the micro, meso, and macro level in an interdisciplinary context.

References

Bacow L., Wheeler M. (1984) *Environmental Dispute Resolution*, Plenum, New York.

Bill R., Fritsch D. (1991) *Grundlagen der Geoinformationsysteme,* Band 1, 2, Wichman-Verlag, Karlsruhe.

Bingham G. (1984) *Resolving Environmental Disputes. A Decade of Experience*, Foundation, Washington D.C.

Haag G. (1986) Stochastic Theory for Residential and Labour Mobility including Travel Networks, in Nijkamp P. (ed.) *Technological Change, Employment and Spatial Dynamics, Lecture Notes in Economics and Mathematical Systems*, 270, Springer-Verlag, Berlin, 340-357.

Haag G. (1990) Housing 2 - A Master Equation Approach and Services 2 - A Master Equation Approach for Labour Mobility, in Bertuglia C.S., Leonardi G., Wilson A.G. (eds.) *Urban Dynamics: Designing an Integrated Model*, Routledge, London, 184-236.

Haller M. (1990) Risiko-Management und Risiko-Dialog, in Schüz M. (ed.) *Risiko und Wagnis. Die Herausforderung der industriellen Welt,* Band 1, Neske Pfullingen, 229-236.

Hübler K.H., Otto-Zimmermann K. (eds.) (1989) *Bewertung der Umweltverträglichkeit Bewertungsmaßstäbe und Bewertungsverfahren für die Umweltverträglichkeitsprüfung*, Blottner-Verlag, Taunusstein.

Kliemstedt H., Fürst D. (1990) Umweltqualitätsziele - Diskussionsstand und Perspektiven für die ökologische Orientierung von Planungen, Schriftenreihe des Fachbereichs Landespflege der Universität Hannover, Beiträge zur räumlichen Plannung, 27.

Renn O., Webler T., Rakel H., Dienel P., Johnson B. (1993) Public Participation in Decision Making: A Three Step Procedure, *Policy Sciences, 26*, 189-214.

Vogel H., Fiedler F., Vogel B. (1992) Berechnung der Ausbreitung von chemisch reaktiven Luftverunreinigungen für vorgegebene Emissionsszenarien und meteorologische Ausbreitungsbedingungen, 8, Statuskolloquium des PEF (April 1992) KfK-PEF, 104.

Subject Index

Author Index

List of Contributors

Peter M. Allen
International Ecotechnology Research Centre
Cranfield University
Bedford, MK43 OAL
United Kingdom

Cristoforo Sergio Bertuglia
Dipartimento di Scienze e Tecniche per i Processi di Insediamento
Politecnico di Torino
Viale Mattioli 39
10125 Torino
Italia

Francesca Bertuglia
Borsista CNR
Viale Manzoni 30
00185 Roma
Italia

Giuliano Bianchi
Rete dell'Alta Tecnologia
Regione Toscana
Via Ciro Menotti 6
50136 Firenze
Italia

Federico M. Butera
Dipartimento di Disegno Industriale e Tecnologia dell'Architettura
Politecnico di Milano
Via Bonardi 3
20133 Milano
Italia

Roberto Camagni
Dipartimento di Economia e Produzione
Politecnico di Milano
Piazza Leonardo da Vinci 32
20133 Milano
Italia

Rainer Carius
Center of Technology Assessment in Baden-Württemberg
Industriestr. 5
D-70569 Stuttgart
Germany

Ennio Cascetta
Dipartimento di Ingegneria dei Trasporti
Università degli Studi Federico II
Via Cesare Battisti 15
80134 Napoli
Italia

Valter Cavallaro
Dipartimento Interateneo Territorio
Politecnico di Torino
Viale Mattioli 39
10125 Torino
Italia

Marco Dellasette
Dipartimento di Idraulica, Trasporti e Infrastrutture Civili
Politecnico di Torino
Corso Duca degli Abruzzi 24
10129 Torino
Italia

Dimitrios S. Dendrinos
Urban and Transportation Dynamics Laboratory
Transportation Center
2011 Learned Hall
The University of Kansas
Lawrence
Kansas 66045-2962, USA

Frank C. Englmann
Institut für Sozialforschung
Keplerstr. 17, 9. OG
70174 Stuttgart
Germany

Sandro Fabbro
Dipartimento di Ingegneria Civile
Università di Udine
Via delle Scienze 208
33100 Udine
Italia

Andreas Faludi
Planologisch en Demografisch Instituut
Universiteit van Amsterdam
Nievwe Prinsengracht 130
1018 vz Amsterdam
The Netherlands

Fiorenzo Ferlaino
IRES
via Bogino 21
10123 Torino
Italia

Luigi Fusco Girard
Dipartimento di Conservazione dei Beni Architettonici ed Ambientali
Università degli Napoli "Federico II"
Via Cesare Battisti 15
80134 Napoli
Italia

Francesco Indovina
Dipartimento di Analisi Economica e Sociale del Territorio
Istituto Universitario di Architettura
Ca' Tron
S. Croce 1957
30135 Venezia
Italia

Sabine Köberle
Center of Technology Assessment in Baden-Württemberg
Industriestr. 5
D-70569 Stuttgart
Germany

Agostino La Bella
Dipartimento di Informatica, Sistemi e Produzione
Università di "Tor Vergata"
Via della Ricerca Scientifica
00187 Roma
Italia

Tunney F. Lee
Department of Architecture
The Chinese University of Hong Kong
Shatin, New Territories
Hong Kong

Silvana Lombardo
Dipartimento di Pianificazione Territoriale e Urbanistica
Università di Roma "La Sapienza"
Via Flaminia 70
00195 Roma
Italia

Giuseppe Longhi
Dipartimento di Urbanistica
Istituto Universitario di Architettura
Ca' Tron
S. Croce 1957
30135 Venezia
Italia

Lorenza Lucchi Basili
Via Paolo Dore 3
40135 Bologna
Italia

Dino Martellato
Dipartimento di Scienze Economiche
Cannaregio
Fond.ta S. Giobbe 873
Università di Venezia
30121 Venezia
Italia

Luigi Mazza
Dipartimento di Scienze del Territorio
Politecnico di Milano
Via Bonardi 3
20133 Milano
Italia

Alfredo Mela
Dipartimento di Scienze e Tecniche per i
Processi di Insediamento
Politecnico di Torino
Viale Mattioli 39
10125 Torino
Italia

Peter Nijkamp
Free University
Department of Regional Economics
De Boelelaan 1105
1081 HV Amsterdam
The Netherlands

Sylvie Occelli
IRES - Istituto di Ricerche
Economico-Sociali del Piemonte
via Bogino 21
10123 Torino
Italia

Bettina Oppermann
Center of Technology Assessment
in Baden-Württemberg
Industriestr. 5
D-70569 Stuttgart
Germany

Giorgio Preto
Dipartimento di Scienze e Tecniche
per i Processi di Insediamento
Politecnico di Torino
Viale Mattioli 39
10125 Torino
Italia

Denise Pumain
Université de Paris - 1
INED - Institut National d'Études
Démographiques
27, Rue du Commandeur
75014 Paris
France

Giovanni Rabino
Dipartimento di Ingegneria
dei Sistemi Edilizi e Territoriali
Politecnico di Milano
Piazza Leonardo da Vinci 32
20133 Milano
Italia

Aura Reggiani
Dipartimento di Scienze Economiche
Università di Bologna
Piazza Scaravilli 2
40126 Bologna
Italia

Ortwin Renn
Center of Technology Assessment
in Baden-Württemberg
Industriestr. 5
D-70569 Stuttgart
Germany

Walter Scheuerer
Institute of Landscape Planning and Ecology
University of Stuttgart
Keplerstr. 11
D-70174 Stuttgart
Germany

Vittorio Silvestrini
Dipartimento di Fisica
Università di Napoli "Federico II"
Piazzale Tecchio 80
80125 Napoli
Italia

Carlo Socco
Dipartimento Interateneo Territorio
Politecnico di Torino
Viale Mattioli 29
10125 Torino
Italia

Angela Spence
Dipartimento di Scienze e Tecniche per i Processi di Insediamento
Politecnico di Torino
Viale Mattioli 39
10125 Torino
Italia

Roberto Tadei
Dipartimento di Automatica e Informatica
Politecnico di Torino
Corso Duca degli Abruzzi 24
10129 Torino
Italia

Maria Tinacci Mossello
Dipartimento di Scienze Economiche
Università di Firenze
Via Curtatone 1
50126 Firenze
Italia

Mariolina Toniolo
Assessorato all'Urbanistica
Amministrazione Comunale di Venezia
Palazzo Costa
Cannaregio 2396
30121 Venezia
Italia

Marina van Geenhuizen
Free University
Department of Regional Economics
De Boelelaan 1105
1081 HV Amsterdam
The Netherlands

MIX
Papier aus verantwortungsvollen Quellen
Paper from responsible sources
FSC® C105338

If you have any concerns about our products,
you can contact us on
ProductSafety@springernature.com

In case Publisher is established outside the EU,
the EU authorized representative is:
Springer Nature Customer Service Center GmbH
Europaplatz 3, 69115 Heidelberg, Germany

Printed by Libri Plureos GmbH
in Hamburg, Germany